Lecture Notes in Physics

For information about Vols. 1–23, please contact your bookseller or Springer-Verlag.

Vol. 24: R. F. Snipes, Statistical Mechanical Theory of the Electrolytic Transport of Nonelectrolytes. V, 210 pages. 1973.

Vol. 25: Constructive Quantum Field Theory. The 1973 "Ettore Majorana" International School of Mathematical Physics. Edited by G. Velo and A. Wightman. III, 331 pages. 1973.

Vol. 26: A. Hubert, Theorie der Domänenwände in geordneten Medien. XII, 377 Seiten. 1974.

Vol. 27: R. K. Zeytounian, Notes sur les Ecoulements Rotationnels de Fluides Parfaits. XIII, 407 pages. 1974.

Vol. 28: Lectures in Statistical Physics. Edited by W. C. Schieve and J. S. Turner. V, 342 pages. 1974.

Vol. 29: Foundations of Quantum Mechanics and Ordered Linear Spaces. Advanced Study Institute, Marburg 1973. Edited by A. Hartkämper and H. Neumann. VI, 355 pages. 1974.

Vol. 30: Polarization Nuclear Physics. Proceedings 1973. Edited by D. Fick. IX, 292 pages. 1974.

Vol. 31: Transport Phenomena. Sitges International Schools of Statistical Mechanics, June 1974. Edited by G. Kirczenow and J. Marro. XIV, 517 pages. 1974.

Vol. 32: Particles, Quantum Fields and Statistical Mechanics. Proceedings 1973. Edited by M. Alexanian and A. Zepeda. V, 132 pages. 1975.

Vol. 33: Classical and Quantum Mechanical Aspects of Heavy Ion Collisions. Proceedings 1974. Edited by H. L. Harney, P. Braun-Munzinger, and C. K. Gelbke. VII, 311 pages. 1975.

Vol. 34: One-Dimensional Conductors GPS Summer School Proceedings, 1974. Edited by H. G. Schuster. VII, 371 pages. 1975.

Vol. 35: Proceedings of the Fourth International Conference on Numerical Methods in Fluid Dynamics, 1974. Edited by R. D. Richtmyer. V, 457 pages. 1975.

Vol. 36: R. Gatignol, Théorie Cinétique des Gaz à Répartition Discrète de Vitesses. II, 219 pages. 1975.

Vol. 37: Trends in Elementary Particle Theory. Proceedings 1974. Edited by H. Rollnik and K. Dietz. V, 472 pages. 1975.

Vol. 38: Dynamical Systems, Theory and Applications. Proceedings 1974. Edited by J. Moser. VI, 624 pages. 1975.

Vol. 39: International Symposium on Mathematical Problems in Theoretical Physics. Proceedings 1975. Edited by H. Araki. XII, 562 pages. 1975.

Vol. 40: Effective Interactions and Operators in Nuclei. Proceedings 1975. Edited by B. R. Barrett. XII, 339 pages. 1975.

Vol. 41: Progress in Numerical Fluid Dynamics. Proceedings 1974. Edited by H. J. Wirz. V, 471 pages. 1975.

Vol. 42: H II Regions and Related Topics. Proceedings 1975. Edited by D. Downes and T. L. Wilson. XII, 488 pages. 1975.

Vol. 43: Laser Spectroscopy. Proceedings 1975. Edited by S. Haroche, J. C. Pebay-Peyroula, T. W. Hänsch, and S. E. Harris. X, 466 pages. 1975.

Vol. 44: R. A. Breuer, Gravitational Perturbation Theory and Synchrotron Radiation. VI, 196 pages. 1975.

Vol. 45: Dynamical Concepts on Scaling Violation and the New Resonances in e^+e^- Annihilation. Edited by B. Humpert. VII, 248 pages. 1976.

Vol. 46: E. J. Flaherty, Hermitian and Kählerian Geometry in Relativity. VIII, 365 pages. 1976.

Vol. 47: Padé Approximants Method and Its Applications to Mechanics. Edited by H. Cabannes. XV, 267 pages. 1976.

Vol. 48: Interplanetary Dust and Zodiacal Light. Proceedings 1975. Edited by H. Elsässer and H. Fechtig. XII, 496 pages. 1976.

Vol. 49: W. G. Harter and C. W. Patterson, A Unitary Calculus for Electronic Orbitals. XII, 144 pages. 1976.

Vol. 50: Group Theoretical Methods in Physics. 4th International Colloquium. Nijmegen 1975. Edited by A. Janner, T. Janssen, and M. Boon. XIII, 629 pages. 1976.

Vol. 51: W. Nörenberg und H. A. Weidenmüller. Introduction to the Theory of Heavy-Ion Collisions. IX, 273 pages. 1976.

Vol. 52: M. Mladjenović, Development of Magnetic β-Ray Spectroscopy. X, 282 pages. 1976.

Vol. 53: D. J. Simms and N. M. J. Woodhouse, Lectures on Geometric Quantization. V, 166 pages. 1976.

Vol. 54: Critical Phenomena. Sitges International School on Statistical Mechanics, June 1976. Edited by J. Brey and R. B. Jones. XI, 383 pages. 1976.

Vol. 55: Nuclear Optical Model Potential. Proceedings 1976. Edited by S. Boffi and G. Passatore. VI, 221 pages. 1976.

Vol. 56: Current Induced Reactions. International Summer Institute, Hamburg 1975. Edited by J. G. Körner, G. Kramer, and D. Schildknecht. V, 553 pages. 1976.

Vol. 57: Physics of Highly Excited States in Solids. Proceedings 1975. Edited by M. Ueta and Y. Nishina. IX, 391 pages. 1976.

Vol. 58: Computing Methods in Applied Sciences. Proceedings 1975. Edited by R. Glowinski and J. L. Lions. VIII, 593 pages. 1976.

Vol. 59: Proceedings of the Fifth International Conference on Numerical Methods in Fluid Dynamics. 1976. Edited by A. I. van de Vooren and P. J. Zandbergen. VII, 459 pages. 1976.

Vol. 60: C. Gruber, A. Hintermann, and D. Merlini, Group Analysis of Classical Lattice Systems. XIV, 326 pages. 1977.

Vol. 61: International School on Electro and Photonuclear Reactions I. Edited by C. Schaerf. VIII, 650 pages. 1977.

Vol. 62: International School on Electro and Photonuclear Reactions II. Edited by C. Schaerf. VIII, 301 pages. 1977.

Vol. 63: V. K. Dobrev et al., Harmonic Analysis on the n-Dimensional Lorentz Group and Its Application to Conformal Quantum Field Theory. X, 280 pages. 1977.

Vol. 64: Waves on Water of Variable Depth. Edited by D. G. Provis and R. Radok. 231 pages. 1977.

Vol. 65: Organic Conductors and Semiconductors. Proceedings 1976. Edited by L. Pál, G. Grüner, A. Janossy and J. Sólyom. 654 pages. 1977.

Vol. 66: A. H. Völkel, Fields, Particles and Currents. VI, 354 pages. 1977.

Vol. 67: W. Drechsler and M. E. Mayer, Fiber Bundle Techniques in Gauge Theories. X, 248 pages. 1977.

Lecture Notes in Physics

Edited by J. Ehlers, München, K. Hepp, Zürich
R. Kippenhahn, München, H. A. Weidenmüller, Heidelberg
and J. Zittartz, Köln
Managing Editor: W. Beiglböck, Heidelberg

94

Group Theoretical Methods in Physics

Seventh International Colloquium and Integrative Conference on Group Theory and Mathematical Physics, Held in Austin, Texas, September 11–16, 1978

Edited by
W. Beiglböck, A. Böhm and E. Takasugi

Springer-Verlag
Berlin Heidelberg GmbH 1979

Editors
Wolf Beiglböck
Institut für Angewandte Mathematik
Im Neuenheimer Feld 5
D-6900 Heidelberg

Arno Böhm
Eiichi Tagasugi
University of Texas at Austin
Department of Physics
Austin, TX 78712/USA

ISBN 978-3-540-09238-4 ISBN 978-3-540-35345-4 (eBook)
DOI 10.1007/978-3-540-35345-4

Originally published by Springer-Verlag Berlin Heidelberg New York in 1979

2153/3140-543210

PREFACE

The VIIth International Group Theory Colloquium, organized as an Integrative Conference on Group Theory and Mathematical Physics, took place at The University of Texas at Austin from September 11-16, 1978. It brought together scientists working in traditionally different subfields of physics as well as mathematicians and chemists whose work is related to the general theory of groups and group representations. The main purpose of the conference was to provide an opportunity of intercommunication between the traditional subfields for scientists who are using similar or related mathematical methods.

The conference was subdivided into two parts of equal importance: 1) Review talks of such a nature that they made new developments accessible to a wide audience of non-specialists, and 2) talks of a technical nature which were presented in special topics sessions or in poster sessions. This subdivision is reflected in the Proceedings. The review talks are usually printed in their full length, followed by summaries (prepared by the authors) of the technical talks.

The Proceedings are subdivided into chapters along the lines of the sessions, each chapter containing the review talks and the technical papers. The various chapters have been edited by the session organizers and the editors of the Proceedings.

We would like to express our gratitude

To the session organizers for their help with the Proceedings and the organization of the sessions;
To the members of the International Advisory Committee for suggesting topics, organizers and speakers;
To the sponsors for providing the financial foundation for the conference; and
To Janee Trybyszewski for her administrative assistance which brought order out of chaos.

Austin, Texas
October, 1978

The editors:

W. Beiglböck
A. Böhm
E. Takasugi

SPONSORS:

International Union of Pure and Applied Physics (IUPAP)
International Association of Mathematical Physics (AMP)
United States National Science Foundation (NSF)
United States Department of Energy (DOE)
American Physical Society (APS)
The University of Texas at Austin
- The College of Natural Sciences (Dr. A. R. Schrank, Dean)
- The Graduate School (Dr. Irwin C. Lieb, Vice-President and Dean of Graduate Studies)
- Office of the President (Dr. Lorene L. Rogers, President)

INTERNATIONAL ADVISORY COMMITTEE:

H. Bacry (Marseille)
A. Bohr (Niels Bohr Institute)
I. M. Gel'fand (Moscow)
M. Hamermesh (Minnesota)
E. İnönü (Istanbul)
A. Janner (Nijmegen)
P. Kramer (Tübingen)
A. Lichnerowicz (College de France)
M. Moshinsky (Mexico)
Y. Ne'eman (Tel-Aviv and Texas)
L. O'Raifeartaigh (Dublin)
H. Sato (Kyoto)
E. M. Stein (Princeton)
S. Sternberg (Harvard and Tel-Aviv)
E.C.G. Sudarshan (Bangalore and Texas)
R. B. Teese (MPI Munich)
J. Werle (Warsaw)
J. A. Wheeler (Texas)
E. P. Wigner (Princeton)
P. Winternitz (Montreal)

CONFERENCE ORGANIZER: A. Böhm

CO-ORGANIZER: F. A. Matsen

SCIENTIFIC SECRETARY: E. Takasugi

TABLE OF CONTENTS

Officers and Contributors of the Group Theory and Fundamental Physics Foundation....1

The Wigner Medal....3

Tribute to Eugene P. Wigner
A. BÖHM 3

Tribute to Valentine Bargman
E.P. WIGNER....5

GAUGE GROUPS AND SOLITONS
organized by L. O'Raifeartaigh

D. CAMPBELL: Solitons....7

R. JACKIW: Semi-Classical Results in Quantum Field Theory....16

H. BACRY and J. NUYTS: Some Classical Solutions of SU(2) Yang-Mills Equations....25

P.B. BURT: Collapsons: New Nonlinear Excitations....27

P. KERNER: The Excited States of the Prasad-Sommerfield Soliton and its SU(3)-Generalization....29

F. LUND: Geometry for Solitons and Inverse Scattering....31

C. MEYERS, M. DE ROO and P. SORBA: Instantons and Embeddings....33

W.M. DE MUYNCK: A Consistent Quantization Procedure for Nonlinear Problems....35

D. VILLARROEL: A Lienard-Wiechert Type of Solution in SU(2)....37

UNITARY GROUP, ATOMIC, MOLECULAR AND SOLID-STATE PHYSICS
organized by A. Janner, P. Kramer and F.A. Matsen

J.D. LOUCK: Application of the Boson Polynomials of U(n) to Physical Problems....39

J. PALDUS: Unitary Group Approach to Molecular Electronic Structure....51

A.M. BINCER: Shift Operators for the Classical Groups....66

B.L. DAVIES and A.P. CRACKNELL: Computer Programs for the Reduction of Kronecker Products and Symmetrized Kronecker Powers of Space Group Irreducible Representations....67

R. DIRL: A New Method for Calculating Clebsch-Gordan Coefficients................70

G.L. FINDLEY and S.P. McGLYNN: A Biological Field Theory of Evolution............71

S. FLODMARK and P.O. JANSSON: General Flodmark-Blokker Method for Irreducible Representatives 73

C. GRUBER and J.R. FONTAINE: Group Structure for General Lattice Systems and Surface Tension 74

N.J. GÜNTHER: Dynamical Symmetries of the Time-Dependent Lewis-Riesenfeld Harmonic Oscillator 75

W.G. HARTER and C.W. PATTERSON: Induced Representations and Spontaneous Molecular Symmetry Breaking 77

A. JANNER and T. JANNSEN: "Phase Ordering" and Symmetry in $Hg_{3-\delta}AsF_6$............82

M.V. JARIĆ: On the Application of Group Theory to the Renormalization-Group Method 83

M.V. JARIĆ and J.L. BIRMAN: Renormalization Group Theory of Structural Phase Transitions in A-15:$Pm3n$-O_h^3 85

L. MICHEL, J. MOZRZYMAS and B. STAWSKI: Structure of the Images of the Irreducible Linear Representations of the Crystallographic Little Space Groups..........86

C.J. NELIN and F.A. MATSEN: Unitary Group Formulation of Hartree-Fock Theory 87

A. RIECKERS: On Symmetries of Infinite Macrosystems in a Standard Representation 89

D.J. ROWE and G. ROSENSTEEL: Geometry of Collective Motion......................91

J.J. SULLIVAN: On Labeling the Basis in Gln and Identifying the Racah Algebra with S_N 95

C.E. WULFMAN: Dynamical Groups of Parametrized Systems: Atomic Supermultiplets..........97

FIBER BUNDLES AND EXTENDED PARTICLE STRUCTURE
organized by A. Böhm

W. DRECHSLER: Geometrically Formulated Gauge Dynamics for Extended Hadrons 98

V.G. KADYSHEVSKY: Fundamental Length Hypothesis in a Gauge Theory Context 114

F. MANSOURI: Superunified Theories in Superspace..............................125

R.N. SEN: Relativity Principles and Fibre Bundles 134

A.O. BARUT: Dynamical Groups for the Motion of Relativistic Composite Systems ..144

E. GUTKIN: Clebsh-Gordon Coefficients for the Holomorphic Discrete Series.......145

J.F. POMMARET: Deformation Cohomology of the Primitive Infinite Lie Algebras.......147

H.M. RUCK: Interactions in a Model of Extended Particles.......148

E. SEILER: Rigorous Results in Lattice and Continuum Gauge Quantum Field Theories.......151

L.P. STAUNTON: A Unitary Relativistic Wave Equation Exhibiting Extended Particle Structure.......152

P.B. YASSKIN, J. ISENBERG and P.S. GREEN: Line Space Construction of Non-Self-Dual Yang-Mills Fields.......154

GROUP THEORY IN QUANTUM MECHANICS
organized by A.O. Barut, M. Moshinsky, E.P. Wigner and P. Winternitz

H. van DAM and L.C. BIEDENHARN: An Explicit Model Exhibiting Mass and Spin Mixing.......155

M. MOSHINSKY and T.H. SELIGMAN: The Ambiguity Group for Canonical Transformations in Classical Mechanics and its Role in their Representation in Quantum Mechanics.......164

J. PATERA and R.T. SHARP: Generating Functions for Characters of Group Representations and their Applications.......175

S.T. ALI and E. PRUGOVECKI: Phase Space Representations of the Poincaré Group and their Applications to Relativistic Particle Dynamics.......184

J. BECKERS: On Subgroups of Physical Symmetry Groups and their Invariant "Electromagnetic" Fields and Potentials.......186

U. CATTANEO: Covariant Observables and Instruments.......187

P. COMBE, R. RODRIGUEZ, M. SIRIGUE-COLLIN and M. SIRUGUE: Quantum Spin Systems, Fermi Systems and Group Extensions.......189

H. EKSTEIN: Review of Research on Presymmetry.......192

J.P. GAZEAU: On the Four Euclidean Conformal Group Structure of the Sturmian Operator.......194

J.P. GAZEAU, A. RONVEAUX and M.C. DUMONT-LEPAGE: An SL(2,$\mathbb{R}$) Approach to Schrödinger Spectral Problem with Pseudosingular Potential.......195

B. KÜMMERER: Ergodic Theory in von Neumann Algebras.......196

S. MALIN: SU(2) Harmonic Analysis as a Basis for Quantization.......197

A. SCHOBER: Some New Aspects of Cayley-Quantum Theory.......200

R.K.P. ZIA and D.M. KAPLAN: Application of Quasiclassical Methods to Uniaxial Spin Systems.......201

GROUPS AND SEMIGROUPS IN THE DESCRIPTION OF DECAYING SYSTEMS
organized by L.P. Horwitz

H. BAUMGÄRTEL: Mathematical Remarks to Resonances and their Eigenfunctionals in Decay-Scattering Systems, Demonstrated by Means of Friedrich's Model....202

L. FONDA, G.C. GHIRARDI, A. RIMINI and T. WEBER: Quantum Decay Processes and Quantum Dynamical Semigroups....212

V. GORINI and G. PARRAVICINI: Unstable Quantum States and Rigged Hilbert Spaces....219

A.P. GRECOS and I. PRIGOGINE: Irreversible Processes in Quantum Theory....229

W.C. SCHIEVE and T. BAILEY: Analytic Continuation in Decay-Scattering Systems....239

H. BAUMGÄRTEL: On the Inverse Problem of the Abstract Relativistic Scattering Theory....244

A. BÖHM: The Rigged Hilbert Space and Decaying States....245

L. DAVIDOVICH and H.M. NUSSENZVEIG: Excitation and Decay of a Multilevel Atom....250

G.N. FLEMING: Dynamical Spin Spreading of Unstable Particles in Four Models....251

L. FONDA, G.C. GHIRARDI, C. OMERO, A. RIMINI and T. WEBER: Sequential Decays of Open Quantum Systems....252

N. GIOVANNINI and C. PIRON: Quantization via the Imprimitivity Systems and Superselection Rules....254

A. GROSSMANN and R. HOEGH-KROHN: A Class of Soluble One Particle Models....255

A. GROSSMANN: Resonances in Quantum Mechanics over Phase Space....255

J. LUKIERSKI: Complex Mass and Field Operator for Unstable Particle....256

G. VITIELLO: A Canonical Description of Semigroup Law for Unstable Systems....258

SYPMPLECTIC STRUCTURE AND GEOMETRIC QUANTIZATION
organized by K. Bleuler and A. Lichnêrowicz

F. BAYEN: Deformation of Symplectic Structure and Quantization....260

M.J. GOTAY and J.M. NESTER: Presymplectic Hamilton and Lagrange Systems, Gauge Transformations and the Dirac Theory of Constraints....272

A. LICHNEROWICZ: Deformation Theory and Quantization....280

J. CZYZ: On Certain Events in Geometric Quantization....290

M.J. GOTAY and J. ISENBERG: Kostant-Souriau Quantization of Robertson-Walker Cosmologies with a Scalar Field....293

R.V. MENDES: Infinite-Dimensional Lie Groups and Quantum Dynamical Systems 296

G. ROSENSTEEL and E. IHRIG: Geometric Quantization of Nuclear Collective Models 298

SPECTRUM GENERATING GROUPS
organized by J. Werle

A. BÖHM and R.B. TEESE: Idea and Application of Spectrum-Generating SU(3) and SU(4) 301

A. GARCIA: Baryon Semileptonic Decays-Spectrum Generating SU(3) and the Cabibbo Model 311

P. KIELANOWSKI: Spectrum-Generating SU(3) and Local Current Algebra 317

M.M. NIETO: The Effect of the Choice of Wave Functions on Theoretical Predictions for Symmetry Breaking Processes: A View from the DKP Formalism 324

S. ONEDA: Algebraic Approach to Hadrons without Seeing Quarks 334

R.L. INGRAHAM: Gauge Invariance with Massive Gauge Bosons 341

Y.S. KIM and M.E. NOZ: Group Theoretical Interpretation of Relativistic Hadronic Structures 345

M.D. SLAUGHTER: A Simple Broken-Symmetry Treatment of the Ground and Radially Excited State Mesons and the $\Delta^+ \to P\gamma$ Decay 346

ALFRED SCHILD MEMORIAL SESSION ON GROUP THEORY IN GENERAL RELATIVITY
organized by J. Plebanski and J.A. Wheeler

J.N. GOLDBERG: Self-Dual Gauge Fields and Space Times 349

A.J. HANSON and T. REGGE: Torsion and Quantum Gravity 354

C. TEITELBOIM: Surface Deformations, their Square Root and the Signature of Spacetime 362

F.J. ERNST: The Role of Group Theory in the Quest for Exact Solutions of Einstein's Field Equations: Some Recent Developments 372

L. HALPERN: On Group Covariant Physical Laws and Gravitation 379

S. HORI: On the Exact Solutions of Tomimatsu-Sato Family for Stationary Axial-Symmetric Gravitational Fields 382

I. OZSVÁTH: Remarks to Homogeneous Solutions of Einstein's Field Equations 383

G.E. *TAUBER and* Y. FELDMAN: Description of Paricles with Internal Structure in General Relativity 391

GROUP THEORY IN NUCLEAR PHYSICS
organized by A. Arima and L.C. Biedenharn

J.B. FRENCH and J.P. DRAAYER: Symmetries and Statistical Behavior in Fermion Systems 394

K.T. HECHT and W. ZAHN: SU_3 Symmetry and Integral Kernels for Nuclear Cluster Problems 408

F. IACHELLO: Dynamical Symmetries in Nuclei 420

L. WEAVER: Recent Work on Collective Motion 430

P.H. BUTLER: The Calculation of 6j Symbols for Compact Groups 438

G. JOHN, P. KRAMER and K.D. HETZEL: Application of Group Theory to Composite Nuclear Particle Interactions 439

SUPERGRAVITY AND GRADED LIE ALGEBRAS
organized by Y. Ne'eman and S. Sternberg

V.G. KAC: Contravariant Form for Infinite-Dimensional Lie Algebras and Superalgebras 441

Y. ILAMED: Dual Sets in Associative Algebras and Generalized Lie Algebras 446

N.R. RANGANATHAN and J.S. PRKASH: Molien Function for a Symmetric Group 448

M. SCHEUNERT: Generalized Lie Algebras 450

J. LUKIERSKI: Superconformal Group and Quark-Like Fermionic Coordinates 451

M. OMOTE and S. KAMEFUCHI: Parafields and Supergroup Transformations 452

M. BATCHELOR: Supermanifolds 458

C.P. BOYER: Graded G-Structures 466

S. DESER: Space-Time Properties of Supergravity 468

T. REGGE: The Gauge Group and Geometry of Supergravity 477

M. KAKU: Recent Developments in Superconformal Gravity 480

P. NATH and R. ARNOWITT: Geometrization in Superspace and Local Supersymmetry 483

W. SIEGEL: Superfield Supergravity 487

GROUP THEORY IN PARTICLE PHYSICS
organized by J. Pasupathy and E.C.G. Sudar

A.O. BARUT: How to Build Hadron Multiplets from Stable Particles 490

S.K. BOSE: Structure of the Multiquark Meson States 499

F. GÜRSEY: Octonionic Structures in Particle Physics........................... 508

C.C. CHIANG: Multilocal Field Theory 522

H. HØGAASEN and P. SORBA: Multiquark States 523

M. JASPERS and J. BECKERS: On the Eriksenlike Form of the Melosh Transformation and its Group Theoretical Interpretation 525

P. JASSELETTE: Working with H.A. Kramers in SU(4) 527

C.S. KALMAN: Dynamical Groups and the Quarkonium Problem 528

R.C. KING and A.H.A. QUBANCHI: Weight Multiplicities of the Exceptional Lie Groups 530

H. REEH: Conserved Currents and Symmetries of the S-Matrix 532

Z. STIPČEVIĆ: Group Theoretical Aspects of More-Than-Four Quark Models 534

I. SZCZYRBA: Infinite Unitary Group, Quarks, Unitary Symmetry 535

LIST OF PARTICIPANTS 537

THE GROUP THEORY AND FUNDAMENTAL PHYSICS FOUNDATION

Chairman: Arno Böhm, Physics Department
The University of Texas at Austin

Vice-Chairman: Morton Hamermesh, Physics Department
The University of Minnesota

Treasurer: John D. Dollard, Mathematics Department
The University of Texas at Austin

PATRONS:

Hamad A. Alhendi
L. C. Biedenharn
Sharifa Böhm
H.B.G. Casimir
John D. Dollard
Lawrence Dresner
F. J. Dyson
Einstein Centennial Celebration Committee at Southern Illinois University
J. B. Garg
Marvin Goldberger
Morton Hamermesh
Richard Haymaker
Yehiel Ilamed
Erdal Inonu
Y. S. Kim
E. J. Konopinski
James D. Louck
W. M. MacDonald
Kazumi Maki
F. A. Matsen
Bernd T. Matthias
Esperanza de Moshinsky
Marcos Moshinsky
Tadao Nakano
Pedro L. Garcia Perez
T. O. Philips
Physical Chemistry Laboratory, ETH Zurich, Switzerland
Humitaka Sato
Marlan Scully
Frederick Seitz
The Solvay Institute, Brussels
SKM, Inc.
Edward Teller
Texas Nuclear
Tracor, Inc.
C. W. Ufford
Instituto de Fisica, Universidad Nacional Autonoma de Mexico
Janette Hegner Wheeler
John Archibald Wheeler

SPONSORS:

R. K. Adair
Norman Austern
V. Bargmann
Joseph Birman
D. Allan Bromley
Harvey Brooks
Philip B. Burt
E.G.D. Cohen
B. L. Davies
R. H. Dicke
G. and S. Domokos
Leonard Eisenbud
Arnold M. Feingold
Stig Flodmark
Stephen A. Fulling
Rudolph Haag
Christopher J. Henrich
Emmett L. Hudspeth
J. H. Irving
John R. Klauder
Wolfgang Ludwig
Louis Michel
R. Nottrot
Robert F. O'Connell
S. Okubo
L. C. Papaloucas
Constantin Piron
Luigi A. Radicati
Frederick Reines
Fritz Rohrlich
Paul Roman
Frank Stern
Marshall H. Stone
Eric W. Vogt
Steven Weinberg
V. F. Weisskopf
A. S. Wightman
Pavel Winternitz
Chen N. Yang

CONTRIBUTORS:

Julius Ashkin
A. Bohm
Burt V. Bronk
Morrel H. Cohen
Volker Enss
U. Fano
William Faris
Hans Frauenfelder
R. H. Gluckstern
Robert Hofstadter
B. R. Judd
S. T. Kuroda
Francis J. Narcowich
Enrico Onofri
W. Opechowski
Herbert Piller
Arlan B. Ramsay
Martin G. Redlich
Peter Rejto
Yitzhak J. Sharon
L. C. Shepley
Eric Sheldon
A. Shimony
N. H. Zebouni

THE WIGNER MEDAL

The high point of the VIIth International Group Theory Colloquium was the first presentation of The Wigner Medal. Two medals were presented, the first to Eugene P. Wigner and the second to Valentine Bargmann. At the presentation, addresses were given by A. Böhm and E. Wigner. These addresses are reproduced below.

The Wigner Medal is administered by the Group Theory and Fundamental Physics Foundation, a publicly supported organization. Donations are tax deductible as provided in Section 170 of the Internal Revenue Code.

Tribute to Eugene P. Wigner by A. Böhm

It is a great honor for me, an honor which I could not have conceived of even a few months ago, to make this presentation of The Wigner Medal to Eugene P. Wigner.

Yesterday Professor Wigner told some of us that his 1939 paper on the representations of the Poincare group had originally been rejected by the journal on the grounds that the results were not of interest. I learned particle physics many years later, but for me, and many of my contemporaries, these representations of Wigner have always been the basis for understanding of the elementary particles. And as far as I know, the 1939 paper was also the starting point of a new area of mathematics: the theory of induced representations of non-compact groups. This theory was worked out in full generality by George Mackey, whom we are also happy to have among us today.

But this is, of course, only one of Professor Wigner's many contributions. The description of symmetries by group representations has its basis in Wigner's work. And he has given many applications of this basic concept which are not only of theoretical beauty but also of very practical use. To mention just one: The Wigner-Eckart Theorem has been of great use in atomic physics, where it was first applied, and it has remained useful for nuclear physics. It is one of the few reliable tools for the calculation of experimental numbers in particle physics.

Though he is the father of group-theoretical methods in quantum physics, the father of us all, he has made many contributions in other areas. Even in quantum mechanics books which make little use of group theory Wigner's name is frequently quoted.

The formula which is perhaps used more often than any other formula in experimental physics is also connected with his name. It is the Breit-Wigner formula which has been applied in all areas of quantum physics. And I could go on in this way.

You all know who Professor Wigner is, and I really did not have to say even this much. But most of you are not so familiar with The Wigner Medal, and I shall, therefore, tell you a little bit about it.

The Group Theory Colloquia were started about seven years ago by Bacry and Janner and have grown ever since. The people who come to the group theory colloquia work in the different conventional branches of physics, and perhaps in each of these conventional fields they are in a minority. There has been no adequate body to represent them. Our conference here in Texas is the first Group Theory Conference--and also the first conference on Mathematical Physics in general--which re-

ceived IUPAP sponsorship. Due to the lack of a formal Group Theory organization, this sponsorship was not easy to obtain. It appeared to me that the group theoreticians needed a central point around which they could unite. In the absence of a formal organization, I thought that a medal might best serve this purpose.

The support for a Wigner award was tremendous. There was, however, opposition from Wigner himself, who said that one should not name an award after a living person. But, fortunately, we succeeded in changing his mind.

The Wigner Medal is now permanently endowed from contributions by scientists throughout the world. The Medal will be awarded approximately every two years on the occasion of an international conference. The Medal is administered by the Group Theory and Fundamental Physics Foundation, and the Standing Committee of the Group Theory Colloquia has acted as the first selection committee. We intend that The Wigner Medal shall be a truly international award, and that the selection committee shall always have international character.

The inscription on the first Wigner Medal reads: "Awarded to Eugene Wigner in recognition of his contributions to the understanding of Physics through group theory. September 1978." Professor Wigner has honored us by lending us his name for this medal, and I hope we will be able to honor him by making this one of the most prestigious awards in science.

Tribute to Valentine Bargmann by Eugene Wigner

It is a pleasure to say a few words about my esteemed colleague, Valya Bargmann, but it would be an even greater pleasure to have him here with us and have him participate in our discussions and to hear his comments, always very pertinent. Unfortunately, he was unable to come.

Let me say a few words about his past life. He was born in Berlin, from Russian Parents, in 1908. Valya went to school, first in Germany, but later moved to Switzerland to escape, this time Hitler's terrible government. Valya obtained his doctor's degree in physics from the University of Zürich in 1936. He was 28 years old then, but he had published 3-1/2 papers even before his doctorate. Soon after that he moved to Princeton's Institute for Advanced Studies and was with us there until 47, the last year as Visiting lecturer at the University. He became a valued member of the theoretical physics community soon after his arrival, and we owe many important remarks to him, made in the course of many personal discussions, a good number of them with Einstein, starting soon after his arrival at Princeton - which, incidentally, coincided in time with my return there from Wisconsin. In 47 he moved for a year to Pittsburgh, but then returned to Princeton as a member of the University, and he is still with us there, still a member of the community, whom we consult on many questions frequently even though, officially, he retired a couple of years ago.

It is difficult to describe any colleague's scientific work in a few words, but one should try. Let me begin by saying that, except for Von Neumann, I never met anyone with greater versatility, with wider interests, than Bargmann. He wrote interesting and significant papers on general relativity, in the domain of quantum mechanics, in the classical mechanics of shock waves with which he got into contact as a result of his wartime research. He contributed significantly to group theory and its role in physics, to the problems of symmetry in general, to scattering and collision theory.

Let me go a bit more into detail. His contribution to the general theory of relativity was directed, principally, toward an incorporation of quantum theoretical concepts into that theory, in particular, the general relativistic description of the spin. But he also collaborated with Einstein and Peter Bergmann on the effort to provide a common basis for gravitational and electromagnetic interaction. His possibly best known paper in the domain of elementary quantum mechanics was his early paper in which, following Fock, he discovered a new symmetry of the Schrödinger equation of the hydrogen atom and derived from it the "accidental degeneracy" in the spectrum of this atom. This article appeared, actually, in the same year in which he received his doctor's degree in Zürich and was, probably, at least in part, responsible for his invitation to Princeton.

The article I last mentioned was already in the area of group theory and its use in quantum mechanics. But his later contribution to this subject was even more significant. After his wartime work, concerned with shock waves and classical mechanics, he turned wholeheartedly toward this subject. He determined the unitary representations of some non-compact groups, in particular, also of the homogeneous Lorentz group, and this became the origin of a great deal of research in pure mathematics. Many colleagues consider his research in this area his most important contribution to mathematical physics. I might mention in this connection also his rigorous proof of the unitary-antiunitary theorem, and there are several other very interesting articles of a very general nature that he contributed to "group theory and quantum mechanics or to the abstract theory of groups.

Bargmann's contributions to scattering theory are also remarkable. The article that is most generally known shows that the energy dependence of the amplitude of the spherically symmetric part of the scattered wave does not give a unique determination of the scattering potential. There are several different potentials which produce the same spherically symmetric part (s-wave) of the scattered wave - they differ though in the number and position of the bound states that they produce.

Let me mention, finally, an article which impressed me as much as any of the other articles I mentioned and the content of which was fully unexpected to me. It is a replacement of our universally used complex Hilbert space by a Hilbert space of analytic functions. This is a very interesting discovery, the details of which I cannot communicate right now. Perhaps I can mention that the scalar product of two analytic functions is defined as the integral over the whole complex plane of their product - the imaginary part of one with the opposite sign - multiplied by $e^{-|z|^2}$, the z being the complex variable.

I spoke at some length about Valya's publications, that is, about his public and direct contributions to science, but I consider his private, that is, personal, contributions at least equally remarkable. His willingness and ability to help, to clarify scientific problems and also results is simply unique. When I had the last few days problems with the mathematical definition of active and passive transformations, I naturally wanted to turn to him for help - as I often did before, and, I am sure, as many others did also. I should have mentioned also his teaching skill - he was one of the most popular teachers and his clarity, not only in class, but also in public addresses, is simply unsurmountable. I hope he will be with us soon again, and it was a true and deep pleasure to learn that he would receive a reward for his abilities, helpfulness and warmth, and that I am appointed to transmit it.

SOLITONS

David Campbell
Theoretical Division, Los Alamos Scientific Laboratory
University of California, Los Alamos, New Mexico 87545

I. INTRODUCTION

Although the term "soliton" is little more than a decade old,[1] the concepts which it reflects have already proved fruitful in a number of problems in physics and mathematics, and the excitement surrounding these often bizarre objects has in recent years exceeded that in any other subject that touches both these disciplines. In this very brief review I can present only a soliton "sampler", touching on but a few of the most interesting results in this area. Given the nature of the conference I have chosen to include some recent, tentative observations on the possible relation of group theory - specifically, non-Abelian gauge group theory - to solitons. Since this material resides at the frontier of soliton research, I will have to pass very quickly through the more familiar territory. Hopefully, the references will be sufficient to fill in the many gaps I am forced to leave.[2]

II. SOLITONS: WHAT, WHERE, AND WHY?

Consider classical wave motion in one space dimension and time. Let $u(x,t)$ be the function describing this motion and thus obeying a wave equation (not necessarily linear) in the independent variables x and t. A "travelling wave" solution, $u_T(x,t)$, to this equation depends on x and t only through $\xi = x - vt$, for a constant velocity v. A "solitary wave " is a travelling wave corresponding to a disturbance that is localized in ξ. Note that this definition allows different values of u_T as $\xi \to \pm\infty$, provided that the physical disturbence corresponding to u_T can be regarded as localized. We shall see explicit examples of this point later. A "soliton" is defined as a solitary wave which asymtotically (in time) preserves its amplitude, shape, and velocity - in short, emerges unscathed - after collisions with other solitary waves. In our discussion, we shall use this strict definition of "soliton", but readers should be aware that in the literature the term "soliton" is occasionally used to describe any solitary wave, independent of its behavior in interactions.

The simplest example of a soliton occurs in the linear, dispersionless wave equation,

$$\phi_{tt}(x,t) - c^2\phi_{xx}(x,t) = 0 \tag{1}$$

where c is a constant. This equation is manifestly linear; the term "dispersionless" refers to the algbraic "dispersion relation" between ω, the Fourier transform variable conjugate to t, and k, that conjugate to x. Equation (1) implies $\omega = \pm ck$ and hence the group velocity, $d\omega/dk$, equals $\pm c$, a constant independent of k. Thus, for waves travelling in a given direction, all Fourier modes move with the same group velocity, and the pulses do not disperse. It is trivial to construct solitary wave solutions to (1): an example is

$$\phi_s(x,t) = e^{-(x-ct)^2} \quad . \tag{2}$$

Further, since Eq. (1) is linear, it is easy to see that this is indeed a soliton, for, using superposition, the solution

$$\phi_{ss}(x,t) = e^{-(x-ct)^2} + e^{-(x+ct)^2} \tag{3}$$

describes two solitary waves of the form (2), which are widely separated for large negative time, come together as $t \to 0$, and separate again - retaining their shape, amplitude, and velocity - for $t \to +\infty$. Of course, this example is trivial, and if solitons existed only in the linear wave equation, then the concept would be empty. The excitement surrounding solitons arises from their quite unexpected appearance in non-linear, dispersive wave equations, for which the above simple construction of solitons fails totally. Actually, with hindsight it is possible to give an intuitive argument for the possible existence of solitons in such equations.[2] Dispersion implies that different Fourier components travel with different velocities and hence that an initial solitary wave pulse should break up. But non-linearity implies that the different Fourier modes are coupled, and thus, if the coupling can counteract the dispersion, solitons can result from the balance between non-linearity and dispersion.

Of the many physically interesting non-linear, dispersive equations containing solitons, three are particularly useful as examples for our later discussion. First, the "Korteweg-de Vries" (KdV) equation,[2,3]

$$u_t(x,t) + u(x,t)u_x(x,t) + u_{xxx}(x,t) = 0 \tag{4.a}$$

for which the one soliton solution is

$$u_s(x,t) = 3v\ \mathrm{sech}^2\left(\frac{\sqrt{v}}{2}(x-vt)\right) \quad . \tag{4.b}$$

Note that the non-linear nature of this wave is manifested by the relations between the amplitude (A) and the velocity ($A \propto v$) and between the width (w) and the velocity ($w \propto 1/v^{1/2}$).

Second, the "non-linear Schrödinger" (NLS) equation,[2,4]

$$i\psi_t(x,t) + \psi_{xx}(x,t) + k|\psi(x,t)|^2\psi(x,t) = 0 \tag{5.a}$$

which has the one soliton solution[2,5]

$$\psi(x,t) = \psi_0 \frac{\exp\left[\frac{iv_1}{2}(x-v_2t)\right]}{\cosh\left[\sqrt{\frac{k}{2}}\,\psi_0(x-v_1t)\right]} \quad , \qquad \psi_0 = \left(\frac{v_1(v_1-2v_2)}{2k}\right)^{1/2} \tag{5.b,c}$$

where v_1 is the envelope and v_2 the carrier velocity. Again the non-linear nature of the wave is shown by the dependence of the amplitude, ψ_0, on the other parameters.

Third, the "sine-Gordon" equation

$$\phi_{tt}(x,t) - \phi_{xx}(x,t) + \sin\phi(x,t) = 0 \tag{6.a}$$

for which the soliton solution[2]

$$\phi_s(x,t) = 4\tan^{-1}\left(\exp\left[\frac{x-vt}{\sqrt{1-v^2}}\right]\right) \tag{6.b}$$

provides an example of a soliton having different asymptotic values as $\xi = x - vt$.

For contrast, let me mention a non-linear, dispersive equation which has a solitary wave solution which is not a soliton: the "non-linear Klein-Gordon" (or "ϕ^4 field theory") equation,[6]

$$\phi_{tt}(x,t) - \phi_{xx}(x,t) + \phi^3(x,t) - \phi(x,t) = 0 \tag{7.a}$$

which has the solitary wave but non-soliton solution

$$\phi(x,t) = \tanh\left(\frac{x - vt}{\sqrt{1 - v^2}}\right) \tag{7.b}$$

the form of which is quite similar to the soliton (6.b).

At this point you are hopefully all intrigued as to how I know that the solutions (4.b), (5.b) and (6.b) are solitons whereas that in (7.b), despite its similarity to (6.b), is not. Recall that to answer this question one must be able to construct a solution corresponding to two waves interacting; since these non-linear equations do not permit superposition, it is difficult to see how to do this analytically. In fact, in recent years, an elegant analytic technique has been developed[2,5,7,8] which, for equations (4.a), (5.a), and (6.a), does permit one to verify the soliton nature of these solutions. This "inverse scattering transform"[2,5,7,8] is the subject of the next section. But, and this is an important historical lesson, prior to the discovery of this analytic technique, computer experiments, in which initially widely separated solitary waves were scattered numerically, had already "proved" that these objects were solitons. Indeed, the word "soliton" itself was coined to describe the results of computer studies of the KdV equation,[1] and from the earliest days of solitons,[9] computer simulations have played an absolutely central and usually leading role[10] in the development of the subject. At present, computer simulations are still the only systematic method for studying solitons in two or more spatial dimensions. Thus it is important to recognize that this form of "experimental mathematics" is here to stay.[11]

III. ARE SOLITONS MATHEMATICALLY RESPECTABLE?

The remarkable properties of solitons remained frustrating curiosities of "experimental mathematics" until the development of a systematic procedure for solving a wide class of non-linear wave equations in one spatial dimension.[2,5,7,8] This method, called the "inverse scattering transform" - "IST" for short - solves a non-linear problem by a sequence of linear manipulations in the manner shown in Figure 1.

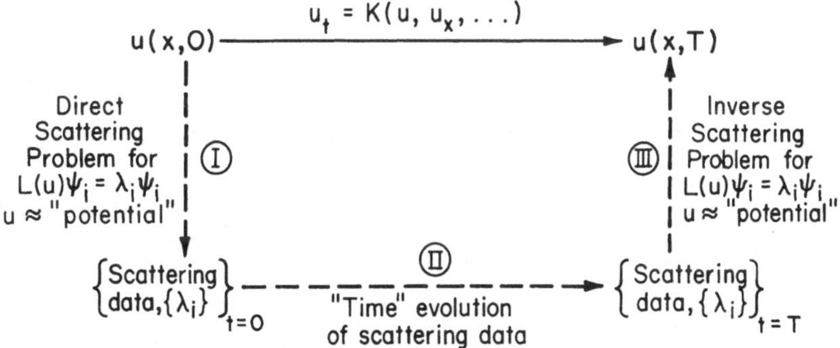

FIGURE 1. A Schematic Diagram of the Inverse Scattering Transform

The IST is conceptually analogous to the Fourier transform method for solving a linear wave equation. As an example, consider a "linear analog" of the KdV equation,

$$u_t(x,t) + u_{xxx}(x,t) = 0 \quad . \tag{8}$$

By introducing the Fourier transform - $\hat{u}(k,t)$ - of u with respect to x (the analog of step I in Figure 1) one obtains a diagonalized time evolution for $\hat{u}(k,t)$:

$$\hat{u}_t(k,t) = -ik^3\hat{u}(k,t) \quad . \tag{9}$$

Solving this equation for $\hat{u}(k,T)$ given $\hat{u}(k,0)$ is clearly the analog of step II in Figure 1. Finally, inverting the Fourier transform to find $u(x,T)$ corresponds to step III in Figure 1.

This analogy is useful, but it leaves one absolutely crucial question completely unanswered: what linear operator, $L(u)$, should one associate with a given non-linear equation, $u_t = K(u,u_x,\ldots)$? In the first application of the IST, the appropriate operator was found by an inspired leap of insight, and although there has been some progress in understanding the relation between K and L,[12-17] there is at present no constructive method for finding L given K. Thus the best way to understand the IST in more detail is to plunge into an example. Consider the KdV equation in the form

$$u_t \equiv K(u,u_x,\ldots) = -uu_x - u_{xxx} \quad . \tag{10}$$

It has been shown[2,7] that the appropriate associated linear eigenvalue problem for this equation is, in physicists' language, the Schrödinger equation (!) with $u(x,t)$ as a potential: namely,

$$L(u)\psi = \lambda\psi \quad , \quad \text{where } L(u) \equiv -\partial_x^2 - \frac{1}{6}u(x,t) \quad . \tag{11.a,b}$$

As far as (11.a) is considered, the "time" coordinate in $u(x,t)$ is simply a parameter: that is, the operator L contains "time" dependence only through u, and this dependence is in turn described by $K(u,u_x\ldots)$ in (10). The spectrum of L is the set of eigenvalues $\{\lambda_i\}$ of (11.a); from elementary quantum mechanics, one knows that the spectrum is related to the scattering data - i.e., the set of bound state energies and the reflection coefficient, $R(\lambda)$ - associated with L. In principle both $\{\lambda_i\}$ and the corresponding set of eigenfunctions $\{\psi_i\}$ will depend on "time" as a parameter via (11.b). But if we wish to have "simple" time evolution, then a natural requirement is that the spectrum of L remain unchanged in time; let us study the consistency and implication of this "isospectral flow" condition. Clearly the (point spectrum[18]) $\{\lambda_i\}$ is invariant if there exists a unitary operator $U(t)$ such that

$$L(t) \equiv U(t)L(0)U^+(t) \tag{12}$$

since then, indicating explicitly only the parametric time dependence,

$$L(0)\psi_i(0) = \lambda_i(0)\psi_i(0) \tag{13.a}$$

or

$$U(t)L(0)U^+(t)U(t)\psi_i(0) = \lambda_i(0)U(t)\psi_i(0) \quad . \tag{13.b}$$

Thus with

$$\psi_i(t) \equiv U(t)\psi_i(0) \quad , \quad L(t)\psi_i(t) = \lambda_i(0)\psi_i(t) \quad , \tag{14.a,b}$$

which implies

$$d\lambda_i/dt = 0 \quad . \tag{15}$$

Since U is a unitary operator depending on a single parameter t, there exists an hermitian operator B such that

$$iU_t = BU \quad . \tag{16}$$

Then by direct calculation from (12) and 15.a), we find

$$iL_t = [B,L] \quad \text{and} \quad i\psi_t = B\psi \quad . \tag{17.a,b}$$

But recalling (11.b), we see that

$$iL_t = -\frac{i}{6}u_t = \frac{i}{6}(uu_x + u_{xxx}) \tag{18}$$

by (10). Hence consistency requires that the "Lax pair"[2,12] L and B satisfy

$$[B,L] = \frac{i}{6}(uu_x + u_{xxx}) \quad . \tag{19}$$

At the moment it is not at all clear what these formal manipulations signify. First, L and B are not specified by K, except through the constraint (19); we have asserted the form of L, can we find B? Second, although $d\lambda_j/dt = 0$, clearly (17.b) suggests $(d\psi_j/dt) \neq 0$; then how can the time evolution of the scattering data be simple? Let me treat the former question first: the solution for B - and I ask you to verify that it does satisfy (19) - is

$$B = -4i\partial_x^3 - iu(x,t)\partial_x - \frac{1}{2}u_x(x,t) \quad . \tag{20}$$

Now rephrase the latter question; since B is such a complicated operator, how can the scattering data evolve simply? From (17.b) and (20) it is clear that calculating $\psi_j(x;t)$ is difficult, but the scattering data are defined for $|x| \to \infty$ and we expect - it is certainly true for solitary wave solutions - that $u(x,t) \to 0$ as $|x| \to \infty$. Then as $|x| \to \infty$, B approaches the simple asymptotic form $-4i\partial_x^3$, and the scattering data will evolve simply in time. Explicitly, for bound states with $\lambda_j \equiv -\kappa_j^2$, the wave function $\psi_j(x;t)$ will asymptote, as $x \to +\infty$,[19] to

$$\psi_j(x,t) \underset{x \to +\infty}{\simeq} c_j(t)\exp[-\kappa_j x] \tag{21}$$

where κ_j is independent of t by isospectrality and where (17.b) implies, using the asymptotic form of B, that the normalization coefficient $c_j(t)$ in (21) obeys a trivial first order differential equation in time with solution

$$c_j(t) = c_j(0)\exp(4\kappa_j^3 t) \quad . \tag{22}$$

For scattering states, $\lambda = k^2 > 0$, the asymptotic forms are

$$a(t)e^{+ikx} + b(t)e^{-ikx} \underset{-\infty \leftarrow x}{\simeq} \psi_k(x;t) \underset{x \to +\infty}{\simeq} c(t)e^{+ikx} + d(t)e^{-ikx} \quad , \tag{23}$$

where k does not depend on t. Again using (17.b) with the asymptotic form of B yields the simple results

$$a(t) = a(0)e^{+4ik^3t} \quad , \quad c(t) = c(0)e^{+4ik^3t} \tag{24.a}$$

and

$$b(t) = b(0)e^{-4ik^3t} \quad , \qquad d(t) = d(0)e^{-4ik^3t} \tag{24.b}$$

so that, choosing $d(0) = 0$ and normalizing the incident flux, we find

$$R(t) \equiv b(t)/a(t) = R(0)\exp[-\,i8k^3t] \quad \text{and} \quad T(t) \equiv c(t)/a(t) = T(0) \tag{25.a,b}$$

where R and T are the standard reflection and transmission coefficients. Hence the scattering data do indeed have simple time evolution; given these data at t=0, it is trivial to calculate them at t = T. This completes step II of Figure 1. There remains only step III, the inversion of the transform; this is precisely the inverse scattering problem for the one-dimensional Schrödinger equation, which is solvable by the Gelfand-Levitan Marchenko linear integral equation[2,20,21] technique; about this I can say only that the scattering data required to reconstruct the potential are κ_i and c_i for each bound state and $|R(k)|$ and arg $R(k)$ for all real k.[22] Our analysis has shown that κ_i and $R(k)$ are constant in time, while c_i and arg $R(k)$ have simple time evolution. Thus the construction of $u(x,T)$ -- and hence the solution of the KdV equation -- by linear means is in principle complete.

Among the virtues of the IST is a direct explanation for the mysterious stability of the solitons of KdV. Suppose we consider the Schrödinger potential corresponding to the single soliton solution (4.b) at t = 0 . From (11.b) we have

$$V_{SCH}(x) = -\frac{1}{6}\,u(x,0) = -\frac{v}{2}\,\text{sech}^2\left(\frac{\sqrt{v}}{2}\,x\right) \tag{26}$$

so that V_{SCH} is one of the special potentials with <u>no</u> reflection at any k - $R(k;0) = 0$ -- and precisely one bound state at $\lambda \equiv -\kappa^2 = -\,v/4$.[23] Since $R(k;0) = 0$, $R(k;t) = 0$ for all t by (25.a) and the "potential" always remains reflectionless and with precisely one bound state. This suggests (the correct) association of bound states in V_{SCH} with solitons in $u(x,t)$; if $u(x,0)$ corresponds to a V_{SCH} with N bound states (at λ_i, $i=1,\ldots N$) , N solitons with velocities $v_i = -4\lambda_i (> 0)$ will emerge asymptotically. A less obvious (but also correct) association is that a non-zero reflection coefficient in V_{SCH} implies the existence of "radiation" - i.e., oscillations which spread out in time - in $u(x,t)$; thus an arbitrary initial pulse in KdV will break up into a definite number of solitons plus radiation as $T \to \infty$.

When the IST solution to the KdV equation appeared in 1967,[7] it was hailed as an elegant achievement, but the relevance of the IST as a method was questioned, since it applied to one equation only. For several years researchers struggled to find which linear eigenvalue problem -- i.e., which L -- could be associated with the non-linear evolution operator, K, for any of the other soliton equations. The key to solving this problem came from turning it around; namely, instead of going from K to L, pick a "natural" L and see which non-linear evolution equations leave the spectrum of L invariant in time (isospectrality). Since the Schrödinger equation worked for KdV, a natural generalization is to look at a Dirac-like linear eigenvalue problem.[5,8] In one space dimension a Dirac-like system consists of two coupled equations, which can be expressed in a 2 x 2 matrix formalism in which the linear operator L and eigenfunctions v are

$$L_2 \equiv \begin{pmatrix} i\partial_x & -\,iq(x,t) \\ ir(x,t) & -\,i\partial_x \end{pmatrix} \quad \text{and} \quad v = \begin{pmatrix} v_1 \\ v_2 \end{pmatrix} \tag{27}$$

so that the eigenvalue equation reads, for eigenvalue "ξ" ,

$$L_2 v = \xi v \quad . \tag{28}$$

This eigenvalue problem is known as the "Zakharov-Shabat"[5] or "AKNS"[8] system. The two potentials $q(x,t)$ and $r(x,t)$ are the analogs of $u(x,t)$ in (13.b) and thus will satisfy some non-linear evolution equation. Of course, r and q need not be independent; when $r = q^*$, for example, the resulting eigenvalue problem is appropriate to the case of the non-linear Schrödinger equation.[5,8] From our discussion of the KdV case, the method of attack is clear: we should demand $d\xi/dt = 0$ and seek a (2 x 2) operator B such that[24] $v_t = Bv$. Without giving any details, we simply note that B's have been found such that all of the soliton equations (4.a) - (6.a) can be solved by the Z-S/AKNS system. Thus it incorporates much of what we know analytically about solitons and is therefore a good place to search for the more general structure which must underlie these remarkable equations.

Several recent articles have suggested that this hidden structure underlying the soliton equations is related to group theory.[14,16,25-29] To understand this suggestion, rewrite (27) as in the form $v_x = Av$, where the form of A follows from that of L_2 in (27). Demanding equality of mixed second derivatives - $v_{tx} = v_{xt}$ - leads to the "compatibility" conditon on A and B

$$A_t - B_x + [A,B] = 0 \quad . \tag{30}$$

Now A follows from (27) and B is as yet undetermined; thus

$$A = \begin{pmatrix} -i\xi & q \\ r & +i\xi \end{pmatrix} \quad \text{and} \quad B = \begin{pmatrix} \alpha & \beta \\ \gamma & \delta \end{pmatrix} \quad . \tag{31}$$

One can check that the compatibility condition and isospectrality imply $\delta = -\alpha$.[8] Thus A and B are both 2 x 2 traceless matrices and can be represented by the Pauli matrices, which suggests an underlying SU(2) or SO(3) algebra and, hence group. In fact, more detailed analysis indicates that the appropriate group is actually SL(2,R).[25,27-28] If we expand A and B in terms of the three matrices forming the basis for the SL(2,R) algebra, so that

$$A = \sum_{i=1}^{3} a_i X_i \quad \text{and} \quad B = \sum_{i=1}^{3} b_i X_i \tag{32}$$

where

$$[X_k, X_\ell] = c_{k\ell}{}^i X_i \tag{33}$$

with $c_{k\ell}{}^i$ the structure coefficients for SL(2,R),[27] then Eq. (30) can be written as

$$a_t^i - b_x^i + c_{k\ell}{}^i a^k b^\ell = 0 \quad . \tag{34}$$

Defining a Yang-Mills vector potential A_μ^i by $A_0^i = a^i$, $A_1^i = b^i$ where "0" is the time and "1" the space component, we see that (34) implies the vanishing of the non-abelian field strength tensor,

$$0 = f_{\mu\nu}^{\ i} \equiv \partial_\mu A_\nu^i - \partial_\nu A_\mu^i + c_{k\ell}^{\ i} A_\mu^k A_\nu^\ell \qquad \begin{matrix} i,k,\ell = 1,2,3 \\ \mu,\nu = 0,1 \end{matrix} \quad . \tag{35}$$

Thus, in a sense, <u>all</u> soliton equations which can be studied by the ZS/AKNS system are related to the gauge group "vacuum", $f_{\mu\nu}^i = 0$; hence, in principle, it appears possible to transform one equation to another via SL(2,R) gauge transformations.[28,29] Although this approach seems to offer a theoretical unification, it remains to be seen whether it has any calculational utility. Note that, in modern differential geometric/fiber bundle language, $f_{\mu\nu}^1$ is the curvature tensor on the manifold $R^2 \times SL(2,R)$; this suggests a possible role for differential geometry in soliton equations.[14,17,25]

IV. WHAT'S LEFT TO BE DONE?

I would be remiss if I did not at least mention three important additional topics related to solitons. First, for all the soliton equations we have studied there exist certain non-linear composition laws - called "Bäcklund transformations"[30] - that allow one to go from an explicit N soliton solutions to an N + 1 soliton solution. The role of these transformations in an underlying "theory" of soliton equations is at present only partially clarified.[28,30,31] Second, soliton - or, more generally, solitary wave - solutions to the partial differential equations corresponding to the "classical" limit of the equations of quantum field theories have been used to calculate, in various semi-classical schemes,[6,32] the bound states of these quantum theories. In the case of the sine-Gordon system, for example, these semi-classical soliton methods[33] are now known[34] to give the exact quantum spectrum. Finally, it has now been established[35] that soliton equations are infinite dimensional, completely integrable Hamiltonian systems. In fact, the transformation to the scattering data is precisely the canonical transformation that diagonalizes the Hamiltonian.[35,36] Complete integrability implies that the motion in phase space will follow closed orbits (tori) and will not be ergodic. This recent interpretation explains[37] one of the first striking results found in the field: namely, the "recurrences" of the initial state, indicating non-thermalization and hence non-ergodic behavior, observed by Fermi, Pasta, and Ulam in numerical studies of a non-linear lattice.[9]

Since this subject has a history of presenting profound puzzles which are eventually resolved in unanticipated ways, it is appropriate for me to close by mentioning some of the present open problems. As I have already indicated, the first important problem is to find a deep unification - be it through a group theoretic,differential geometric, or other approach - of soliton theory. In part, this search is motivated by the desire to solve a second -- and perhaps the outstanding -- problem in soliton physics: namely, to construct soliton solutions in two and more space dimensions. Here, although there has been considerable numerical and some analytic progress,[11,38] there is still no procedure analogous to the IST for solving even a limited number of non-linear problems constructively. Finally, a third important problem is that of perturbation on soliton systems. How much of the exotic and striking behavior of these systems remains if the system is modified slightly? How close, for example, is the ϕ^4 field theory to the sine-Gordon equation? These, and a wide variety of other interesting questions, guarantee that the subject of solitons will continue to fascinate, and frustrate, mathematicians and physicists for many years to come.

ACKOWLEDGEMENTS

It is a pleasure to thank Jim Corones, Andy Hanson, Dave McLaughlin, and Simon Ruijsenaars for many valuable discussions and comments and to ackowledge the hospitality of the Aspen Center for Physics, where part of this work was done.

REFERENCES

1. N. J. Zabusky and M. D. Kruskal, Phys. Rev. Lett. 15, 240-243 (1965).
2. The best general introduction to solitons remains the classic review by A. C. Scott, F. Y. F. Chu, and D. W. McLaughlin, in Proc. IEEE, 61 1443-1483 (1973).
3. D. J. Korteweg and G. de Vries, Phil. Mag. 39, 422-443, (1895).
4. V. I. Talanov, Zh. Eksp. Teor. Fiz. Pis. Red. 2, 218-222 (1965); trans JETP Lett. 2, 138-141 (1965); P. L. Kelley, Phys. Rev. Lett. 15, 1005-1008 (1965).
5. V. E. Zakharov and A. B. Shabat, Zh. Eksp. Teor. Fiz. 61, 118-134 (1971); trans. Soviet Physics JETP 34, 62-69 (1972).
6. For a good review see R. Rajaraman, Phys. Lett. 21C, 229-313 (1975).
7. C. S. Gardner, J. M. Greene, M. D. Kruskal, and R. M. Miura, Phys. Rev. Lett. 19 1095-97 (1967).
8. M. J. Ablowitz, D. J. Kaup, A. C. Newell, and H. Segur, Stud. In App. Math. 53 249-315 (1974).
9. E. Fermi, J. Pasta, and S. Ulam, "Studies of Nonlinear Problems. I", Los Alamos Scientific Laboratory Report, LA-1940 (1955).

10. A notable exception to this rule is the sine-Gordon equation, where the analytic two soliton solution was found prior to the numerical experiments which showed the soliton nature of the solution in (6.b). Compare A. Seeger, M. Donth, and A. Kochendörfer, Z. Phys. 134, 173-193 (1953) to J. K. Perring and T. H. R. Skyrme, Nuc. Phys. 31, 550-555 (1962).
11. For a recent review of the status of this "experimental mathematics," see V. G. Makhankov, Phys. Lett. 35C, 1-128 (197).
12. P. Lax, Comm. Pure and App. Math., 21, 467-490 (1968).
13. J. Corones and D. McLaughlin, Phys. Rev. A10, 2051-2062 (1974).
14. H. Wahlquist and F. Estabrook, J. Math. Phys. 16, 1-7 (1975).
15. G. Lamb, Phys. Rev. Lett. 37, 235-237 (1976).
16. J. Corones, B. Markovsky, and V. Rizov, J. Math. Phys. 18, 2207-2213 (1977).
17. F. Lund and T. Regge, Phys. Rev. D 14, 1524-1535 (1976); F. Lund, Phys. Rev. Lett. 38, 1175-1178 (1977).
18. More precisely, we shall consider only the time independence of the point spectrum of L; for the continuum, it is easiest to work directly with $R(\lambda)$.
19. Studying the asymptotic behavior as $x \to -\infty$ is equally simple but yields no new information; see ref. (2).
20. I. M. Gelfand and B. M. Levitan, Izvest. Akad. Nauk. SSSR Ser. Mat. 15 309-360 (1951) [Am. Math. Soc. Trans. 1, 253-304 (1953)].
21. V. A. Marchenko, Dokl. Akad. Nauk. SSR 72, 457-460 (1950); 104, 695-698 (1955); Z. S. Agranovich and V. A. Marchenko, The Inverse Problem of Scattering Theory (Gordon and Breach, New York, 1963).
22. V. Bargman, Rev. Mod. Phys. 21, 488-493 (1949); R. Jost and W. Kahn; Det. Kgl. Dan. Vidensk. Selsk. Mat. - Fys. Medd. 27, No. 9 (1953).
23. See, for example, P. M. Morse and H. Feshbach, Methods of Theoretical Physics, Part II, pp. 1649-1659 (McGraw-Hill, New York, 1953).
24. The absence of a factor of "i" here is purely conventional and is for notational convenience only.
25. R. Hermann, Phys. Rev. Lett. 36, 835-836 (1976).
26. J. Corones, J. Math. Phys. 18, 163-164 (1977).
27. M. Crampin, F. Pirani, P. Robinson, Lett. Math. Phys. 2, 15-19 (1977).
28. M. Crampin, Phys. Lett. 66A, 170-172 (1978).
29. J. Corones, "An Illustration of the Lie Group Framework for Soliton Equations: Generalizations of the Lund-Regge Model", J. Math. Phys., to appear.
30. Bäcklund Transformations, The Inverse Scattering Method, Solitons, and Their Applications, Robert M. Miura, ed., Lecture Notes in Mathematics 515, A. Dold and B. Eckmann, eds. (Springer Verlag, Berlin, 1976).
31. M. Wadati, H. Sanuki, K. Konno, Prog. Theor. Phys. 53, 419-436 (1975).
32. See, for example, R. Dashen, B. Hasslacher, and A. Neveu, Phys. Rev. D 10, 4114-4129, 4130-4138 (1975); J. Goldstone and R. Jackiw, Phys. Rev. D 11, 1486-1498 (1975); A. Neveu and J. Gervais, eds, Phys. Lett. 23C, 237-374 (1976).
33. R. Dashen, B. Hasslacher, and A. Neveu, Phys. Rev. D 11, 3424-3450 (1975).
34. A. Luther, Phys. Rev. B. 14, 2153-2159 (1976).
35. V. Zakharov and L. Faddeev, Funkt. Anal. i Ego. Pril. 5, 18-27 (1971); (Transl.) Funct. Anal. and its Applic. 5, 280-287 (1972).
36. For a pedagogical presentation, see the article by H. Flaschka and D. McLaughlin, in ref. [30].
37. V. Zakharov, Zh. Eksp. Teor. Fiz. 65, 219-225 (1973); (Transl.) Sov. Phys. JETP 38, 108-110 (1974).
38. F. Tappert, private communication; see also J. C. Goldstein, Los Alamos Informal Report, LA 6833-MS (1977).

SEMI-CLASSICAL RESULTS IN QUANTUM FIELD THEORY*

R. Jackiw

Center for Theoretical Physics
Laboratory for Nuclear Science and Department of Physics
Massachusetts Institute of Technology
Cambridge, Massachusetts 02139

During the last few years, approximation techniques have developed for obtaining accurate results about quantum field theory without use of the perturbative expansion. Rather, we have solved field equations as classical non-linear, partial differential equations, and then extracted from the classical description of the dynamical system a semi-classical, quantal picture. The findings have been startling; effects are exposed which were not anticipated from the earlier, perturbative studies, and the semi-classical method yields qualitative and occasionally even numerical accounts of the various processes. Moreover, once the semi-classical description of a new phenomenon is established, it sometimes happens that a careful reexamination of the complete quantal model exhibits the effect without recourse to any approximation — with the hindsight of semi-classical knowledge some oversights have been corrected.

In the short time allotted to me and in front of this general audience, I can give only a very brief outline of the topic, which is quite broad and frequently technical. For further descriptions, please see the several reviews which are now available.[1]

Two categories of solutions to classical field equations have thus far been utilized as probes of the quantum theory. For the first kind, we seek stable, non-dissipative, finite-energy solutions. Frequently stability is insured by non-trivial topological structure of the field configurations at spatial infinity; in that case the solutions are called "topological solitons" — a term borrowed from engineers and applied mathematicians, but here used in a broader sense.[2]

It has been shown that these classical solutions indicate that in the corresponding quantum theory there are particle states which are neither the "elementary" particles of the model, nor bound states of a finite number of these. Rather, the quantum solitons are bound states of an infinite number of elementary particles; they are coherent, collective excitations, and as such they exhibit many remarkable properties. They carry quantum numbers, whose conservation insures the solitons' stability against decay; but these conserved quantities need not arise from a Noether symmetry; instead they can be manifestations of the non-trivial topological structure. The quantum solitons sometimes are fermions with half-integer spin even though the underlying field theory is constructed from bosons with integral spin.[3]

Soliton solutions with interesting topological properties have thus far been found only in field theories which are not realistic models for elementary particle physics. The models live in unphysical dimensionality of space; one-dimensional solitons are plentiful, but examples of two- and three-dimensional ones make use of unphysical internal symmetry groups. (In branches of physics not concerned with elementary particles some of these models have practical significance, and the soliton phenomena are experimentally verified.) Thus soliton investigations have been more important theoretically than phenomenologically; they provide theoretical laboratories where semi-classical procedures can be tested.

An example of an exact result which emerged from soliton investigations is a fascinating "duality" of description: the same physical model has two field theoretical versions. In the first, there are "elementary" particles associated with small fluctuations of the fundamental fields — the conventional states — also there are soliton particles. The second, dual description makes use of quite a different Lagrangian, and now the solitons are associated with the fields of this Lagrangian, while the previously elementary particles emerge as bound states. This is the situation with the sine-Gordon and massive Thirring models[4] — unphysical models in one spatial dimension, to be sure; but it has been conjectured that something similar may happen in a realistic model, and that confinement of quarks may be understood through these ideas.[5]

The second type of classical solution useful for understanding the quantum theory is obtained by continuing the equations to imaginary time ($t \to -i\tau$). When such solutions have finite action, they are called "instantons" or "pseudoparticles" and they are interpreted as evidence that in the quantum theory there are motions which are forbidden for classical, real-time, physics, but are allowed quantum mechanically. This interpretation is based on the well-known fact that the semi-classical wave-function describing tunnelling contains an exponent which is the action for the tunnelling motion, proceeding classically but for imaginary time.[6]

Instanton solutions occur in Yang-Mills theory, and the consequent semi-classical analysis has been principally carried out for that model, since it is the centerpiece of all contemporary descriptions of elementary particle phenomena. Moreover, we have realized that older discussions of the quantal Yang-Mills theory overlooked an important effect.

The question is how do quantum states respond to gauge transformations. Conventionally it was said that they are gauge invariant. Indeed the time component of the Yang-Mills field equation — the non-Abelian version of Gauss' law — implies that the generator of infinitesimal gauge transformations annihilates physical states. It then follows that any gauge transformation which can be constructed by iterating an infinitesimal one must indeed leave the state invariant. However, it has been now realized that in a non-Abelian gauge theory there are gauge transformations which cannot be continuously deformed to the identity. They are called "large" gauge transformations, in contrast to the "small" ones, which are deformable to the identity. Physical states need not be invariant under the large transformation; gauge invariance of the physical description is insured provided states are phase-invariant.

$$|\Psi\rangle \xrightarrow[\text{large gauge transformation}]{} e^{i\theta}|\Psi\rangle$$

In this way we have learned that quantum Yang-Mills theory is described by an angle, whose existence had been missed earlier.

It can be shown that regions in gauge-potential function-space which are gauge copies, with a large gauge transformation connecting them, are separated by a finite energy barrier. Therefore, one may expect that in the quantum theory there will be tunnelling between them. Indeed the classical instanton solution describes an evolution (for imaginary time) between two such gauge-potential configurations.[6] Explicit, numerical estimates of tunnelling amplitudes are obtained by describing the theory with a functional integral, continuing the integral to Euclidean space (imaginary time), and evaluating it approximately by an expansion around the instanton.[7] But, as we have seen, the circumstances which give rise to tunnelling may be recognized without any approximation.

One can wonder whether this phenomenon may be a gauge artifact, and would disappear if a choice of gauge were made which removes the freedom of performing gauge transformations; for example the Coulomb gauge, as in electrodynamics. However, it was noted that the transversality condition on the vector potential as well as additional boundary conditions, which conventionally define the Coulomb gauge, fail to specify the gauge uniquely in a non-Abelian theory.[8] Moreover, it was realized that this circumstance is related to the presence of precisely the large gauge transformations which distinguish Yang-Mills theory from an Abelian gauge theory.[9] A choice of gauge which does not run afoul of these difficulties has been recently given; in it one clearly sees the emergence of the non-trivial angle.[10]

The discovery of tunnelling in Yang-Mills theory has a phenomenologically important consequence. The theory of strong interactions built around the Yang-Mills model — quantum chromodynamics — seems to possess an unwanted symmetry, a chiral U(1) invariance. This symmetry is not realized in nature, even approximately, since it predicts the existence of a particle, approximately degenerate in mass with the pions. Since no such particle is known, a mechanism must be found which interferes with the action of this symmetry. The tunnelling phenomenon, combined with the axial-vector current anomaly, works exactly in the desired direction to remove the chiral U(1) symmetry from the model. The phenomenological significance of the angle θ is not clear. Indeed, since it is a CP violating parameter, its magnitude must be consistent with observed CP violation. At present we know of no theoretical way to determine θ; in the absence of a compelling model for CP violation, we also cannot say how large θ should be to agree with experiment.

The instanton has properties which have interested both physicists and mathematicians. Thus we are participating in a welcome conjunction of research between physics and mathematics. The instanton satisfies the field equations by being self-dual.

$$F_a^{\mu\nu} = {}^*F_a^{\mu\nu}$$

(Here $F_a^{\mu\nu}$ is the gauge field, obtained from the gauge potential A_a^{μ}.

$$F_a^{\mu\nu} = \partial^\mu A_a^\nu - \partial^\nu A_a^\mu + g\varepsilon_{abc} A_b^\mu A_c^\nu$$

The coupling constant is g; the gauge group is SU(2) with ε_{abc} its structure constants. The dual field ${}^*F_a^{\mu\nu}$ is constructed with the antisymmetric tensor.

$$*F^{\mu\nu}_{a} = \tfrac{1}{2}\varepsilon^{\mu\nu\alpha\beta} F_{\alpha\beta}$$

Since the dual gauge field always satisfies the Bianchi identities

$$\partial_{\mu} *F^{\mu\nu}_{a} + g\varepsilon_{abc} A_{b\mu} *F^{\mu\nu}_{c} = 0$$

the field equations

$$\partial_{\mu} F^{\mu\nu}_{a} = g\varepsilon_{abc} A_{b\mu} F^{\mu\nu}_{c} = 0$$

are necessarily satisfied by self-dual fields.) The non-trivial topological structure of the instanton is reflected in the fact that its Pontryagin index is non-zero. The Pontryagin index q is a topological invariant which characterizes 4-dimensional gauge fields.

$$q = \frac{1}{32\,\pi^2} \int d^4x \; *F^{\mu\nu}_{a} F_{a\mu\nu}$$

For a non-singular potential, which tends to a non-singular pure gauge at infinite x, q takes on integer values, and for the instanton solution which was found first $q=1$.[11] Subsequent analysis, both by physicists and mathematicians, categorized all self-dual instantons. It was shown that for $q=n$, the most general self-dual instanton depends on $8|n|-3$ parameters (for SU(2); analogous formulas hold for other groups).[12] Moreover, a $5|n|+4$ family of solutions was explicitly found by physicists,[13] while mathematicians gave a constructive procedure for determining the most general $8|n|-3$ parameter solution.[14]

Another point of contact between physicists and mathematicians arises when one examines the coupling of an external, linear system to the instanton; for example one studies the Dirac equation in the field of an instanton. (Small deformations of instantons satisfy equations of this type as well.) Very frequently these equations possess zero-eigenvalue modes, which are described by a famous mathematical result — the Atiyah-Singer index theorem. Equivalent to this theorem is a result familiar to physicists — the axial-vector current anomaly.[15]

This activity has shed much light on the mathematical structure of classical Yang-Mills theory, but the complete physical import of all these results has not yet been established. Only the original solution with unit Pontryagin index has played a role in practical, physical calculations of tunnelling; solutions with larger index appear to give only negligibly small corrections. Thus one open question concerns the proper role in physical theory of instantons with large Pontryagin index. Beyond this, one would like to know how to utilize other solutions of the classical Yang-Mills theory, both in Euclidean space and in Minkowski space, as well as "solutions" which fail to satisfy the equations at isolated singularities.[16]

References

1. S. Coleman, "Classical Lumps and Their Quantum Descendants" in New Phenomena in Sub-Nuclear Physics, A. Zichichi, ed., Plenum Press, New York (1977); S. Coleman, "Uses of Instantons" in N.N., A. Zichichi, ed., Plenum Press, New York (in press); R. Jackiw, "Quantum Meaning of Classical Field Theory", Reviews of Modern Physics, 49, 681 (1977); R. Jackiw, C. Nohl and C. Rebbi, "Classical and Semi-Classical Solutions of the Yang-Mills Theory" in Particles and Fields, D. Boal and A. Kamal, eds., Plenum Press, New York (1978).
2. For a review of engineering and applied mathematics uses of solitons see A. Scott, F. Chu and D. McLaughlin, Proc. IEEE 61, 1443 (1973).
3. R. Jackiw and C. Rebbi, Phys. Rev. Lett. 36, 1116 (1976); P. Hasenfratz and G. 't Hooft, Phys. Rev. Lett. 36, 1119 (1976);

A. Goldhaber, Phys. Rev. Lett. 36, 1122 (1976).

4. S. Coleman, Phys. Rev. D11, 2088 (1975).

5. G. 't Hooft, "Extended Objects in Gauge Field Theories" in Particles and Fields, D. Boal and A. Kamal, eds., Plenum Press, New York (1978), and Nucl. Phys. B138, 1 (1978).

6. R. Jackiw and C. Rebbi, Phys. Rev. Lett. 37, 172 (1976); C. Callan, R. Dashen and D. Gross, Phys. Lett. 63B, 334 (1976).

7. G. 't Hooft, Phys. Rev. Lett. 37, 8 (1976) and Phys. Rev. D14, 3432 (1976); F. Ore, Phys. Rev. D16, 2577 (1977).

8. S. Mandelstam, Lecture at American Physical Society Meeting, Washington DC (1977); V. Gribov, Lecture at 12th Winter School, Leningrad (1977), and Nucl. Phys. B139, 1 (1978).

9. R. Jackiw, I. Muzinich and C. Rebbi, Phys. Rev. D17, 1576 (1978); I. Singer, Comm. Math. Phys. 60, 7 (1978).

10. J. Goldstone and R. Jackiw, Phys. Lett. 74B, 81 (1978); V. Baluni and B. Grossman (unpublished); L. Faddeev, A. Izergin and V. Korepin (unpublished).

11. A. Belavin, A. Polyakov, A. Schwartz and Y. Tyupkin, Phys. Lett. 59B, 85 (1975).

12. A. Schwarz, Phys. Lett. 67B, 172 (1977); R. Jackiw and C. Rebbi, Phys. Lett. 67B, 189 (1977); M. Atiyah, N. Hitchin and I. Singer, Proc. Nat. Acad. Sci. USA 74, 2662 (1977).

13. R. Jackiw, C. Nohl and C. Rebbi, Phys. Rev. D15, 1642 (1977).

14. M. Atiyah, N. Hitchin, V. Drinfeld and Y. Manin, Phys. Lett. 65A, 185 (1978).

15. S. Coleman (unpublished); A. Schwarz, Phys. Lett. 67B, 172 (1977); J. Kiskis, Phys. Rev. D15, 2329 (1977); L. Brown, R. Carlitz and C. Lee, Phys. Rev. D16, 417 (1977); R. Jackiw and C. Rebbi, Phys. Rev. D16, 1052 (1977); N. Nielsen and B. Schroer, Nucl. Phys. B127, 493 (1977).

16. Some of these solutions and speculations concerning their physical significance are found in the last review cited in Ref. 1.

* This work is supported in part through funds provided by the U.S. DEPARTMENT OF ENERGY (DOE) under contract EY-76-C-02-3069.

SOME CLASSICAL SOLUTIONS OF SU (2) YANG-MILLS EQUATIONS

H. BACRY and J. NUYTS
Université d'Aix-Marseille II
Centre de Physique Théorique, CNRS, Marseille

1. Given a classical Maxwell field $f_{\mu\nu} = (\vec{e}, \vec{b})$ created by localized magnetic sources and satisfying $\vec{e}.\vec{b} = 0$, it is possible to associate with it solutions of the SU(2) - Yang-Mills equations in the following way :

Find real functions ω and g such that

$$f_{\mu\nu} = \partial_\mu \omega \, \partial_\nu g - \partial_\nu \omega \, \partial_\mu g \tag{1}$$

the Yang-Mills field associated with it is given by

$$F_{\mu\nu} = f_{\mu\nu} G \tag{2}$$

where

$$G = \begin{vmatrix} g & \sqrt{1-g^2}\, e^{-i\omega} \\ \sqrt{1-g^2}\, e^{i\omega} & -g \end{vmatrix} \tag{3}$$

$F_{\mu\nu}$ satisfies the uncoupled Yang-Mills equation and derives from the potential

$$A_\mu = -i \left[G , \partial_\mu G \right] \tag{4}$$

The Maxwell field $f_{\mu\nu}$ is gauge invariant.

2. As simplest examples, we have those corresponding to SO(2) - invariant magnetic sources for which ω can be chosen to be

$$\omega = \tan^{-1} \frac{x_2}{x_1} \tag{5}$$

Among them, the 't Hooft monopole corresponds to

$$g = \frac{x_3}{r} \qquad (r^2 = x_1^2 + x_2^2 + x_3^2) \tag{6}$$

The magnetic dipole is given by

$$\vec{b} = \left(\frac{3x_1x_3}{r^5} , \frac{3x_2x_3}{r^5} , \frac{3x_3^2 - r^2}{r^5} \right) \tag{7}$$

$$g = -\frac{\lambda}{r^3} \tag{8}$$

a result which is easily generalized to cylindrical multipoles.

3. Set of n monopoles. The result is simple : just add the Maxwell field associated with each monopole and add the corresponding G's . The Yang-Mills field is

$$F_{\mu\nu} = \left(\sum_i f_{\mu\nu}^{(i)}\right)\left(\sum_j G^{(j)}\right) \tag{9}$$

4. The ephemeron

Non static cases can also be investigated. Among them, the dipole ephemeron corresponds to the magnetic source

$$j_1 = 0 \quad , \quad j_2 = 0 \quad , \quad j_3 = -\partial_4\left[\delta^{(4)}(x)\right] \quad , \quad j_4 = \partial_3\left[\delta^4(x)\right] \tag{10}$$

a conserved current with cylindrical symmetry. The Yang-Mills field associated with the problem is given by (5) and

$$g = -\frac{\lambda}{\pi^2}\,\frac{x_1^2 + x_2^2}{x_1^2 + x_2^2 + x_3^2 + x_4^2}$$

For other details and examples, the reader is referred to :

H. Bacry and J. Nuyts Phys. Lett., to appear

H. Bacry and J. Nuyts Nucl. Phys. ,to appear

COLLAPSONS: NEW NONLINEAR EXCITATIONS

PHILIP B. BURT

Clemson University, Clemson, S.C. 29631

In the development of quantum field theory one of the principal ideas has been that the interaction can be considered small in some sense. In particular, scattering processes are represented by three elements-fields describing noninteracting particles in the distant past and future related by interpolating fields. In this way the assumption of smallness enters when the interpolating fields(or their matrix elements) approach the asymptotic fields(matrix elements) describing the noninteracting particles. By assumption, attention is confined to fields and interactions for which a portion of the interaction vanishes for asymptotically large times while a residual portion dresses the noninteracting particles, giving them their physical masses. This assumption is a modified expression of the classical idea of a controllable interaction. However, physical systems, through virtual processes, experience persistent self interactions at all times. Incorporation of this idea into the basic assumption of quantum field theory can be expected to lead to a broader perspective for the theory. The perturbation theory associated with switchable interactions, while not supplanted by this approach, becomes more peripheral.

A field theory with persistent self interactions must be intrinsically nonlinear. That is, if only one type of field is considered, the field equations must have interactions(nonlinearities) present at all times. This type of theory can then describe the persistent self interaction-hence ever present virtual processes-without reference to other fields. Only if the interaction strength is rigorously zero does the field equation become a free field equation. Such field theories already exist in classical physics. The solitary wave is an example of an excitation of a system which contains the interaction for all times. It represents an intrinsically nonlinear mode of the system. In relativistic quantum field theories there is an analogue of the classical solitary waves which has the properties described above(1,2). It is a solution of the nonlinear field equation independent of perturbative solutions and it is persistently self interacting. The basic field equations are nonlinear and dispersive but, like its classical counterpart, the new excitation, viewed as a wave, has constant phase velocity. The wave shape is independent of time. Unlike the classical solitary wave, some of these generalized solitary wave solutions reduce to free field solutions in the absence of self interactions. Even so, they are nonperturbative. The generalized solitary waves, since they are independent modes of the system, contribute new terms to the propagator. The amplitude for the system to make a transition from spacetime points $\check{y}$ to $\check{x}$ consists of a sum of terms

$$P(\check{x}-\check{y}) = P_{pert}(\check{x}-\check{y}) + P_{col}(\check{x}-\check{y}) \qquad (1)$$

where P_{pert} is the perturbation theory term and P_{col} is intrinsically nonlinear. A striking property of P_{col} is that the mass spectrum, obtained by finding the poles of P_{col} in momentum space, is, in most cases, independent of the self interaction coupconstants. The generalized solitary wave field operators describe a coherent many particle system. Thus, scattering amplitudes contain new, nonlinear contributions arising from states with mass $(2pn+1)m$, where n is an integer, m is the mass of the noninteracting field (or the physical mass if perturbation effects are included) and p is determined by the form of the interaction. The coupling constants enter P_{col} only in such a way that the relative probability for excitation of a given mass state depends on the self interaction strength. The fields are special limiting cases of nonlinear excitations which I have called collapsons(3,4). The collapson-a wave solution of a nonlinear, dispersive field equation-contains a nonlinear superposition of waves described by distinct wave vectors. These wave vectors can be classified into subsets which have the property that all vectors in a subset are either timelike or spacelike. If the wave vectors are timelike or if their space components are all parallel there is effectively only one wave vector in the subset. The multiple waves in the subset collapse to a single wave. Collapsons exist for both polynomial and non polynomial interactions. As an example, double collapsons of sine Gordon are

$$\partial_\mu \partial^\mu \phi + m^2 \sin\phi = 0 \; (m^2>0), \qquad \phi = 4\tan^{-1}\alpha(1-uv)(u-v)^{-1}, \tag{2}$$

$$u = \sum_i u_i \exp(\beta_i \check{k}_i \cdot \check{x}), \qquad v = \sum_i v_i \exp(\eta_i \check{p}_i \cdot \check{x}) \tag{3}$$

where u_i and v_i are arbitrary constants, the sum is arbitrary, $\check{k}\cdot\check{x} = k_0 x_0 - \vec{k}\cdot\vec{x}$ and

$$\beta_i = (-m^2/\check{k}_i^2)^{1/2}, \quad \eta_i = (-m^2/\check{p}_i^2)^{1/2}, \quad \check{p}_i^2 \neq 0, \quad \check{k}_i^2 \neq 0, \tag{4}$$

$$\beta_i \beta_j \check{k}_i \cdot \check{k}_j = -m^2 = \eta_i \eta_j \check{p}_i \cdot \check{p}_j, \quad \beta_i \eta_j \check{k}_i \cdot \check{p}_j = -(1-\alpha^2) m^2 (1+\alpha^2)^{-1} \tag{5}$$

Eq.(5) restricts the subsets as discussed above. The vectors $\vec{k}_i$ must be coplanar as must $\vec{p}_i$. There are three special cases to consider: $\check{k}_i, \check{p}_i$ spacelike; in a particular coordinate system $\beta_i \check{k}_i = (\omega_i, m, \omega_i, 0)$, $\eta_i \check{p}_i = (\Omega_i, m(1-\alpha^2)(1+\alpha^2)^{-1}, \Omega_i, 2\alpha m(1+\alpha^2)^{-1})$ for arbitrary $\omega_i, \Omega_i, \alpha$; thus for $|\alpha|>0$ the solutions are 3+1 dimensional multisolitons: $\check{k}_i$, $\check{p}_i$ timelike; let $\check{p}^2 = \check{k}^2 = m^2$; in the coordinate system in which $\beta\check{k} = (m,0,0,0)$, $\check{p}$ can be written $\eta\check{p} = (m(1+\xi^2)(1-\xi^2)^{-1}, \pm 2\xi(1-\xi^2)^{-1}, 0, 0)$, $\xi = i\alpha \neq 1$; these solutions have direct application in quantum field theory: $\check{k}_i$ timelike, $\check{p}_i$ spacelike; for $\beta\check{k} = (m,0,0,0)$ the unique $\eta\check{p} = (im(1-\alpha^2)(1+\alpha^2)^{-1}, m, im(1-\alpha^2)(1+\alpha^2)^{-1}, 0)$, α arbitrary. Other collapsons have been found with similar properties and similar applications(6).

References: (1) P.B. Burt, Acta Physica Polon. B7, 617(76); (2) P.B. Burt, "Nonperturbative Self Interactions, Solitary Waves and Others", NATO Adv. Stud. Inst., Istanbul 77 (A.O. Barut, ed, D. Reidel, 78); (3) P.B. Burt, Lett. Nuovo Cim. 13, 26(75); P.B. Burt and M. Sebhatu, Lett. Nuovo Cim. 13, 1o4(75); M. Sebhatu, Nuovo Cim. 33A, 568(76); B.A.P. Taylor, Clemson dissertation(78, unpublished); (4) P.B. Burt, Proc. Roy. Soc. (London) A359, 479(78) Lett. Nuovo Cim. 20, 501(77); P.B. Burt and B.A.P. Taylor, Acta Phys. Polon. B9, 335(78); (5) G.B. Whitham, preprint; (6) P.B. Burt, "Multidimensional Solutions of the Sine Gordon Equation (in press).

THE EXCITED STATES OF THE PRASAD-SOMMERFIELD SOLITON AND ITS SU(3)-GENERALIZATION

Richard KERNER

I.M.T.A.,Université P. et M. Curie, Paris, France.

The aim of this paper is to investigate the asymptotic behaviour of an SU(3) monopole-like soliton ([1]),([2]) in the Prasad-Sommerfield limit ([3]). We find a series of asymptotic expansions at infinity, which can be identified with the excited spherically symmetric states of the monopole. By matching as smoothly as possible these asymptotic solutions somewhere in between r = 0 and r = ∞, we get an approximation of a global solution. Then the energy levels of the corresponding excited states can be obtained by feeding in these solutions to the energy integral.

The Lagrangian of the model, proposed by 't Hooft ([4]), is :

$$\mathcal{L} = -\frac{1}{4}\,\mathrm{Tr}\,F_{\mu\nu}F^{\mu\nu} - \frac{1}{2}\,\mathrm{Tr}\,(D_\mu\phi)(D^\mu\phi) - \frac{\Lambda}{4}(\mathrm{Tr}\,\phi^2 - v^2)^2 \tag{1}$$

where

$$F_{\mu\nu} = \partial_\mu A_\nu - \partial_\nu A_\mu + \frac{e}{2i}\left[A_\mu, A_\nu\right] \quad , \quad D_\mu\phi = \partial_\mu\phi + \frac{e}{i}\left[A_\mu, \phi\right] \tag{2}$$

The fields A_μ and ϕ are Lie-algebra-valued; they can be expressed in some basis as $A_\mu = A^a_\mu L_a$, $\phi = \phi^a L_a$, with $|L_a, L_b| = C^c_{ab} L_c$ being the commutation relations of the Lie algebra of the gauge group. For SU(3) we tale $L_a = \lambda_a$, the Gell-Mann matrices, a=1,2,...8, and we seek a static solution in the form

$$A_o = 0 \quad A_k = \sum_{i,j=1}^{3} \frac{\epsilon_{kij}\,\lambda_i\,x_j}{er^2}\left[K(r)-1\right] \; , \; \phi = a_s \sum_{k=1}^{3} \frac{\lambda_k x_k}{er^2} H(r) + b_s \lambda_8 \frac{G(r)}{er^2} \tag{3}$$

here, $r = \sqrt{x^k x_k}$, i,j = 1,2,3; s=1,2,3; $a_s^2 + b_s^2 = 1$ and $a_1 = 1$, $a_2 = \frac{1}{2}$, $a_3 = \frac{\sqrt{3}}{2}$.

The three different values of a_s correspond to different embeddings of SU(2) into SU(3).(for details see([1])).

The equations resulting from the variational principle applied to the Lagrangian density (1) are, in the limit when $\Lambda \to 0$, the following :

$$r^2K'' = K\left(K^2 + a_s^2\,H^2 - 1\right) \quad , \quad r^2H'' = 2HK^2 \quad , \quad r^2G'' = 0 \tag{4}$$

(here $K' = \frac{dK}{dr}$, etc).

We want to find the solutions of the system (4) giving <u>finite</u> value to the energy (Lagrangian) integral :

$$L = -\frac{4\Pi}{e^2}\int_0^\infty dr\left[K'^2 + \frac{(1-K^2)^2}{2r^2} + \frac{a_s^2(rH'-H)^2}{2r^2} + \frac{b_s^2(rG'-G)^2}{2r^2} + \frac{a_s^2\,H^2K^2}{r^2}\right] \tag{5}$$

At $r \to 0$ the only behaviour giving no infinities in (5) is $K \underset{r\to 0}{\to} 1 + Ar^2$, $H \underset{r\to 0}{\to} Br^2$, whereas $G = G_o r$ and does not contribute to the energy integral at all.

A and B are arbitrary constants up to now. At infinity we have $K \underset{r\to\infty}{\to} 0(\frac{1}{r})$ and $H \underset{r\to\infty}{\to} C_r + D + 0(\frac{1}{r})$. It is important to note that in 't Hooft's case D = 0, whereas in Prasad-Sommerfield limit it is arbitrary. If we put now this asymptotic form of H(r) into the first equation of (4), we get up to higher orders, a linear equation valid for $r \to \infty$:

$$K'' = K\left[a_s^2 C^2 + \frac{2a_s^2 C D}{r} + \frac{a_s^2 D^2 - 1}{r^2}\right] \tag{6}$$

which can be identified as Whittaker's equation of the second limit if we put $2a_s Cr = z$, $\lambda = -a_s D$, $\mu^2 = a_s^2 D^2 - \frac{3}{4}$, so that

$$\frac{d^2K}{dz^2} = K\left[\frac{1}{4} - \frac{\lambda}{z} + \frac{\mu^2 - \frac{1}{4}}{z^2}\right] \tag{7}$$

The solutions of this equations are well known ([5]) :

$$K_1(\lambda,\mu,z) = z^{\mu+1/2} e^{-z/2}\Phi(\mu - \lambda + \tfrac{1}{2}, 2\mu+1; z) \quad ; \quad K_2(\lambda,\mu;z) = K_1(\lambda,-\mu;z) \tag{8}$$

where $\Phi(\alpha,\beta;z) = 1 + \frac{\alpha}{\beta}z + \frac{\alpha(\alpha+1)}{\beta(\beta+1)}\frac{z^2}{2!} + \ldots$ is the confluent hypergeometric function. It is easy to show that only one of the two independent solutions (8) gives no infinity in the energy integral, subject to the condition $\mu - \lambda + \frac{1}{2} = -n$, a negative integer. This gives $-\lambda = a_s D = -\frac{n^2+n+1}{2n+1}$ and $\mu = -\frac{n^2+n-1/2}{2n+1}$.

It has to be inderlined that though these solutions are only asymptotic, the argument leading to the quantization of the constant D is an exact statement, because it holds to any degree of approximation.

Having at our disposal the asymptotic forms of the solutions near $r = 0$ and $r = \infty$ we can match them together in between in order to obtain an approximate global solution and evaluate the energy level.

If we match in a continuous (C°-class) way these asymptotic forms of our solutions at some finite points, say $r = R_K$ for $K(r)$ and $z = R_H$ for $H(r)$, the energy integral will depend only on R_K and R_H. There exists a unique choice of the parameters R_K and R_H giving the minimum value of the energy integral; this gives the best approximation possible to the exact solution. We have been able to perform this programme for the first three levels; here are the results (in units $\frac{4\pi C}{e^2}$) :

a_s \ n	1	2	3
1	1.037	3.018	5.420
1/2	0.676	2.333	3.943
$\sqrt{3}/2$	0.919	2.790	4.928

These are, of course, the upper bounds for the exact values. If we visualize this table on a graph, it looks like if the energy was a linear function of n, with the value of a_s fixing different slopes.

References

1. A. Sinha, Phys. Rev. D, Vol. 14 N°8 (1976

2. A. Chakrabarti, Ann. Inst. H. Poincaré, Vol. 23, N°2, (1975)

3. Prasad and Sommerfield, Phys. Rev. Lett. 35, 151 (1975)

4. 't Hooft G., Nucl. Phys. B, 79, 276 (1974)

5. Gradshtein and Ryzhik, Tables of Integrals, Pergamon Press, London 1965.

GEOMETRY FOR SOLITONS AND INVERSE SCATTERING

Fernando Lund

Departamento de Física, Universidad de Chile,
Santiago, Chile.

There have been a number of developments [1] in the past couple of years that seem to indicate that a geometric, as a complement to an analytic, way of looking at non-linear problems can be very fruitful. These developments have to do mainly with dynamical systems that are completely integrable and they relate mostly to the so-called Inverse Scattering Method in which a non-linear equation is regarded as the compatibility condition of a linear system of equations.

One of the several geometric paths that can be taken is the following:[2] consider a two-dimensional surface isometrically embedded in a three dimensional euclidean space. The surface is equipped with an orthonormal tangent frame consisting of two (unit) tangent vectors and the unit normal at each point. This tangent frame satisfies an overdetermined linear system of equations, the equations of Gauss-Weingarten. The coefficients of this linear system depend on the metric and extrinsic curvature (also called second fundamental form) of the surface. The compatibility conditions that the overdetermined system of Gauss-Weingarten must satisfy are in general a set of non-linear equations in which t the unknowns are the metric and extrinsic curvature. In this manner one obtains in a natural way a non linear equation and an associated linear system in one package.

The simplest example occurs under certain simplifying conditions which can be described, very roughly, as saying that the surface has constant curvature. In this case both the metric and the extrinsic curvature can be described by one function of two variables, say θ (x,t), and the non-linear compatibility conditions reduce to the sine-Gordon equation. The Gauss-Weingarten equations are nothing else than the linear equations found by Ablowitz <u>et al</u>[3] to solve the Sine-Gordon system through the inverse scattering method.

Relaxing the condition of constant curvature leads to a generalization of the Sine-Gordon equation described by the Lagrangian

$$\mathcal{L} = \frac{1}{2} (\partial_\mu \theta)^2 - \frac{1}{2} \sin^2 \theta + \frac{1}{2} \cot^2 \theta \, (\partial_\mu \lambda)^2 \qquad (1)$$

in which the dynamical objects are two scalar fields in two-dimensional spacetime. From what has been said above, there is a linear eigenvalue problem of which the equations of motion following from (1) are the integrability condition. This eigenvalue problem can be treated [4] according to the standard scattering techniques, leading to a solution of the initial value problem for (1). This system, apart from being of interest in its own right, arises in the study of the O(4) nonlinear σ-model[5], in a

U(2) theory of massless fermions with chiral invariance[6], and in a theory of strings interacting through a scalar field.[7]

The idea of using the equations obeyed by the tangent frame to a surface in order to determine a linear problem associated with a nonlinear equation has been used also by Maison[8] in connection with a stationary, axially symmetric gravitational field obeying the Einstein equations and by Jevichy and Papanicolaou[9] in a study of the continuous Heisenberg spin chain.

ACKNOWLEDGMENTS

A travel grant from the University of Chile is gratefully acknowledged.

REFERENCES

1. See for instance the Proceedings of the International Symposium of Nonlinear Evolution Equations Solvable via the Inverse Scattering Transformation, Accademia dei Lincei, Rome, June 15-18, 1977, F. Calogero Ed. Also L.Faddeev, talk given at Princeton, Spring 1978. G.E. Lamb Jr. Phys. Rev.Lett. 37 , 235 (1976); J.Math.Phys. 18 , 1654 (1977).

2. F.Lund, "Solitons and Geometry", Proceedings of the NATO ASI on "Nonlinear Equations in Physics and Mathematics; A.O. Barut Ed. Reidel-Dordrecht 1978.

3. M.Ablowitz, D.J. Kanp, A.C. Newell and H.Segur Phys.Rev.Lett 30 (1973) and Stud.Appl. Math. 53 , 249 (1974).

4. F.Lund, Phys.Rev.Lett. 38 , 1175 (1977). "Classically Solvable Field Theory Model", to be published in Annals of Physics.

5. K. Pohlmeyer, Commun.Math.Phys. 46 , 207 (1976) F.Lund, Phys.Rev. D15 , 1540 (1977).

6. A.Neven and N.Papanicolaou, Commun. Math.Phys. 58 , 31 (1978).

7. F.Lund and T.Regge, Phys.Rev. D14 , 1524 (1976).

8. D.Maison, Max Planck preprints, April 1978.

9. A. Jevichs and N.Papanicolaou, Princeton preprint, July 1978.

INSTANTONS AND EMBEDDINGS

C. Meyers, M. de Roo and P. Sorba.

The set of gauge field configurations of a gauge group G contains, as particular cases, gauge field configurations of subgroups of G. In certain applications it is important to know whether or not a gauge field corresponding to a group G is reducible to some smaller group included in G. In particular, problems of this kind have been encountered in the construction of instantons (solutions of the Euclidean equations of motion of pure Yang-Mills theories) for gauge groups other than SU(2). We shall use group-theoretical methods to study various aspects of the problem of embeddings of gauge groups.

1. DEFINITION AND GENERAL PROPERTIES OF EMBEDDINGS

The mathematical problem of embedding a simple algebra $\tilde{G}$ in a simple algebra G has been discussed by Dynkin (Math. Sb 30 (1952) 349, Amer. Math. Soc. Transl., Sec. 2, Vol. 6, p. III).

A faithful embedding of an algebra $\tilde{G}$ in an algebra G is defined by an injective mapping f of $\tilde{G}$ into G: $\tilde{X} \to f(\tilde{X}) \in G$ for every $\tilde{X} \in \tilde{G}$ such that

$$f\left([\tilde{x},\tilde{y}]\right) = [f(\tilde{x}), f(\tilde{y})]$$

We have the relation

$$\left(f(\tilde{x}), f(\tilde{y})\right) = j_f\,(\tilde{x},\tilde{y}) \qquad \tilde{x},\tilde{y} \in \tilde{G}$$

in which $(\tilde{X},\tilde{Y}) = \mathrm{Tr}\, \mathrm{ad}\tilde{X}\, \mathrm{ad}\tilde{Y}$ is the value of the Killing form for $\tilde{X}$ and $\tilde{Y}$. This relation determines a scalar factor j_f independent of $\tilde{X},\tilde{Y}$ which is called the Dynkin index of the embedding.

Now let us define what we mean by an embedding of a gauge field. Let $A_\mu = A_\mu^\alpha(x)X_\alpha$ be a gauge potential on the Euclidean space E^4, defined by analytic functions $A_\mu^\alpha(x)$, where X_α are the generators of a compact group G. Then the gauge potential $A_\mu(x)$ of G is an embedding of $\tilde{G}$ in G (or is reducible to $\tilde{G}$) if there exists a gauge in which $A_\mu(x)$ belongs to the algebra of the proper subgroup $\tilde{G}$ of G.

If the gauge field A_μ is an embedding of $\tilde{G}$ in G the Dynkin index of the embedding, $j_{\tilde{G}/G}$, plays an important role in the calculation of the topological charge. The charge will be $q = j_{\tilde{G}/G}\, q_1$, where q_1 is the charge of A considered as a $\tilde{G}$ gauge field (K.M. Bitar and P. Sorba, Phys. Rev. D16 (1977) 431; C. Meyers, M. de Roo and P. Sorba, CERN preprint TH.2562 (1978)).

GENERAL CHARACTERIZATION OF EMBEDDINGS

Embeddings can be characterized by the following theorem (C. Meyers, M. de Roo and P. Sorba, Nuclear Phys. B140 (1978) 533).

Theorem: Let $\mathfrak{F}$ be a G gauge field, where G is a compact group, on a simply connected open set of E^4 with an analytic gauge potential $A_\mu(x)$. If the field strength tensor $F_{\mu\nu}$ and its covariant derivatives $D_\rho F_{\mu\nu}$, $D_\rho D_\sigma F_{\mu\nu}$, ... can be written in terms of a proper subalgebra $\mathcal{M}$ of $\mathcal{G}$, the Lie Algebra of G, then there exists a gauge in which $A_\mu(x)$ can be written in terms of $\mathcal{M}$, i.e., the field $\mathfrak{F}$ is an embedding of the subgroup M of G in this open set.

The theorem is general in the sense that it refers to quite arbitrary gauge field configurations. We have shown that the conditions of the theorem stated above are in fact sufficient to generalize it to all of S^4: on S^4 it is therefore sufficient to consider F and its gauge derivatives in one simply connected open region.

PRACTICAL CRITERIA FOR THE REDUCIBILITY OF GAUGE FIELD CONFIGURATIONS

In the above-mentioned paper, we gave necessary and sufficient conditions for the irreducibility of an SU(3) gauge field configuration. This result was subsequently applied to prove that certain SU(3) instanton solutions were indeed irreducible (F.A. Bais and H.A. Weldon, Univ. Pennsylvania Preprints, 1978). Although the complete generalization of these SU(3) results to arbitrary groups are complicated, some of these properties can be extended (C. Meyers, M. de Roo and P. Sorba, CERN preprint TH.2562 (1978)).

DESCRIPTION OF EMBEDDINGS IN THE PARAMETER SPACE OF INSTANTONS

A method for constructing all self-dual solutions to the Euclidean Yang-Mills equations for compact Lie groups has been given by Atiyah et al. (Phys. Letters 65A (1978) 185). The matrix formulation of this construction can be given which emphasizes the potential properties of the parameter space (C. Meyers and M. de Roo, CERN preprint TH.2543 (1978)). In this approach any G instanton solution is considered as an embedding of an SU(n) instanton solution. It appears that in this general construction of self-dual solutions the reducibility of the gauge field implies certain symmetry conditions on the parameter space. We have determined these conditions for Sp(n), O(n) and SU(p) × SU(q) embeddings (C. Meyers, M. de Roo and P. Sorba, CERN Preprint TH.2562 (1978)).

A CONSISTENT QUANTIZATION PROCEDURE FOR NON-LINEAR PROBLEMS.

Willem M. de Muynck, Eindhoven University of Technology, Eindhoven, The Netherlands.

In order to overcome the ordering problem in the quantization of the Hamiltonian of a non-linear system, several authors [1,2] have proposed to remove ordering ambiguities by quantizing the Hamiltonian together with the generators of the symmetry group which leaves this Hamiltonian invariant. Although this method has been very successful, it was realized that it only works, provided the system possesses sufficient symmetry. The quantization method we propose here, is independent of the existence of symmetry in the Hamiltonian, since it essentially depends on the simultaneous quantization of the position observable and the generator w of a transformation group (the covariance group) which changes the Hamiltonian.

We choose the group of canonical point transformations as a covariance group. The one parameter group of classical point transformations, generated by the observable $w(q,p) = p_\ell \, \Lambda^\ell(q)$, $(q = (q^1, .., q^n)$, $p = (p_1,\dots,p_n)$; summation over ℓ is understood) can be written down explicitly as

$$q_\tau^k(q,p) = e^{\tau\Lambda^\ell(q)\,\partial_\ell}\, q^k, \quad p_{k_\tau}(q,p) = \frac{\partial q^j}{\partial q_\tau^k}\, p_j, \tag{1}$$

in which $-\infty < \tau < \infty$ and $\dfrac{\partial q^j}{\partial q_\tau^i}\, \Lambda^i(q_\tau) = \Lambda^j(q)$.

The circumstance that q_τ^k is depending on q only, suggests a quantization procedure in which there is a one-to-one correspondence between the family of classical observables $q_\tau^k(q)$ and a family of functions of the quantum mechanical position operator, obtained by replacing q^k by the operator Q^k. It is known already for a long time [3] that there exists a group of unitary transformations generated by the operator $W = \frac{1}{2}\{\Lambda^\ell(Q), P_\ell\}_+$, such that $Q_\tau^k = e^{i\tau W}\, Q^k\, e^{-i\tau W}$ precisely represents this family. Moreover the generator W is unique up to a gauge transformation. Under the transformation generated by W the position operators Q^k and the momentum operators P_k transform as

$$\begin{aligned} Q_\tau^k &= e^{i\tau W}\, Q^k\, e^{-i\tau W} = e^{\tau\Lambda^\ell(Q)\,\partial/\partial Q^\ell}\, Q^k \\ P_{k\tau} &= e^{i\tau W}\, P_k\, e^{-i\tau W} = \tfrac{1}{2}\{P_j, \frac{\partial Q^j}{\partial Q_\tau^k}\}_+ = \frac{\partial Q^j}{\partial Q_\tau^k}\, \hat{g}_\tau^{\frac{1}{4}}\, P_j\, \hat{g}_\tau^{-\frac{1}{4}} \\ \hat{g}_\tau^{k\ell} &= \sum_i \frac{\partial Q^k}{\partial Q_\tau^i}\, \frac{\partial Q^\ell}{\partial Q_\tau^i}, \quad \hat{g}_\tau = (\det \hat{g}_\tau^{k\ell})^{-1} \end{aligned} \tag{2}$$

If we now apply this quantization scheme to the free particle Hamiltonian by requiring that the quantization of q,p and w is covariant under the group of point transformations, we get the following scheme, which can be considered as a quantization scheme for the non-linear system described by the classical Hamiltonian $H_\tau^{cl} = \frac{1}{2}\, \hat{g}_\tau^{k\ell}(q)\, p_k p_\ell$:

$$\begin{array}{ccc} H_0^{cl}(q,p) = \frac{1}{2}\sum_i p_i p_i & \xrightarrow{\;w = p_\ell\Lambda^\ell(q)\;} & H_\tau^{cl}(q,p) = \frac{1}{2}\hat{g}_\tau^{k\ell}(q)\, p_k p_\ell \\ \downarrow & & \downarrow \\ H_0^{qm} = \frac{1}{2}\sum_i P_i P_i & \xrightarrow{\;W\;} & H_\tau^{qm} = \frac{1}{2}\hat{g}_\tau^{-\frac{1}{4}}\, P_k\, \hat{g}_\tau^{\frac{1}{4}}\, \hat{g}_\tau^{k\ell}(Q)\, \hat{g}_\tau^{\frac{1}{4}}\, P_\ell\, \hat{g}_\tau^{-\frac{1}{4}} \end{array} \tag{3}$$

The result, H_{τ}^{qm}, obtained in this way, strongly resembles the Hamiltonian derived by Podolsky[4] using the coordinate transformation $\bar{q}^k = \bar{q}^k(q)$ from Carthesian coordinates q to curvilinear coordinates $\bar{q}$, followed by a change of wave function according to $\psi(q(\bar{q})) \to \bar{\psi}(\bar{q}) = \bar{g}^{\frac{1}{4}}\ \psi(q(\bar{q}))$, in which $\bar{g}(\bar{q})^{\frac{1}{2}}$ is the Jacobian of the coordinate transformation. Indeed it is possible to demonstrate the relation between the Podolsky Hamiltonian and our consistent Hamiltonian H_{τ}^{qm}. Putting in the Podolsky Hamiltonian $\bar{q}^k = q_{\tau}^k$, $q_{\tau}^k(q)$ being given by (1), and performing subsequently a mere change of notation: $q_{\tau}^k \to q^k$ leads to our Hamiltonian $H_{-\tau}^{qm}$, apart from the fact that from the quantization scheme (3) it follows that in $H_{-\tau}^{qm}$ the momentum operator $P_k = -i\ \partial/\partial q^k$ refers to a Carthesian coordinate q. Also the Hilbertspace representation of the covariance group, given by

$$\psi_{-\tau}(q) = e^{-i\tau W}\ \psi(q) = \hat{g}_{-\tau}^{\frac{1}{4}}(q)\ \psi(q_{-\tau}(q)) \tag{4}$$

clearly corresponds with the above-mentioned Podolsky transformation.
For suitable functions $\Lambda^{\ell}(q)$ the point transformations constitute a group G of automorphisms of configuration space R^n. This group can be implemented into a system of imprimitivity, based on configuration space R^n, consisting of the spectral measure $E^Q(\Delta) = \int_{\Delta} dE^Q$, $\Delta \subset R^n$, corresponding with the operators Q_{τ}^k (2) according to

$$Q_{\tau}^k = \int_{R^n} q_{\tau}^k(q)\ dE^Q \tag{5}$$

and satisfying the equality

$$e^{-i\tau W}\ E^Q(\Delta)\ e^{i\tau W} = E^Q(\Delta_{\tau}),\ q_{\tau} \in \Delta_{\tau}\ \text{iff}\ q \in \Delta \tag{6}$$

We shall call (5) and (6) a configuration space system of imprimitivity. The requirement that the quantization of the position variable is covariant under point transformations is equivalent with the existence of this configuration space system of imprimitivity. So we might start from this imprimitivity system, satisfying (6), in order to define a position observable [5]. It is interesting to note that the transformations (1) do not define an analogous imprimitivity system, related to the momentum observable. So our quantization procedure provides the position observable with an exceptional status.

References.

1. J.M. Charap, Journ.Phys. A 6, 393 (1973)
2. M. Omote, H. Sato, Progr.Theor.Phys. 47, 1367 (1972)
3. B.S. De Witt, Phys.Rev. 85, 653 (1952), Rev.Mod.Phys. 29, 377 (1957)
4. B. Podolsky, Phys.Rev. 32, 812 (1928)
5. C. Piron, Foundations of Quantum Physics, Benjamin, Inc. 1976, chapt. 5

A LIENARD-WIECHERT TYPE OF SOLUTION IN SU(2)

D.Villarroel

Departamento de Física, Universidad de Chile
Casilla 5487, Santiago, Chile.

The Yang-Mills equation for the SU(2) group in Minkowski space-time are

$$\partial^{\mu} F^{a}_{\mu\nu} + \varepsilon^{abc} A^{b\mu} F^{c}_{\mu\nu} = 0 \quad , \qquad (1)$$

where

$$F^{a}_{\mu\nu} = \partial_{\mu} A^{a}_{\nu} - \partial_{\nu} A^{a}_{\mu} + \varepsilon^{abc} A^{b}_{\mu} A^{c}_{\nu} \; . \qquad (2)$$

Greek indexes refer to space-time and take the values 0,1,2,3. The SU(2) indexes are denoted by a,b,c and run from 1 to 3. We also introduce indexes r,s,t running from 1 to 3 , to denote the space indexes of space-time. The metric is $g_{\mu\nu}$ = diag (-,+,+,+).

In order to find a class of solution of (1), we introduce the following symbol $\eta'^{a}_{\mu\nu}$: $\eta'^{a}_{rs} = \varepsilon_{ars}$, $\eta'^{a}_{or} = -i\delta_{ar}$, $\eta'^{a}_{ro} = i\delta_{ar}$, $\eta'^{a}_{oo} = 0$. Then putting $A^{a}_{\mu} = \eta'^{a}_{\mu\nu} \partial^{\nu}\phi$, eq. (1) is written as

$$\eta'^{a}_{\nu\sigma} \{ \partial^{\sigma} (\partial^{2}\phi - \phi_{\omega}\phi^{\omega}) + 2\phi^{\sigma} (\partial^{2}\phi - \phi_{\omega}\phi^{\omega}) \} = 0 \qquad (3)$$

where , $\partial^{2} = \partial_{\mu}\partial^{\mu}$, $\phi_{\omega} = \partial_{\omega}\phi$, $\phi_{\mu\nu} = \partial_{\mu}\partial_{\nu}\phi$. Introducing now a new scalar ψ by mean of $\psi = e^{-\phi}$, it is a simple matter to see that every solution of

$$\partial^{2}\psi = c\psi^{3} \quad , \qquad (4)$$

for arbitrary constant c , generates a complex solution of the Yang-Mills equations (1).

Because of the close analogy of the Yang-Mills fields with ordinary electrodynamics, it is natural to ask about the existence of a singular solution in SU(2) , which would be the analog of the Lienard-Wiechert potentials of the electron. Let us consider a solution of this type in the present formalism : Let z (τ) be an arbitrary timelike world-line parametrized by its proper time τ. Let x be an arbitrary point and $z_{\mu}(\tau)$ its corresponding retarded position, i.e. the point where the wordline cuts the null cone drawn from x into the past. Let v_{μ} be the velocity at the retarded point and $\rho \; (x_{\mu} - z_{\mu}) v^{\mu}$, then the scalar

$$\psi = \frac{1}{\rho} \qquad (5)$$

is a solution of (4) with c=0. This solution has the pathological property of giving rise to an identically vanishing rate of radiated energy. In fact, the energy-momentum

$$T_{\mu\nu} = F^{a}{}_{\mu}{}^{\gamma} F^{a}{}_{\gamma\nu} + \frac{1}{4} g_{\mu\nu} F^{a}{}_{\sigma\omega} F^{a\sigma\omega} , \tag{6}$$

is in the present formalism given by

$$T_{\mu\nu} = 2\phi_{\mu}\phi_{\nu} (\phi_{\omega}\phi^{\omega} - \partial^{2}\phi) + 2\phi_{\mu\nu}(\phi_{\omega}\phi^{\omega} - \partial^{2}\phi)$$
$$- \frac{1}{2} g_{\mu\nu}(\phi_{\omega}\phi^{\omega} - \partial^{2}\phi)(\phi_{\sigma}\phi^{\sigma} + \partial^{2}\phi) , \tag{7}$$

and it turns out to vanish identically for (5). In order to obtain a solution with $T_{\mu\nu} \neq 0$ we must choose a solution of (4) with $c \neq 0$.

However, it can be shown that a scalar solution of (4) with $c \neq 0$ that depends locally on retarded quantities does not exist.

Let us remark that the Lienard- Wiechert type of solution found by Kaku [1] in SU(4) , has also a vanishing energy-momentum tensor. This follows from the fact that Kaku's solution is an imbedding in SU(4) of the present one.

ACKNOWLEDGMENTS

The author acknowledges the financial support from the Vicerrectoría de Asuntos Académicos de la Universidad de Chile (SDCCA).

REFERENCES

[1] M.Kaku, Phys. Rev. D13 , 2881 (1976).

APPLICATION OF THE BOSON POLYNOMIALS OF U(n) to PHYSICAL PROBLEMS†

J. D. Louck
Theoretical Division, Los Alamos Scientific Laboratory, Los Alamos, New Mexico 87545

I. *Introduction*

The use of boson operators as a method for studying the representations of the unitary groups is well known from the original work of Jordan [1] and Schwinger [2]. The effectiveness of this technique is found in four basic properties of boson operators under unitary transformations: (i) the number operator is invariant; (ii) the boson operator commutation relations are form invariant; (iii) polynomial forms are mapped into polynomial forms; and (iv) there is a natural invariant inner product defined for pairs of polynomials.

The purpose of this talk is (i) to review the properties of a general class of polynomials in the boson operators which have been found useful for obtaining the explicit unitary irreducible representations (irreps) of the unitary group itself; and (ii) to show how these same polynomials provide a unified approach for obtaining the explicit solutions to several classic problems in physics and chemistry.

The literature relating to (i) above and to the problems alluded to in (ii) is enormous, and space permits us to list only selected references where further sources may be found. In keeping with the general review nature of this talk, no detailed proofs are given and no specific credits cited. (Confer, however, the listed references.)

II. *The* U(n) * U(n) *boson polynomials*

Let A denote the n x n matrix of boson operators, $A = (a_i^j)$, $(i,j = 1,2,\ldots,n)$, and $\bar{A}$ the matrix of conjugate bosons, $\bar{A} = (\bar{a}_i^j)$. The boson operators in each of the sets $\{a_i^j\}$ and $\{\bar{a}_i^j\}$, respectively, commute, and otherwise they satisfy the commutation relations $[\bar{a}_i^j, a_k^\ell] = \delta_{j\ell}\delta_{ik}$. In terms of this notation, it is the conjugate boson operators which annihilate the "vacuum" state $|0\rangle$, that is, $\bar{a}_i^j|0\rangle = 0$ all i,j.

We denote polynomials in the boson operators a_i^j and over the complex numbers by the notation P(A), and call such a polynomial a *boson polynomial*. The inner product of two such polynomials is written in two alternative forms, (P,P') or $\langle P|P'\rangle$, and is defined by

$$(P,P') = \langle P|P'\rangle \equiv \langle 0|P^*(\bar{A})P'(A)|0\rangle , \tag{1}$$

where P^* denotes the complex conjugate polynomial to P.

In applications to physical problems, it is often the case that the superscripts $j = 1,2,\ldots,n$ and the subscripts $i = 1,2,\ldots,n$ labelling the boson operators $\{a_i^j\}$ have distinct rôles, e.g., indices labelling particles and indices labelling components of operators relative to a laboratory frame. It is with this possibility in mind that we are led to consider unitary transformations of the boson matrix A of the form

$$A \to \tilde{U}AV \ , \ U, \ V\varepsilon \ U(n) \ . \tag{2}$$

Thus, for $V = I_n$, each column of A undergoes the same transformation $\tilde{U}$, while, for $U = I_n$, each row of A undergoes the same transformation V. The transposed matrix $\tilde{U}$ of U is used in Eq. (2) so that when the transformation (2) is followed by $A \to \tilde{U}'AV'$, the composed transformation is $A \to \widetilde{(U'U)}A(V'V)$.

We seek now to classify the set of all polynomials $\{P(A)\}$ according to their transformation properties under the linear substitution of boson operators given by Eq. (2). As a first step in this classification we observe that the mapping (2) carries homogeneous polynomials into homogeneous polynomials. Thus, the problem is reduced to one of classifying all polynomials which are homogeneous of degree N, $P(\lambda A) = \lambda^N P(A)$, where N is an arbitrary nonnegative integer.

The set of polynomials which are homogeneous of degree N in the n^2 bosons $\{a_i^j\}$ defines a vector space $V^{[N]}$ of dimension $\dim V^{[N]} = (n^2+N-1)!/(n^2-1)!N!$. Furthermore, $V^{[N]}$ is the carrier space of irrep $[N] = [N0\ldots0]$ of the unitary group $U(n^2)$. This representation of $U(n^2)$ is obtained from the linear transformation of polynomials of $V^{[N]}$ given by $p(\underset{\sim}{a}) \to p(\tilde{W}\underset{\sim}{a})$, each $W\varepsilon U(n^2)$, where $\underset{\sim}{a}$ denotes the column vector of boson operators: $\underset{\sim}{a} = \text{col}\ (a_1^1\ldots a_1^n;\ldots;a_n^1\ldots a_n^n)$. Observing that the transformation (2) may be written in the alternative form: $\underset{\sim}{a} \to (\tilde{U}\otimes\tilde{V})\underset{\sim}{a}$, where $\otimes$ denotes the matrix direct product, we may now state in group theoretical terms the classification problem for boson polynomials under the transformation (2): *Split the carrier space* $V^{[N]}$ *of irrep* $[N]$ *of* $U(n^2)$ *into a direct sum of carrier spaces of irreps of the direct product group* $U(n)\times U(n)\subset U(n^2)$.

In order to state the solution to this classification problem, let us first recall that the single-valued irreps of U(n) are in one-to-one correspondence with the n-tuples $[m_{1n}m_{2n}\ldots m_{nn}]$ in which the m_{in} are integers (positive, zero, or negative) which satisfy $m_{1n}\geqslant m_{2n}\geqslant\ldots\geqslant m_{nn}$. Thus, the inequivalent unitary matrix irreps of U(n) may be denoted by $D^{[m]} = \{D^{[m]}(U)|U\varepsilon U(n)\}$, where $[m] = [m_{1n}\ldots m_{nn}]$ runs over all n-tuples of ordered integers.

An abstract characterization of the splitting problem posed above is contained in the result: *Each irrep of* U(n)×U(n) *of the form* $D^{[m]}\otimes D^{[m]} = \{D^{[m]}(U)\otimes D^{[m]}(V)\,|\,U,\ V\varepsilon U(n)\}$, *where each set of irrep labels satisfies* $m_{1n}\geqslant\ldots\geqslant m_{nn}\geqslant 0$ *and* $m_{1n}+\ldots+m_{nn} = N$, *occurs exactly once in the restriction of irrep* [N] *of* $U(n^2)$ *to the subgroup* U(n)×U(n).

The result stated above implies that the space $V^{[N]}$ splits into a direct sum of (perpendicular) subspaces $H^{[m]}$, that is, $V^{[N]} = \sum_{[m]} \oplus H^{[m]}$, where $H^{[m]}$ is the carrier space of irrep $D^{[m]}\otimes D^{[m]}$ of U(n)×U(n). Our problem of classifying boson polynomials under the transformation (2) has now been reduced to determining those polynomials which belong to the space $H^{[m]}$. We call these polynomials U(n)*U(n) *boson polynomials*, the * designating that the irrep of U(n)×U(n) carried by the space $H^{[m]}$ "shares" the U(n) irrep label [m]. The remainder of Sec. II is devoted to a description of the properties of U(n)*U(n) boson polynomials.

We first describe two alternative notations for enumerating a set of basis vectors of the space $H^{[m]}$.

Gel'fand pattern scheme. The basis vectors of $H^{[m]}$ are in one-to-one correspondence with the set of *double Gel'fand patterns:*

$$\left\{ \begin{pmatrix} (m') \\ [m] \\ (m) \end{pmatrix} \middle| \begin{matrix} (m') \text{ is a Gel'fand pattern} \\ \\ (m) \text{ is a Gel'fand pattern} \end{matrix} \right\} . \tag{3}$$

A Gel'fand pattern

$$\begin{pmatrix} [m] \\ (m) \end{pmatrix} = \begin{pmatrix} m_{1n} & & m_{2n} & & \cdots & & m_{n-1\,n} & & m_{nn} \\ & m_{1n-1} & & m_{2n-1} & & & & m_{n-1\,n-1} & \\ & & \ddots & & & & \cdot^{\cdot^{\cdot}} & & \\ & & & m_{12} & & m_{22} & & & \\ & & & & m_{11} & & & & \end{pmatrix} \tag{4}$$

is a triangular array of nonnegative integers containing n rows in which [m] is the set of ordered integers denoting an irrep of U(n) and the integers m_{ij}, i, j = 1,2,..., n-1 in (m) may assume all sets of values consistent with the "betweenness" conditions":

$$m_{ij+1} \geqslant m_{ij} \geqslant m_{i+1j+1} \tag{5}$$

$\begin{pmatrix} [m] \\ (m') \end{pmatrix}$ is an array of the same type as (4) which for notational convenience is inverted over $\binom{[m]}{(m)}$ in (3), the common irrep labels [m] being written only once.

Example. For n = 3 and $[m_{13}m_{23}m_{33}] = [210]$ there are eight Gel'fand patterns

$$\begin{pmatrix} 2 & 1 & 0 \\ & 2 \;\; 1 & \\ & 2 & \end{pmatrix} \begin{pmatrix} 2 & 1 & 0 \\ & 2 \;\; 1 & \\ & 1 & \end{pmatrix} ; \begin{pmatrix} 2 & 1 & 0 \\ & 2 \;\; 0 & \\ & 2 & \end{pmatrix} \begin{pmatrix} 2 & 1 & 0 \\ & 2 \;\; 0 & \\ & 1 & \end{pmatrix} \begin{pmatrix} 2 & 1 & 0 \\ & 2 \;\; 0 & \\ & 0 & \end{pmatrix}$$
$$\begin{pmatrix} 2 & 1 & 0 \\ & 1 \;\; 0 & \\ & 1 & \end{pmatrix} \begin{pmatrix} 2 & 1 & 0 \\ & 1 \;\; 0 & \\ & 0 & \end{pmatrix} ; \begin{pmatrix} 2 & 1 & 0 \\ & 1 \;\; 1 & \\ & 1 & \end{pmatrix} \qquad (6)$$

and sixty-four double Gel'fand patterns.

The significance of the betweenness conditions (5) and of the integers in the Gel'fand pattern (4) is readily understood as a geometrical realization of the Weyl branching law for the chain of unitary subgroups $U(n) \supset U(n-1) \supset \ldots \supset U(2) \supset U(1)$.

The *weight* or *content* of a Gel'fand pattern is the row vector $W = (w_1, w_2, \ldots, w_n)$ where w_j is defined to be the sum of the entries in row j minus the sum of the entries in row j-1: $w_j = \sum_i m_{ij} - \sum_i m_{ij-1}$ $(w_1 \equiv m_{11})$.

Weyl tableau scheme. The basis vectors of $H^{[m]}$ are in one-to-one correspondence with the set of *double standard Weyl tableaux* of shape $[\lambda_1\lambda_2\ldots\lambda_n]$ $(\lambda_i \equiv m_{in}; m_{nn} \geqslant 0)$:

$$\left(\begin{array}{llll|llll} i_{11} & i_{12} & \cdots & i_{1\lambda_1} & j_{11} & j_{12} & \cdots & j_{1\lambda_1} \\ i_{21} & i_{22} & \cdots & i_{2\lambda_2} & j_{21} & j_{22} & \cdots & j_{2\lambda_2} \\ \vdots & & & & \vdots & & & \\ i_{n1} & i_{n2} & \cdots & i_{n\lambda_n} & j_{n1} & j_{n2} & \cdots & j_{n\lambda_n} \end{array}\right) . \qquad (7)$$

For completeness we recall that a *Young frame* of shape $[\lambda_1\lambda_2\ldots\lambda_n]$ has λ_1 boxes (nodes) in row 1 (top row), λ_2 boxes in row 2,... λ_n boxes in row n. A *standard Weyl tableau* is a Young frame which has been "filled in" with integers selected from 1,2,...,n in such a way that the sequence of integers in each row is nondecreasing as read from left to right and the sequence of integers in each column is strictly increasing as read from top to bottom. The *weight* or *content* W of a standard Weyl tableau is the row vector $W = (w_1, w_2, \ldots, w_n)$, where w_k equals the number of times k appears in the tableau. [A *standard Young tableau* is a standard Weyl tableau such that $\lambda_1+\ldots+\lambda_n = n$ and $W = (1,1,\ldots,1)$.]

Example. The standard Weyl tableaux corresponding to the Young frame [210] are:

$$\begin{array}{|c|c|}\hline 1 & 1 \\ \hline 2 & \\ \cline{1-1}\end{array}\quad \begin{array}{|c|c|}\hline 1 & 2 \\ \hline 2 & \\ \cline{1-1}\end{array}\ ;\quad \begin{array}{|c|c|}\hline 1 & 1 \\ \hline 3 & \\ \cline{1-1}\end{array}\quad \begin{array}{|c|c|}\hline 1 & 2 \\ \hline 3 & \\ \cline{1-1}\end{array}\quad \begin{array}{|c|c|}\hline 2 & 2 \\ \hline 3 & \\ \cline{1-1}\end{array}\ ;$$
$$\begin{array}{|c|c|}\hline 1 & 3 \\ \hline 3 & \\ \cline{1-1}\end{array}\quad \begin{array}{|c|c|}\hline 2 & 3 \\ \hline 3 & \\ \cline{1-1}\end{array}\ ;\quad \begin{array}{|c|c|}\hline 1 & 3 \\ \hline 2 & \\ \cline{1-1}\end{array}\ . \tag{8}$$

Remark. There is a one-to-one correspondence between Gel'fand patterns (4) having fixed labels [m] and the standard Weyl tableaux of shape [m] ($m_{nn}\geqslant 0$). It is this result which allows one to go back and forth between the notations (3) and (7) for the basis vectors of $\mathcal{H}^{[m]}$. Row j of the standard Weyl tableau corresponding to the Gel'fand pattern (4) is

$$\underbrace{\boxed{j}\,\cdots\,\boxed{j}}_{m_{jj}}\,\underbrace{\boxed{j+1}\,\cdots\,\boxed{j+1}}_{m_{jj+1}-m_{jj}}\,\cdots\,\underbrace{\boxed{n}\,\cdots\,\boxed{n}}_{m_{jn}-m_{jn-1}}\ . \tag{9}$$

Observe then that the weight of a Gel'fand pattern and that of the corresponding standard Weyl tableau agree.

Let us now describe the U(n)*U(n) polynomials which span the space $\mathcal{H}^{[m]}$. We begin with the description of the simplest polynomials which are those corresponding to the Young frame having 1 row and p boxes so that $[m] = [p0\ldots0] = [p\dot{0}]$:

$$B\begin{pmatrix}(m')\\ [p\ \dot{0}]\\ (m)\end{pmatrix}(A) = \left[\prod_{i=1}^{n}(w_i)!\,(w'_i)!\right]^{1/2}\sum_{\boxed{\alpha}}\ \prod_{i,j=1}^{n}(a_i^j)^{\alpha_i^j}\big/(\alpha_i^j)!\ , \tag{10}$$

where W and W′ are the weights of the lower and upper Gel'fand patterns, respectively, and $\boxed{\alpha}$ denotes a square array of the nonnegative integers (α_i^j) in which the entries in each row i must sum to w_i and the entries in each column j must sum to w'_j. These "magic square" constraints are symbolized by the notation:

$$\boxed{\alpha} = \begin{array}{|cccc|l}\hline \alpha_1^1 & \alpha_1^2 & \cdots & \alpha_1^n & w_1\\ \alpha_2^1 & \alpha_2^2 & \cdots & \alpha_2^n & w_2\\ \vdots & & & & \\ \alpha_n^1 & \alpha_n^2 & \cdots & \alpha_n^n & w_n\\ \hline \multicolumn{1}{c}{w'_1} & w'_2 & & \multicolumn{1}{c}{w'_n} & \end{array}\ . \tag{11}$$

The summation in Eq. (10) is over all (α_i^j) which satisfy these constraints.

The general U(n)*U(n) boson polynomial is only slightly more complicated in appearance than (10):

$$B\begin{pmatrix}(m')\\ [m]\\ (m)\end{pmatrix}(A) = M^{1/2}([m]) \sum_{\boxed{\alpha}} C\begin{pmatrix}(m')\\ [m]\\ (m)\end{pmatrix}(\alpha) \prod_{i,j=1}^{n} (a_i^j)^{\alpha_i^j} / [\alpha_i^j)!]^{1/2}, \qquad (12)$$

where the significance of $\boxed{\alpha}$ is as before [Eq. (11)]. $M[m])$ is a normalization factor, $M([m]) = \prod_i (m_{in}+n-i)!/\prod_{i<j}(m_{in}-m_{jn}+j-i)$, and the C-coefficient is a Wigner coeficient for the subduction $[N]\downarrow U(n)\times U(n)$ where $N = \sum_i m_{in}$. These coefficients are given explicitly by the matrix element

$$C\begin{pmatrix}(m')\\ [m]\\ (m)\end{pmatrix}(\alpha) = C\begin{pmatrix}(m)\\ [m]\\ (m')\end{pmatrix}(\tilde{\alpha}) = \left\langle \begin{pmatrix}[m]\\ (m')\end{pmatrix}\right| \left\langle\begin{matrix}(\Gamma_n)\\ [w_n \quad \dot{0}]\\ (\alpha_n)\end{matrix}\right\rangle \cdots \left\langle\begin{matrix}(\Gamma_2)\\ [w_2 \quad \dot{0}]\\ (\alpha_2)\end{matrix}\right\rangle \left\langle\begin{matrix}(\Gamma_1)\\ [w_1 \quad \dot{0}]\\ (\alpha_1)\end{matrix}\right\rangle \left| (0) \right\rangle, \qquad (13)$$

where the symbol

$$\left\langle\begin{matrix}(\Gamma_k)\\ [w_k \quad \dot{0}]\\ (\alpha_k)\end{matrix}\right\rangle = \left\langle\begin{matrix} & & \gamma_{11} & & \\ & & \cdot\gamma_{12} \quad 0 & & \\ & \gamma_{1k-1} & \cdot & \cdot & \\ \alpha_k^1+\ldots+\alpha_k^n & & \cdot & & \cdot \\ \alpha_k^1+\ldots+\alpha_k^n & 0 & \cdots & & 0 \\ & \ddots & & \cdot & \\ & \alpha_k^1+\alpha_k^2 & & 0 & \\ & & \alpha_k^1 & & \end{matrix}\right\rangle, \quad \gamma_{1i} \equiv \sum_{j=1}^{i}(m_{jk}-m_{jk-1}), \qquad (14)$$

denotes a U(n) Wigner operator. We shall say more about the coefficients (13) later. For now it suffices to observe that the coefficients (13), while complicated, are completely known.

The double Gel'fand pattern boson polynomials possess a number of important properties which are summarized below:

(i) Pairs of polynomials having distinct Gel'fand patterns are orthogonal [scalar product given by (1)].

(ii) The polynomials of given weight (W,W') are a basis of the vector space spanned by all monomials in the (a_i^j) which contain w_i occurrences of the subscript i and w'_j occurrences of the supercript j.

(iii) The polynomials corresponding to all partitions [m] of N and all patterns (m′) and (m) are a basis for all homogeneous polynomials of degree N in the (a_i^j).

(iv) Replacing the boson matrix A by a unitary matrix $U\varepsilon U(n)$ yields the unitary irrep [m] of U(n), that is,

$$D^{[m]}_{(m)(m')}(U) = B\begin{pmatrix}(m')\\ [m]\\ (m)\end{pmatrix}(U). \qquad (15)$$

(v) The polynomials corresponding to fixed [m] are a basis of the carrier space of irrep $D^{[m]}\otimes D^{[m]}$ of U(n)×U(n), that is,

$$B\begin{pmatrix}(m')\\ [m]\\ (m)\end{pmatrix}(\tilde{U}AV) = \sum_{(\mu)(\mu')} D^{[m]}_{(\mu)(m)}(U)\,D^{[m]}_{(\mu')(m')}(V)\,B\begin{pmatrix}(\mu')\\ [m]\\ (\mu)\end{pmatrix}(A). \tag{16}$$

Example. As a simple, nontrivial, example of the boson polynomials, we obtain the unitary irreps of SU(2) in the form:

$$D^{j}_{mm'}(U) = N^{1/2}_{jm} \sum_{\boxed{\alpha}} \frac{(u^1_1)^{\alpha^1_1}(u^1_2)^{\alpha^1_2}(u^2_1)^{\alpha^2_1}(u^2_2)^{\alpha^2_2}}{(\alpha^1_1)!(\alpha^1_2)!(\alpha^2_1)!(\alpha^2_2)!}, \tag{17}$$

where $N_{jm} = (j+m)!(j-m)!(j+m')!(j-m')!$ and the sum is over all non-negative (α^j_i) such that $\alpha^1_1+\alpha^1_2=j+m'$, $\alpha^2_1+\alpha^2_2=j-m'$, $\alpha^1_1+\alpha^2_1=j+m$, $\alpha^1_2+\alpha^2_2=j-m$.

Remark. In an alternative theory to that above, one associates to each pair of corresponding columns in the double standard tableau (7) a *determinantal boson:* $\{(i_{1k}i_{2k}\cdots i_{\lambda'_k k})\ ;\ (j_{1k}j_{2k}\cdots j_{\lambda'_k k})\} \to a^{j_{1k}\cdots j_{\lambda'_k k}}_{i_{1k}\cdots i_{\lambda'_k k}}$. Forming the product over all column pairs then yields a *Weyl boson polynomial.* These polynomials are, in general, nonorthogonal, but possess properties (i)-(v) above when "orthogonal" is replaced by "linear independent" and "unitary" is dropped.

III. *Applications to physical problems*

1. *The Yamanouchi real, orthogonal representations of* S_n. Let $\{I_P | P\varepsilon S_n\}$ denote the Cayley representation of the symmetric group S_n by n x n permutation matrices:

$$I_P = [e_{i_1} e_{i_2} \cdots e_{i_n}],\quad P = \begin{pmatrix}1 & 2 & \cdots & n\\ i_1 & i_2 & \cdots & i_n\end{pmatrix}, \tag{18}$$

where e_i denotes the unit column vector, e_i = col [0...010...0] (1 in position i). We now replace the boson matrix A by I_P in Eq. (12) and restrict the Gel'fand patterns (m) and (m′) to those having weight W = W′ = (1,1,...,1). The result simplifies to [cf. Eq. (15)]

$$D^{[m]}_{(m)(m')}(I_P) = \left[\frac{n!}{\dim [m]}\right]^{1/2} \left\langle \begin{pmatrix}[m]\\ (m)\end{pmatrix} \right| \left\langle [1\,\dot{0}]^{\gamma_n}_{i_n} \right\rangle \cdots \left\langle [1\,\dot{0}]^{\gamma_n}_{i_2} \right\rangle \left\langle [1\,\dot{0}]^{\gamma_1}_{i_1} \right\rangle \left| \begin{matrix}[\dot{0}]\\ (0)\end{matrix} \right\rangle, \tag{19}$$

where: (i) dim [m] denotes the dimension of irrep [m] of S_n; (ii) $\langle 1^{\gamma}_{i}\dot{0}\rangle$ denotes a fundamental U(n) Wigner operator in which $\begin{pmatrix}1 & \dot{0}\\ & i\end{pmatrix}$ and $\begin{pmatrix}& \gamma\\ 1 & \dot{0}\end{pmatrix}$ (inverted) denote n-rowed patterns having weights (0...010...0) with 1 in position i and γ, respectively; and (iii) the sequence of integers $(\gamma_n,\ldots,\gamma_2,\gamma_1)$ is the Yamanouchi symbol of the Gel'fand pattern $\begin{pmatrix}[m]\\ (m')\end{pmatrix}$, W′ = (1,...,1). ($\gamma_s$ is the number of the row in which integer s appears in the standard tableau.)

Let $D^{[m]}(P)$ denote the matrix of dimension, dim[m], with elements in row (m) and column (m′) given by Eq. (19). Then the result we have obtained may be summarized as: $\{D^{[m]}(P)|P\varepsilon S_n\}$ *is the Yamanouchi real, orthogonal representation of* S_n.

Remarks. While each γ_k appearing in Eq. (19) may assume the values 1,2,...,n, the matrix element of the string of fundamental Wigner operators is automatically zero unless $(\gamma_n,\ldots,\gamma_2,\gamma_1)$ is the Yamanouchi symbol described above. The matrix elements of the fundamental U(n) Wigner operators are completely known so that Eq. (19) is a completely general and explicit result yielding all irreps of S_n.

2. *Explicit N-particle states transforming irreducibly under* S_N *and* SU(n). We consider that N⩾n (the case N⩾n may be treated similarly) and specialize the polynomials (12) to those having irrep labels of the form $[m_{1n}m_{2n}\ldots m_{Nn}0\ldots0]$, $\sum_{i=1}^{N} m_{in}=N$, and upper Gel'fand patterns (m′) having the weight $W'=(1,\cdots,1,0,\ldots,0)$ containing N ones and n-N zeroes. We obtain:

$$(M[m_{1n}\ldots m_{Nn}\dot{0}])^{-1/2}\, B\begin{pmatrix} (m') \\ [m_{1n}\ldots m_{Nn}\dot{0}] \\ (m) \end{pmatrix}(A) \tag{20}$$

$$= \sum_{k_1\ldots k_N\,=\,1}^{n} \left\langle \begin{pmatrix} [m_{1n}\ldots m_{Nn}\dot{0}] \\ (m) \end{pmatrix} \middle| \left\langle \begin{matrix} [1^{\gamma_N}\dot{0}] \\ k_N \end{matrix} \right\rangle \cdots \left\langle \begin{matrix} [1^{\gamma_1}\dot{0}] \\ k_1 \end{matrix} \right\rangle \middle| \begin{pmatrix} [\dot{0}] \\ (0) \end{pmatrix} \right\rangle a^1_{k_1} a^2_{k_2} \cdots a^N_{k_N}\,,$$

where $(\gamma_N,\ldots,\gamma_2,\gamma_1)$ is the Yamanouchi symbol of the standard Young tableau corresponding to the N-rowed Gel'fand pattern $\begin{pmatrix} [m_{1n}\ldots m_{Nn}] \\ (m') \end{pmatrix}$, $W'=(1,\ldots,1)$.

The transcription of the boson polynomials (20) to basis vectors of the Hilbert space H of the union of N physical systems, considered as a single system, is accomplished by the correspondence $a^1_{k_1}a^2_{k_2}\ldots a^N_{k_N} \leftrightarrow |k_1\rangle\otimes|k_2\rangle\otimes\ldots\otimes|k_N\rangle$, where $\{|k\rangle\,|k=1,2,\ldots,n\}$ is the (orthonormal) basis of the Hilbert space H_n of single-particle states, and H is the tensor product space $H = H_n\otimes\ldots\otimes H_n$ (N times; dim $H = n^N$). This result follows by considering the unitary transformations of single-particle states given by U: $|i\rangle \rightarrow \sum_j u_{ji}|j\rangle$, each $U\varepsilon U(n)$, and the permutations of identical particles which induce the transformations P: $|k_1\rangle\otimes|k_2\rangle\otimes\ldots\otimes|k_N\rangle \rightarrow |k_{i_1}\rangle\otimes|k_{i_2}\rangle\otimes\ldots\otimes|k_{i_N}\rangle$ of the tensor product space, where P is the rearrangement $1\rightarrow i_1$, $2\rightarrow i_2,\ldots,N\rightarrow i_N$ of the subscripts 1,2,...,N. The resulting transformation of the basis vector $|k_1\rangle\otimes\ldots\otimes|k_N\rangle$ is then exactly the same as that of the boson product $a^1_{k_1}\ldots a^N_{k_N}$ under $A\rightarrow \tilde{U}AI_P$ where A now denotes the nxN boson matrix (a^j_i), i=1,...,n;j=1,...,N. (Since the boson operators having j>N do not appear in the right-hand side of (20), we can replace the nxn matrix boson by the nxN matrix boson.) We thus obtain an explicit expression for the *orthonormal* N-particle states which transform irreducibly under U(n) and S_N:

$$\left|(T_{U_n}|T_{S_N})\right\rangle = \sum_{k_1\ldots k_N} {}_1^n \left\langle \begin{pmatrix}[m]\\(m)\end{pmatrix} \middle| \left\langle [1_{k_N}^{\gamma_N}\dot{0}] \right\rangle \ldots \left\langle [1_{k_1}^{\gamma_1}\dot{0}] \right\rangle \middle| \begin{pmatrix}[\dot{0}]\\(0)\end{pmatrix}\right\rangle$$

$$\times\ |k_1\rangle \otimes |k_2\rangle \otimes \ldots \otimes |k_N\rangle\ . \tag{21}$$

In stating this result, we have replaced the somewhat cumbersome (because of repetition of labels) Gel'fand pattern notation by the double tableau notation: T denotes a standard tableau of shape $[m_{1n}\ldots m_{nn}]$, $\Sigma m_{in}=N$; T_{U_n} denotes the standard Weyl tableau corresponding to $\begin{pmatrix}[m]\\(m)\end{pmatrix}$; and T_{S_N} denotes the standard Young tableau which has the Yamanouchi symbol $(\gamma_N,\ldots,\gamma_2,\gamma_1)$. (There is no restriction between N and n in the final result (21) other than $\sum_{i=1}^{n} m_{in}=N$.)

In terms of the double tableau notation, the orthonormality and transformation properties of the basis vectors (21) take the forms:

$$\left\langle (T'_{U_n}|T'_{S_N}) \middle| (T_{U_n}|T_{S_N})\right\rangle = \delta(T'_{U_n},T_{U_n})\ \delta(T'_{S_N},T_{S_N})\ , \tag{22}$$

where $\delta(T',T) = 0$ for distinct tableaux and $\delta(T',T) = 1$ for identical tableaux.

$$U:\ \left|(T_{U_n}|T_{S_N})\right\rangle \rightarrow \sum_{(m')} D^{[m]}_{(m')(m)}(U) \left|(T'_{U_n}|T_{S_N})\right\rangle\ , \tag{23}$$

$$P:\ \left|(T_{U_n}|T_{S_N})\right\rangle \rightarrow \sum_{(\gamma'_N\ldots\gamma'_1)} D^{[m]}_{(\gamma'_N\ldots\gamma'_1),(\gamma_N\ldots\gamma_1)}(P) \left|(T_{U_n}|T'_{S_N})\right\rangle\ .$$

3. *Spin states for N particles of spin-$\frac{1}{2}$ which transform irreducibly under* S_N. These states are obtained as a special case of Eq. (21), namely, $n = 2$, $m_{12}+m_{22}=N$, $2S = m_{12}-m_{22}$, so that $m_{12} = \frac{N}{2}+S$, $m_{22} = \frac{N}{2} -S$, where the total spin S may assume the values $S = \frac{N}{2}, \frac{N}{2} -1,\ldots, \frac{1}{2}$ or 0. Thus, the tableau T has the shape

$$T = \begin{array}{|c|c|c|c|c|c|}\hline \ & \ & \cdots & \ & \cdots & \ \\ \hline \ & \ & \cdots & \ \\ \cline{1-4}\end{array} \quad \begin{array}{l}\frac{N}{2}+S\\ \frac{N}{2}-S\end{array} \tag{24}$$

and the orthonormalized basis vectors are:

$$\left|(T_{U_2}|T_{S_N})\right\rangle = \left|(\gamma_N\ldots\gamma_2\gamma_1);\ SM_S\right\rangle \tag{25}$$

$$= \sum_{\sigma_1\ldots\sigma_N} {}_1^2 \left\langle \begin{pmatrix}\frac{N}{2}+S & \frac{N}{2}-S\\ \frac{N}{2}+M_S\end{pmatrix} \middle| \left\langle [1_{\sigma_N}^{\gamma_N}0] \right\rangle \ldots \left\langle [1_{\sigma_1}^{\gamma_1}0] \right\rangle \middle| \begin{pmatrix}0 & & 0\\ & 0 & \end{pmatrix}\right\rangle$$

$$\times\ |\sigma_1\rangle \otimes |\sigma_2\rangle \otimes \ldots \otimes |\sigma_N\rangle\ ,$$

where $|1\rangle = |\frac{1}{2},\frac{1}{2}\rangle$, $|2\rangle = |\frac{1}{2},-\frac{1}{2}\rangle$ denote the two single-particle spin states. [Observe that σ_i,γ_i=1 and 2 correspond, respectively, to the patterns $\begin{pmatrix}1 & 0\\ & 1\end{pmatrix}$ and $\begin{pmatrix}1 & 0\\ & 0\end{pmatrix}$.]

The evaluation of the matrix elements of the U(2) Wigner operators in Eq. (25) is discussed below.

4. *Equivalent electron configurations* ℓ^N. Let the single-electron states $|n\ell m\rangle$ of fixed energy E_n and fixed angular momentum ℓ be denoted by $|k\rangle = |\ell-m+1\rangle$ so that $k=1,2,\ldots,2\ell+1$, and let the two spin states be denoted by $|\sigma\rangle = |\frac{3}{2} - \mu\rangle \equiv |\frac{1}{2},\mu\rangle$ so that $\sigma=1,2$. Thus, the set of single-particle states of the electron is $\{|k\sigma\rangle = |k\rangle \otimes |\sigma\rangle \mid k=1,\ldots,2\ell+1;\ \sigma=1,2\}$.

The Slater determinantal states for the configuration ℓ^N are given by

$$|(k_1\sigma_1)(k_2\sigma_2)\ldots(k_N\sigma_N)\rangle = \frac{1}{\sqrt{N!}} \sum_P (-1)^p P(|k_1\sigma_1\rangle \otimes \ldots \otimes |k_N\sigma_N\rangle). \qquad (26)$$

If we order the pairs of integers (k,σ) by the rule $(1,1)<(2,1)<\ldots<(2\ell+1,1)<(1,2)<(2,2)<\ldots<(2\ell+1,2)$, then an orthonormal basis of the space $V_A(\ell^N)$ of Slater states is obtained from the states (26) by imposing $(k_1\sigma_1)<(k_2\sigma_2)<\ldots<(k_N\sigma_N)$. Hence, the space has dimension, dim $V_A(\ell^N) = \binom{4\ell+2}{N}$.

A basic problem in the LS-coupling scheme is to introduce a new basis into the space $V_A(\ell^N)$ such that the total orbital angular momentum $\underline{L}$ and the total spin $\underline{S}$ of the N spin-$\frac{1}{2}$ particles have the standard irreducible action on the new basis (L^2, L_3, S^2, S_3 diagonal).

This problem is partially solved by the following well known procedure: We introduce the tableau $\tilde{T}$ which is conjugate (or dual) to the tableau T given by (24), and we use the fact that the antisymmetric irrep [1 1...1] of S_N occurs only in the direct product of conjugate irreps, and then exactly once. Furthermore, the coefficients for this reduction are given by the simple formula $C(T_{S_N}) = \sigma(T_{S_N})/\dim[\frac{N}{2}+S,\frac{N}{2}-S]$, where $\sigma(T_{S_N})$ is the signature of the tableau T_{S_N}. (The signature of a standard tableau is the signature of the permutation $1\to t_1, 2\to t_2, \ldots, N\to t_N$ where $t_1 t_2 \ldots t_N$ are the entries in row 1, followed by those in row 2,... .) Carrying out this procedure, using the vectors (25) and the vectors (21) for the tableau conjugate to (24), we obtain:

$$\left|(\tilde{T}_{U_{2\ell+1}} | T_{U_2})\right\rangle = \sum_{T_{S_N}} C(T_{S_N}) \left|(\tilde{T}_{U_{2\ell+1}} | \tilde{T}_{S_N})\right\rangle \otimes \left|(T_{U_2} | T_{S_N})\right\rangle , \qquad (27)$$

where the summation is over all standard Young tableaux of shape T [Eq. (24)].

The vectors (27) *corresponding to all* U_2 *Weyl tableaux of shape* T [Eq. (24)], *to all* $U_{2\ell+2}$ *Weyl tableaux of shape* $\tilde{T}$, *and to all shapes given by* $S = \frac{N}{2}, \frac{N}{2} - 1, \ldots, \frac{1}{2}$ *or* 0 *are an orthonormal basis of the space* $V_A(\ell^N)$, *and the total spin* $\dot{S}$ *has the standard irreducible action* (S^2, S_3 diagonal) *on this basis.*

The result (27) does not solve the problem of constructing states of good orbital angular momentum (LM_L). This requires reducing irrep $[\dot{2}\dot{1}] \equiv [2^{\frac{N}{2}-S}\, 1^{2S}]$ of $U(2\ell+1)$ into irreps of the rotation group. More precisely, since the single-particle states undergo the unitary transformation $V \equiv D^{2\ell+1}(U) \in U(2\ell+1)$, each $U \in SU(2)$, the problem is to reduce the representation $D^{[\dot{2}\dot{1}]}(V)$ of the rotation group into irreducible constituents $D^L(U)$. Alternatively, when expressed in terms of the Lie algebras $\{L_i | i=1,2,3\}$ of $SU(2)$ and $\{E_{ij} | i,j=1,2,\ldots,2\ell+1\}$ of $U(2\ell+1)$, the problem is to determine the linear combinations of the vectors $\left|(\tilde{T}_{2\ell+1} | T_{U_2})\right\rangle$ on which $L_+ = \sum_m [(\ell-m)(\ell+m+1)]^{1/2} E_{\ell-m,\ell-m+1}$ $L_- = \sum_m [(\ell+m)(\ell-m+1)]^{1/2} E_{\ell-m+2,\ell-m+1}$, $L_3 = \sum_m m\, E_{\ell-m+1,\ell-m+1}$ have the standard irreducible action. Since this reduction entails summing only over the tableaux $\tilde{T}_{U_{2\ell+1}}$, the antisymmetry and spin properties of the states (27) are preserved, and the states (27) retain their importance in the LS-coupling problem. [Equivalent electron configurations j^N may also be obtained from Eq. (20) by identifying the states $|k_i\rangle$ with the ℓs-coupled states $|jm\rangle$, choosing $n = 2j+1$ and $[m] = [1\ldots10\ldots0]$ (N ones, $2j+1-N$ zeroes), $\gamma_k = k$, and reducing irrep $[1^N]$ of $U(2j+1)$ into irreps of the rotation group.]

IV. *Evaluation of the coefficients.* The matrix elements of the Wigner operators which appear in Eqs. (18), (20), and (24) may be evaluated using the rules of the pattern calculus (Ref. 12). These rules are simple to apply and give the following form for the nonzero matrix elements of a fundamental Wigner operator between arbitrary Gel'fand states (for specific evaluations, it is easier to apply the pattern calculus rules directly than to specialize the general result below):

$$\left\langle \begin{pmatrix} [m] \\ (m) \end{pmatrix} \gamma_n \cdots \gamma_i \,\middle|\, \left\langle \begin{matrix} [1^{\gamma}\dot{0}] \\ i \end{matrix} \right\rangle \,\middle|\, \begin{pmatrix} [m] \\ (m) \end{pmatrix} \right\rangle = \tag{28}$$

$$\prod_{k=i}^{n} S(\gamma_{k-1}-\gamma_k) \left[\prod_{\substack{s=1 \\ s\neq\gamma_k}}^{k} \frac{(p_{\gamma_{k-1}k-1} - p_{sk} + 1)}{(p_{\gamma_k k} - p_{sk})} \prod_{\substack{t=1 \\ t\neq\gamma_{k-1}}}^{k-1} \frac{(p_{\gamma_k k} - p_{tk-1})}{(p_{\gamma_{k-1}k-1} - p_{tk-1} + 1)} \right]^{1/2},$$

where: (i) the final pattern is obtained from the initial one by shifting row j ($j=n,n-1,\ldots,i$) to $[m_{1j}+\delta_{1\gamma_j}, \ldots, m_{jj}+\delta_{j\gamma_j}]$ ($\gamma_j = 1,\ldots,j$), $\gamma_n \equiv \gamma$; (ii) $S(j-i) = +1$ for $j \geqslant i$ and -1 for $j < i$; (iii) the $k=i$ factor in the product has sign $S(i-\gamma_i)$ and all factors containing $p_{\gamma_{i-1}i-1}$ are to be omitted; and (iv) $p_{ij} \equiv m_{ij}+j-i$.

†Work performed under auspices of the USERDA.

REFERENCES

1. P. Jordan, Z. Physik 94 (1935), 531-535.

2. J. Schwinger, "On Angular Momentum," [Reprinted in "Quantum Theory of Angular Momemtum," ed. by L. C. Biedenharn and H. van Dam, Academic Press, New York, 1965, 229-279.]

3. H. Weyl, "The Theory of Groups and Quantum Mechanics," translated by H. P. Robertson, Methuen, London, 1931; reissued Dover, New York, 1949.

4. G. Racah, Phys. Rev. 76 (1949), 1352-1365.

5. G. Racah, Group theory and spectroscopy, Engeb. Exakt. Naturw. 37 (1965), 28-84.

6. M. Hamermesh, "Group Theory and its Applications to Physical Problems," Addison-Wesley Publ. Co., Reading, Massachusetts, 1962.

7. B. R. Judd, "Operator Techniques in Atomic Spectroscopy," McGraw-Hill Book Co., New York, 1963.

8. M. Moshinsky, J. Math. Phys. 4 (1963), 1128-1139.

9. G. E. Baird and L. C. Biedenharn, J. Math. Phys. 4 (1963), 1449-1466.

10. J. D. Louck, J. Math. Phys. 6 (1965), 1786-1804.

11. M. Moshinsky, J. Math. Phys. 7 (1966), 691-698.

12. L. C. Biedenharn and J. D. Louck, Commun. Math. Phys. 8 (1968), 80-131.

13. P. Kramer and M. Moshinsky, Group theory of harmonic oscillators and nuclear structure, in "Group Theory and its Applications," ed. by E. M. Loebl, Academic Press, New York, 1968, 339-468.

14. W. A. Goddard, Phy. Rev. 157 (1967), 73-80; ibid, 81-93.

15. P. Kramer, Z. Phys. 216 (1968), 68.

16. J. D. Louck, Am. J. Phys. 38 (1970), 3-42.

17. W. J. Holman, Nuovo Cimento 4A (1971), 904-931.

18. E. Chacón, M. Ciftan, and L. C. Biedenharn, J. Math. Phys. 13 (1972), 577-589.

19. T. H. Seligman, J. Math. Phys. 13 (1972), 876-879.

20. J. D. Louck and L. C. Biedenharn, J. Math. Phys. 14 (1973), 1336-1357.

21. P. Doubilet, G.-C. Rota, and J. Stein, Studies in Appl. Math., Vol. LIII, No. 3, 1974, 185-216.

22. T. H. Seligman, "Double Coset Decompositions of Finite Groups and the Many Body Problem," Burg Verlag, Basel, 1975.

23. W. G. Harter and C. W. Patterson, "A Unitary Calculus for Electronic Orbitals," Springer-Verlag, Berlin, 1976.

24. C. W. Patterson and W. G. Harter, J. Math. Phys. 17 (1976), 1125-1136; 17, 1137-1142.

UNITARY GROUP APPROACH TO MOLECULAR ELECTRONIC STRUCTURE*

Josef Paldus

Department of Applied Mathematics, University of Waterloo

Waterloo, Ontario, Canada, N2L 3G1.**

I. INTRODUCTION

The symmetry (invariance) groups of atomic, molecular or solid state systems, which leave the pertinent Hamiltonians invariant, have been used in quantum mechanical studies of these systems for a long time. In contrast, the class of Lie groups (algebras) that may be regarded as dynamical groups (algebras) of these systems [1,2], have only been exploited in similar studies relatively recently. Indeed, even though it was already observed by Jordan [3] that the Hamiltonians of many-body systems are expressible in terms of the generators (infinitesimal operators) of $GL(\infty,\mathbf{C})$ or $U(\infty)$, it was not until some rather recent developments in the representation theory of classical groups [4] that the exploitation of these observations became feasible. This feasibility was first noted by Moshinsky [5], who showed how the unitary group $U(n)$ can be conveniently used in nuclear shell-model calculations. This approach enabled the efficient construction of both spin and isospin adapted N-particle bases, as well as the construction of pertinent matrices. The same formalism can be readily adapted to N-electron problems [6], in which case it can be considerably simplified [7-13]. This approach provides not only a very elegant and compact spin-free formalism [13], which has been advocated in quantum chemistry for a long time by Matsen [14], but also very efficient and versatile algorithms suitable for computer implementation [8-12]. In addition, other symmetries may conveniently be accounted for in this formalism [9,15], as well as other than shell-model approaches [16].

This development led in turn to some new advances based on the pertinent symmetry group formalism, namely the symmetric group $S(N)$ and spin group(s) $SU(2)$ [or $SU(2)^{\otimes N}$]; [17,18]. The relationship with the unitary group approach offers not only new insight, but also new algorithms, which would be hard - if not

---

* Invited talk at the Integrative Conference on Group Theory and Mathematical Physics, The University of Texas in Austin, September 11-16, 1978.

** Also at Department of Chemistry, University of Waterloo and Guelph-Waterloo Center for Graduate Work in Chemistry.

impossible - to find solely on the basis of the symmetry groups of the system, and vice versa.

In this review we shall attempt to give a brief outline of these developments and to establish the relationships among various approaches. We shall base our presentation on the unitary group approach to the many-body problem as developed by Moshinsky [5]. We shall then indicate the simplifications which are possible for many-electron systems, where considerable development has occurred during the past few years, as well as some current and future developments.

II. PROBLEM FORMULATION

1. In a quantum mechanical determination of the electronic structure of atoms, molecules or solids we are faced with the problem of finding the eigenvalues and eigenvectors of the non-relativistic N-electron Hamiltonian $\hat{H}$, involving at most two body forces,

$$\hat{H} = \sum_{A,B} \langle A|z|B\rangle X_A^{\dagger}X_B + \tfrac{1}{2}\sum_{A,B,C,D} \langle AB|v|CD\rangle X_A^{\dagger}X_B^{\dagger}X_D X_C \ , \quad (1)$$

where $X_A^{\dagger}(X_A)$ are creation (annihilation) operators defined on some complete orthonormal set of single particle states $|A\rangle$, spanning the one-particle Hilbert space $H_{(1)}$, and z and v designate the one-particle (kinetic and external potential energy) and two-particle (Coulomb potential) operators, respectively. The Hamiltonian (1) is particle number conserving and is thus defined on the N-particle component $\overline{H}_{(N)}$ of the Fock space, which may be regarded as a completion of the pre-Hilbert space $H_{(N)}$, defined as the N-th rank tensor product of one-particle Hilbert spaces $H_{(1)}$; $H_{(N)} = H_{(1)}^{\otimes N}$.

2. The above problem can be simplified by using various invariance properties of $\hat{H}$, Eq. (1), in particular those arising from

(i) indistinguishability of electrons,
(ii) spin-independence of forces involved (approximate),
(iii) spatial symmetry of the external potential field.

The first property requires that we restrict $\hat{H}_{(N)}$ to its totally antisymmetric component ${}^{A}\overline{H}_{(N)}$ (Pauli principle). The antisymmetric states spanning ${}^{A}\overline{H}_{(N)}$ are referred to as configurations and the problem II.1 as complete configuration interaction (CI).

The spin-independent Hamiltonian has the form

$$\hat{H} = \sum_{a,b} \langle a|z|b\rangle \sum_{\sigma} X^{\dagger}_{a\sigma} X_{b\sigma} + \tfrac{1}{2} \sum_{a,b,c,d} \langle ab|v|cd\rangle \sum_{\sigma,\tau} X^{\dagger}_{a\sigma} X^{\dagger}_{b\tau} X_{d\tau} X_{c\sigma} , \qquad (2)$$

where we assumed that $|A\rangle = |a\rangle|\sigma\rangle$, $|a\rangle$ forming a complete orthonormal basis of a spinless one-particle Hilbert space and $|\sigma\rangle$, (σ=1,2), designating the spin-up ($|1\rangle$) and spin-down ($|2\rangle$) states of a two-dimensional one-electron spin space. Since $\hat{H}$, Eq. (2), commutes with both the square of the total spin operator $\hat{S}^2$ and its z-component $\hat{S}_z$, we can use their eigenvalues to label uniquely the pertinent invariant subspaces. The basis vectors of such subspaces are referred to as <u>spin-adapted configurations</u> and the problem II.1 as <u>spin-adapted CI</u>.

Finally, one can symmetry-adapt the N-electron basis to the operations of various point or space groups which leave $\hat{H}$ invariant. This is essential for atomic and solid state systems but much less so for the molecular ones: not only do many molecules not possess any spatial symmetry whatsoever, but even for symmetric ones the symmetry is of little use in studying problems like chemical or photochemical reactions, where non-equilibrium geometries predominate.

3. Even with the simplifications just outlined, the problem II.1 cannot be solved for systems with $N > 2$. We thus restrict outselves to the corresponding <u>model problem</u> defined in some <u>finite</u> dimensional subspace ${}^{A}V_{(N)}$ of ${}^{A}\overline{H}_{(N)}$, defined as the totally antisymmetric component of the N-th rank tensor product of a <u>finite</u> dimensional one-particle (orbital) space $V_{(1)}$, $\dim V_{(1)} = n$; $V_{(N)} = V_{(1)}^{\otimes N}$. The pertinent <u>model Hamiltonian</u> $\hat{H}$ is again given by Eqs. (1) and (2), where now the summations over the orbital (a,b,...) or spinorbital (A,B,...) labels are finite summations over n or $2n$ terms, respectively. The problem II.1 of constructing the matrix representative of $\hat{H}$ in ${}^{A}V_{(N)}$ (the so called <u>CI matrix</u>) and of finding its eigenvalues and eigenvectors is referred to as a (spin and/or symmetry adapted) <u>full CI</u> problem.

4. Often even the model problem II.3 cannot be handled by present day computing technology [even for very simple molecule like H_2O $(N = 10)$ the dimension of the full CI problem exceeds 10^7 when a reasonable one-electron basis $(n=20)$ is used]. One then has to make further approximations by restricting the model space ${}^{A}V_{(N)}$ to some proper subspace ${}^{A}W_{(N)}$. This approach is referred to as a <u>limited CI</u> method.

Even though the choice of $V_{(1)}$ and ${}^{A}W_{(N)}$ is crucial for the success of the model problem, we cannot elaborate on it here except to mention that $V_{(1)}$ is usually spanned by the set of Hartree-Fock orbitals (occupied and virtual). In selecting ${}^{A}W_{(N)}$ one tries to include configurations which describe the relevant dissociation and/or excitation processes, and which give significant energy contributions (cf., for example, Ref. [20]). In any case, the CI expansions are slowly convergent and, consequently, large CI problems arise (presently matrices of dimension 10^3-5.10^5 are used).

5. As follows from II.2, a significant reduction in size of the CI problem is achieved when spin-adapted configurations are used. There exist many methods for the construction of spin-adapted states and for the evaluation of pertinent matrix elements of $\hat{H}$ (cf., for example, Ref. [20]). In view of the slow convergence of CI expansions, II.4), it is essential that the computing labor in constructing the matrix representatives of $\hat{H}$ be minimal. It is becoming increasingly evident that such a technique is provided by the unitary group approach (cf., for example, Ref. [21]).

III. UNITARY GROUP AND THE MANY-BODY PROBLEM

1. In 1935, Jordan [3] had already observed that the Hamiltonian (1) can be expressed in terms of the operators e_{AB} ,

$$e_{AB} = X_A^\dagger X_B \ , \tag{3}$$

as follows

$$\hat{H} = \sum_{A,B} \langle A|z|B\rangle\, e_{AB} + \tfrac{1}{2} \sum_{A,B,C,D} \langle AB|v|CD\rangle\, (e_{AC}e_{BD} - \delta_{BC}e_{AD}) \ . \tag{4}$$

The operators (3) can be regarded as generators of $GL(\infty,\mathbb{C})$ or $U(\infty)$, since they satisfy the following relations

$$[e_{AB},e_{CD}] \equiv e_{AB}e_{CD} - e_{CD}e_{AB} = \delta_{BC}e_{AD} - \delta_{AD}e_{CB} \ , \tag{5a}$$

$$e_{AB}^\dagger = e_{BA} \ . \tag{5b}$$

The same applies to the corresponding model Hamiltonian (cf., II.3) where the e_{AB} form generators of $GL(2n,\mathbb{C})$ or $U(2n)$. Thus, knowing the irreps of $U(2n)$ we can construct the corresponding matrix representatives of $\hat{H}$. In this case, however, only the totally antisymmetric representations are admissible (Pauli principle).

2. It has been shown by Moshinsky [5] that the above formalism is particularly useful for spin-independent Hamiltonians (2). In this case one easily finds that in addition to the operators (3) ,

$$e_{a\sigma,b\tau} = X_{a\sigma}^\dagger X_{b\tau} \ , \tag{3'}$$

their partial traces

$$E_{ab} = \sum_{\sigma} e_{a\sigma,b\sigma} = \sum_{\sigma} X^{\dagger}_{a\sigma} X_{b\sigma} , \tag{6a}$$

$$\mathcal{E}_{\sigma\tau} = \sum_{a} e_{a\sigma,a\tau} = \sum_{a} X^{\dagger}_{a\sigma} X_{a\tau} , \tag{6b}$$

also satisfy the relations (5). Moreover,

$$[E_{ab}, \mathcal{E}_{\sigma\tau}] = 0 . \tag{7}$$

These partial traces E_{ab} and $\mathcal{E}_{\sigma\tau}$ may thus be regarded as generators of U(n) and U(2), respectively. Since, however, $\hat{H}$, Eq. (2), is spin-independent, it is expressible solely through the orbital generators E_{ab} of U(n) :

$$\hat{H} = \sum_{a,b} \langle a|z|b\rangle E_{ab} + \tfrac{1}{2} \sum_{a,b,c,d} \langle ab|v|cd\rangle (E_{ac}E_{bd} - \delta_{bc}E_{ad}) . \tag{8}$$

3. For spin-independent $\hat{H}$, Eqs. (2) and (8), we can thus symmetry adapt the N-electron states to the subgroup chain

$$U(2n) \supset U(n) \otimes U(2) . \tag{9}$$

Labeling the irreps of U(k) by their maximal weights $(\underline{m}) \equiv (m_1, m_2, \ldots, m_k)$; $m_1 \geq m_2 \geq \ldots \geq m_k$, $m_i \in \mathbf{Z}$; or equivalently by the corresponding Young tableaus {m} (cf., for example, Ref. [22]), we note that the physically admissible totally antisymmetric irreps $\Gamma(\{1^N\})$ of U(2n) decompose as follows when subduced to U(n) ⊗ U(2)

$$\Gamma^{U(2n)}(\{1^N\}) \downarrow U(n) \otimes U(2) = \sum_{m} \oplus \, \Gamma^{U(n)}(\{m\}) \otimes \Gamma^{U(2)}(\{\tilde{m}\}) , \tag{10}$$

where $\{\tilde{m}\}$ is the Young tableau conjugate to {m} and the direct sum extends over all partitions of N . In view of this conjugacy relationship, the irreps of U(n) are uniquely determined by the irreps of U(2) and vice versa, while the irreps of U(2) are uniquely labeled by the total spin quantum number S . Thus the only physically allowed irreps for U(n) have maximal weights $(\underline{m})$ with $m_i = 2$ for $i = 1,\ldots,a$; $m_i = 1$ for $i = a + 1,\ldots,a + b$ and $m_i = 0$ for $i = a + b + 1,\ldots,n$, where $a = N/2 - S$ and $b = 2S$, since the total particle number must be conserved, $\sum m_i = N$. We can choose their carrier spaces as the desired spin-adapted subspaces of $V_{(N)}$ for the CI approach. Since the model Hamiltonian (2) is expressed through the generators of U(n) , Eq. (8), we can completely ignore the U(2) subgroup once the total spin S is specified.

4. Summary: Consider the model spin-independent Hamiltonian (2) describing an N-electron system and defined on the N-th rank tensor product space whose

orbital part is $V^{\otimes N}$, $\dim V = n \geq N$. Then as the spin-adapted CI spaces, characterized by the total spin quantum number S , we can use the carrier spaces of the irreps $\Gamma(\{2^{N/2-S}\,1^{2S}\})$ of $U(n)$. Further, we can choose such spin-adapted configurations for the CI approach, which are identifiable with the vectors of the Gelfand-Tsetlin [4,22] canonical basis. The latter are uniquely labeled by the Gelfand tableaus [m]. Knowing the representation matrices of the generators E_{ab} we can obtain the matrix representative of $\hat{H}$, Eq. (8), in the same basis, thus obtaining the desired CI matrix.

IV. N-electron formalism

1. Gelfand tableaus (GT) pertinent to an N-electron problem are referred to as electronic Gelfand tableaus (EGT). Since the highest weight components m_i can only equal 0,1 or 2 (cf., III.3), we have that $2 \geq m_{ij} \geq 0$ for an arbitrary entry m_{ij} of any EGT. We can thus more efficiently characterize EGT's by an $n \times 3$ tableau $[x_{ij}]$, $(i=1,\ldots,n;\ j=1,2,3)$, [23] (called earlier [8,9,11] an ABC tableau), whose entries x_{ij} give the number of 2's, 1's and 0's in the corresponding row of the EGT [m] ,

$$x_{ij} = \sum_{k=1}^{i} \delta(j-1, m_{ki}) \ . \tag{11}$$

The top row (i=n) defines the irrep $\Gamma(x) \equiv \Gamma(\{2^a 1^b\})$ under consideration,

$$x_{n3} \equiv a = N/2 - S \ , \quad x_{n2} \equiv b = 2S \ , \quad x_{n1} \equiv c = n - N/2 - S \ , \tag{12}$$

$$\operatorname{Dim} \Gamma(x) = \frac{n+1-a-c}{n+1} \binom{n+1}{a} \binom{n+1}{c} \ , \tag{13}$$

where $\binom{m}{n}$ is the usual binomial coefficient. Since, further, the canonical basis is adapted to the chain $U(n) \supset U(n-1) \supset \ldots \supset U(1)$, we have

$$\sum_{j=1}^{3} x_{ij} = i \ , \tag{14}$$

so that any $n \times 2$ subtableau also uniquely determines a given EGT. It is often convenient to choose columns 3 and 1, which form a so called AC tableau or simply a tableau [8,9,11,24] .

2. The lexicality conditions require that the tableau columns form nondecreasing finite integer sequences, whose adjacent members differ by at most unity, i.e.

$$1 \geq x_{i+1,j} - x_{i,j} \geq 0 \ , \ j = 1 \text{ or } 3 \ . \tag{15}$$

3. This suggests in turn that we introduce the difference tableaus Δ^k ,

(k=1,2) with entries

$$\Delta^k x_{ij} = \Delta^{k-1} x_{ij} - \Delta^{k-1} x_{i-1,j} \qquad (j=1,3;\ i=k,k+1,\ldots,n) \tag{16}$$

where

$$\Delta^0 x_{ij} = x_{ij}\ ,\ x_{oj} = 0 \qquad (j=1,3)\ . \tag{17}$$

Clearly

$$x_{ij} = \sum_{k=1}^{i} \Delta^1 x_{kj} \quad \text{and} \quad \Delta^1 x_{kj} = 0 \ \text{or}\ 1\ . \tag{18}$$

Thus, the Δ^1 tableau again fully determines a given EGT and can be efficiently represented by binary strings in digital computers. One can also easily formulate the lexicality conditions for them [11,24]. The possible rows of a Δ^1 tableau are (01), (00), (11) and (10), which correspond to the occupation numbers (diagonal matrix elements of weight generators E_{aa}) 0,1,1 and 2, respectively; or, equivalently, to the (at most) four possibilities which arise at each level of canonical basis generation (cf., [8,9]).

4. The tableau structure described in IV.1-3 permits a very compact computer storage of a canonical basis, since the number of distinct rows for a given i $(i=1,\ldots,n)$ is rather limited even for $i \approx [n/2]$, and each given i-th row can be followed by at most four distinct (i-1)-th rows (cf., IV.3) [8,9,12]. For example, the canonical basis for the case of $N = 10$, $n = 20$ and $S = 1$ has 99,419,400 vectors but only 355 distinct rows so that it can be stored in less than $(3+2.4).355 \approx 4$ kbytes [12]. The explicit expressions for the number of distinct rows were derived by Shavitt [12], who also introduced a simple graphical representation of the canonical basis (cf., also [9]) which nicely illustrates its basic structure and is helpful in many derivations (cf., V.2).

5. The matrix elements of elementary generators are given by the simple formula

$$\begin{aligned} &<[\Delta^1 x_{kj}^{(i)}]\ |E_{i-1,i}|\ [\Delta^1 x_{kj}]> = \\ &= \delta(\Delta^2 x_{ij}, j-2)\ [h_i/(h_i-\Delta^2 x_{ij})]^{r_i} \end{aligned} \tag{19}$$

$$(j=1,3;\ i=1,\ldots,n)\ ,$$

where

$$r_i = \Delta^2 x_{i3}\ \Delta^2 x_{i1}/2\ , \tag{20}$$

$$h_i = i - \sum_{k=1}^{i-1} (\Delta^1 x_{k3} + \Delta^1 x_{k1}) > \Delta^2 x_{ij}\ . \tag{21}$$

The entries of the bra-tableau $[\Delta^1 x_{kj}^{(i)}]$ are the same as the entries of the ket-tableau $[\Delta^1 x_{kj}]$ except for $k = i - 1$ and $k = i$, when

$$\Delta^1 x_{kj}^{(i)} = \delta(\Delta^1 x_{kj};\ 0) \ ; \ (k=i-1,i) \ . \tag{22}$$

Note that any row (column) of an elementary generator matrix representative has <u>at most two</u> nonvanishing entries and that formula (19) enables one to determine simultaneously a given row or column for all elementary raising (lowering) generators. The values of $\Delta^2 x_{ij}$ are either 0 or ± 1, so that r_i is either 0 or $\pm\frac{1}{2}$, while h_i is simply obtained from the digital sum of appropriate tails of binary strings forming the columns of Δ^1 tableaus. Thus, the values of matrix elements (19) are either 1 or the positive squareroot of a ratio of two integers differing by 1.

Equivalent but less compact formulas employing the Weyl tableau labeling of canonical basis vectors were given by Harter and Patterson[10], whose derivation relied basically on the symmetric group formalism. The equivalence of both approaches and simple rules relating our tableaus with Weyl tableaus were given in Ref. [24].

6. In evaluating the non-elementary generator matrix elements one can either use (i) recursive relations based on (5a), namely

$$E_{ij} = [E_{i,j-1}, E_{j-1,j}] \ , \text{ or} \tag{23}$$

(ii) direct formulas or algorithms, based on (i) [9] or other approaches (cf., V.2).

The first approach is probably the most efficient one if full CI is considered. However, in limited CI the second approach is to be preferred if it can be formulated so as to avoid the matrix multiplication implied by the rhs of (23) [even though only at most two nonvanishing terms are involved in each sum, cf. IV.5]. Such algorithms have been recently developed [12,17,18], (cf., V.2). Analogous techniques may also be used to evaluate the matrix representatives of generator products, which appear in the two-electron part of $\hat{H}$ [12,18].

7. We must also mention that the unitary group approach as outlined here should also be very useful in so called <u>direct</u> or <u>vector method</u> (VM) CI approaches [25-27]. Since many modern diagonalization algorithms for large sparce matrices are essentially based on Krylov-type sequences [28], the basic step in each iteration is the evaluation of the state $|\Phi_{i+1}\rangle = \hat{H}\,|\Phi_i\rangle$, where $|\Phi_i\rangle$ is a given linear combination of chosen N-electron configurations (a trial eigenvector). This step can in fact be carried out without ever actually constructing the CI matrix of $\hat{H}$ (which proves to be not only a time-consuming step but the storage problems become insurmountable). Thus, in the direct approaches the two-electron integrals

are accessed one by one and used to generate $|\Phi_{i+1}\rangle = \hat{H}\,|\Phi_i\rangle$ from $|\Phi_i\rangle$ in each iteration.

It should be obvious from the structure of the unitary group formalism that it can be conveniently employed in the direct approaches [9,11,12]. Particularly for the VM approach [27] the so called harmonic excitation diagrams [11], further pursued by Robb et al. [29], seem to be very appropriate. However, the same facility is provided by the compactly stored canonical basis (IV.4) as discussed in more detail by Shavitt [12].

V. Relationship with other approaches

The close relationship between the representation theories of the symmetric group $S(N)$ and the general linear groups $GL(n,\mathbb{C})$ or unitary groups $U(n)$ is well known. In view of the conjugation property as expressed in Eq. (10) one can also see the relationship with the simple spin-coupling approach within the $SU(2)$ formalism. Both the symmetric group $S(N)$ and the Racah $SU(2)$ formalism were used extensively in the past to achieve the spin-adaptation of the CI problem. Recently, however, these approaches received a new stimulus from the unitary group formulation and were in turn very helpful in devising improved computational algorithms for the unitary group approach. It is impossible to indicate even briefly all the facets of this interrelationship between the various approaches in the limited space available to us here, and we shall thus restrict ourselves to mentioning the most important advances achieved in this respect.

1. It has been observed by Moshinsky and Seligman [6] that the electronic Gelfand-Tsetlin canonical states may be identified (up to a phase) with the Yamanouchi-Kotani (Y-K) [30] basis of the $S(N)$ representation theory. The latter basis is also easily related to the $SU(2)$ basis obtained by an appropriate sequential coupling of the electron spins. Using our notation, the occupation numbers n_i are simply given by the Δ^1 tableau entries, while the intermediate spin quantum numbers S_i are given by the middle column entries x_{i2} of the ABC tableaus, $S_i = x_{i2}/2$. Thus, the electronic Gelfand state characterized by the tableau $[x_{ij}]$ can be written (up to a phase) as follows

$$|[x_{ij}]\rangle = \sum_{m_1,\ldots,m_{\tilde{n}}} \sum_{M_1,\ldots,M_{\tilde{n}}} \Big[\prod_{i=1}^{n} \langle s_i m_i S_{i-1} M_{i-1} | S_i M_i \rangle \Big]$$

$$\times \Big[\prod_{j\in I_2} 2^{-\frac{1}{2}} \sum_{\sigma_j',\sigma_j''} \langle \tfrac{1}{2}\sigma_j' \; \tfrac{1}{2}\sigma_j'' | 00\rangle \Big] \Big[\prod_{i=1}^{n} \mathbf{X}_i^{\dagger}(n_i) \Big] |0\rangle \;, \quad (\tilde{n}=n-1) \qquad (24)$$

where $\langle a\alpha\ b\beta \mid c\gamma \rangle$ is the usual Clebsch-Gordan coefficient for SU(2), I_2 is the index set of doubly occupied orbital labels,

$$I_2 = \{i \mid i=1,\ldots,n;\ n_i = 2\} \ , \tag{25}$$

$$n_i = 1 + \Delta^1 x_{i3} - \Delta^1 x_{i1} \ , \tag{26}$$

$|0\rangle$ is the physical vacuum state (basis vector of the zero-particle component $\overline{H}_{(0)}$ of the Fock space: the one-dimensional Hilbert space of complex numbers), and

$$\mathbf{X}_i^{\dagger}(n_i) = \begin{cases} I & \text{if} \quad n_i = 0 \\ X^{\dagger}_{im_i} & \text{if} \quad n_i = 1 \\ X^{\dagger}_{i\sigma_i'} X^{\dagger}_{i\sigma_i''} & \text{if} \quad n_i = 2 \ , \end{cases} \tag{27}$$

where I is the identity operator. The resulting total spin quantum number S and its projection M are given by S_n and M_n, respectively. Finally, $s_i = \frac{1}{2}$, $m_i = \pm\frac{1}{2}$ if $n_i = 1$ and $s_i = m_i = 0$ if $n_i = 0$ or 2.

2. The states (24) span the carrier spaces either for the irreps of U(n), when regarded as n-orbital states or, when the orbital occupancies are fixed (i.e., when regarded as N-electron spin functions) they also span carrier spaces (or their subspaces) for the irreps of S(N), with basis states adapted to the Y-K chain (when no double occupancies occur, we get a standard Young-Yamanouchi chain). In the second case the E_{ij} matrix elements are given (up to a simple phase and normalization factor) as the N-electron spin function matrix elements of the appropriate "line-up" permutations, which can be shown to be the cyclic permutations $(r_i, r_i+1, \ldots, r_j)$ is $r_i < r_j$ or $(r_i, r_i-1, \ldots, r_j)$ if $r_i > r_j$, where $r_i(r_j)$ is the first occurrence of the orbital $i(j)$ in the bra (ket) [18,31,32].

The U(n) oriented approach using states (24) has been exploited in the work by Gouyet et al. [17], while the S(N) oriented approach has been used by Drake and Schlesinger [18]. In both cases these authors represented the pertinent states and evaluated the matrix elements using the graphical techniques of spin algebras [33]. With this technique the relationship of S(N) and U(n) based approaches becomes particularly transparent, since the spin diagrams used in the S(N) approach are simply obtained by deleting zero angular momentum lines, associated with unoccupied orbitals, in graphs used in the U(n) approach. This technique provides not only the expressions for the elementary generator matrix elements, Eq. (19), (possibly up to a phase), but also convenient formulas for the nonelementary generators, given as a single product of simple (two angular momenta equal to ½) 6j coefficients [17,18].

It is remarkable that the same types of formulas have been recently obtained independently by Shavitt [12], relying solely on the U(n) formalism (IV.1-5). These formulas enable an efficient computation of generator matrix elements even with the truncated canonical bases employed in limited CI approaches (cf., IV.5).

3. A similar problem is faced in the calculation of the two-electron part of the limited CI matrix, involving products of two generators. Again, some intermediate states $|[m'']\rangle$ involved in the evaluation of the generator product matrix elements,

$$\langle[m']|E_{ij}E_{k\ell}|[m]\rangle = \sum_{m''} \langle[m']|E_{ij}|[m'']\rangle \langle[m'']|E_{k\ell}|[m]\rangle , \tag{28}$$

will lie outside the set of states chosen for a limited CI. Shavitt [12] presented an algorithm which recursively generates these states, and involves at most three non-vanishing terms in each step in the "overlapping" part [34] of the generator product. However, an even simpler algorithm is obtained following Drake and Schlesinger [18]: the spin diagram for the "overlapping" part of the generator product is given by a 3nj symbol of the first or second kind, which in turn is expressible as a sum (containing only two terms) of products of simple 6j coefficients. The U(n) version of this algorithm is also readily obtained [35a].

4. It should be mentioned here that the approach based on graphical methods of spin algebras is very general and has considerable potential for obtaining the irreps of S(N) in bases adapted to various chains of subgroups and for establishing transformations between such bases (cf., for example, Ref. [36]). With an appropriate generalization of the SU(2) graphical techniques to non-simply reducible groups [37] one can use this technique to obtain an arbitrary irrep of S(N) , as well as of other groups. We also found this technique very valuable in formulating the unitary group approach in the hole-particle formalism [38], (cf. VI).

VI. CONCLUSIONS

We have indicated how the U(n) formalism provides an efficient algorithm for the molecular shell model (CI) calculations. Even though several programs exploying this scheme [8,21,29,35] already exist, much better ones should be available in the near future, particularly those based on the direct CI ideas. We feel that particularly in the development of efficient codes there is still plenty of room for improvement due to the great flexibility and versatility of the unitary group formalism.

To conclude this review, we mention at least a couple of basic problems, whose solution would be desirable for future development:

(i) an efficient treatment of the non-abelian geometrical symmetry of the molecular nuclear framework within the unitary group formalism, and

(ii) formulation of the unitary group approach in the hole-particle formalism.

The first problem requires a general formalism for the handling of irreps of $U(n)$ subduced to the subgroup chain

$$U(n) \supset M_1 \supset M_2 \supset \dots \supset M_k , \tag{29}$$

where M_i $(i=1,\dots,k)$ are various non-abelian [39,40] point groups, while in the second problem the following chain is involved

$$U(m+n) \supset U(m) \otimes U(n) , \tag{30}$$

as well as the mixed co- and contra-variant representations. There exist a number of papers concerning the chain (29), [41], but none of these approaches seem to be simple and universal enough to apply in electronic structure calculations. Ideally, one would only add some additional labels to those of the EGT, to identify uniquely the irreps and basis vectors of their carrier spaces, while preserving the simple algorithms for matrix element evaluation. For the chain (30) a unique labeling of the canonical basis vectors was given by Mickelsson [42], but as far as we know there are no formulas available for the matrix elements of the generators. A basic unitary group formulation of the particle-hole approach was given by Flores and Moshinsky [5,43]; however, there are still problems with the efficient handling of states with lower than maximal multiplicities as well as with the evaluation of matrix elements of particle-hole generators (i.e., tensor operators) connecting different irreps, associated with different excitation levels in the hole-particle formalism As mentioned in V. 4 we have recently overcome [38] these difficulties using the graphical techniques of spin algebras [33]. It remains to be seen, however, if competitive algorithms can be obtained using only group theoretical methods.

ACKNOWLEDGEMENTS

The author wishes to acknowledge a continued support by the National Research Council of Canada, which made it possible to carry out the research reported in this review. I am also very much obliged to my colleagues Drs. B.G. Adams and P.E.S. Wormer, Mr. M.J. Boyle and Professor J. Čížek for countless discussions and friendly help. Finally, my sincere thanks are due to Helen Warren and Debbie Gooding-Mustin for the immaculate typing of the manuscript.

REFERENCES

[1] While a symmetry group leaves the Hamiltonian invariant, this is not necessarily the case for a dynamical group [2a], in which case we only require that the Hamiltonian is expressible through its generators. In addition, one sometimes requires [2b] that all of the dynamics of the system is contained in a single irreducible representation (irrep) of the dynamical group. However, in the case of U(n) different irreps are needed for different multiplets of the model systems studied.

[2] (a) W. Miller, Jr., Symmetry Groups and Their Applications (Academic Press, N.Y., 1972), p. 400.
(b) Cf., for example, C.E. Wulfman, in Group Theory and Its Applications, Vol. 2, E.M. Loebl, Ed. (Academic Press, N.Y., 1971), p. 145.

[3] P. Jordan, Z. Phys. 94, 531 (1935).

[4] I.M. Gelfand and M.E. Tsetlin, Dokl. Akad. Nauk SSSR 71, 825, 1070 (1950); G.E. Baird and L.C. Biedenharn, J. Math. Phys. 4, 1449 (1963); 5, 1723, 1730 (1964), 6, 1847 (1965); M. Moshinsky, J. Math. Phys. 4, 1128 (1963).

[5] M. Moshinsky, J. Math. Phys. 7, 691 (1966); Group Theory and the Many-Body Problem (Gordon and Breach, N.Y., 1968).

[6] M. Moshinsky and T.H. Seligman, Ann. Phys. (N.Y.) 66, 311 (1971).

[7] W.G. Harter, Phys. Rev. A8, 2819 (1973).

[8] J. Paldus, J. Chem. Phys. 61, 5321 (1974); Intern. J. Quantum Chem., Symp. 9, 165 (1975).

[9] J. Paldus, in Theoretical Chemistry: Advances and Perspectives, Vol. 2, H. Eyring and D.J. Henderson, Eds. (Academic Press, N.Y., 1976), p. 131.

[10] W.G. Harter and C.W. Patterson, Phys. Rev. A13, 1067 (1976); A Unitary Calculus for Electronic Orbitals (Springer-Verlag, Berlin, 1976).

[11] J. Paldus, in Electrons in Finite and Infinite Structures, P. Phariseau and L. Scheire, Eds. (Plenum Publ. Co., N.Y., 1977), p. 411.

[12] I. Shavitt, Intern. J. Quantum Chem., Symp. 11, 131 (1977); Symp. 12 (in press).

[13] F.A. Matsen, Intern. J. Quantum Chem., Symp. 8, 379 (1974); 10, 525 (1976).

[14] F.A. Matsen, Adv. Quantum Chem. 1, 59 (1964), J. Amer. Chem. Soc. 92, 3525 (1970).

[15] F.A. Matsen, T.L. Welsher and B. Yurke, Intern. J. Quantum Chem. 12, 985 (1977); F.A. Matsen and T.L. Welsher, 12, 1001 (1977).

[16] (a) P.E.S. Wormer and A. van der Avoird, J. Chem. Phys. 57, 2498 (1972); Intern. J. Quantum Chem. 8, 715 (1974); P.E.S. Wormer, Ph.D. Thesis, University of Nijmegen, The Netherlands, 1975.
(b) F.A. Matsen, Advan. Quantum Chem. 11 (in press), Intern. J. Quantum Chem., Symp 12 (in press); Accounts of Chem. Research (in press).

[17] J.-F. Gouyet, R. Schranner and T.H. Seligman, J. Phys. A: Math. Gen. 8 285 (1975); J.-F. Gouyet, Rev. Mex. Fis (in press).

[18] J. Drake, G.W.F. Drake and M. Schlesinger, J. Phys. B8, 1009 (1975); G.W.F. Drake and M. Schlesinger, Phys. Rev. A 15, 1990 (1977).

[19] A.A. Cantu, D.J. Klein, F.A. Matsen and T.H. Seligman, Theoret. Chim. Acta 38, 341 (1975).

[20] I. Shavitt, in Methods of Electronic Structure Theory, H.F. Schaefer III, Ed. (Plenum Publ. Co., N.Y., 1977), p. 189.

[21] G.A. Segal, R.W. Wetmore and K. Wolf, Chemical Physics 30, 269 (1978).

[22] J.D. Louck, Amer. J. Phys. 38, 3 (1970).

[23] We call these matrices tableaus because of the unconventional numbering of rows and columns used: the rows are numbered from the bottom upwards and the columns from right to left in order to achieve a simple correspondence with the EGT's.

[24] J. Paldus, Phys. Rev. A14, 1620 (1976).

[25] T. Sebe and J. Nachamkin, Ann. Phys. (N.Y.) 51, 100 (1969); R.R. Whithead, Nucl. Phys. A 182, 290 (1972).

[26] B. Roos, Chem. Phys. Letters 15, 153 (1972); B.O. Roos and P.E.M. Siegbahn, in Methods of Electronic Structure Theory, H.F. Schaefer III, Ed. (Plenum Publ. Co., N.Y., 1977), p. 277.

[27] R.F. Hausman, S.D. Bloom and C.F. Bender, Chem. Phys. Letters 32, 483 (1975), R.H. Hausman and C.F. Bender, in Methods of Electronic Structure Theory, H.F. Schaefer III, Ed. (Plenum Publ. Co., N:Y., 1977), p. 319, C.F. Bender, J. Comput. Phys. (in press).

[28] Cf., for example, Ref. [20] and C. Lanczos, J. Res. Nat. Bur. Stand. 45, 255 (1950); E.R. Davidson, J. Comput. Phys. 17, 87 (1975).

[29] M.J. Downward and M.A. Robb, Theoret. Chim. Acta 46, 129 (1977); D. Hegarty and M.A. Robb, "Direct CI Calculations using the Calculus of the Unitary Group", preprint. (Unfortunately, this work does not fully exploit the simple structure and sparseness of generator matrices.)

[30] M. Kotani, A. Amemiya, E. Ishiguro and T. Kimura, Tables of Molecular Integrals, (Maruzen Co., Ltd., 2nd edition, Tokyo, 1963).

[31] P.E.S. Wormer, unpublished results.

[32] C.R. Sarma and S. Rettrup, Theoret. Chim. Acta 46, 63 (1977); S. Rettrup and C.R. Sarma, Theoret. Chim. Acta 46, 73 (1977), cf., also, C.R. Sarma and K.C. Dinesha, J. Math. Phys. 8, 1662 (1978).

[33] A.P. Jucys, I.B. Levinson and V.V. Vanagas, Mathematical Apparatus of the Theory of Angular Momentum (Israel Program for Scientific Translations, Jerusalem, 1962, and Gordon and Breach, N.Y., 1964); A.P. Jucys and A.A. Bandzaitis, The Theory of Angular Momentum in Quantum Mechanics, (Institute of Physics and Mathematics of the Academy of Science of the Lithuanian S.S.R., Mintis Vilnius, 1965, in Russian); E. El Baz and B. Castel, Graphical Methods of Spin Algebras (M. Dekker, N.Y., 1972); D.M. Brink and G.R. Satchler, Angular Momentum, 2nd edition (Clarendon Press, Oxford, 1968).

[34] The "overlapping" part in the product $E_{ij}E_{k\ell}$ consists of rows $k, k+1,\dots,j$ if $i \le k < j \le \ell$ (product of two raising generators) or of rows $\ell, \ell+1,\dots,j$ if $i \le \ell < j \le k$ (product of a raising and a lowering generator).

[35] (a) J. Paldus, unpublished results,
(b) J. Downing, unpublished results.

[36] J. Paldus and P.E.S. Wormer, Phys. Rev. A (in press); cf. also J. Paldus, B.G. Adams and J. Čížek, Intern. J. Quantum Chem. 11, 813 (1977).

[37] V.P. Karasev, in Group Theoretical Methods in Physics, D.V. Skobel'tsyn, Ed. (Proc. of the P.N. Lebedev Physics Institute, Vol. 70, Consultants Bureau, N.Y., 1975), p. 141; V.K. Agrawala and J.G. Belinfante, Ann. Phys. (N.Y.) 49, 130 (1968); P.A.M. Guichon, Report LYCEN 7570, University of Lyon, France (1975); M. Kibler and E. El Baz, Report LYCEN 7872, University of Lyon, France (1978).

[38] J. Paldus and M.J. Boyle, unpublished results.

[39] Abelian point groups may be handled quite easily [9], even with an efficient canonical basis storage (IV.4), [40].

[40] I. Shavitt, unpublished results.

[41] Cf., for example, B.R. Judd, W. Miller, Jr., J. Patera and P. Winternitz, J. Math. Phys. 15, 1787 (1974).

[42] J. Mickelsson, J. Math. Phys. 11, 2803 (1970), 12, 2378 (1971).

[43] J. Flores and M. Moshinsky, Nucl. Phys. A93, 81 (1967).

SHIFT OPERATORS FOR THE CLASSICAL GROUPS*

Adam M. Bincer, Physics Department, University of Wisconsin, Madison, Wisconsin 53706

Shift operators S were first introduced and applied to U(n) by Nagel and Moshinsky. By definition, the action of S_k (resp. S^k) on a semimaximal vector produces another semimaximal vector with the k-th weight component shifted down (resp. up) by unity. Thus these S provide an explicit algorithm for the construction of a basis in the irreducible representation space. By using the concept of tensor operators due to Louck and Biedenharn I obtain simple expressions for the S for all classical groups. The importance of adequate notation cannot be over-emphasized. I use the notation of Nwachuku and Rashid and extend their unified treatment of symplectic and orthogonal groups to the unitary groups by simply requiring the symbol ε_a to vanish and the indices to range from 1 to n for U(n). References to all authors are given in my paper, J. Math. Phys. 19, 1173 (1978).

Of those vectors contained in the representation space (specified by the highest weight f) of the group the maximal vector (specified by the highest weight $\tilde{f}$) of a subgroup is called the semimaximal vector $|f;\tilde{f}\rangle$. The group and subgroup are respectively $U(n)\supset U(n-1)$, $O(2n+1)\supset O(2n)$, $Sp(2n)\supset Sp(2n-2)$ and $O(2n)\supset O(2n-1)$. The U(n-1) subgroup is obtained by omitting one from the range of indices; the O(2n) subgroup by omitting zero; the Sp(2n-2) subgroup by omitting ±1; the O(2n-1) subgroup by omitting ±1 and reinstating zero via $G_b^0 = (G_b^1+G_b^{-1})/\sqrt{2}$. The shift operators may be defined by (plus for $S = S^k$, minus for $S = S_k$; all indices in subgroup range)

$$[G_a^a,S]|f;\tilde{f}\rangle = \pm\delta_k^a S|f;\tilde{f}\rangle,\ a>0 \text{ and } [G_b^a,S]|f;\tilde{f}\rangle = 0,\ a>b.$$

My result for S is (all indices, displayed or suppressed, in subgroup range)

$$S_k = \{V \prod_{j=k_{min}}^{k-1} (G-g_j\mathbb{1})\}_k\ ,\quad S^k = \{[\prod_{j=k+1}^{k_{max}} (G-\hat{g}_j\mathbb{1})]V\}^k,\qquad V^d = V_d^{\dagger},$$

where $k_{min(max)}$ is the minimum (maximum) value of indices in subgroup and V are essentially those generators of the group that are not in the subgroup: $V_d = G_d^1$ for U(n) and Sp(2n), $= G_d^0$ for O(2n+1), $= G_d^1-G_d^{-1}$ for O(2n). Lastly g_j is the partial hook of Louck and Biedenharn (ε = 0, 1, -1 for U, O and Sp respectively):

$$g_j = \tilde{f}_j + \sum_{d<j} (1-\varepsilon\delta^d_{-j}),\ \hat{g}_j = g_j + 1 - \delta_j^0 .$$

By proceeding in this fashion down the natural subgroups chain one obtains a basis in the representation space for the unitary and orthogonal groups. Since, in contrast to the U and O groups, the restriction $Sp(2n)\downarrow Sp(2n-2)$ is not multiplicity-free, additional labels are needed to specify the semimaximal vector uniquely. I have constructed shift operators for these additional labels but must omit them here for lack of space. It is noteworthy that the resultant basis is orthogonal.

*Work supported in part by the Wisconsin Alumni Research Association.

Computer programs for the reduction of Kronecker products and symmetrized Kronecker powers of space group irreducible representations.

B.L. DAVIES[1] and A.P. CRACKNELL[2]

[1]School of Mathematics and Computer Science, University College of North Wales, BANGOR LL57 2UW, Wales, U.K.

[2]Carnegie Laboratory of Physics, University of Dundee, DUNDEE DD1 4HN, Scotland, U.K.

The reduction of the Kronecker product of the irreducible representations $(\Gamma_p^{\underset{\sim}{k}_i} \uparrow G)$ and $(\Gamma_q^{\underset{\sim}{k}_j} \uparrow G)$ of a space group G into its irreducible components $(\Gamma_r^{\underset{\sim}{k}_\ell} \uparrow G)$ is determined by the values of the Clebsch-Gordan (C-G) series coefficients $C_{pq,r}^{\underset{\sim}{k}_i \underset{\sim}{k}_j, \underset{\sim}{k}_\ell}$ in the expression

$$(\Gamma_p^{\underset{\sim}{k}_i} \uparrow G) \boxtimes (\Gamma_q^{\underset{\sim}{k}_j} \uparrow G) \equiv \sum_{\ell,r} C_{pq,r}^{\underset{\sim}{k}_i \underset{\sim}{k}_j, \underset{\sim}{k}_\ell} (\Gamma_r^{\underset{\sim}{k}_\ell} \uparrow G) , \qquad (1)$$

where $(\Gamma_p^{\underset{\sim}{k}_i} \uparrow G)$ is a (single or double valued) irreducible representation of G induced from the small (or allowed) representation $\Gamma_p^{\underset{\sim}{k}_i}$ of the little group $G^{\underset{\sim}{k}_i}$, etc. and $\underset{\sim}{k}_i, \underset{\sim}{k}_j, \underset{\sim}{k}_\ell$ are restricted to lie in the representation domain Φ of G. At last year's Colloquium in Tübingen we reported on the Wave Vector Selection Rule (WVSR) and Kronecker Product (KP) programs we have developed to evaluate the coefficients in equation (1) for any choice of vectors $\underset{\sim}{k}_i$ and $\underset{\sim}{k}_j$ in Φ for any of the 230 space groups [1,2] . At that time our programs had run successfully for more than 150 groups and since then we have completed the remainder and the results are in the press [3] .

The Kronecker n^{th} power of an induced representation has been shown by Gard [4] to be capable of reduction in terms of the symmetric group of degree n , so that within each symmetry class, the reduction is expressed as a sum of induced representations. This is the generalization of the work of Mackey [5] and Bradley and Davies [6] for the case $n = 2$ to which we shall confine ourselves in the following.

The calculation to be performed is the evaluation of the coefficients $C(\pm)_{p,r}^{\underset{\sim}{k}_i, \underset{\sim}{k}_\ell}$ in the expansion of the symmetrized and antisymmetrized Kronecker square respectively:

$$[(\Gamma_p^{\underset{\sim}{k}_i} \uparrow G) \boxtimes (\Gamma_p^{\underset{\sim}{k}_i} \uparrow G)] \equiv \sum_{\ell,r} C(+)_{p,r}^{\underset{\sim}{k}_i, \underset{\sim}{k}_\ell} (\Gamma_r^{\underset{\sim}{k}_\ell} \uparrow G) , \qquad (2)$$

$$\{(\Gamma_p^{\underset{\sim}{k}_i} \uparrow G) \boxtimes (\Gamma_p^{\underset{\sim}{k}_i} \uparrow G)\} \equiv \sum_{\ell,r} C(-)_{p,r}^{\underset{\sim}{k}_i,\underset{\sim}{k}_\ell} (\Gamma_r^{\underset{\sim}{k}_\ell} \uparrow G) \ . \tag{3}$$

The coefficients in equation (1) when $\underset{\sim}{k}_i = \underset{\sim}{k}_j$ and $p = q$ are related to those in equations (2) and (3) by

$$C_{pp,r}^{\underset{\sim}{k}_i\underset{\sim}{k}_i,\underset{\sim}{k}_\ell} = C(+)_{p,r}^{\underset{\sim}{k}_i,\underset{\sim}{k}_\ell} + C(-)_{p,r}^{\underset{\sim}{k}_i,\underset{\sim}{k}_\ell} \ . \tag{4}$$

The first stage in the calculation is to evaluate the WVSRs for the Kronecker square which determine the values of $\underset{\sim}{k}_\ell \in \Phi$ which may appear on the right hand sides of equations (1) - (3) . Each WVSR

$$R_\lambda \underset{\sim}{k}_i + R_\mu \underset{\sim}{k}_i \cong \underset{\sim}{k}_\ell \tag{5}$$

is determined by the pair (R_λ, R_μ) where $\{R_\lambda | \underset{\sim}{v}_\lambda\}$ and $\{R_\mu | \underset{\sim}{v}_\mu\}$ form a very restricted subset of the left coset representatives of $G^{\underset{\sim}{k}_i}$ with respect to G [7]. Furthermore, the set of WVSRs is in one-to-one correspondence with the set of double cosets $G^{\underset{\sim}{k}_i} : G^{\underset{\sim}{k}_i}$ of G :

$$G = \sum_\nu G^{\underset{\sim}{k}_i}\{R_\nu | \underset{\sim}{v}_\nu\} G^{\underset{\sim}{k}_i} \tag{6}$$

where

$$R_\nu = R_\mu^{-1} R_\lambda \ , \tag{7}$$

and (R_λ, R_μ) defines the WVSR in equation (5). Thus, each WVSR may be labelled self-inverse or non-self-inverse according as the corresponding double coset is self-inverse or non-self-inverse respectively. It is convenient to classify the WVSRs as belonging to type 0 , 1 or 2 according as the corresponding double coset is self-inverse and identical to $G^{\underset{\sim}{k}_i}$, self-inverse but not identical to $G^{\underset{\sim}{k}_i}$ or non-self-inverse, respectively. It can then be shown from Bradley and Davies [6] that the coefficients in equations (2) and (3) are given by

$$C(\pm)_{p,r}^{\underset{\sim}{k}_i,\underset{\sim}{k}_\ell} = \sum_{\lambda,\mu} D^\pm(\lambda,\mu \ ; \ p,r)\,\delta(R_\lambda \underset{\sim}{k}_i + R_\mu \underset{\sim}{k}_i - \underset{\sim}{k}_\ell) \tag{8}$$

where

$$\delta(\underset{\sim}{k}) = \begin{cases} 1 & \text{if } \underset{\sim}{k} \text{ is a reciprocal lattice vector} \\ 0 & \text{otherwise} \end{cases} \tag{9}$$

and $D^\pm(\lambda,\mu \ ; \ p,r)$ is defined as follows.

<u>WVSR TYPE 0</u>

$$D^\pm(\lambda,\mu \ ; \ p,r) = \frac{|T|}{2|N_{\lambda\mu}|} \sum_{\{S|\underset{\sim}{w}\} \in N_{\lambda\mu}/T} \chi_{\lambda p}^{\underset{\sim}{k}_i}(\{S|\underset{\sim}{w}\})^2 \pm \chi_{\lambda p}^{\underset{\sim}{k}_i}(\{S|\underset{\sim}{w}\}^2)\ \chi_r^{\underset{\sim}{k}_\ell *}(\{S|\underset{\sim}{w}\}) \tag{10}$$

WVSR TYPE 1

$$D^{\pm}(\lambda,\mu\ ;\ p,r) = \frac{|T|}{2|N_{\lambda\mu}|}\sum_{\{S|\underset{\sim}{w}\}\in N_{\lambda\mu}/T}[\chi_{\lambda p}^{\underset{\sim}{k}_i}(\{S|\underset{\sim}{w}\})\chi_{\mu p}^{\underset{\sim}{k}_i}(\{S|\underset{\sim}{w}\})\chi_r^{\underset{\sim}{k}_\ell *}(\{S|\underset{\sim}{w}\}) \pm$$

$$\chi_{\lambda p}^{\underset{\sim}{k}_i}((\{A|\underset{\sim}{a}\}\{S|\underset{\sim}{w}\})^2)\chi_r^{\underset{\sim}{k}_\ell *}(\{A|\underset{\sim}{a}\}\{S|\underset{\sim}{w}\})] \tag{11}$$

and $\{A|\underset{\sim}{a}\}\ \varepsilon\ (\{R_\lambda|\underset{\sim}{v}_\lambda\}G^{\underset{\sim}{k}_i}\{R_\mu|\underset{\sim}{v}_\mu\}^{-1}) \cap (\{R_\mu|\underset{\sim}{v}_\mu\}G^{\underset{\sim}{k}_i}\{R_\lambda|\underset{\sim}{v}_\lambda\}^{-1})$

WVSR TYPE 2

$$D^{\pm}(\lambda,\mu\ ;\ p,r) = \frac{|T|}{2|N_{\lambda\mu}|}\sum_{\{S|\underset{\sim}{w}\}\in N_{\lambda\mu}/T}\chi_{\lambda p}^{\underset{\sim}{k}_i}(\{S|\underset{\sim}{w}\})\chi_{\mu p}^{\underset{\sim}{k}_i}(\{S|\underset{\sim}{w}\})\chi_r^{\underset{\sim}{k}_\ell *}(\{S|\underset{\sim}{w}\}) \tag{12}$$

In equations (10) - (12)

$$N_{\lambda\mu} = G^{R_\lambda \underset{\sim}{k}_i} \cap G^{R_\mu \underset{\sim}{k}_i} \cap G^{\underset{\sim}{k}_\ell}\ , \tag{13}$$

$$\chi_{\nu p}^{\underset{\sim}{k}_i}(\{S|\underset{\sim}{w}\}) = \chi_p^{\underset{\sim}{k}_i}(\{R_\nu|\underset{\sim}{v}_\nu\}^{-1}\{S|\underset{\sim}{w}\}\{R_\nu|\underset{\sim}{v}_\nu\})\ ,\ (\nu = \lambda,\mu) \tag{14}$$

T is the invariant subgroup of translations, and $\chi_p^{\underset{\sim}{k}_i}$, $\chi_r^{\underset{\sim}{k}_\ell}$ are the characters of the small representations $\Gamma_p^{\underset{\sim}{k}_i}$, $\Gamma_r^{\underset{\sim}{k}_\ell}$ of $G^{\underset{\sim}{k}_i}$, $G^{\underset{\sim}{k}_\ell}$ respectively. In equation (8) the summation over λ,μ is taken over all WVSRs (R_λ,R_μ) given by equation (5). Equation (8) has been incorporated into the KP program which has run successfully for all special $\underset{\sim}{k}$ vectors $\underset{\sim}{k}_i\ \varepsilon\ \Phi$ for all 230 space groups.

One of us (BLD) is grateful to Professor G.W. Mackey for a most stimulating discussion during the poster session at the conference.

1. A.P. Cracknell and B.L. Davies, 1977, Lecture Notes in physics, 79, Group theoretical methods in physics, Sixth International Colloquium, Tübingen, 1977, 405 - 407 (Springer, Berlin).

2. B.L. Davies and A.P. Cracknell, 1977, Lecture notes in physics, 79, Group theoretical methods in physics, Sixth International Colloquium, Tübingen, 1977, 408 - 410 (Springer, Berlin).

3. A.P. Cracknell and B.L. Davies, Kronecker product tables (Plenum, New York) (in press).

4. P. Gard, 1973, J. Phys. A : Math. Nucl. Gen., 6, 1807 - 1828.

5. G.W. Mackey, 1953, Am. J. Math., 75, 387 - 405.

6. C.J. Bradley and B.L. Davies, 1970, J. Math. Phys., 11, 1536 - 1552.

7. C.J. Bradley, 1966, J. Math. Phys., 7, 1145 - 1152.

A NEW METHOD FOR CALCULATING CLEBSCH-GORDAN COEFFICIENTS

R. Dirl

Institut für theoretische Physik, TU Wien
A-1040 Wien, Karlsplatz 13; Austria

The present paper deals with a new method for calculating CG-coefficients for finite (or compact continuous) groups. This method consists of considering the columns of the CG-matrices as G-adapted vectors of a Euclidean space.

$$\mathbf{D}^{\alpha\beta}(x)\,\vec{C}_j^{\alpha\beta;\gamma w} = \sum_k \mathbf{D}^{\gamma}_{kj}(x)\,\vec{C}_k^{\alpha\beta;\gamma w} \tag{1}$$

Thereby $\mathbf{D}^{\alpha\beta}(x) = \mathbf{D}^{\alpha}(x) \otimes \mathbf{D}^{\beta}(x)$ denotes projective matrix representations of a Kronecker product of G, and $\{\vec{C}_j^{\alpha\beta;\gamma w}\}_{pr} = C^{\alpha\beta}_{pr;\gamma wj}$ the matrix elements of the corresponding CG-matrix $C^{\alpha\beta}$. In order to obtain unitary CG-matrices it suffices to require, that the vectors $\vec{C}_j^{\alpha\beta;\gamma w}$ are orthonormal with respect to all indices γwj. Orthogonality of these vectors with respect to γj follows immediately from the transformation properties (1) by applying the usual projection method, whereas the orthogonality with respect to the "multiplicity index" w can be achieved in different ways. This can be done in any way by Schmidt's procedure or by applying the corresponding projection operator to elements of a special orthonormalized basis of the underlying Euclidean space. In doing so we obtain special vectors which are (apart from the normalization) already a part of the desired columns of the CG-matrix, if a special orthogonality relation is satisfied. Hence the present method allows to identify in a systematic way the multiplicity index w in terms of special column indices $(q_v s_v)$ of the original matrix $\mathbf{D}^{\alpha\beta}$.

The utility of the present method [1] has been demonstrated recently as CG-coefficients for non-symmorphic space groups were computed [2,3,4]. Without reference to a special space group simple defining equations for the multiplicity index were derived. The reason for obtaining these simple equations arises from the generalized permutation structure of space group unirreps due to the method (induction) by which they are usually found. For a detailed discussion the reader is referred to Refs.1-4.

1 R. Dirl: Clebsch-Gordan Coefficients: General Theory (J.Math.Phys., in press)
2 R. Dirl: Multiplicities for Space Group Representations (J.Math.Phys., in press)
3 R. Dirl: Clebsch-Gordan Coefficients for Space Groups (J.Math.Phys., in press)
4 R. Dirl: Clebsch-Gordan Coefficients for Pn3n (J.Math.Phys., in press)

A BIOLOGICAL FIELD THEORY OF EVOLUTION*

G. L. Findley and S. P. McGlynn

Choppin Chemical Laboratories
The Louisiana State University
Baton Rouge, Louisiana 70803

INTRODUCTION: Watson and Crick [1] suggested that the information content of a gene, at the molecular level, resides in the linear arrangement of the nucleic acid bases T, C,A,G (T ≡ thymine, C ≡ cytosine, A ≡ adenine, G ≡ guanine) in DNA (deoxyribonucleic acid). Since then, extensive chemical and biological investigations have greatly expanded our knowledge of the fundamental processes which comprise molecular genetics [2]. Molecular genetics is concerned with investigations into the physico-chemical basis for heredity. Such research has lead to a theory of molecular genetics of which the cornerstone is the so-called "central dogma" [2]: that is, that DNA replication, DNA transcription into m-RNA (messenger ribonucleic acid) and m-RNA translation into protein serve as the molecular basis for the hereditary process.

Perhaps one of the most exciting prospects to arise from the advances in molecular genetics is the possibility of investigating biological evolution through a study of changes in the linear base sequence of DNA. The purpose of the present work is to provide a mathematical framework descriptive of the time evolution of the information content of DNA.

FORMALISM: Space limitations preclude presentation of details [3]. Thus, we confine ourselves to a qualitative discussion of Evolutionary Field Theory (EFT).

The mathematical foundation of EFT resides in a 65-dimensional differentiable manifold in a coordinatization such that each manifold point represents (i), the number of DNA codons [4] of each specific type in a physically-realizable or non-physically-realizable DNA; and (ii), the evolutionary time of the DNA. This manifold is termed the informational space-time manifold, M. Specific considerations of the general biological problem lead one to conclude that M cannot be Euclidean but must be taken, at least conditionally, to be Riemannian. Such a distinction leads to a decomposition of evolution into (i), permutational evolution, PE (i.e., that which proceeds via a permutation of codon order in a DNA); and (ii), substitutional evolution, SE (i.e., that which proceeds via an overall change in the number of codons of a specific type in a DNA). Biological evolution, at the molecular level, is a composite of PE and SE.

Curves in M are interpreted to represent the evolutionary progress of DNA. Evolutionary motions in M are taken to be geodesics, and this stipulation restricts the intrinsic structure of M to be that which accurately reflects the nature of biological evolution. It follows, then, that the curvature of M determines a biological _evolutionary field_, and _evolutionary equations-of-motion_ result from the geodesic equations. Finally, it follows that the solution to those evolutionary questions which are formulated at the DNA level resides, in principle, in knowledge of the intrinsic structure of M: that is, knowledge of the biologically correct _genetic-cosmology_.

DISCUSSION: This work, formal though it be, represents a significant simplification: namely, attention is redirected from the complex physico-chemical processes involved in evolution (as mediated through natural selection) to the totally geometric concept of an evolutionary field produced by the curvature of M.

The distinction between PE and SE is trenchant. This distinction, in fact, serves as a critical test of EFT, since the essential difference between PE and SE appears not to have been noticed before. Additionally, a recent extension of EFT [5] indicates the possibility of taking SE to be determinative of the curvature of M, while treating PE as a gauge field.

REFERENCES: 1. J. D. Watson and F. H. Crick, _Nature_ _171_, 964 (1953).

2. J. D. Watson, _Molecular Biology of the Gene_, 3rd ed., W. A. Benjamin, Inc., Menlo Park, Calif. (1976).

3. G. L. Findley, "A Riemannian-Geometric Realization of Molecular Genetics," Ph.D. Thesis, LSU (1978); G. L. Findley and S. P. McGlynn, "Riemannian Geometry and Molecular Genetics," _J. Math. Phys._, submitted for publication.

4. A DNA codon is a triple of the DNA bases {T,C,A,G}.

5. G. L. Findley and S. P. McGlynn, in preparation.

*Research supported by USDOE.

GENERAL FLODMARK-BLOKKER METHOD FOR IRREDUCIBLE REPRESENTATIVES

by

Stig Flodmark and Per-Olof Jansson

Institute of Theoretical Physics, University of Stockholm

Vanadisvägen 9, S-113 46 Stockholm, Sweden

In dealing with the crystallographic space groups a computer program IRREP for constructing irreducible representations of finite groups exists |1|. The theory is now generalized and programmed. Using Dixon's method for irreducible characters |2| the projection operator (in the regular representation) for a space containing only the jth irrep (l_j times) is found. If the group contains an element without degeneracy in the irrep its eigenstate in this l_j^2-dimensional space transforms irreducibly |3|. Otherwise we proceed as follows:

(i) Find a common eigencolumn C belonging to the jth irrep of a complete (degeneracy removing) set of commuting regular matrix representatives.

(ii) Find l_j linearly independent columns according to $V(A) = \Gamma^{reg}(A)\ C\ (A \in G)$ and orthogonalize these columns.

(iii) Find the irreps as ${}^j\Gamma(S) = V^{\dagger}\Gamma^{reg}(S)V$, where V is a rectangular matrix made up by the l_j orthonormal columns in (ii).

Method to find a common vector space $V_A \cap V_B$ of the degenerate eigenspaces V_A, V_B of A and B: $V_A + V_B = (V_A \cap V_B) \oplus (V_A^{\perp} + V_B^{\perp})$. The intersection $V_A \cap V_B$ is found by orthogonalizing vectors in the union $V_A + V_B$ towards $V_A^{\perp} + V_B^{\perp}$. This procedure is used repeatedly in (i).

To prove the generality of our method we state the following theorem: Any matrix M with the same dimension as the irrep ${}^j\Gamma$ can be written as a linear combination over the irreps of the elements of the group: $M = \Sigma_{A \in G}\ C_A\ {}^j\Gamma(A)$. This theorem is a consequence of the orthogonality relations $N_{rs}^{pq} = (l_j/g)\ \Sigma_{A \in G}\ {}^j\Gamma^*_{pq}(A)\ {}^j\Gamma_{rs}(A) = \delta_{pr}\ \delta_{qs}$. Defining a matrix N^{pq} with these elements, thus $M = \Sigma_{p,q}\ M_{pq}\ N^{pq}$ (where M_{pq} are arbitrary complex numbers) is an arbitrary matrix. Comparing the two expressions for M above, we obtain $C_A = (l_j/g)\ \Sigma_{p,q}\ M_{pq}\ {}^j\Gamma^*_{pq}(A)$, which proves the stated theorem.

In case of degeneracy for certain group elements there is always some other group element diminishing this degeneracy. As a matter of fact, we can choose the matrix M above as a block matrix with the first diagonal block diagonal with different diagonal elements. The degeneracy of a group element in the corresponding subspace can obviously be removed by some other group element, due to the form of M. Thus there is always a complete set in (i) above.

1. S. Flodmark, J. Comp. Phys., 25, 314 (1977).

2. S. Flodmark, E. Blokker, in 'Group Theory and its Applications' 2, Acad. Press NY, 1971.

3. SF, EB, Int. J. Quantum Chem. 1S (1967), 703; 4 (1971), 463; 5 (1971), 569; 6 (1972), 925.

GROUP STRUCTURE FOR GENERAL LATTICE SYSTEMS AND SURFACE TENSION

Ch. Gruber, EPFL, Lausanne, Switzerland

J.R. Fontaine, Université de Louvain, Louvain-La-Neuve, Belgique

1. Abstract We suggest to introduce the surface tension as definition of a phase transition. We then give a geometrical criteria to prove the existence or absence of such a phase transition using the group structure associated with lattice systems.

2. Group Structure Let $\mathcal{L}$ be a countable set of points in $\mathbb{R}^\nu$; with each x in $\mathcal{L}$ is associated a L.C.A. group $\mathcal{G}_x = \{\theta_x\}$ and $\mathcal{G}_{\mathcal{L}} = \prod_{x\in\mathcal{L}} \mathcal{G}_x = \{\underline{\theta}\}$. Let $\mathcal{B}$ be a countable set of indices; with each b in $\mathcal{B}$ is associated a L.C.A. group $\hat{\mathcal{G}}_b = \{\vartheta_b\}$ and $\hat{\mathcal{G}}_{\mathcal{B}} = \prod_{b\in\mathcal{B}} \hat{\mathcal{G}}_b = \{\underline{\vartheta}\}$. Finally $\gamma : \underline{\theta} \mapsto \gamma(\underline{\theta})$ is a group homomorphism of $\mathcal{G}_{\mathcal{L}}$ onto $\hat{\mathcal{G}}_{\mathcal{B}}$ and for each b in $\mathcal{B}$, V_b is a real function on $\hat{\mathcal{G}}_b$. With each b in $\mathcal{B}$ it is then possible to associate $B_b \subset \mathcal{L}$.

Then for each finite $\Lambda \subset \mathcal{L}$ and $\underline{\theta}^0 \in \mathcal{G}_{\mathcal{L}}$, $\mathcal{B}_\Lambda = \{b \in \mathcal{B}; B_b \cap \Lambda \neq \phi\}$ and the partition function $Z(\Lambda; \underline{\theta}^0)$ is defined by :

$$Z(\Lambda;\underline{\theta}^0) = \int_{\mathcal{G}_\Lambda} \prod_{x\in\Lambda} d\theta_x \prod_{b\in\mathcal{B}_\Lambda} \exp V_b[\gamma_b(\underline{\theta}+\underline{\theta}^0)]$$

3. Surface tension Let $\mathcal{L} = \mathcal{L}^u \cup \mathcal{L}^d$, $\mathcal{L}^u = \{x \in \mathcal{L};\ x^\nu > 0\}$; with $\underline{\theta}^1, \underline{\theta}^2$ in $\mathcal{G}_{\mathcal{L}}$ define $\underline{\theta}^{12} = \underline{\theta}^1_{\mathcal{L}^u} + \underline{\theta}^2_{\mathcal{L}^d}$. With Λ parallelipiped with sides $(L,\dots,L,M)$ and $\underline{\theta}^1, \underline{\theta}^2$ in $\mathcal{G}_{\mathcal{L}}$ we define the surface tension

$$\tau^{12} = -\lim_{L\to\infty}\lim_{M\to\infty} \frac{1}{L^{\nu-1}} \log \frac{Z(\Lambda;\underline{\theta}^{12})}{Z(\Lambda;\underline{\theta}^1)}$$

Theorem 1 $|\tau^{12}| \leq 2\,K\,C_{12}$ where $K = \sup_b \sup_{\vartheta_b} |V_b[\vartheta_b]|$

$$C_{12} = \lim_{L\to\infty} \frac{1}{L^{\nu-1}} \sup_{\underline{\theta}\in\mathcal{G}_\Lambda} |\gamma(\underline{\theta}+\underline{\theta}^2_{\mathcal{L}^d}-\underline{\theta}^1_{\mathcal{L}^d})|$$

$|\underline{\vartheta}|$ = cardinality of the set of b such that $\vartheta_b \neq 0$.

Theorem 2 ($\mathcal{G}_x$ discrete; $\underline{\theta}^1=0$, $\underline{\theta}^2 \in$ Kernel γ) If the system is such that there exists a LT-HT dual and if $V_b[0] > V_b[\vartheta_b]$ for all $\vartheta_b \neq 0$, $\exp V_b[\ ]$ is in $\mathcal{L}^1[\hat{\mathcal{G}}_b]$ then there exists T_0 such that for $T < T_0$

$$2\,K\,C_{12} \geq |\tau^{12}| \geq C\cdot C_{12} \qquad \text{with } C > 0$$

Theorem 3 ($\mathcal{G}_x$ compact; $\underline{\theta}^2$ related to $\underline{\theta}^1$ by internal symmetry). If $\mathcal{F}\exp V_b[\ ]$ is in $\mathcal{L}^1[\mathcal{G}_b]$ then there exists T_1 such that $\tau^{12} = 0$ for all $T > T_1$.

4. Conclusion For finite spin, ferromagnetic systems, there will be a phase transition associated with surface tension if and only if $C_{12} \neq 0$. C_{12} is a geometrical constant which can be computed by means of duality transformation.

DYNAMICAL SYMMETRIES OF THE TIME-DEPENDENT LEWIS-RIESENFELD HARMONIC OSCILLATOR.

Neil J. Günther

Physics Department, University of Southampton, Hampshire, England, U.K. S09 5NH.

Symmetry aspects of the Lewis-Riesenfeld oscillator Hamiltonian

$$H(t) = \frac{1}{2}\sum_{i}^{n}(p_i p_i + \omega^2(t) q_i q_i) \qquad (1)$$

have been studied in detail, [1,2,3]. For the time-independent oscillator the usual dynamical symmetry associated with the extra ("accidental") degeneracy of states is contained in the group SU(n). This same group was shown to be a minimal non-invariance of the Hamiltonian given by (1) provided one defined a symmetric tensor invariant,

$$A_{ij} = \frac{1}{2}\{\rho^{-2} q_i q_j + (\rho p - \dot{\rho} p)_i (\rho p - \dot{\rho} q)_j\} \qquad (2)$$

where $\rho = \rho(t)$ is related to the arbitrary time-dependent function, $\omega(t)$ through the nonlinear differential equation [4]

$$\rho^3(\ddot{\rho} + \omega^2(t)\rho) = \lambda^2. \qquad (3)$$

λ is a real valued constant associated with the angular momentum of auxiliary motion in a 4-dimensional hyperspace, [1].

Although A_{ij} does not commute with (1), its total time derivative vanishes

$$[A_{ij}, H(t)] = i\, \partial A_{ij}/\partial t \qquad (4)$$

so confirming it as a bona fide dynamical invariant. An algebra isomorphic to SU(n) can be shown to close under commutation with TrA_{ij}, [1]. It is in this sense that the Lewis-Riesenfeld Hamiltonian admits exact solutions analogous to those of the time-independent Hamiltonian, so that recourse to perturbative methods is rendered unnecessary.

In particular when n = 3, the traceless tensor B_{ij} has five independent components associated with the coset space SU(3)/SO(3) and together with the generators of SO(3) furnishes the regular representation for the noninvariance algebra. Sp(6) and SU(4) can be shown to play the role of higher dimensional noninvariance algebras, [2] and the respective SU(3) multiplet structure was obtained using established branching rules, [5].

The introduction of an arbitrary time-dependency into the frequency of the oscillator Hamiltonian (n = 3) has the effect of reducing the invariance symmetry from SU(3), in the time-independent case, to SO(3) for the Lewis-Riesenfeld case, (1). The form of the generators responsible for breaking the SU(3) invariance can be given a dynamical realisation, [3]. The "sourceless" form of (1) is rewritten in the interaction picture $H(t) = H_0 + H_1(t)$ by imposing the transformation $\omega^2(t) = \omega_0^2 + \Omega^2(t)$. Then H_0 is the usual time-independent oscillator Hamiltonian invariant under SU(3) while $H_1(t)$ is invariant only under the subgroup of SO(3) transformations. Although the interaction term is not a generator of SU(3) it can be shown to possess well defined transformation properties, [3]. $H_1(t)$ can be identified with the trace of a canonical bilinear form associated with the generators of $Sp(6,\mathbb{R})$.

More precisely, the bilinear $\frac{1}{2}\sum_k^3 (q_k^2 - p_k^2)$ transforms exactly as a 6 according to the commutation relations amongst the SU(3) generators

$$\frac{1}{2}\sum_k \left[J_i, (q_k^2 - p_k^2)\right] = 0$$

$$\frac{1}{2}\sum_{ij} \left[B_{ij}, \left[B_{ij}, (q_k^2 - p_k^2)\right]\right] = \frac{10}{3}\sum_k (q_k^2 - p_k^2)$$

In this normalisation 10/3 is the eigenvalue of the quadratic Casimir corresponding to the U.I.R., 6. One could anticipate this result on the basis of the branching rule $Sp(6)\downarrow SU(3)$: $\underline{21} \supset \underline{8} + \underline{6} + \overline{\underline{6}} + \underline{1}$ confirming that any symplectic form must lie in $\underline{6} + \overline{\underline{6}}$.

The class of Hamiltonians represented in (1) maybe of importance in Relativity Theory [6] through their transformation properties [1]. The relevance of the associated dynamical symmetries for quantum mechanics [7] and field theory [2,8] remain to be investigated further.

References:

[1] N.J.Günther and P.G.Leach, J.M.P.18, 572(1977).

[2] N.J.Günther, J.M.P. (To be published).

[3] N.J.Günther, J.M.P. (Submitted for publication).

[4] C.J.Eliezer and N.J.Günther, La Trobe Univ. Report(1976).

[5] R.C.King, Unpublished lecture notes - Univ. of Southampton(1977).

[6] J.A.Wheeler, private communication.

[7] P.W.Higgs, (To be published).
H.I.Leemon, J.Phys.A(Submitted for publication.)

[8] G.'t Hooft, Phys.Rev.D14, 3432(1976).
F.R.Ore Jr., Phys.Rev.D15, 470(1977).

INDUCED REPRESENTATIONS AND SPONTANEOUS MOLECULAR SYMMETRY BREAKING

William G. Harter
School of Physics
Georgia Institute of Technology
Atlanta, GA 30332

Chris W. Patterson
Univ. of Calif. - LASL
Theoretical Division T-7
Los Alamos, NM 87545

The energy levels of symmetric systems are generally associated with irreducible representations (irreps) of their symmetry groups according to the earliest ideas of WIGNER[1] The states belonging to each degenerate level are "partners which span an irrep basis. Generally higher symmetries tend to have dimensional irreps and correspondingly higher degeneracy. The term symmetry breaking refers to the results of putting a system of $\mathcal{G}$ symmetry into an externally applied perturbation of lower symmetry $\mathfrak{g}$. Then each energy level E^A of degeneracy ℓ^A, splits into sublevels E^a, E^b,... and E^c of degeneracy ℓ^i where: $\ell^a + \ell^b + \ell^c = \ell^A$, according to reduction of irrep D^A as follows:

$$D^A(\text{of } \mathcal{G}) \downarrow \mathfrak{g} \sim D^a + D^b + \ldots + D^c \qquad (1)$$

into irreps D^i of broken symmetry $\mathfrak{g}$.

Now recent results involving molecular spectra have shown clearly that another type of symmetry breaking exists which tends to increase rather than reduce degeneracy. For lack of a better name one may call this spontaneous symmetry breaking since it is analogous to effects of the same name in solid state and high energy physics. The increased extraordinary near-degeneracies in laser spectra can be associated (HARTER AND PATTERSON (2,3)) with induced representation (indreps), which reduce as follows

$$D^a(\text{of } \mathfrak{g}) \uparrow \mathcal{G} \sim D^A + D^B + \ldots D^C \qquad (2)$$

into irreps of the larger symmetry $\mathcal{G} \subset \mathfrak{g}$. As a system "falls" into lower symmetry $\mathfrak{g}$, the levels associated with $\mathcal{G}$ irreps D^A, D^B... contained in the induced representation become degenerate.

The simplest example of indreps in molecules may be in a model of inversion-doublets in NH_3 and related XY_3 molecules. As shown in Fig. 1a the states |X up> and |X down> are mixed by an off-diagonal tunneling amplitude (-S) in the model Hamiltonian matrix <H>. The eigenstates in |+> and |-> correspond to the wavefunctions and eigenvectors depicted in Fig. 1b. They are even and odd, respectively, to horizontial reflections, i.e. they transform according to (±) irreps of the "global symmetry" $\mathcal{G} = C_h$. However, if the tunneling splitting (2S) or frequency (2S/h) between these states approaches zero then the original states |X up> or |X down> become eigenstates as indicated in Fig. 1c. Then the horizontial reflection symmetry of the molecule is broken down to local nonsymmetry C_1 and the inversion doublet becomes degenerate.

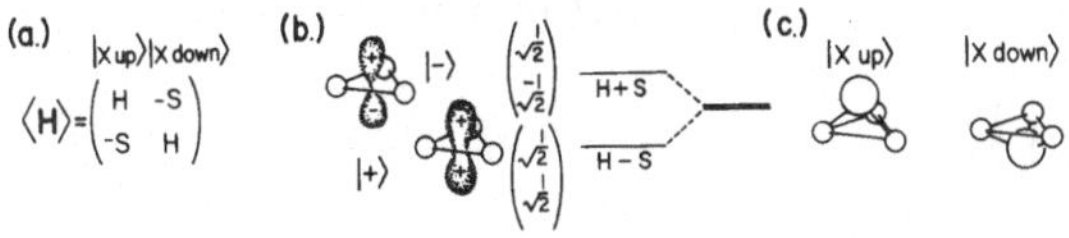

Fig. 1. Model of Inversion doubling in Ammonia (NH_3) Molecules.

It is easy to prove that inversion doublets correspond to the induced representation $D^a(\text{of } C_1) \uparrow C_h$ by the Frobenius Reciprocity relation. The reciprocity relation

$$f^A(D^a(\text{of } \mathfrak{g}) \uparrow \mathcal{G}) = f^a(D^A(\text{of } \mathcal{G}) \downarrow \mathfrak{g}) \qquad (3)$$

equates the frequency f^A which D^A appears in the indrep $D^a \uparrow \mathcal{G}$ to the more commonly derived frequency **fA** of D^a in irrep D^A reduced to $\mathcal{g}$. Since $\mathcal{g} = C_1$ has only one irrep D^a which is equal **to** $D^{\pm} \downarrow C_1$, it follows that $D^a \uparrow C_h \simeq D^+ + D^-$.

Analogous indreps appear in the theory of rovibrational spectra of octahedral (SF_6) and tetrahedral (CF_4, SiF_4, etc.) molecules. (HARTER AND PATTERSON (2)) The Hamiltonian for these molecules has the form

$$H = \nu + BJ^2 + \ldots + gT^4 \qquad (4)$$

where the interesting spectra effects come from 4th rank tensor terms T^4. For pure rotational excitations T^4 has the form of a centrifugal distortion operator

$$T^4 = \frac{7}{12} T^4_o + \frac{5}{24} (T^4_4 + T^4_{-4}) = (J^4_x + J^4_y + J^4_z - \frac{3}{5} J^4) \qquad (5)$$

where T^k are irreducible tensors and J_ν are angular momentum operators defined in the molecular frame. The function $r=[1 + gT^4]$ is plotted in $\{J_xJ_yJ_z\}$ space in Fig. 2 subject to the constraint J^2=(constant). The (g=1.2) surface is shown in Fig. 2a corresponds to the distortion properties of the octahedral SF_6 molecule; i.e., highest energy or lowest distortion for $\vec{J}$ along the 4-fold x,y, or z axes, and lowest energy and highest distortion for $\vec{J}$ along the 3-fold {1,1,1} axes. The (g=-1,2) case shown in Fig. 2b corresponds "opposite" distortion properties of tetrahedral CF_4 or cubic (XY_8) molecules

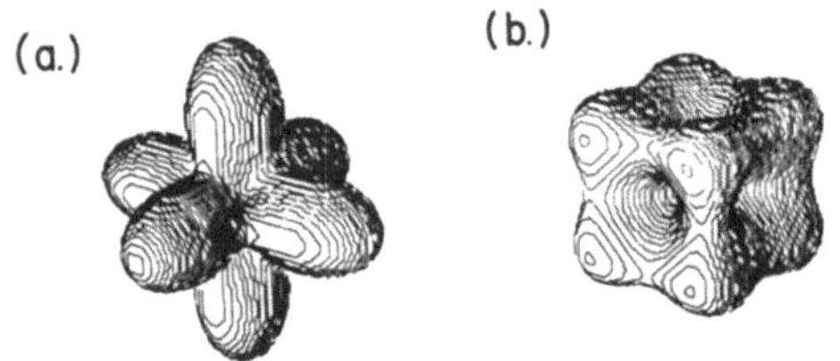

Fig. 2. Molecular Centrifugal Distortion Potential in $\{J_xJ_yJ_z$ Space (a) Octahedral (XY_6) Molecules (b) Tetrahedral XY_4 and Cubic (XY_8) Molecules.

An example of laser spectra of SF_6 shown in Fig. 3 shows lines belonging to clusters of cubic irreps A_1 or A_2 (dimension=1), E(dimension=2), and $T_1 \equiv F_1$ or $T_2 \equiv F_2$ (dimension=3). The clusters correspond to induced representations listed in the columns of the correlation tables between irreps of cubic (O) and cyclic (C_3) or (C_4) symmetry.

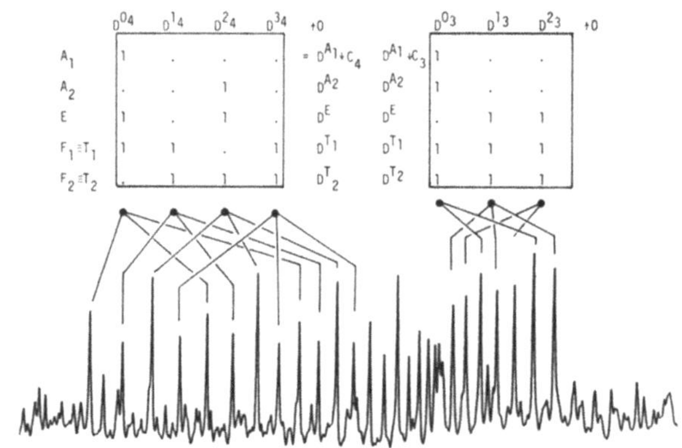

Fig. 3. ν_4 P(88) IR Spectra of SF_6 (K.C. Kim, W.B. Person, D. Seitz, and B.J. Krohn; to be published in J. Mol. Spec.) Each well defined peak corresponds to a cluster of irreps as indicated. Peak height ratios can be computed using Table 1-b.

The clusters $D^a 3{\uparrow}O = T_1 + E + T_2, A_1 + T_1 + T_2 + A_2$, etc. correspond to molecules "stuck" rotating on their three-fold axes while $D^a 4{\uparrow}O = A_1 + T_1 + E, T_1 + T_2$, etc. correspond to four-fold rotation. The rotational doublet, triplet, and quartet clusters are analogous to the ammonia inversion doublets arising when the X-atom is stuck in the "x-up" or "x-down" potential valley. To visualize the rotational sticking it helps to think of states of quantum angular momentum as cones stuck in one of the hills or valleys of Fig. 2. Let the expectation values $\langle J_z \rangle = M$ and $\langle J \rangle = \sqrt{J(J+1)}$ define the altitude and slant-height of the angular momentum cone as shown in Fig. 4a. Then the cluster energies are given to a good approximation (for high J and M) by the radial distance to the intersection of (J,M) cones and the potential hill or valley (viz. (Fig. 2) in which they are centered. (See Fig. 4b). This follows from an approximation (EDMONDS(4)) shown in Fig. 4c.

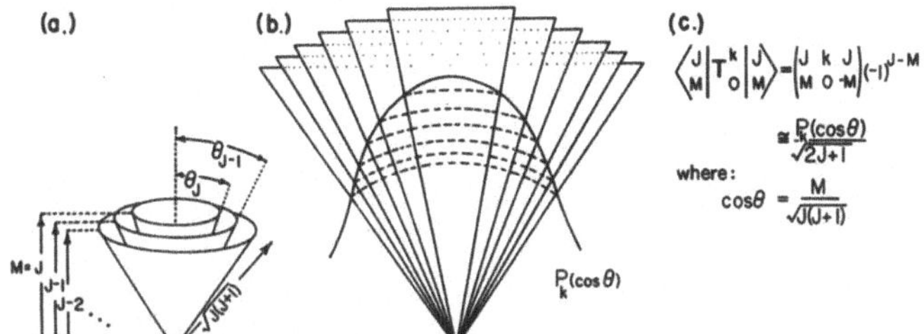

Fig. 4. Geometrical Approximation for Cluster Energies.
(a) Angular Momentum Cones
(b) Cluster energy levels at cone-valley intersections
(c) Approximation for Wigner Coefficients

Therefore clustering is due to two things: localization and symmetry. First, high momentum states $|{}^J_M\rangle$ with M nearly equal to J have momentum localized in the direction of the axis of quantization. $|{}^J_M\rangle$ states belonging to different axes, i.e., different but equivalent valleys, do not overlap appreciably and may therefore be good approximate eigenstates. Second, the symmetry of the potential makes a set of six $|{}^J_M\rangle$ states localized on six respective four-fold axes have degenerate zeroth order energies; and the same for a set of eight states localized on respective three-fold axes. (The cluster orbital degeneracies of six and eight were first noticed in computer calculations by LEA, LEASK, AND WOLF (4,5) and by DORNEY AND WATSON (6)).

Tunneling Hamiltonian matrices analogous to Fig. 1a may be defined for each cluster. For example the (-S) in Fig. 5a is the small overlap or tunneling amplitude between neighboring 3-fold valleys. The eigensolutions in Fig. 5b correspond to irreps contained in $(O_3{\uparrow}O)$, and their ordering and spacing agrees with an unidentified quartet seen in very high resolution SiF_4 spectra in Fig. 5c.

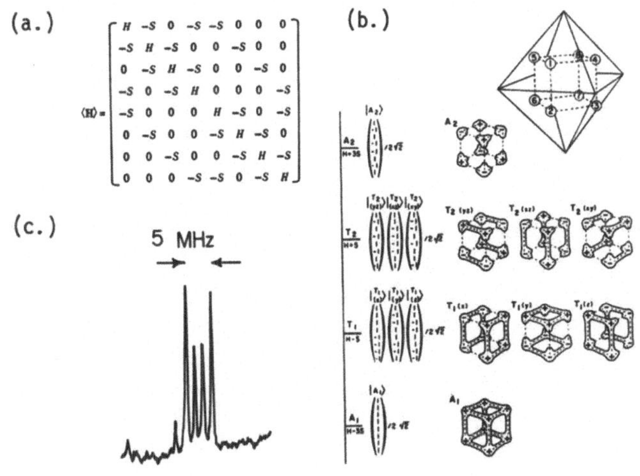

Fig. 5. $O_3{\uparrow}O$ Octahedral Cluster Splitting (a) Tunneling matrix. (b) Predicted splitting of irrep levels. (c) observed superfine spectra of SiF_4 by H.P. Layer and F.R. Petersen NBS (Communicated by J.T. Hougen)

The peak heights in Fig. 5c correspond to the spin-½ (SU_2) nuclear spin degeneracie of the A species (□□□□, Total spin I=2) for which 2I + 1 = 5 and the T species (□□□ over □, total spin I=1) for which 2I + 1 = 3. It is instructive to correlate the permutation tableaus of S_4 with tetrahedral symmetry of XY_4 molecules (table 1a), the tableaus of S_6 and S_8 with cubic symmetry of XY_6 and XY_8 molecules, respectively (Table b and c) for spin ½ Y nuclei. The columns in tables b and c correspond to "permutation-clusters" which are degenerate as long as XY_6 and XY_8 molecules are "stuck" in cubic subgroups of S_6 or S_8. Also it is easy to prove that E-species of XY_4 molecules belong to "inversion clusters" which are doublets like Fig. 1. However, since it is not feasible for normal $XY_{4,6,8}$ molecules to invert or permute their nuclei, the permutation or inversion cluster splitting will normally be infinitely smaller than the inversion doubling in NH_3 spectra.

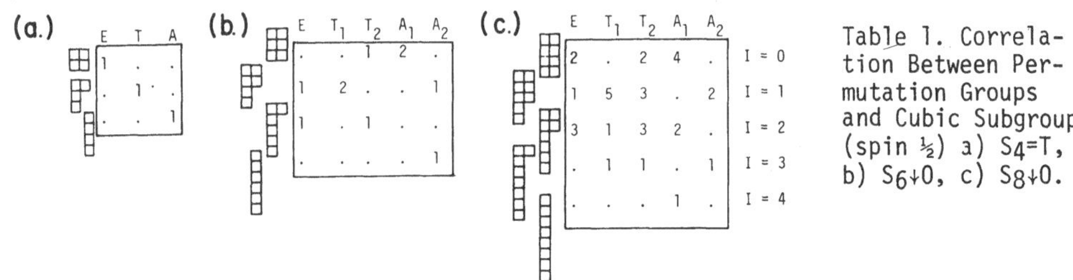

(a.)

Tableau	E	T	A
[2,2]	1	.	.
[2,1,1]	.	1	.
[1,1,1,1]	.	.	1

(b.)

Tableau	E	T_1	T_2	A_1	A_2
[2,2,2]	.	.	1	2	.
[2,2,1,1]	1	2	.	.	1
[2,1,1,1,1]	1	.	1	.	.
[1,1,1,1,1,1]	.	.	.	.	1

(c.)

Tableau	E	T_1	T_2	A_1	A_2	
[2,2,2,2]	2	.	2	4	.	I = 0
[2,2,2,1,1]	1	5	3	.	2	I = 1
[2,2,1,1,1,1]	3	1	3	2	.	I = 2
[2,1,1,1,1,1,1]	.	1	1	.	1	I = 3
[1,1,1,1,1,1,1,1]	.	.	.	1	.	I = 4

Table 1. Correlation Between Permutation Groups and Cubic Subgroup (spin ½) a) S_4=T, b) $S_6 \downarrow O$, c) $S_8 \downarrow O$.

Nevertheless, it may be possible to see permutation-inversion structure under so-called Case 2 conditions when superfine (rotational cluster) splitting becomes less than hyperfine splitting. (HARTER AND PATTERSON, 7) As A, T, or E levels become more nearly degenerate these states can be more easily mixed by nuclear spin-rotation or other hyperfine interactions, and they are perturbed as shown in Fig. 6.

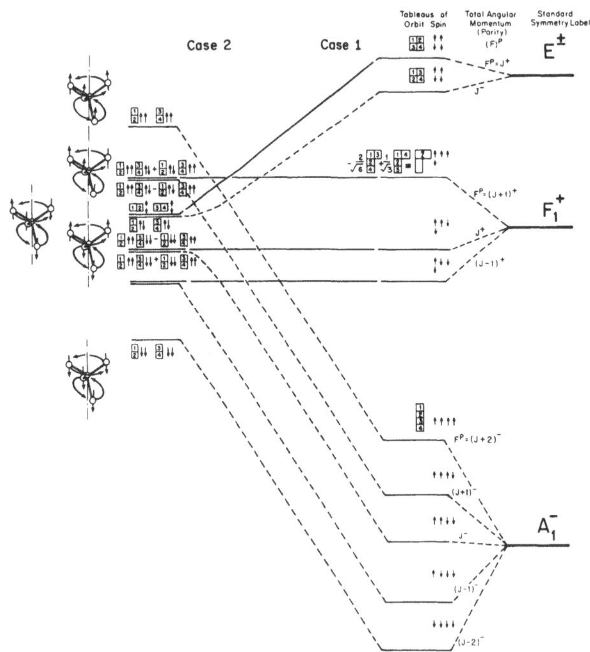

Fig. 6. $O_4 \uparrow O$ Cluster Level Correlation Between Case 1 Case 2.

The cluster splitting and hyperfine mixing is a function of how deeply the momentum cone in Fig. 4 intersects the valley; tunneling amplitudes in a cluster spectrum such as Fig. 3 vary over fifteen or twenty orders of magnitude. (See also Ref. 7). The splitting is least and the mixing is greatest for clusters on the limbs of the pattern since the momentum for these is most nearly localized to an internal symmetry axes. Most SF_6 or CF_4 molecules at room temperature may well be in nearly degenerate Case 2 cluster states for which the hyperfine mixing is important.

This mixing marks the final stage of the symmetry breaking. The nuclear spins align themselves to one internal rotation axis. Tunneling or tumbling onto other symmetry axes is unfeasible for the molecule. This implies that some "rotational" permutations become unfeasible, too. Then labeling of the mixed-species states can be done using broken tableaus as shown by the case (2) side of the $O_4{\uparrow}O$ cluster in Fig. 6. At this point the permutation or inversion of nuclear spin states rather than the nuclei themselves may become important. For example the inversion doublet (E-$\begin{smallmatrix}1&2\\3&4\end{smallmatrix}$, $\begin{smallmatrix}1&3\\2&4\end{smallmatrix}$) in Fig. 6 will be split by odd permutations of nuclear spin states or of nuclei. While the latter is impossible in ground states of stable molecules, the former is possible through hyperfine interactions whose effects become important as Case 2 is approached.

In conclusion, the formation of degenerate rotational clusters ($C_4{\uparrow}O$, $C_3{\uparrow}O$,...) may be accompanied by the splitting of permutation-inversion clusters ($A{\uparrow}S_n$, $E{\uparrow}S_n$,...) due to spin exchange. The study of spontaneous symmetry breaking promises many interesting physical and chemical effects based on the theory of cluster spectra.

1. E.P. Wigner, Gruppen theorie und ihre Anwendung anf die Quantenmechanik de Atomspektren, Vieweg Braumschweig (1931) (Trans. by J.J. Griffin, Academic Press, 1959).
2. W.G. Harter and C.W. Patterson, Phys. Rev. Lett. 38, 224 (1977); J. Chem. Phys. 66, 4872 (1977); Int. J. Quantum Chem., Symposium 11, 479 (1977).
3. W.G. Harter, C.W. Patterson, and F.J. Paixao, Rev. Mod. Phys. 50 37 (1978).
4. A.R. Edmonds, Angular Momentum in Quantum Mechanics, Princeton Univ. Press (1957), p. 122.
5. K.R. Lea, M.J.M. Leask, and W.P. Wolf, J. Phys. Chem. Sol. 23, 1381 (1962).
6. A.J. Porney, and J.K.G. Watson, J. Mol. Spec. 42, 1 (1972).
7. W.G. Harter and C.W. Patterson, in Advances in Laser Chemistry III, Ed. A.H. Zewail, Springer-Verlag, (1978) p. 455-463.

"Phase ordering" and symmetry in $Hg_{3-\delta}AsF_6$

A. Janner and T. Janssen, Institute for Theoretical Physics, University of Nijmegen, Nijmegen, NL.

1. General Considerations. According to J.P. Pouget et al.[1)] the compound $Hg_{3-\delta}AsF_6$ has a composite crystal structure which at $120^{\circ}K$ undergoes a phase ordering phase transition. In this structure one distinguishes 3 subsystems: one is formed by the AsF_6 octahedra (building the host lattice) and the two others consist of Hg chains parallel to the a- and b-axis of the host lattice, respectively. Each subsystem has space group symmetry (disregarding modulations), but since corresponding lattice constants are incommensurable, there is no 3-dim. space group description for the whole system. At room temperature the relative phases of the Hg chains are at random, but this is no longer the case below $120^{\circ}K$. The structure then has long range order and the superspace group approach developed by the authors [2)] can be applied.

2. Summary of the experimental data. The AsF_6 subsystem has a $G_1 = I4_1/amd$ symmetry; the Hg(1) one (formed by chains along the a-direction) has $G_2 = A2/m$, and the Hg(2) subsystem (chains along the b-direction) $G_3 = B2/m$ symmetry. The corresponding Bragg reflections (expressed in the AsF_6 reference system) are indexed by $(hk\ell)$, $(h(3-\delta),\ k-h\delta,\ell)$ and $(h-k\delta,\ k(3-\delta),\ell)$ respectively with $h+k+\ell = 2n$. Furthermore the phase ordering is revealed by m=0 extinction and m=1 enhancement of reflections of the type $(m(3-\delta),\ m(3-\delta),0)$. The δ-parameter is related to the ratio of the Hg and AsF_6 lattice constants and to the relation between the origins of the two Hg subsystems (at $(x\ \frac{1}{4}\ \frac{1}{4})$ and $(0\ y\ 0)$ respectively) with respect to the AsF_6 lattice. One has: $2(x-y+\frac{1}{4}) = (3-\delta)^{-1} = a_{Hg}/a_{AsF_6}$.

3. The superspace group approach. The system of Bragg reflections is described by a 4-dim. Z-module generated by the reciprocal vectors (011), (101), (110) and $(\delta\delta 0)$. After imbedding in a 4-dim. space, such that the above vectors appear as 3-dim. projection of 4 linearly independent vectors, a 4-dim. periodic structure is obtained whose set of equivalent positions has the form: $P_{\nu j} = (\vec{r}_{\nu j} + \vec{n}_{\nu} - \pi_{\nu}\vec{t},\vec{t})$, where $\nu = 1,2,3$ labels the subsystems, j the inequivalent atoms in the unit cell of the ν-th lattice Λ_{ν}; $\vec{n}_{\nu} \in \Lambda_{\nu}$; π_{ν} are suitably defined projections, and $\vec{t}$ is here the 4th coordinate. The symmetry transformations are of the form $g = (g_E,g_I) \in E(3) \times E(1)$ and are indicated here in the usual notation for euclidean transformations: $g_E = \{R_E|\vec{a}_E\}$, $g_I = \{R_I|\vec{a}_I\}$. The symmetry conditions for the sets $P_{\nu j}$ then become:
$\{R_E|\vec{a}_E + \pi_{\nu}\vec{a}_I\} \in G_{\nu}$ and $\pi_{\nu}R_I\vec{t} = R_E\pi_{\nu}\vec{t}$.

4. Superspace symmetry of $Hg_{3-\delta}AsF_6$. One gets $G = P^{I2/m}_{\bar{1}\,\bar{1}\,s}$ as a 4-dim. superspace group generated by the lattice translations $(\bar{\tfrac{1}{2}}\tfrac{1}{2}\tfrac{1}{2}0)$, $(\tfrac{1}{2}\bar{\tfrac{1}{2}}\tfrac{1}{2}0)$, $(\tfrac{1}{2}\tfrac{1}{2}\bar{\tfrac{1}{2}}\delta)$ and (0001) and by the following two elements: $g=(\{2_z|(0\tfrac{1}{2}0)\},\{\bar{1}|0\})$ and $g'=(\{m_z|(0\tfrac{1}{2}0)\},\{1|\tfrac{1+\delta}{2}\})$. These two last generators are present as a consequence of the observed phase ordering and are responsible for the corresponding systematic extinctions.

5. References.
1) J.P. Pouget et al. Brookhaven National Lab. preprint BNL-24306. 2) A. Janner and T. Janssen, Phys.Rev. B15 (1977)643; A. Janner in Group Theoretical Methods in Physics, R.T. Sharp and B. Kolman ed., Academic Press (1977), p.9; T. Janssen in Electron-Phonon interactions and Phase Trans., T. Riste ed., Plenum Press, (1977), p.172.

On the application of group theory to the renormalization-group method*

Marko V. Jarić

University of California, Physics Department, Berkeley, Calif. 94720

Abstract

We have summarized the results of an investigation into the symmetry properties of renormalization-group (RG) transformations. We also formulated problems for further research: (i) classification of equivalent Hamiltonians and corresponding RG transformations for l-component fields; (ii) generalizations to lattice systems; (iii) Further study of the appearence of continious symmetry groups of transformations commuting with RG.

The purpose of the present paper is to summarize the results of the investigations into symmetry properties of the renormalization-group (RG) transformations as well as to formulate some of the questions that remain unanswered.

Jona-Lasinio[1] has been interested in studying classes of equivalent (i.e. related by difeomorphisms) RG transformations. He has established several important theorems concerning such RG transformations. Wegner[2] and Green[3] have also established several invariance theorems for RG transformations generated by a flow vector.

The connection between classes of equivalent Hamiltonians and classes of RG transformations was also studied. Korzhenevskii[4] has studied the connection between the properties of the RG transformation and the symmetry of the Hamiltonian. He considered an example of a two-component field with cubic symmetry. Zia and Wallace[5] have previously analyzed a general case of two- and three-component fields. In both cases Landau-Ginzburg-Wilson Hamiltonians were considered.

The author[6] has studied a general l-component field ψ. He has considered the LGW Hamiltonian with m linearly independent, D-invariant, quartic polynomials in ψ ($D \subseteq O(l)$). Corresponding m coupling constants form a parameter space Π on which RG transformations R act. It was

*Supported in part by grants: CUNY-FRAP; NSF-DMR 76-20641-A01; and by the Miller Institute for Basic Researsh in Science.

shown that a group G_g of all orthogonal transformations on Ψ which leave the Hamiltonian form-invariant is a normalizer of D in O(l):

$$G_g=\{g:\ g^{-1}Dg=D,\ g\ O(l)\} \quad . \tag{1}$$

G_g induces on Π a group G_T of orthogonal transformations which commute with R's. G_T is an orthogonal representation of G_g. Expansion of R into covariants of G_T was introduced and properties of fixed points were studied. In the case when

$$D=\Delta\oplus\Delta^* \quad , \tag{2}$$

and Δ has at least one quartic invariant, and when

$$\Delta \mathcal{Q} \Delta^* \tag{3}$$

(i.e. Δ and Δ^* are quasi-equivalent), then

$$O(2)\subseteq G_T \quad . \tag{4}$$

Thus in this case lines of fixed points, generated by O(2), can appear.

It appeares important to study complex representations Δ in order to answer the question as to whether Δ has quartic invariants. A more intriguing question is: under which conditions are Δ and Δ^* quasi-equivalent? Another significant problem would be to analyze a general Hamiltonian with $\frac{1}{4!}l(l+1)(l+2)(l+3)$ quartic terms for an l-component field. The classes of equivalent Hamiltonians and corresponding R's should be deduced using a subduction of the fourth symmetrized power of O(l). This question was partialy answered in Ref.5.

As a final remark, there remains a question of the generalization of the work of Ref.6 to spin lattice systems. One would think that group-theoretical methods would be more readily applicable to such systems. Unfortunately until the present such applications have been very limited.

References: [1] G. Jona-Lasinio in Collective Properties of Physical Systems, Nobel Symposium Vol.24, edited by B. and S. Lundquist (Academic, New York, 1973); [2] F.J. Wegner, J.Phys. C7, 2098 (1974) and in Phase Transitions and Critical Phenomena, Vol.6, edited by C. Domb and M.S. Green (Academic, London, 1976); [3] M.S. Green, Phys, Rev. B15, 5367 (1977); [4] A.L. Korzhenevskiǐ, Sov. Phys.-JETP, 44, 751 (1976); [5] R.K.P. Zia and D.J. Wallace, J. Phys. A8, 1089 (1975); [6] M.V. Jarić, Phys. Rev. B18, xxx and xxx (Sep. 1, 1978).

Renormalization Group Theory of Structural Phase Transitions in A-15: Pm3n-O_h^3 †

Marko V. Jarić* and Joseph L. Birman
Physics Department, City College, C.U.N.Y., New York, N.Y. 10031
* Presently: Dept. of Physics, University of California, Berkeley, CA 94720.

Abstract

Results of a RG theory of phase transitions in A-15, or Pm3n structure are discussed and compared to the Landau theory results.

Space group Pm3n-O_h^3 is of interest owing to the occurrence of high T_c superconductors (V_3Si; Nb_3Sn and others) in this structure as well as owing to a structural phase transition (Pm3n → tetragonal group) which occurs in some systems at a proximate temperature T_m. Basis functions for the irreducible representations of Pm3n are likely candidates for "order parameters" involved in these transitions: they may be interpreted using electronic, "soft phonons", band Jahn-Teller or other pictures. High interest attaches in particular to the multidimensional irreducible space group representations of Pm3n, especially those with wave vectors $\Gamma:(0,0,0)$; $X:(\pi/a,0,0)$; $R:(\pi/a, \pi/a, \pi/a)$.

Previously we analyzed possible second order "Landau" phase transitions: Pm3n → G, where G is a subspace group, "driven" by order parameters of types $\vec{\Gamma}(j^{\pm})$, $j=1,\ldots 5$; $*\vec{X}(j)$, $j=1,\ldots 4$; and $*\vec{R}(j) = \vec{R}(j)$, $j=1,2,3$. Note the little group of $\vec{R}$ is m3n, so $*\vec{R}=\vec{R}$; and representation R(3) is six dimensional - the highest allowed for any space group. In that analysis, Landau, and Lifschitz, and subduction (and chain subduction) criteria were used.

In the present work the Wilson-Fisher RG method, and "ε expansion" was used to investigate fixed points of Landau-Ginzburg-Wilson Hamiltonians

$$H = -\frac{1}{2}\int dv\ [\ \gamma\psi_\alpha^2 + \nabla\psi_\alpha\cdot\nabla\psi_\alpha - u_k\ I_k(\{\psi_\alpha\})] \qquad (1)$$

where $\{\psi_\alpha\}$ are the set of basis functions of the irreducible representation; I_k are independent quartic polynomial invariants composed from $\{\psi_\alpha\}$; γ, and u_k are parameters; and sum convention is used. All the above-mentioned $\vec{\Gamma}(j^{\pm})$, $*\vec{X}(j)$, $\vec{R}(j)$ were analyzed.

We find that only 1,2, and 3 dimensional irreducible representations give stable fixed points ; we obtain the critical exponents by calculation or by gauge transformation of (1) to equivalent solved Hamiltonian. A new result is the critical exponents for the RG and Landau allowed second order transition based on R(1); the RG theory gives critical exponents, and agrees with Landau theory for $\Gamma(j^{\pm})$; neither any $*X(j)$ nor $R(j)$ $j = 2,3$ give second order transitions. All the RG results are given in more detail elsewhere: MVJ & JLB, Phys. Rev. B17, 4368 (1978).

† Supported in part by: PSC-BHE grant #12343, NSF-DMR78-12399 and the Miller Institute for Basic Research in Science.

"Structure of the Images of the Irreducible Linear Representations of the Crystallographic Little Space Groups"

Louis Michel, IHES, 91440 Bures-sur-Yvette, France
Jan Mozrzymas, Institute for Theoretical Physics, Wrocław University, Poland
Boguslaw Stawski, Technical University, Wrocław, Poland

Many problems in solid state physics require the use of the irreps (= irreducible linear representations) of the space groups G_j (j = 1 to 230). A representation of a group is determined by its kernel and its image. Most of its properties depend only on the image (e.g. the invariants). An irrep of G_j is characterized by a wave vector k and is induced from an irrep of $G_{j,k}$, the little group of k in G_j. There is an infinite set of irreps of the $G_{j,k}$'s. We have established that the image of each one is generated by one of <u>only 25 images</u> of irreps of a Schur covering $S(P_k)$ of the little point group P_k and a set of phases forming a U(1) subgroup. In the most physically important casesof finite image, these phases form a cyclic group of order n, this integer depending on k. We have characterized these 25 images.

We can also say that the images of the irreps of $G_{j,k}$ are also the images of the irreps of P(j,k), central extensions of P_k by Imk (i.e., $P_k = P(j,k)/\mathrm{Imk}$, and Imk $\subset$ center of P(j,k)) where k maps the translation subgroup τ of G_j into U(1) $t \mapsto e^{-ik\cdot t}$. Note that P_k and Imk are independent of "j" in any of the 73 arithmetic classes and we proved that they are so related:

$P_k = C_i, C_{2h}, C_{4h}, D_{2h}, S_6, C_{6h}, D_{3d}, D_{4h}, D_6, D_{6h}, T_h, O, O_h$	$\mathrm{Imk} = \{1\}, Z_2$
$P_k = D_2, D_{2d}, S_4, D_4, T, T_d$	$\mathrm{Imk} = \{1\}, Z_2$ or Z_4
$P_k = D_3, C_{3h}, D_{3h}$	$\mathrm{Imk} = \{1\}, Z_2, Z_3$ or Z_6

P_k =	C_6, C_{6v}	C_3, C_{3v}	C_2, C_{2v}, C_4, C_{4v}	C_s	C_1	
Imk =	Z, Z_m	$Z \times Z_3, Z, Z_m$	$Z \times Z_2, Z, Z_m$	$Z^2 \times Z_2, Z^2, Z \times Z_m, Z_m$	$Z^3, Z^2 \times Z_m, Z \times Z_m, Z_m$	$m \geq 1$

The G_j's are extensions corresponding to the cohomology group $H^2_\Delta(P,\tau)$, $P \xrightarrow{\Delta} \mathrm{Aut}\,\tau$ similarly the $G_{j,k}$'s correspond to $H^2_{\Delta'}(P_k,\tau)$ where $\Delta' = \Delta|_{P_k}$ and the P(j,k) correspond to $H^2_o(P_k,\mathrm{Imk})$ where o means trivial action. The P(j,k) are defined by the functorial homomorphisms: $H^2_\Delta(P,\tau) \xrightarrow{a_k} H^2_{\Delta'}(P_k,\tau) \xrightarrow{k} H^2_o(P_k,\mathrm{Imk})$. For the 10 cyclic P_k's the P(j,k) are Abelian. For $P_k = C_1, C_i, S_4, S_6, C_{3h}, D_3$, $P(j,k) = \mathrm{Imk} \times P_k$. Finally all P(j,k) are of the form A(j,k) x Q(j,k) where A(j,k) is Abelian. We will give in a forthcoming publication the list of the 27 (?) nonisomorphic Q(j,k) and of all Abelian A(j,k)'s.

Unitary Group Formulation of Hartree-Fock Theory†

C. J. Nelin and F. A. Matsen
Departments of Chemistry and Physics, University of Texas, Austin, Texas 78712 USA

The unitary group many-particle Hamiltonian is

$$H = \sum_{rs}^{\rho}\sum h_{rs}E_{rs} + \tfrac{1}{2}\sum_{rstu}^{\rho}\sum\sum\sum v_{rs,tu}(E_{rs}E_{tu} - \delta_{s,t}E_{ru}) \tag{1}$$

where ρ is the number of orbitals, h_{rs} is a one-particle matrix element, $v_{rs,tu}$ is a two particle matrix element, and the E_{rs} are the infinitesimal generators of $U(\rho)$, which satisfy the usual commutation relation. The Hilbert space of this Hamiltonian is

$$V_\rho = \sum_{[\lambda]} \oplus V_\rho^{[\lambda]} \tag{2}$$

where $[\lambda]$ is a partition of a positive integer, N, (For fermions $[\lambda] = [1^N]$, for ordinary spin-free electrons and isospin free nucleons $[\lambda] = [2^\rho, 1^{N-2\rho}]$, $\rho = \frac{N}{2}$ -S,I and for superfreons $[\lambda] = [4^{\lambda_4}, 3^{\lambda_3}, 2^{\lambda_2}, 1^{\lambda_1}]$) and $V_\rho^{[\lambda]}$ is an invariant space with respect to the enveloping algebra of $U(\rho)$. A basis state, $|(m)\rangle$ has a weight,

$$\{m\} = \{g_r, g_{r-1}, \cdots\} \tag{3}$$

where g_r is an occupation number. The basis states are eigenvectors to the diagonal generators, whereas, non-diagonal generators raise or lower the weight of the state. Among these basis states there exists a unique highest weight state, $|(m)^*\rangle = |\rangle$ which is the state of lowest zero-order energy. A singly excited state is defined as

$$|SE\rangle = B\, E_{xy}|\rangle \qquad g_x < g_y \tag{4}$$

where B is a normalization.

To develop a generalized restricted Hartree-Fock theory we follow work of Rowe[1] and Thouless.[2] The Hartree-Fock orbitals are those orbitals which give an extremum of the energy of the highest weight state of $V_\rho^{[\lambda]}$. We carry out the variation of the orbitals by a unitary transformation,

$$|v\rangle = e^Z|\rangle \tag{5}$$

where

$$Z = \sum_{\substack{x \\ g_x<g_y}}\sum_{y} (Z_{xy}E_{xy} - Z^*_{xy}E_{yx}) \tag{6}$$

and Z is skew-Hermitian. To calculate the energy we use the Baker-Campbell-Hausdorff expansion

$$E(v) = \langle v|H|v\rangle = \langle|e^{-Z}He^{Z}|\rangle = \langle|H|\rangle + \langle|[H,Z]|\rangle + \frac{1}{2!}\langle|[[Z,H]Z^+]|\rangle + \cdots \tag{7}$$

For the energy to be an extremum we require that terms to first order in the variation are zero. This implies n extremum conditions of the form

$$\langle|[E_{xy},H]|\rangle = 0 \tag{8}$$

where n is the number of single excitations. By substitution of the Hamiltonian (1) and (8) we obtain the extremum conditions and from there the Fock operators.

For a close shell system ($g_p = 0$ and $g_c = 1$ for fermions, $g_c = 2$ for freons) we let $x = p, y = c$, which leads to one class of extremum conditions

$$\langle|[E_{pc},H]|\rangle = h_{cp} + \sum_{c''}(g_c v_{cp,c''c''} - v_{cc'',c''p}) = 0 \tag{9}$$

and one Fock operator

$$F = \sum_{ij}\sum(h_{ij} + \sum_{c''}(g_c v_{ij,c''c''} - v_{ic'',c''j}))E_{ij} \tag{10}$$

For details see Matsen.[3] A detailed treatment of an open shell system ($g_c = 2$, $g_o = 1$, $g_p = 0$) is given by Matsen and Nelin.[4] For this system there are three classes of extremum conditions and three Fock operators.

The Hartree-Fock stability condition is

$$\langle|[[Z,H],Z^+]|\rangle \geqslant 0 \tag{11}$$

The resulting matrix equation is directly applicable in the Random Phase Approximation. (See Matsen[3] and Matsen and Nelin[4])

References

1. D. J. Rowe, Nuclear Collective Motion, Methuen and Co. Ltd., London (1970).
2. D. J. Thouless, The Quantum Mechanics of Many-Body Systems, Academic Press, New York and London (1961).
3. F. A. Matsen, Adv. Quantum Chem., in press.
4. F. A. Matsen and C. J. Nelin, Intern. J. Quant. Chem., submitted for publication.

†Supported by the Robert A. Welch Foundation of Houston, Texas.

On Symmetries of Infinite Macrosystems in a Standard Representation

A. Rieckers
Institut für Theoretische Physik
der Universität Tübingen
D-7400 Tübingen, Germany

For the dynamical treatment of infinite quantum systems - one of the great challenges for present day theoretical physics - one has developed the so-called algebraic approach with its basic notion of a quasilocal (simple) C^*-algebra $\mathcal{A}$ [1]. One of the most agreeable features of an infinitely extended dynamical system is that the intrinsic symmetries of the interaction forces are not destroyed by the accidental shape of a finite container. This gives rise to the hope that in the algebraic formulation the dynamical symmetries of many body systems can be fully exploited. The main difficulty is, however, to find the appropriate mathematical realization of the symmetry transformations. For, model studies of many body systems ([2] , [3] , [4]) have led to the conclusion that time translations cannot, in general, be realized by $*$-automorphisms of $\mathcal{A}$ but are of a more general kind which is most conveniently described in the Schrödinger picture. As transformations in the state space $\mathcal{S}(\mathcal{A})$ they are merely defined on a certain subset $\mathcal{S}_0 \subset \mathcal{S}(\mathcal{A})$ of distinguished states. And for $\mathcal{S}_0$ one should only assume the properties of convexity, norm-closedness, and invariance under local excitations. Since the dynamical symmetries commute by definition with the time translations they have to leave $\mathcal{S}_0$ invariant, too. Dynamics and dynamical symmetries being special cases of structural symmetries (corresponding to the Wigner symmetries of traditional quantum mechanics) we have generally to connect them with the selected subset $\mathcal{S}_0$. More specifically we define:

(i) The group $G_{\boxminus}$ of all structural symmetries of a class $\boxminus := (\mathcal{A}, \mathcal{S}_0)$ of physical systems consists of all affine invertible mappings μ of $\mathcal{S}_0$ onto $\mathcal{S}_0$.

(ii) If ν denotes a weakly continuous representation of the additive group $\mathbb{R}$ in $G_{\boxminus}$, the triple $\Sigma := (\mathcal{A}, \mathcal{S}_0, \nu)$ is called a dynamical system.

(iii) The dynamical symmetry group G_{Σ} of Σ is the set of all structural symmetries which commute with all ν_t , $t \in \mathbb{R}$.

It should be observed that we use here a considerable generalization of Kadison's symmetry concept [5] , since fullness of $\mathcal{S}_0$ and

$\sigma(\mathcal{S}_0, \mathcal{A})$ -continuity of μ have been dropped. Thus it is very interesting that a version of the KMS-condition is also applicable to this case, as has been shown in [6] . Hence we have the notion of a thermodynamic equilibrium state of Σ and can investigate its symmetry properties. In this note, however, we communicate only the main tool for such an analysis being the content of the theorem below.

Let π_φ be the GNS - representation of $\mathcal{A}$ corresponding to the state φ and Ω_φ the cyclic vector. We introduce $\pi := \oplus_{\varphi \in \mathcal{S}_0} \pi_\varphi$ and $\mathcal{B} := \pi(\mathcal{A})''$. Every $\varphi \in \mathcal{S}_0$ gives rise to a $\tilde{\varphi} \in \mathcal{N}(\mathcal{B})$, the set of all normal states on $\mathcal{B}$, by setting $\langle \tilde{\varphi}, A \rangle :=$ $(\Omega_\varphi, A\Omega_\varphi)$, $A \in \mathcal{B}$. One can show that this correspondence is a homeomorphism of $\mathcal{S}_0$ onto $\mathcal{N}(\mathcal{B})$ in the norm topology. Thus, we have for every $\mu \in G_\square$ a well defined $\tilde{\mu}$ acting in $\mathcal{N}(\mathcal{B})$, especially $\tilde{\nu}_t$ and the dual mappings $\tau_t := \tilde{\nu}_t^*$, which act in $\mathcal{B}$. With this in mind we have

Theorem: Let Σ possess a factorial time invariant state. Then the triple $\tilde{\Sigma} := (\mathcal{B}, \mathcal{N}(\mathcal{B}), \tau)$ associated with Σ by the above definitions is a W^* -system, where for every $t \in \mathbb{R}$, τ_t is a $\sigma(\mathcal{B}, \mathcal{B}_*)$ -continuous $*$-automorphism. G_Σ is in one-to-one correspondence with the set of all $\sigma(\mathcal{B}, \mathcal{B}_*)$ -continuous Jordan automorphisms of $\mathcal{B}$ which commute with τ_t , $\forall\, t \in \mathbb{R}$. $\varphi \in \mathcal{S}_0$ is β -KMS for Σ iff $\tilde{\varphi}$ is β -KMS for $\tilde{\Sigma}$.

[1] Emch, G.G.: "Algebraic Methods in Statistical Mechanics and Quantum Field Theory", Wiley, New York (1972)

[2] Emch, G.G. and H.J. Knops: J. Math. Phys. 11, 3008 (1970)

[3] Dubin, D.A. and G. Sewell: J. Math. Phys. 11, 2990 (1970)

[4] Jelinek, F.: Commun. Math. Phys. 9, 169 (1968)

[5] Kadison, R.V.: Topology 3, Suppl. 2, 177 (1965)

[6] Roos, H.: "KMS condition without time automorphisms in the observable algebra", preprint, University of Göttingen, Germany, (July 1978)

Geometry of Collective Motion

D.J. Rowe, Department of Physics, University of Toronto,
Toronto, Ontario M5S 1A7, Canada

and G. Rosensteel, Department of Physics, Tulane University,
New Orleans, Louisiana, 70118, U.S.A.

Two major objectives in many-body physics are: to extract a collective Hamiltonian subdynamics from the microscopic Hamiltonian equations and to realize collective states in many-particle Hilbert space. Based on the several investigations of these problems by ourselves and others [1-10], we present a systematic geometric approach to the achievement of both objectives.

The first step is to decompose N-particle configuration space, $\mathbb{R}^{3N}$, into orbits of a kinematical collective group, e.g. SO(3) or $GL_+(3,\mathbb{R})$, and a smooth transversal. Collective motion is defined to be motion on an orbit surface and intrinsic motion, motion along a transversal.

The kinetic energy for N particles is proportional to Δ, the LBO (Laplace-Beltrami operator) on $\mathbb{R}^{3N}$. Let $p_{ni} = -i\hbar\partial/\partial x^{ni}$, $n=1,\ldots,N, i=1,2,3$ denote the usual single-particle momenta and $\pi_\nu = \pi_\nu^{ni}(x)p_{ni}$, $\nu=1,\ldots,3N$, a set of vector fields on $\mathbb{R}^{3N}$ that form a basis of tangent vectors at each point. Then, if $g_{\mu\nu} = \sum_{n,i}\pi_\mu^{ni}(x)\pi_\nu^{ni}(x)$ is the metric and $g^{\mu\nu}$ its inverse, one can show [11] that $\Delta = \sum_{n,i} p_{ni}^2$ can be expressed in the form

$$\Delta = (\pi_\mu - i\hbar\partial\pi_\mu^{ni}/\partial x^{ni})g^{\mu\nu}\pi_\nu. \tag{1}$$

Hence, to decompose Δ into collective and intrinsic components, one has simply to select a set of vector fields $\{\pi_\nu\}$ which separates into a subset $(X_1,X_2,\ldots)$ of collective momenta, tangent to the orbits of the chosen kinematical group, and a complementary set of intrinsic momenta $(\Pi_1,\Pi_2,\ldots)$, orthogonal to the orbit surfaces. Then $g(X_\mu,\Pi_\sigma) = 0$, all μ,σ, i.e. $(g^{\mu\nu})$ is block diagonal, and Δ separates

$$\Delta = \Delta_{coll} + \Delta_{intr} \tag{2}$$

with no cross terms.

For example, if for collective quadrupole dynamics one considers the kinematical collective group $GL_+(3,\mathbb{R})$ and takes as collective momenta the basis $\tau_{ij} = x^{ni}p_{nj}$ $i,j=1,2,3$ of a realization of its Lie algebra, one immediately obtains

$$\Delta = \sum_{i,j,k}(\tau_{ij} - i\hbar N\delta_{ij})Q_{ik}^{-1}\tau_{kj} + \Delta_{intr} \tag{3}$$

*This paper has been added to the proceedings because Dr. Rowe could not attend the conference.

where (Q^{-1}_{ij}) is the inverse of the quadrupole tensor $Q_{ij} = \sum_n x^{ni} x^{nj}$.

The collective component of Δ can be further decomposed and expressed in terms of generators of more physically interesting collective motions [6] by observing that any x on a $GL_+(3,\mathbb{R})$ orbit containing a fixed point $x_0 \in \mathbb{R}^{3N}$ can be expressed

$$x = r_1 \cdot S \cdot r_2 \cdot x_0, \quad r_1, r_2 \in SO(3), \; S \in \mathbb{R}^3_+, \tag{4}$$

where $\mathbb{R}^3_+$ is the group of diagonal positive definite real 3×3 matrices and the action on $\mathbb{R}^{3N}$ is given by the natural action on N copies of $\mathbb{R}^3$. Generators of motion in r_1, S and r_2 are respectively the vector fields

$$L_A = \sum_{B,C} \varepsilon_{ABC} \sum_{i,j} R_{Bi} R_{Cj} \tau_{ij} \tag{5}$$

$$t_A = \sum_{i,j} R_{Ai} R_{Aj} \tau_{ij} \tag{6}$$

$$\mathcal{L}_A = \sum_{B,C} \varepsilon_{ABC} \frac{\lambda_C}{\lambda_B} \sum_{i,j} R_{Bi} R_{Cj} \tau_{ij} \tag{7}$$

(A=1,2,3), where $R(x) \in SO(3)$ is the rotation matrix that diagonalizes (Q_{ij}) at $x \in \mathbb{R}^{3N}$. Specifically, if λ_1, λ_2, λ_3 are the principal quadrupole moments, t_A generates vibrations in λ_A and L_A and $\mathcal{L}_A$ generate r_1- and r_2-rotations, respectively, about the principal axis A. The latter have sometimes been called 'vortex' rotations [6,8]. In terms of these vector fields, one readily evaluates eq.(3) to obtain

$$\Delta_{coll} = \sum_A [t_A \frac{1}{\lambda_A^2} - i\hbar(N-3)\frac{1}{\lambda_A^2} - 2i\hbar \sum_{B\neq A} \frac{1}{\lambda_A^2-\lambda_B^2}]t_A$$

$$+ \sum_A [\frac{\lambda_B^2+\lambda_C^2}{(\lambda_B^2-\lambda_C^2)^2} (L_A^2+\mathcal{L}_A^2) - \frac{4\lambda_B\lambda_C}{(\lambda_B^2-\lambda_C^2)} L_A\mathcal{L}_A] \tag{8}$$

(A,B,C, cyclic).

As intrinsic momenta we seek a set of vector fields orthogonal to the $GL_+(3,\mathbb{R})$ orbit surfaces. If x_0 is some fixed non-zero vector in $\mathbb{R}^{3N}$, it can be shown [12] that any other non-zero $x \in R^{3N}$ can be expressed

$$x = r_1 \cdot S \cdot R \cdot x_0, \quad r_1 \in SO(3), \; S \in \mathbb{R}^3_+, \; R \in SO(N). \tag{9}$$

Thus it follows that $\mathcal{L}_A$ generates rotations of an SO(3) subgroup of SO(N). In fact one can readily show that

$$-\mathcal{L}_A \equiv \mathcal{J}_{BC} = \sum_{m,n} \mathcal{D}_{Bm} \mathcal{D}_{Cn} j_{mn} \qquad \text{(A,B,C, cyclic)} \tag{10}$$

where

$$j_{mn} = \sum_{i=1}^{3} (x^{mi} p_{ni} - x^{ni} p_{mi}) \tag{11}$$

is an SO(N) angular momentum and $\mathcal{D}$ is the SO(N) rotation matrix that diagonalizes $\mathcal{Q}_{mn} = \sum_{i=1} x^{mi} x^{ni}$. A complementary set of orthogonal intrinsic momenta is given by the SO(N) angular momenta

$$J_{\alpha A} = \sum_{m,n} \mathcal{D}_{\alpha m} \mathcal{D}_{An} j_{mn}, \quad \alpha=4,\ldots,N,\ A=1,2,3. \tag{12}$$

Inserting these vector fields into eq.(3), along with the collective momenta (5)-(7) immediately gives

$$\Delta = \Delta_{coll} + \sum_{A=1}^{3} \sum_{\alpha=4}^{N} \frac{1}{\lambda_A^2} J_{\alpha A}^2 . \tag{13}$$

To construct a collective Hamiltonian subdynamics we must now introduce a set of position observables whose values distinguish points (not necessarily uniquely) on the collective submanifold. The position observables should be such that, together with the momenta, they close on a Lie algebra of observables. Furthermore, they should form a sufficient set for the description of collective phenomena. If the collective submanifold is the generic $GL_+(3,\mathbb{R})$ orbit, a suitable set is given by the six quadrupole moments Q_{ij}, whose values specify the orientation and quadrupole shape of the system. If we adjoin these observables to a basis for the Lie algebra $gl_+(3,\mathbb{R})$ of $GL_+(3,\mathbb{R})$, as momenta, we obtain the Lie algebra of observables $cm_+(3) \sim [\mathbb{R}^6]gl_+(3,\mathbb{R})$. This algebra has been considered previously as a spectrum generating algebra for interesting collective Hamiltonians.

The collective Hilbert space is the space of all square integrable functions on the collective submanifold with respect to the measure induced from $\mathbb{R}^{3N}$. This space carries a unitary representation of $cm_+(3)$ in which the collective momenta act as differential operators and the position observables as multiplicative operators. The full N-particle Hilbert space is the space of square integrable functions on the generic $SO(3) \otimes \mathbb{R}_+^3 \otimes SO(N)$ orbit. Now the square integrable functions on SO(N) carry a representation of the symmetric group S_N. Furthermore, since $SO(3) \otimes \mathbb{R}_+ \otimes SO(N) \supset GL_+(3,\mathbb{R})$ one is enabled to decompose N-particle Hilbert space into irreducible subspaces with respect to both $cm_+(3)$ and S_N. The latter is important because it enables one to take fully into account the particle statistics. In particular, for fermions, one can combine collective subspaces of given permutation symmetry to spin or spin-isospin spaces of contragredient symmetry to construct product spaces of fully anti-symmetric states. Details of this decomposition will be given in a paper to follow [12].

To summarize, we have shown how to decompose N-particle configuration space $\mathbb{R}^{3N}$ into collective submanifolds and smooth transversals, how to introduce a Lie algebra of collective observables and how to decompose N-particle Hilbert space into irreducible collective subspaces of definite permutation symmetry. The realization of collective states in microscopic terms now reduces to the problem of finding the eigenstates of the corresponding collective Hamiltonian within the given irreducible collective subspaces. This is just the problem that is solved in diagonalizing a phenomenological collective Hamiltonian except that, in the latter case, it is usual to restrict consideration to vortex free (i.e. L=0) subspaces. Finally, a collective Hamiltonian may be derived from the microscopic N-particle Hamiltonian by adjoining the collective kinetic energy, from eq.(8), to the potential energy obtained by restricting the microscopic potential to a collective subspace.

References

1. F.M.H. Villars, Nucl. Phys. 3 (1957) 240; Ann. Rev. Nucl. Sci. 7 (1957) 185.
2. D.J. Rowe, Nucl. Phys. A152 (1970) 273.
3. W. Zickendraht, Journ. Math. Phys. 12 (1971) 1663.
4. E.V. Vanagas and R.K. Kalinauskas, Yad. Fiz. 18 (1973) 768.
5. R.M. Asherova, V.A. Knyr, Y.F. Smirnov and V.N. Tolstoy, Yad. Fiz. 21 (1975) 1126.
6. P. Gulshani and D.J. Rowe, Can. Journ. Phys. 54 (1976) 970.
7. G. Rosensteel and D.J. Rowe, Ann. Phys. 96 (1976) 1.
8. O.L. Weaver, R.Y. Cusson and L.C. Biedenharn, Ann. Phys. 102 (1976) 493.
9. G. Rosensteel and D.J. Rowe,"The SP(3,$\mathbb{R}$) model of nuclear collective motion", in Group Theoretical Methods in Physics (Proc. 5th Int. Colloquium, ed. R.T. Sharp and B. Kolman, Acad. Press, 1977).
10. D.J. Rowe and G. Rosensteel, "The nuclear collective model and the symplectic group", in Group Theoretical Methods in Physics (Proc. 6th Int. Colloquium, ed. P. Kramers and A. Rieckers, Springer-Verlag, 1977).
11. D.J. Rowe and G. Rosensteel, Journ. Math. Phys. (to be published).
12. D.J. Rowe and G. Rosensteel, to be published.

On Labeling the Basis in Gln and Identifying the Racah Algebra with S_N

J. J. Sullivan, Physics Dept., U. New Orleans, New Orleans, LA 70122

The Schur-Weyl construction of Nth rank tensor irreps of Gln and its subgroups Un and SUn results in basis kets $|\lambda mM\rangle$ labeled by an index m under action of the group S_N and an index M under action of the group Gln. The irrep lable λ is common to both groups S_N and Gln. Usually tensor coupling is considered only within the Racah algebra of a single group. In a series of papers[1,2] we have exploited the duality of coupling between S_N and Gln to establish nontrivial identifications and orthogonality properties for the various coupling coefficients (n-j symbols, isoscalar factors and CFP) that occur in the generalized Racah algebra of Un. Two principal physical applications utilizing such an algebra are a) elementary particles and fields and b) (atomic and nuclear) shell theories. The usual subgroup sequences which specify the precise meaning of the index M are different for the two applications. It is important to clarify the labeling involved, because isoscalar factors and CFP depend on the subgroup specification.

The nontrivial aspects of the label M are essentially combinatorial. In a generalized Gel'fand scheme as used in elementary particle theories the index M expands to specify the weight of the state (W) and the combinatorial prescription m' according to which the totally symmetric component weights $W=\{w_i\}$ are to be coupled (in accord with the Littlewood-Richardson rules) to form a tensor of type λ. The labeling is explicit when one considers the action of a matrix basis projector $(\underline{\lambda|mm'})$ of the group S_N applied to a primitive (i.e., completely symmetric in the component weights) weight ket $\otimes|w_i\rangle$, where $\sum_{i=1}^{n} w_i=N$. The resulting (unnormalized) ket is labeled by $|\lambda m,m'(W)\rangle$. The generalized Gel'fand scheme corresponds to the subgroup decomposition $Gln_1+n_2/Gln_1\otimes Gln_2$ in which the label m'(W) is expanded to $\lambda_1 m_1(W_1)\lambda_2 m_2(W_2)$ according to the generalized branching rule.

In shell theory applications one forms symmetrized tensors of a basic defining irrep that can itself be considered as a symmetrized tensor irrep of a more fundamental defining space. For usual angular momentum applications the defining tensor irrep can be considered to be the (2j+1) dimensional totally symmetric [2j] rank tensor irrep of SU_2. In S_N the construction of symmetrized powers of an irrep corresponds to symmetrized outer products, a process when applied to Schur functions Littlewood calls an outer plethysm. If $[2j]\otimes[\lambda] = \sum_J [jN+J,jN-J]$ designates the bi-partition decomposition of the symmetrized outer product, the resulting ket is labeled by $|\lambda m,J(M)\rangle$, where M is the usual weight specification in SU_2. For the dimension n=2j+1, there is a unitary transformation $\langle\lambda m,m'(W)|\lambda'm'',J(M)\rangle=\delta^{\lambda\lambda'}\delta^{mm''}$ $(m'(W)|\lambda Mj|J)$ between the two equivalent basis sets in which λM and j act as fixed parameters. Multiplicity labels such as seniority numbers may be needed and can simply be appended to the label J. The SU_2 weight M is a fixed parameter since a given weight (W) determines M. For (α,β) signifying the defining SU_2 irrep the (2j+1) states of [2j] correspond to weights $(\alpha^{2j-k}\beta^k)$ and weight $(W)=\{w_i\}$ has

$M = \Sigma(j-k)w_k$. For example j=1 and N=3 (three equivalent p orbitals) the basis transformation blocks as

M	$(w_0, w_1\ w_2)$	dimension	$(m(w)\|\lambda Mj=1\|J)$				
3	(3, 0, 0)	1	1	-	-	-	-
2	(2, 1, 0)	3	1	-	1	-	-
1	(2, 0, 1)	3	$(1/5)^{1/2}$	$-(4/5)^{1/2}$	$1/\sqrt{2}$	$-1/\sqrt{2}$	-
	(1, 2, 0)	3	$(4/5)^{1/2}$	$(1/5)^{1/2}$	$1/\sqrt{2}$	$1/\sqrt{2}$	-
0	(1, 1, 1)	6	$(3/5)^{1/2}$	$-(2/5)^{1/2}$	(*)		1
	(0, 3, 0)	1	$(2/5)^{1/2}$	$(3/5)^{1/2}$	-	-	-
			[6]	[4,2]	[5,1]	[4,2]	$[3^2]$	[3+J,3-J]
			[3]		twice [2,1]		$[1^3]$	λ

total dimension 27 = 7 + 3 + 2 (5 + 3) + 1

(including M < 0)*, λ = [2,1] , M = 0 separates into symmetric and antisymmetric parts

$$\begin{matrix} & [5,1] & [4,2] \\ (1,1,1)_A & 1 & 0 \\ (1,1,1)_S & 0 & 1 \end{matrix}\ .$$

Values of the matrix elements have been obtained by identification within the symmetric group. This identification and identification of 9-j coupling coefficients, isoscalar factors and CFP with their equivalents in S_N in both schemes discussed here are considered in detail in refs. 2.

1. J. J. Sullivan, J. Math. Phys. 14, 387 (1973); 16, 756 (1975); 16 1707 (1975); and Proceedings Int. Symp. on Math. Phys., Mexico City, Jan. 5-8, 1976, Vol I, 253.
2. J. J. Sullivan, J. Math. Phys. 19 1674, 1681 (1978).

DYNAMICAL GROUPS OF PARAMETRIZED SYSTEMS: ATOMIC SUPERMULTIPLETS

CARL E. WULFMAN
Department of Physics
University of the Pacific
Stockton, California, 95211

Abstract

Lie groups that allow transformation of parameters in a Hamiltonian are investigated. Stability considerations suggest treating the parameters as dynamical variables. This yields generalized Hamiltonian systems in an enlarged classical phase space, and "third quantized" Hamiltonians in quantum mechanics. These provide a conceptually and mathematically convenient tool for uncovering relationships between the properties of related physical systems.

For atoms, allowing the nuclear charge Z to become a dynamical variable with discrete spectrum yields a third quantized Hamiltonian $H(p_i, q_i, Z_{op})$ whose set of eigenstates

$$H\Psi_{EZ} = E\Psi_{EZ},\ Z_{op}\Psi_{EZ} = Z\Psi_{EZ} \tag{1}$$

is Barut's atomic supermultiplet.[1] This system which we term a Baruton, admits subgroups of $Sp(8) \sim SO(5,4)$ as degeneracy and spectrum generating groups.

The 1/Z expansion method for N-electron atoms is formulated in terms of operations of the adjoint group of $(Sp(8)_{orb} \times SU(2)_{spin} \times)^N$ acting in a finite second-quantized representation with hydrogenic basis $\{\phi\}$. Variationally determined effective nuclear charges ζ are conceived of as taking the place of Z:

$$Z_{op}\phi_\zeta = \zeta\phi_\zeta \quad . \tag{2}$$

In this finite basis SO(9) is also a dynamical group of the Baruton, as defined by (1).

A semi-empirical investigation of the relationship between one-electron properties (ionization potentials, electron affinities) of different atoms leads to the identification of the periodic chart as a representation table for an SO(7) subgroup of the SO(9) dynamical group.

Reference:

1. A.O. Barut in Proc. Rutherford Centennial Symposium, University of Canterbury, Christchurch, N.Z., 1971) P. 126.

Geometrically Formulated Gauge Dynamics for Extended Hadrons

W. Drechsler

Max-Planck-Institut für Physik und Astrophysik, Munich, Fed. Rep. Germany

Abstract:

A de Sitter gauge invariant set of field equations is investigated as a possible basis for a gauge description of extended hadrons. The formalism uses an underlying geometric structure given by a fiber bundle over space-time with Cartan connection possessing as fiber a 4-dimensional space of constant curvature characterized by a curvature radius R chosen to be of the order of a Fermi. The constant R represents an elementary length parameter of geometric origin associated with strong interaction physics. A curvature is induced on the bundle space through a hadronic matter distribution described by a generalized bilocal wave field $\psi(x,\xi)$ where x denotes a point in the base space (space-time) and ξ varies in the local fiber. An expansion of the internal motion associated with the variable ξ is given in terms of "de Sitter plane waves", i.e. the so-called horospherical waves, which are the analogue of the usual plane waves in flat Minkowski space-time. In this context the harmonic analysis of scalar and spinor fields in (4,1) de Sitter space is discussed and its relevance to the SO(4,1) gauge theory is pointed out.

I. Introduction

To gain an understanding of the intriguing problem of hadron structure and the hadronic mass spectrum it seems essential to incorporate into the theoretical framework the fact that the hadrons observed in nature are extended objects. We want to take this property from the very beginning into consideration by introducing at a fundamental level of the description an elementary length parameter R of the order of one Fermi. To define a geometric stratum on which extended hadrons could manifest themselves as extended entities we shall assume that in the small the space to be used in the formulation of a hadron dynamics possesses a richer structure than the one given by a four dimensional flat space-time manifold. In fact, we shall assume that the basic geometry to be used in representing hadronic phenomena is provided by a fiber bundle with Cartan connection raised over space-time and possessing the SO(4,1) de Sitter group as structural group. Without repeating the motivation for this choice in detail we mention four relevant points*):

i) An elementary length is built into the underlying geometric structure in characterizing the fiber of the bundle by a length parameter R. As fiber we take a 4-dimensional space-time of constant curvature, V'_4, with curvature radius R (being, as mentioned, of the order of a Fermi) on which the SO(4,1) de Sitter

*) Compare in this context the arguments presented in refs. 1-3.

Presented at the Integrative Conference on Group Theory and Mathematical Physics, Austin, Texas 11-16 September 1978 group*) acts as a group of motion. V_4' is isomorphic to the noncompact coset space SO(4,1)/SO(3,1).

ii) The Cartan nature of the bundle implies that the fiber over each space-time point x, i.e. the local de Sitter space $V_4'(x)$, is tangent to space-time at x**).

iii) The suggested geometric framework allows a hadronic matter distribution, represented by a bilocal wave field $\psi(x,\xi)$ with $x \in V_4$***) and $\xi \in V_4'(x)$, to react back on the geometric stratum by inducing a curvature on the bundle space which is determined by a hadronic current associated with the field $\psi(x,\xi)$.

iv) The Casimir operators of the de Sitter group correlate mass and spin leading to a rotator spectrum of states similar to a Regge spectrum with the higher mass states possessing larger spins[(4)] ****).

The geometrically motivated formalism leads to a gauge description for extended yet elementary hadrons[(1-3)(7)]. There are no constituents in the true sense of this word[(8)] introduced in this description. Hadronic matter is represented, as already mentioned, by a generalized bilocal wave field $\psi(x,\xi)$ with x labelling a space-time point and the internal coordinates ξ varying in the local fiber, $V_4'(x)$, the latter being isomorphic to a copy of an abstract (4,1) de Sitter space sitting over the point x. Technically speaking $\psi(x,\xi)$ is a cross section on a vector bundle associated with the de Sitter frame bundle, the latter being a principal fiber bundle over space-time with the fiber and structural group being identical to the SO(4,1) de Sitter group. The section $\psi(x,\xi)$ will carry representation character with respect to the local Lorentz group operating in T_x, the tangent space to V_4 at x, and with respect to the internal or gauge group SO(4,1). We shall factor $\psi(x,\xi)$ into $\psi(x,\xi) = \varphi(x)\cdot\Phi_x(\xi)$ where our present knowledge requires - as we shall briefly indicate below - to regard the spacetime part $\varphi(x)$ as a local q-number field of conventional quantum field theory, whereas the de Sitter part $\Phi_x(\xi)$ is a c-number field describing the motion in the fiber over x. $\Phi_x(\xi)$ is a quantity varying smoothly in ξ and x on the bundle space.

The use of a generalized wave operator $\psi(x,\xi)$ with specific q-number and c-number content amounts to the description of hadronic phenomena in terms of a bilocal field (or generalized wave function in a one-particle theory) being observable only modulo the action of a gauge group G which for physical and geometrical reasons is here taken to be the noncompact ten parameter group SO(4,1). The freedom of changing the local

---

*) We take SO(4,1) in favour of SO(3,2) since the associated (4,1) de Sitter space is compact in its spacial extensions and noncompact in time, while the (3,2) de Sitter space has closed timelike geodesics.

**) The tangent space T_x to space-time at x and the tangent space $T_{\overset{\circ}{\xi}}$ to $V_4'(x)$ at $\overset{\circ}{\xi}$ are identified by an isomorphism (soldering condition). $\overset{\circ}{\xi}$ is the point left invariant by the Lorentz subgroup SO(3,1) of SO(4,1).

***) V_4 denotes a generally curved Riemannian space-time. For most of the subsequent discussion we shall, however, disregard long range gravitational fields and assume the base space to be flat Minkowski space-time M_4.

****) Compare also ref. 5 and 6 in this context.

cross section with which the hadronic matter field $\psi(x,\xi)$ is identified corresponds to the freedom of performing de Sitter gauge transformations. The basic dynamical equations have, of course, to be invariant against such SO(4,1) gauge transformations. To be definite we shall choose in the formulation of our basic equations presented in the next section for the de Sitter part of $\psi(x,\xi)$ describing a hadron the lowest dimensional spinor representation of SO(4,1), i.e. the defining four-dimensional representation of the covering group USp(2,2)*) of SO(4,1). The de Sitter part $\Phi_x(\xi)$ represents thus - as far as the ξ-dependence is concerned - a wave function for a spin 1/2 object moving in a space-time of constant curvature. This bares a certain resemblence to the choice of spin 1/2 for the quarks in the more conventional constituent models. We point out in passing that with these finite dimensional nonunitary representations associated with the internal degrees of freedom no usual quantum mechanical probability interpretation for the motion in the fiber in terms of supposedly existing constituents or subunits of hadrons emerges[(10)]**).

We point out once again in concluding this introduction that our generalized hadronic matter field $\psi(x,\xi)$ is a mixture of a q-number and a c-number quantity. This property seems to be a more general feature of various formalisms investigated in recent times: They are constructed neither with purely second quantized i.e. canonical operator fields nor with purely classical c-number fields despite the starting point of various authors from one side or the other. The interplay of q-number and c-number properties for field quantities introduced to describe extended objects in hadron physics seems to us to be of a fundamental importance. We should, moreover, point out in this context that in the extended particle formalism based on differential geometric concepts and reasoning, which will be reviewed in more detail below, no Higg's algorithm is involved to produce arbitrary mass shifts for particles or fields as is done in theories based on a certain Lagrangian. The expectation in this bilocal bundle description is rather that for the interpretation of the solutions to the basic nonlinear equations written down in Sect. II below and the emergence of a hadronic mass spectrum no such ad hoc mass-giving mechanism is in fact needed. The general physical picture in the proposed fiber bundle formalism concerning the mass problem is qualitatively the following: Due to the coupling between translational motions (taking place on the base space) and internal motions (taking place in the fiber) which is inherent in the equations of motion the inertia of a certain hadronic state against changes of its translational motion depends on its internal state of motion, hence, is different for different internal excitations. We think that the in-

---

*) USp(2,2) is the intersection of the groups U(2,2) and Sp(4,C). It can also be regarded as a subgroup of SL(2,Q)[(9)].

**) To allow such a probability interpretation one would have to represent the internal motion of hadrons in terms of unitary but infinite dimensional representations of the (non-compact) de Sitter group rendering thereby the formulation of the internal motion much more complicated without any apparent gain in physical insight.

corporation of an elementary length of the order of a Fermi into the basic geometric stratum on which the dynamics of hadrons is formulated and the abandoning of the extreme idealization of regarding the underlying hadron dynamics as a dynamics of point-like objects or local constituent fields possessing no intrinsic scale of length is the first step into a direction which, indeed, takes seriously the extension of hadronic objects as experimentally observed in nature. This, together with the gauge interaction concept, we think will open new prospects to the solution of the hadron mass problem.

II. SO(4,1)-Gauge Invariant Equations of Motion

The de Sitter space was originally proposed as a possible cosmological space-time where local inhomogeneities are smoothed out. The geometry in de Sitter space is the intrinsic geometry of a hyperboloid which can be embedded into a 5-dimensional pseudo-Riemannian space which we call E_5. In E_5 the (4,1) de Sitter space is given by a one-shell hyperboloid defined by the quadric*)

$$V_4' \quad : \quad [\xi\xi] \quad = \quad \xi^a \xi^b \eta_{ab} \quad = \quad -R^2 \tag{2.1}$$

with η_{ab} = diag (1, -1, -1, -1, -1). The group of hyperbolic rotations in E_5 leaving the form $[\xi\xi] = -R^2$ invariant is the SO(4,1) de Sitter group. SO(4,1) acts transitively on the hyperboloid (2.1) as well as on the cone $[\zeta\zeta] = 0$. One can identify V_4' with the noncompact coset space SO(4,1)/SO(3,1) where the Lorentz group SO(3,1) is the stability subgroup of the point $\overset{\circ}{\xi}{}^a = (0, 0, 0, 0, -R)$. Following Gelfand, Graev and Vilenkin(11) we shall, moreover, identify antipodal points on the hyperboloid turning it into a so-called imaginary Lobachevsky space**). As was explained in the introduction we consider here a space-time of constant curvature, V_4', with curvature radius $R \lesssim 10^{-13}$ cm as <u>the fiber</u> of a bundle of Cartan type***) raised over space-time. We denote this Cartan bundle over V_4 by $T^R(V_4)$ and refer to it as to the de Sitter bundle over space-time given by

$$T^R(V_4) \quad = \bigcup_{x \in V_4} V_4'(x) \quad . \tag{2.2}$$

Mathematically $T^R(V_4)$ is a fiber bundle****) E(B,F,G,P) associated to the de Sitter frame bundle $P = P(B{=}V_4,\ G{=}SO(4,1)$ defined by

$$E\ (B = V_4,\ F = SO(4,1)/SO(3,1),\ \ G = SO(4,1),\ P)\ . \tag{2.3}$$

---

*) We adopt the summation convention and sum identical covariant and contravariant indices a,b,c ... over 0, 1, 2, 3, 5.

**) Compare in this context also E. Schrödinger(12)

***) For a detailed mathematical investigation of these fiber bundles, which are usually omitted in text books on differential geometry, see Ch. Ehresmann(13).

****) B: base space; F: fiber type; G: structural group; P: principal bundle.

The bundle $T^R(V_4)$ can be viewed as a generalized tangent bundle. It reduces to the affine tangent bundle, $T(V_4)$, over space-time in the limit $R \to \infty$[14]. This limit corresponds to the Inönü-Wigner contraction limit for the structural groups of the respective bundles i.e. $SO(4,1) \xrightarrow{R\to\infty} ISO(3,1)$. However, we shall take R in the following to be fixed and finite, in fact, treat R as a fundamental constant of nature of the order of a Fermi*). Neglecting from now on long range gravitational fields and focussing the attention on the strong interaction i.e. fiber degrees of freedom we shall assume space-time to be Minkowskian and thus base the following discussion on the de Sitter bundle over Minkowski space, $T^R(M_4)$.

Now we consider a hadronic matter field on $T^R(M_4)$ which we shall take to be a bilocal bispinor object

$$\psi(x,\xi) = \{\psi^{AA'}(x,\xi);\ A,\ A' = 1,2,3,4\} = \{\varphi^A(x)\Phi^{A'}{}_x(\xi);\ A,\ A' = 1,2,3,4\} \tag{2.4}$$

with A being a usual Dirac spinor index (for a spin 1/2 baryon) and A' being an internal i.e. de Sitter spinor index. The representation of a hadronic matter field in a certain gauge can now be viewed as a cross section on the vector bundle

$$E\ (B=M_4,\ F=C^4,\ G=USp(2,2),\ P) \tag{2.5}$$

associated with the de Sitter frame bundle $P=P(M_4,\ SO(4,1))$ over M_4. The fiber of (2.5) is a 4-dimensional complex representation space for de Sitter spinors,and the structural group of the bundle is the covering group USp(2,2) of SO(4,1)**).

The gauge transformation property of $\psi(x,\xi)$ is expressed through the following equations (with $\hat{S}(x)$ acting on the primed spinor index)

$$\psi(x,\ \xi') = \hat{S}(x)\ \psi\ (x,\xi), \tag{2.6}$$

with

$$\xi'^a = [A(x)]^a_b\ \xi^b, \tag{2.7}$$

where $A(x) \in SO(4,1)$, $\hat{S}(x) \in USp(2,2)$, and with $\gamma^a[A^{-1}(x)]^b_a = \hat{S}(x)\gamma^b\hat{S}^{-1}(x)$ defining the homomorphism $\hat{S}(x) \to A(x)$ of USp(2,2) onto SO(4,1). Here γ^a; $a = 0,1,2,3,5$ denotes a set of five Dirac matrices $\gamma^a = (\gamma^\mu;\ \mu = 0,1,2,3,\ \ \gamma^5 = \gamma^0\gamma^1\gamma^2\gamma^3)$ obeying $\{\gamma^a,\ \gamma^b\} = 2\eta^{ab}\cdot 1$, and $\gamma^{a\dagger} = \gamma^0\gamma^a\gamma^0$.
The invariance of (2.1) under hyperbolic rotations, A(x), requires the pseudoorthogonality condition $A^{-1}(x) = \eta A^T(x)\eta$ where A^T denotes the transposed matrix and η is the 5x5 matrix with the elements η_{ab}. Furthermore, $\hat{S}^{-1}(x)$ is given by $\hat{S}^{-1}(x) = \gamma^0\ \hat{S}^\dagger(x)\gamma^0$ implying that the quantity $\psi^\dagger(x,\xi)\gamma^0$ transforms inversely under de Sitter gauge transformations. For later convenience we define a double barred spinor field by taking the adjoint with respect to the Dirac and the de Sitter indices and multiply

---

*) The precise value of R will have to be fixed in comparing the hadronic masses appearing in this formalism with the experimentally observed mass values.

**) The expression (2.5), of course, implies the existence of a SO(4,1) spinor structure over space-time.

each of them by γ^o which is, in an apparent direct product notation,

$$\bar{\bar{\psi}}(x,\xi) = \psi^\dagger(x,\xi)\,\gamma^o \otimes \gamma^o . \tag{2.8}$$

After these preliminaries we define the gauge dynamics on the de Sitter bundle $T^R(M_4)$ by the following sets of equations[(2)(3)*)]:

$$\gamma^\mu[\partial_\mu + i\Gamma_\mu^{\ R}(x)]\,\psi(x,\xi) = -i\, m\, \psi(x,\xi) \tag{2.9}$$

$$\frac{1}{2R}\,\gamma^a\gamma^b\, L_{ab}(\xi)\,\psi(x,\xi) = (\mu + \frac{2i}{R})\,\psi(x,\xi) \tag{2.10}$$

$$D^\mu\, R^R_{\mu\nu ab}(x) = \bar{\kappa}\, J_{\nu ab}(x) \tag{2.11}$$

$$D_\kappa R^R_{\mu\nu ab}(x) + D_\mu R^R_{\nu\kappa ab}(x) + D_\nu R^R_{\kappa\mu ab}(x) = 0 \tag{2.12}$$

Eq. (2.9) is the Dirac equation for the translational motion on $T^R(M_4)$ with m denoting the mass of the hadron described by $\psi(x,\xi)$. $D_\mu = \partial_\mu + i\Gamma_\mu^{\ R}(x)$ represents the operator for the gauge invariant derivative of a de Sitter spinor quantity defined over $T^R(V_4)$ where the 4x4 matrix valued connection coefficients, $\Gamma_\mu^{\ R}(x)$, are given by

$$\Gamma_\mu^{\ R}(x) = \frac{1}{2}\,\Gamma^R_{\mu ab}(x)\, S^{ab}, \tag{2.13}$$

with

$$S^{ab} = \frac{i}{4}\,[\gamma^a, \gamma^b] , \tag{2.14}$$

providing a 4x4 matrix representation for the Lie algebra of SO(4,1)[**)]. The 40 quantities $\Gamma^R_{\mu ab}(x)$ are connection coefficients specifying the geometry in the de Sitter bundle[***)]. In eq. (2.9) they play the role of 40 external gauge potentials provided by the geometry which are, however, coupled to the matter distribution immersed into the geometry in a non-linear way expressed by eq. (2.11) as will be explained below. Eq. (2.10) is the Dirac equation[****)] for the motion in the fiber i.e. in the variable ξ. The operators $L_{ab}(\xi)$ generating hyperbolic rotations in E_5 are defined by

---

*) Matrices γ^μ act on the Dirac spinor index, matrices γ^a act on the de Sitter spinor index of $\psi(x,\xi)$.

**) The Lie algebra of SO(4,1) is defined by: $i[M^{ab},M^{cd}] = \eta^{ac}M^{bd}+\eta^{bd}M^{ac}-\eta^{ad}M^{bc}-\eta^{bc}M^{ad}$.

***) More exactly, they are the coefficients of the connection on the bundle of de Sitter frames, $P(M_4,\ SO(4,1))$, which was introduced above.

****) Compare in this context Dirac's paper of 1935[(15)] as well as that of Gürsey and Lee[(16)].

$$L_{ab}(\xi) = i\,(\xi_a \partial_b - \xi_b \partial_a) \tag{2.15}$$

with $\xi_a = \eta_{ab}\xi^b$ and $\partial_a = \partial/\partial\xi^a$. The parameter μ in eq. (2.15) is an internal mass parameter characterizing further the de Sitter representation character of the internal spin 1/2 modes. We shall come back to the range of values for μ when we discuss solutions of eq. (2.10) in the next section.

The basic equations defining the way how a matter distribution reacts back onto the underlying geometry are provided by eqs. (2.11) which were called the <u>current-curvature-equations (c.c.-equations)</u>[2]. They define the strong fiber dynamics (SFD) generating an internal i.e. de Sitter curvature from a hadronic matter distribution. $R^R_{\mu\nu ab}(x)$ are the components of the de Sitter curvature tensor derived from the connection coefficients $\Gamma^R_{\mu ab}(x)$ according to (we leave out the arguments x for simplicity):

$$R^R_{\mu\nu ab} = \partial_\mu \Gamma^R_{\nu ab} - \partial_\nu \Gamma^R_{\mu ab} + \Gamma^R_{\mu ac}\,\Gamma^{Rc}_{\nu b} - \Gamma^R_{\nu ac}\,\Gamma^{Rc}_{\mu b} \;. \tag{2.16}$$

The $R^R_{\mu\nu ab}(x)$ are antisymmetric in μ and ν as well as in a and b. The l.h.-side of eq. (2.11) represents the covariant divergence of the curvature tensor with respect to the bundle connection $\Gamma^R_{\mu ab}(x)$ where the covariant derivative, $D_\kappa R^R_{\mu\nu ab}(x)$, is defined by

$$D_\kappa R^R_{\mu\nu ab} = \partial_\kappa R^R_{\mu\nu ab} - \Gamma^{Rc}_{\kappa a} R^R_{\mu\nu cb} - \Gamma^{Rc}_{\kappa\, b} R^R_{\mu\nu ac} \;. \tag{2.17}$$

The r.h.-side of eq. (2.11) is determined by the source current of the hadronic matter distribution which is obtained by integrating a bilinear density in the fields $\bar{\bar{\psi}}(x,\xi)$ and $\psi(x,\xi)$ in an invariant manner over the local fiber to obtain a local antisymmetric (in a,b) tensor gauge current defined by

$$J_{\nu ab}(x) = \int_{V'_4(x)} \bar{\bar{\psi}}(x,\xi)\,(\gamma_\nu \otimes M_{ab})\,\psi(x,\xi)\,d\mu(\xi) \tag{2.16}$$

where again we have used an obvious direct product notation for Dirac and de Sitter matrix products. $d\mu(\xi)$ is the SO(4,1) invariant measure on the de Sitter hyperboloid given by

$$d\mu(\xi) = \frac{R}{|\xi^5|}\, d\xi^0 \wedge d\xi^1 \wedge d\xi^2 \wedge d\xi^3 \;. \tag{2.19}$$

For the integral (2.18) to exist $\psi(x,\xi)$ is required to be of compact support on $V'_4(x)$. Furthermore, $\bar{\kappa}$ is a coupling constant of dimension $[\text{length}^{-3}]$. One could use the length parameter R to introduce a dimensionless coupling constant g according to $\bar{\kappa} = g/R^3$. g measures the strength of the coupling between the hadronic matter field, on the one side, and the curvature on the underlying de Sitter bundle, on the other side, where the curvature is to be associated with strong interaction effects (representing a halo of gauge field strengths). Thus g plays the role of a strong interaction coupling constant.

The last set of equations above, eqs. (2.12), are Bianchi identities representing integrability conditions for the de Sitter curvature tensor $R^R_{\mu\nu ab}(x)$.

As a consequence of the equations of motion the source current $J_{\mu ab}(x)$ is seen to be covariantly conserved i.e.

$$D^\mu \ J_{\mu ab}(x) = 0 \ . \tag{2.20}$$

The equations (2.9) - (2.12) describing the mutual interplay between hadronic matter, on the one side, and the underlying fiber bundle geometry, on the other side, are de Sitter gauge covariant. This is at once seen from eqs. (2.6) and (2.7) and the associated transformation law for the de Sitter spinor connection $\Gamma^R_\mu(x)$ given by

$$\hat{\Gamma}^R_\mu(x) = \hat{S}(x)\Gamma^R_\mu(x)\ \hat{S}^{-1}(x) - i\ \hat{S}(x)\,\partial_\mu\ \hat{S}^{-1}(x)\ . \tag{2.21}$$

The gauge transformation law of the $R^R_{\mu\nu ab}(x)$ is that of a covariant second rank tensor, i.e.

$$\hat{R}^R_{\mu\nu ab}(x) = [A^{-1}(x)]^c_a\ [A^{-1}(x)]^d_b\ R^R_{\mu\nu cd}(x) \tag{2.22}$$

and similarly for $J_{\mu ab}(x)$ as a result of eqs. (2.6), (2.7) and (2.18) and the invariance of the measure (2.19) i.e. $d\mu(A(x)\xi) = d\mu(\xi)$.

In closing this section on the gauge invariant equations of motion we like to add two comments: One refering to the description of leptons in this formalism the other relating to the question of antimatter. As regards the latter point we mention that it is possible to define a matter-antimatter conjugation, $\hat{C}$, leaving the eqs. (2.9) - (2.12) invariant.[2] $\hat{C}$ turns out to be a discrete symmetry combining an ordinary charge conjugation for the Dirac space-time part $\varphi(x)$ of $\psi(x,\xi)$ with an internal $\hat{P}\hat{T}$-transformation*) performed globally on the bundle in all the fibers. For consistency the Dirac part - or point spinor part - $\varphi(x)$ of $\psi(x,\xi)$ is required, as mentioned in the introduction, to possess q-number anticommutation character leading thus to a source current given by

$$J_{\mu ab}(x) = j^D_\mu(x) \int_{V'_4(x)} \bar{\Phi}_x(\xi)\ M_{ab}\Phi_x(\xi)\ d\mu(\xi) \tag{2.17}$$

where $j^D_\mu(x)$ is the usual Dirac current for a local canonical fermion field $\varphi(x)$ with $j^D_\mu(x)$ being odd under ordinary charge conjugation C, i.e.

$$j^D_\mu(x) = \frac{1}{2}\,[\bar{\varphi}(x)\gamma_\mu\varphi(x) - \tilde{\varphi}(x)\ \tilde{\gamma}_\mu\tilde{\bar{\varphi}}(x)] \tag{2.18}$$

where ~ denotes the transpose. The integral in eq. (2.17) is a c-number quantity, $F_{ab}(x)$, describing form factor effects. (For a detailed discussion of discrete symmetries on the de Sitter bundle space see ref. 2).

---

*) I.e. internal space and time reflexion $\xi^\mu \to -\xi^\mu$; $\mu = 0,1,2,3$, $\xi^5 \to \xi^5$.

As regards the representation of leptons in a theory based on a geometric stratum identified with the de Sitter bundle $T^R(M_4)$ we observe that one has to represent leptons as internal i.e. de Sitter scalars. Then the dualism between matter and geometry which we called SFD above and which is expressed through the c.c.-equations (2.11) is suspended for leptons. There are no effects mediated through the fibers for leptons represented as internal scalars. Thus, with such an assignment, leptons would develop neither extension nor strong interaction.

III. Search for Solutions - The de Sitter Plane Waves

There are no solutions of the set of equations (2.9) - (2.12) known yet. Especially the current-curvature-equations (2.11) in the presence of a matter source current still require careful investigation. Our strategy now is to determine first the basic solutions describing the motion in the fiber. To this end the plane wave solutions of the basic SO(4,1) invariant field equations for scalar, spinor and vector fields in a space-time of constant curvature have been determined(17). Mathematically this relates to the problem of the harmonic analysis of functions defined on the homogenous space SO(4,1)/SO(3,1) and the characterization of the basic modes for the motion in the fiber by a set of generalized Fourier coefficients. We thus first have to solve eq. (2.10) for a fixed fiber (i.e. with x kept constant*)). As a preparation to this let us in a first step determine the plane wave solutions of the analogue of the Klein-Gordon equation in de Sitter space reading in the form suggested by Penrose(18) and Chernikov and Tagirov(19):

$$(\Box + \frac{2}{R^2} + \mu^2)\ \Phi(\xi) = 0 \quad . \tag{3.1}$$

Here $\Box = \frac{1}{\sqrt{-g}}\partial_\mu \sqrt{-g}\ g^{\mu\nu}\partial_\nu$ is the Laplace-Beltrami operator on the de Sitter hyperboloid *) being identical to the second order Casimir operator $I_1 = \frac{1}{2R^2} L^{ab}(\xi)L_{ab}(\xi)$ in the scalar case **), μ is the mass of the scalar field $\Phi(\xi)$, and $\frac{2}{R^2}$ is the quantity identical with $\frac{1}{6}$ times the scalar curvature of the de Sitter space rendering eq. (3.1) conformally invariant in the massless case. As eigenvalue equation for the Casimir operator I_1 eq. (3.1) assumes the form

---

*) We, therefore, suppress the variable x in parts of the discussion in this section.

**) The Casimir operators of SO(4,1) are $I_1 = (1/2R^2)M_{ab}M^{ab}$ and $I_2 = W_a W^a$ with $W_a = (1/8R)\varepsilon_{abcde}M^{bc}M^{de}$ where ε_{abcde} is the Levi-Civita tensor in E_5. I_2 has zero eigenvalue in the scalar case. In general $M^{ab} = L^{ab}(\xi) + S^{ab}$ where S^{ab} is a set of spin matrices.

$$\frac{1}{2R^2} L^{ab}(\xi)\, L_{ab}(\xi)\, \Phi_\kappa(\xi) = \frac{\kappa(\kappa+3)}{R^2}\Phi_\kappa(\xi) \qquad (3.1')$$

where we have called the eigenfunctions $\Phi_\kappa(\xi)$ and denoted the eigenvalues by $\frac{1}{R^2}\kappa(\kappa+3)$. It is easy to verify that the functions*)

$$\Phi_\kappa(\xi) = |[\xi\zeta]|^\kappa \qquad ((3.2)$$

with ζ being a vector on the cone in the embedding space E_5, i.e. $[\zeta\zeta] = 0$, is a solution of eq. (3.1'). Comparing with the work of Kuriyan, Mukunda and Sudarshan[20] based on the eigenvalue spectrum of the Casimir operators of SO(4,1) it is seen that for $\kappa = -\frac{3}{2} + i\rho;\ 0 \leq \rho < \infty$ one obtains the continuous series (or principal series, p.s.) of unitary irreducible representations (u.i.r.), for $\kappa = -\frac{3}{2} + r;$ $0 < r < \frac{1}{2}$ one obtains the exceptional series (or supplementary series, s.s.) of u.i.r., and for $\kappa = e-1$, $e = 0,1,2, \ldots$ the discrete series (d.s. of u.i.r.**). The symmetry $\kappa \to -\kappa - 3$ leads to u.i.r. equivalent to the ones given above. Comparing eqs. (3.1) and (3.1') it results that for the continuous series characterized by ρ the relation

$$\mu^2 = \frac{\rho^2}{R^2} + \frac{1}{4R^2} \qquad (3.3)$$

holds showing that $\frac{\rho^2}{R^2} + \frac{1}{4R^2}$ plays the role of a mass squared for a scalar particle in de Sitter space.

We, furthermore, normalize the null vector ζ appearing in eq. (3.2) such that $\zeta^5 = \frac{1}{R}$***) i.e. we introduce the 5-vector

$$\zeta^{(\pm)} = (\zeta^{(\pm)\mu}, \frac{1}{R}) \qquad (3.4)$$

with $\zeta^{(\pm)\mu} = \pm(\sqrt{\frac{1}{R^2} + \vec{\zeta}^2}, \vec{\zeta})$ obeying $\zeta^{(\pm)\mu}\zeta^{(\pm)\nu}\eta_{\mu\nu} = \frac{1}{R^2}$. The superscript $(\pm)$ stands for the sign of the zeroth component of $\zeta^{(\pm)}$. Each null vector in E^5 determines what is called a <u>horosphere</u> or <u>horocycle of the first kind</u> in de Sitter space. This is a surface in V_4' characterized by the equations[11]****)

$$|[\xi\zeta]| = 1\ ; \quad [\zeta\zeta] = 0\ . \qquad (3.5)$$

---

*) Remember that $[\xi\zeta] = \xi^a\zeta^b\eta_{ab}$.

**) For details see refs. (17) and (20).

***) Due to the identification of antipodal points on V_4' it suffices to consider null vectors ζ with $\zeta^5 > 0$.

****) A horosphere of the second kind characterized by ζ is determined by $[\xi\zeta] = 0$; $[\zeta\zeta] = 0$.

Due to the particular form (3.4) for ζ the equation $|[\xi\zeta^{(\pm)}]| = 1$ defines a horosphere through the point $\overset{\circ}{\xi}$ (the origin in stereographic coordinates) characterized by $\zeta^{(\pm)}$. If one denotes by $<\xi,\zeta^{(\pm)}>$ the quantity $\log |[\xi\zeta^{(\pm)}]|$ representing the distance of an arbitrary point $\xi \in V_4'$ from the horosphere passing through $\overset{\circ}{\xi}$ determined by $\zeta^{(\pm)}$ one sees that the eigenfunctions (3.2) can be written (changing slightly the notation

$$\Phi_\kappa(\xi) = \Phi_\kappa(\xi,\zeta^{(\pm)}) = e^{\kappa<\xi,\zeta^{(\pm)}>} \quad . \tag{3.6}$$

These are the plane wave solutions - or, more exactly, the horospherical wave solutions - of the scalar wave equation in de Sitter space. They are eigenfunctions of the Laplace-Beltrami operator on V_4' being constant on each horosphere parallel to the one through $\overset{\circ}{\xi}$ characterized by $\zeta^{(\pm)}$. Because of eq. (3.4) the quantity $<\xi,\zeta^{(\pm)}>$ is invariant under transformations of the stability subgroup SO(3,1) of the point $\overset{\circ}{\xi}$. This property corresponds to the Lorentz-invariance of the phase of the Minkowski plane wave solutions. Moreover, the solutions (3.6) have the property that $\Phi_\kappa(\overset{\circ}{\xi},\zeta^{(\pm)}) = 1$.

It is not difficult to show using stereographic projection coordinates that for the principal series the functions (3.6) go over in the limit $R \to \infty$, with $\frac{\rho^2}{R^2}$ kept fixed (see eq. (3.3)), into the usual positive ($e^{-ip.x}$) and negative ($e^{ip.x}$) frequency plane wave solutions of the Klein-Gordon equation in Minkowski space-time:

$$\Phi_{-\frac{3}{2}+i\rho}(\xi,\zeta^{(\pm)}) = e^{(-\frac{3}{2}+i\rho)<\xi,\zeta^{(\pm)}>} \xrightarrow[\frac{\rho^2}{R^2}=\mu^2]{R\to\infty} e^{\pm ip.x} \quad . \tag{3.7}$$

Here we have denoted the scalar product in Minkowski space by a dot. The phase p.x is constant on a plane with normal p having the distance p.x from the origin implying a completely analogous interpretation as for the horospherical solutions (3.6) in a space-time of constant curvature which was mentioned above. From this discussion and eqs. (3.3) and (3.4) it is suggestive to introduce for the case of the p.s. a four vector $\rho\zeta^{(\pm)\mu}$ of length squared $\frac{\rho^2}{R^2}$ having the property that in the flat space limit, with V_4' going over into M_4, it goes over into the momentum vector, $\pm p^\mu$, with $p^2 = \mu^2$, i.e.

$$\lim_{R\to\infty} \rho\zeta^{(\pm)\mu} = \pm\, p^\mu = \pm\,(\sqrt{\mu^2+\vec{p}^2},\ \vec{p}) \quad . \tag{3.8}$$

After these remarks one can now define a general principal series solutions of eq. (3.1), remembering eq. (3.3), by*)

$$f_{\kappa=-\frac{3}{2}+i\rho}(\xi) = \int_B \hat{f}(\rho,\zeta)\ e^{(-\frac{3}{2}+i\rho)<\xi,\zeta>} db \quad . \tag{3.9}$$

---

*) Compare in this context the work of S. Helgason[(21),(22)].

Here $\rho \geq 0$ characterizes the representation, B denotes the part of the boundary of the de Sitter space, i.e. the cone in E_5 given by $[\zeta\zeta] = 0$, over which the integration takes place and

$$db = \delta([\zeta\zeta])d^4\zeta = [\Theta(\zeta^o - \sqrt{\frac{1}{R^2} + \vec{\zeta}^2}) + \Theta(\zeta^o + \sqrt{\frac{1}{R^2} + \vec{\zeta}^2})] \frac{d^3\zeta}{2|\zeta^o|} d\zeta^o \qquad (3.10)$$

represents the associated measure. Inserting the expression (3.10) and integrating over ζ^o eq. (3.9) decomposes into the following positive (-) and negative (+) frequency contributions

$$f_{\kappa=-\frac{3}{2}+i\rho}(\xi) = \int_B [\hat{f}(\rho,\zeta^{(-)})e^{(-\frac{3}{2}+i\rho)<\xi,\zeta^{(-)}>} + \hat{f}(\rho,\zeta^{(+)})\, e^{(-\frac{3}{2}+i\rho)<\xi,\zeta^{(+)}>}]\frac{d^3\zeta}{2\zeta^o} \qquad (3.11)$$

where $d^3\zeta/2\zeta^o$ is the measure on the hyperboloid $\zeta^{(\pm)\mu}\zeta^{(\pm)\nu}\eta_{\mu\nu} = \frac{1}{R^2}$ being the intersection of the cone $[\zeta\zeta] = 0$ with the plane $\zeta^5 = \frac{1}{R}$.

For an arbitrary scalar function $f(\xi)$ on V_4' one can write down a spectral decomposition in terms of functions $\hat{f}(\kappa,\zeta^{(\pm)})$ involving besides the integration over db appearing in eq. (3.9) an additional integration over the u.i.r. characterized by κ with a Plancherel measure $d\mu(\kappa)$. For the group SO(4,1) the Plancherel measure will be nonzero for the principal and the discrete series of u.i.r.[23]. Thus the Fourier analysis on the hyperboloid V_4' isomorphic to the homogenous space SO(4,1)/ SO(3,1) is given by the formula[21],[22],[23]

$$f(\xi) = f_{d.s.}(\xi) + \int_{R^+}\int_B \hat{f}(\rho,\zeta)\, e^{(-\frac{3}{2}+i\rho)<\xi,\zeta>}\sigma(\rho)d\rho db \quad . \qquad (3.12)$$

Here $f_{d.s}(\xi)$ denotes the contribution of the discrete series (which we did not write in expanded form), $\sigma(\rho)d\rho$ is the Plancherel measure for the principal series with the integration over ρ extending over the positive real axis. The generalized Fourier coefficient or Fourier transform, $\hat{f}(\rho,\zeta)$, is obtained by integrating $f(\xi)$ with the SO(4,1) invariant measure $d\mu(\xi)$ against a plane wave, i.e. (for the p.s. part*))

$$\hat{f}(\rho,\zeta) = \int_{V_4'} f(\xi)\, e^{(-\frac{3}{2}-i\rho)<\xi,\zeta>} d\mu(\xi) \quad . \qquad (3.13)$$

The proof of the inversion (3.13) of the expansion (3.12) involves the use of the following transformation rule for the distance $<\xi,\zeta>$ with $\zeta = \zeta^{(\pm)}$ as given by eq. (3.4) and $A \in SO(4,1)$:

$$< A\xi, A\zeta > = <\xi,\zeta > + < A\mathring{\xi}, A\zeta > \quad . \qquad (3.14)$$

For A being an element Λ of the Lorentz subgroup, with $\Lambda\mathring{\xi} = \mathring{\xi}$ and $< \mathring{\xi},\zeta > = <\mathring{\xi},\Lambda\zeta > = 0$, eq. (3.14) expresses the Lorentz invariance of the distance $< \xi,\zeta >$. The addition of the term involving $A\,\mathring{\xi}$ in (3.14) is due to the shift of the

---

*) The d.s. part can be treated similarly. It requires, however, considerably more discussion.

origin*). Furthermore, one needs the following relation for the change of variables on B

$$\left|\frac{d(A\zeta)}{db}\right| = e^{-3\langle A^{-1}\overset{o}{\xi},\zeta\rangle} = e^{3\langle A\overset{o}{\xi},A\zeta\rangle} . \tag{3.15}$$

As regards the representation character of the solutions (3.9) or (3.11) it is shown in ref. 21 by a generally valid argument which we mention here only briefly in its application to the de Sitter group that the mapping

$$f_{-\frac{3}{2}+i\rho}(\xi) \rightarrow T_A f_{-\frac{3}{2}+i\rho}(\xi) = f_{-\frac{3}{2}+i\rho}(A^{-1}\xi) \tag{3.16}$$

with $A \in SO(4,1)$, provides a unitary irreducible representation of SO(4,1) in the Hilbert space $\mathcal{H}_{-\frac{3}{2}+i\rho}$ of functions (3.9) for which $\hat{f}(\rho,\zeta)$ is square integrable on B. From the self-adjointness of the operator $\square$ for u.i.r. (i.e. restricting the κ-values to the range mentioned above) it, moreover, follows that the $f_\kappa(\xi)$ are orthogonal for different eigenvalues κ, i.e.

$$\int_{V'_{4}(x)} f^*_\kappa(\xi)\, f_{\kappa'}(\xi)\, d\mu(\xi) = 0 \quad \text{for } \kappa \neq \kappa' . \tag{3.17}$$

Having treated the scalar case we are now in the position to write down the horospherical wave solutions of the Dirac equation (2.15) on V'_4. Suppressing again the variable x and calling the four component de Sitter spinor $\psi(\xi)$ we solve the equation

$$\frac{1}{2R}\,\gamma^a\gamma^b L_{ab}(\xi)\,\psi(\xi) = (\mu + \frac{2i}{R})\,\psi(\xi) . \tag{3.18}$$

It was shown previously(1)**) that in eq. (3.18) $\mu = \pm\tilde{\mu}$, where $\tilde{\mu}$ is a positive fermionic mass parameter. Considering the ansatz

$$\psi_\kappa(\xi,\zeta^{(\pm)}) = |[\xi\zeta]|^\kappa\,\omega(\zeta^{(\pm)}) \tag{3.19}$$

with $\zeta^{(\pm)}$ as given by eq. (3.4) and with $\omega(\zeta^{(\pm)})$ denoting a constant four spinor solving in analogy to the usual Dirac momentum spinors $u(\pm p)$ the equation

$$(\zeta^a\gamma_a)\,\omega(\zeta^{(\pm)}) = [\zeta^{(\pm)\mu}\gamma_\mu - \frac{1}{R}\gamma^5]\,\omega(\zeta^{(\pm)}) = 0 , \tag{3.20}$$

---

*) In flat space-time the analogue of eq. (3.14) is the Poincaré transformation of the phase (x.p) of a plane wave solution. With $\bar{g}x = \Lambda x+a$ and $\bar{g}p=\Lambda p$, where $\bar{g}\in ISO(3,1)$, one has $(\bar{g}x.\bar{g}p)=(x.p)+(a.\Lambda p)$, with $a=\bar{g}.0$ being the shift of the origin.

**) Compare also ref. 17

it is easy to see that (3.19) is, indeed, an eigenfunction of the operator $D = \frac{1}{2R}\gamma^a\gamma^b L_{ab}(\xi)$ with eigenvalue $-i\kappa/R$. Comparison with eq. (3.18) leads to the relation

$$-i\,\frac{\kappa}{R} = \mu + \frac{2i}{R} = \pm\tilde{\mu} + \frac{2i}{R} \quad . \tag{3.21}$$

Relating this to the analysis of ref. 20 based on the Casimir invariants of SO(4,1)*) one obtains the following values for κ: $\kappa = -2 + i\rho$; $0 < \rho < \infty$ for the continuous series (principal series, p.s.), and $\kappa = 0,1,2, \ldots$ for the discrete series (d.s.). There is no supplementary series in this case. The mirror symmetry leaving the eigenvalue of the Casimir operators invariant is now given by $\kappa \to -\kappa - 4$ which amounts to the transition to representations equivalent to the ones mentioned above. Writing down eq. (3.21) for the p.s. one has $\mu = \rho/R$. Thus, for $\kappa = -2 + i\rho$, $\rho > 0$, one finds $\rho = \mu R$, while for the equivalent representation $\kappa = -2 - i\rho$, $\rho > 0$, one finds $\rho = -\mu R$, i.e.

$$\frac{\rho}{R} = \tilde{\mu} \quad . \tag{3.22}$$

As in the scalar case one shows that letting ρ grow with R such that $\tilde{\mu}$ in eq. (3.22) stays constant in taking the flat space limit the functions

$$\tilde{\psi}_{\kappa=-2+i\rho}(\xi,\zeta^{(\pm)}) = e^{(-2+i\rho)<\xi,\zeta^{(\pm)}>}\,(1 - \frac{\hat{\xi}}{R})\,\omega(\zeta^{(\pm)}) \tag{3.23}$$

where $\hat{\xi} = \xi^a\gamma_a$, reduce for $R \to \infty$ to the usual plane wave solution $\psi_p(x) = u(\pm p)\cdot e^{\mp ip.x}$ of the free Dirac equation in Minkowski space of a particle with mass $\tilde{\mu}$ *), i.e.

$$\gamma^\mu \partial_\mu \psi_p(x) = -i\tilde{\mu}\psi_p(x) \tag{3.24}$$

where

$$e^{(-2+i\rho)<\xi,\zeta^{(\pm)}>}(1-\frac{\hat{\xi}}{R})\,\omega(\zeta^{(\pm)}) \xrightarrow[\frac{\rho}{R}=\tilde{\mu}]{R\to\infty} u(\mp p)e^{\pm ip.x} \quad . \tag{3.25}$$

In analogy to eq. (3.9) one can now write down a general principal series solution of eq. (3.18) for four component spinors as a superposition of horospherical waves belonging to a definite value for $\rho = \mu R > 0$:

$$\begin{aligned}\psi_{\kappa=-2+i\rho}(\xi) = \int_B [\hat{g}(\rho,\zeta^{(-)})\omega(\zeta^{(-)})\, e^{(-2+i\rho)<\xi,\zeta^{(-)}>} + \\ + \hat{g}(\rho,\zeta^{(+)})\omega(\zeta^{(+)})\, e^{(-2+i\rho)<\xi,\zeta^{(+)}>}]\frac{d^3\zeta}{2\zeta o} \quad .\end{aligned} \tag{3.26}$$

---

*) Compare ref. 17.

After this rather lengthy discussion concerning the mathematics of the de Sitter plane or horospherical waves of scalar and spinor type we return to our gauge theory for extended particles which was based on a fiber bundle over space-time possessing a standard fiber identical to a (4,1) de Sitter space. Knowing the solutions of eq. (3.18) we now expand the de Sitter part $\Phi_x(\xi)$ of $\psi(x,\xi)$ according to eq. (3.26) with the Fourier coefficients $\hat{g}(\rho,\zeta^{(\pm)})$ being allowed to depend on $x \in M_4$, i.e. we write in a short hand notation for $\kappa = -2+i\rho$ (p.s.)

$$\Phi_x(\xi) = \int_B \hat{g}(\rho,\zeta;x)\ \omega(\zeta)\ e^{(-2+i\rho)<\xi,\zeta>} db \quad . \tag{3.27}$$

Entering with this expression into eqs. (2.9) and (2.11), remembering eq. (2.17) yields a coupled system of equations in which internal degrees of freedom contained in expressions like $\omega(\zeta')\ M_{ab}\omega(\zeta)$ and translational degrees of freedom are intricately mixed. In order to construct a solution of these equations one would now proceed by restricting the internal motion to certain very simple modes, choose special gauges, restrict the description for stable hadronic objects to the maximal compact subgroup SO(4) of SO(4,1) etc. etc. . Needless to say, so far no explicit solution to these nonlinear inhomogencus equations have been found. What is known from further work of Sasaki[24] are solutions to the free SO(4,1) gauge theory defined by eqs. (2.11) for a vanishing source current. Again these solutions could be expressed in terms of special de Sitter plane waves. Because of lack of space we, however, cannot review these results here in detail.

IV. Concluding Remarks

Our starting point was the supposition that there exists in nature a parameter R of the order of a Fermi which provides a scale for the hadronic masses. Usually it is thought that such a scale would appear automatically in the theory as a result of renormalization. In the presented description, however, such a length parameter is built into the basic geometric structure on which the hadron dynamics is formulated. The proposal is to use as the basic underlying geometry a fiber bundle of Cartan type over space-time with structural group SO(4,1). Not a fully, in all aspects, second quantized gauge theory is aimed at here requiring probably for its rigorous mathematical definition an underlying lattice structure with a lattice spacing determined by a parameter like R destroying thereby the Lorentz invariant of the theory. Instead we use continuum mathematics and differential geometry on a higher dimensional manifold (the de Sitter bundle) representing the basic hadronic matter field by a bilocal spinor quantity, $\psi(x,\xi)$, possessing both q-number and c-number properties.

References

(1) W. Drechsler, Fortschr.Phys. 23, 607 (1975).

(2) W. Drechsler, Found. of Phys. 7, 629 (1977).

(3) W. Drechsler, Nuovo Cimento, 41A, 597 (1977).

(4) A. Bohm, The Dynamical Group of a Simple Particle Model, in Lectures in Theoretical Physics, Vol. IXB. Editors: A.O. Barut and W.E. Brittin Gordon and Breach, 1967, p. 327.

(5) A. Bohm, Phys.Rev. 175, 1767 (1968).

(6) A. Bohm, Relativistic Rotators - A Quantum Mechanical de Sitter Bundle, in Proc. of the Int. Symposion on Mathematical Physics, Mexico City 1976, Vol. 2, p. 377.

(7) W. Drechsler, Wave Equation for Extended Hadrons, in Group Theoretical Methods in Physics, Editors. A. Janner, T. Janssen, and M. Boom, Lecture Notes in Physics, Vol. 50, Springer Verlag, 1976, p. 37.

(8) W. Heisenberg, Die Naturwissenschaften 63, 1 (1976).

(9) R. Takahashi, Bull.Soc.Math. France 91, 289 (1963).

(10) W. Drechsler, Phys. Letters, 66B, 439 (1977).

(11) I.M. Gelfand, M.I. Graev, and N.Ya. Vilenkin, Generalized Functions, Vol. 5, Chapter V. Academic Press, London 1966.

(12) E. Schrödinger, Expanding Universe, Cambridge University Press 1956.

(13) Ch. Ehresmann, Colloque de Topologie (espaces fibrés), Bruxelles 1950, p. 29.

(14) W. Drechsler, J.Math.Phys. 18, 1358 (1977).

(15) P.A.M. Dirac, Ann.Math. 36, 657 (1935).

(16) F. Gürsey, and T.D. Lee, Proc.Nat.Acad.Sci. 49, 179 (1963).

(17) W. Drechsler, and R. Sasaki, Nuovo Cimento 46A, 527 (1978).

(18) R. Penrose, Proc.Roy.Soc. (London) 284, 159 (1965).

(19) N.A. Chernikov, and E.A. Tagirov, Ann.Inst. Henri Poincaré 19, 109 (1968).

(20) J.B. Juriyan, N. Mukunda, and E.C.G. Sudarshan, Commun.Math.Phys. 8, 204 (1968).

(21) S. Helgason, Lie Groups and Symmetric Spaces, in Battelle Rencontres, Editors: C.M. De Witt, and J.A. Wheeler, Benjamin Inc. New York 1968, p. 1.

(22) S. Helgason, Functions on Symmetric Spaces, in Homogenous Analysis on Homogenous Spaces. Ann.Math.Soc., Providence, Rhode Island, 1973.

(23) R.S. Strichartz, Journ.of Functional Analysis 12, 341 (1973).

(24) R. Sasaki, Some Classical Solutions of the Sourceless SO(4,1) Gauge Field Equations, Preprint MPI-PAE/PTh 24/78, June 1978.

Fundamental Length Hypothesis in a Gauge Theory Context†

VLADIMIR G. KADYSHEVSKY*
Fermi National Accelerator Laboratory, Batavia, Illinois 60510

ABSTRACT

The new gauge formulation of the electromagnetic interaction theory, containing the "fundamental length" ℓ as a universal scale like $\hbar$ and c, is worked out. A key part belongs to the 4-dimensional de Sitter p-space, with the curvature radius $\hbar/\ell c$. In the new approach the electromagnetic potential becomes a 5-vector associated with de Sitter group O(4.1). The extra fifth component, called the τ-photon, similar to scalar and longitudinal photons, does not correspond to an independent dynamical degree of freedom. Respectively, the new local gauge group is larger than the ordinary one and depends intrinsically on the fundamental length ℓ.

The gauge invariant equations of motion, replacing the Dirac-Maxwell equations, are set up. The new formulation is minimal with respect to the 5-potential but is not so in terms of the usual 4-potential. As a result, the underlying physics looks much richer than the ordinary electromagnetic phenomena. The new scheme predicts the existence of the electric dipole moments for charged particles, leading to a direct violation of P- and CP-symmetries, and the new universal correction to the (g-2)-anomaly. Further, some new group of internal symmetry, $SU_\tau(2)$, arises that can be used to describe the μe-symmetry of the electromagnetic interactions. It turns out that $SU_\tau(2)$-symmetry is violated by the 4-fermion type interaction, induced by τ-photons, with associated coupling constant $\sim\alpha\ell^2$. This novel interaction might give rise to the μe-mass difference and processes like $\mu \to 3e$, $\mu \to e\gamma$, etc.

In the limit $\ell \to 0$, the new field equations turn into the Dirac-Maxwell equations for the electron, muon and electromagnetic fields. So, one may consider this approach as a generalization in a profound way of the standard theory of electromagnetic interactions at small distances $\lesssim \ell$ (high energies $\gtrsim 1/\ell$).

The upper bound for the fundamental length ℓ is discussed taking into account the various experimental data.

1. FUNDAMENTAL LENGTH AND DE SITTER MOMENTUM SPACE

In the present talk we shall discuss a new gauge formulation of the electromagnetic interaction theory which is based on a concept of fundamental length. This new hypothetical constant we denote as ℓ. Together with $\hbar$ and c it is expected to regulate all microscopic phenomena. The quantity

$$M = \hbar/\ell c \tag{1.1}$$

is called the fundamental mass.

† Contribution to the Integrative Conference on Group Theory and Mathematical Physics, Austin, Texas, September 1978.

*Permanent address: Joint Institute for Nuclear Research, Dubna, USSR.

The idea of an existence of a new universal length, and therefore mass, that would fix a scale in 4-dimensional space-time, and therefore 4-dimensional momentum-energy space, was discussed in literature of the last four decades in different contexts.[1-14]

Most of the people who tried to introduce a fundamental length into field theory, pursued a quite clear and practical goal: to cure the theory from the ultraviolet divergences. But it turned out that a theory can survive with this chronic disease, and work as a quantitative scheme, if it possesses genetically a renormalizability property. Nowadays, the principle of renormalizability has imperceptibly become one of the corner stones of the quantum field theory. As a result, interest in a fundamental length has almost died out (see, however, refs. 15-18).

The greatest triumph of the renormalization approach to the formulation of the quantum field theory is certainly quantum electrodynamics (QED). The predictions of QED agree with a number of highly precise experiments. The upper bound for the magnitude of the fundamental length, established in experiments in the test of QED at high energies, now is given by

$$\ell \lesssim 10^{-15}\,\mathrm{cm} \quad . \tag{1.2}$$

The harmony and elegance of QED make an impression which cannot be darkened even by the obviously algorithmic character of the renormalization procedure. It should be clear that the fundamental length hypothesis is first of all a challenge to contemporary QED. In other words, this hypothesis can survive only if it will lead naturally to modifying QED in a profound way.

The crucial advantage of QED is that the form of the interaction in this theory is dictated by gauge symmetry arguments. It is called the minimal interaction principle and is symbolized by the following substitution law ($\hbar = c = 1$)

$$p_\mu \rightarrow p_\mu - e_0 A_\mu(x) \quad . \tag{1.3}$$

This substitution leads one to the inhomogeneous Dirac-Maxwell equations for "bare" fields

$$(i\partial\!\!\!/ - e_0 A\!\!\!/(x) - m_0)\psi(x) = 0 \tag{1.4a}$$

$$\frac{\partial F^{\mu\nu}(x)}{\partial x^\nu} = e_0 \bar{\psi}(x)\gamma^\mu \psi(x) \tag{1.4b}$$

where the field strengths are defined as $F^{\mu\nu}(x) = \frac{\partial A^\mu(x)}{\partial x_\nu} - \frac{\partial A^\nu(x)}{\partial x_\mu}$.

Let us mention, to be complete, that local gauge transformations of the fields $\psi(x)$, $\bar{\psi}(x)$ and $A(x)$, leaving Eqs. (1.4a)-(1.4b) invariant, are given by*:

$$\psi(x) \rightarrow e^{ie_0\lambda(x)}\psi(x) \quad , \quad \bar{\psi}(x) \rightarrow e^{-ie_0\lambda(x)}\bar{\psi}(x) \tag{1.5a}$$

$$A_\mu(x) \rightarrow A_\mu(x) - \frac{\partial\lambda(x)}{\partial x^\mu} \quad ; \quad \lambda^\dagger(x) = \lambda(x) \quad . \tag{1.5b}$$

*In p-representation the hermiticity condition of λ-function, evidently, becomes

$$\lambda^\dagger(p) = \lambda(-p) \quad . \tag{1.6}$$

The rule (1.3) does not contain any scale like ℓ or M and for this reason is universally applied to all space-time intervals and to all values of 4-momenta. Therefore, if one adopts the fundamental length hypothesis it means that the substitution law (1.3) and its consequences are probably invalid or incomplete in the domains

$$|x| \lesssim \ell \qquad (1.7a)$$

$$|p| \gtrsim M \quad . \qquad (1.7b)$$

Let us consider just one consequence of (1.3). Choosing $e_o A_\mu = \text{const} = k_\mu$ we obtain obviously, zero field strengths: $F_{\mu\nu}(x) = 0$. But corresponding substitution (1.3) is not yet an identity transformation, namely

$$p_\mu \rightarrow p_\mu - k_\mu \quad . \qquad (1.8)$$

This is a pseudoeuclidean parallel shift transformation of the 4-dimensional p-space, testifying that a geometry of this space is a Minkowskian one.

So we may conclude that our fundamental length hypothesis challenges the Minkowskian structure of the momentum 4-space in the region (1.7b). But if the momentum 4-space is not everywhere pseudoeuclidean then what is a reasonable alternative? According to a general geometrical classification, the (pseudo)euclidean spaces are those with zero curvature. Their closest neighbors are spaces with non-zero constant curvature. In the present 4-dimensional case, these curved spaces are so-called "de Sitter spaces."

Let us try to impose on 4-momentum space the de Sitter geometry realized on the one-sheeted 5-hyperboloid

$$p_0^2 - p_1^2 - p_2^2 - p_3^2 - M^2 p_4^2 = -M^2 \quad . \qquad (1.9)$$

The curvature radius M we identify with the fundamental mass (1.1), assuming that this quantity is large enough (cf. (1.2)). Note that Eq. (1.9) places no constraint on timelike 4-momenta, and it is therefore not in conflict with the construction of Fock space and Poincaré invariance of the S-matrix. Besides (1.9), there exists only one more de Sitter space satisfying the correspondence principle at $M \rightarrow \infty$, namely

$$p_0^2 - p_1^2 - p_2^2 - p_3^2 + M^2 p_4^2 = M^2 \quad .$$

But in this geometry we are faced with the universal upper bound for time-like momenta $p_0^2 - \vec{p}^2 \leq M^2$, which is inconsistent with the implementation of a unitary representation of the Poincaré group on Fock space.

Since the mass shell hyperboloids $p^2 = m_1^2$, $p^2 = m_2^2$,... can be equally well embedded into de Sitter p-space (1.9) or in flat Minkowskian p-space free physical particles cannot distinguish between these two geometries. Actually, only virtual (interacting) particles can probe the geometrical structure of 4-momentum space.

In the "flat limit," i.e. in the region of small virtual momenta

$$|p| \ll M$$

one can neglect the curvature of de Sitter p-space, and therefore the new formalism reduces to the ordinary theory.

For virtual momenta belonging to the region (1.7b), the curvature of de Sitter p-space becomes a crucial factor. It means that the old (Minkowskian) and new (de Sitterian) formalisms should lead to quite different descriptions of particle interactions at small space-time intervals.

A general approach to the construction of quantum field theory on a base of de Sitter p-space has been put forward and investigated in Refs. 19-29.* The concept of local gauge transformations and gauge vector field was transferred to the new geometrical arena in Ref. 30.

2. NEW CONCEPT OF LOCAL GAUGE TRANSFORMATION AND GAUGE VECTOR FIELD

It is clear, independently of arguments connected with p-space geometry, that in a theory based on the fundamental length hypothesis, the notion of a local gauge group should be revised or generalized in some nontrivial manner. However, geometrical or group theoretical arguments allow it to be done in an essentially unique way.

Indeed, one can realize that in de Sitter p-space (1.9), λ-functions parametrizing the gauge transformation in question may be written as follows:

$$\lambda(p_o, \vec{p}, p_4) = \delta(p_o^2 - \vec{p}^2 - M^2 p_4^2 + M^2)\tilde{\lambda}(p_o, \vec{p}, p_4) \tag{2.1}$$

with the hermiticity condition inherited from (1.7)

$$\lambda^\dagger(p, p_4) = \lambda(-p, p_4) \quad . \tag{2.2}$$

The next step is connected with the following observation: if

$$\lambda(p, p_4) = \int e^{-ipx-ip_4\tau} \lambda(x, \tau)d^4x d\tau \tag{2.3}$$

then

$$(\Box - M^2 \frac{\partial^2}{\partial\tau^2} - M^2)\lambda(x, \tau) = 0$$

$$\lambda(x, \tau)^\dagger = \lambda(x, -\tau) \quad . \tag{2.4}$$

So in the new scheme, the gauge functions λ may be treated as local functions of five variables (x^μ, τ), with the obligatory constraints (2.4). The extra space-like variable τ can be

---

*The use of such a momentum space in a field theory was pioneered in Refs. 4 and 8. The list of other papers on this subject can be found in Ref. 20. In Ref. 10, the field theory with the momentum space of variable curvature was discussed. We should point out that in all previous attempts to employ a non-euclidean p-space, Poincaré invariance of the theory was not maintained. The concept of nonlocal electromagnetic field based on ideas which were close to a non-euclidean momentum space hypothesis, holding Poincaré invariance, was developed many years ago in Ref. 3.

interpreted, due to its commutativity with the Poincaré group generators, as some internal parameter of the theory. All λ-functions which parametrize the conventional gauge transformations (1.6a)-(1.6b) can be found among functions $\lambda(x, 0)$.

Since the localization of the new gauge group happened to be connected with the configurational 5-space, the relevant gauge vector field, i.e. the electromagnetic potential, has to be a 5-vector. Let us denote it as

$$A_M(x, \tau) = \left(A_\mu(x, \tau), A_4(x, \tau)\right) \qquad \mu = 0, 1, 2, 3 \quad . \tag{2.5}$$

The extra component $A_4(x, \tau)$ we call the τ-photon. The neutrality of the electromagnetic field gives rise to the relation

$$A_M(x, -\tau) = \left(A_\mu^{\dagger}(x, \tau), -A_4^{\dagger}(x, \tau)\right) \quad . \tag{2.6}$$

The equations of motion for all five components (2.5) are of the form (we put $\hbar = c = \ell = M = 1$)[30]

$$\left\{\begin{array}{l} A_\mu(x, \tau) + i\dfrac{\partial A_\mu(x, \tau)}{\partial \tau} - i\dfrac{\partial A_4(x, \tau)}{\partial x^\mu} = 0 \\[2ex] A_4(x, \tau) - i\dfrac{\partial A_4(x, \tau)}{\partial \tau} + i\dfrac{\partial A_\nu(x, \tau)}{\partial x_\nu} = 0 \\[2ex] \left(\Box - \dfrac{\partial^2}{\partial \tau^2} - 1\right) A_\mu(x, \tau) = 0 \end{array}\right. \quad . \tag{2.7}$$

It is readily verified that Eqs. (2.7) are invariant under the following gauge transformation:

$$\begin{aligned} A_\mu(x, \tau) &\to A_\mu(x, \tau) - \frac{\partial \lambda(x, \tau)}{\partial x^\mu} \\ A_4(x, \tau) &\to A_4(x, \tau) + i\,\lambda(x, \tau) - \frac{\partial \lambda(x, \tau)}{\partial \tau} \end{aligned} \tag{2.8}$$

where $\lambda(x, \tau)$ satisfies the relations (2.4). Further, due to (2.6), Eqs. (2.7) remain unaltered after the transformation $\tau \to -\tau$ combined with hermitian conjugation. We shall refer to this property as the τ-invariance.

As was shown in Ref. 30 the τ-photon does not correspond to an independent dynamical degree of freedom and can be excluded by an appropriate gauge transformation (2.8). But, similar to scalar and longitudinal photons, it plays the important mediating role in an interaction.[31]

Putting

$$B_M(x, \tau) = e^{-i\tau} A_M(x, \tau) \tag{2.9}$$

one can rewrite (2.8) as follows

$$B_M(x, \tau) \to B_M(x, \tau) - \frac{\partial}{\partial x^M}\left(e^{-i\tau}\lambda(x, \tau)\right) \quad M = 0, 1, 2, 3, 4 \quad . \tag{2.10}$$

The 5-potential $B_M(x, \tau)$ has a clear geometrical meaning in the fiber bundle theory context.* Introducing the 5-dimensional field strengths

---

*A generalization of (2.10) to the non-abelian case is straightforward.[30]

$$F_{MN}(x,\tau) = \frac{\partial B_M(x,\tau)}{\partial x^N} - \frac{\partial B_N(x,\tau)}{\partial x^M} \qquad (2.11)$$

we can conclude from this definition and Eqs. (2.7) that

$$F_{\mu 4}(x,\tau) = 0 \;, \quad \frac{\partial F^{\mu\nu}(x,\tau)}{\partial \tau} = 0 \quad ; \qquad (2.12)$$

$$\frac{\partial F^{\mu\nu}(x,0)}{\partial x^\nu} = 0 \quad . \qquad (2.13)$$

Eq. (2.13) coincides with the standard Maxwell equation. This prompts that $B_\mu(x, 0)$ can be identified with the ordinary electromagnetic 4-potential $A_\mu(x)$:

$$B_\mu(x,0) \equiv A_\mu(x) \quad . \qquad (2.14)$$

A formulation of the free Dirac theory in the 5-dimensional (x, τ)-space is described in Ref. 31. The important feature of that scheme is a use of the 8-component de Sitter spinors

$$\Psi(x,\tau) = \begin{pmatrix} \psi_1(x,\tau) \\ \psi_2(x,\tau) \end{pmatrix} \qquad (2.15)$$

where $\psi_a(x, \tau)$ $(a = 1, 2)$ are the conventional 4-component spinors. This is connected with the requirement of the τ-invariance.

3. GENERALIZED PRINCIPLE OF MINIMAL ELECTROMAGNETIC INTERACTION AND NEW FIELD EQUATIONS

The next natural step is to assume that the new theory of electromagnetic interaction has to be invariant under the following group of the local gauge transformations:

$$\Psi(x,\tau) \rightarrow e^{ie_o e^{-i\tau}\lambda(x,\tau)}\,\Psi(x,\tau)$$

$$B_M(x,\tau) \rightarrow B_M(x,\tau) - \frac{\partial}{\partial x^M}\left(e^{-i\tau}\lambda(x,\tau)\right) \qquad M = 0,1,2,3,4 \qquad (3.1)$$

where $\lambda(x, \tau)$ satisfies, as before, the relations (2.4) and e_o is the electric charge. The appropriate generalization of the <u>minimal principle of electromagnetic interaction</u> can be given by the substitution law

$$i\frac{\partial}{\partial x^M} \rightarrow i\frac{\partial}{\partial x^M} - e_o B_M(x,\tau) \qquad M = 0,1,2,3,4 \quad . \qquad (3.2)$$

In the new approach we obtain the following field equations, substituting the Dirac-Maxwell equations (1.4a)-(1.4b) (the units are $\hbar = c = 1$)[31]

$$(i\partial\!\!\!/ - e_o A\!\!\!/ - m_o)\psi_1(x) = \frac{ie_o \ell \cos\theta_o}{4}\gamma^5\sigma^{\mu\nu}F_{\mu\nu}(x)\psi_1(x) -$$

$$- \frac{e_o \ell \sin\theta_o}{4}\sigma^{\mu\nu}F_{\mu\nu}(x)\psi_1(x) - \frac{e_o^2\ell^2}{8}\left(\bar\psi_1(x)\gamma_\mu\psi_2(x) + \bar\psi_2(x)\gamma_\mu\psi_1(x)\right)\gamma^\mu\psi_1(x) \quad , \qquad (3.3a)$$

$$(i\not{\partial} - e_o\not{A} - m_o)\psi_2(x) = \frac{ie_o\ell \cos\theta_o}{4}\gamma^5\sigma^{\mu\nu}F_{\mu\nu}(x)\psi_2(x) -$$

$$- \frac{e_o\ell\sin\theta_o}{4}\sigma^{\mu\nu}F_{\mu\nu}(x)\psi_2(x) + \frac{e_o^2\ell^2}{8}(\overline{\psi}_1(x)\gamma_\mu\psi_2(x) + \overline{\psi}_2(x)\gamma_\mu\psi_1(x))\gamma^\mu\psi_2(x) \quad , \quad (3.3b)$$

$$\frac{\partial F^{\mu\nu}}{\partial x^\nu} = e_o \sum_{a=1,2}\Big[\overline{\psi}_a(x)\gamma^\mu\psi_a(x) - \frac{i\ell}{2}\cos\theta_o\frac{\partial}{\partial x^\nu}(\overline{\psi}_a(x)\gamma^5\sigma^{\mu\nu}\psi_a(x)) +$$

$$+\frac{\ell}{2}\sin\theta_o\frac{\partial}{\partial x^\nu}(\overline{\psi}_a(x)\sigma^{\mu\nu}\psi_a(x))\Big] \quad , \quad (3.3c)$$

where $\sigma^{\mu\nu} = \frac{i}{2}(\gamma^\mu\gamma^\nu - \gamma^\nu\gamma^\mu)$ and

$$\sin\theta_o = \frac{\sqrt{1+m_o^2\ell^2}-1}{m_o\ell} \quad , \quad \cos\theta_o = \sqrt{\frac{2}{1+\sqrt{1+m_o^2\ell^2}}} \quad . \quad (3.4)$$

Eqs. (3.3) contain as the conventional minimal electromagnetic interaction as the <u>non-minimal</u> terms:

i) electric dipole moment (EDM) interaction $\sim \ell\cos\theta_o$;

ii) magnetic dipole moment (MDM) interaction $\sim \ell\sin\theta_o$, called also the Pauli interaction;

iii) τ-photon interaction $\sim e_o^2\ell^2$. Such a name is chosen because this interaction survives even if there are no "ordinary" photons ($A_\mu(x) = 0$) and, therefore, it could be induced only by the τ-photons.

Neglecting the τ-photon interaction in Eqs. (3.3a)-(3.3b), we obtain that the resulting set of equations, together with Eq. (3.3c), becomes invariant under some SU(2)-group operating in (ψ_1, ψ_2)-space. This group will be denoted as $SU_\tau(2)$.

We would like to stress that <u>all</u> interaction terms in Eqs. (3.3), minimal and non-minimal, are originated by the <u>minimal</u> interaction in terms of the 5-potential $B_M(x,\tau)$.

4. INTERPRETATION AND DISCUSSION

We suggest to consider the $SU_\tau(2)$-symmetry, deeply ingrained in our scheme, as a manifestation of the μe-symmetry of the electromagnetic interactions. So, for instance,*

$$\psi_1 = e$$
$$\psi_2 = \mu \quad . \quad (4.1)$$

The τ-photon interaction, violating $SU_\tau(2)$-symmetry, presumably gives rise to the muon-electron mass difference and the processes $\mu \to 3e$, $\mu \to e\gamma$, etc. Analyzing the experimental data, concerning these decays, one obtains the following upper bound for the fundamental length ℓ:

$$\ell \lesssim 3\cdot 10^{-18}\text{ cm} \quad . \quad (4.2)$$

Next, let us consider Eqs. (3.3a)-(3.3b) in the Pauli approximation. Assuming that

$$A^o(x) = \phi(r) = \text{arbitrary} \quad , \quad \vec{A}(x) = \tfrac{1}{2}[\vec{H}\times\vec{r}] \quad ; \quad \vec{H} = \text{const.} \quad , \quad (4.3)$$

and expanding Eqs. (3.3a)-(3.3b) in powers of $1/c$, one finds the following generalized Pauli equation:

$$i\hbar\frac{\partial}{\partial t}\phi_a(\vec{r},t) = \left[\frac{\vec{p}^2}{2m_o} - e_o\phi(r) - e_o\frac{(\vec{L}+g\vec{S})\cdot\vec{H}}{2m_oc} + \frac{e_o\cos\theta_o}{Mc}(\vec{S}\cdot\vec{E})\right]\phi_a(\vec{r},t) \quad (a=1,2) \tag{4.4}$$

where

$$\vec{p} = -i\hbar\frac{\partial}{\partial\vec{r}}\ ;\ \vec{L} = [\vec{r}\times\vec{p}]\ ;\ \vec{S} = \frac{\hbar\vec{\sigma}}{2}\ ;$$

$$\vec{E} = -\frac{\partial}{\partial\vec{r}}\phi(r)\ ;\ g = 2\sqrt{1+\frac{m_o^2}{M^2}}\quad .$$

The quantity

$$\frac{g-2}{2} = \sqrt{1+\frac{m_o^2}{M^2}} - 1 \tag{4.5}$$

defines the intrinsic anomalous magnetic moment of the Dirac particle in our approach. Writing the EDM-interaction in the canonical form $U_{EDM} = (\vec{d}\cdot\vec{E})$ one obtains the expression for the intrinsic electric dipole moment of the same particle

$$\vec{d} = -e_o\frac{\cos\theta_o}{Mc}\vec{S} = -e_o\ell\frac{\cos\theta_o}{2}\vec{\sigma}\quad . \tag{4.6}$$

Thus, one can conclude:

i) the Dirac particle in this scheme is an extended object from the very beginning;

ii) the theory developed predicts P- and CP-violations in the electromagnetic interaction, since charged particles possess the EDM.

Eq. (4.4) holds in the general case of arbitrary ratio between the particle mass m_o and the fundamental mass M. For leptons, evidently, $m_o/M \ll 1$ and therefore

$$\left(\frac{g-2}{2}\right)_{lepton} \simeq \frac{m_o^2}{2M^2} \tag{4.7}$$

$$|\vec{d}|_{lepton} \simeq \frac{e_o\ell}{2} \tag{4.8}$$

Comparing (4.7) with the current theoretical and experimental uncertainties in this quantity we obtain one more upper bound for the fundamental length*:

$$\ell \lesssim 2.6\cdot 10^{-17}\ \text{cm} \tag{4.9}$$

that is not very far from (4.2).

Using (4.2), (4.8) and (4.9) we can conclude that the upper bound for leptonic EDM, in order of magnitude, is at least

$$|\vec{d}|_{lepton} \lesssim e_o\,(10^{-17} \div 10^{-18})\ \text{cm} \tag{4.10}$$

---

*To our knowledge, it is the first time a highly precise experiment, designed for a test of a validity of QED, is used to estimate the fundamental length.

This is consistent with the old experimental data on a direct measurement of the electron and muon EDM,[32] with observed shifts of atomic levels,[32] with parity violation effects in atoms,[33] with the recent search for the parity violation in the polarized electron scattering.[34]

On the other hand, a number of experiments were performed on indirect estimation of the electron EDM through the measurements of EDM that it induces in atoms. From the result obtained[35–38] and (4.8) one might conclude that

$$\ell \lesssim 10^{-24} \div 10^{-23}\ \mathrm{cm} \quad . \qquad (4.11)$$

Let us emphasize once more that the existence of non-zero EDM for elementary particles should lead to the violation of the CP-symmetry.[39] Thus, according to our approach, the mechanism of the CP-violation may be purely electromagnetic if the theory of the electromagnetic interactions is based on the fundamental length hypothesis.

It does not need any comment that the experimental discovery of the particle EDM would be of great importance for the present theory. Concerning leptons, which have not undergone strong interactions, one should realize that, due to (4.8), the measurement of their EDM is the straight measurement of the fundamental length.

Coming back to the particle MDM we can rewrite this quantity as follows

$$\vec{\mu} = \frac{e_0 g}{2m_0 c}\vec{S} = \frac{e_0}{Mc}\sqrt{\frac{m^2 + M^2}{m^2}}\,\vec{S} \quad . \qquad (4.12)$$

So $e_0\hbar/2MC \simeq e_0\ell/2$ plays the role of a minimal magneton attainable only when $m_0 \gg M$. Hence, superheavy Dirac particles, if such objects somewhere exist,* should serve not only as sources of the static Coulomb field but also as sources of the static magnetic field, produced by the magnetic dipole moment $(e_0\hbar\vec{\sigma})/(2Mc)$.

I am sincerely grateful to Professor Arno Bohm for his kind invitation extended to me to attend this conference.

REFERENCES

1 G.V. Watagin, Zc. Phys. 88, 92 (1934).

2 W. Heisenberg, Zc. Phys. 101, 533 (1936); W. Heisenberg, Introduction to the Unified Field Theory of Elementary Particles, Inters. Publ. 1966.

3 M.A. Markov, JETP 10, 1311 (1940).

4 H. Snyder, Phys. Rev. 71, 38 (1947); 72, 68 (1947).

5 C.N. Yang, Phys. Rev. 72, 814 (1947).

6 M.A. Markov, Nucl. Phys. 10, 140 (1958); A.A. Komar and M.A. Markov, Nucl. Phys. 12, 190 (1959); M.A. Markov, Hyperons and K-mesons, GIFML, Moscow, 1958.

---

*Cf. the "maximon" considered by Markov.[52]

[7] D.I. Blokhintsev, UFN 61, 137 (1957).

[8] Yu. A. Gol'fand, JETP 37, 504 (1959); 43, 256 (1962); 44, 1248 (1963).

[9] V.G. Kadyshevsky, JETP 41, 1885 (1961), AN USSR Doklady, 147, 588, 1336 (1962).

[10] I.E. Tamm, Proceedings of XII International Conference on High Energy Physics, vol. II, p. 229, Atomizdat, Moscow (1964); Proceedings of Inter. Confer. on Elementary Particles, Kyoto (1965).

[11] R.M. Mir-Kasimov, JETP 49, 905, 1161 (1965); 52, 533 (1967).

[12] D.A. Kirzhnits, UFN 90, 129 (1966).

[13] A.N. Leznov, JINR preprint P2-3590, p. 52 (1967).

[14] D.I. Blokhintsev, "Space and Time in Microworld," Moscow, Nauka, 1970.

[15] G.V. Efimov, Particles and Nuclei, 1, No. 1, 256 (1970); 5, No. 1, 223 (1974).

[16] M.A. Solov'ev and V. Ya. Feinberg, in "Non-local, non-linear and non-renormalizable theories," D2-9788, JINR, Dubna (1976).

[17] S. Fubini, CERN preprint TH 2129-CERN (1976).

[18] J.D. Bjorken, Proceedings of B. Lee Memorial Conference (to be published).

[19] V.G. Kadyshevsky, JINR preprint P2-5717, Dubna (1971).

[20] V.G. Kadyshevsky, in the book "Problems of Theoretical Physics" dedicated to the memory of I.E. Tamm, Moscow, Nauka, 1972; in "Non-local, Non-linear and Non-renormalizable Theories," D2-7161, Dubna (1973).

[21] A.D. Donkov, V.G. Kadyshevsky, M.D. Mateev and R.M. Mir-Kasimov, Bulgar. Journ. of Physics 1, 58, 150, 233 (1974); 2, 3 (1975); Proceedings of Internat. Conference on Mathemat. Problems of Quantum Field Theory and Quantum Statistics, pp. 85-129, Moscow, Nauka (1975); JINR preprint E2-7936, Dubna (1974).

[22] R.M. Mir-Kasimov, "Axiomatic Quantum Field Theory and de Sitter momentum space," in P1,2-7642, JINR, Dubna (1973).

[23] V.G. Kadyshevsky, M.D. Mateev and R.M. Mir-Kasimov, JINR preprints: E2-8892, P2-8877, Dubna (1975).

[24] V.G. Kadyshevsky, "Fundamental length as a new scale in quantum field theory," in D1,2-9342, Dubna (1975).

[25] A.D. Donkov, V.G. Kadyshevsky, M.D. Mateev and R.M. Mir-Kasimov, in "Non-local, Non-linear and Non-renormalizable theories," D2-9788, Dubna (1976); Proceedings of the XVIII International Conference on High Energy Physics, Tbilisi, p. A5-1, D1,2-10400, Dubna (1977).

[26]M.D. Mateev, "Processes at High Energies and Fundamental Length Hypothesis," in D2-10533, p. 257, Dubna (1977).

[27]I.P. Volobuyev, TMF 28, 331 (1976).

[28]R.M. Mir-Kasimov, I.P. Volobuyev, Acta Physica Polonica B9, 2 (1978).

[29]V.G. Kadyshevsky, M.D. Mateev, R.M. Mir-Kasimov and I.P. Volobuyev, JINR preprint, E2-10860 (1977).

[30]V.G. Kadyshevsky, Fermilab-Pub-78/22-THY, Nuclear Physics (in press).

[31]V.G. Kadyshevsky, Fermilab-Pub-78/70-THY, Annals of Physics (in press).

[32]F.L. Shapiro, UFN, 95, 145 (1968).

[33]L. Barkov, Parity Violation in Bi-atoms, XIX Int. Conference on High Energy Physics, Tokyo, 1978.

[34]R.E. Taylor, Parity Violation in Polarized eD-Scattering, XIX Int. Conf. on High Energy Physics, Tokyo, 1978.

[35]M.C. Weisskopf, et al., Phys. Rev. Lett. 21, 1645 (1968); T. S. Stein, et al., Phys. Rev. 186, 39 (1969).

[36]M.A. Player and P.G.H. Sandars, J. Phys. B3, 1620 (1970).

[37]P.G.H. Sandars and R.M. Sternheimer, Phys. Rev. A11, 473 (1975).

[38]B.V. Vasiliev, E.V. Kolycheva, JINR preprint, P14-10948 (1977).

[39]L.D. Landau, JETP, 32, 405 (1957).

[40]M.A. Markov, JETP 51, 878, 1966; Progress of Theor. Phys., H. Yukawa Suppl., p. 85 (1965).

Superunified Theories in Superspace*

Freydoon Mansouri**

Department of Physics, Yale University, New Haven, Conn. 06520

ABSTRACT

The structure of superunified theories in superspace is discussed from the point of view of local gauge invariance. It is shown that the same two requirements which lead to the correct description of gravity determine the structure of these theories. In particular, all the transformation laws of the fields follow from geometry.

---

* Research (Yale Report COO-3075-213) supported by the U. S. Department of Energy.

** Invited talk given at Integrative Conference on Group Theory and Mathematical Physics, Sept., 1978, Austin, Texas, U.S.A.

I. Introduction

In a superunified theory based on the geometry of local gauge invariance, the primary objectives are two: (i) To describe the theory in terms of fields naturally arising in the geometry; (ii) To obtain the transformation laws of the fields directly from geometry. Once these are achieved, one would then have to impose additional physical requirements to uniquely determine the theory of interest. The fact that the first objective can be attained is by now well-known[1]. But a completely satisfactory solution to the second objective has been harder to come by: Various ways of obtaining such transformation laws have either been model dependent or have involved separate assumptions with no clear realtion to geometry[2]. One aim of this talk is to show that once the geometry is fully specified by the physics it describes, the transformation laws do indeed follow from geometry. This specification of geometry is based on two requirements: (i) In contrast to Yang-Mills theory, gravity is described by a non-linear realization of the space-time gauge symmetry[1,3,4,5]; (ii) When the gauge group involves a space-time symmetry, horizontal and vertical transformations in the corresponding fiber bundle must be properly "interlocked"[6,7]. As will be seen below, to implement these ideas one is naturally led to work in superspace. Then to make contact with the physical world, a physical interpretation of superspace is suggested.

II. Geometry of Gravity

A comparison of Einstein's field equations with those of Yang-Mills shows that these two sets of equations are structurally different[1,3,4]. Moreover, these differences are not due to the specific form of Einstein-Hilbert action. So any geometrical theory which has both gravity and Yang-Mills sectors must account for this difference. From the point of view adopted here, there are two reasons for the underlying difference:

(i) Contrary to Yang-Mills, gravitation is described by a non-linear realization of gauge symmetry[1,3,5]. In connection with internal symmetries, there is ample literature on non-linear realizations. Here I want to formulate the problem more generally so that it is directly applicable to gravity. Let G be a group and H one of its subgroups. Then the generators of G are

$$\{X_{\hat{A}}\} = \{T_{\hat{a}},\ S_{\hat{i}}\} \tag{2.1}$$

where $\{T_{\hat{a}}\}$ generate H and $\{S_{\hat{i}}\}$the coset G/H. Consider the Lie algebraic object

$$\begin{aligned}\hat{D}_\mu &= \partial_\mu + h_\mu^{\hat{A}} X_{\hat{A}} \\ &= \partial_\mu + H_\mu^{\hat{a}} T_{\hat{a}} + K_\mu^{\hat{i}} S_{\hat{i}} = D_\mu + K_\mu \end{aligned} \tag{2.2}$$

If we only demand that $\hat{D}_\mu$ as a whole transform covariantly under G, one obtains linear gauge fields of Yang-Mills theory. If, on the other hand, one demands that D_μ and K_μ separately be covariant under G, then one obtains a non-linear realization of the gauge symmetry G, which is linear with respect to H. If G is a space-time gauge group this allows one to understand how a gravitation theory can be covariant despite the explicit appearance of the gauge fields K_μ^i in the action. But this is not enough because internal symmetries can also be realized non-linearly. How can we tell that we are dealing with a space-time symmetry?

(ii) To ensure that the gauge group in question describes not an internal but a space-time symmetry we must further require that the horizontal and vertical transformation be "interlocked"[6,7]. To be explicit, suppose G = SO(3,2), and H = SO(3,1) and compare the parallel transport of vector fields in space-time along a curve with the lift of that curve in the fiber bundle[6]. This uniquely relates the connection in space-time to the gauge potentials of the subgroup H of the fiber

$$H_\mu^{ij} = \eta^{ik} K_k^\lambda K_\nu^j \Gamma_{\mu\lambda}^{\ \nu} \tag{2.3}$$

Then one completes the interlocking by extending the connection from H to G. A natural way to do this[8] is to go to a "local" horizontal basis $\{X_i\}$ such that

$$X_i = K_i^\mu D_\mu \ ; \quad D_\mu = K_\mu^i X_i \tag{2.4}$$

$$K_\mu^i K_j^\mu = \delta_j^i \tag{2.5}$$

The identification is then made by setting $K_\mu^i X_i$ equal to K_μ in (2.2)[6]. In contrast to internal symmetry, it is the interlocking feature which provides the constraint for a non-linear realization of space-time symmetries.

With

$$\left[D_\mu, D_\nu \right] = -R^{\hat{i}\hat{j}}_{\mu\nu} X_{\hat{i}\hat{j}} \tag{2.6}$$

one finds

$$\left[X_i, X_j \right] = -R^{ij}_{\mu\nu} X_{\hat{i}\hat{j}} - T_{\mu\nu}^{\ \ i} X_i \tag{2.7}$$

Just as in Yang-Mills theory under an infinitesimal gauge transformation we have

$$D_\mu \rightarrow (1 + \varepsilon^{\hat{A}} X_{\hat{A}}) D_\mu (1 - \varepsilon^{\hat{A}} X_{\hat{A}}) \tag{2.8}$$

Taking the interlocking into account, this gives for local translations

$$\delta K^i_\mu(x) = -D_\mu \varepsilon^j - \varepsilon^j K^\lambda_j T_{\lambda\mu}^{\ \ i} \tag{2.9}$$

$$\delta H^{ij}_\mu(x) = \varepsilon^j K^\lambda_j R^{ij}_{\lambda\mu} \tag{2.10}$$

Note that just as in Yang-Mills theory these transformations follow from the geometry. Hence, they are compatible with <u>any</u> invariant Lagrangian and close "off-shell".

In summary, gravity can be described in terms of a non-linear realization of a gauge symmetry subject to the "interlocking" constraint.

As an immediate test of validity of the above transformation laws consider the gravity-Yang-Mills system, so that $G = G_{gravity} \otimes G_{internal}$ and $H = SO(3,1) \otimes G_{internal}$. Then following the same steps as in pure gravity we have

$$D_\mu = \partial_\mu + H^{ij}_\mu X^{ij} + A^a_\mu X_a = K^i_\mu X_i = K_\mu \tag{2.11}$$

$$\left[X_i, X_j \right] = -K^\mu_i K^\nu_j \left[R^{mn}_{\mu\nu} X_{mn} + F_{\mu\nu}^{\ \ a} X_a + T_{\mu\nu}^{\ \ k} X_k \right] \tag{2.12}$$

Under local translations; one obtains, in addition to (2.9) and (2.10)

$$\delta A^a_\mu = \varepsilon^j K^\lambda_j F_{\lambda\mu}^{\ \ a} \tag{2.13}$$

III. Geometry of Supergravity Theories

The formalism of section II has a straight-forward extension to supergravity theories since any such theory must have gravity as a subtheory. The group G for supergravity theories contains both the internal and space-time symmetries. For definiteness consider OSp(N;4). If we require that internal symmetry and only the SO(3,1) part of space-time symmetry be realized linearly, the linear subgroup is reduced to $H = SO(3,1) \otimes SO(N)$. So the correct description of gravity excludes supersymmetry from H in this class of gauge groups[3]. The G/H part of the group is realized non-linearly. A natural way of implementing our interlocking requirement is to again identify the base manifold with the local G/H[8]. But to do so one must enlarge the base manifold into a superspace[9].

Consider a fiber bundle over a base manifold S with coordinates $\{Z^I\} = \{(x^i, \theta^\alpha)\}$ and fiber G with generators

$$\{X_{\hat{A}}\} = \{\hat{X}_{\hat{I}}, \hat{X}_{\hat{i}\hat{j}}, \hat{X}_a\} \; ; \text{ "a"} \Longrightarrow \text{internal symmetry} \tag{3.1}$$

$$\{I,J\} = \{i, j, \ldots(\text{even}); \alpha, \beta, \mu, \ldots(\text{odd})\} \tag{3.2}$$

On the Lie algebraic quantity

$$\hat{D}_I = \partial_I + H_I^{\hat{i}\hat{j}} \hat{X}_{\hat{i}\hat{j}} + H_I^a \hat{X}_a + K_I^{\hat{J}} \hat{X}_{\hat{J}} \tag{3.3}$$

$$= D_I + K_I$$

We impose our two requirements which specify the connection in the bundle: First we require that D_I and K_I separately transform covariantly under G, so that G/H part of the group is realized non-linearly. Then we proceed to interlock the horizontal and vertical transformations. As a result the local basis $\{X_{\hat{I}}\}$ defined by

$$\hat{X}_{\hat{I}} = K_{\hat{I}}^J D_J \; ; \; D_I = K_I^{\hat{J}} X_{\hat{J}} \tag{3.4}$$

$$K_{\hat{M}}^I K_I^{\hat{J}} = \delta_{\hat{M}}^{\hat{J}} \; ; \; K_M^{\hat{I}} K_{\hat{I}}^J = \delta_M^J \tag{3.5}$$

allows us to identify $K_I^{\hat{J}} X_{\hat{J}}$ with K_I in (3.3). From the definition of curvature and the global OSp(N;4) algebra we have

$$\left[D_I, D_J \right\} = -R_{IJ}^{\hat{i}\hat{j}} X_{\hat{i}\hat{j}} - F_{IJ}^a X_a \tag{3.6}$$

where $$R_{IJ}^A = (-)^{\sigma_I \sigma_J} \left[H_{I,J}^A - (-)^{\sigma_I \sigma_J} H_{J,I}^A + f_{BC}^A H_J^B H_I^C \right] \tag{3.7}$$

One also finds

$$\left[X_{\hat{I}}, X_{\hat{J}} \right\} = -(-)^{\sigma_I(\sigma_J + \sigma_J)} K_{\hat{J}}^J K_{\hat{I}}^I \left[R_{IJ}^{\hat{i}\hat{j}} X_{\hat{i}\hat{j}} + F_{IJ}^a X_a + T_{IJ}^{\hat{M}} X_{\hat{M}} \right] \tag{3.8}$$

where

$$T_{IJ} = T_{IJ}^{\hat{M}} X_{\hat{M}} = D_I K_J - (-)^{\sigma_I \sigma_J} D_J K_I \tag{3.9}$$

Gauge transformations have exactly the same form as in (2.8). In particular, for local infinitesimal parameters of G/H we get

$$\delta K_I^{\hat{J}} = -D_I \varepsilon^{\hat{J}} - \varepsilon^{\hat{M}} K_{\hat{M}}^{L} T_{LI}^{\hat{J}} \tag{3.10}$$

$$\delta H_I^{\hat{i}\hat{j}} = \varepsilon^{\hat{M}} K_{\hat{M}}^{J} R_{JI}^{\hat{i}\hat{j}} \tag{3.11}$$

$$\delta H_I^{\hat{a}} = \varepsilon^{\hat{M}} K_{\hat{M}}^{J} F_{JI}^{a} \tag{3.12}$$

$$\begin{aligned} \delta R_{IJ}^{mn} = (-)^{\sigma_I \sigma_J} \quad & \times \\ \times \quad & [(D_J \varepsilon^{\hat{M}}) K_{\hat{M}}^{N} R_{NI}^{\hat{m}\hat{n}} + (-)^{\sigma_M \sigma_J} \varepsilon^{\hat{M}} D_J K_{\hat{M}}^{N} R_{NI}^{\hat{m}\hat{n}} \\ & -(-)^{\sigma_I \sigma_J} D_I \varepsilon^{\hat{M}} K_{\hat{M}}^{N} R_{NJ}^{\hat{m}\hat{n}} - (-)^{\sigma_I(\sigma_J + \sigma_M)} \varepsilon^{\hat{M}} D_I K_{\hat{M}}^{N} R_{NJ}^{\hat{m}\hat{n}} \\ & +(-)^{\sigma_N \sigma_J} \varepsilon^{\hat{M}} K_{\hat{M}}^{N} D_J R_{NI}^{\hat{m}\hat{n}} - (-)^{\sigma_I(\sigma_J + \sigma_N)} \varepsilon^{\hat{M}} K_{\hat{M}}^{N} D_I R_{NJ}^{\hat{m}\hat{n}}] \end{aligned} \tag{3.13}$$

Note the structural similarity of the transformation laws (3.11) and (3.12): with respect to $\varepsilon^{\hat{M}}$, each gauge field of the linear subgroup H undergoes a variation proportional to its own curvature. This is to be contrasted with the transformation law for δK_I^J. As can be seen from (3.10) the fields K_I^J transform covariantly, and their variation with respect to $\varepsilon^{\hat{M}}$ is more like that of curvature components given above than the bona fide gauge fields of the subgroup H.

We have thus achieved our objective of deriving a set of model-independent transformation laws for all the fields in the theory. These geometrical transformation laws are "off-shell" and not tied down to any particular Lagrangian. For illustrative purposes we have taken the gauge group to be OSp(N;4). Our method is quite general, however, and is applicable to any gauge group within which the gravity sector can be described correctly. In particular, conformal and superconformal theories based on the gauge groups SU(N;4) can be treated with relatively minor changes. One would only have to note that to carry out our interlocking which ensures the correct description of gravity, the base manifold is not put into one-to-one correspondence with G/H but with an appropriate fiber F. With this taken into account, one can then repeat for SU(N;4) the steps enumerated above for OSp(N;4). The details will be given elsewhere[8].

The price we had to pay for a natural enforcement of our interlocking requirement and a systematic derivation of the transformation laws was to work in a superspace. Since one is then forced to take the idea of a superspace seriously, one may wonder about the immediate practical advantages of such a formulation, which complement its elegance and generality. To begin with we note that all the fields in such a theory directly enter the geometry. Then remembering that all such theories must contain pure gravity and Yang-Mills theories as sub-theories and that we know the physically interesting action forms of such theories, we can look for similarly natural actions in superspace with the hope that their reduction to space-time would result in a highly structured theory of physical interest. To this end a general action of the form

$$I(\Omega) = \int d^{1+m+n}_{V}\, \Omega^{IJ}_{A}\,(x,\theta) R^{A}_{IJ}(x,\theta) \tag{3.14}$$

was written down in reference 1, where dV is the invariant volume element of a superspace with m+1 Bose and n Fermi coordinates, and Ω^{IJ}_{A} is a covariant object determined by the <u>additional</u> physical requirements of the given theory. Recently, an explicit action of the form (3.14) has been found by Gell-Mann et al[10] for simple supergravity, thus lending support to the usefulness of the superspace approach.

We must now look for a way of translating the results obtained in superspace into the space-time language.

IV. A Physical Interpretation of Superspace

The formalism presented in section III is independent of any specific interpretation of the superspace. But the very idea of working in superspace creates a number of conceptual difficulties which must be dealt with if one is to make contact with the real world. To see this consider a point (x^i,θ^α) in superspace. Under a supersymmetry transformation represented by the parameter ε^α one arrives at the point

$$(x'^i,\ \theta'^\alpha) = (x^i + \delta x^i,\ \theta^\alpha + \varepsilon^\alpha)$$

Even when x^i, i = 0, ..., 3, are c-numbers, the quantities x'^i are not since the components $\delta x^i = \varepsilon^\alpha f^i_{\alpha\beta}\theta_\beta$ are not real numbers. So one must decide in what way the coordinates $\{x^i\}$ represent ordinary space-time. An obvious possibility is to regard $\{x^i\}$and $\{\theta^\alpha\}$ as, respectively, even and odd elements of a Grassman algebra. But then ordinary space-time would also have to be specified by non-c-number coordinates. Moreover, the parameters of the Lie subgroups of the relevant supergroups would, in general, not be c-numbers. Whether this point of view can lead to a consistent description of space-time is at present not known.

Let us then consider the possibility of retaining the c-number nature of space-time coordinates and see if it can still be embedded in a superspace. Consider the points $(x_c^i, \varepsilon^\alpha)$, where x_c^i 's are c-number coordinates, and ε^α is an arbitrary but fixed Grassman number for all x_c. Clearly any two points belonging to this set have c-number separations. Next we enlarge this set to an orbit, the physical orbit, which includes all the points in superspace related to (x_c, ε) by a supersymmetry transformation. Thus the coordinates of a point on the physical orbit would have the form $(x_c + \delta x, \theta + \varepsilon)$ where $\delta x^i = \varepsilon^\alpha f^i_{\alpha\beta}\theta_\beta$. By a suitable choice of metric on the physical orbit the line element can be made svpersymmetrically invariant, so that the points on this orbit will have c-number separations. Therefore the embedding of space-time manifold in this orbit is free of the usual ambiguities.

If one restricts the (x,θ) dependence of our fields and their transformation laws to the physical orbit, their θ-dependence can be made superfluous. Then taking our interlocking requirement into account, we can compare the resulting expressions with those obtained in the 4-dimensional formulation of supergravity theories. For simple and SO(2)-extended supergravities we have verified that this leads to correct field structure and transformation laws for latter theories. The generality of the method leads us to believe that it should work for other supergravity theories as well.

A more detailed exposition of the contents of this section will be given elsewhere.

It is a pleasure to acknowledge fruitful discussions with F. Ardalan, G. Domokos, T. Eguchi, S. Kövesi-Domokos, and C. Schaer.

References

(1) See, e.g., F. Mansouri, Phys. Rev. D16, 2456 (1977). Our earlier works can be traced from this article.

(2) For other attempts, see Y. Neeman and T. Regge, Phys. Lett. 74B, 54 (1978); S. W. MacDowell, unpublished.

(3) L. N. Chang and F. Mansouri, Phys. Rev. D17, 3168 (1978); Phys. Lett. B, in press.

(4) F. Mansouri, Proceedings of Johns Hopkins workshop, G. Domokos and S. Kovesi-Domokos Editors, April, 1978.

(5) See also F. Gürsey and L. Marchildon, Phys. Rev. D17, 2038 (1978).

(6) L. N. Chang, K. Macrae, and F. Mansouri, Phys. Rev. D13, 235 (1976); F. Mansouri and L. N. Chang, Phys. Rev. D13, 3192 (1976).

(7) A. Trautman, Ann. N. Y. Acad. Sci. 262, 241 (1975).

(8) For other ways of enforcing interlocking see reference 1 and F. Mansouri and C. Schaer, in preparation.

(9) For other superspace formulations see, e.g., J. Wess and B. Zumino, Phys. Lett. 66B, 361 (1977); and L. Brink *et al*, Phys. Lett. 74B, 336 (1977); W.Siegal and J. Gates, Harvard preprint, 1978; R. Arnowitt and P. Nath, Northeastern preprint, 1978.

(10) M. Gell-Mann, *et al*, Caltech preprint, Sept., 1978.

(11) A. Salam and J. Strathdee, Phys. Rev. 11D, 1521 (1975).

RELATIVITY PRINCIPLES AND FIBRE BUNDLES

R.N. Sen
Department of Mathematics
Ben Gurion University of the Negev
Beersheba 84120, Israel

INTRODUCTION

In this talk I would like to discuss some of the things which we have learnt, and may hope to learn, by combining the relativity principles of physics with the topological and geometrical structures called *fibre bundles*. For the moment, the term "relativity principles" will mean only Galilean and Special Relativity, and despite the seeming generality of the title, we shall not be intentionally concerned with gauge theories.

The first topic which we shall discuss is the theory of symmetry of *infinite* quantum-mechanical systems. By the latter we mean exclusively systems with infinitely many massive particles. The question was: do the relativity principles apply *in full* to such systems? The usual answer was *no*, because the state space of quantum mechanics is some sort of a Hilbert space, and even if one can construct a Hilbert space for an infinite system as defined above, one cannot find unitary operators upon it which will implement the boosts in a continuous fashion. Physically, no unitary operator can supply an infinite amount of energy and momentum. Therefore one said that the boost symmetries were "broken".

We were not satisfied with this answer for several reasons [1, 2] of which we mention only one. To us it seemed that to boost an infinite system was equivalent to passing to a different inertial frame, and that this equivalence ought to be formalizable [3]. Now it is obvious that to do so *one has to change the state space*. New classes of state spaces have been introduced twice in this century. First, with the discovery of quantum mechanics as opposed to classical mechanics (Hilbert space, Fock space). Second, with the discovery of methods for coping with systems with *infinitely many* as opposed to finitely many degrees of freedom (states of positive linear functionals over C^*-algebras of quasi-local observables). The reason why we want to introduce yet another class of state spaces for infinite systems is that we are interested in the global implications of symmetries rather than in local dynamics. Thus our approach may be said to complement the algebraic approach.

Let us now return to the problem of formalizing the notion of distinct inertial frames in the nonrelativistic quantum mechanics of infinite systems. Let $\vec{w}$ denote a boost velocity, $H = H_0$ the Hilbert space in the rest-frame $\vec{w} = 0$, and $H_{\vec{w}}$ the Hilbert space in the inertial frame $\vec{w} \neq 0$. Now we should expect the different $H_{\vec{w}}$ to be carbon copies of each other, and therefore of *the* Hilbert space H. Nevertheless, the Hilbert spaces $H_{\vec{w}}$ should be totally disjoint, because the superposition of vectors from Hilbert spaces belonging to different inertial frames does not admit of any physical interpretation. Finally, application of the Galilei boost $\vec{v}$ should take $H_{\vec{w}}$ to $H_{\vec{w}+\vec{v}}$, space and time translations should act entirely within a given inertial frame, whereas the rotations should act both in $H_{\vec{w}}$ and upon the parameter $\vec{w}$. Now $\vec{w}$ which labels an inertial frame can be thought of as a point on the quotient space (left-cosets) $M = G/T \times E$, where G, T and E stand respectively for the inhomogeneous Galilei, time translation and Euclidean groups. Then the family of Hilbert spaces $\{H_{\vec{w}}\}$ can be thought of as the set of ordered pairs $(\vec{w}, \phi)$, where $\vec{w} \in M$ and $\phi \in H$. Call it B, and look upon $H_{\vec{w}}$ as a subset of $B : H_{\vec{w}} = \{(\vec{w}, \phi);\ \vec{w}$ fixed, $\phi \in H\}$. Then there exists a natural definition of $(\vec{w}, \phi) + (\vec{w}, \psi) = (\vec{w}, \phi + \psi)$, of $\alpha(\vec{w}, \phi) = (\vec{w}, \alpha\phi)$ for any complex number α, and of $((\vec{w}, \phi), (\vec{w}, \psi)) = (\phi, \psi)$ where (ϕ, ψ) is the inner product in H. However, there exists no natural definition of either $(\vec{v}, \phi) + (\vec{w}, \psi)$ or of $((\vec{v}, \phi), (\vec{w}, \psi))$ when $\vec{v} \neq \vec{w}$. So we may regard B as *a bundle of Hilbert spaces*, $B = M \times H$, based on M, with the natural projection $\pi : B \to M$. We already have some information about the action of G on B, and this is enough to lead us to the general problem which is formulated and solved in the next section.

BUNDLE REPRESENTATIONS

Let M be a topological space, L a topological vector space, $B = M \times L$ the topological product considered as a bundle over M with the projection $\pi : B \to M$, and G a T_0, second-countable locally compact topological group (generally Lie). A *bundle representation* $\mathcal{D}$ of G on B is defined to be a continuous map $\mathcal{D} : G \times B \to B$ with the following properties: (i) $\mathcal{D}(g, b) = h(g)b$, where $g \in G$ and $b \in B$, and $h(g)$ is an invertible bundle map $h(g) : B \to B\ \forall\, g \in G$ such that $h(g)^{-1} = h(g^{-1})$. Then $h(e) = id_B$. (ii) $h(g_1g_2, b) = h(g_1, g_2b) \circ h(g_2, b)$ $\forall\, g_1, g_2 \in G$, $b \in B$. Clearly such a family of bundle maps $\{h(g)\}$ induces a family of base maps $\{\bar{h}(g)\}$, $\bar{h}(g) : M \to M$, $\pi h(g)b = \bar{h}(g)\pi(b)$ which are homeomorphisms and satisfy $\bar{h}(g_1g_2, x) = \bar{h}(g_1, g_2x)\bar{h}(g_2, x)$ $\forall\, g_1, g_2 \in G$, $x \in M$. (iii) $\mathcal{D}$ preserves the linear structure on L.

Since $\mathcal{D}$ is a map into a product space, the continuity of $\mathcal{D}$ is equivalent to the continuity of the two projections $\Pi_M \circ \mathcal{D} : G \times B \to M$ and $\Pi_L \circ \mathcal{D} : G \times B \to L$. The first of these is the same as the continuity of the map $\bar{\mathcal{D}} : G \times M \to M$ defined by $\bar{\mathcal{D}}(g, x) = \bar{h}(g)x$. To spell out the second in detail, let us take the case of $L = H$, a separable Hilbert space with the strong topology. Then (iii) should be amended to read "$\mathcal{D}$ preserves the inner product in H". We now pass to a subclass of bundle representations.

We define a *canonical* bundle representation of G to be a representation on $B = M \times H$, where (i) $M = G/H$ (left-cosets), topologized by the quotient topology, and H is a *closed* subgroup of G, and (ii) H is a separable Hilbert space, so that G acts on H by unitary operators. There exists a natural projection

$p : G \to G/H = M$ which sends every element of G into its left-coset in H, and p is a continuous open map. There exists also a continuous natural action of G on M given by $gx = p(gp^{-1}(x))$, where $x \in M$ and $p^{-1}(x)$ denotes the inverse image, i.e. the fibre over x. With such a choice of base space for the bundle B, the action of G upon it is automatically continuous. Now write $b = (x, \phi)$, where $b \in B$, $x \in M$ and $\phi \in \mathcal{H}$, and $h(g)b = gb = g(x, \phi) = (gx, u(g, x)\phi)$. Here u has to be a unitary operator on $\mathcal{H}$ and has to satisfy (i) $u(e, x) = 1 \ \forall x \in M$, (ii) $u(g_1g_2, x) = u(g_1, g_2x)u(g_2, x) \ \forall g_1, g_2 \in G, x \in M$. To solve for such u's we first choose a *Borel section* of H in G, i e. a Borel map $\eta : M \to G$ which satisfies $p \circ \eta = id_M$. Such maps exist in great profusion. Then with this η we construct a (G, M, H) cocycle, using the famous formula

$$(1) \qquad k(g, x) = \eta(gx)^{-1} g\eta(x) .$$

It is readily seen that k is a Borel map of $G \times M$ *into* H. By imposing a non-triviality condition, e.g. $k(h, x_0) = h \ \forall h \in H$, where $x_0 = H$ on M, we can make k surjective. This nontriviality condition is surely satisfied if $\eta(x_0) = e$, which is not really restrictive. Finally we choose a continuous unitary representation D of H on $\mathcal{H}$ and set $u(g, x) = D(k(g, x))$. Then the second half of the continuity requirement, i.e. $\Pi_{\mathcal{H}} \circ \mathcal{D} : G \times B \to \mathcal{H}$, is met because k is a Borel function. The explicit formula for the canonical bundle representation is

$$(2) \qquad g(x, \phi) = (gx, D(k(g, x))\phi) .$$

This representation is determined by the unitary representation D of H and the cocycle k (eqn. (1) has the *unique* solution $\eta(x) = ak(a, x_0)^{-1}$, where a is *any* element of G such that $x = ax_0$), but the latter can be eliminated by passage to an equivalence class. A *coordinate transformation* on the bundle representation $\mathcal{D} = \{h(g), D\}$ is defined to be an invertible bundle map Ω which takes $\{h(g), D\}$ to $\{h'(g), D\}$, where $h'(g) = \Omega^{-1} \circ h(g) \circ \Omega \ \forall g$. A change of Borel section is easily seen to be a coordinate transformation Ω which induces different unitary transformations on different fibres. Thus $\{h(g), D\}$ is determined by D but not by k. Finally, as the action of G on $M = G/H$ is transitive, questions concerning reducibility and irreducibility of the representation $\{h(g), D\}$ amount exactly, by the Whitney sum process, to the corresponding questions concerning the unitary representation D of H on $\mathcal{H}$.

This concludes our extremely condensed survey of bundle representations. For further details see [3, 5]. Although here we have considered only canonical bundle representations, these suffice to include all cases of interest [5] except *ergodic bundle representations*, in which G acts ergodically on the base space. We have not treated such representations at all.

APPLICATIONS OF BUNDLE REPRESENTATIONS

To ascertain whether or not bundle representations of the Galilei and the Poincaré groups correspond to physical objects in a systematic rather than a sporadic or accidental fashion, we must first determine a large class of such representations and then examine them individually. This is time-consuming, and we give below only a few main results and "spinoffs" of this analysis. One general point should be borne in mind: we call a system quantum-mechanical only if the time-translations are represented on the fibre (this condition is not sufficient; see [3, 5]).

Consider first the case of the Galilei group G and its central extension $\tilde{G}$. Bundle representations of these groups based on the boosts appear to correspond to stable (i.e. zero-temperature limit) nonrelativistic zero and nonzero mass, quantum-mechanical particle-like elementary exictations with arbitrary helicity in a homogeneous medium. The base space can be extended to include space translations and/or rotations. In this case the base space appears to act as a *source* or *sink* of linear and/or angular momentum. The excitations still remain particle-like (i.e. localizable) when the rotations are included in the base space, but not when the space translations are so included. There exist bundle representations in which the boosts are represented not on the base space but on the fibre; these representations appear to lack physical meaning because there seem to be no quantum-mechanical observables corresponding to the boosts. In all bundle representations of $\tilde{G}$ in which the subgroup Θ extending G is represented on the fibres, a generalization of Bargmann's superselection rule holds, irrespective of whether the space translations and/or the boosts are represented on the base space or the fibre.

It appears to be a general feature that *irreducible* bundle representations of the above kind are devoid of physical interest. They contain only one energy state, and do not contain localizable states when they should. Elementary excitations correspond to very special *reducible representations*. The possibility of a momentum-energy or spin-energy *dispersion law* $E = E(\vec{p})$, $E = E(j)$ arises from the reducibility of the representation and is not dictated by the symmetry of the problem. This reflects the fact that dispersion laws of elementary excitations are related to the detailed microstructure and not the global symmetry of the system. *The correspondence between physical elementarity and mathematical irreducibility which obtains in particle physics breaks down completely in the physics of elementary excitations.*

So far it appears that nonrelativistic zero-mass quantum-mechanical systems cannot be described by any formalism other than that of bundle representations. Moreover, such excitations may occur in a wide variety of homogeneous media other than superfluid ^{4}He, e.g. liquid lead [6]. Physical considerations make it appear highly unlikely that these exictations are directly responsible for the *hydrodynamic* behaviour of superfluid helium. Looking around for other (London-type) explanations, one is led to the idea that there might be a *vector potential interaction* between atoms [5, 7] in addition to the usual van der Waals one. If so, it could explain, on the one hand, the anomalous temperature dependence of the roton dip [8] as well as contribute towards a theory of dislocations and melting.

Turning now to the case of phonons and finite-mass excitations in perfect crystals, it appears that these too can be identified with *appropriate bundle representations of the full Galilei or Poincaré group*. The base spaces in these cases are $V \times (E/\Gamma)$, where $V(=\mathbb{R}^3)$ is the space of the boosts, E is the full Euclidean group, Γ a space group, $\times$ means topological product and $=$ means homeomorphism. If Γ contains no rotation/reflection elements the results are easiest to compute and to interpret. Owing to the presence of the crystal lattice, the free evolution in time behaves in an unexpected fashion. If Γ is symmorphic the situation does not worsen significantly *in principle*, but 72 of the 73 such cases have not yet been computed. If Γ is nonsymmorphic then the determination of the homogeneous spaces E/Γ is a nontrivial problem which has only recently been solved by J. Wolf [9]. His results have not yet been physically interpreted, but I would venture to guess that many strange phenomena in crystals would seem much less strange when properly interpreted in terms of "Euclidean space forms".

Similarly, one may consider bundle representations of the Poincaré group. If one wishes to talk about spin rather than helicity - and the Fermi sea would seem to leave us no alternative - one has to include space reflections. Then it is tempting to pass to local fields in x-space instead of those in momentum space. This can be done, at least in the cases of spin ½ and 1. The two most important features which emerge seem to be: (i) positive and negative $(\text{mass})^2$ appear to be on a similar footing. (ii) The two spin directions are *completely decoupled* from each other (this would also be true in the Galilei group), which gives rise to the possibility of an electronic analogue of the (optical) Faraday effect. This effect has not yet been observed; the "samples" required are not known.

To conclude this section, let us consider an application of the bundle representation formalism to a semiclassical situation. Let P and L be the Poincaré and the Lorentz groups respectively. Then $P/L = M_0$ is space-time, homeomorphic to $\mathbb{R}^4$, and in bundle representations of P based on M_0 the time translations are implemented on the base space and not on the fibre. Thus we are free to consider finite-dimensional nonunitary representations $D^{(i,j)}$ of L on vector spaces $V^{(i,j)}$, i, j being 0, ½, 1, ... Writing $x \in M$, $\Lambda \in L$ and $\{a, \Lambda\} \in P$, we have the natural action $\{a, \Lambda\}x = a + \Lambda x$ of P on M, and there exists a natural continuous cross-section $\eta(x) = \{x, 1\}$ and a canonical cocycle $k(\{a, \Lambda\}, x) = \{0, \Lambda\}$. Then the bundle representation determined by the nonunitary representation $D^{(i,j)}$ is, explicitly

$$\{a, \Lambda\}\ (x, \phi) = (a + \Lambda x,\ D^{(i,j)}(\Lambda)\phi) \tag{3}$$

where $\phi \in V^{(i,j)}$. Now it is obvious that a cross-section of the bundle $M \times V^{(i,j)}$

is a *tensor field* of type (i, j), and the bundle representation formula (3) is none other than the relativistic transformation law for such fields. Specializing (3) to $i = j = \frac{1}{2}$, $\phi = A_\mu$, we may write it in the more familiar form

(4) $$A_\mu(x) \xrightarrow{\{a,\Lambda\}} \Lambda_\mu{}^\nu\, A_\nu(a + \Lambda x) \ .$$

The point of this discussion is that finite-dimensional nonunitary representations of noncompact groups *are not excluded for classical fields*. Therefore if one wants to admit a noncompact internal symmetry group G *without admitting infinite multiplets*, one should first construct bundle representations of $P \times G$ based on space-time and finite-dimensional representations of $L \times G$ on the fibre, pass to classical fields by taking cross-sections of the bundles $M \times V$, and then quantize these fields. With this we terminate our incomplete survey of the applications of bundle representations.

CATEGORIES OF PHYSICAL SYSTEMS AND INTERACTIONS BETWEEN THEM

Bundle representations can be generalized in two directions. One is that we consider wider classes of state spaces. We may, in fact, define a *category* of model systems to be a triplet $\{K, G, \Delta\}$ where K is a category of mathematical objects, G a relativity group of physics, and Δ the class of admissible actions or representations of G on K. From the physical point of view, we shall be interested in very few mathematical categories K, very few groups G and, for any such pair $\{K, G\}$, very few representations $\mathcal{D} \in \Delta$.

I would like to know what is the state space of a classical infinite *compressible* fluid and how the Galilei group acts upon it. The reason is simple, and anyone familiar with the Galilei group will recognize it immediately. In this case we shall have an *internal energy density*

(5) $$I(\vec{x}, t) = E(\vec{x}, t) - p(\vec{x}, t)^2/2m(\vec{x}, t)$$

which, because of its dependence on $\vec{x}$ and t, *cannot be legislated away*. So we may have a problem with the free time evolution. Moreover, owing to the presence of the internal energy density one might expect to make some contact with thermodynamics.

The second generalization is that we can find a framework for describing the interaction - or part of it - between different categories of physical systems which share the same relativity group. The two following observations provide us with our point of departure: (1) The bundles we have used are themselves *products* $M \times H$ of objects from different categories. (2) The group action automatically defines a free time evolution, i.e. classifies noninteracting states. Now suppose that we have two model systems $(S_1, G, \mathcal{D}_1)$ and $(S_2, G, \mathcal{D}_2)$ belonging to the categories $\{K_1, G, \Delta_1\}$ and $\{K_2, G, \Delta_2\}$ respectively. We can then define a product category $K = K_1 \times K_2$, have a class of admissible actions $\Delta = \Delta_1 \times \Delta_2$ of G upon it, and form the category of model systems $\{K, G, \Delta\}$. The states in this category are free states of a composite system. We now adopt an S-matrix type viewpoint, with the (possible) difference that we identify $K_{in} = K_{out} = K$. Then we may partially describe some interactions between these two model systems by an S-operator, i.e. a map $S : K \to K$ which commutes with $G : gSg^{-1} = S \ \forall g \in G$, and which may be interpreted as a transition from an initial to a final state.

Such a framework might provide a model for describing *a quantum-mechanical measurement process* in which the initial and the final states are prepared and registered by purely classical devices.

STRUCTURE OF THE TOTAL SPACE OF HIGH-ENERGY PHYSICS

Now I would like to discuss whether or not there is any connection between general relativity and the internal symmetries of high-energy physics within the framework of relativity principles and fibre bundles. This question arises in the following manner. The symmetry group of high-energy physics is $P \times G$, where P is Poincaré and G an internal symmetry group. This group may be looked upon as a transformation group of an underlying space $B_0 = M_0 \times G$, where M_0 is space-time and the space G is the group manifold itself of the group G. Now the physical distinction between an internal and a space-time symmetry is that a space-time symmetry operation can be *performed* in some sense in which an internal symmetry operation cannot. This distinction is between the *algebraic operations* of P and those of G. Now, if we "forget" the algebraic structure and look upon $B_0 = M_0 \times G$ as only a geometric object, a manifold, the following is a very reasonable question: can we factorize B_0 *uniquely* into $M_0 \times G$? The answer is obviously *yes*, because B_0 is *defined* to be the product $M_0 \times G$. Therefore we change the question slightly: instead of B_0, we consider another manifold B which is homeomorphic to B_0, i.e. to $M_0 \times G$, and we ask: is every such B *naturally homeomorphic* to $M_0 \times G$, or do we have to use a coordinate system to describe such a homeomorphism. The physical translation of this question is the following: is there a natural, unique factorization of the total space of high energy physics into "space-time" and an "internal space", or do we have some arbitrariness in performing this factorization? The answer is the following:

If M_0 is flat in the Riemannian (not *pseudo-Riemannian*) sense, then $M_0 = \mathbb{R}^4$, i.e. M_0 is naturally homeomorphic to $\mathbb{R}^4$. Then B can be viewed as a fibre bundle over $\mathbb{R}^4$ with the identity as the group of the bundle, i.e. B is naturally homeomorphic to $\mathbb{R}^4 \times G$. However, if we replace M_0 by any 4-manifold $\mathbb{R}^4$ which is homeomorphic, *but not naturally homeomorphic* to $\mathbb{R}^4_0$ (geometrically this means that M is curved or twisted, but still contractible), then it is fairly obvious that a manifold B which is homeomorphic to $M \times G$ is not generally naturally homeomorphic to the latter. One needs a coordinate system to describe this homeomorphism, and therefore there arises a group of admissible coordinate transformations.

Now let us assume that, under certain reasonable conditions "C"

(6) FR ⇒ FPR

where FR stands for flat Riemannian and FPR for flat pseudo-Riemannian (of the Minkowski signature). Then, again under "C", we have

(7) *Not* FPR ⇒ *Not* FR

⇒ B is not *naturally* homeomorphic to $M \times G$

= there is a certain arbitrariness in the separation between *space-time* and the *space of internal symmetries*.

Note that in the second implication in (7) the emphasis is on *naturality*; so far we are not questioning that B *is* homeomorphic to $M \times G$.

This is the overall picture. We shall try to clarify it somewhat more in the following section.

G-STRUCTURES AND PRINCIPAL BUNDLES

Let $M = (X, A)$ be a real, connected, orientable, smooth n-manifold. The atlas A is maximal with respect to the property that for any $U, V \in A$, $U \cap V \neq \emptyset$, the map $U \cap V \to GL(n, \mathbb{R})$ given by the Jacobian matrix of the transition map is smooth. Furthermore, the Jacobian matrix of the composition of two transition maps is the matrix product of the Jacobian matrices of the two maps. Now if there exists a sub-atlas A_G of A such that the corresponding transition matrices belong to the subgroup G of $GL(n, \mathbb{R})$, and A_G is maximal with respect to this property, then A_G is said to define a G-*structure* on M [10]. A manifold M may simultaneously have more than one G-structure. For example, if $X = \mathbb{R}^4$ and A the natural atlas on X, then $M = (\mathbb{R}^4, A)$ has at least two distinct and physically interesting G-structures: the homogeneous Galilei structure and the Lorentz structure, hereafter called the G_0- and L-structure respectively. The latter is the same as a pseudo-Riemannian $O(3, 1)$ structure. A Riemannian structure on an n-manifold is an $O(n)$-structure. A manifold M with a G-structure can conveniently be regarded as a principal bundle over M with group and fibre G [11]. Since M is paracompact, the group of the bundle can always be reduced to a maximal compact subgroup of it [12], from which it follows that every real n-manifold has a Riemannian structure. A general n-manifold M with its $GL(n, \mathbb{R})$ structure, regarded as a principal bundle over M, is sometimes called a *frame bundle*.

A mathematically interesting class of G-structures are the *almost-complex structures*. Here $n = 2m$ and the subgroup of $GL(2m, \mathbb{R})$ is $GL(m, \mathbb{C})$. In this case the "square root of minus one", i.e. a 2n dimensional real linear transformation whose square is -I, varies from point to point on M. The only *spheres* which admit an almost complex structure are S^2 and S^6.

It would be of interest to know whether or not there exist connected orientable 4-manifolds M, except the trivial case $M = \mathbb{R}^4$ mentioned earlier, which admit a homogeneous Galilei structure. That is, we are looking for a principal bundle $\mathcal{B} = [B, M, \pi, G_0, G_0]$ where the group and fibre G_0 are the homogeneous Galilei group, and which admits a nontrivial G_0-invariant connection. In particular, it would be interesting to know whether or not S^4 admits a G_0-structure.

With this preparation we can make the question posed in the previous section more precise.

Regarding internal symmetries, we have a principal bundle $\mathcal{B} = [B, M, \pi, G, G]$ where M is a connected orientable 4-manifold and G is an internal symmetry group. We assume that M is contractible, so that $\mathcal{B}$ has a cross-section and is homeomorphic to $M \times G$. However, this homeomorphism is not generally natural and one says that the bundle $\mathcal{B}$ is G-*equivalent* to the product. That is, M is a nonflat Riemannian manifold.

Interpreting the principal bundle $\mathcal{B}$ as a G-structure on M, we see that G should be a subgroup of $GL(4, \mathbb{R})$. So we should encounter no difficulties as long as we are interested in the internal symmetry group $SU(2) \times SU(2)$. However, in order to admit SU(3) or SU(4), there remains the further problem of devising a "unitary trick" to pass from $SL(4, \mathbb{R})$ to SU(4).

Regarding general relativity, the principal bundle is $\mathcal{B}' = [B', M, \pi', L, L]$, where L is the Lorentz group and we have a pseudo-Riemannian $O(3, 1) = L$ structure on M, and again this structure is not trivial.

In order to arrive at physically significant conclusions, we have to proceed as follows. (1) Define a notion of *compatibility* between different G-structures defined on the same manifold. This notion should be such that the flat L-structure be compatible with the flat O(4) structure, as well as the flat G_0-structure on $M_0 = \mathbb{R}^4$. (2) Determine whether or not *nonflat* L-structures on M are compatible with a flat O(4) structure on a manifold which is homeomorphic to $\mathbb{R}^4$. If the answer to (2) is in the affirmative, there is no connection between internal symmetries and general relativity in the sense outlined above. If the answer to (2) is in the *negative*, then (3) devise an acceptable method for converting the SL(4) structure on M into an SU(4) structure on it. If this can be done, we can probably say that SU(4) arises out of the Cartan-Ambrose-Singer holonomy theorem [13, 14].

My conjecture is that the answer to (2) is generally in the *negative*, and that the trick (3) can be devised.

ACKNOWLEDGEMENTS

I would like to acknowledge many useful conversations with my colleague H. Gauchman, and would like to thank Mrs. Yael Ahuvia for preparing this camera-ready typescript.

REFERENCES

1. H.J. Borchers and R.N. Sen, *Comm. math. Phys.* 42 (1975) 101.
2. R.N. Sen, *Nonrelativistic Zero-mass Systems*, Lecture Notes, Göttingen (1974).
3. ———, *Theory of Symmetry in the Quantum Mechanics of Infinite systems*, I; to appear in *Physica* A.
4. ———, *Theory of Symmetry in the Quantum Mechanics of Infinite Systems*, II; to appear in *Physica* B.
5. ———, *Bundle Representations and their Applications*, Lecture Notes (in preparation).
6. K.S. Singwi, *Atomic Motions in Liquids and Neutron Scattering*, in: *Theory of Condensed Matter* (ICTP, Trieste, Winter 1967), IAEA, Vienna (1968), and the references cited therein.
7. J.M. Jauch, *Helv. Physica Acta* 37 (1964) 284.
8. D.G. Henshaw and A.D. Woods, in: *Proc. 7th Int. Conf. on Low Temperature Physics*, edited by G.M. Graham and A.C. Hollis Hallett, Univ. of Toronto Press, Toronto (1961) p. 539.
9. J.A. Wolf, *Spaces of Constant Curvature*, 3rd edition, Publish or Perish, Inc., Boston, Mass. (1974).
10. S.S. Chern, *The Geometry of G-Structures*, *Bull. Amer. Math. Soc.* 72 (1966) 167.
11. S. Sternberg, *Lectures on Differential Geometry*, Prentice-Hall, Englewood Cliffs, N.J. (1964).
12. N. Steenrod, *The Topology of Fibre Bundles*, Princeton (1951); with an appendix added November 1956.
13. S. Kobayashi and K. Nomizu, *Foundations of Differential Geometry*, Vol. I, Interscience, New York (1963).
14. R.L. Bishop and R.J. Crittenden, *Geometry of Manifolds*, Academic Press, New York (1964).

DYNAMICAL GROUPS FOR THE MOTION OF RELATIVISTIC COMPOSITE SYSTEMS

A. O. Barut
Department of Physics, University of Colorado, Boulder, Colorado 80309

A general study is presented to treat composite systems as "elementary" relativistic objects endowed with internal degrees of freedom. The emphasis is on the global quantum numbers and observed spectral properties of the system; it is "inverse" spectral method. The general postulates are given. The mathematical tools are the theory of induced representations and finite or infinite component wave equations. The induction process from the so-called dynamical groups is introduced. The electron-positron complex is the simplest prototype of the method, and the role of the dynamical group O(4,2) for this system provides a new treatment of the electron-positron complex. The generalizations to more complex systems are introduced in an inductive way based on the example of the two-body problem. Various realizations of the conformal dynamical algebra are introduced. Next the principles underlying the choice of the infinite-component wave equations are discussed and various examples are discussed. The electromagnetic interactions of the composite systems are treated on the basis of the minimal coupling of the infinite-component wave equation. Finally, a scattering theory of relativistic composite systems is outlined.

The full text will appear as a chapter in the book "Group Theory and Many-Body Physics" edited by P. Kramer and H. Stumpf, Vieweg Verlag, 1978.

Clebsch-Górdon coefficients for the holomorphic discrete series

E. Gutkin, University of Utah, Salt Lake City, Utah 84112
(partially supported by the NSF-Grant 78 - 01826)

Let G be the connected component of identity of the group of automorpisms of an Hermitian symmetric domain X, let K be a maximal compact subgroup of G and H a Cartan subgroup of K. It is known by Harish-Chandra (I) that G has the holomorphic discrete series of representations and that these representations have some properties analogous to those of finite dimensional representations. In the present work I study tensor products of representations of the holomorphic discrete series and the theorems I prove confirm this analogy.

Let $\mathfrak{g}, \mathfrak{k}, \mathfrak{h}$ be the complex Lie algebras of G, K, H respectively and let $\langle , \rangle$ be the Cartan form of $\mathfrak{g}$. A root of $\mathfrak{g}$ relatively to $\mathfrak{h}$ is called compact if the corresponding root vector belongs to $\mathfrak{k}$ and noncompact otherwise. Fix an ordering of roots and let ρ be the half sum of the positive roots. Representations of the holomorphic discrete series are parametrized by the set Λ of K-dominant weights such that for $\lambda \in \Lambda$ and for any positive noncompact root α, $\langle \lambda + \rho , \alpha \rangle < 0$. For any $\lambda \in \Lambda$ denote by T_λ the corresponding representation of the holomorphic discrete series and by π_λ the irreducible representation of K with the highest weight λ. For a unitary representation T denote by T* its contragredient representation.

Theorem I. Let $\lambda, \mu \in \Lambda$. Then the tensor product $T_\lambda \otimes T^*_\mu$ is induced by the representation $\pi_\lambda \otimes \pi^*_\mu$ of K.

Theorem 2. Let $\lambda, \mu \in \Lambda$. There exists an integer valued function $C(\lambda,\mu,\nu)$ on Λ such that $T_\lambda \otimes T_\mu = \sum_\nu C(\lambda,\mu,\nu) T_\nu$.

In view of the analogy with the finite dimensional case , I call numbers $C(\lambda,\mu,\nu)$ the Clebsch-Gordon coefficients fir the holomorphic discrete series. The next theorem gives an expression of $C(\lambda,\mu,\nu)$ through the usual Clebsch-Gordon coefficients.

Theorem 3. Let $\mathfrak{g} = \mathfrak{p}_+ \oplus \mathfrak{k} \oplus \mathfrak{p}_-$ be the Harish-Chandra decomposition and consider the symmetric algebra $S(\mathfrak{p}_-)$ as a K-module. Then for any $\lambda, \mu \in \Lambda$

$$\pi_\lambda \otimes \pi_\mu \otimes S(\mathfrak{p}_-) = \sum_\nu C(\lambda,\mu,\nu) \pi_\nu \qquad \text{(I)}$$

The next theorem gives an assymptotic formula for $C(\lambda,\mu,\nu)$.

Assume for the convenience of exposition that X is an irreducible symmetric domain and let $\alpha_1, \dots, \alpha_\ell$ be the maximal strongly orthogonal system of $(\mathfrak{g}, \mathfrak{k})$. Set $\beta_1 = \alpha_1, \beta_2 = \alpha_1 + \alpha_2, \dots, \beta_\ell = \alpha_1 + \alpha_2 + \dots + \alpha_\ell$ and let B be the semigroup generated by $\beta_1, \dots, \beta_\ell$. For a weight λ and a representation π of K we set $\lambda >> \pi$ if for any weight ν of π, $\lambda + \nu$ is a K-dominant weight. Denote by $\langle \nu, \pi \rangle$ the multiplicity of the weight ν in the representation π.

Theorem 4. Let $\lambda, \mu \in \Lambda$. For any $\nu \in \Lambda$ such that $\nu >> \pi_\lambda \otimes \pi_\lambda^* \otimes \pi_\mu \otimes \pi_\mu^*$

$$C(\lambda,\mu,\nu) = \sum_{\beta \in B} \langle \nu + \beta, \pi_\lambda \otimes \pi_\mu \rangle \qquad (2)$$

Formulas (I) and (2) permit to obtain explicit expressions for the Clebsch-Gordon coefficients for the conformal group SU(2,2). These formulas and the proofs of the theorems I - 4 will be published elsewhere.

References

(I) Harish-Chandra. Representations of semisimple Lie groups I , , I. Amer.J.Math. 77(I955), 743 - 777, 78(I956), I -4I, 564 - 628.

DEFORMATION COHOMOLOGY OF THE PRIMITIVE INFINITE LIE ALGEBRAS

by J.F.Pommaret , Collège de France , Paris

The theory of deformation of Lie algebras,created by M.Gerstenhaber in 1964 (ref 11),has been initiated by Segal,Inonü and Wigner in 1953 (ref 18),while looking for a link between the commutation$ laws of the Poincaré algebra and those of the Galilée algebra.

From the work of D.C.Spencer in 1962 (ref 45) one might conjecture the existence of a link between deformations of algebraic structures and deformations of geometric structures (ref 31).

We have solved this conjecture in 1977,introducing new cohomological methods for the study of Lie equations and transitive infinite Lie algebras.More precisely,we have shown that the generalization of the Chevalley-Eilenberg cohomology for finite Lie algebras (well known in that case because of the Whitehead lemmas) was not a similar one for infinite Lie algebras.In fact,such a later one cannot be computed in general,as it follows from the work of D.S.Rim in 1966 (ref 40).

On the contrary,we proved that this Chevalley-Eilenberg cohomology was a particular case of our deformation cohomology,finding again by this way all the classical definitions of the deformation theory of finite Lie algebras,both with their interpretation in terms of cohomological arguments,but within a quite different background.

Our purpose,in the present work,is to apply these methods to the six primitive infinite Lie algebras of E.Cartan's classification,the use of which is fundamental in physics today.As a by-product,we prove that these algebras are rigid by computing explicitely their groups of deformation cohomology.

The idea is to construct,when $\mathcal{D}$ is an involutive Lie operator, the Janet sequence: $0 \longrightarrow \Theta \longrightarrow T \xrightarrow{\mathcal{D}} F_0 \xrightarrow{\mathcal{D}_1} F_1 \xrightarrow{\mathcal{D}_2} \cdots \xrightarrow{\mathcal{D}_n} F_n \longrightarrow 0$ which is a resolution of the solution sheaf Θ of $\mathcal{D}$ .The problem is then to determine the cohomology at $\mathcal{T}_n$ of the purely algebraic induced sequence: $0 \longrightarrow Z(\Theta) \longrightarrow C(\Theta) \xrightarrow{\mathcal{D}} \mathcal{T}_0 \xrightarrow{\mathcal{D}_1} \mathcal{T}_1 \xrightarrow{\mathcal{D}_2} \cdots \xrightarrow{\mathcal{D}_n} \mathcal{T}_n \longrightarrow 0$ where $Z(\Theta)$ and $C(\Theta)$ are respectively the center and the centralizer of Θ .

We hope that these methods,based on the jet theory and the use of finite length differential sequences,will give a new impetus to the theory of infinite Lie algebras and its application to physics.

Nota:the references are taken from the book:
J.F.Pommaret:Systems of partial differential equations and Lie pseudogroups (Gordon and Breach,New York,London,Paris,1978,410p).

INTERACTIONS IN A MODEL OF EXTENDED PARTICLES

Herbert M. Ruck
Physics Department, Duke University
Durham, North Carolina 27706, USA

I study some physical aspects of interactions in a relativistic model of extended particles defined by Dirac [1] and further developed in [2]. The wave function for bosons with spin s and magnetic number m, $\varphi^{s,m}(x,q)$, (x is the position of the center of mass of the particle in Minkowski space, $q\in \mathbb{R}^2$ are internal coordinates) is defined by linear differential constraints. In addition the wave function satisfies the higher-order differential equation

$$(\Box - M^2(V_0))\varphi^{s,m}(x,q)=0 \quad . \tag{1}$$

M^2 is a function of the operator V_0, which acts on the internal degrees of freedom $V_0\varphi^{s,m} =(s+1)\varphi^{s,m}$, such that an arbitrary relation between the spin- and mass spectrum $M^2= \alpha + \beta s + \ldots$ is possible. A multi - spinorial representation $\varphi_{\alpha\beta\ldots}(x,q)$ has been found in [3].

I consider (1) as a dynamical equation for elastic interactions, where the spins and masses of the particles involved do not change. Eq. (1) reduces to the Klein-Gordon equation. I assume a coupling of the electromagnetic potential $A_\mu(x)$ to the conserved current $j^s_\mu(x,q)$ ($\varphi^s= \Sigma\, c_m\, \varphi^{s,m}$)

$$j^s_\mu(x,q)= \tfrac{1}{2}e\, [\, \partial_\mu\bar{\varphi}^s(x,q)\varphi^s(x,q) - \bar{\varphi}^s(x,q)\partial_\mu\varphi^s(x,q)\,] \, , \tag{2}$$

with
$$\partial_\mu j^s_\mu(x,q) = 0 \, . \tag{3}$$

The electric charge of the particle is e. Due to (3) the coupling $j^s_\mu(x,q)A_\mu(x)$ is gauge invariant. The scattering of the extended particle defined by φ^s with a pointlike particle determines the elastic electromagnetic formfactor F in momentum space. In Born approximation the differential cross-section factorizes:

$$\frac{d\sigma}{d\Omega} = \frac{d\sigma}{d\Omega}\bigg|_{\text{pointlike}} F^2(\hat{p}',\hat{p}) \quad . \tag{4}$$

with
$$F^2(\hat{p}',\hat{p}) = \int d^2q\, \bar{\varphi}^s(\hat{p}',q)\varphi^s(\hat{p},q), \tag{5}$$

p',p are the scattered respective incident momentum and $\hat{p}'=p'/M, \hat{p}=p/M$. The normalized formfactor of a spin zero particle is a function of momentum transfer in the scattering plane defined by $p=(p_0,0,0,p_3)$ and $p'=(p_0,p_1',0,p_3')$:

$$F(\hat{p}',\hat{p}) = \left(1 + \frac{4}{M^2} (\vec{p}' - \vec{p})^2\right)^{-1} , \tag{6}$$

which corresponds to a Yukawa I- type charge distribution in position space [4]

$$\rho(r) = 6 \frac{a}{r} \exp(- \sqrt{6} \frac{r}{a}) , \tag{7}$$

with $a^2= \frac{3}{2} M^{-2}$. Such a hard core structure is not observed in nature.

An example of inelastic interactions of electromagnetic origin , with change of spin and mass, is the one photon decay induced by the coupling

$$\bar{\varphi}^{s',m'}(x,q)S_{\mu\nu}(x)F_{\mu\nu}(x)\varphi^{s,m}(x,q) \tag{8}$$

where $S_{\mu\nu}$ is the "spin-part" of the generators of the Poincaré group and $F_{\mu\nu}$ the electromagnetic field tensor. $|s'-s| = 1$, $|m'- m| \leq 1$.

The decay width Γ for the special case $s'=1$, $s=0$ ($1 \to 0 + \gamma$) is linear in M_1 and monotone decreasing in $\epsilon=M_0/M_1$, $\epsilon \in [0,1]$:

$$\Gamma = M_1 \; g(\epsilon). \tag{9}$$

On a linear Regge trajectory ($M^2 = \alpha + \beta s$) $\epsilon^2= \alpha/(\alpha + \beta)$. Therefore the instability of the particles against γ-decay increases with the slope parameter β.

Inelastic interactions without exchange of any bosons are given by polynomial interactions e.g. the φ^4 interaction:

$$\bar{\varphi}_1(x,q)\bar{\varphi}_2(x,q)\varphi_3(x,q)\varphi_4(x,q) \tag{10}$$

where the indices denote spin and mass. The coupling (10) is not relativistic covariant. In spite of this we may remark some physical properties of the amplitude defined by (10), which may persist in a covariant model of polynomial interactions. (10) simulates a scattering at small scattering parameters (central collisions). The dynamics depends on the velocity of the particles involved. This kind of scaling behavior is supported by high energy experiments. The ratio between the forward and backward amplitude is of the order of unity (energy dependent), a feature not observed in experiments.

This work has been supported by the Deutsche Akademische Austauschdienst (DAAD)

[1] P.A.M. Dirac,Proc. Roy. Soc. A322,435 (1971), A328, 1 (1972), and "Directions in Physics", John Wiley & Sons,N.Y.(1978)
[2] H.van Dam and L.C. Biedenharn, Phys. Rev. D14, 405 (1976) and ref. therein

[3] H.van Dam and L.C. Biedenharn, "An Explicit Model Exhibiting a Generalization of Supersymmetry", Preprint (1978)

[4] R. Herman and R. Hofstadter, "High Energy Electron Scattering Tables", Stanford University Press, Stanford, California (1960).

RIGOROUS RESULTS IN LATTICE AND CONTINUUM GAUGE QUANTUM FIELD THEORIES

Erhard Seiler, Department of Physics, Princeton University, Princeton, N.J.08540

While gauge field theories have long been recognized as fundamentally important for the description of the basic interactions in nature, a rigorous mathematical investigation of their properties has only begun rather recently. It seems to be very important to find out how much of the existing lore can really be proven (or maybe disproven) mathematically and thereby to separate fact from fiction.

In a joint work with K. Osterwalder [8] the general structure of the theory was investigated on a lattice; this is the only known way of introducing cutoffs without destroying gauge invariance. Many physical properties can be established already at this level; in particular physical positivity, the existence of an infinite volume limit with a finite mass gap and the mass generation through the Higgs mechanism (see also [3,5,7]). Information on the confinement of quarks can also be obtained at this level [1,2,4,6,8,9].

Work in progress with D. Brydges and J. Fröhlich [1] further analyzes the lattice theories; in particular we establish very general "diamagnetic inequalities" and correlation inequalities.

These inequalities are essential tools for studying the continuum and infinite volume limits of the easiest nontrivial gauge theory, the two-dimensional Higgs model. They also give structural information, including a confinement result for fractional charges. For this model the methods of constructive field theory are sufficiently powerful to prove stability.

Finally, a limiting case of this model (essentially equivalent to the Massive Sine-Gordon model) allows a convergent expansion which can be interpreted as the activity expansion for an instanton (vortex) gas. The analogue of this expansion for the general case is under investigation.

References

[1] Brydges, D., Fröhlich, J., and Seiler, E., "On the Construction of Quantized Gauge Fields I: General Results", IHES (Bures-sur-Yvette) preprint, August 1978, and papers in preparation.

[2] Challifour, J.L. and Weingarten, D., J. Math. Phys. 19 (1978) 1134-6.

[3] DeAngelis, G.F., deFalco, D., Guerra, F., Phys. Lett. 68B (1977) 255-7; Phys. Rev. D17 (1977) 1624-8; Lett. Nuovo Cim. 18 (1977) 536-8, 19 (1977) 55-8, and Salerno preprints.

[4] Gawedski, K., "On Confinement of Fermions in Strongly Coupled Lattice Gauge Theory", Max-Planck Institut preprint 1977.

[5] Glimm, J., Jaffe, A., Phys. Lett. 66B (1977) 67-69; Commun. Math. Phys. 56 (1977) 195-212.

[6] Israel, R., Nappi, C., "Quark Confinement in the Two-Dimensional Lattice Higgs-Villain Model", Harvard preprint 1978.

[7] Lüscher, M., Commun. Math. Phys. 54 (1977) 283-292.

[8] Osterwalder, K., Seiler, E., Ann. Phys. (N.Y.) 110 440-471; "Lattice Gauge Theories" in: Springer Lecture Notes in Physics 80, G. dell'Antonio, S. Doplicher, G. Jona-Lasinio, eds.

[9] Seiler, E., Phys. Rev. D18 (1978) 482.

A UNITARY RELATIVISTIC WAVE EQUATION EXHIBITING EXTENDED PARTICLE STRUCTURE

L. P. Staunton
Drake University
Des Moines, Iowa 50311

Lorentz invariant relativistic wave equations which contain operators realizing unitary representations of the Lorentz group[1] are known to have spacelike and lightlike solutions, and frequently exhibit unphysical mass-spin spectra. Unitary models of a purely algebraic nature[2] permit consideration of physical states only, but preclude the use of a gauge invariance principle to effect coupling.

In contrast, consider the covariant, 4-Hamiltonian wave equation[3] $P_\mu\psi=(m\Gamma_\mu+iS_{\mu\nu}P^\nu)\psi$, where the operators Γ_μ and $S_{\mu\nu}$ form a unitary representation of the Lie algebra of SO(3,2) with the properties that $S_{\mu\nu}S^{\mu\nu}=-3/2, \epsilon^{\mu\nu\alpha\beta}S_{\mu\nu}S_{\alpha\beta}=0$, and $\Gamma^\mu\Gamma_\mu=-1/2$, i.e., operators realizing the Majorana representation of the Lorentz group. This wave equation imposes four constraints upon the single function ψ which are not all equivalent. These may be shown to lead uniquely to a set of simultaneous, Lorentz invariant wave equations[4], viz. $(P^2-m^2)\psi=0, W^2\psi=-3/4m^2\psi$, and $(\Gamma^\mu P_\mu-m)\psi=0$. It follows that ψ is the massive, spin 1/2, timelike solution of Majorana's equation, so that the covariant, 4-Hamiltonian wave equation defines a projection upon the solution space of Majorana, one which eliminates the unphysical spacelike and lightlike solutions, and avoids as well the familiar problem of a descending mass-spin spectrum by limiting consideration to a single mass-spin state. Further, since a single covariant equation defines the theory, gauge principles may be used to introduce couplings.

As is well known, the spectrum of the unitary operator Γ_o is strictly positive, so that only positive-energy solutions are present. Consequently, ψ represents a state which cannot, a priori, be localized, but is an extended object which exhibits a constituent substructure. The use of infinite dimensional matrices makes identification of this substructure difficult, but if a realization[5] of the Majorana operators in terms of two dimensionless variables, q_1, and q_2, with conjugate operators η_1 and η_2 is employed, then e.q., $\Gamma_o=1/4(q_1^2+q_2^2+\eta_1^2+\eta_2^2)$, and the single component wave function solution $\psi(\chi^\mu,q_1,q_2)$ is the first occupied state of the relativistic two dimensional null-plane harmonic-oscillator problem[6]. The two constituents are therefore permanently bound and "asymptotically free".

A manifestly covariant Minkowski space analysis of the constituent structure may be obtained[7] via consideration of the Heisenberg picture generated by the generalized, scalar "Hamiltonian" $\Phi=m-\Gamma^\mu P_\mu$, annihilating the states, $|>$, in terms of an associated scalar parameter, τ, whose role is that of the proper time, i.e., $d\chi^\mu/d\tau=-i[\chi^\mu,\Phi]$. The resulting operator differential equations yield a solution system isomosphic to a quantized, covariant dual string containing only a single normal mode. The Virasoro[8] gauge conditions are identities in this model, not additional constraints, as a consequence of the unitary representation employed, i.e., $\Gamma^\mu\Gamma_\mu$ is identically a c-number. The normal mode operator commutation relations obtained form, therefore, an informative subset of those necessary for a properly quantized string in 4-dimensions, and suggest the origin of many of the difficulties usually encountered[9], e.g., the dual string operator usually denoted a_o^μ and identified with the 4-momentum on the classical level, is here found to have the form $a_o^\mu = M^{-2}p^\mu(m-\Phi)$, so that its matrix elements are p^μ/m, while its commutation relations are quite different. The 4-Hamiltonian wave equation defining the theory here imposes the relation $M^2|\rangle = m^2|\rangle$, and requires as well that the negative frequency normal mode operator annihilate the states.

A non-quantum but relativistic limit theory may be obtained[10] by straightforward replacement of commutators with Poisson brackets. This procedure is not formal, but well defined since the unitary operators Γ_μ, $S_{\mu\nu}$ have classical definitions as functions of two continuous variables and associated conjugates with respect to which partial derivatives may be taken. The differential equations so obtained reveal a Minkowski space rotational motion in the hyperplane orthogonal to p_μ. Further, since the classical functions Γ_μ are the covariant 4-velocity, and satisfy identically the relation $\Gamma^\mu\Gamma_\mu=0$ (Note: a <u>different</u> relation) as classical functions, the rotation is a Zitterbewegung. This result characterizes the two constituents as massless, and the model as a limit of the classical Nambu string[11], with explicitly solved constraints.

The first step toward an extention of this model to incorporate several constituents (normal modes), that of exhibiting a null-plane, continuous variable realization of the Lie algebra of Sp(8,R), has recently been reported.[12]

1. E.Majorana, Nuovo Cimento 9,335(1932); Y.Nambu, Prog. Theor. Phys. Supp. 37, 38, 368(1966), Phys. Rev. 160,1171(1967); C. Fronsdal, Phys. Rev. 156, 1665(1967); A.O.Barut and H.Kleinert, Phys. Rev. 156,1546(1967); A.O.Barut, D. Corrigan, and H.Kleinert, Phys. Rev. 167,1527(1968).
2. A.Bohm, Phys., Rev. 175, 1767(1968) and refs. therein.
3. L.P. Staunton, Phys. Rev. D10,1760(1974).
4. W^μ is the Pauli-Lubanski operator. $W^2=-s(s+1)m^2$ on states.
5. P.A.M.Dirac, J. Math. Phys. 4,901(1963).
6. L.C.Biedenharn, M.Y.Han and H.van Dam, Phys. Rev. D8 1735(1973).
7. L.P.Staunton, Phys. Rev. D13,3269(1976).
8. M.A.Virasoro, Phys. Rev. D1, 2933(1970).
9. R.Marnelius, Nucl. Phys. B104,477(1976), and refs. therein.
10. L.P.Staunton and S. Browne, Phys. Rev. D12,1026(1975).
11. Y.Nambu, Copenhagen Lectures, 1970 (unpublished).
12. L.P.Staunton, J. Math. Phys. 19,1471(1978).

LINE SPACE CONSTRUCTION OF NON-SELF-DUAL YANG-MILLS FIELDS*

PHILIP B. YASSKIN, Dept of Physics, Harvard University, Cambridge, MA 02138

JAMES ISENBERG, Dept. of Physics, University of Maryland, College Park, MD 20742

PAUL S. GREEN, Dept. of Mathematics, University of Maryland, College Park, MD 20742

Recently, Ward [1] has developed a twistorial description of self-dual Yang-Mills fields on pieces of conformally completed complex Minkowski space, CM. The solutions covering S^4(conformally completed real Euclidean space) are the instantons of interest in quantum field theory. We have modified Ward's construction to describe non-self-dual fields and have found currentless and reality conditions [2,3]. A similar modification has been investigated independently by Witten [4].

Notationally, if S is any subset of CM, we let $P_\alpha(S)$, $P_\beta(S)$ and L(S) denote the sets of α-planes, β-planes, and null lines which intersect S. Since every null line is the intersection of a unique α-plane and a unique β-plane, there are projections $L(S) \to P_\alpha(S)$ and $L(S) \to P_\beta(S)$ and an injection $L(S) \to P_\alpha(S) \times P_\beta(S)$.

Ward has shown that any self-dual gauge field with gauge group, G, on a "nice" region, $S \subset CM$, corresponds to a principle G-bundle over $P_\alpha(S)$ which is trivial over $P_\alpha(p)$ for all $p \in S$. A similar statement relates anti-self-dual fields to bundles over $P_\beta(S)$. We have proved that there is a 1-1 correspondence between (1) analytic, principal G-bundles over S with connection, A, and curvature, F; and (2) analytic, principal G-bundles over L(S) which are trivial over L(p) for all $p \in S$.

Unlike self-dual fields, non-self-dual fields do not automatically satisfy the currentless Yang-Mills equations. We prove (a) the gauge field (A and F) satisfies the currentless Yang-Mills equations iff the bundle over L(S) has a third order extension to a neighborhood of L(S) within $P_\alpha(S) \times P_\beta(S)$. We also prove the successively more restrictive conditions, (b) the gauge field satisfies the currentless Yang-Mills equations and has commuting self-dual and anti-self-dual parts of its holonomy group iff the bundle over L(S) has an extension to a neighborhood of L(S) within $P_\alpha(S) \times P_\beta(S)$; (c) the gauge field is self-dual iff the bundle over L(S) is the pull back of a bundle over $P_\alpha(S)$ along the projection $L(S) \to P_\alpha(S)$; and (d) the gauge field is flat iff the bundle over L(S) is trivial.

Also, unlike self-dual fields, more general fields may be "real" on real Minkowski space. We prove (e) when restricted to real Minkowski space (or real Euclidean space) the gauge field (A and F) is reducible to a real gauge group iff there exists an "appropriate" conjugation on the bundle over L(S).

A more detailed statement of our results may be found in [2] where we also outline the constructions needed to demonstrate the correspondence between (1) and (2) above. The proof of the most important result, (a) above, may be found in [3].

We suggest four ways in which our construction may be useful in the future. (i) Our construction provides a new tool for trying to prove the conjecture that any currentless Yang-Mills solution, regular on all of S^4, must have commuting self-dual and anti-self-dual parts of its holonomy group.(ii) Since non-self-dual fields need not be currentless, it may be possible to self-consistently couple non-self-dual fields to source fields. (iii) Since our construction produces all gauge fields, (not just those satisfying the currentless Yang-Mills equations), it may be possible to quantize using path integrals over all bundles on L instead of over all guage fields on CM. (iv) Our construction should generalize to describe Yang-Mills fields on curved spacetimes where L becomes the set of null geodesics, but in general, P_α ceases to exist.

*Supported in part by the National Science Foundation under Grant PHY-76-20029.

[1] R.S. Ward, Phys. Lett. 61A (1977), 81-82.

[2] J. Isenberg, P. B. Yasskin and P. S. Green, Phys. Lett. __B(1978), to be pub.

[3] J. Isenberg and P. B. Yasskin, in Proc. of NSF Summer Workshop on Applications of Complex Manifold Techniques to Problems in Theoretical Physics, 1978, ed. by D. Lerner and P. Sommers, to be published.

[4] E. Witten, Phys. Lett. __B (1978), to be published.

AN EXPLICIT MODEL EXHIBITING MASS AND SPIN MIXING

by

H. van Dam
Physics Department, University of North Carolina
Chapel Hill, North Carolina 27514

and

L.C. Biedenharn*
Physics Department, Duke University
Durham, North Carolina 27706

I. *Introduction*

The research discussed in this paper originated with Dirac's analysis of a remarkable representation of SO(3,2)[1], and with the "new Dirac equation" for particles of spin zero[2]. We shall of necessity focus on the most recent developments, but we must mention first, several interesting developments following Dirac's work. One of these is Staunton's equation for spin-1/2, another concerns the classical limit, where one finds, in the center of mass frame, a single constituent circling the center at the velocity of light.[3] We also want to point out that many of the developments in this area were foreseen by A. Bohm[4], but not fully appreciated at the time. The fiber bundle viewpoint, discussed below, should be distinguished from the much more general Cartan bundles advocated by Drechsler[5].

In this paper we shall limit ourselves to a discussion of recent results concerning Poincaré covariant models incorporating (discrete) mass and spin mixing. A series of theorems[6] of increasing generality have been proven which deny the existence of any such model. To circumvent such theorems one need only avoid one of the (possibly implicit) hypotheses; supersymmetry, as is now well known, achieves this circumvention by adjoining Grassmann elements affecting Bose-Fermi transformations[7]. It is generally believed [8] that this particular realization[6] of supersymmetry (which cannot incorporate mass mixing in the algebra) is the *only* possible way to circumvent the no-go theorems. There exists, however, a different structure achieving this goal[9]. We shall review this structure here and put it in an algebraic form which facilitates comparison with supersymmetry. We find a generalization of the standard supersymmetry[7,8] and of the version developed by J. Schwinger[10].

---

*Supported in part by the National Science Foundation.

II. *A Remarkable Representation of* SO(3,2)

The generators of this group are represented over a digenerate pair of boson operators[1]: $[\xi_i,\pi_j] = i\delta_{ij}$, $[\xi_i,\xi_j] = 0 = [\pi_i,\pi_j]$, $i,j = 1,2$. In the standard manner one can construct creation and annihilation operators which satisfy $[a_i,a_j^\dagger] = \delta_{ij}$.

With the column vector Q_α = column $(\xi_1,\xi_2,\pi_1,\pi_2)$, one can write the commentation relations[2]

$$[Q_\alpha,Q_\beta] = i\,\gamma^0_{\alpha\beta}, \text{ where} \qquad (1)$$
$$\gamma^0 = \begin{pmatrix} 0 & 1\!\!1 \\ -1\!\!1 & 0 \end{pmatrix},$$

in which $1\!\!1$ is the 2 dimensional unit matrix. For any symmetric matrix $\gamma^0 N$ one finds, with (1)

$$[Q^\dagger\,\gamma^0\,N\,Q,\,Q_\alpha] = -2i\,N_{\alpha\beta}\,Q_\beta\;. \qquad (2)$$

If, in addition, N is real, one obtains a Hermitian operator $Q^\dagger\gamma^0 NQ$; there are ten linearly independent operators of this type. These can be chosen as

$$S^{\mu\nu} = \frac{1}{8}\,Q^\dagger\gamma^0\,\{\gamma^\mu,\gamma^\nu\}\,Q\;, \qquad (3)$$
$$V^\mu = \frac{1}{2}\,Q^\dagger\gamma^0\gamma^\mu Q\;,$$

where the matrices γ^μ must be real (Majorana representation) and must include the γ^0 of eq. (1). The choice of representation is otherwise free.

Note that:

$$V^0 = \frac{1}{2}\,(\pi_1^2 + \xi_1^2 + \pi_2^2 + \xi_2^2) = a_1a_1^\dagger + a_2a_2^\dagger + 1, \qquad (4)$$

is a positive definite Hermitian operator.

One finds that the ten generators defined in (3) generate a unitary representation $U(\Sigma)$ of the covering group of the connected part of the SO(3,2) de Sitter group[1,2]. Also, with (2), one finds an associated nonunitary representation by 4x4 matrices:

$$U(\Sigma)\,Q_\alpha\,U^{-1}(\Sigma) = \mathcal{D}^{-1}(\Sigma)_{\alpha\beta}\,Q_\beta\;. \qquad (5)$$

III. *Quasi-Newtonian Form for Generators of the Poincare´ Group (Thomas Form)*

This form was given in the last two citations of reference 9. The operators for P_μ, $M^{\mu\nu}$ are defined on functions $\psi(\vec{p},\xi_1,\xi_2)$ with inner product $(\phi,\chi) = \int d\vec{p}\, d\xi_1 d\xi_2\, \phi^*\chi$. The operators are

$$
\begin{aligned}
P_i &= p_i \quad, \quad i=1,2,3, \\
P_0 &= +[\vec{P}^2+M^2]^{1/2} \quad, \\
M_{ij} &= \varepsilon_{ijk}\,(\vec{X}x\vec{P})_k + S_{ij} \quad, \\
M_{0i} &= \frac{1}{2}\,(X_iP_0+P_0X_i) + tP_i/P_0 + [P_0+M]^{-1}(\varepsilon_{ijk}P_jJ_k) \quad,
\end{aligned}
\qquad (6)
$$

where $X_i = -i\,\frac{\partial}{\partial p_i}$, $S_{ij} = \varepsilon_{ijk}J_k$ and $M = f(V_0)$ in which f is a positive monotonically increasing function. In (6) the symbols S_{ij} and V_0 are defined by equations (3) and (4). The generators (6) are an easy generalization of those given by Thomas, and by Bacry for a particle with spin s[11]. The construction of the generators (6) is quite general. All one needs to get a unitary representation of the Poincare´ group is a special "internal" Hilbert space *H*. This Hilbert space is only special in the sense that it must contain 3 Hermitian operators S_{ij} with the commutation relations of angular momentum and, further, a V_0 (having a positive definite spectrum) which commutes with the S_{ij}. One can adjoin this *H*, in direct product fashion, to the space of functions of $\vec{p}$ (scalar particles) as in (6). In this way one can combine internal symmetry groups with the Poincare´ group in a nontrivial manner. Examples, a relativistic SU(6) and SL(3,R), are given in reference 9. These more general aspects will be discussed briefly in Section V.

The generators (6) describe a Regge band with each spin s contained once, the mass being given by $m^2=f(s)$. The ground state of the degenerate oscillator gives the spin zero state, the two first excited states the spin 1/2 state, etc. The operators $a_i^\dagger$ increase the spin, the a_i lower the spin, suggesting a connection to supersymmetry (cf. Section V).

The system given by (6) is Poincare´ invariant, but it is not manifestly invariant in the sense that it is not given in terms of four vectors, four spinors, etc. Therefore, it is not suitable for introducing interactions.

IV. *Explicitly Poincaré Invariant Formulation*

This formulation is obtained by a transformation analogous to an inverse Foldy-Wouthuysen transformation. One now considers functions $\psi(\vec{p},p^0,\xi_1,\xi_2) \equiv \psi(p,\xi_1,\xi_2)$ and puts in the mass spectrum by an equation of motion. Note that the generators (6) allow only solutions for p in the forward lightcone. The key is the introduction of ground states and of creation and annihilation operators which parametrically contain the unit "four velocity" $\hat{p}$ in the direction of p (operators aligned with p).

For an arbitrary p in the forward lightcone the "aligned" annihilation operators are

$$T_\alpha = \frac{1}{2}\,(1\!\!1 + i\,\hat{p}_\mu\gamma^\mu)_{\alpha\beta}\,Q_\beta\ , \tag{7}$$

with Q_β from Section II. The projection operator on the right-hand side of (7) implies that T_α has only two independent components. For $p=(0,0,0,m)$ one finds that these components are just the destruction operators, a_1 and a_2. The creation operators aligned with p are given by using the projection operator with the minus sign on the right-hand side of (7), or, alternatively, by

$$\overline{T}_\alpha = T^\dagger_\beta\,\gamma^0_{\beta\alpha}\ . \tag{8}$$

It is useful to note that the T_α, $\overline{T}_\alpha$ obey the relation:

$$[T_\alpha,\overline{T}_\beta] = (-i)(\tfrac{1}{2})(1\!\!1 + i\,\hat{p}_\mu\gamma^\mu)_{\alpha\beta}\ .$$

The ground state "aligned with p" is given by

$$\tfrac{1}{2}(1\!\!1 + i\,\hat{p}_\mu\gamma^\mu)_{\alpha\beta}\,Q_\beta\,e_0(\hat{p},\xi_1,\xi_2) = 0\ ,$$

$\alpha = 1,2,3,0$. This is the new Dirac equation[2] and $e_0(\hat{p},\xi_1,\xi_2)$ is the solution to that equation. This solution describes a spin zero particle as it is invariant under the little group of p. The four functions $\overline{T}_\alpha\,e_0(\hat{p},\xi_1,\xi_2)$ describe, similarly, a spin 1/2 particle, etc. The functions obtained in this manner form a complete set. Out of a general $\psi(p,\xi_1,\xi_2)$, with p in forward lightcone, one can project a spin 0 wave function by

$$\psi_0(p) = \int d\xi_1 d\xi_2\; e_0{}^*(\hat{p},\xi_1,\xi_2)\,\psi(p,\xi_1,\xi_2)\ , \tag{9}$$

a spin 1/2 wave function by

$$\psi_\alpha(p) = \int d\xi_1 d\xi_2\,(\overline{T}_\alpha\;e_0(\hat{p},\xi_1,\xi_2))^*\,\psi(p,\xi_1,\xi_2)\ , \tag{10}$$

and, similarly, multispinor wave functions $\psi_{\alpha\beta\ldots}(p)$ for higher spin.

The mass spectrum of (6) must now be imposed as a subsidiary condition. For $f(V^0) = V^0$ in (6) one has

$$(P^2 - M_{0p}^2)\psi(p,\xi_1,\xi_2) = 0 \quad , \qquad (11)$$

with $M_{0p}^3 = M_0^{\ 2}\ P\cdot V^{[12]}$, where M_0 is a constant and V is defined in (3). Note that, as P·V is Hermitian, this equation has no space-like solutions.

V. *An Algebra which Extends the Poincaré Algebra and Contains Operators for Raising and Lowering Mass and Spin*

To start, we restrict our discussion to the solutions of equation (11). On that space the operators $\Pi_\pm \equiv \frac{1}{2}(1 \pm i\ \gamma\cdot P\ M_{op}^{-1})$ act as a pair of orthogonal projection operators. To proceed further in the way suggested by (7), we must not only change the spin, as in (7), but also scale the four-vector p to the new mass value. Notice also that $[M_{op},Q_\alpha] \neq 0$, so that $[\Pi_\pm,Q_\alpha] \neq 0$. Let us introduce the scaling generators s, defined by: $[s,P] = i\ P$; $[s,X] = -i\ X$. Using this scaling generator we define the (Lorentz invariant) scaling operator: $S \equiv \exp\ [-i\ (\ln M_{op})s]$, such that: $S^{-1}\ P\ S = M_{op}^{-1}\ P$. (This uses $[P,M_{op}] = 0$).

The desired global four-component spinorial destruction operator is then defined by:

$$S_\alpha \equiv (\Pi_+)_{\alpha\beta}\ S^{-1}\ Q_\beta\ S. \qquad (12)$$

The conjugate (raising) operator $\bar{S}_\alpha$ is defined by: $\bar{S}_\alpha = S_\beta^+(\gamma_o)_{\beta\alpha}$.

The spinorial operators satisfy the following commutation rules:

$$[S_\alpha,S_\beta] = [\bar{S}_\alpha,\bar{S}_\beta] = 0;\ [S_\alpha,\bar{S}_\beta] = (-i)(\tfrac{1}{2})(1+i\gamma\cdot P\ M_{op}^{-1})_{\alpha\beta} \quad . \qquad (13)$$

These algebraic relations are to be contrasted with the analogous (standard) supersymmetry algebra, where commutation is replaced by anti-commutation, and the "mass term" (the unit operator in Π_+) is missing.

The spinorial operators obey a useful identity relating them to the mass operators, $m_o^{\ 2}\ \sum_\alpha \{\bar{S}_\alpha,S_\alpha\} = 8\ M_{op}^2$. We remark that the specific choice of mass operator used in (11) (linear Regge band) is not required; any positive monotonic function of spin can be achieved as in (6).

The S_α (and $\bar{S}_\alpha$) transform as spinors under the Lorentz group. That is: $[M_{\mu\nu},S_\alpha] = (\sigma_{\mu\nu})_{\beta\alpha}S_\beta$, and analogously for $\bar{S}_\alpha$.

The most interesting commutation relations involve the translation operator:

$$[P_\mu, S_\alpha] = (M_{op} S_\alpha M_{op}^{-1} - S_\alpha) P_\mu \quad , \tag{14}$$

and the "velocity" operator $P_\mu M_{op}$:

$$[P_\mu M_{op}^{-1}, S_\alpha] = 0 \quad . \tag{15}$$

Remarks: (*i*) The relation given by Eq. (14) is in sharp contrast to the standard supersymmetry relation (having the right-hand side zero). However, if one seeks to unite multiplets of *different* mass then necessarily a relation of this form must obtain.

(*ii*) It will be noticed at once that the Poincaré algebra extended with S_α, $\overline{S}_\alpha$ *does not close.* It is in this way that the construction avoids the no-go theorems, since there is, in fact, no (finite rank) Lie group associated to the resulting algebra. The structure is, however, physically well-defined; the Lie group requirement is for mathematical convenience only.

(*iii*) The relation given by Eq. (15) is very important, for it shows that the algebra generated by: $\{P_\mu M_{op}^{-1}, M_{\mu\nu}, S_\alpha, \overline{S}_\alpha\}$ *does indeed close.* (The projection operator Π_+ commutes with these generators.) From a physical point of view this result is quite natural. Consider a system, in its rest frame, that can exist in several discrete mass states; a boost transformation to a moving frame changes *velocity*, not momentum, so that all the different mass states have, in the moving frame, the same velocity. The possibility of forming coherent combinations of wave functions with differing mass and spin (subject to superselection rules) thus exists in each frame indexed by a unit four-velocity.

(*iv*) The algebraic Lie group associated to the algebra of the previous remark defines an interesting physical structure. One may view this structure geometrically in terms of the three-dimensional surface of a unit-mass hyperboloid in the forward light cone. For every four-velocity, there is a corresponding point of the surface; over each point we may associate a set of different mass-spin states. The vector space of these states is precisely the fiber of a principle fiber bundle[5], whose base space consists of the points of the hyperbolic surface. The isotropy group of a given point (the transformations leaving a given four-velocity invariant) is isomorphic to the group SU2 extended by the shift operators $S_\alpha, \overline{S}_\alpha$, adapted to this velocity.

To establish the relationship between the present model and the standard supersymmetry structure, we decompose any generic solution

$\psi(p,\xi_1,\xi_2)$ of Eq. (11) into a sum of wave functions of definite spin, as discussed at the end of Section IV, (9) and (10). The variation $\delta_+\psi(p,\xi_1,\xi_2) \equiv \eta_\alpha \bar{S}_\alpha\ \psi(p,\xi_1,\xi_2)$ translates onto these wave functions $\psi_{\alpha\beta\ldots}$ of definite spin as the action:

$$\delta_+\psi_o = 0;\ \delta_+\psi_\alpha = 1/2\ (\frac{\gamma\cdot p}{m_{1/2}} - i)_{\alpha\beta}\ \eta_\beta\psi_o(+)\ ; \tag{16}$$

$$\delta_+\psi_{\alpha\beta} = 1/2\ [(\frac{\gamma\cdot p}{m_1} - i)_{\alpha\gamma}\ \eta_\gamma\psi_\beta(+) + (\frac{\gamma\cdot p}{m_1} - i)_{\beta\gamma}\eta_\gamma\psi_\alpha(+)];\ldots,$$

where the plus sign on the right-hand side means that the mass has been scaled up from m_j to m_j+1/2.

For the variation defined by $\bar{\eta}_\beta\ S_\beta$ one finds in a similar way:

$$\delta_-\psi_o = \bar{\eta}_\beta\psi_\beta(-);\quad \delta_-\psi_\alpha = \bar{\eta}_\beta\psi_{\alpha\beta}\ (-);\ \ldots\ . \tag{17}$$

Let us emphasize that the actions represented by eqs. (16) and (17) map solutions of eq. (11) into solutions; the parameters η are c-numbers at this stage.

Next we consider the second-quantized form of the solutions to eq. (11); this can be done by second-quantizing the wave functions $\psi_{\alpha\beta\ldots}$ which now become free field operators. The spin-statistics theorem implies that field operators with an odd number of Dirac indices anticommute; this requires that the parameters η in eqs. (16) and (17) must be Grassmann variables anti-commuting with themselves and all odd-indexed field operators.

Using the operators S_α, $\bar{S}_\beta$ one can construct "supersymmetry" operators which connect only a finite number of states. For example, the supersymmetry operators given by Schwinger[10] for two adjacent levels j, j+1/2 are easily obtained by using δ_+ on the level j and δ_- on the level j+1/2. This composite supersymmetry operation is found to satisfy *commutation* relations, whereas the basic δ_+ and δ_- satisfy *anti-commutation* relations.

This reversal from anticommutation (for the operation δ_+ and δ_-) to commutation relations (for standard supersymmetry) necessarily occurs whenever one restricts the action to a *finite* number of levels (but mixing integer and half-integer spin).

Remarks: (*i*) The use of spinorial operators S_α, $\bar{S}_\beta$ for constructing the algebra is not essential. Instead one may use four-vector shift operators, tensorial shift operators,..., as one chooses. Products of spinorial operators automatically yield higher rank shift operators. It is of interest to note that for vectorial shift operators there exists a variant distinct from that given by the product of two spinorial operators. This variant is characterized by: $[K_\mu, \bar{K}_\nu] = (g_{\mu\nu} - P_\mu P_\nu M_{op}^{-2})$.

(ii) The construction of an explicitly Poincaré invariant model realizing "relativistic SU6" with both mass and spin mixing is now straightforward[9]. One defines an SU3 triplet of spinorial generators: S_α^i (i = 1,2,3) and generalizes eq. (13) to read:

$$i[S_\alpha^i, \bar{S}_\beta^j] = \delta_{ij} \, (\Pi_+)_{\alpha\beta} \quad . \tag{18}$$

The multiplets associated to this symmetry are the totally symmetric irreps [n0] of SU6; that is: 1,6,21,56,... . The baryon 56-multiplet is thereby realized; mass-splitting can be incorporated.

(iii) Since the existence of relativistic SU6 has been regarded as problematic, let us discuss the geometric meaning of our construction. The symmetry group SU6 is the maximal compact subgroup of the algebraic Lie group generated by the "velocity-Poincaré" generators, extended by S_α^i and $\bar{S}_\beta^j$ *(iii)*. There have been difficulties in realizing "relativistic SU6" within the standard supersymmetry framework. The incorporation of mass by spontaneous symmetry breaking is a complicated construction[13]. More serious is the fact that anti-symmetric SU(6) multiplets necessarily occur. Both difficulties disappear in the realization of generalized supersymmetry given above.

We thank P. van Nieuwenhuizen for a discussion which led to this work.

References

[1] P.A.M. Dirac, J. Math. Phys. 4, 901 (1963).
[2] P.A.M. Dirac, Proc. Roy. Soc. A322, 435 (1971), A328, 1 (1972).
[3] L.P. Staunton, Phys. Rev. D10, 1760 (1974), Phys. Rev. D13, 3269 (1976); Cf. also L.C. Biedenharn and H. van Dam in "Many Degrees of Freedom in Particle Theory", edited by H. Satz (Plenum Press, N.Y., 1978) p. 19.
[4] A. Bohm, Phys. Rev. 175, 1767 (1968).
[5] W. Drechsler, M.E. Mayer, "Fiber Bundle Techniques in Gauge Theories", Lect. Notes in Phys. 67, (Springer, Berlin) 1977.
[6] L. O'Raifeartaigh, Phys. Rev. 139, (1965) B1052; S. Coleman, J. Mandula, Phys. Rev. 159, (1967) 1252.
[7] J. Wess, B. Zumino, Nucl. Phys. B70, (1974) 39; the review of Fayet, et al. [3] cites further reference.
[8] S. Ferrara, B. Zumino, Nucl. Phys. B85, (1974) 413; P. Fayet, S. Ferrara, Physics Report 32, (1976) 249; A. Salam, J. Strathdee, "Supersymmetry and Superfields", (ICTP, Trieste) IC/76/12; R. Haag, J.T. Lopuszanski, M. Sohnius, Nucl. Phys. B88 (1975) 257.
[9] L.P. Staunton, H. van Dam, Lett. N.C. 7, (1973); H. van Dam, L.C. Biedenharn, Phys. Lett. 62B, (1976) 190; ibid. Phys. Rev. D14 (1976) 405.
[10] J. Schwinger, preprint UCLA/78/TEP/11, (May, 1978).
[11] L.H. Thomas, in "Quantum Theory of Atoms, Molecules and the Solid State" (a tribute to Professor J.C. Slater, edited by Per-Olov Löwdin), Academic Press (N.Y.) 1966; cf. pps. 93-96; H. Bacray,

"Lecons sur la theorie des groupes et des particules elementaires" (Gordon and Breach, N.Y., 1966), p. 309; Cf. also J. Schwinger, "Particles, Sources and Fields" (Addison Wesley, Reading, 1970) p. 19. The representations given by these authors, are those found by E.P. Wigner for particles of mass m and spin s, cf. Ann. Math. 40, 149 (1939).

[12] A.O. Barut, I.H. Duru, Lett. N.C. 8, (1973) 768.

[13] P. van Nieuwenhuizen, to be published.

THE AMBIGUITY GROUP FOR CANONICAL TRANSFORMATIONS IN CLASSICAL MECHANICS AND ITS ROLE IN THEIR REPRESENTATION IN QUANTUM MECHANICS

M. Moshinsky* and T.H. Seligman

Instituto de Física, Universidad de México (UNAM)

Apdo. Postal 20-364, México 20, D.F.

Abstract

We discuss, through the example of the canonical transformations leading to action and angle variables of the two dimensional isotropic oscillator, the representation in quantum mechanics of non-bijective canonical transformations. We stress in particular the importance of a classical concept, the ambiguity group, both in suggesting the quantization rules for the example just mentioned as well as in achieving the representation in quantum mechanics through the corresponding concept of ambiguity spin.

The problem of the representation in quantum mechanics of classical canonical transformations has had a long and distinguished history[1,2], but even today is not fully understood[3]. In particular the subtle aspects related to the representation of non-bijective, i.e. not one to one onto, canonical transformations, have only recently come to the attention of physicists[4-7].

It is our purpose to review the latter developments, which include such concepts as "ambiguity groups" in the phase space transformations and "ambiguity spins" in the corresponding Hilbert spaces. We

*Member of the Instituto Nacional de Energía Nuclear and El Colegio Nacional.

shall do this through the detailed discussion of a single example, the two dimensional isotropic harmonic oscillator, though we shall also indicate at the end of this paper other problems of importance to which a similar reasoning was applied.

The dynamical group of the two dimensional oscillator is the semi-direct product of a four dimensional symplectic group and the Weyl group i.e. $Sp(4) \wedge W(2)$, whose Lie Algebra generators are

$$Sp(4):\ q_i q_j, (1/2)(q_i p_j + p_j q_i), p_i p_j \ ; \tag{1a}$$

$$W(2):\ q_i, p_i, 1;\ i,j=1,2\ . \tag{1b}$$

Still dealing with <u>classical</u> mechanics, we introduce the creation and annihilation observables by the definitions

$$\eta_j=(1/\sqrt{2})(q_j - ip_j),\ \ \xi_j=(1/\sqrt{2})(q_j+ip_j), \tag{2}$$

$$\eta_\pm=(1/\sqrt{2})(\eta_1 \pm i\eta_2),\ \xi_\pm=(1/\sqrt{2})(\xi_1 \mp i\xi_2), \tag{3}$$

where we employ units in which $\hbar$, the mass of the particle and the frequency of the oscillator are taken equal to 1. The Lie Algebra (1) can then also be written in terms of linear and quadratic expressions in $\eta_\pm, \xi_\pm$.

We shall consider a subalgebra of four elements of the above algebra

$$H=\eta_+\xi_+ + \eta_-\xi_-, M=\eta_+\xi_+ - \eta_-\xi_-;\ N=\eta_+,\ R=\eta_+\eta_-\ , \tag{4}$$

where we check trivially that we have the Poisson bracket relations

$$\begin{aligned}&\{H,M\}=0,\ \{H,N\}=-iN,\ \{H,R\}=-2iR\\ &\{M,N\}=-iN,\ \{M,R\}=0,\ \{N,R\}=0.\end{aligned} \tag{5}$$

This subalgebra contains the Hamiltonian H and the angular momentum M.

We can now pass to the enveloping algebra of (4) to get a Weyl Lie Algebra defined by

$$|Q_1|=H-|M|,\ Q_2=M \tag{6a,b}$$

$$P_1 Q_1/|Q_1|=(i/2)\ln\left[2R(H^2-M^2)^{-1/2}\right],\ P_2=i\ln\left[\sqrt{2}N(H+|M|)^{-1/2}\right], \tag{7a,b}$$

where from (5) we immediately check that

$$\{Q_i,P_j\}=\delta_{ij} \quad i,j=1,2 \ . \tag{8}$$

As $|Q_1|$, Q_2 are clearly the actions associated with the two dimensional oscillator, then P_1,P_2 are one possible form of the corresponding angles and from (3,4) we easily see that they take only real values. As $H\geq 0$ and larger than $|M|$, we denote the first action variable by the absolute value $|Q_1|$ rather than by Q_1 itself and thus the corresponding canonically conjugate variable will be $P_1Q_1/|Q_1|$ rather than P_1 [5].

The relations (7) can be exponentiated to give

$$\exp(-2iP_1Q_1/|Q_1|)=2R(H^2-M^2)^{-1/2} \ , \tag{9a}$$

$$\exp(-iP_2)=\sqrt{2}\ N(H+|M|)^{-1/2} \ , \tag{9b}$$

where the appearance of the algebraic expressions $(H^2-M^2)^{-1/2}$, $(H+|M|)^{-1/2}$ on the right hand side will not bother us as they will be applied later to quantum mechanical basis in which H,M are diagonal and thus we can replace them there by their eigenvalues.

We now proceed to analyze the ambiguity group implicit in equations (6,9), which while being associated with a completely classical problem will give us a hint about the quantization rules. We note that if we carry out any of the following canonical transformations in the action-angle phase space

$$Q_1\to -Q_1, P_1\to -P_1 \ ; \ Q_1\to Q_1, \ P_1\to P_1+s_1\pi \ ; \tag{10a,b}$$

$$Q_2\to Q_2, \ P_2\to P_2+2s_2\pi \ ; \tag{10c}$$

(where $s_1,s_2=0,\pm1,\pm2,\ldots.$) the left hand side of (6),(9) remains invariant. Thus we conclude that there is an ambiguity group of canonical transformations defined by (10) that connects all the points of the action-angle phase space that are mapped on a single point in the original phase space.

The explicit appearance of H,M,R,N of (4) in (6),(9) indicates on the other hand that two different points in the original phase space are mapped on <u>different</u> points in the new phase space. Thus the full ambiguity group includes only the transformations (10) in the new phase space. We shall denote this group by

$$(T_1 \wedge I)\times T_2 \tag{11}$$

where $T_1 \wedge I$ is the semidirect product of translations by π of P_1, and inversions as indicated in (10a,b) while T_2 is associated with the translations by 2π of P_2 of (10c); as the two groups are independent we have the direct product indicated in (11).

We now turn our attention to the irreducible representations of the ambiguity group $(T_1 \wedge I)\times T_2$ which in an abstract two dimensional space (x_1,x_2) can be defined by the operations

$$I\begin{bmatrix}x_1\\x_2\end{bmatrix}=\begin{bmatrix}-x_1\\x_2\end{bmatrix},\; T_{1s}\begin{bmatrix}x_1\\x_2\end{bmatrix}=\begin{bmatrix}x_1+s\pi\\x_2\end{bmatrix},\; T_{2s}\begin{bmatrix}x_1\\x_2\end{bmatrix}=\begin{bmatrix}x_1\\x_2+2s\pi\end{bmatrix}, s=0,\pm1,\pm2\ldots . \tag{12}$$

In this space the complete set of plane waves

$$\exp\left[i(\nu_1x_1+\nu_2x_2)\right] , \tag{13}$$

forms a basis for a representation, albeit a reducible one, of the group $(T_1 \wedge I)\times T_2$. We note though that it is possible to write

$$\nu_1=2\sigma'(n+\lambda');\; n=0,1,2,\ldots ;\; 0\le\lambda'<1,\; \sigma'=\pm1 ; \tag{14a}$$

$$\nu_2=m+\tau';\; m=0,\pm1,\pm2,\ldots,\; 0\le\tau'<1 , \tag{14b}$$

and then the subsets of plane waves

$$\exp\{i\left[2\sigma'(n+\lambda')x_1+(m+\tau')x_2\right]\} \tag{15}$$

for the values $\sigma'=\pm1$ are basis for a two dimensional irreducible representation of the group (11) characterized by (λ',τ') i.e. we have the correspondances

$$I\rightarrow\begin{bmatrix}0&1\\1&0\end{bmatrix},\; T_{1s}\rightarrow\begin{bmatrix}\exp(i2\pi\lambda's)&0\\0&\exp(-i2\pi\lambda's)\end{bmatrix},\; T_{2s}\rightarrow\begin{bmatrix}1&0\\0&1\end{bmatrix}\exp(i2\pi\tau's) . \tag{16}$$

We proceed to make use of the results of the previous paragraph to derive a representation in an appropriately generalized Hilbert space of the canonical transformation (6),(9). For this we require first the normalized eigenstates $\psi_{nm}(q_1',q_2')$ of H,M in (4), where we use Dirac's notation[2] q_1',q_2' for the c-numbers associated with the observables q_1,q_2. From the definitions of $\eta_\pm$ in (2,3), considered now as operators, we immediately obtain

$$\psi_{nm}(q_1',q_2')=[(n+|m|)!\,n!]^{-1/2}\begin{cases}\eta_+^{n+m}\eta_-^{n}\psi_{00}(q_1',q_2')\text{ for } m\geq 0\\ \eta_+^{n}\eta_-^{n+|m|}\psi_{00}(q_1',q_2')\text{ for } m<0\end{cases}\qquad(17)$$

and from (4), in our units, they correspond to the following eigenvalues of the Hamiltonian H and the angular momentum M

$$H\psi_{nm}=(2n+|m|)\psi_{nm},\quad M\psi_{nm}=m\psi_{nm}\ .\qquad(18)$$

In (17) the explicit form of $\psi_{00}(q_1',q_2')$ is

$$\psi_{00}(q_1',q_2')=\pi^{-1/2}\exp\{-(1/2)(q_1'^2+q_2'^2)\}\ .\qquad(19)$$

We turn now our attention to the action observables Q_1,Q_2 whose eigenstates are given by

$$\Psi_{\nu_1\nu_2}(Q_1',Q_2')=\delta(Q_1'-\nu_1)\delta(Q_2'-\nu_2),\ -\infty\leq\nu_1,\nu_2\leq\infty\qquad(20)$$

where again we employ Dirac's notation[2] for the c-numbers Q_1',Q_2' associated with the observables Q_1,Q_2. The corresponding eigenvalues are then of course

$$Q_i'\Psi_{\nu_1\nu_2}(Q_1',Q_2')=\nu_i\Psi_{\nu_1\nu_2}(Q_1',Q_2'),\ i=1,2\ .\qquad(21)$$

We now note that in the abstract space (x_1,x_2) introduced above, these variables have the transformation properties (12) identical to those of (P_1,P_2) in (10). If we replace (x_1,x_2) by (P_1',P_2'), the c-numbers associated with (P_1,P_2), and pass from momentum to configuration space, the expression (13) will become then the wave function (20). But the analysis given above indicates that we must decompose the set of plane waves (13) into the subsets (15) associated with irreducible representations of the ambiguity group $(T_1\wedge I)\times T_2$. Thus, correspondingly, the set of eigenstates (20) of the action variables

Q_1, Q_2 can be decomposed into the subsets

$$\psi_{nm}^{\lambda'\tau'\sigma'}(Q_1', Q_2') = \delta[Q_1' - 2\sigma'(n+\lambda')]\,\delta[Q_2' - (m+\tau')] \ , \tag{22a}$$

$$\sigma' = \pm 1;\ 0 \leq \lambda', \tau' < 1;\ n = 0,1,2,\ldots\ ;\ m = 0, \pm 1, \pm 2, \ldots , \tag{22b}$$

characterized by the indices (λ', τ') that determine the irreducible representation of the group (11) and the index $\sigma' = \pm 1$ associated with the row of the two dimensional representation (16).

It is now possible to pass from the full set of states (17) in the original Hilbert space to those in the subsets mentioned above through the matrix of an operator $\mathbf{U}$ given by

$$\begin{aligned} \langle q_1' q_2' | U_{\lambda'\sigma'\tau'} | Q_1' Q_2' \rangle &= \sum_{n,m} \psi_{nm}(q_1', q_2')\, \psi_{nm}^{\lambda'\sigma'\tau' *}(Q_1', Q_2') = \\ &\equiv (q_1'\ q_2', \lambda'\tau'\sigma' | U | Q_1'\ Q_2' \rangle \ , \end{aligned} \tag{23}$$

which on the left hand side is characterized by the indices associated with the irreducible representations of the ambiguity group. On the other hand on the right hand side we have redefined the bras to include in them the indices $\lambda'\tau'\sigma'$. This was done for the following reason: The new phase space $(Q_1, Q_2; P_1, P_2)$ maps, because of (10), on an infinite number of sheets in the original phase space $(q_1, q_2; p_1, p_2)$. The presence of the ambiguity group, through its irreducible representations, suggests then that we introduce in the original Hilbert space an "ambiguity spin" characterized by $0 \leq \lambda', \tau' < 1$ and $\sigma' = \pm 1$ to have in it the equivalence of the many sheeted structure of the original phase space.

The expression (23) provides a unitary or, if we wish to be more exact, an isometric operator $\mathbf{U}$ as the bras in the original and the kets in the new Hilbert space are distinct. Thus while (23) relates the kets $|q_1' q_2', \lambda'\sigma'\tau')$ with $|Q_1' Q_2'\rangle$, the inverse operator $\mathbf{U}^{-1} = \mathbf{U}^{\dagger}$ is given by

$$\langle Q_1' Q_2' | U^{\dagger} | q_1' q_2', \lambda'\tau'\sigma') = (q_1' q_2', \lambda'\tau'\sigma' | U | Q_1' Q_2' \rangle^{*} \tag{24}$$

where the symbol † stands for hermitian conjugate and * for conjugate.

From (23),(24) it is easy to check that

$$<Q'_1 Q'_2|U^{\dagger}U|Q''_1 Q''_2>=\delta(Q'_1-Q''_1)\delta(Q'_2-Q''_2) \ , \tag{25a}$$

$$\begin{aligned}(q'_1 q'_2,\lambda'\sigma'\tau'|UU^{\dagger}|q''_1 q''_2,\lambda''\sigma''\tau'')= \\ =\delta(q'_1-q''_1)\delta(q'_2-q''_2)\delta(\lambda'-\lambda'')\delta(\tau'-\tau'')\delta_{\sigma'\sigma''} \ ,\end{aligned} \tag{25b}$$

where in (25a) we integrate over the intermediate q'_1,q'_2 as well as λ',τ' and sum over σ', while in (25b) we integrate over the intermediate Q'_1,Q'_2.

We now proceed to find the operator relations in quantum mechanics corresponding to the relations (6),(9) that define the canonical transformation to action and angle variables in classical mechanics.

For any classical observable in the action-angle phase space $F(Q_1,Q_2;P_1,P_2)$ we have a quantum mechanical operator

$$F(Q'_1,Q'_2;-i\partial/\partial Q'_1,-i\partial/\partial Q'_2)\delta(Q'_1-Q''_1)\delta(Q'_2-Q''_2) \ , \tag{26}$$

in the Hilbert space spanned by the kets $|Q'_1 Q'_2>$. We wish then to find the corresponding operators in the original Hilbert space given symbolically by

$$\mathbf{f}(q_i,p_j)= \mathbf{U}F(Q_i,P_j)\mathbf{U}^{\dagger} \ , \tag{27}$$

where F, **f** stand now for operators; $\mathbf{U},\mathbf{U}^{\dagger}$ are given by (23), (24), and we use a bold face letter for **f** to indicate that besides its operator character with respect q'_1,q'_2 and their derivatives, it is also a matrix in the indices λ',τ',σ' as shown, for example, by (25b) in the case when F is the unit operator.

From (25a) we also see that (27) can be rewritten as

$$\mathbf{f}(q_i,p_j)\mathbf{U}=\mathbf{U}F(Q_i,P_j)=[F^{\dagger}(Q_i,P_j)\mathbf{U}^{\dagger}]^{\dagger} \tag{28}$$

and we shall use this last formula to obtain the correspondance between operators in the original and new Hilbert spaces.

We shall consider only two types of observables in the phase space $(Q_1,Q_2;P_1,P_2)$whose corresponding operators in the new Hilbert space we wish to translate to the original one. The first will be the left hand side of (6) i.e. $|Q_1|,Q_2$ that become the operators

$$|Q_1'|\delta(Q_1'-Q_1'')\delta(Q_2'-Q_2''),Q_2'\delta(Q_1'-Q_1'')\delta(Q_2'-Q_2'') \ , \qquad (29a,b)$$

which remain unchanged when we take their hermitian conjugate required in (28).

The second will be the left hand side of (9a,b) where the hermitian conjugate of the corresponding operators become

$$\exp(2\partial/\partial|Q_1'|)\delta(Q_1'-Q_1'')\delta(Q_2'-Q_2'') \ , \qquad (30a)$$

$$\exp(\partial/\partial Q_2')\delta(Q_1'-Q_1'')\delta(Q_2'-Q_2'') \ , \qquad (30b)$$

where these expressions follow from the fact that

$$i(Q_1/|Q_1|)P_1\to(Q_1'/|Q_1'|)\partial/\partial Q_1'=\partial/\partial|Q_1'| \ , \ iP_2\to\partial/\partial Q_2' \ . \qquad (31)$$

Applying then (28) to the operators (29),(30) we get straightforwardly[5-7] the following correspondances of operators

$$|Q_1|\leftrightarrow\| (H-|M|+\lambda')\delta(\lambda'-\lambda'')\delta(\tau'-\tau'')\delta_{\sigma'\sigma''}\| \ , \qquad (32a)$$

$$Q_2\leftrightarrow\| (M+\tau')\delta(\lambda'-\lambda'')\delta(\tau'-\tau'')\delta_{\sigma'\sigma''}\| \ , \qquad (32b)$$

$$\exp(-2iP_1Q_1/|Q_1|)\leftrightarrow$$

$$2\eta_+\eta_-[(H+2)^2-M^2]^{1/2} \| \delta(\lambda'-\lambda'')\delta(\tau'-\tau')\delta_{\sigma'\sigma''}\| \ , \qquad (33a)$$

$$\exp(-iP_2)\leftrightarrow$$

$$[\sqrt{2}(H+|M|)^{-1/2}\eta_+]^{(|M|+M)/2|M|}[\sqrt{2}\xi_-(H+|M|)^{-1/2}]^{(|M|-M)/2|M|} \times$$

$$\times \| \delta(\lambda'-\lambda'')\delta(\tau'-\tau')\delta_{\sigma'\sigma''}\| \qquad (33b)$$

where H,M are given in (4) and $\eta_\pm$ in (3).

The expressions (32),(33) are the quantum mechanical equivalent of the classical expressions (6),(9) that define the canonical transformation. We must now check whether they reduce to them in the classical limit i.e. when H,|M| take large integer values. Clearly in that case

λ',τ' can be disregarded in (32) and thus, except for the unit matrix $\|\delta(\lambda'-\lambda'')\delta(\tau'-\tau'')\delta_{\sigma'\sigma''}\|$ we recover the classical relation (6). For (33) the situation is more complex as we see that in the classical limit, where the order of the observables is no longer relevant and we can disregard 2 as compared to H we recover, up to the unit matrix just mentioned, the classical relations

$$\exp(-2iP_1Q_1/|Q_1|)=2\eta_+\eta_-(H^2-M^2)^{-1/2}\ , \qquad (34a)$$

$$\exp(-iP_2)=\eta_+^{(|M|+M)/2|M|}\xi_-^{(|M|-M)/2|M|}\sqrt{2}(H+|M|)^{-1/2} \qquad (34b)$$

The expression (34a) is identical to (9a) but (34b) coincides with (9b) only when M is positive when η_+ appears and not for M negative when ξ_- appears. Thus the operator correspondances (33) are not associated with the classical relations (9) but rather with (34) where the latter are a kind of symmetrized form in η_+,ξ_- of the former. It is easy to check though that (6),(34) still define a canonical transformation for which $\{Q_i,P_j\}=\delta_{ij}$.

We have completed the discussion of the representation in quantum mechanics of the canonical transformation leading to action and angle variables of the two dimensional oscillator.

Interesting conclusions can be drawn from this problem which, as we shall later indicate, hold also for several other problems of this type that we have discussed[4-7]. To begin with we stress that the ambiguity group is a <u>purely classical</u> concept, yet from its irreducible representations (whose derivation is a problem of mathematics and not of physics) we were led to reclassifying the full set of states $\delta(Q_1'-\nu_1)\delta(Q_2'-\nu_2),-\infty\leq\nu_1,\nu_2\leq\infty$ into subsets for which ν_1,ν_2 are given by (14). This decomposition immediately recalls the <u>quantum mechanical</u> spectra of the operators $H-|M|,M$ in (18) displaced by the values λ',τ' associated with the irreducible representation of the ambiguity group. Thus it strongly suggests that quantum mechanical spectra or, more modestly, the Bohr-Sommerfeld quantization rules, are in some way implicit in the classical problem.

Another point is that when we take the classical limit of the operator correspondances (32),(33) we get the expressions (6),(34) but multiplied by $\|\delta(\lambda'-\lambda'')\delta(\tau'-\tau'')\delta_{\sigma'\sigma''}\|$. This indicates that even classically it is important to endow phase space with a much more complex structure, if we want to translate unambiguously the algebra of observables in the (q_i,p_j) phase space to the one of (Q_i,P_j) and viceversa.

Finally we note that once action variables such as $|Q_1|,Q_2$ of (6) are introduced, the angle variables P_1,P_2 can be defined in a variety of ways and still keep the Poisson bracket relation $\{Q_i,P_j\}=\delta_{ij}$. The choice we made in (9), or rather its symmetrized form in η_+,ξ_- given in (34), was deliberate, so that with the unitary representation (23) we get an operator correspondance (33) that in the classical limit reduces to (34). Had we made another choice of P_1,P_2 we would have to change the definition (23) introducing in it an appropriate phase $\exp(i\phi_{nm})$ so that the correspondance of operators gives in the classical limit the new choice of the canonical transformation.

All the properties mentioned hold for other problems of representations of canonical transformations we have analyzed: Transformations that take us from an oscillator of unit frequency to one of frequency κ^{-1} where κ is an integer[4)]; transformations taking us from (q,p) to the Hamiltonian and its appropriately conjugate variable for the free particle and the repulsive oscillator[5,6)]; transformations taking us from (q,p) to action and angle variables of the Coulomb problem for negative energies or, in an appropriate generalization of the concept of action, for positive energies[7)].

The above examples seem to indicate that we have a procedure for finding the unitary representation in quantum mechanics of a wide class of non-bijective canonical transformations, through the use of the concept of ambiguity group and ambiguity spin.

REFERENCES

1. M. Born, W. Heisenberg and P. Jordan, Z. Phys. 35, 557 (1926).

2. P.A.M. Dirac, "The principles of quantum mechanics", Oxford at the Clarendon Press, First Edition 1930, Third edition 1947 pp. 103-107.

3. P.A. Mello and M. Moshinsky, J. Math. Phys. 16, 2017 (1975).

4. P. Kramer, M. Moshinsky and T.H. Seligman, J. Math. Phys. 19, 683 (1978).

5. M. Moshinsky and T.H. Seligman, Annals of Physics (N.Y.)(in press, 1978).

6. M. Moshinsky and T.H. Seligman, Proceedings of the 1977 Bonn Conference on "Differential Geometric Methods in Mathematical Physics" (Springer-Heidelberg 1978).

7. M. Moshinsky and T.H. Seligman, Annals of Physics (submitted for publication).

GENERATING FUNCTIONS FOR CHARACTERS OF GROUP REPRESENTATIONS AND THEIR APPLICATIONS*

J. Patera, Université de Montréal, Montréal, Québec, Canada

and

R.T. Sharp, McGill University, Montreal, Quebec, Canada

ABSTRACT

A new method in group representation theory is explained with the help of a number of examples of typical problems.

I. INTRODUCTION

Generating functions provide a new tool in the representation theory of continuous and discrete groups which is in many aspects superior to existing techniques. Its distinguishing feature is that it solves an infinity of problems of a given type at the same time. Thus a generating function for Clebsch-Gordan series of a group G is equivalent to a table of decomposition of all direct products of all irreducible representations of the group.

The first particular generating function for group representations was written in 1897 by T. Molien[1]. For many years the generating functions were used only in the context of discrete groups and for generation of invariants (identity representations)[2]. Generating functions for continuous groups are rather recent. At first they were again used only for a study of scalars[3]. The genuine versatility of this mathematical tool started to emerge only after first computations involving generating functions for representations different from the trivial one demonstrated examples for continuous[4] as well as finite groups[5].

In this paper we first list some typical problems solved using generating functions (Sec. II), then we describe with simple examples how generating functions provide answers to these problems (Sec.III). In Sec. IV we define the character generator, which is a special case of a generating function for characters, and derive a suitable formula for it. In Sec. V the generating function for characters of semisimple Lie groups and for finite groups is defined and its conversion into a generating function for representations is demonstrated. Sec. VI contains brief explanations of how the generating functions are set up for the relevant typical problems of Sec. II.

II. TYPICAL PROBLEMS SOLVED BY GENERATING FUNCTIONS

Any of the following problems is solved by providing a *single* generating function (all representations are of finite dimension):

1. Calculation of weights of all irreducible representations of a given semisimple Lie group G.

* Work supported in part by the National Research Council of Canada and by the Ministère de l'Education du Québec.

2. Determination of dimensions of all irreducible representations of G.
3. Reduction of all irreducible representations of a given semisimple Lie group G to representations of a given Lie subgroup H (Branching rules calculation).
4. Reduction of all irreducible representations of a given semisimple Lie group G to any representation of a given finite subgroup of G.
5. Decomposition of all tensor products of any two representations of a given semisimple Lie group G into a direct sum of irreducible representations.
6. Multiplicities of polynomial G-tensors of all degrees. Components of such a tensor span a representation space for a given group G (finite or continuous) and are homogeneous polynomials in components of another given G-tensor.
7. Calculation of all plethysms based on a given representation of a continuous or finite group G.
8. Structure of the universal enveloping algebra of any semisimple group G as a reducible representation space for G.
9. Integrity bases related to problems 3-8 above.

III. EXPLOITATION OF A GIVEN GENERATING FUNCTION

A derivation of a suitable expression of a generating function for a given problem is often a lengthy computational task, which is of little interest to someone who wants to use a known generating function in order to get an asnwer to a specific question. Therefore we describe how generating functions can be used on some examples of simple cases before explaining their derivation.

A generating function is a rational expression whose numerators and denominators are polynomials in auxiliary variables with integer coefficients. When expanded into a power series in all variables it contains only terms with positive integer coefficients.

In general, generating functions are used in two different ways:
(i) As power series; then the required information is obtained from the presence of various terms and their coefficients (multiplicities).
(ii) Information about the corresponding integrity basis is inferred from a special rational form. Such a generating function has only positive terms in its numerators, and its denominators have the form of a product of terms (1-X), where X is a product of powers of auxiliary variables.

In order to explain how a generating function is used for problems of Sec.II, we consider now some simple examples.

1. The generating function for weights of SU(2) representations:

$$\frac{1}{(1-A\alpha)(1-A\alpha^{-1})} = 1 + A(\alpha+\alpha^{-1}) + A^2(\alpha^2+1+\alpha^{-2}) + \ldots \qquad (1)$$

An exponent of a power A^a refers to the irreducible representation of SU(2) of dimension $a+1$, i.e. of angular momentum $a/2$. The exponents of α in

$(\alpha^a+\alpha^{a-2}+\ldots+\alpha^{-a})$ constitute the weight system of that representation. The multiplicity of a weight b is the coefficient of α^b. A similar interpretation is easily given to the generating functions for SU(3) weights,

$$\frac{1}{(1-A\alpha\beta)(1-A\alpha^{-1}\beta)(1-B\alpha\beta^{-1})(1-B\alpha^{-1}\beta^{-1})}\left(\frac{1}{1-A\beta^{-2}}+\frac{B\beta^2}{1-B\beta^2}\right), \qquad (2)$$

and O(5) weights,

$$\frac{1}{(1-A\alpha^{-1})(1-A\beta^{-1})(1-B\alpha\beta)(1-B\alpha^{-1}\beta)}\left(\frac{1+B}{(1-B\alpha\beta^{-1})(1-B\alpha^{-1}\beta^{-1})}+\frac{(1+B)A\alpha}{(1-A\alpha)(1-B\alpha\beta^{-1})}+\frac{A\beta}{(1-A\alpha)(1-A\beta)}\right). \qquad (3)$$

2. Generating function for dimensions of irreducible rerpesentations of

$$SU(2): \frac{1}{(1-A)^2} = 1+2A+3A^2+\ldots \qquad (4)$$

$$SU(3): \frac{1}{(1-A)^2(1-B)^2}\left(\frac{1}{1-A}+\frac{B}{1-B}\right) = 1+3A+3B+6A^2+6B^2+8AB+\ldots \qquad (5)$$

$$O(5): \frac{1}{(1-A)^2(1-B)^2}\left(\frac{1+B}{(1-B)^2}+\frac{(1+B)A}{(1-A)(1-B)}+\frac{A}{(1-A)^2}\right) = 1+4A+5B+\ldots \qquad (6)$$

Obviously the exponents of the auxiliary variables label irreducible representations while the coefficients are the dimensions.

3. The generating function for the $SU(3) \supset SU(2)\times U(1)$ branching rules:

$$[(1-ATY)(1-AY^{-2})(1-BTY^{-1})(1-BY^2)]^{-1} = 1+A(TY+Y^{-2})+B(TY^{-1}+Y^2)+\ldots \qquad (7)$$

Here A^pB^q specifies the irreducible representation (p,q) of SU(3) of dimension $\frac{1}{2}(p+1)(q+1)(p+q+2)$, the power T^t specifies SU(2) representation of dimension t+1, and the exponent of Y provides the one dimensional representation of U(1). Alternately, the series (7) can be reorganized to the form

$$\sum_{k=0}^{\infty}(AB)^k + TY\sum_{j=0}^{\infty}A^{1+j}B^j+\ldots \qquad (8)$$

which indicates all the SU(3) representations containing a given representation of $SU(2)\times U(1)$.

4. Generating function for the branching rule $O(3) \supset$ tetrahedral group. For simplicity we write only the generating function which gives the multiplicity of the tetrahedral representation Γ_1 in O(3) representations.

$$\frac{1+A^6}{(1-A^3)(1-A^4)} = 1+A^3+A^4+2A^6+\ldots \qquad (9)$$

Presence of a term cA^a indicates that the (2a+1)-dimensional representation of O(3) contains the representation Γ_1 exactly c times.

5. The generating function for SU(2) Clebsch-Gordan series.

$$[(1-A_1A)(1-A_2A)(1-A_1A_2)]^{-1} = 1+A_1A+A_2A+A_1A_2(1+A^2)+\ldots \qquad (10)$$

A term $A_1^aA_2^bA^c$ indicates that the tensor product of two SU(2) representations of dimensions a+1 and b+1 contains the representation of dimension c+1. One interprets similarly the generating function for SU(3) Clebsch-Gordan series:

$$\frac{1}{(1-A_1A)(1-B_1B)(1-A_2A)(1-B_2B)(1-A_1B_2)(1-B_1A_2)}\left(\frac{1}{1-A_1A_2B}+\frac{B_1B_2A}{1-B_1B_2A}\right). \qquad (11)$$

Here A_1,B_1 and A_2,B_2 correspond to the first and second factor representations.

6. Multiplicities and degrees of the SU(2) tensors whose components are polynomials in components of a tensor transforming according to the 5-dimensional irreducible representation of SU(2):

$$\frac{1+U^3A^3}{(1-U^2)(1-U^3)(1-UA^2)(1-U^2A^2)} = 1+UA^2+U^2(1+A^2+A^4)+\ldots \qquad (12)$$

A term cU^aA^b denotes that there are c linearly independent tensors of degree a transforming according to the SU(2) representation of dimension 2b+1.

7. Generating function for plethysms based on SU(2) representations of dimensions 2 and 3. A plethysm is a term in a tensor product of representations with the symmetry of a given Young tableau. To calculate a plethysm means to decompose such a product into a direct sum. The two generating functions are

$$[(1-AL)(1-B)]^{-1} = 1+B+AL+\ldots \qquad (13)$$

$$[(1-AL)(1-A^2)(1-BL)(1-B^2)(1-C)]^{-1}(1+ABL) = 1+C+AL+BL+\ldots \qquad (14)$$

The coefficient of A^aB^b in (13) or $A^aB^bC^c$ in (14), a polynomial in L, gives the plethysms $(\frac{1}{2})^{\{ab\}}$ and $(1)^{\{abc\}}$ respectively. A power L^d indicates the presence of an SU(2) representation of dimension d+1 in (13) and of dimension 2d+1 in (14). Also

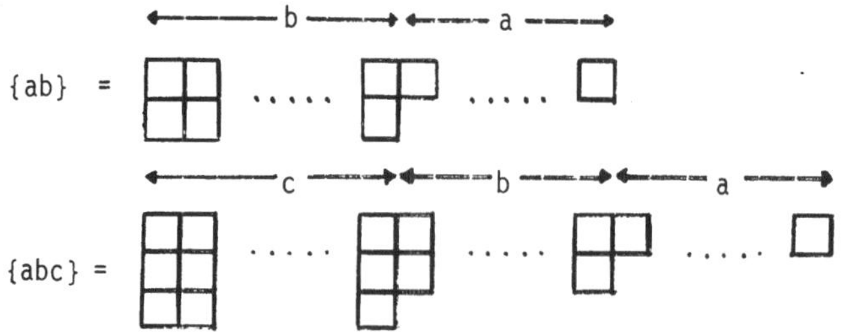

8. Generating functions for the structure of the universal enveloping algebra of SU(2) and SU(3). One has respectively

$$[(1-U^2)(1-UL)]^{-1} = 1+UL+U^2(1+L^2)+\ldots \tag{15}$$

$$\frac{1+U^2AB+U^4A^2B^2}{(1-U^2)(1-U^3)(1-UAB)(1-U^3A^3)(1-U^3B^3)} = 1+UAB+U^2(1+AB+A^2B^2)+\ldots \tag{16}$$

where the terms U^dL^a and $U^dA^pB^q$ indicate that the reducible space spanned by the elements of degree d contains an SU(2) irreducible subspace of dimension $2a+1$ and SU(3) irreducible subspace of dimension $\frac{1}{2}(p+1)(q+1)(p+q+2)$ respectively.

9. The integrity basis for tetrahedral scalars contained in SU(2) representations

$$[(1-A_1^6)(1-A_2^8)]^{-1}(1+A_3^{12}) = 1+A_1^6+A_2^8+A_1^{12}+A_3^{12}+\ldots \tag{17}$$

The existence of three elements I_1, I_2 and E_3 of the integrity basis for tetrahedral scalars is inferred from the left side of (17). The elements are of degree 6, 8 and 12 respectively in the components of SU(2) spinor. A term $A_1^{6a}A_2^{8b}A_3^{12\varepsilon}$ on the right side of (17) indicates that the SU(2) representation of dimension $3a+4b+6\varepsilon+1$ contains the tetrahedral scalar composed from the elements of the integrity basis as $I_1^aI_2^bE_3^\varepsilon$ where a and b are non-negative integers while $\varepsilon = 0$ or 1.

IV. CHARACTER GENERATOR

The character generator for a group G, apart from providing the characters, or weights, of all irreducible representations, turns out to be a convenient starting point for the calculation of the generating functions for representations which are the solutions of problems 3-9.

We define the character generator for irreducible representations of a semisimple Lie group G as

$$X_A(\alpha) = \sum_\lambda \chi_\lambda(\alpha)A^\lambda, \tag{18}$$

where the summation extends over all finite irreducible representations λ of G, $\chi_\lambda(\alpha)$ is the character, $A^\lambda = \prod_{i=1}^{\ell} A_i^{\lambda_i}$, where ℓ is the rank of G and the non-negative integers λ_i label the irreducible representation. To effect the sum in (18) and thereby to evaluate $X_A(\alpha)$ it is convenient to use Weyl's formula for the character in terms of the characteristic function $\xi_\lambda(\alpha)$

$$\chi_\lambda(\alpha) = \xi_\lambda(\alpha)/\Delta(\alpha); \qquad \Delta(\alpha) \equiv \xi_0(\alpha). \tag{19}$$

We introduce the generating function for the characteristics of irreducible representations

$$\Xi_A(\alpha) = \sum_\lambda \xi_\lambda(\alpha)A^\lambda. \tag{20}$$

Recall Weyl's formula for $\xi_\lambda(\alpha)$:

$$\xi_\lambda(\alpha) = \sum_S (-1)^S \Big[\prod_{j=1}^{\ell} \Big(\prod_{i=1}^{\ell} \alpha_i^{(S\gamma_j)_i}\Big)^{\lambda_i}\Big]\Big[\prod_{k=1}^{\ell} \alpha_k^{(SR)_k}\Big]. \tag{21}$$

The S-summation extends over the elements of the Weyl group, γ_j is the highest weight of the j'th fundamental representation, $(S\gamma_j)_i$ is the i'th component of the vector $S\gamma_j$; R is half the sum of the positive roots of G. It is important to notice that the exponents of α_i's depend linearly on the λ_j's; moreover a standard choice of scalar product in the Lie algebra makes the coefficients of the λ_j's integers. Hence, when (21) is substituted into (20) the λ-sums are infinite geometric series and can be simply evaluated. We find

$$\Xi_A(\alpha) = \sum_S (-1)^S \frac{\prod_{k=1}^{\ell} \alpha_k^{(SR)_k}}{\prod_{j=1}^{\ell} \Big(1 - A_j \prod_{i=1}^{\ell} \alpha_i^{(S\gamma_j)_i}\Big)}, \tag{22}$$

and finally

$$\chi_A(\alpha) = \Xi_A(\alpha)/\Delta(\alpha). \tag{23}$$

This computation of $\chi_A(\alpha)$ consists of straightforward elementary steps; however the details become lengthy for higher rank groups.

As an illustration consider the case of SU(2) group. One has

$$\xi_\lambda(\alpha) = \alpha^{\lambda+1} - \alpha^{-\lambda-1}, \qquad \Delta(\alpha) = \alpha - \alpha^{-1},$$

$$\chi_\lambda(\alpha) = \alpha^{\lambda} + \alpha^{\lambda-2} + \ldots + \alpha^{-\lambda} = \frac{\alpha^{\lambda+1} - \alpha^{-\lambda-1}}{\alpha - \alpha^{-1}},$$

$$\Xi_A(\alpha) = \sum_\lambda \xi_\lambda(\alpha) A^\lambda = \frac{\alpha}{1-A\alpha} - \frac{\alpha^{-1}}{1-A\alpha^{-1}} = \frac{\alpha - \alpha^{-1}}{(1-A\alpha)(1-A\alpha^{-1})},$$

$$\chi_A(\alpha) = \Xi_A(\alpha)/\Delta(\alpha) = \frac{1}{(1-A\alpha)(1-A\alpha^{-1})}.$$

The character generator $\chi_A(\alpha)$ solves the problem 1 of Sec. II (and also the problem 2 after all the variables α are set equal to unity). The character generator is a particular case of the generating function for characters which we discuss next.

V. CONVERSION OF A GENERATING FUNCTION FOR CHARACTERS INTO A GENERATING FUNCTION FOR REPRESENTATIONS

In each of problems 3-8 one can write down fairly easily a generating function $f(U,\alpha)$ for weights. The variables α carry the weights; the information carried by the variables U varies from problem to problem. In principle, $f(U,\alpha)$ can be expanded in characters of the group in question:

$$f(U,\alpha) = \sum_n U^n \sum_\lambda m_{n\lambda} \chi_\lambda(\alpha). \tag{24}$$

The multiplicities $m_{n\lambda}$ are usually difficult to calculate directly from (24). It is much easier to get the $m_{n\lambda}$ from the corresponding generating function for irreducible representations:

$$F(U,A) = \sum_n U^n \sum_\lambda m_{n\lambda} A^\lambda. \tag{25}$$

Moreover the form of F (a rational function of U and A of rather special type) contains information about the integrity basis for the corresponding tensors.

Our problem here is to convert $f(U,\alpha)$ of (24) into $F(U,A)$ of (25). How to find $f(U,\alpha)$ is explained in Sec. VI. Assuming G to be a semisimple Lie group, the first step is to multiply (24) by $\Delta(\alpha)$,

$$\Delta(\alpha)f(U,\alpha) = \sum_n U^n \sum_\lambda m_{n\lambda} \xi_\lambda(\alpha); \tag{26}$$

$m_{n\lambda}$ is just the coefficient of the lowest weight term of the characteristic which has the form

$$\prod_{i=1}^{\ell} \alpha_i^{-M_i(\lambda)} ;$$

the exponents $M_i(\lambda)$ are linear functions of the λ_i's with non-negative integer coefficients. Hence to find $m_{n\lambda}$, one has only to multiply (26) by

$$\prod_{i=1}^{\ell} \alpha_i^{M_i(\lambda)-1}$$

and take residues with respect to the α_i's for poles inside the unit circles (assuming $|U| < 1$). To keep track simultaneously of all irreducible representations we also multiply (26) by

$$\prod_{j=1}^{\ell} A_j^{\lambda_j}$$

and sum over λ_j's from 0 to ∞ before taking the α-residues. The infinite λ-sums are geometric series. The result is a rational expression for $F(U,A)$, which now can be expanded in powers of the auxiliary variables U and A to determine the $m_{n\lambda}$.

For a finite group G there is a similar procedure. The generating function for characters is now

$$f_s(U) = \sum_n U^n \sum_r m_{nr} \chi_{rs}; \tag{27}$$

r refers to the r'th irreducible representation Γ_r, s refers to the s'th class of elements. We want to convert $f_s(U)$ into the corresponding generating function $F_r(U)$ defined by

$$F_r(U) = \sum_n U^n m_{nr} \tag{28}$$

For this we use the orthogonality of characters and get

$$F_r(U) = \frac{1}{N}\sum_s N_s f_s(U)\chi^*_{rs}. \qquad (29)$$

Here N is the order of G, N_s is the number of elements in the class s and * denotes complex conjugation.

VI. GENERATING FUNCTIONS FOR CHARACTERS FOR TYPICAL PROBLEMS

The remaining task is to write down the generating functions $f(U,\alpha)$ and $f_s(U)$ for the problems 3-8.

3. To calculate the generating functions for irreducible representations of a Lie subgroup H contained in irreducible representations of a group G, one starts with the character generator $X_A(\alpha)$ of G. By a straightforward projection one substitutes for the variables α in terms of (possibly fewer) new variables β which carry the subgroup weights; the character generator $X_A(\alpha)$ thus becomes the generating function $f(A,\beta)$ for subgroup characters.

4. To find the generating function for irreducible representations of a finite group H contained in a Lie group G, we start again from the character generator of G. For each class s of H, one substitutes for the variables the numerical values corresponding to that class. Then the character generator $X_A(\alpha)$ becomes the generating function $f_s(A)$ for subgroup characters.

5. The Clebsch-Gordan series problem is a particular case of problem 3. The group is now GXG and the subgroup is G.

6. In this problem we are given a representation λ of G. When G is a semisimple Lie group the generating function for characters is

$$f(U,\alpha) = [\prod_w(1-U\alpha^w)]^{-1}. \qquad (30)$$

Here the product is over the weights w of λ, and α^w means $\prod_i \alpha_i^{w_i}$, where w_i are the integer components of the weight w. When G is finite

$$f_s(U) = [\det(1-UA_s)]^{-1}, \qquad (31)$$

where A_s is the matrix which represents an element of the class s in the representation Γ .

7. This problem is an application of 3 or 4. To construct the plethysm $(\lambda)^{\{a\}}$, where λ is an n-dimensional representation of a group G, one considers the branching rule problem for $U(n) \supset G$.

8. This problem is a special case of problem 6, where the starting representation λ is the adjoint representation of the Lie group G.

REFERENCES

1. T. Molien, Sitzungsber. König. Preuss. Akad. Wiss. 1152 (1897).

2. B. Meyer, Can.J.Math. 6, 135 (1954); A.G. McLellan, J.Phys. C7, 3326 (1974); N.J.A. Sloane, Amer.Math.Month. 84, 82 (1977); M.W. Jaric, J.L.Birman, J.Math. Phys. 18, 1456, 1459, 2085 (1977).

3. J. Igusa, Amer.J.Math. 89, 817 (1967); B.R. Judd, W. Miller Jr., J. Patera, P. Winternitz, J.Math.Phys. 15, 1787 (1974); W.McKay, J.Patera, R.T.Sharp, Comp. Phys.Comm. 10, 1 (1975); R.T. Sharp, J.Math.Phys. 16, 2050 (1975); C.Quesne, J. Math.Phys. 17, 1452 (1976), 18, 1210 (1977); R.P. Stanley, "Combinatorics and Invariant Theory", M.I.T. preprint.

4. R. Gaskell, A. Peccia, R.T. Sharp, J.Math.Phys. 19, 727 (1978).

5. J.Patera, R.T.Sharp, P.Winternitz, J.Math.Phys. 19, xxx (1978); P.Desmier, R.T. Sharp, J.Math.Phys. 19, xxx (1978); J.Patera, R.T. Sharp "Generating Function Techniques Pertinent to Spectroscopy and Crystal Physics", Proceedings of the NATO Advanced Study Institute, St.Francis Xavier University, Antigonish, Nova Scotia, Canada, August 1978 (to be published by Plenum Corp.).

6. Explicit projection matrices for a large number of group-subgroup pairs are foundin Tab.IV (pp. 260-262) of the article by W.McKay, J.Patera, D.Sankoff in Computers in Nonassociative Rings and Algebras, ed. R.E.Beck and B.Kolman, Academic Press, N.Y. 1977.

PHASE SPACE REPRESENTATIONS OF THE POINCARÉ GROUP & THEIR APPLICATIONS TO RELATIVISTIC PARTICLE DYNAMICS

S. Twareque Ali
Department of Mathematics
University of Prince Edward Island
Charlottetown, P.E.I., Canada

E. Prugovečki
Department of Mathematics
University of Toronto
Toronto, Ontario, Canada

Phase space representations of quantum mechanics are extremely useful, both theoretically - in understanding the relationship of quantum to classical mechanics- as well as computationally [c.f., e.g., Refs. 1 and 2 and related work cited therein]. We study, therefore, how far phase space ideas - both classical and quantum - can be viewed relativistically, and in particular, if one can overcome thereby some of the traditional problems of relativistic quantum mechanics. In this regard we construct here some representations of the Poincaré group $P_{\uparrow}^{+} = \mathbb{R}^4 \circledS SL(2,\mathbb{C})$ on Hilbert spaces of functions of the phase space variables $\underline{q}$ and $\underline{p}$.

Let $L^2(\Gamma)$ be the Hilbert space of all complex-valued, square integrable (w.r.t. $d^3q\, d^3p$) functions f of $\underline{q}$ and $\underline{p}$. Let $\mathcal{K}^j$ be a $2j+1$ dimensional spinor space ($j=0, 1/2, 1, \ldots$). The following is a (reducible) unitary representation of $P_{\uparrow}^{+}$ on $\mathcal{K}^j \otimes L^2(\Gamma)$, corresponding to a classical particle of mass m and spin-j:

$$(U(a,\Lambda)f)(\underline{q},\underline{p}) = \exp\left[-\frac{im^2c^2}{\hbar}(\Lambda^{-1})^0{}_\nu(\underline{q}-a)/(\Lambda^{-1})^0{}_\nu p^\nu\right] \times L^j(\Lambda^{-1}_p \Lambda \Lambda_{\Lambda^{-1}p}) f(\{a,\Lambda\}^{-1}(\underline{q},\underline{p})), \tag{1}$$

$\{a,\Lambda\} \in P_{\uparrow}^{+}$, U its unitary operator representative and L^j is a unitary irreducible representation of $SU(2)$ on $\mathcal{K}^j$. The representation in Eq. (1) can be canonically induced [3] from a unitary irreducible representation of the subgroup $T \otimes SU(2)$ of $P_{\uparrow}^{+}$ (T= time translations). Let

$$f(\underline{q},\underline{p},t) = (U((-ct,\underline{0}),I)f)(\underline{q},\underline{p}) \quad \text{and} \quad \rho = \|f(\underline{q},\underline{p},t)\|^2.$$

Then,

$$p_\mu \hat{P}^\mu \rho(\underline{q},\underline{p},t) = 0, \qquad \hat{P}^\mu = i\hbar \frac{\partial}{\partial q_\mu}. \tag{2}$$

The current, $j_\mu(q) = \int p_\mu \rho(\underline{q},\underline{p},t)\frac{d^3p}{p_0}$ satisfies the continuity equation [cf. Ref. 3]:

$$\partial_\mu j^\mu = 0. \tag{3}$$

The quantum mechanical representations (mass m, spin-j) are also obtained on $\mathcal{K}^j \otimes L^2(\Gamma)$ as

$$\begin{aligned}(U(a,\Lambda)f)(\underline{q},\underline{p}) &= \exp\left[-\frac{im^2c^2}{\hbar}(\Lambda^{-1})^0{}_\nu(\underline{q}-a)/(\Lambda^{-1})^0{}_\nu p^\nu\right] \\ &\times\left\{\exp\left[-\frac{im}{\hbar}(\Lambda^{-1})^0{}_\nu(\underline{q}-a)/(\Lambda^{-1})^0{}_\nu p^\nu\right](H-H_c) \right.\\ &\left.\times L^j(\Lambda^{-1}_p \Lambda \Lambda_{\Lambda^{-1}p})f\right\}(\{a,\Lambda\}^{-1}(\underline{q},\underline{p})),\end{aligned} \tag{4}$$

where, $H = c[\hat{\underline{P}}^2 + m^2c^2]$, $H_c = \frac{c}{p_0}[\underline{p}\cdot\hat{\underline{P}} + m^2c^2]$. This time $f(\underline{q},\underline{p},t)$ satisifes the Klein-Gordon equation

$$\left(\Box_q - \frac{m^2c^2}{\hbar^2}\right)f(\underline{q},\underline{p},t) = 0, \qquad \Box_q = -\frac{1}{c^2}\frac{\partial^2}{\partial t^2} + \underline{\nabla}_q^2. \tag{5}$$

On certain irreducible subspaces the current conservation law of Eq. (3) is again satisfied - contrast the usual theory where no such positive probability density $\|f\|^2$, with positive energy and satisfying current conservation, exists. These representations have been used to develop a completely consistent description of a relativistic particle interacting with an electromagnetic field [cf. Refs. 4 and 5,& related papers cited therein]. The dynamics is non-local and has to be understood in the sense of stochastic phase spaces [4]. It is gauge invariant and can be formulated in a manifestly covariant way.

The projection $\mathbb{P}_e$ onto the irreducible subspaces of the representation in Eq. (4) is obtained as

$$(\mathbb{P}_e f)(\underline{q},\underline{p}) = h^{-3/2}\int_\Gamma \overline{e(\underline{q}',\underline{p}')}\,(U^*((0,\underline{q}),\Lambda_p)f)(\underline{q}',\underline{p}')\,d^3\underline{q}'\,d^3\underline{p}' , \tag{6}$$

where e is a complex valued function satisfying

$$\int_\Gamma |e(\underline{q},\underline{p})|^2\, d^3\underline{q}\, d^3\underline{p} = mc , \qquad e(R\underline{q}, R\underline{p}) = e(\underline{q},\underline{p}), \tag{7}$$

(for all $R \in SU(2)$), and the reproducibility condition

$$\begin{aligned} e_{\underline{q},\underline{p}}(\underline{q}',\underline{p}') &= \int_\Gamma \overline{e_{\underline{q}',\underline{p}'}(\underline{q}'',\underline{p}'')}\, e_{\underline{q},\underline{p}}(\underline{q}'',\underline{p}'')\, d^3\underline{q}''\, d^3\underline{p}'' \\ &= \exp\left[-\frac{i m^2c^2}{\hbar^2}(\Lambda_p^{-1})^0_k(\underline{q}'-\underline{q})^k/(\Lambda_p^{-1})^0_\nu p'^\nu\right] e(\{(0,\underline{q}),\Lambda_p\}^{-1}(\underline{q}',\underline{p}')). \end{aligned} \tag{8}$$

If $\mathcal{K}^j \otimes L^2(\Gamma_e) = \mathbb{P}_e[\mathcal{K}^j \otimes L^2(\Gamma)]$, then $\mathcal{K}^j \otimes L^2(\Gamma_e)$ carries a unitary irreducible representation of $\mathcal{P}_\uparrow^+$ which is equivalent [3] to the standard (Wigner) representation $\tilde{U}$ on $\mathcal{K}^j \otimes L^2(\mathbb{R}^3, \frac{d^3k}{k_0})$ for mass m and spin-j:

$$(\tilde{U}(a,\Lambda)\psi)(k) = \exp\left[-\frac{i}{\hbar}ka\right] L^j(\Lambda_k^{-1}\Lambda\Lambda_{\Lambda^{-1}k})\,\psi(\Lambda^{-1}k). \tag{9}$$

The fact that Eq. (1) corresponds to a classical representation is to be inferred from its associated canonical system of imprimitivity [3]. Furthermore, Eq. (2) is a relativistic generalization of the Liouville equation in classical statistical mechanics.

The quantum representation in Eq. (4) does not admit a canonical system of imprimitivity. However on the subspaces $\mathcal{K}^j \otimes L^2(\Gamma_e)$ the normalized positive operator valued measure

$$a(E) = \int_E |e_{\underline{q},\underline{p}}\rangle\langle e_{\underline{q},\underline{p}}|\, d^3\underline{q}\, d^3\underline{p} , \qquad a(\Gamma) = I , \tag{10}$$

does define a generalized system of imprimitivity in the sense of Ref. 6, which leads in turn to the stochastic interpretation of the phase space Γ in the quantum case.

REFERENCES

1. S. T. Ali and E. Prugovečki, Physica 89A, 501 (1977).

2. S. T. Ali and E. Prugovečki, International J. Theor. Phys. 16, 689 (1977).

3. S. T. Ali, 'On some representations of the Poincaré group on phase space', University of Toronto preprint (1978).

4. E. Prugovečki, J. Math. Phys. 19, (1978), to appear.

5. E. Prugovečki, Phys. Rev. D18, (1978), to appear.

6. S. T. Ali and E. Prugovečki, J. Math. Phys. 18, 219 (1977).

ON SUBGROUPS OF PHYSICAL SYMMETRY GROUPS AND THEIR INVARIANT "ELECTROMAGNETIC" FIELDS AND POTENTIALS

J. BECKERS,
Physique Théorique et Mathématique, Institut de Physique,
Université de Liège, Liège, Belgique

Recent studies have determined all the maximal subalgebras of the conformal algebra when conjugacy classes are studied under conformal transformations(1) or under Poincaré transformations(2). These maximal subalgebras and other non-maximal ones have been studied(3)(4) as parts of a general program for subalgebra structure analyses.

Fields and potentials, invariant under one of the above substructures, have been discussed in recent works(2)(5)(6)(7)(8)(9).

In the context of differential geometry, the determination of 2-forms (fields F) and 1-forms (potentials A) invariant under some Lie groups of infinitesimal transformations is equivalent to the vanishing of the Lie derivatives (L) of these differential forms with respect to the vector fields (X) induced by the one-parameter subgroups. The corresponding formulae are :

$$L_X F = 0 \quad \leftrightarrow \quad F_{\mu\nu,\alpha}\xi^\alpha + F_{\mu\alpha}\xi^\alpha{}_{,\nu} + F_{\alpha\nu}\xi^\alpha{}_{,\mu} = 0 \qquad (1)$$

and

$$L_X A = 0 \quad \leftrightarrow \quad A_{\mu,\alpha}\xi^\alpha + A_\alpha\xi^\alpha{}_{,\mu} = 0 \qquad (2)$$

where, for a general infinitesimal conformal transformation

$$x^\mu \to x'^\mu = (\delta^\mu{}_\nu + \omega^\mu{}_\nu)x^\nu + a^\mu + \rho x^\mu + 2x^\mu(c\cdot x) - c^\mu(x)^2 \qquad (3)$$

$$= x^\mu + \xi^\mu , \quad (\mu=0,1,2,3), \qquad (4)$$

the vector fields X are defined by

$$X = \xi^\mu \partial_\mu . \qquad (5)$$

All the conformal substructures including the Poincaré- and E(3)-ones can now be used in order to determine their invariant F- and A-tensors. The results(11) have to be discussed in comparison with other studies(2)(5)(6)(8)(10) when Maxwell theory is taken into account.

REFERENCES :

(1) J. PATERA et al., Montreal Preprint CRM-697 (1977).
(2) J. BECKERS et al., Journ.Math.Phys. (September 1978).
(3) H. BACRY, Ph. COMBE & P. SORBA, Rep.Math.Phys. 5,145,361 (1974).
(4) J. PATERA et al., Journ.Math.Phys. 15,1378,1932 (1974).
(5) H. BACRY, Ph. COMBE & J.L. RICHARD, Nuov.Cim. 67A,267 (1970).
(6) Ph. COMBE & P. SORBA, Physica 80A,271 (1975).
(7) J. BECKERS & G. COMTE, Bull.Soc.Roy.Sc.Liège 45,279 (1976).
(8) J. BECKERS et al., Journ.Math.Phys. 18,72 (1977).
(9) J. BECKERS & M. JAMINON, Physica A (November 1978).
(10) J. BECKERS, M. JAMINON & J. SERPE, to be published (1979).
(11) J. BECKERS & G. COMTE, to be published (1979).

COVARIANT OBSERVABLES AND INSTRUMENTS *)

U. CATTANEO

Institut de Physique, Université de Neuchâtel
CH-2000 Neuchâtel (Switzerland)

The axiomatic approach to statistical physical theories proposed by Davies and Lewis [1] is sufficient to describe all the models so far considered for quantum and classical probability theories. For instance, the suggested mathematical framework can be realized by taking as state space the real Banach space $\mathcal{T}(\mathcal{H})_a$ of all self-adjoint trace class operators in a separable complex Hilbert space $\mathcal{H}$, with the trace norm. The set of all positive trace class operators in $\mathcal{H}$ is then a cone $\mathcal{T}(\mathcal{H})_a^+$ generating $\mathcal{T}(\mathcal{H})_a$ and the <u>states</u> are the elements of $\mathcal{T}(\mathcal{H})_a^+$ of trace one, namely, they are the states of the conventional description of an irreducible quantum mechanical system without superselection rules. In this model, the observables and instruments of Davies and Lewis are defined as follows. Let $\mathcal{B}_X$ be a σ-field of subsets of a set X, let $\mathcal{L}(\mathcal{H})$ be the set of all bounded (linear) operators in $\mathcal{H}$, and let $\mathcal{L}(\mathcal{T}(\mathcal{H})_a)$ be the set of all bounded operators in $\mathcal{T}(\mathcal{H})_a$. An <u>observable</u> on X acting in $\mathcal{H}$ is a measure $M:\mathcal{B}_X \to \mathcal{L}(\mathcal{H})$ which is countably additive in the ultraweak topology and is such that, for each $B \in \mathcal{B}_X$, M(B) is a positive operator satisfying $M(B) \leq M(X) = Id_{\mathcal{H}}$. If M is projection-valued, we say that it is a <u>decision observable</u> [2]. An <u>instrument</u> on X acting in $\mathcal{H}$ is a measure $\mathcal{E}: \mathcal{B}_X \to \mathcal{L}(\mathcal{T}(\mathcal{H})_a)$ which is countably additive in the strong operator topology, satisfies $Tr(\mathcal{E}(X)\rho) = Tr(\rho)$ for all $\rho \in \mathcal{T}(\mathcal{H})_a$, and is such that, for each $B \in \mathcal{B}_X$, $\mathcal{E}(B)$ maps $\mathcal{T}(\mathcal{H})_a^+$ into itself. Given $\mathcal{E}$, there exists a unique observable $M_{\mathcal{E}}$ on X acting in $\mathcal{H}$ which satisfies

$$Tr(M_{\mathcal{E}}(B)\rho) = Tr(\mathcal{E}(B)\rho)$$

for all $B \in \mathcal{B}_X$ and all $\rho \in \mathcal{T}(\mathcal{H})_a$.

The definition of an observable just given is different from the usual one in two respects: firstly, since the value space X is not necessarily a subset of $\mathbb{R}$ and secondly, since an observable is not necessarily a projection-valued measure (as in the case of customary observables, via the spectral theorem). Using this definition, it is possible to construct a joint observable for position and momentum ([3], 3.4) which is obviously not a decision observable and can be interpreted as giving the probabilities of approximate simultaneous position and momentum measurements. The notion of an instrument gives a mathematical form to the physical idea of a measurement.

*) Supported by the Swiss National Science Foundation

Now let G be a topological group and let the value space X be a G-space. This situation occurs when the observables considered are symmetric under the action of G. The observable M (resp. the instrument $\mathcal{E}$) defined above is said to be G-covariant with respect to a strongly continuous unitary representation U of G on $\mathfrak{H}$ if

$$M(g.B) = U(g)M(B)U(g)^{-1}$$
$$(\text{resp. } \mathcal{E}(g.B) = U(g)\circ\mathcal{E}(B)\circ U(g)^{-1})$$

for all $g \in G$ and all $B \in \mathfrak{B}_X$.

Using Naimark's dilation theorem we can prove the following

Proposition. [4] Let $\mathfrak{H}$ be a separable complex Hilbert space, let G be a second countable locally compact group, let X be a countably generated Borel G-space with Borel structure $\mathfrak{B}_X$, and let M be an observable on X acting in $\mathfrak{H}$. Suppose that M is G-covariant with respect to a strongly continuous unitary representation U of G on $\mathfrak{H}$. Then there exist a separable complex Hilbert space $\mathfrak{H}_e$ containing $\mathfrak{H}$ as a closed subspace, a strongly continuous unitary representation U_e of G on $\mathfrak{H}_e$, and a decision observable M_e on X acting in $\mathfrak{H}_e$ which is G-covariant with respect to U_e satisfying

$$WU(g) = U_e(g)W \qquad \text{for all } g \in G,$$
$$WM(B) = M_e(B)W \qquad \text{for all } B \in \mathfrak{B}_X,$$

where W is the canonical injection of $\mathfrak{H}$ into $\mathfrak{H}_e$, such that the set

$$\{M_e(B)W\psi \mid B \in \mathfrak{B}_X \text{ and } \psi \in \mathfrak{H}\}$$

is total in $\mathfrak{H}_e$. The triple $(\mathfrak{H}_e, U_e, M_e)$ is unique up to unitary equivalence. ∎

This result and Mackey's Imprimitivity Theorem can be applied to give a classification of all G-covariant observables and instruments on transitive standard Borel G-spaces [5].

References

[1] E.B. Davies and J.T. Lewis: Commun.math.Phys. 17, 239-260 (1970).

[2] G. Ludwig: "Deutung des Begriffs "physikalische Theorie" und axiomatische Grundlegung der Hilbertraumstruktur der Quantenmechanik durch Hauptsätze des Messens", Lecture Notes in Physics 4. Berlin: Springer-Verlag 1970.

[3] E.B. Davies: "Quantum Theory of Open Systems". New York: Academic Press 1976.

[4] U. Cattaneo: "On Mackey's Imprimitivity Theorem" (preprint).

[5] U. Cattaneo: To appear.

QUANTUM SPIN SYSTEMS, FERMI SYSTEMS AND GROUP EXTENSIONS

Ph. COMBE* R. RODRIGUEZ*

M. SIRUGUE-COLLIN** M. SIRUGUE

C.N.R.S. - LUMINY - Case 907
Centre de Physique Théorique
F-13288 MARSEILLE CEDEX 2 (France)

In the framework of the Weyl quantization procedure we examine the structure of Fermi systems and quantum spin systems. First of all we study the central extension of the abelian group of the classical spin system and show that there exist only two such extensions which satisfy natural physical restrictions. Moreover we prove that both extensions lead to the same C^*-algebra up to an isomorphism which has a nice structure. Finally we show that Fermi operators and spin operators can be seen as quantized of functions on phase space which is actually a compact group.

Let Λ be a lattice whose cardinality is at most countable, $\{\sigma_i^x, \sigma_i^y, \sigma_i^z\}$ the 2x2 Pauli spin matrices on the lattice point i. Let

$$\Sigma(x,y) = i^{-|x\cap y|} \prod_{i\in x} \sigma_i^x \prod_{j\in y} \sigma_j^y \qquad (x,y \text{ finite subsets of } \Lambda).$$

These operators are unitary, of square one, and moreover satisfy the following relation

$$\Sigma(x,y)\,\Sigma(x',y') = i^{-|x\cap y| - |x'\cap y'| + |(x\Delta x')\cap(y\Delta y')|}\,(-1)^{|x'\cap y|}\,\Sigma(x\Delta x', y\Delta y')$$

where $|x|$ denotes the cardinality of x, and $x\Delta y$ the symmetric difference between x and y. If we denote by $\mathcal{F}_\Lambda$ the abelian group of finite subsets of

---

* U.E.R. Scientifique de Luminy, Université d'Aix-Marseille II

** Université de Provence, Centre Saint-Charles, Marseille

with the symmetric difference as group law, quantum spin algebra appears as the linear span of an unitary representation of the central extension of $\mathcal{P}_\Lambda \times \mathcal{P}_\Lambda$ by the multiplier [1]

$$\xi^{S}((x,y);(x',y')) = i^{-|x\cap y| - |x'\cap y'| + |(x \Delta x')\cap(y \Delta y')|}\,(-1)^{|x'\cap y|}$$

Conversely one can look at the other central extensions of $\mathcal{P}_\Lambda \times \mathcal{P}_\Lambda$ and if we notice that such an extension is completely characterized by the bicharacter of $\mathcal{P}_\Lambda \times \mathcal{P}_\Lambda$, $b_\xi((x,y);(x',y')) = \xi((x,y);(x',y'))\,\overline{\xi((x',y');(x,y))}$ the physically relevant ones are given by the

Theorem : [2]

There are only two bicharacters b of $\mathcal{P}_\Lambda \times \mathcal{P}_\Lambda$ which for every cardinality of have the following properties

i) Symmetry : $b((x',y'),(x,y)) = b((x,y),(x',y'))$;

ii) $b((x,y),(x,y)) = 1$;

iii) Non degeneracy, i.e. $b(xy,x'y') = 1 \quad \forall\ x',y' \in \mathcal{P}_\Lambda \Rightarrow x = y = \phi$

iv) $b((x,y),(x',y')) = b((y,x),(y',x'))$;

v) invariance under the finite permutation of points in Λ, they are

$$b^{S}((x,y),(x',y')) = (-1)^{|x\cap y'| + |x'\cap y|}$$

$$b^{F}((x,y),(x',y')) = (-1)^{|x\cap x'| + |y\cap y'| + (|x|+|y|)(|x'|+|y'|)}$$

If we construct on these two extensions the C^*-algebras which are the strict analogues of the C^*-algebra of canonical commutation relation which is built on the central extension of the abelian group R^{2n} by the multiplier $\xi((x,p),(x',p')) =$ $= e^{i\frac{\hbar}{2}(xp' - x'p)}$ we get respectively the U.H.F. algebra and the Clifford algebra of canonical anticommutation relations.

Consequently we can quantize "à la Weyl" functions on the "classical phase space" $\mathcal{S}_\Lambda \times \mathcal{S}_\Lambda$ which is a compact space ($\mathcal{S}_\Lambda \times \mathcal{S}_\Lambda$ is the dual of $\mathcal{P}_\Lambda \times \mathcal{P}_\Lambda$), according to the following theorem.

Theorem : [3]

Let $\mathcal{B}_\Lambda$ be the algebra of functions on $\mathcal{S}_\Lambda \times \mathcal{S}_\Lambda$ which are Fourier transforms of ℓ_1 functions on $\mathcal{P}_\Lambda \times \mathcal{P}_\Lambda$, then

$$Q(f) = \sum_{x,y\in\mathcal{P}_\Lambda} \{\mathcal{F}f\}(x,y)\ \delta_{x,y}$$

where $\mathcal{F}f$ is the Fourier transform of f, defines an element of the

U.H.F. algebra (resp. Clifford algebra) and $\delta_{x,y}$ are the generators of the U.H.F. algebra (resp. Clifford algebra). Moreover

$$\mathcal{G}(f)^* = \mathcal{G}(f^*) \qquad f^*(x,y) = \overline{f(x,y)}$$

$$\mathcal{G}(f_1)\,\mathcal{G}(f_2) = \mathcal{G}(f_1 \circ f_2) \qquad f_1 \circ f_2 = \mathcal{F}(\mathcal{F}f_1 \times \mathcal{F}f_2)$$

and
$$g_1 \times g_2(x,y) = \sum_{x',y' \in \mathcal{P}_\Lambda} \xi((x,y);(x',y'))\, g_1(x',y')\, g_2(x \Delta x', y \Delta y')$$

$$\mathcal{G}(1) = 1$$

$$\mathcal{G}(\chi_{x,y}) = \delta_{x,y} \qquad x, y \in \mathcal{P}_\Lambda$$

$$\chi_{x,y}(x',y') = (-1)^{|x \cap y'| + |x' \cap y|}$$

The well known isomorphism between U.H.F. algebra and Clifford algebra is given by an automorphism of the "classical phase space" which preserves the Haar measure.

REFERENCES

[1] MACKEY G.W., Acta Math. 99, 265 (1958).

[2] COMBE Ph., RODRIGUEZ R., SIRUGUE-COLLIN M., and M. SIRUGUE, An uniqueness Theorem for Anticommutation Relations and Commutation Relations of Quantum Spin Systems. To appear in Comm. math. Phys. .

[3] COMBE Ph., RODRIGUEZ R., SIRUGUE-COLLIN M., and M. SIRUGUE, On the Quantization of Spin Systems and Fermi Systems. To appear in J.M.P. .

REVIEW OF RESEARCH ON PRESYMMETRY

Hans EKSTEIN

Centre National de la Recherche Scientifique
Centre de Physique Théorique, MARSEILLE

The motivation for this inquiry is the absence of a well-founded kinematics in relativistic quantum mechanics in contrast to Galilean quantum mechanics with its canonical commutation relations. A better understanding of the geometric origin of the latter promises to suggest kinematic principles in relativistic theory, thereby restricting the field of possible specific assumptions - much as Einstein's principle of equivalence narrowed the set of possible theories of gravitation.

The first step is a more explicit physical interpretation of theory. Another procedure is defined as a pair $(b,\{P_n\})$ where b is a collection of blueprints for making observation instruments, and $\{P_n\}$ a set of space-time points which are to coincide with marks on the instruments, times at which buttons are to be pressed, etc. A motion is defined as a permutation : $(b,\{P_n\}) \rightarrow (b,\{gP_n\})$ of the set of observation procedures where $g: P \rightarrow gp$ is a transformation of space-time.

The theory of presymmetry introduces a new structure: the algebra of observation procedures. Strictly speaking, an observable (e.g. a self-adjoint operator) is the mathematical image of an equivalence class of observation procedures, and presymmetry takes these equivalence classes apart. Physical motions of measuring instruments induce automorphisms of the algebra of observation procedures, and their group includes accelerations of all degrees. The analysis of group realizations by automorphisms leads to Newton's law for the two cases: Galilean quantum mechanics and classical relativistic field theory. Newton's second law is taken to be the following testable statement: the knowledge of expectation values of the elements of the canonical subalgebra at one instant (generated by the momenta, positions and spins for quantum mechanics or by fields and their first time-derivatives for classical field theory) provides full predictive power.

The canonical commutation relations $[q_i,p_k] = i\ \delta_{i_k}$ are obtained as a theorem for multiparticle systems.

A strengthened form of the Einstein equivalence principle is obtained by applying presymmetry principles to the acceleration group.

Work on a presymmetry foundation of relativistic quantum mechanics is in progress.

The philosophical implications of presymmetry concerning the uniqueness or conventionality of space-time are explored in a paper to appear (Ref. 7).

References

1) H. Ekstein, Phys. Rev. 184, 1315 (1969)

2) Y. Avishai, H. Ekstein and J. E. Moyal, J. Math. Phys. 13, 1139 (1972)

3) H. Ekstein, Phys. Rev. D 5, 4 (1971)

4) Y. Avishai, H. Ekstein, Foundational Phys. 2, 257 (1972)

5) Y. Avishai and H. Ekstein, Phys. Rev. D 7, 983 (1973)

6) Y. Avishai and H. Ekstein, Comm. Math. Phys. 37, 193 (1974)

7) H. Ekstein, Is Physical Space Unique or Optional? to appear in Proceedings of the Boston Colloquium in the Philosophy of Science.

ON THE FOUR EUCLIDEAN CONFORMAL GROUP STRUCTURE OF THE STURMIAN OPERATOR

J.P. GAZEAU
Laboratoire de Chimie Physique
11, rue Pierre et Marie Curie
75 005 PARIS - FRANCE -

The Sturmian method (1) or equivalently the "symmetrized kernel" method (2) consists in replacing the eigenvalue problem for the interacting particle Hamiltonian : $(H_0 + V - E)\psi = 0$, $E < 0$, $\psi \in L^2(\mathbb{R}^3)$, by the spectral problem for the so-called Sturmian operator $\mathcal{N} \equiv \mathcal{N}(E)$, $(\mathcal{N} - \nu)\phi = 0$, ϕ in some Hilbert space $\mathcal{H}_E$. The energy $E < 0$ is regarded as a parameter such that the eigenvalues n of $\mathcal{N}$ are compelled to verify the equation $n = n(E) = \nu(E)$, ν being some ratio of physical entities. This method is interesting because the operator $\mathcal{N}$, the classical equivalent of which is always the ratio $V^{-1}(H_0 - E)$, owns a discrete spectrum for a large class of interactions V.

The main result of this work (3) lies in the disclosure of the algebraic nature of the operator $\mathcal{N}^{-1}$, as a "linear superposition" of representation operators $\mathcal{T}(g)$, for g belonging to a so-called dynamical semi-group $G_< \subset G = SU^*(4) \simeq Spin(1,5)$, the four Euclidean Conformal Group. When restricted to Sp(1,1) $\subset$ spin(1,4) $\subset G_<$, the representation $\mathcal{T}$ is equivalent to one complementary series unitary irreducible representation (4).

Explicitely, there exists a $\mathcal{H}_E \simeq L^2(SU(2))$-valued distribution (4) T on G , with support in $G_<$ completely determined by V , such that

$$\mathcal{N}^{-1} = \int_{G_<} dT(g)\,\mathcal{T}(g)$$

(at least on a dense subset of $\mathcal{H}_E$)

The only hypotheses are the usual Galilean pseudoinvariance of V and the self adjointness of $\mathcal{N}^{-1}$.

The matrix elements of $\mathcal{T}$ with respect to two different basis of $\mathcal{H}_E$ have been calculated (5) as well as the "character" $\chi = \mathrm{Tr}\,\mathcal{T}$.

A large field of applications is allowed by this group theoretical understanding of the Sturmian operator : calculation of traces, norms, resolvents, transition matrix elements. An illustration is given by a new and effective approach to the bound state problem for a Yukawian potential and by a straightforward calculation of the successive terms of the Born Series for the H-atom in a perturbative electromagnetic field.

(1) M. ROTENBERG, Ann. Phys. (N.Y.) 19, 262 (1962).

(2) M. SCADRON, S. WEINBERG and J. WRIGHT, Phys. Rev. 135,B.202(1964). B. SIMON, Commun. Math. Phys. 21, 192 (1971).

(3) J.P. GAZEAU, Thesis Paris (1978).

(4) G. WARNER "Harmonic Analysis on Semi-simple Lie Groups" (Springer Verlag, Berlin (1972)).

(5) J.P. GAZEAU, J. Math. Phys. 19, 1041 (1978).

AN SL(2,$\mathbb{R}$) APPROACH TO SCHRÖDINGER SPECTRAL PROBLEM WITH PSEUDOSINGULAR POTENTIAL

J.P. Gazeau*, A. Ronveaux**, M.Cl. Dumont-Lepage**

* Laboratoire de Chimie Physique de l'Université de Paris VI, Rue Pierre et Marie Curie, 11, F-75005 Paris, France.

** Département de Physique, Facultés Universitaires Notre-Dame de la Paix, Rue de Bruxelles, 61, B-5000-Namur, Belgium.

The integral Schrödinger equation [1] for central potential can be reduced [2], using a "Fourier-Fock transformation", to the eigenvalue problem for the so-called sturmian operator η_ℓ^{-1}, acting on the Hilbert space $\mathcal{H}_\ell$ of the square integrable function $a(\alpha)$ $0 \leqslant \alpha \leqslant \pi$ with the measure $d\alpha \sin^{\ell+1}\alpha$.

For each ℓ, the eigenvalue equation for the bound state E is replaced by : $(I - \nu\eta_\ell^{-1})\, a = 0$ with

$$\nu = \frac{mg}{\sqrt{-2mE\hbar^2}} .$$

For the pseudosingular potential $V(r) = -gr^{-(s+2)}$, $-1 \leqslant s < 0$; $g > 0$, we show that $\eta_\ell^{-1} \equiv \eta_{s,\ell}^{-1}$ is a "linear superposition" of a local representation $T^{\ell+1}$ with multiplier of SL(2,$\mathbb{R}$) [3] :

$$\eta_{s\ell}^{-1} = \frac{1}{\Gamma(s+1)} \left(\frac{2\cdot\sqrt{-2mE}}{\hbar}\right)^{s+1} \int_0^{+\infty} d\beta\, \beta^s \int_0^{+\infty} dt\; T^{\ell+1}\,(g(\beta)\; d(t))$$

with $g(\beta) = \begin{pmatrix} 1+\beta & \beta \\ -\beta & 1-\beta \end{pmatrix}$ $\begin{pmatrix} e^{t/2} & 0 \\ 0 & e^{-t/2} \end{pmatrix} = d(t)$.

The operator $\eta_{s\ell}^{-1}$ is a compact selfadjoint operator. It can be approximated by finite rank operators and, in the sturmian Coulomb basis of the Gegenbauer polynomials, the bound states spectrum can be obtained very quickly [2] and therefore the (g,s) behaviour of the lowest states.

1. M. Rotenberg, Ann.Phys. (N.Y.), *19*, 262 (1962).
2. J.P. Gazeau, J.Math.Phys., *19*, 1041 (1978), and thesis (Paris VI).
3. W. Miller Jr., *Lie Theory and Special Functions*, Mathematics in Science and Engineering, Vol. 43, Academic Press, 1968.

Ergodic Theory in von Neumann Algebras.

Foundation of thermodynamics demands the existence of equilibrium states which are states invariant under time translation and any state should behave in the middle as one of these invariant states.
Two concepts were developed to describe such a situation: The concept of mean ergodicity (v. Neumann, 1932) and the concept of individual ergodicity (Birkhoff, 1931). They proved ergodic theorems for time translation on $L^2(X,\mu)$ resp. on $L^1(X,\mu)$ induced by some time translation on the phase space X.
Meanwhile, both theorems have been generalized in a considerable way. In modern quantum field theory andquantum statistical mechanics, the mean ergodic theorem for groups of *-automorphisms on v. Neumann algebras obtained by I.Kovacs and J.Szücs (1966) and the individual ergodic theorem for the group generated by one *-automorphism on a v.Neumann algebra (C.Lance, 1977) are of great interest. But in many situations(e.g. in the case of a quantum stochastic process) time development can not be described by *-automorphisms. Instead one uses semigroups of contractions.
Generalizations of both theorems to great classes of contraction semigroups on v.Neumann algebras are proved.

1. Kümmerer,B.;Nagel,R.:Mean Ergodic Semigroups on W^*-Algebras. To appear in Acta Sci. Math..
2. Kümmerer,B.:Mean Ergodic Semigroups of Contractions in W^*-Algebras. Group Theoretical Methods in Physics, Tübingen 1977. Lecture Notes in Physics 79, 479-481. Springer 1978.
3. Kümmerer,B.:A Non-Commutative Individual Ergodic Theorem. Inventiones math. 46, 139-145 (1978).

Burkhard Kümmerer
Mathematisches Institut d. Universität
Auf der Morgenstelle 10
74 Tübingen/Germany

SU(2) HARMONIC ANALYSIS AS A BASIS FOR QUANTIZATION

S. Malin
Department of Physics
Ben-Gurion University
Beer Sheba 84120, Israel
and
Department of Physics and Astronomy
Colgate University
Hamilton, New York 13346, U.S.A.

Abstract. This is a brief review of (i) Carmeli's method of formulation and quantization of fields in terms of functions over the group SU(2); (ii) Work by Barut, Carmeli and the present author on applications of the method to Maxwell's equations, Weyl's equation, Dirac's equation and the linearized equations of General Relativity.

Carmeli's method of formulation[1] and quantization[2] of fields using the group SU(2) is based on the following observation: consider the angular variables θ, ϕ in the usual spherical system of coordinates. The space $0\leq\theta<\pi$, $0\leq\phi<2\pi$ does not constitute the space of a group. If, however, one adds to them a third, superflous, angle ϕ_2, with the range $0<\phi_2<2\pi$, then the space of the three angles $\phi_1 = \pi/2-\phi$, θ, ϕ_2 is precisely the space of the group SU(2). The usual methods of the theory of compact groups can then be applied; the superflous angle ϕ_2 drops out of final expressions which correspond to observed physical quantities.

The application of this idea to Maxwell's Equations, proceeds as follows[1,2]: let us introduce the complex vector field $\vec{V} = \vec{E}+i\,\vec{B}$ and the functions

$$\eta_\pm = -1/\sqrt{2}\,(V_\phi \pm iV_\theta)e^{\pm i\phi_2}\,, \qquad \eta_0 = V_2 \tag{1}$$

The functions $\eta_\pm$, η_0 can be considered, for each value of the time t and the radius r as function of $u \in SU(2)$, with

$$u = \begin{pmatrix} \cos\theta/2\,\exp[i/2(\phi_1+\phi_2)] & i\sin\theta/2\,\exp[-i/2(\phi_1-\phi_2)] \\ i\sin\theta/2\,\exp[i/2(\phi_1-\phi_2)] & \cos\theta/2\,\exp[-i/2(\phi_1+\phi_2)] \end{pmatrix}. \tag{2}$$

It was shown by Carmeli that Maxwell's equations in free space are equivalent to the following set of equations for the η functions:

$$\begin{aligned} &1/\sqrt{2}\;\; 1/r(\partial/\partial r \pm \partial/\partial t)(r^2\eta_0) \mp K_\pm\eta_\mp = 0\,, \\ &(\pm\partial/\partial r + \partial/\partial t)(r\eta_\pm) + 1/\sqrt{2}\;\; K_\pm\eta_0 = 0\,, \end{aligned} \tag{3}$$

where the operators $K_\pm$ are defined by

$$K = e^{\mp i\phi_2}(\pm\cot\theta\,\partial/\partial\phi_2 + i\,\partial/\partial\theta \mp \mathrm{cosec}\theta\,\partial/\partial\phi_1)\,. \tag{4}$$

These operators, along with

$$K_3 = i\,\partial/\partial\phi_2\,, \tag{5}$$

are well known from the theory of representations of SU(2). They satisfy the following relations:

$$K_{\pm}T^{j}_{mn} = [(j \pm m + 1)(j \mp M)]^{\frac{1}{2}} T^{j}_{m\pm 1,n} , \qquad (6)$$

where $T^{j}_{mn}(u)$ are the matrix elements of the irreducible representation of weight j of the group SU(2).

The functions $\eta_{\pm}$ and η_0 can be expanded in the following way:

$$\eta_{\pm}(t,r,u) = \sum_{j=1}^{\infty} \sum_{n=-j}^{j} \alpha^{j}_{\pm 1,n}(t,r) T^{j}_{\pm 1,n}(u) ,$$
$$\eta_{0}(t,r,u) = \sum_{j=0}^{\infty} \sum_{n=-j}^{+j} \alpha^{j}_{0,n}(t,r) T^{j}_{0,n}(u) , \qquad (7)$$

where the coefficients are given by

$$(2j + 1)^{-1} \alpha^{j}_{\pm 1,n}(t,r) = \int \eta_{\pm}(t,r,u) T^{j}_{\pm 1,n}(u) du ,$$
$$(2j + 1)^{-1} \alpha^{j}_{0,n}(t,r) = \int \eta_{0}(t,r,u) T^{j}_{0,n}(u) du , \qquad (8)$$

and $du = (1/16)\pi^{-2} \sin\theta \, d\phi_1 d\theta \, d\phi_2$ is the invariant measure over SU(2), normalized so that $\int du = 1$.

Substituting expansions (7) in Eq. (3) one can obtain a partial differential equation for the $\alpha^{j}_{0,m}$ and express $\alpha^{j}_{\pm 1,m}$ in terms of $\alpha^{j}_{0,m}$, i.e.,

$$(\partial^2/\partial t^2 - \partial^2/\partial r^2)(r^2 \alpha^{j}_{0,m}) + j(j + 1)\alpha^{j}_{0,m} = 0 , \qquad (9)$$
$$j = 0,1,2,\ldots, ; \qquad m = -j,\ldots, + j ,$$

$$\alpha^{j}_{\mp 1,m} = \pm 1/\sqrt{2j(j+1)} \; 1/r \; [\partial/\partial r \pm \partial/\partial t](r^2 \alpha^{j}_{0,m}) , \qquad (10)$$
$$j = 1,2,\ldots; \qquad m = -j,\ldots, j .$$

We thus arrive at the conclusion that the functions $\alpha^{j}_{0,m}(t,r)$ determine $\alpha^{j}_{\mp 1,m}$ completely, through substitution in Eq. (10). The problem of solving Maxwell's equations reduces, therefore, to the solution of Eq. (9) for a single scalar complex function $\eta_0(t,r,u)$.

Denoting $\alpha^{j}_{m} \equiv r^2 \alpha^{j}_{0,m}$, Eq. (9) with α^{j}_{m} substituted for $r^2\alpha^{j}_{0,m}$ can be obtained from the Lagrangian density

$$= \sum_{j} w_j^{-1} [\partial\alpha^{*j}_{m}/\partial x^{\mu} \, \partial\alpha^{j}_{m}/\partial x_{\mu} - j(j+1)/r^2 \; a^{*j}_{m} \, \alpha^{j}_{m}]$$

where α^{*j}_{m} is the complex conjugate of α^{j}_{m} and $w_j = 2j(j+1)(2j+1)$. The α^{j}_{m}, α^{*j}_{m} and the canonical momenta $\pi^{j}_{m} \equiv \partial \;/\dot{\alpha}^{j}_{m} = w_j^{-1} \dot{\alpha}^{*j}_{m}$, $\pi^{*j}_{m} \equiv \partial \;/\partial\dot{\alpha}^{*j}_{m} = w_j^{-1} \dot{\alpha}^{j}_{m}$ are then assumed to satisfy the standard equal-time commutation relations. The commutation relations for the other field quantities follow from Eq. (10). This method of quantization is gauge-free.

When interaction with matter is considered it was shown by Aharonov and Carmi,[3] extending Aharonov and Bohm's classical result,[4] that the vector potential has a local physical meaning in the context of quantum mechanics. As a first step towards applying the present method to quantum electrodynamics the equation for the vector potential in terms of functions over the group SU(2) was derived by Barut and present author.[5]

The present method was also applied to (i) The formulation of both the Weyl and Dirac equations;[6] (ii) The problem of scattering of electromagnetic radiation;[7] the spherical analog of positive and negative helicities of plane waves was obtained; (iii) The linearized equations of General Relativity;[8] the analysis was carried out in the Newman-Penrose formalism and led to a gauge-free quantization of these equations. Further work is now in progress in collaboration with Professor M. Semon.

It is a pleasure to acknowledge the hospitality of Professor Kenneth Greisen and his colleagues at the Center for Radiophysics and Space Research, Cornell University, where this work was done.

References

1. M. Carmeli, J. Math. Phys. 10, 569 (1969).
2. M. Carmeli, Nuovo Cimento B67, 103 (1970).
3. Y. Aharonov and G. Carmi, Found. Phys. 4, 75 (1974); 3, 493 (1973).
4. Y. Aharonov and D. Bohm, Phys. Rev. 115, 485 (1959).
5. A.O. Barut and S. Malin, Found. Phys. 5, 375 (1975).
6. S. Malin, J. Math. Phys. 16, 679 (1975).
7. A.O. Barut, M. Carmeli and S. Malin, Ann. Phys. (NY) 77, 454 (1973).
8. S. Malin, Phys. Rev. D8, 2338 (1974).

Some new aspects of Cayley-Quantum Theory

Anton Schober, Technische Universität Berlin

An eight-dimensional real Clifford algebra is used for coordinatization of the most simple non-trivial sub - CROC's (Fig. 1) of a given CROC (canonical relatively orthocomplemented lattice). By such sub-CROC a projective plane is determined which at least contains seven points (Fig. 2). Specializing the Clifford algebra to the Cayley field these seven points are set into correspondence with the seven subalgebras generated by the real and one imaginary Cayley unit, respectively. (See fig. 3 for a multiplication scheme.) General pure quantums states are assumed to be sequences of Cayley numbers. The imaginary Cayley units operate on these states. By some subsidiary conditions on physical states it is achieved that the commutator algebra of the imaginary Cayley units represent the Lie algebra of the Lorentz group extended by the torus. Skew-hermiticity of the generators restricts physical states to be n-tuples of Cayley numbers, $2 \leq n \leq 8$. Specializing the Clifford algebra otherwise (see fig. 4), a representation of the SO(4) algebra is obtained in a similar way. In that case, no restriction on the length of the sequences representing quantum states occurs.

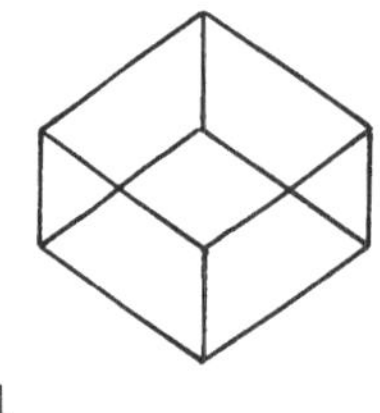

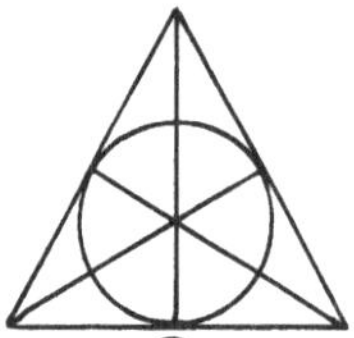

Fig. 1

Fig. 2

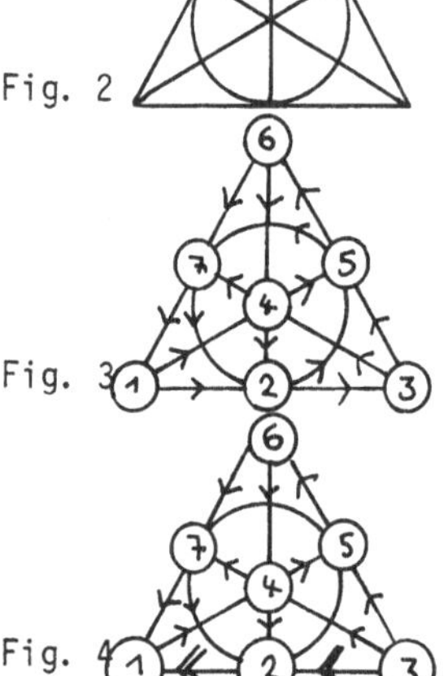

Fig. 3

Fig. 4

APPLICATION OF QUASICLASSICAL METHODS TO UNIAXIAL SPIN SYSTEMS

R.K.P.Zia and D.M.Kaplan, Virginia Polytechnic Institute, Blacksburg, VA., USA

The quasiclassical formalism, to be distinguished from semi-classical analyses, provides a fully quantum mechanical treatment of physical systems. Since Wigner's pioneering work[(1)] the formalism has been extended to include spin.[(2)] The most obvious interest is the possibility of a perturbative approach to the Heisenberg system about the 'classical limit' ($J=\infty$). Here we explore a similar possibility for the uniaxial spin system, the Ising model being the $J=\frac{1}{2}$ case.

The formalism involves mapping (a) the uniaxial spin operator $\underline{J}$ (e.g. diagonal matrix with elements $J, J-1, \ldots, -J$) to the monomial Jx, where $x \in [-1,1]$, (b) the identity operator to the constant 1, (c) powers of $\underline{J}$ to polynomials in x, and so, (d) an arbitrary operator $\underline{A}$ to a function of x, A(x). The trace of $\underline{A}$ is simply given by the integral of A(x) over x, multiplied by $J+\frac{1}{2}$ for normalization.

Thus, thermodynamic quantities of a spin J system are <u>explicit functions of J</u>, allowing the set up of a perturbation theory. For example the Gibbs free energy may be expressed as $F(J) = F(J_o) + (J-J_o)[\partial F(J_o)/\partial J] + \ldots$

Not surprisingly, $\partial F/\partial J$ involves correlation functions[(3)] (of a local cluster of spins, if the interaction is short ranged.) So, if the spin J_o system is known, information on others may be obtained via perturbation theory. In two dimensions, a well studied case is the Ising system. An analysis of the singular structure of F allows us to calculate the critical temperature (as a series in $J-\frac{1}{2}$) and explore universality of critical behaviour. Both aspects have been investigated to first order.[(3)] In particular, $T_c^{-1}\partial T_c/\partial J|_{J=\frac{1}{2}} \simeq 3.6$ and no universality violating amplitudes have been found. Higher order contributions, the $J=\infty$ limit, and implications for renormalization group analyses are being studied.

----

(1) E.P.Wigner, Phys. Rev., 40, 749 (1932).

(2) For Heisenberg spins, see D.M.Kaplan, Transp. Theory & Stat. Phys., 1,81 (1971). For uniaxial spins, see R.L.Bowden and D.M.Kaplan, Journ. Phys., A9,1655 (1976).

(3) R.K.P.Zia and D.M.Kaplan, to be published.

Mathematical remarks to Resonances and their Eigenfunctionals in Decay-Scattering systems, demonstrated by means of Friedrich's model.

H. Baumgärtel

Zentralinstitut für Mathematik und Mechanik
Akademie der Wissenschaften der DDR
Berlin, DDR.

Paper presented at the Integrative Conference on Group Theory and Mathematical Physics,
Austin, Texas, September 11-16, 1978.

Abstract

A finite-dimensional Friedrich's model with a condition of analyticity is considered as a typical decay-scattering system. In particular virtual poles (resonances) are characterized by an eigenvalue problem within the frame-work of Gelfand triples.

§ 1 Decay-Scattering systems. The terminus decay-scattering system is due to L.P. Horwitz and J.-P. Marchand [1]. It denotes "a suitable mathematical model which describes resonance and decay (two aspects of one and the same physical object) in a unified manner".

1.1. Scattering. Let H be a selfadjoint operator on a separable Hilbert space $\mathfrak{H}$ and let H be bounded below (H represents the Hamiltonian of a quantum-mechanical system, $\mathfrak{H}$ its state space). Further let $\mathfrak{A}$ be a linear abelian system of selfadjoint operators on $\mathfrak{H}$, let $1 \in \mathfrak{A}$ and let all $A \in \mathfrak{A}$ with $A \neq \gamma 1$ (γ real) be absolutely continuous ($\mathfrak{A}$ represents the system of "free" Hamiltonians). Each A is an a-priori possible channel Hamiltonian. Let $Q_A^{\pm}$ be the orthogonal projection onto the subspace of all $f \in \mathfrak{A}$ such that $W_{\pm}(H,A)f = s\text{-}\lim_{t \to \pm} e^{itH} e^{-itA} f$ exists. Let $Q_A^+ = Q_A^- = Q_A$. A is called a channel Hamiltonian and Q_A the corresponding channel projection if $Q_A \neq 0$. Hence the set of channels is at most denumerable: $A_1, A_2 \ldots$ As usual one forms the Hilbert space $\mathfrak{H}^o$ as the direct Hilbert sum of all channel subspaces $Q_{A_\rho} \mathfrak{H}$, the "free Hamiltonian" A^o as the direct sum of all channel Hamiltonians A_ρ and the wave operator $W_{\pm}(H, A^o)$ as the compositum of all channel wave operators $W_{\pm}(H, A_\rho) Q_{A_\rho}$. The scattering operator S is defined by $S = W_+(H, A^o)^* W_-(H, A^o)$, it commutes with H^o. If ima $W_{\pm}(H, A^o) = P^{ac}(H)$ where $P^{ac}(H)$ denotes the orthoprojection onto the subspace of absolute continuity of H, the system is called asymptotically complete. In this case S is unitary.

1.2. Decay. It is generally accepted, that scattering should be understood as the creation of an unstable system, a so-called compound system, which subsequently decays. The scattering model of 1.1. doesn't reflect this aspect. In order to connect that model with this aspect, one is guided to introduce a certain "ideal" Hamiltonian H_∞ in $\mathfrak{H}$ or in a suitable subspace of $\mathfrak{H}$. This operator should have the following properties: (I) there is a certain spectral discrete part $H_\infty \restriction P_\infty \mathfrak{H} = \sum_j \lambda_\infty^{(j)} P_\infty^{(j)}$ of H_∞, which corresponds to the "compound system" and it's energy levels, (II) the "channel structure" of H_∞ with regard to $\mathfrak{A}$ is the same as that of H, (III) H_∞ should be a "perturbation" of H in the sense of a decoupling of that part of the interaction which is responsible for the decay of the compound system.

1.3. Remark. In some cases such a decoupling may be generated by a projection P, which is "local" with regard to H (which means that dom $H \cap P\mathfrak{H}$ is dense in $P\mathfrak{H}$ and dom $H \cap P^{\perp}\mathfrak{H}$ is dense in $P^{\perp}\mathfrak{H}$). If one forms the Hamiltonian $H_{\mu} = H + \mu P$, $\mu \geq 0$, and if $\mu \to \infty$, then one obtains a limit H_{∞} (by strong resolvent convergence) which is selfadjoint in $P^{\perp}\mathfrak{H}$ and which is exactly the Friedrich's extension of $H \restriction P^{\perp}\mathfrak{H}$ (see H. Baumgärtel and M. Demuth [2]).

1.4. Unstable embedded eigenvalues. The system $\{H, H_{\infty}\}$ reflects the decay aspect of the compound system, which is stable, if the Hamiltonian H_{∞} is actual, under the influence of the real evolution e^{-itH}.

Hence from the mathematical point of view the theory of a decay-scattering system $\{H, H_{\infty}, \mathfrak{A}\}$ is the theory of unstable (embedded) eigenvalues λ_{∞} of H_{∞} under the influence of the perturbation $H_{\infty} \mapsto H$, the investigation of the connection between unstability of λ_{∞} and the decay of the eigenstates of H_{∞} and the investigation of the influence of λ_{∞} to the scattering, that is the investigation of the behaviour of the cross-sections near λ_{∞} (the so-called "resonance problem" in [1]). Central notions in this connection are the "virtual poles" and the (mathematical) "resonances", which must be carefully defined and investigated.

There are some rigorous mathematical results, which shall be demonstrated in the following only for the example of the well-known Friedrich's model and for a simple special case (for a general and detailed presentation see H. Baumgärtel [3] , H. Baumgärtel, M. Demuth, M. Wollenberg [4] , J.S. Howland [5, 6, 7]).

§ 2 Friedrich's model

2.1. *Definition.* For simplicity the definition of Friedrich's model is given as follows. Let $\mathfrak{K}$ and $\mathfrak{W}$ be Hilbert spaces of finite dimensions $n < \infty$. Let $0 < \lambda_0 < 1$ and $\mathfrak{H} = L^2(E, \mathfrak{K}, d\lambda) \oplus \mathfrak{W}$, where $E = [0,1]$. Now H_0 is defined by $H_0 \{f(\lambda), v\} = \{\lambda f(\lambda), \lambda_0 v\}$, where $f(\cdot) \in L^2(E, \mathfrak{K}, d\lambda)$, $v \in \mathfrak{W}$. The orthogonal projection from $\mathfrak{H}$ onto $\mathfrak{W}$ is denoted by P_0. Let Γ: $\mathfrak{W} \mapsto P_0^{\perp}\mathfrak{H}$ be a (bounded) linear operator with the property dim ima $\Gamma = n$. Then $\{H_0, H_0 + \Gamma + \Gamma^*\}$ is called a Friedrich's model, with one eigenvalue λ_0, embedded into the absolutely continuous spectrum E of H_0, which has the constant multiplicity n. Let $H = H_0 + \Gamma + \Gamma^*$.

The next aim is the formulation of the main assumption. Let G_0 be a fixed region of the complex plane, where $E \subset G_0$. We denote by Φ the set of all holomorphic vector-functions f: $G_0 \mapsto \mathfrak{K}$. As is well-known, with regard to the family of semi-norms $|\varphi|_{\mathfrak{K},K} = \sup_{z \in K} |\varphi(z)|_{\mathfrak{K}}$, where $K \subset G_0$ is compact, Φ is a nuclear space (see for instance I.M. Gelfand and N.J. Wilenkin [8, p. 85]). The "canonical" operator $E(z)$: $\Phi \to \mathfrak{K}$, which is defined by $E(z)\varphi = \varphi(z)$, is holomorphic on G_0. The inclusion $\Phi \subset L^2(E, \mathfrak{K}, d\lambda)$ is true.

ASSUMPTION 1. *Let* ima $\Gamma \subset \Phi$ *and* ima $\Gamma(z) = \mathfrak{K}$ *for all* $z \in G_0$, *where* $\Gamma(z) = E(z)\Gamma$.

If Friedrich's operator Γ depends on a perturbation parameter $\mu \geq 0$, $\Gamma = \Gamma_\mu$, then additionally is imposed

ASSUMPTION 2. *If* $\mu \to 0$, then $\|\Gamma_\mu\| \to 0$.

In this case assumption 1 should be true for all μ .

2.2. Remark. Assumption 1 excludes the existence of real embedded eigenvalues α of H, $0 \leq \alpha \leq 1$. For if $H\{f, \varphi\} = \alpha\{f, \varphi\}$ with $f \in L^2(E, \mathfrak{K}, d\lambda)$, $\varphi \in \mathfrak{H}$, then $H_0\varphi + \Gamma^* f = \alpha\varphi$, $H_0 f + \Gamma\varphi = \alpha f$, that is $x f(x) + \Gamma(x)\varphi = \alpha f(x)$, $0 \leq x \leq 1$, hence $f(x) = \frac{\Gamma(x)\varphi}{\alpha - x}$. Because of assumption 1 we have $|\Gamma(x)\varphi|^2_{\mathfrak{K}} > 0$ for all $x \in E$ and all $\varphi \neq 0$, $\varphi \in \mathfrak{H}$. Hence from $f \in L^2$ immediately $\varphi = 0$ and $f = 0$ follow.

2.3. Virtual poles

2.3.1. The Lifšic-matrix. We put

$$H^+(z) = \{ z - \lambda_0 - \Gamma^*(z - H_0)^{-1}\Gamma \} \restriction P_0\mathfrak{H} \quad , \operatorname{Im} z > 0.$$

$H^+(z)$ is called Lifšic-matrix. It is connected with the partial resolvent $P_0(z - H)^{-1}P_0$ by

$$P_0(z - H)^{-1}P_0 \restriction P_0\mathfrak{H} = H^+(z)^{-1} , \qquad \operatorname{Im} z > 0.$$

Because of assumption 1 $H^+(z)$ is holomorphic continuable, for instance into $\mathbb{C}_+ \cup G_0 - (\mathbb{R} - E)$. We denote the Riemannian surface of $H^+(z)$ by $\mathfrak{R}$ (without boundary and branching points).

DEFINITION 1. The point $\zeta \in \mathfrak{R}$ is called a virtual pole, if $\det H_+(\zeta) = 0$.

One obtains easily: ζ is a virtual pole if and only if ζ is a pole of $H^+(z)^{-1}$.

Hence if ζ is a virtual pole, then $\operatorname{Im}\zeta \leq 0$. In general the real virtual poles form the eigenvalue spectrum of H in $0 < \lambda < 1$ (see for instance J.S. Howland [7], H. Baumgärtel [3]). Hence in the present case there are no real virtual poles.

In the perturbation case because of assumption 2 and $\det H_0^+(z) = (z - \lambda_0)^n$ the existence of n virtual poles (where their multiplicities as roots of $H^+_\mu(z)$ is taken into account) in the neighborhood of λ_0 for sufficiently small μ may be proved by means of Rouche's theorem. In this case the trajectories of the virtual poles tend to λ_0 for $\mu \to 0$. The splitting behavior of λ_0 into the virtual poles is very complicated in general. A simple case (total splitting) is obtained, if the analytic continuation $W^+_\mu(z)$ of $\Gamma^*_\mu(z-H_0)^{-1}\Gamma_\mu$ has a representation of the following form: $W^+_\mu(z) = \sum_{\rho=1}^{n} \alpha_{\mu\rho}(z) P_{\mu\rho}(z)$, where the $P_{\mu\rho}(z)$ form a system of disjunct projections, holomorphic at $z = \lambda_0$, and where $P_{\mu\rho}(z) \to P_\rho$ and $\alpha_{\mu\rho}(z) \to 0$, if $\mu \to 0$, $\sum_{\rho=1}^{n} P_\rho = 1$. In this case the virtual poles may be calculated by iteration from the equations

$$z - \lambda_0 = \alpha_{\mu\rho}(z), \quad \rho = 1, 2, \ldots, n.$$

2.3.2. Eigenfunctionals. Let $\mathcal{D} = \Phi \oplus \mathcal{W}$ and let $\mathcal{D}$ be equipped with the product topology of Φ and $\mathcal{W}$. $\mathcal{D}$ is nuclear (and countably Hilbert) and dense in $\mathfrak{H}$. The scalar product (x, y) of $\mathfrak{H}$ is for $x, y \in \mathcal{D}$ a continuous sesquilinear functional. Let $\mathcal{D}^* = \Phi^* \times \mathcal{W}$ be the linear space of all antilinear functionals on $\mathcal{D}$. Then the tripel $\mathcal{D} \subset \mathfrak{H} \subset \mathcal{D}^*$ is a Gelfand triple (rigged Hilbert space) (see for instance I.M. Gelfand and N.J. Wilenkin [8, p. 104]). An important class of functionals in $\mathcal{D}^*$ is the class of so-called δ-functionals, which are defined by $\langle \delta^*_{\lambda_0,k} | d \rangle = (k, E(\lambda_0)\varphi)_{\mathfrak{K}}$, where $d = \varphi + v$.

It is well-known that for $H = H_0 + \Gamma + \Gamma^*$ in $\mathcal{D}^*$ a complete system of eigenfunctionals for real values exists. Of course, the eigenvalue problem may be solvable also for other (complex) values. We solve the eigenvalue problem for H with regard to the tripel $\mathcal{D} \subset \mathfrak{H} \subset \mathcal{D}^*$ and for $\zeta \in G_0 - \{0,1\}$.

Let $d^* = \{\varphi^*, h\}$, $\varphi^* \in \Phi^*$, $h \in \mathfrak{H}$. The eigenvalue problem is $H^* d^* = \xi d^*$, that is

(1) $\langle d^* | (H_0 + \Gamma + \Gamma^*)(\varphi + h_0)\rangle = \xi\langle d^* | \varphi + h_0\rangle$, $\varphi \in \Phi$, $h_0 \in \mathfrak{H}$.

Hence one obtains two equations

(2) $$\langle \varphi^* | (\bar{\xi} - H_0)\varphi\rangle = (\Gamma h, \varphi)_{P_0^\perp \mathcal{G}}, \quad \varphi \in \Phi,$$

(3) $$\langle \varphi^* | \Gamma h_0\rangle = (\xi - \lambda_0)(h, h_0)_{\mathfrak{H}}, \quad h_0 \in \mathfrak{H}.$$

First (2) is solved, if $h \in \mathfrak{H}$ is given. If $\operatorname{Im} \xi > 0$, then there is only one solution $\varphi_h^*(\xi)$, namely

$$\langle \varphi_h^*(\xi) | \varphi\rangle = \left((\xi - H_0)^{-1}\Gamma h, \varphi\right)_{P_0^\perp \mathcal{G}}, \quad \varphi \in \Phi.$$

By analytic continuation of the right hand side one obtains a solution $\varphi_h^*(\xi)$ for all $\xi \in G_0 - \{0,1\}$. Now we solve the homogenuous equation

(4) $$\langle \varphi^* | (\bar{\xi} - H_0)\varphi\rangle = 0, \quad \varphi \in \Phi.$$

Such a functional φ^* is necessarily a δ - functional, that is $\langle \varphi_k^* | \varphi\rangle = (k, E(\bar{\xi})\varphi)_{\mathfrak{K}}$ and the solutions φ_k^* of (4) may be indexed by the vectors $k \in \mathfrak{K}$. Hence the general solution of (2) is given by

(5) $$\varphi^* = \varphi_h^*(\xi) + \varphi_k^*(\xi), \quad k \in \mathfrak{K}.$$

Now we put (5) into (3) and obtain finally

(6) $$H^+(\xi)h = -\Gamma(\bar{\xi})^* k$$

for the determination of h, where $\Gamma(\bar{\xi})^*$ is the adjoint of $\Gamma(\bar{\xi}): \mathfrak{H} \mapsto \mathfrak{K}$.

If ξ is not a virtual pole, then $h = -H^+(\xi)^{-1}\Gamma(\bar{\xi})^* k$ and the eigenspace F_ξ^* of all eigenfunctionals corresponding to ξ consists of all $d^* = \{ \varphi_h^*(\xi) + \varphi_k^*(\xi), - H^+(\xi)^{-1} \Gamma(\bar{\xi})^* k \}$, where $\langle \varphi_k^*(\xi) | d \rangle = (k, E(\bar{\xi}) \varphi)_{\mathfrak{K}}$ and $\varphi_h^*(\xi)$ is given by (6), that is the ϕ^*-part of a non-zero eigenfunctional always contains a δ-like term, namely $\varphi_k^*(\xi)$.

If ξ is a virtual pole, then Ker $H^+(\xi) \supset \{o\}$. If $h \in$ Ker $H^+(\xi)$, then from (6) and assumption 1 necessarily $k = 0$ follows. Hence in this case $d^* = \{ \varphi_h^*(\xi), h \}$. In the general case $h \in \mathfrak{H}$ one obtains from assumption 1 $k = -(\Gamma(\bar{\xi})^*)^{-1} H^+(\xi) h$ and $d^* = \{ \varphi_h^*(\xi) + \varphi_k^*(\xi), h \}$.

We obtain the following

THEOREM. $\xi \in G_o - \{0;1\}$ *is a virtual pole if and only if the eigenspace* $F_\xi^* \in \mathcal{D}^*$ *of all eigenfunctionals contains a subspace consisting of all* $d^* = \{ \varphi_h^*(\xi), h \}$, *where* $h \in \mathfrak{H}_\xi$ *and* $\mathfrak{H}_\xi$ *is a certain proper non-zero subspace of* $\mathfrak{H}$.

In other words, a virtual pole ξ may be characterized by the condition, that there are eigenfunctionals which do not contain a δ - like term.

2.3.3. *Remark.* The connection between virtual poles and the decay problem may be investigated by means of the partial resolvent (see L.P. Horwitz and J.-P. Marchand [1], see also M. Demuth [9]).

2.4. *Scattering resonances.*

For Friedrich's model the scattering amplitude $T(\lambda): \mathfrak{K} \mapsto \mathfrak{K}$, is a function on E. A well-known formula (see for instance J.S. Howland [7]) connects $T(\lambda)$ with the partial resolvent:

$$T(\lambda) = \Gamma(\lambda) H^+(\lambda)^{-1} \Gamma(\lambda)^* . \tag{7}$$

From (7) immediately follows the analytic continuability of $T(\lambda)$. The Riemannian surface of $T(\lambda)$ is equal that of $H^+(z)$.

The poles of T(z) are called (scattering) resonances. Because of $T(\lambda) = S(\lambda) - 1$, where $S(\lambda)$ is the scattering matrix, there are no real resonances. A point ζ is a resonance if and only if it is a virtual pole.

In other cases, where real virtual poles are possible, the nonreal virtual poles and the resonances coincide.

If $\Gamma = \Gamma_\mu$, then $T(\lambda) = T_\mu(\lambda)$ depends on $\{\lambda, \mu\}$. For $\mu = 0$ one obtains $T_0(\lambda) \equiv 0$. But $T_\mu(\lambda)$ is by no means continuous with regard to μ. The (weakened) problem of resonance scattering in this case consists of the determination of trajectories $\lambda(\mu)$ such that for instance

$$\liminf_{\mu \to 0} \| T_\mu(\lambda(\mu)) \|^2_{2,\mathfrak{K}} \geq \gamma > 0.$$

In the case considered here the real parts of the virtual poles have this property. In particular for the case of total splitting this can be seen easily, because the main part of $T_\mu(\lambda)$ is obtained in the form

$$\sum_{\rho=1}^{n} \frac{1}{\lambda - \zeta_\rho(\mu)} \left\{ \frac{1}{1 - \alpha_\rho'(\lambda, \mu)} \Gamma(\lambda) P_{\rho\mu}(\lambda) \Gamma(\lambda)^* \right\}$$

where the virtual poles are denoted by $\zeta_\rho(\mu)$ and $\alpha_\rho' = \partial\alpha_\rho / \partial\lambda$. Finally let us note how the channels are included in this model. Let A_α be a channel Hamiltonian and let $P_\alpha^\pm$ be the orthogonal projection onto ima $W_\pm(H_0, A_\alpha)Q_\alpha$, which commutes with H_0. Hence $P_\alpha^\pm$ corresponds to projector-functions $P_\alpha^\pm(\lambda): \mathfrak{K} \longmapsto \mathfrak{K}$, defined on E. The partial cross-section for the transition between the channel α and β is given by

$$\sigma_{\alpha\to\beta}(\lambda) = \| P_\beta^+(\lambda)\, T(\lambda)\, P_\alpha^-(\lambda) \|^2_{2,\mathfrak{K}},$$

hence the resonance effects of the virtual poles may be deducted from these formulas.

REFERENCES.

[1] L.P. Horwitz and J.-P. Marchand, The decay-scattering system, Rocky Mountain J. of Math. 1 (1971), 225-253.

[2] H. Baumgärtel and M. Demuth, Decoupling by a projection, Reports on Math. Phys. to appear.

[3] H. Baumgärtel, Resonances of Perturbed Selfadjoint Operators and their Eigenfunctionals, Math. Nachr. 75 (1976), 133-151.

[4] H. Baumgärtel, M. Demuth and M. Wollenberg, On the equality of resonances (poles of the scattering amplitude) and virtual poles, Math. Nachr. to appear.

[5] J.S. Howland, Perturbation of embedded eigenvalues, Bull. Amer. math. Soc. 78 (1972), 280-283.

[6] J.S. Howland, Puiseux series for resonances near an embedded eigenvalue, Pac. J. Math. 55 (1974), 157-176.

[7] J.S. Howland, The Livsic Matrix in Perturbation Theory, J. math. Analysis Appl. 50 (1975), 415-437.

[8] I.M. Gelfand und N.J. Wilenkin, Verallgemeinerte Funktionen (Distributionen) IV, Berlin 1964.

[9] M. Demuth, Pole Approximation and Spectral Concentration, Math. Nachr. 73 (1976), 65-72.

QUANTUM DECAY PROCESSES AND QUANTUM DYNAMICAL SEMIGROUPS

L. Fonda, G.C. Ghirardi
International Centre for Theoretical Physics, Trieste, Italy
and
Istituto di Fisica Teorica dell'Università, Trieste, Italy
A. Rimini and T. Weber
Istituto di Fisica Teorica dell'Università, Trieste, Italy

1. INTRODUCTION

The quantum description of unstable systems has been the subject of many investigations, particularly in recent times, in which several conceptual problems connected with this description have been reconsidered and analyzed[1]. Let us recall some of them:

i) The problem of the exponential or non-exponential nature of the decay probability.

ii) The problem of characterizing in a general way the unstable system. This problem has been dealt with within the S-matrix axiomatic approach by associating unstable systems with analytical properties of the S-matrix or of the resolvent operator, or within the group theoretic approach which associates unstable systems with representations of the time-translation semigroup[2].

iii) The problem of giving a description which is adherent to the actual physical situation and takes into account the fact that unstable systems, which live long enough to merit the investigation of their time evolution, interact with the environment.

To discuss these matters, in particular point iii), one starts with an investigation of the properties of the so-called non-decay probability

$$P(t) = |A(t)|^2, \tag{1.1}$$

$$A(t) = \langle u | e^{-iHt} | u \rangle, \tag{1.2}$$

i.e. the probability of finding the system at time t in the same state in which it has been prepared with certainty at the initial time t=0, if no disturbances from the outside world have occurred in the same interval of time. H is here the self-adjoint Hamiltonian operator.

When an ensemble of identically prepared systems is considered we of course have $N(t) = N(0)\,P(t)$. As well known, the classical theory, based on the assumption that the decay process does not depend on the past history of the individual decaying systems, yields for N(t) the expression

$$N(t) = N(0)\, e^{-t/\tau} \tag{1.3}$$

which defines τ as the life-time of the decaying system.

There are some general properties of P(t) which can be proved completely in general by the standard theory of Fourier transforms:

1. If the Fourier transform of $|u\rangle$ on the improper scattering eigenstates of H has a narrow Breit-Wigner resonance shape for a large energy interval around the mass of the unstable particle , then there is a large interval of time for which P(t) is almost exponential, of the type $\exp(-\gamma t)$, where γ is the width of the resonance.

2. Under the condition that the mean energy in the state $|u\rangle$ be finite, $\langle u|H|u\rangle < \infty$, P(t) has a vanishing derivative at the origin

$$\left.\frac{dP(t)}{dt}\right|_{t=0} = 0 \tag{1.4}$$

The deviations from the exponential decay law occurring at small times, as implied by Eq. (1.4), play a very important role. Calculations on explicit models[3], as well as very simple qualitative arguments, show that deviations extend at least up to times of the order of (we use units $\hbar = c = 1$):

$$t = \gamma/(\Delta E)^2, \tag{1.5}$$

ΔE being the energy spread given by

$$\Delta E = \left[\langle u|H^2|u\rangle - \langle u|H|u\rangle^2\right]^{\frac{1}{2}} \tag{1.6}$$

From (1.5) we see that the greater the "theoretical" life-time $1/\gamma$ of the unstable system the smaller is the time region where deviations from the exponential occur.

3. If the Hamiltonian has a spectrum which is bounded from below, a necessary physical assumption, then P(t) decreases for large times more slowly than any exponential; in practice it vanishes according to an inverse power of the time. The deviations from the purely exponential law become appreciable at times for which

$$e^{-\gamma t/2} \simeq \frac{(2\ell+1)!!}{2^{\ell+2}\sqrt{\pi}} \; \frac{(\gamma t)^{-(\ell+3/2)}}{\left\{\frac{1}{4}+\frac{E_R^2}{2}\right\}^{\frac{1}{2}(\ell+5/2)}} \tag{1.7}$$

ℓ being the wave in which the resonance occurs, and E_R its position on the energy axis.

As shown by explicit calculations, oscillations appear in P(t) in the transition region from the exponential to the power-like behaviour. At first sight this appears to be against common sense, since in certain time intervals the number of undecayed systems appears to be increasing instead of steadily decreasing. As a matter of fact this is a typical quantum mechanical effect arising from a process of reformation of the unstable system. This can be very simply seen by writing the state vector at time t as the following superposition:

$$e^{-iHt}|u\rangle = A(t)\,|u\rangle + |\varphi(t)\rangle, \tag{1.8}$$

where $|\varphi(t)\rangle$ is orthogonal to $|u\rangle$ and contains of course the decay products. Applying to both sides of (1.8) the evolution operator exp $(-i\,H\,t')$ and taking the scalar product with $|u\rangle$, one gets[4]

$$A(t+t') = A(t)\,A(t') + \langle u|e^{-iHt'}|\varphi(t)\rangle. \tag{1.9}$$

We see from this equation that, if the term $\langle u|e^{-iHt'}|\varphi(t)\rangle$ were missing, the function A(t) would be a pure exponential for all times. This term instead provides a regeneration of the unstable system through the rescattering of the decay products among themselves. Since this term cannot be zero, we now perfectly understand why the decay law is not a pure exponential: one is forced to introduce a history -

dependent factor in its evolution equation. There is, however, a way to make the rescattering term disappear, and this is provided by nature. In fact in almost all practical cases the unstable system is not left undisturbed by the experimenter. As a matter of fact, an actual decaying system does not evolve isolated, but interacts repeatedly with its detectors before decaying. In the next Section we shall see how one can account phenomenologically for these interactions. In particular the decay law will turn out to be a pure exponential for all relevant times, yielding the reconciliation between classical and quantum descriptions, and a semigroup law of time evolution for the density operator will emerge very naturally.

2. INTERACTIONS WITH THE ENVIRONMENT

As emphasized by various authors[5], an unstable system which lives long enough unavoidably interacts repeatedly with the measuring apparatuses so that one is actually dealing with a non-isolated(open) quantum system. The recognition of this fact is obvious for almost all experiments devised to measure the non-decay probability $P(t)$ such as those involving track-visualising devices. Also for other kinds of experiments, as those on radioactive materials, analogous considerations can be made.

One must then try to incorporate the effects of the interactions with the surrounding in the dynamics of the unstable system. Now it is easy to recognize that the experimental set-ups are such that the interactions with the surrounding correspond to yes-no experiments investigating whether the system is decayed or not. To be more precise,this information is obtained by means of experiments aimed at ascertainting whether or not the relative distance between the decay fragments is smaller or larger than some distance R which is characteristic of the measuring apparatus. As a very clear example, one can think of a neutral particle decaying into two charged fragments, in the presence of any one of the tracks - visualizing devices used in actual experiments. When the two fragments are very close to each other, they appear as a neutral system; only when their relative distance gets larger than, say, the ionisation distance of the particles of the surrounding medium, this medium can react, identifying them as two distinct charged entities. An analysis of other experimental set-ups allows to draw similar conclusions.

This localization procedure has been exhaustively discussed in ref.6. As discussed by those authors, if P_R is the projection operator on a relative distance of the decay products less than the distance R due to the resonant dynamics of the unstable system, application of P_R on a wave packet overlapping the resonance energy leads essentially to a unique state, provided the time T_R spent by the decay products to travel the distance R be significantly smaller than the quantum mechanical life-time

$$T_R \lesssim \frac{1}{\gamma} \tag{2.1}$$

Another important point concerns the frequency of the interactions with the environment. In actual experiments the system is subjected to interactions occurring at random times during the process. One can think of the case of bubble or cloud chambers or photographic emulsions. Obviously, at least each bubble or activated grain corresponds to an interaction of the unstable system with the measuring apparatus equivalent to a localization of the fragments within a distance R.

What we have then to do now, is to write a dynamical equation for the evolution of a quantum system subjected to the above discussed localization procedures during its evolution. Due to the above analysis, it seems appropriate to schematize phenomenologically these complicated interactions according to the quantum theory of measurement: each interaction with the environment is considered as a yes-no experiment associated with the projection operators P_R and $1 - P_R$. It is very nice that an evolution equation for such a system can be obtained. Of course the equation must involve the density operator, as measurement processes induce transitions from a pure state to a statistical mixture. As it is obvious, one cannot forget about the interactions with the environment when the mean frequency λ of the reduction processes is greater than the width γ of the resonance

$$\lambda \gtrsim \gamma \tag{2.2}$$

i.e. when several reductions take place in a quantum-mechanical life-time. We claim that in most of the experimental situations the two above inequalities (2.1) and (2.2) are well satisfied.

Let us consider experiments in bubble chambers or photographic emulsions. In these cases the velocities of the decaying products are a fraction of the velocity of light. To compute $1/\lambda$ we can then simply evaluate the linear density of emulsion grains or bubbles which are of the order of 6μ or 1 m.m, respectively, yielding $1/\lambda \simeq 2\cdot 10^{-14}$ sec. or $1/\lambda \simeq 3\cdot 10^{-12}$ sec., respectively. This will be an upper bound for $1/\lambda$. To get a lower bound one can evaluate λ by the formula

$$\lambda = \sigma \varrho v , \tag{2.3}$$

where σ is the cross-section for electromagnetic interactions, ϱ the density of scatterers and v the velocity of the fastest of the decay products. According to (2.3) one gets

$$\lambda \simeq 10^{16} \text{ sec}^{-1} . \tag{2.4}$$

From these considerations one can safely state that[7]

$$10^{-16} \text{sec} < \frac{1}{\lambda} < 3\cdot 10^{-12} \text{ sec.} \tag{2.5}$$

For what concerns the localization distance R, since atomic phenomena are involved in the measurement, R turns out to be of the order of atomic or molecular dimensions

$$10^{-8} \text{ cm} < R < 10^{-7} \text{ cm} , \tag{2.6}$$

so that the time of flight T_R turns out to be

$$3\cdot 10^{-19} \text{ sec} < T_R < 3\cdot 10^{-18} \text{ sec.} \tag{2.7}$$

Remembering the life times of the elementary particles for which a determination of P(t) in track experiments is presently performable ($10^{-6} \div 10^{-10}$ sec), one sees that the inequality (2.1) is very well satisfied. Actually it is very well satisfied also for very short living particles such as π°. Comparing the life-times with Eq. (2.5) we see that for all the above mentioned particles, exception made of course for π°, also inequality (2.2) is well satisfied.

For α and β decays one must also take into account that the decay products interact repeatedly with the ensemble of surrounding atoms of the radioactive sample. It turns out that Eqs. (2.1) and (2.2) are even better satisfied in these cases.

3. THE EVOLUTION EQUATIONS

To write an evolution equation for a system subjected to reduction processes is very easy. If this system is left isolated in the time interval (t_o, t) the density operator evolves according to:

$$\varrho(t) = e^{-iH(t-t_o)} \varrho(t_o) e^{iH(t-t_o)} \tag{3.1}$$

The effect of a measurement (at a given time t_1) of the operator whose associated projection operators are P_R and $1 - P_R$ is

$$\varrho(t_1) \longrightarrow P_R\, \varrho(t_1) P_R + (1-P_R)\, \varrho(t_1)(1-P_R). \tag{3.2}$$

According to (3.2) we are considering a non selective measurement, i.e. no matter what result we find, we consider the complete ensemble of systems, composed of the subensembles for which different results have been found. If a selection is made, for instance disregarding the subensemble of the decayed systems, only the first term in (3.2) must appear. This will be our case since we are interested in all systems which have be found undecayed in all measurements up to time t.

From (3.1) and (3.2) one gets[8]

$$i \frac{d\,\varrho(t)}{dt} = \left[H, \varrho(t)\right] - i\lambda \left\{ \varrho(t) - P_R\, \varrho(t) P_R - (1-P_R)\, \varrho(t)(1-P_R) \right\} \tag{3.3}$$

This equation is equivalent to the integral equation

$$\varrho(t) = e^{-\lambda t} e^{-iHt} \varrho(0) e^{iHt} + \lambda \int_0^t d\delta\; e^{-\lambda\delta} e^{-iH\delta} \left\{ P_R \varrho(t-\delta) P_R + (1-P_R)\varrho(t-\delta)\cdot(1-P_R) \right\} e^{iH\delta} \tag{3.4}$$

If one disregards the subensemble of the decayed systems, one gets

$$\varrho(t) = e^{-\lambda t} e^{-iHt} \varrho(0) e^{iHt} + \lambda \int_0^t d\delta\; e^{-\lambda\delta} e^{-iH\delta} P_R \varrho(t-\delta) P_R e^{iH\delta} \tag{3.5}$$

Due to the remark that, when a reduction process takes place and the system is found undecayed, if the condition (2.1) is satisfied the system is left essentially in a unique state $|u\rangle$, we can actually substitute in (3.5) P_R with $|u\rangle\langle u|$ and we can also put $\varrho(0) = |u\rangle\langle u|$, assuming that initially the system has been analogously identified by a reduction process. Eq. (3.5) becomes then

$$\varrho(t) = e^{-\lambda t} e^{-iHt} |u\rangle\langle u| e^{iHt} + \lambda \int_0^t d\delta\; e^{-\lambda\delta} e^{-iH\delta} |u\rangle \langle u| \varrho(t-\delta) |u\rangle \langle u| e^{iH\delta} \tag{3.6}$$

Taking the sandwich of this equation between the state $|u\rangle$ we obtain the basic equation (see the paper by Fonda, Ghirardi, Rimini and Weber under ref. 5):

$$F(t) = e^{-\lambda t} P(t) + \lambda \int_0^t d\delta\; e^{-\lambda\delta} P(\delta)\, F(t-\delta), \tag{3.7}$$

where $F(t) = \langle u|\, \varrho(t)\, |u\rangle$ is the experimental probability of survival of the unstable system up to the time t. Eq. (3.7) is solved by using the Laplace transform, For the transformed f (s) of F(t) one gets

$$f(s) = p\,(s+\lambda)\Big/\Big[1 - p(s+\lambda)\Big] , \tag{3.8}$$

where p(s) is the Laplace transform of P(t). In evaluating the inverse Laplace transform of (3.8), one sees that there is a dominant contribution coming from the pole of the denominator in (3.8) at a value $s=-1/\tau$ such that

$$p\,(\lambda - \frac{1}{\tau}) = \frac{1}{\lambda} \tag{3.9}$$

The remaining contributions to F(t) decrease as $e^{-\lambda t}$ and thus, if condition (2.2) is satisfied, after few reductions, can be disregarded. The resulting expression for F(t) for all relevant times is then[9]:

$$F(t) = e^{-t/\tau} \tag{3.10}$$

Eq. (3.10) shows that, taking into account the interactions with the environment, the non-exponential behaviour of P(t) is completely eliminated. According to Eq. (3.9) one has that the experimental life time τ is given by the equation

$$\lambda \int_0^\infty dt\; e^{-\lambda t}\, e^{t/\tau}\, P(t) = 1 \tag{3.11}$$

which relates it to the quantum mechanical non-decay probability P(t) and to the frequency λ of the reductions. Eq. (3.11) implies a dependence of the experimentally observed life-time τ on the experimental set-ups. If P(t) would be a pure exponential $e^{-\gamma t}$ for all times, one would have $\tau = 1/\gamma$. In particular cases this dependence of τ on λ is very difficult to detect[7] (see also the paper by Omero and Persi under ref. 3).

Before concluding this paper we want to make some general comments. Eqs. (3.3 - 3.6) which govern the process are particular cases of the theory of Quantum Dynamical Semigroups[10]. Actually it has recently been proved[11] that if one looks for the most general one parameter mapping among the set of self-adjoint positive semidefinite operators of unit trace

$$\varrho(t) = \Sigma_t\; \varrho(0) \tag{3.12}$$

satisfying the conditions

i) that Σ_t be linear

ii) that Σ_t be completely positive in the sense of Stinespring (a requirement with a strong physical motivation since it amounts to require that the transformed statistical operator of any larger system containing as a subsystem the system evolving with Σ_t be positive semidefinite).

iii) that the family Σ_t constitutes a one parameter semigroup associated with the forward time translations semigroup

$$\Sigma_t \cdot \Sigma_{t'} = \Sigma_{t+t'}, \quad \forall\, t,t' > 0$$

(from the physical point of view this assumption amounts to assert that the modifications occurring in the environment during the process, do not change the features of its future actions on the same system).

iv) that the generator of the semigroup be a bounded operator,

the evolution equation for $\varrho(t)$ is necessarily of the form

$$\frac{d\,\varrho(t)}{dt} = -i\left[H,\varrho(t)\right] - \tfrac{1}{2}\left\{ \varrho(t), \textstyle\sum_n A_n^+ A_n \right\} + \textstyle\sum_n A_n\,\varrho(t)\,A_n^+ , \qquad (3.13)$$

where H is a self-adjoint operator and $\{ A_n \}$ an arbitrary set of operators such that $\sum_n A_n^+ A_n$ is bounded. Eq. (3.13) preserves the trace

$$\frac{d\,\mathrm{Tr}\,\varrho(t)}{dt} = 0. \qquad (3.14)$$

It has to be remarked that Eq. (3.4) is a particular case of Eq. (3.13) which means, due to ii), that the evolution of the density operator $\varrho(t)$ is associated with the semigroup of time translations. The same can be easily shown to hold for Eq. (3.5) and (3.6) which have a similar structure,apart from the fact that for them the trace is not conserved due to the loss of flux into the decay channels. So, while the old approach of associating unstable particles with the time translation semi-group of operators acting on the Hilbert space of the state vectors was unsuccessful, the same kind of procedure leads to a satisfactory result when it is applied to operators acting on the Banach space of density operators and the actual physical situation is properly taken into account.

A treatment analogous to the one given here for ordinary decay processes, can also be performed for the case of sequential decays[12].

REFERENCES

1. See the review article by L. Fonda, G.C. Ghirardi, A. Rimini,Reports on Progress in Physics 41, 587 (1978).
2. L.P. Horwitz,J.A. La Vita and J.P. Marchand, J. Math.Phys. 12, 2537 (1971).
   K. Sinha, Helv.Phys. Acta 45, 619 (1972).
   D.N. Williams, Commun. Math. Phys. 21, 314 (1971).
3. R.G. Newton, Annals of Physics N.Y. 14, 333 (1961).
   G.N. Fleming, Nuovo Cimento 16A, 232 (1973).
   C. Omero and T. Persi, Nuovo Cimento in press.
4. J. Ersak, Yad.Fiz.9, 468 (1969), Engl.trans.Sov.J.Nucl.Phys.9, 263.
5. J. Rau, Phys. Rev. 129, 1880 (1963).
   A. Beskow and J. Nilsson Ark. Fys. 34, 561 (1967).
   H.D. Zeh, Varenna School IL Corso, 1971.
   H. Ekstein and A.J.F. Siegert, Annals of Physics 68, 509 (1971).
   L. Fonda, G.C. Ghirardi, A. Rimini and T. Weber, Nuovo Cimento 15A, 689 (1973).
6. L. Fonda and G.C. Ghirardi, Nuovo Cimento 67A, 257 (1970) and 6A,553 (1971).
7. A. Degasperis, L. Fonda and G.C. Ghirardi, Nuovo Cimento 21A,471(1974).
8. L. Fonda, G.C.Ghirardi and A.Rimini, Nuovo Cimento 18B,1 (1973).
   E.B. Davies, Helv. Phys. Acta 48, 365 (1975).
9. For a discussion of this point see also: G.N. Fleming, Nuovo Cimento 46A, 579 (1978).
10. S.T.Ali, L. Fonda and G.C. Ghirardi, Nuovo Cimento 25A, 134 (1975).
    E.B. Davies, Quantum Theory of Open Systems, Academic Press, 1976.
11. G.Lindblad, Comm. Math. Phys. 48, 119 (1976).
12. L. Fonda,G.C.Ghirardi,C.Omero,A.Rimini and T.Weber,Phys.Rev. in press and these Proceedings.

UNSTABLE QUANTUM STATES AND RIGGED HILBERT SPACES*

V. Gorini and G. Parravicini**
Physics Department, CPT, The University of Texas at Austin
Austin, Texas 78712†

Abstract

We apply rigged Hilbert space techniques to the quantum mechanical treatment of unstable states in nonrelativistic scattering theory. We discuss a method which is based on representations of decay amplitudes in terms of expansions over complete sets of generalized eigenvectors of the interacting Hamiltonian, corresponding to complex eigenvalues. These expansions contain both a "discrete" and a "continuum" contribution. The former corresponds to eigenvalues located at the second sheet poles of the S matrix, and yields the exponential terms in the survival amplitude. The latter arises from generalized eigenvectors associated to complex eigenvalues on background contours in the complex plane, and gives the corrections to the exponential law.

1. Introduction

When considering unstable states in quantum theory, one is confronted with the following problem (see e.g. [1] and references therein). From the analogy with the classical case, one might believe that the quantum survival probability of the unstable state h, $P(t) = |A(t)|^2$, where $A(t) = (h,\exp(-iHt)h)$ is the survival amplitude, decays exponentially. However, by using some of the Paley-Wiener theorems and the physical hypothesis that the Hamiltonian is bounded below, it is seen that as t approaches infinity, P(t) cannot decrease faster than $\exp(-\alpha t^q)$, $\alpha > 0$, $q < 1$, so that P(t) vanishes slower than $\exp(-it/\Gamma)$ for any positive Γ. Moreover, if one assumes that h is a state with finite expectation value for the energy, it is easily seen that $dP(t)/dt$ vanishes at $t = 0$. These two facts imply that a quantum unstable state cannot have an exact exponential decay, and the need arises for methods allowing for the separation of the exponential term from a background relevant at short and long times only. See however [20].

It has been many times suggested, and in many cases proven, that resonances in scattering theory should be connected either to poles of the analytic continuation of the resolvent of the Hamiltonian, suitably restricted to some manifold, or to poles of the resolvent of certain nonself-adjoint operators associated with the

---

*Invited paper at the VII International Colloquium on Group Theoretical Methods in Physics, Austin, Texas, 1978. Supported in part by NATO Research Grant No. 1380 and by CNR Contratto di Ricerca No. 77.01543.63.

**A Fulbright-Hays Act grant is gratefully acknowledged.

†Permanent address: Istituto di Fisica dell'Universitá, Via Celoria 16, 20133 Milano, Italy, and INFN, Sezione di Milano, Italy.

Hamiltonian (see e.g. [2-5], [6, vol. IV]).

Another recurring argument in the literature is that resonances, i.e. peaks in cross sections, are to be associated with unstable states, as stated for example in any book on quantum scattering theory; for clear formulations of this see [1,7].

Also, from an intuitive point of view, physicists like to think of resonances as associated to complex eigenvalues of the Hamiltonian, the real and negative imaginary part of one such "eigenvalue" being respectively the energy and half-width of the resonance. For a rigorization of this idea along lines different from those we develop here, see [3,8,9].

The aim of this lecture is to show how to give a rigorous ground to the last point above, by the use of analytic eigenfunctionals in the framework of rigged Hilbert spaces (or Gel'fand triplets) [10-12].* This technique is further shown to provide a method, which is intuitively easy, for the analytic continuation of a reduced resolvent of the Hamiltonian across the absolutely continuous (a.c.) spectrum in such a way that its poles are the same as those of the continued S-matrix. As a by-product, the connection between unstable states and resonances is stressed once again. Our method stems from [15], where analytic continuation techniques were used to separate the exponential decay law from the background in the framework of the Friedrichs model [16], and the present lecture and a forthcoming paper [17] are intended to provide a rigorization and generalization of [15]. Our approach is related to that of ref. [8], the main difference being that we allow, and desire, the existence of generalized eigenvectors with complex eigenvalues not restricted to the positions of resonances.

2. Rigged Hilbert spaces and analytic functionals

In this section, we recall some essential features of rigged Hilbert spaces, in connection with the diagonalization of self-adjoint (s.a.) operators. We also review a few facts about analyticity of families of functionals in rigged Hilbert spaces, that depend on a complex parameter z. Finally, we examine the case when such families are families of eigenfunctionals of a s.a. operator, and connect their isolated singularities with eigenfunctionals and associated functionals.

Rigged Hilbert spaces were introduced and studied [10, vol IV; 11,12] in order to give a rigorous and direct meaning to the eigenket and eigenbra formalism developed by Dirac. We shall sketch the formalism by means of an example, the operator $T = -id/dx$ with domain in the Hilbert space $H = L^2(-\infty,+\infty)$ such that T is

*The appearance of "complex eigenvalues" for self-adjoint operators was first noticed and treated by physicists, in connection with the reduction of unitary representations of Lie groups with respect to noncompact subgroups (see e.g., [13]). Rigged Hilbert space techniques in this connection were first introduced, to our knowledge, in [14,21].

s.a. The eigenvalue equation $-idf_p(x)/dx = pf_p(x)$ has as only solutions plane waves $f_p(x) = \exp(ipx)$ that do not belong to H; eigenkets make no sense within H. However, we know that they are associated to Fourier transform, i.e. $\langle\exp(-ipx)|h\rangle \equiv \int_{-\infty}^{+\infty}\exp(-ipx)h(x)dx = \tilde{h}(p)$ can be defined for any h in H to yield a unitary transform from H onto L^2 $(-\infty,+\infty)$ w.r.t. the variable p; then Th is transformed into the function $\langle\exp(-ipx)|Th\rangle = p\langle\exp(-ipx)|Th\rangle = p\tilde{h}(p)$. In general, given any s.a. operator A, there is a unitary operator that diagonalizes A, that is to say that transforms the Hilbert space onto a space of L^2 functions of a real variable λ upon which A acts as the operator of multiplication by λ; such a unitary operator is called the generalized Fourier transform associated to A (see e.g. [6, vol. I]).

Rigged Hilbert space technique consists essentially in finding I) a dense submanifold ϕ II) included in the domain of T, III) invariant with respect to T, endowed with a topology such that IV) the identity embedding of ϕ into H is continuous and V) the action of T from ϕ into ϕ is continuous; finally, things must be arranged in such a way that if $\varphi \in \phi$, $\langle\exp(-ipx)|\varphi\rangle$ defines a linear continuous functional upon ϕ, i.e. it belongs to the dual space ϕ'. It has to be noted that because of I) and IV), H is continuously and densely embedded into ϕ'. Hence, vectors in ϕ' can be approximated by vectors in ϕ, a physically important fact. Then the relationship:

$$\langle\exp(-ipx)|T\varphi\rangle = p\langle\exp(-ipx)|\varphi\rangle \tag{1}$$

valid for almost every p, is to be understood not only as defining the function $p\tilde{\varphi}(p)$ in L^2 $(-\infty,\infty)$, Fourier transformed of $-id\varphi(x)/dx$, but also in the following way. For any p, $\langle\exp(-ipx)|\varphi\rangle$ defines a linear continuous functional upon ϕ, and so does $\langle\exp(-ipx)|T\varphi\rangle$ because T is continuous upon ϕ; therefore T can be seen as acting upon $\langle\exp(-ipx)|$ and transforming it into the new functional $\langle\exp(-ipx)|T$ that acts upon φ according to $\langle\exp(-ipx)|T\varphi\rangle$. Then (1) simply means that the functional $\langle\exp(-ipx)|$ is eigenvector of the extension of T in the space ϕ', with eigenvalue p. In the example at hand, one can choose ϕ to be S, the space of test functions for tempered distributions (whose space is denoted by S') endowed with the usual topology [10 vol. I; 6 vol. I]. We know that the functionals $\langle\exp(-ipx)|$ are in S', so that they are eigenvectors of T in S', even though they do not belong to H. Moreover they form a complete set, in the sense that the scalar product between any two vectors φ and ψ in S is given by:

$$(\varphi,\psi) = \frac{1}{2\pi}\int_{-\infty}^{+\infty}\overline{\langle\exp(-ipx)|\varphi\rangle}\langle\exp(-ipx)|\psi\rangle dp \ , \tag{2}$$

as follows from unitarity of the Fourier transform. Such eigenvectors can be approximated by vectors in H, as is well known. Therefore, if we are looking for a "complete set of eigenvectors" for any given s.a. operator A acting in a Hilbert space H, we may start looking for a submanifold ϕ such that conditions I)-V) are satisfied w.r.t. A. Because of technical reasons, we have also to require that the topology on the submanifold ϕ be such that ϕ is VI) nuclear, VII) barrelled and

VIII) complete. For the definitions we do not give here, see f.i. [10, vol. II; 11]. A triple of spaces $\phi \subset H \subset \phi'$ satisfying conditions I), IV), VI), VII) and VIII) is a rigged Hilbert space (notice that in ref. [10, vol. IV] this term denotes a smaller class of spaces). The triple $S \subset H \subset S'$ is an example of such space. A theorem of Maurin states that a "complete family of eigenvectors" always exists, that is to say the following. Given any s.a. operator A in a Hilbert space H, there exists a rigged Hilbert space $\phi \subset H \subset \phi'$ that satisfies II), III), and V) w.r.t.A. In the dual ϕ' of ϕ there exists a family of eigenvectors $<\lambda|$ of A,

$$<\lambda|A = \lambda<\lambda| \quad \text{or} \quad <\lambda|A\varphi> = \lambda<\lambda|\varphi> \qquad (3)$$

for any $\varphi \in \phi$ and any λ in the spectrum Σ of A. This set is complete, i.e. for any φ and ψ in ϕ it holds:

$$(\varphi,\psi) = \int_\Sigma \overline{<\lambda|\varphi>}<\lambda|\psi>d\mu(\lambda) \qquad (4)$$

μ being some measure on Σ. Otherwise stated, this means that the transform from ϕ into $L^2(\Sigma,\mu)$ given by $|\varphi> \to <\lambda|\varphi>$, where $<\lambda|\varphi>$ is to be considered for fixed φ a function of the parameter λ in Σ, can be extended to the whole of H and then coincides with one of the possible generalized Fourier transforms that diagonalize A [10, vol. IV; 11].

Returning to the example of the derivative operator, we see that here, as well as in the general case, the choice of the rigged Hilbert space is not unique: for example, instead of S we could have chosen for ϕ the space $\mathcal{D}$ of test functions with compact support [10 vol. I; 6 vol. I]. Of course, also this choice is such that Maurin theorem is satisfied, and indeed we know that the plane waves $<\exp(-ipx)|$ are distributions in $\mathcal{D}'$ as well. However, here another interesting feature arises: for any complex z, the functions exp(-ipz) are distributions in $\mathcal{D}'$, that is to say the integrals

$$<z|\varphi> \equiv \int_{-\infty}^{+\infty}\exp(-izx)\varphi(x)dx \qquad (5)$$

converge for any test function with compact support and are continuous w.r.t. the topology of $\mathcal{D}$. Moreover, these distributions are such that, for any complex z and for any φ in $\mathcal{D}$, $<z|T\varphi> = z<z|\varphi>$, i.e. the eigenvalue equation has solutions also for eigenvalues that are not in the spectrum of T; however, the associated eigenvectors in $\mathcal{D}'$ do not play any role in Maurin theorem. This is a general feature: solutions of the eigenvalue problem may arise for eigenvalues in the spectrum as well as for eigenvalues outside the spectrum, that do not play any role in Maurin theorem [13,14].

Another important feature arises in this example: for any fixed φ in $\mathcal{D}$, $<z|\varphi>$ as given in (5) is an analytic function of z, because one can differentiate under the integral. We will say that a family of vectors $<G(z)|$ in ϕ' depending on the complex parameter z is analytic in a region Ω if for any φ in ϕ the ordinary function of z given by $<G(z)|\varphi>$ is analytic in Ω [10, vol. I]. An isolated singularity for an analytic family $\langle G(z)|$ is a point z_0 in the complex plane such that, for at least some φ in ϕ, $\langle G(z)|\varphi\rangle$ is singular at z_0 and the family $\langle G(z)|$ is analytic in some punctured disc about z_0 [10, vol. I]. The integral $\int_\gamma \langle G(z)|dz$ has to be understood as that particular vector in ϕ', call it $\langle c|$, such that for any φ in ϕ, it holds $\langle c|\varphi\rangle = \int_\gamma \langle G(z)|\varphi\rangle dz$ where, for fixed φ, the r.h.s. is an ordinary Riemann integral [10, vol. I]; it can be shown that, under assumptions milder than I)-VIII), if γ is a rectifiable path, $\int_\gamma \langle G(z)|dz$ always converges to a unique $\langle c|$ in ϕ' whenever $\langle G(z)|$ is analytic on γ. Also, with the same assumptions, Cauchy theorems for integrals of ordinary analytic functions hold true for integrals of analytic families in ϕ'. The same is true for Taylor expansions or Laurent expansions about an isolated singularity z_0, i.e.:

$$\langle G(z)| = \sum_{n=-\infty}^{+\infty} \langle c_n(z_0)|(z-z_0)^n \tag{6}$$

$$\langle c_n(z_0)| = \frac{1}{2\pi i}\int (z-z_0)^{-n-1}\langle G(z)| \ , \tag{6'}$$

the series converging in a suitable sense in ϕ' [10, vol. I]. Moreover, the principle of identity of analytic functions, or principle of analytic continuation, holds true in our setting. We are interested in the case when a family of eigenvectors $\langle G(\lambda)|$ of a s.a. operator A, $\langle G(\lambda)|A = \lambda\langle G(\lambda)|$, is the restriction to the spectrum Σ of A of a family $\langle G(z)|$ analytic in some region $\Omega \supset \Sigma$. Then, provided there is in Σ a proper accumulation point which is also in Ω, the family $\langle G(z)|$ satisfies $\langle G(z)|A = z\langle G(z)|$ everywhere in Ω by the principle of identity of analytic functions. Next, consider the coefficients $\langle c_n(z_0)|$ of the Laurent expansion about an isolated singularity z_0 for such an analytic family of eigenvectors. From (6') it is readily seen that [17]:

$$\langle c_n(z_0)|A = z_0\langle c_n(z_0)| + \langle c_{n-1}(z_0)| \ . \tag{7}$$

Therefore, if $\langle c_{n-1}(z_0)|$ vanishes, $\langle c_n(z_0)|$ is an eigenvector with eigenvalue z_0, otherwise it is an associated vector. If the singularity is a pole of order N (or a regular point) then $\langle c(z_0)|$ (or the first nonvanishing coefficient) is an eigenvector with eigenvalue z_0.

3. Singularities of the continued reduced resolvent and analytic families of eigenvectors of the Hamiltonian

Here we envisage a s.a. operator H with suitably good properties and show how the singularities of the analytic continuation of its resolvent, properly reduced, across the spectrum, are associated with corresponding singularities of analytic families of eigenvectors of H. In the case where H is interpreted as the Hamiltonian, these singularities give rise to exponentially decaying discrete contributions to survival amplitudes of nonstationary states.

Suppose that there exists a rigged Hilbert space $\phi \subset H \subset \phi'$ that satisfies conditions I)-VIII) w.r.t. H and that the complete family of eigenvectors $\langle G(\lambda)|$ of H in ϕ' which appear in Maurin theorem is the restriction to the spectrum Σ of H of a family $\langle G(z)|$ analytic in some region $\Omega \supset \Sigma$. For the sake of clarity, assume Σ to be simple and an interval, possibly infinite, of the real axis, and that formula (4) can be written with μ the Lebesgue measure. Throughout this section, we consider valid these assumptions. From sec. 2, we know that $\langle G(z)|H = z\langle G(z)|$ holds everywhere in Ω. Consider also the space ϕ^x of antilinear continuous functionals on ϕ If $\langle G(\bar{z})|$ is in ϕ' then the functional $|F(z)\rangle$ defined by:

$$\langle\varphi|F(z)\rangle = \overline{\langle G(\bar{z})|\varphi\rangle} \tag{8}$$

is in ϕ^x. If $\langle G(\bar{z})|A = \bar{z}\langle G(\bar{z})|$ we have $H|F(z)\rangle = z|F(z)\rangle$. The family $|F(z)\rangle$ is of course analytic in $\bar{\Omega}$, the complex conjugate of the region Ω; for such families of antilinear eigenvectors everything can be stated as it was in sec. 2 for $\langle G(z)|$. The introduction of such antilinear functionals could be avoided, but we will not dwell upon this here [15,17].

Let f(z) be an entire analytic function whose restriction to Σ is bounded, and consider the following identity for any φ and ψ in ϕ:

$$(\varphi,f(H)\psi) = \int_\Sigma \overline{\langle G(\lambda)|\varphi\rangle} f(\lambda)\langle G(\lambda)|\varphi\rangle d\lambda \ . \tag{9}$$

If Γ is a path inside $\Omega \cap \bar{\Omega}$, obtained by deforming Σ, with both end points in Σ, and $z_1,z_2,\ldots,z_n$ are the singular points of $\langle G(z)|$ and/or $|F(z)\rangle$ in the region Λ enclosed by Σ and Γ, Cauchy theorem yields (see figure)

$$(\varphi,f(H)\psi) = \int_\Gamma \langle\varphi|F(z)\rangle f(z)\langle G(z)|\psi\rangle dz - 2\pi i\Sigma_i \text{Res}_{z_i}[\langle\varphi|F(z)\rangle f(z)\langle G(z)|\psi\rangle] \ , \tag{10}$$

the last term at the r.h.s. being the sum over the residues. If, say, $|F(z)\rangle$ is regular in Λ, whereas $\langle G(z)|$ is singular only at z_0, a simple pole, (10) becomes:

$$(\varphi,f(H)\psi) = \int_\Gamma \langle\varphi|F(z)\rangle f(z)\langle G(z)|\psi\rangle - 2\pi i\langle\varphi|F(z_0)\rangle f(z_0)\langle c_{-1}(z_0)|\psi\rangle \ , \tag{11}$$

$\langle c_{-1}(z_0)|$ being an eigenvector of H in ϕ' with eigenvalue z_0, defined by (6) and (6'). In the general case of a higher order singularity, there would appear various coefficients of the Laurent expansions of $\langle G(z)|$ and $|F(z)\rangle$, coefficients which we know to be either eigenvectors or associated vectors with eigenvalues given by the location of the singularities. Therefore, we have expanded scalar products in terms of a "complete" family of eigenvectors and associated vectors of H with complex eigenvalues; singularities of the analytic family of eigenvectors give a discrete

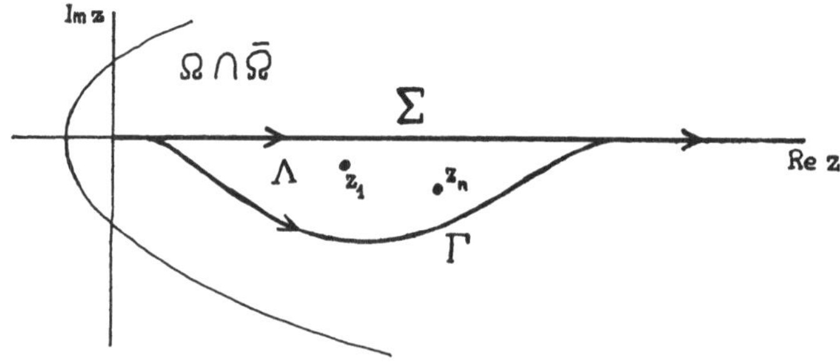

contribution to such expansions. Let us stress here that such eigenvectors can always be approximated by means of vectors in H. Now let $f(z) = \exp(-izt)$ and interpret H as the Hamiltonian of some quantum system. In the case that leads to (11), we have for the survival amplitude $A(t)$ of a state φ in ϕ:

$$A(t) = \int_\Gamma \langle\varphi|F(z)\rangle\langle G(z)|\varphi\rangle \exp(-izt)dz - 2\pi i \exp(-iz_0 t)\langle\varphi|F(z_0)\rangle\langle c_{-1}(z_0)|\varphi\rangle . \quad (11')$$

Assuming Γ lies in the lower halfplane of the complex energy, so that $\mathrm{Im}\, z_0 < 0$, we have associated with the pole at z_0 two eigenvectors, $\langle c_{-1}(z_0)|$ and $|F(z_0)\rangle$, and a time behaviour $\exp(-iz_0 t)$, i.e. exponential decay, explicitly separated from a background given by the integral over Γ. In the case of a higher order pole at z_0, the discrete contribution arising from the singularity would be of the form $P(t)\exp(-iz_0 t)$, with $P(t)$ a polynomial.

Next, let us consider the resolvent $R(w) = (H-w)^{-1}$. Whenever w is not in Σ, $R(w)$ is in the space $L(\phi,\phi^x)$ of continuous operators from ϕ into ϕ^x, where ϕ^x is endowed with the so-called weak topology [10, vol. II; 11]; as such, it has a continuation across Σ, $R^{II}(w)$, which is also in $L(\phi,\phi^x)$ (but not in $L(H,H)$!), its singularities being the same as those of the families $|F(z)\rangle$ and $\langle G(z)|$. Let us remark that only matrix elements between vectors in ϕ make sense for an operator in $L(\phi,\phi^x)$. To see that $R(w)$ is continuable, consider, for any φ and ψ in ϕ, the identity: $(\varphi,R(w)\psi) = \int_\Sigma \overline{\langle G(\lambda)|\varphi\rangle}\langle G(\lambda)|\psi\rangle(\lambda-w)^{-1}d\lambda$. Application of Plemelj formulas, together with a suitable definition of convergence for sequences of operators in $L(\phi,\phi^x)$, and the fact that $|F(z)\rangle\langle G(z)|$ is in $L(\phi,\phi^x)$, yield that:

$$R^{II}(z) = R(z) + 2\pi i|F(z)\rangle\langle G(z)| \quad (12)$$

is the analytic continuation of $R(w)$ across Σ from the upper halfplane into the second Riemann sheet. Clearly its singularities are those of $|F(z)\rangle$ and $\langle G(z)|$.

4. A critical comment

Even though we have been able to connect to each other singularities of the continued resolvent, discrete contributions to the survival amplitude with exponential decay law and eigenvectors (or associated vectors) of the Hamiltonian with complex eigenvalues, we are not in general justified in considering these objects associated with true physical resonances. Indeed, the analytic structure of the eigenvectors critically depends on the choice of ϕ. As a matter of fact, given any z_0 with $\mathrm{Im}\, z_0 < 0$, we can find a ϕ such that the corresponding family of eigenvectors has a simple pole at z_0. Correspondingly, poles for the continued resolvent can be located arbitrarily and therefore, for a Hamiltonian with at least part of the spectrum as in sec. 3, unstable states can be found with arbitrary energy and lifetime. In order to be able to select the true resonances in any particular case, we need additional input from the physics of the problem. We do this in the next section by using a second Hamiltonian and scattering theory.

5. Connection with the second sheet poles of the S matrix

A suitable way, though not the only possible one [1,4,5,6,18], to select physical resonances and to associate them to singularities of analytic families of eigenvectors of the Hamiltonian H, may consist in the introduction of a suitable "unperturbed" Hamiltonian H^0 and in assuming that resonances are connected to the singularities of the continued S matrix. Accordingly, assume we have two Hamiltonians, an "unperturbed" one H^0 and a "perturbed" one H in the Hilbert space $\mathcal{H}$, and let P^0 and P be the orthogonal projections on the absolutely continuous (a.c.) subspaces of H^0 and H respectively (for the definition, see f.i. [6, vol. I]). Assume the existence of the Møller wave operators W_+ and W_- for the pair (H^0,H), defined by $\lim_{t\to\mp\infty}\|\exp(-iH^0t)P^0h - \exp(-iHt)W_\pm P^0h\| = 0$, where h is any vector in $\mathcal{H}$. The S operator is then defined as $S = W_-^* W_+$ and it is unitary in $P^0\mathcal{H}$ (the space of scattering asymptotes). We recall, because it will be important later, the meaning in time independent scattering theory of the wave operators. There is a generalized Fourier transform F^0 that diagonalizes H^0, which transforms the vector h in $\mathcal{H}$ into the function $[F^0(h)]_i(E)$, where the subscript i takes into account degeneracy, say angular momentum, and the vector H^0h into $E[F^0(h)]_i(E)$; moreover there are two generalized Fourier transforms F^+ and F^- that diagonalize H, i.e. $h \to [F^+(h)]_i(E)$, or $h \to [F^-(h)]_i(E)$, such that Hh is transformed by them into $E[F^+(h)]_i(E)$ and $E[F^-(h)]_i(E)$ respectively and such that

$$[F^+(W_+P^0h)]_i(E) = [F^0(P^0h)]_i(E) \tag{13}$$

and

$$[F^-(W_-P^0h)]_i(E) = [F^0(P^0h)]_i(E) \,. \tag{13'}$$

Let $S_{ij}(E)$ be the diagonalization of the S operator w.r.t. H^0, i.e. $[F^0(SP^0h)]_i(E) = \Sigma_j S_{ij}(E)[F^0(P^0h)]_j(E)$ (in the formal theory of scattering, one has: $\langle E,i|S|E'j\rangle = \delta(E-E')S_{ij}(E)$). Then it also holds

$$[F^-(Ph)]_i(E) = \Sigma_j S_{ij}(E)[F^+(Ph)]_j(E) \,. \tag{14}$$

Now we assume that the S matrix elements $S_{ij}(E)$ can be continued from above into the unphysical sheet and we stick to the usual assumption that resonances are to be associated with poles of such continuation. Next, we assume that a single rigged Hilbert space $\phi \subset \mathcal{H} \subset \phi'$ exists such that assumptions II), III), V) of sec. 2 are satisfied w.r.t. both the pairs H^0, P^0 and H, P. Let us remark that we do not assume that the same be true for $W_\pm$ and S, and indeed in general it is not true. Moreover, assume that the generalized Fourier transforms F^0 and F^+ above are implemented by complete families of eigenvectors of H^0 and H, $\langle G_i^0(E)|$ and $\langle G_i^+(E)|$ respectively. Then it is easily seen that

$$\langle G_i^-(E)| \equiv \Sigma_j S_{ij}(E)\langle G_j^+(E)| \tag{15}$$

is a complete family of eigenvectors of H in ϕ' that implements F^-. The families $\langle G_i^+(E)|$ and $\langle G_i^-(E)|$ are the "in" and "out" states of the formal theory of scattering, here in a rigorous setting. Finally, let Σ be the a.c. spectrum of both Hamiltonians and consider the restrictions of the family $\langle G_i^0(E)|$ to $P^0\phi$ and of the families $\langle G_i^{\pm}(E)|$ to $P\phi$. From now on the symbols $\langle|$ and $|\rangle$ will denote such restrictions for the respective families. Assume that for such restrictions the following hold: a) the family $\langle G_i^0(E)|$ is the restriction to Σ of a family $\langle G_i^0(z)|$ analytic in a simply connected region $\Omega_0 \supset \Sigma$ in the plane of the complex energy z; b) the families $\langle G_i^+(E)|$ and $|F_i^-(E)\rangle$ are continuous boundary values on Σ of families $\langle G_i^+(z)|$ and $|F_i^-(z)\rangle$ analytic in some simply connected region Ω_1 in the lower halfplane, $\Omega_1 \subset \Omega_0$. It follows that $\langle G_i^-(E)|$ is analytic in $\bar{\Omega}_1$ and can be continued into Ω, the intersection of Ω_1 with the region of analyticity of the continued S matrix, because of (15) (provided the series converges suitably). The singularities in Ω of the continued family $\langle G_i^-(z)|$ are then exactly the same as those of the continued S matrix (apart from pathological cases in which zeros of $\langle G_i^+(z)|$ cancel poles of $S_{ij}(z)$). Now apply formula (12) to the resolvent of the unperturbed Hamiltonian by means of the eigenvector families $\langle G^0(z)|$ and $|F^0(z)\rangle$; no singularities arise for this operator in Ω_1, because Ω_0 is simply connected and includes Ω_1. Remark that (12) is valid here and in the following for the resolvents restricted to ϕ, only when suitable conditions on the not a.c. parts of the spectra hold; otherwise, they are valid for the respective restrictions to $P\phi$ and $P^0\phi$. Then, apply to the perturbed Hamiltonian formulas (10), (12) and (11') (the last one in case S has only one simple pole in Ω), by inserting the family $\langle G_i^-(z)|$ and letting Γ to be a path in Ω. These formulas show that resonances, i.e. singularities z_i in the analytic S matrix continued from above into the unphysical sheet are associated with: 1) eigenvectors and associated vectors of H with eigenvalues z_i, that can be approximated by vectors in $\mathcal{H}$ just as plane waves can be; 2) singularities, of the same kind, of the continued resolvent of H, whereas the continued resolvent of H_0 is regular; 3) discrete contributions, that decay exponentially, to the survival amplitude of vectors in $P\phi$. Moreover, the exponential contributions of resonances to the survival amplitude are straightforwardly separated from the background.

6. The example of the Friedrichs model

The Friedrichs model [16], equivalent to the Lee model in the first sector with an unstable V particle, provides a simple example in which the theory sketched above can be implemented. Let $\mathcal{H} = C \oplus L^2(0,+\infty)$, so that a vector in $\mathcal{H}$ is a pair $(h^0;h(E))$ where h^0 is a complex number and h(E) is in $L^2(0,+\infty)$. H^0 acts according to $H^0h = (Mh^0;Eh(E))$, with $M > 0$. H acts according to $Hh = (Mh^0 + g\int_0^\infty \bar{f}(E)h(E)dE;Eh(F) + h^0gf(E))$, where f(E) is a suitable cut-off function and g is a coupling constant. If we take f(E) to be the restriction to $[0,+\infty)$ of a function in Z, the space of Fourier transforms of test functions with compact support [10, vol. I], and if we assume that $f(0) = 0$, $f(E) \neq 0$ for $E > 0$, and $g^2 < M/\int_0^\infty |f(E)|^2E^{-1}dE$, we have the following. Choose $\phi = C + Z$: then all the assumptions of sec. 5 are satisfied and

P is the identity. In particular, as $P^0\phi = Z$, it holds $\langle G^0(E)|\varphi\rangle = \varphi(E)$ and $\langle G^0(z)|\varphi\rangle = \varphi(z)$, so $\langle G^0(z)|$ is analytic entire. If we define: $\alpha(z) = z - M - g^2\int_0^\infty |f(E)|^2 (z-E)^{-1} dE$, we have:

$$\langle G^{\pm}(E)|\varphi\rangle = gf(E)[\alpha(E\mp i0)]^{-1}\{\varphi^0 + g\int_0^\infty \overline{f(E')}\varphi(E')(E\mp i0-E')^{-1}dE'\} + \varphi(E)\ .$$

The S operator on $P^0 H$, diagonalized w.r.t. H^0, is $S(E) = \alpha(E-i0)[\alpha(E+i0)]^{-1}$, with continuation into the unphysical sheet: $S(z) = \alpha(z)[\alpha(z) + 2\pi i g^2 \overline{f(\bar{z})} f(z)]^{-1}$. For g small enough and nonzero there is a resonance pole at a point $z_0 = E_0 - i(\Gamma_0/2)$, $\Gamma_0 > 0$, close to M, solution of the equation $\alpha(z) + 2\pi i g^2 \overline{f(\bar{z})} f(z) = 0$ [17].

The theory sketched in sec. 5 holds also for degenerate perturbations; we refer to [19] for other interesting models where this theory applies. For another approach see [22].

References

[1] L. Fonda, G. C. Ghirardi, A. Rimini: Rep. Progr. Phys. 41, 587 (1978).
[2] J. Schwinger: Ann. Phys. (N.Y.) 9, 169 (1960).
[3] A. Grossmann: J. Math. Phys. 5, 1025 (1964).
[4] J. S. Howland: Amer. J. Math. 91, 1106 (1969); Bull. Am. Math. Soc. 78, 280 (1972); Trans. Am. Math. Soc. 162, 141 (1971); J. Math. Anal. and Appl. 50, 415 (1975).
[5] J. Aguilar, J. M. Combes; Commun. Math. Phys. 22, 269 (1971). E. Balslev, J. M. Combes: Commun. Math. Phys. 22, 280 (1971). R. A. Weder: J. Math. Phys. 15, 20 (1974).
[6] M. Reed, B. Simon: Methods of Modern Mathematical Physics (Academic Press New York).
[7] L. P. Horwitz, J. P. Marchand: Rocky Mountain J. Math. 1, 225 (1971).
[8] H. Baumgärtel: Math. Nachr. 75, 133 (1976).
[9] L. P. Horwitz, I. M. Sigal: Tel Aviv University Preprint, 1976.
[10] I. M. Gel'fand, et al.: Generalized Functions (Academic Press, New York, 1964).
[11] K. Maurin: General Eigenfunction Expansions and Unitary Representations of Topological Groups (PWN, Warsaw, 1968).
[12] J. E. Roberts: Commun. Math. Phys. 3, 98 (1966); J. P. Antoine: J. Math. Phys. 10, 53 (1969); O. Melsheimer: I-II, J. Math. Phys. 15, 902 (1974); A. Böhm: Boulder Lectures in Theoretical Physics, Vol. 9A, 255 (1966); The Rigged Hilbert Space and Quantum Mechanics, Lecture Notes in Physics, vol. 78 (Springer, New York, 1978).
[13] J. G. Kuriyan, N. Mukunda, E.C.G. Sudarshan: J. Math. Phys. 9, 2100 (1968).
[14] G. J. Iverson: Unitary Adjoint Representations of the Lorentz Groups I, II, University of Adelaide (1967); Phys. Lett. 26B, 229 (1968).
[15] E.C.G. Sudarshan, C. B. Chiu, V. Gorini: Phys. Rev. D, in press.
[16] K. O. Friedrichs: Commun. Pure and Appl. Math. 1, 361 (1948).
[17] G. Parravicini, V. Gorini, E.C.G. Sudarshan, C. B. Chiu: in preparation.
[18] E. B. Davies: Lett. Math. Phys. 1, 31 (1975).
[19] T. K. Bailey, W. C. Schieve: Complex Energy Eigenstates in Quantum Decay Models, Nuovo Cimento, in press.
[20] A. P. Grecos, I. Prigogine: Irreversible Processes in Quantum Theory, Contribution in these Proceedings, and references quoted therein.
[21] G. Lindblad, B. Nagel: Ann. Inst. H. Poincaré, 13A, 27 (1970).
[22] A. Böhm: Quantum Mechanics (Springer, New York, 1979).

IRREVERSIBLE PROCESSES IN QUANTUM THEORY

A.P. Grecos and I. Prigogine*

Faculté des Sciences, Université Libre de Bruxelles,
Campus Plaine, CP. 231, 1050 Brussels, Belgium.

and *Center for Statistical Mechanics and Thermodynamics,
The University of Texas at Austin, Austin, Texas 78712, U.S.A.

One of the basic problems of non-equilibrium statistical mechanics is the derivation and the interpretation of equations describing dissipative phenomena in dynamical systems. Assuming the validity of reversible (classical or quantal) dynamical laws we look for conditions which permit the description of the evolution in terms of manifestly irreversible "kinetic" equations. The fact that these two modes of description - the dynamical and the kinetic one - have different symmetries with respect to time reversal, is the main conceptual difficulty in formulating and interpreting a dynamical theory of irreversible processes.

In this paper we examine certain aspects of the structure of kinetic equations as well as some physical and mathematical questions of the "problem of irreversibility". Our considerations will be based on the ideas and methods developed during the past few years in Brussels[(1)]. The formalism of subdynamics and the theory of non-unitary transformations to which it leads, provide an appropriate framework in investigating the status of irreversible processes in dynamical systems. To simplify the presentation we have introduced several assumptions, some of which, as one may notice, can be relaxed. Most of the discussion applies to both classical and quantum systems. However, having in mind the problem of the description of decaying states in quantum theory, we generally stress the quantum mechanical aspects of the formalism.

1. <u>Asymptotic Master Equations</u>. A well known example of an irreversible "kinetic" equation is that proposed by Pauli[(2)] for the evolution of the diagonal elements $\rho_0(k)$ $\left[=\langle k|\rho|k\rangle\right]$ of the density matrix ρ, in the case of a "weakly coupled" quantum system. We write the Hamiltonian

in the form $H = H_o + \lambda V$ and we denote by $\{|k>\}$ and $\{\omega^o_k\}$ the eigenvectors and eigenvalues of H_o, i.e. $H_o|k> = \omega^o_k|k>$. Then assuming that the coupling parameter λ is "small", the Pauli master equation reads

$$i\partial_t \rho_o(k) = i\pi\lambda^2 L^{-3}\sum_l |<l|V|k>|^2 \delta(\omega^o_k - \omega^o_l)\left[\rho_o(l) - \rho_o(k)\right] \tag{1.1}$$

Its r.h.s. represents a "collision term" with transition probabilities evaluated at the Born approximation.

Using this equation we may formulate a microscopic theory of transport phenomena in several cases, such as anharmonic solids. Also the study of dissipative phenomena in systems involving light-matter interaction, such as lasers, is based on equations of this type. Thus our interest on the derivation and generalization of the master equation arises from actual needs of modern theoretical physics.

Equation (1.1) may be written in a somewhat more abstract form, namely

$$i\partial_t \rho_o = \Psi_{(2)}(+i0)\rho_o \tag{1.2}$$

where ρ_o is the diagonal part of ρ with respect to the H_o-representation. The operator $\Psi_{(2)}(+i0)$, defined by comparison with the r.h.s. of eq. (1.1), is the second order approximation of a more general expression, the asymptotic collision operator, that we shall introduce in the following.

A basic property of the Pauli equation is that it admits an H-function. This means that the functional $H(t)$ of ρ_o, $H(t)=\Sigma\rho_o(k)\ln\rho_o(k)$, decreases monotonically in time ($dH/dt \leq 0$) . Note that other choices for the H-function, e.g. $\Sigma|\rho_o(k)|^2$, are possible(3). Consequently, assuming that $\rho_o(k)$ = constant is the only stationary solution of eq. (1.2), its validity implies, at least for ρ_o , a monotonic approach to microcanonical equilibrium. Moreover, because of the "H-theorem", $S = -kH$ may be used as a microscopic definition of non-equilibrium entropy for a weakly coupled system.

The main assumptions used in deriving eq. (1.1) are : (i) off-diagonal elements of ρ have been neglected, (ii) only asymptotic effects of the interactions have been retained, leading to a "markovian" equation, (iii) the collision term is evaluated to second order in the coupling parameter. Note also that the system is considered to be enclosed in a box of volume L and, eventually, the thermodynamic limit must be taken.

To discuss the status of Pauli-type equations, a convenient starting point is the so-called generalized master equation(4). This is an exact equation for ρ_o that is obtained from the von Neumann equation

$$i\partial_t \rho = L\rho \quad (=[H\ ,\ \rho]) \tag{1.3}$$

by analyzing its formal solution

$$\rho(t) = \exp\{-iLt\}\ \rho(0) = (2\pi i)^{-1} \int_{\vec{C}} dz\ \exp\{-izt\}\ (L-z)^{-1}\ \rho(0) \tag{1.4}$$

The first derivations were based on perturbation methods but it was soon realized that considerable simplification occurs by appealing to projection methods[(5)].

Suppose that we have a projection P such that

$$\rho_o = P\rho\ ,\quad P = P^2\ (= P^+\ ),\quad Q = I - P, \tag{1.5}$$

and this is in fact the case when ρ_o is the diagonal part of ρ . Then the resolvent $(L-z)^{-1}$ may be decomposed as

$$(L - z)^{-1} = \left[P\ + C(z)\right]\left[PLP + \Psi(z) - z\right]^{-1}\left[P + \mathcal{D}(z)\right] + (QLQ - z)^{-1} Q \tag{1.6}$$

where we have introduced the <u>collision operator</u>

$$\Psi(z) = -PLQ\ (QLQ - z)^{-1}\ QLP\ , \tag{1.7}$$

the <u>destruction operator</u>

$$\mathcal{D}(z) = -PLQ\ (QLQ - z)^{-1}, \tag{1.8}$$

and the <u>creation operator</u>

$$C(z) = -(QLQ - z)^{-1}\ QLP \tag{1.9}$$

(for the terminology see ref. 4). It follows from eqs. (1.4) and (1.6) that $\rho_o(t)$ satisfies the equation

$$i\partial_t \rho_o(t) = PLP\ \rho_o(t) + \int_0^t d\tau\ G(\tau)\ \rho_o(t-\tau) + F(t)\ \rho_c(0) \tag{1.10}$$

Here $G(t)$ and $F(t)$ are the inverse Laplace transforms of $\Psi(z)$ and $\mathcal{D}(z)$ respectively, and $\rho_c(0) = Q\rho(0)$.

By choosing suitable projections we can apply the formalism of generalized master equations to other situations than N-particle systems. We mention the theory of "open systems"(6,7), where $P\rho$ is assumed to describe the state of a "small" (sub)system coupled to a thermal reservoir. Note that we suppose that the decomposition (P,Q) is given and we do not discuss the difficult question of its justification on physical grounds. Eq. (1.10) may be written for any linear system, independently of whether dissipative processes take place or not. Thus its existence does not solve the problem of the relation of irreversibility to dynamics.

Clearly the continuity of the spectrum of L is a necessary condition for a consistent theory of irreversibility. Therefore, quantum systems should be considered in some infinite limit, e.g. the thermodynamic limit ($N \to \infty$, $L^3 \to \infty$, NL^{-3} : finite). For this reason a rigorous discussion is rather difficult and plausible, but nevertheless heuristic, arguments are often used. However, certain aspects may be formulated precisely in simple cases where the Hamiltonian has ab initio a continuous spectrum. In this respect, models of the Friedrichs type(8), that are frequently used to discuss quantum unstable states, are interesting examples.

Equation (1.2) may be viewed as an approximation to eq. (1.5). Assuming that $\rho_c(0) = 0$ and approximating the convolution term by $\left[\int_0^\infty d\tau\, G(\tau)\right] \rho_o(t)$, we get

$$i\partial_t \rho_o(t) \simeq \left[PLP + \Psi(+i0)\right] \rho_o(t) \tag{1.11}$$

The <u>asymptotic collision operator</u> $\Psi(+i0) = \lim_{z\to+i0} \Psi(z)$ has the important property of being dissipative ($\text{Im}\Psi(+i0) \leqslant 0$). Moreover, when considered as a functional of L, we have $\text{Im}\,\Psi(L, +i0) = \text{Im}\,\Psi(-L, +i0)$. Thus when the <u>condition of dissipativity</u> $\text{Im}\,\Psi(+i0) \neq 0$ is satisfied, eq. (1.11) has a different symmetry from the von Neumann equation. The Pauli equation is the first non-trivial approximation of eq. (1.11).

To justify the Pauli equation the "$\lambda^2 t$-approximation" ($\lambda \to 0$, $t \to \infty$, $\lambda^2 t$: finite) is often used (cf. ref. 6). But this argument cannot be applied to strongly coupled systems (finite λ). A detailed analysis(9) has shown that the proper formulation of an <u>asymptotic master equation</u> does not lead to eq. (1.11) but rather to

$$i\partial_t \tilde{\rho}_o(t) = \Theta\, \tilde{\rho}_o(t) \;; \qquad \Theta = \Omega\, \Psi(+i0) \tag{1.12}$$

where Ω takes into account the finite duration of the interactions. The structure of Θ will be indicated in the next section. From this point of view, eq. (1.12) is exact, and eq. (1.2) an approximation to it (and not to eq. (1.10)). Then its solution must be a particular contribution, denoted by $\tilde{\rho}_o(t)$, to the P-component of the state. We need now to determine the relation between $\tilde{\rho}_o(t)$ and $\rho_o(t)$, and specify the appropriate initial condition for eq. (1.12).

2. The Concept of Subdynamics. By hypothesis the spectrum of L (and of QLQ) is real. Therefore the singularities of $[PLP + \Psi(z) - z]^{-1} = P(L-z)^{-1}P$ lie on the real axis. When L has a continuous spectrum, an analytic continuation in some domain below the axis of the branch of $\Psi(z)$ defined for Im $z > 0$ might be possible. The operator Θ is determined by choosing a class of singularities of the analytic continuation of $[PLP + \Psi(z) - z]^{-1}$ in the lower half-plane. Of course, the specific structure of Θ depends on the nature (poles, cuts) of the singularities under consideration. Here we briefly discuss the simplest case where we choose a set of poles arising from the solution of the equations

$$[PLP + \Psi(z_n) - z_nP]\,|u_n> = 0 \tag{2.1a}$$

$$<v_n|\,[PLP + \Psi(z_n) - z_nP] = 0 \tag{2.1b}$$

Note that the "left" $\{<v_n|\}$ and "right" $\{|u_n>\}$ eigenvectors of this "nonlinear eigenvalue problem" are, in general, neither identical nor bi-orthogonal. We suppose that they are complete in the P-subspace.

From the formal solution of the von Neumann equation in terms of the resolvent of L, we conclude that the solutions of eqs. (2.1) give rise to a contribution $\tilde{\rho}_o(t)$ to $\rho_o(t)$ which reads

$$\tilde{\rho}_o(t) = \sum_n \exp\{-iz_nt\}\,(1 - \mu_n')^{-1}\,|u_n>\,<v_n|\;\rho_o(0) \tag{2.2}$$

Here $\mu_n' = <v_n|\,\Psi'(+i0)|\,u_n>$ and the initial condition has no Q-component ($\rho_c(0) = 0$). It is an easy matter to show that $\tilde{\rho}_o(t)$ satisfies eq. (1.12) with

$$\Theta = \sum_n z_n|u_n>\,<\bar{v}_n| \tag{2.3}$$

Here, the set $\{|\bar{u}_n>\}$ is bi-orthogonal to $\{<v_n|\}$ and similarly we denote by $\{<\bar{v}_n|\}$ a set bi-orthogonal to $\{|u_n>\}$. For t = 0 eq. (2.2) implies that

$$\tilde{\rho}_o(0) = A\,\rho_o(0) \quad ; \quad A = \sum_n (1 - \mu_n')^{-1}\,|u_n><v_n| \tag{2.4}$$

Thus the "pseudo-markovian" equation of the previous section, eq.(1.12), describes the evolution of particular contributions to the P-component of the state, characterized by a specific set of relaxation times (or lifetimes in the case of unstable states).

The essential idea of the theory of subdynamics[10,11] is to consider together with $\tilde{\rho}_o(t)$ the contribution $\tilde{\rho}_c(t)$ to the Q-component of the state that has the same time dependence. Let us first examine the effect of $\rho_c(0)$ to $\tilde{\rho}_o(t)$. If we restrict our attention to initial conditions such that $\mathcal{D}(z)\rho_c(0)$ admits an analytic continuation below the real axis with no singularities at $\{z_n\}$, then the previous considerations remain valid but the initial condition given by eq. (2.4) should be replaced by

$$\tilde{\rho}_o(0) = A\left[\rho_o(0) + D\,\rho_c(0)\right] \quad ; \quad D = \sum_n |\bar{u}_n\rangle \langle v_n|\,\mathcal{D}(z_n) \tag{2.5}$$

On the other hand a formal calculation shows that $\tilde{\rho}_c(t)$ is a functional of $\tilde{\rho}_o(t)$, namely

$$\tilde{\rho}_c(t) = C\,\tilde{\rho}_o(t) \quad ; \quad C = \sum_n \mathcal{C}(z_n)\,|u_n\rangle \langle \bar{v}_n| \tag{2.6}$$

Let us denote by $\tilde{\rho}(t)$ the contribution to $\rho(t)$ with exponential time dependence determined by the poles $\{z_n\}$. We have

$$\tilde{\rho}(t) = \left[P + C\right] \exp(-i\Theta t)\, A \left[P + D\right] \rho(0) \tag{2.7}$$

or

$$\tilde{\rho}(t) = \Pi\rho(t) \quad \text{with} \quad \Pi = \left[P + C\right] A \left[P + D\right] \; ; \; \Pi = \Pi^2 (\neq \Pi^+) \tag{2.8}$$

In other words $\tilde{\rho}(t)$ is obtained formally from $\rho(t)$ by a non-orthogonal projection (assuming that some poles z_n have a nonvanishing imaginary part). The details of the calculations may be found in refs. 12 and 13.

There are several problems that need a deeper study. Already questions such as the existence of the asymptotic collision operator $\Psi(+i0)$ or the possibility of an analytic continuation of $\Psi(z)$ are difficult to answer in a general manner, e.g. when the P-subspace is infinite dimensional. However, these type of problems, although mathematically difficult, are not always of basic importance for the physical interpretation of the formalism. On the other hand the precise meaning of the operators C and D, and thus of Π, as well as the structure of the space in which they act, needs further investigation. The formal construction of Π implies the analytic continuation of (classes of matrix elements of) the resolvent of L. Such an operation cannot be performed in a normed space because the continuous spectrum of the operator is a natural boundary for $(L-z)^{-1}$. Several investigations in this direction have already appeared in the literature(14). A study of Friedrichs type models indicates that for a consistent interpretation of the formalism of subdynamics, we may need the introduction of spaces with non-hermitian metric.

3. Non-Unitary Transformations. The asymptotic master equation, eq. (1.12) is obviously irreversible, because $\Theta \neq \Theta^+$ and its spectrum lies in the lower half-plane. Nevertheless, difficulties arise in proving an "H-theorem" given that Θ, contrary to $\Psi(+i0)$, is not necessarily a dissipative operator. From the structure of Θ , eq.(2.3), one may easily see that there exists in the P-subspace an invertible transformation χ , such that

$$\Phi_o = \chi^{-1}\Theta\,\chi \quad ; \quad (2i)^{-1}\left[\Phi_o - \Phi_o^+\right] \leqslant 0 \tag{3.1}$$

For example, a possible expression for χ is

$$\chi = \sum_n b_n\,|u_n\rangle\langle e_n| \qquad\qquad \chi^{-1} = \sum_n b_n^{-1}\,|e_n\rangle\langle\bar{v}_n| \tag{3.2}$$

where $\{b_n\}$ is an arbitrary set of non-vanishing constants and $\{|e_n\rangle\}$ an arbitrary set of orthonormal vectors that is complete in the P-subspace. Of course the form of χ is not uniquely defined by eq. (3.1). Certain symmetries, that we do not discuss here (cf. refs. 11, 13), restrict to some extent the admissible transformations.

Any transformation χ such that Φ_o is dissipative, may be used to construct a functional of $\tilde{\rho}_o$ that decreases in time. An obvious choice is the quadratic form

$$\Omega_o\{\tilde{\rho}_o\} = tr(\chi^{-1}\tilde{\rho}_o)^+(\chi^{-1}\tilde{\rho}_o) \tag{3.3}$$

Because of the second relation in eq. (3.1), it follows that $d\Omega_o/dt \leqslant 0$. However, this form of χ is not unique. The χ transformation is the restriction to the P-subspace of a more general transformation Λ ($\chi = P\Lambda P$) that relates the projections Π and P

$$\Pi = \Lambda P \Lambda^{-1} \tag{3.4}$$

Of course eq. (3.4) is not sufficient to determine Λ and one needs to analyze in more detail all possible contributions to the solution of the von Neumann equation[11].

To conclude this section let us present a few general remarks on non-unitary transformations and their implication in the theory of dissipative phenomena. Independently of the preceeding discussion, where we have insisted on the formulation of a theory linked to the kinetic description, we may ask for conditions that insure the existence of an invertible transformation Λ that transforms L to a dissipative operator, i.e.

$$\Phi = \Lambda^{-1} L \Lambda \quad ; \qquad (2i)^{-1}\left[\Phi - \Phi^{+}\right] \leqslant 0 \tag{3.5}$$

Such a transformation introduces a new representation, called sometimes the "physical representation", in which the evolution is given by a manifestly irreversible equation, namely

$$\overset{(p)}{\rho} = \Lambda^{-1}\rho \quad ; \qquad i\partial_t \overset{(p)}{\rho} = \Phi \overset{(p)}{\rho} \tag{3.6}$$

It may be shown easily that if the spectrum of L is discrete, Λ is unitary and Φ is hermitian. Therefore, for the existence of a non-trivial Λ, the continuity of the spectrum of L is a necessary condition. Moreover it turns out that a Lebesgue spectrum is a sufficient condition (more precise statements may be found in ref. 15).

Here also, eq. (3.5) does not specify a unique transformation and supplementary conditions must be imposed. Our aim is to describe a situation where we have well defined entities (i.e. "physical" particles) in interaction. We refer to a recent paper on the Friedrichs model[16] where a discussion is presented of the possibility of associating the energies of unstable particles as eigenvalues of the (super) operator $\mathcal{H} = \frac{1}{2}[H,\,]_{+}$ $(\mathcal{H}\rho = \frac{1}{2}(H\rho+\rho H))$ and ask then for a Λ which diagonalizes this operator. Thus in the "physical representation" we may associate an entropy production to the process of the decay and at the same time define the "physical" particles.

4. Concluding Remarks. It is impossible in this short note to examine carefully several subtle questions that may be raised in connection with the description of irreversible processes and the formalism that we have sketched. Concerning unstable states we have indicated that their definition should be given in terms of non-orthogonal projection (in appropriate spaces). This means that for the description of decay phenomena it is not sufficient to analyze what happens in some suitable subspace characterized by an orthogonal projection. We have seen that,

in general, the complementary subspace should be taken into account. An illustration of this point is provided by the microscopic derivation of linearized hydrodynamics in a simple fluid(17). There, the (damped) sound, shear and thermal modes are the exact analogue of exponentially decaying states.

We have used throughout this article the von Neumann equation. When we deal with the approach to equilibrium in general terms, mixed states are the rule and pure states the exception. Note also that a Hilbert space, where states and observables are Hilbert-Schmidt operators cannot be a proper framework for a rigorous formulation of such a theory. The Hamiltonian has a continuous spectrum and consequently observables represented by arbitrary bounded operators should be taken into consideration.

One should observe that we have not made any statements about the dominance of the exponentially varying terms in some time interval (e.g. short, intermediate or long times). In our opinion the interpretation of unstable states as exponentially decaying terms in the solution of the von Neumann equation is not related to the possible dominance of these terms. Non-exponential terms simply represent the evolution of the field that is always present.

Acknowledgements. This work has been partially supported by the Belgian Government "Actions de Recherche Concertées" Conv. n° 76/81-II.3. One of us (A.G.) wishes to thank the N.R.C. "Demokritos" in Athens and the "Center for Statistical Mechanics and Thermodynamics" of the University of Texas at Austin for their hospitality during the summer 1978, as well as the "Fonds National de la Recherche Scientifique" (Belgium) for a travel grant.

References.

1. For a comprehensive review see : I. Prigogine and A.P. Grecos, in G. Toraldo di Francia (Ed.), "Problems in the Foundations of Physics", North-Holland, Amsterdam (to appear).
2. W. Pauli, in P. Debye (Ed.), "Probleme der Modernen Physik", Verlag S. Hirzel, Leipzig, 1928.
3. A. Uhlmann in "Proceedings of the International Colloquium on the Role of Mathematical Physics in the Development of Science", UNESCO, Paris, 1978.
4. I. Prigogine, "Non-Equilibrium Statistical Mechanics", Interscience, New York, 1962.
5. R. Zwanzig, J. Chem. Phys. 33, 1388 (1960) ; Physica 30, 1109 (1964).
6. F. Haake, in "Springer Tracts in Modern Physics" 66, 98 (1973).
7. E.B. Davies, "Quantum Theory of Open Systems", Academic Press, New York, 1976.
8. K.O. Friedrichs, Commun. Pure Applied Math., 1, 361 (1948).
9. I. Prigogine and P. Résibois, Physica 27, 629 (1961).
10. I. Prigogine, C. George and F. Henin, Physica 45, 418 (1969).
11. I. Prigogine, C. George, F. Henin and L. Rosenfeld, Chemica Scripta 4, 5 (1973).

12. A.P. Grecos, T. Guo and W. Guo, Physica 80A, 421 (1975).
13. A.P. Grecos and M. Theodosopulu, Acta Phys. Polonica A50, 749 (1976).
14. A. Grossmann, J. Math. Phys. 5, 1025 (1964) ; J.-M. Combes, in "Proceedings of the International Congress of Mathematicians", Vancouver, 1974 ; see also the contributions by V. Gorini and G. Paravacini, H. Baumgartel and W.C. Schieve in these Proceedings.
15. B. Misra, Proc. Nat. Acad. Sci. (U.S.A.) 75, 1627 (1978).
16. C. George, F. Henin, F. Mayné and I. Prigogine, Hadronic J. 1, 520 (1978).
17. M. Theodosopulu, A.P. Grecos and I. Prigogine, Proc. Nat. Acad. (U.S.A.) 75, 1632 (1978) ; M. Theodosopulu and A.P. Grecos, Physica to appear.

ANALYTIC CONTINUATION IN DECAY-SCATTERING SYSTEMS

W. C. Schieve and T. Bailey
University of Texas at Austin

Singularities of the analytically continued Hamiltonian resolvent have long served as the mathematical basis for modeling decay and resonance scattering phenomena. The central idea is well known. A pole at $z = E_R - i\Gamma/2$ ($\Gamma > 0$) lies near the real axis on an "unphysical" sheet of the Riemann surface characterizing the function $(H - z)^{-1}$. This pole leads to the Breit-Wigner form for the scattering cross-section, i.e. $\sigma(E) \sim [(E - E_R)^2 + (\Gamma/2)^2]^{-1}$. If the binding potential of the interaction Hamiltonian is increased, the resonance pole can move into the physical sheet to a point $z = E_B$ with E_B negative real, i.e. the energy of a bound state. The occupation probability of this prepared bound state obeys the Wigner-Weisskopf decay law when the system evolves in time according to the Hamiltonian with the resonant pole in its resolvent. There is a "pole contribution" proportional to $\exp(-\Gamma t)$ and a "threshold contribution" resulting from the branch point of $(H - z)^{-1}$ at the scattering threshold [1].

Several interesting means of calculating this pole contribution have been proposed in recent years. Each involves the use of analytic continuations to define inner products involving resonant solutions to the Schrödinger equation.

In coordinate space calculations [2, 3, 4] the difficulty is that an inner product such as

$$\langle E_m|E_n\rangle = \int d^3r \langle E_m|r\rangle\langle r|E_n\rangle \tag{1}$$

is not in general well-defined when either wavefunction in the integrand is analytically continued to unphysical energy. This is because $\langle r|E_n\rangle \sim \exp(ip_n r)$ for large r, and $p_n = (2mE_n)^{1/2}$ has a negative imaginary part for a resonance momentum. There have been several attempts to avoid this divergence. A suggestion due to Romo [2] is to analytically continue the value of the integral (redefined in terms of a modified bound state wavefunction) rather than the integrand. Berggren [3] redefines the inner product of eq. 1 by inserting a convergence factor $\exp(-\lambda r)$ in the integrand and then taking the limit $\lambda \to 0$ following the integration. This method is not entirely successful since the regularization fails for products of resonance and scattering states.

Berggren also indicates that resonant momentum states may be incorporated in a complete set of eigenvectors with continuum eigenvalues forming a contour deformed into the unphysical region of the complex momentum plane. Thus resonances contribute in the same manner as bound states when the closure relation is used to expand an inner product. For example, the unstable state occupation probability includes oscillating terms resulting from bound states and exponential decay terms associated with resonant states. Sudarshan et al [5] have recently obtained similar results for the simple Lee model from field theory.

In this work we suggest that the existence of a complete set of complex energy eigenvectors follows from very simple properties of real energy eigenstates. We choose a representation which does not exhibit the divergence problems inherent in the coordinate representation, and illustrate the approach with a solvable model.

First we show that the structure of the Hamiltonian resolvent can be used to establish a closure relation involving both bound state and resonant solutions of the Schrödinger equation. We consider a Hamiltonian H with a set of discrete eigenvalues E_n and continuum eigenvalues E which lie on the real axis contour σ. Bound states are denoted by $|E_n\rangle$ and continuum states $|E\beta\rangle$ where β indexes degeneracies in E. Assuming that these eigenvectors form a complete set, the inner product $\langle\psi|\omega\rangle$ can be written

$$\langle\psi|\omega\rangle = \sum_n \langle\psi|E_n\rangle\langle E_n|\omega\rangle + \int_\sigma dE \sum_\beta \langle\psi|E\beta\rangle\langle E\beta|\omega\rangle \ . \tag{2}$$

From the definition of the resolvent $R_{\psi\omega}(z) = \langle\psi|(H-z)^{-1}|\omega\rangle$ it follows that the integrand in eq. (2) can be expressed as

$$\sum_\beta \langle\psi|E\beta\rangle\langle E\beta|\omega\rangle = \frac{1}{2\pi i}\left[R_{\psi\omega}(E+i0) - R_{\psi\omega}(E-i0)\right] . \tag{3}$$

We assume that the vectors $|\psi\rangle$ and $|\omega\rangle$ are chosen such that eq. (3) can be analytically continued from its region of definition, $E \in S$, to the lower half plane ($\mathrm{Im}\, E < 0$). Then the resolvent matrix element $R_{\psi\omega}(E+i0)$ may have simple poles at points z_i in its continuation through the cut σ to the "unphysical" region. The other term on the left side of eq. (3), $R_{\psi\omega}(E-i0)$, is analytic for $\mathrm{Im}\, E < 0$ because the continuation is done without crossing the real axis cut.

While holding the endpoints of σ fixed, we may deform the integration contour of eq. (2) into the lower half plane. If the deformed contour Γ crosses resonant poles, but avoids other singularities of the analytically continued integrand, then the residue theorem implies

$$\langle\psi|\omega\rangle = \sum_n \langle\psi|E_n\rangle\langle E_n|\omega\rangle + \int_\Gamma dz \sum_\beta \langle\psi|z\beta\rangle\langle z\beta|\omega\rangle + \sum_i \mathrm{Res}\, R^{(+)}_{\psi\omega}(z)\big|_{z=z_i} \tag{4}$$

where the (+) superscript in the final sum indicates that the residue is to be calculated on the branch of $R_{\psi\omega}(z)$ located by continuing from the physical branch through the cut to points z such that $\mathrm{Im}\, z < 0$. It is natural to choose a contour Γ such that the only poles between σ and Γ are those associated with bound states of some unperturbed Hamiltonian.

Since bound states of H satisfy the relation

$$\langle\psi|E_n\rangle\langle E_n|\omega\rangle = \lim_{z\to E_n} (E_n - z)\, R_{\psi\omega}(z) \tag{5}$$

it is clear that the last term in eq. (4) is related to the first by an analytic continuation. Of course only a discrete set of points E_n satisfy the bound state eigenvalue condition, but we may still define $<\psi|E_n><E_n|\omega>$ at other points in the physical sheet, and then continue the function to resonance solutions in the unphysical sheet.

Eqs. (2) through (5) show that for suitably chosen $<\psi|$ and $|\omega>$, an alternate way to define the inner product is

$$<\psi|\omega> = \sum_n <\psi|E_n><E_n|\omega> + \int_\Gamma dz \sum_\beta <\psi|z\beta^R><z\beta^L|\omega> + \sum_i <\psi|z_i^R><z_i^L|\omega> \tag{6}$$

with the right eigenvectors $|z_i^R>$ and $|z\beta^R>$ defined as the analytic continuations of the kets $|E_n>$ and $|E\beta>$ respectively, and left eigenvectors defined by continuing $<E_n|$ and $<E\beta|$.

A natural representation in which to calculate these eigenvectors is the spectral representation of a Hamiltonian H_0 having bound state eigenvalues ε_i which become the resonances z_i when the perturbing potential $V = H - H_0$ is added. If H_0 has a single bound state $|\varepsilon_1>$ and a non-degenerate continuum of scattering states $|\varepsilon>$ for ε an element of the real axis contour S, then the closure relation

$$1 = |\varepsilon_1><\varepsilon_1| + \int_S d\varepsilon \; |\varepsilon><\varepsilon| \tag{7}$$

can be used to expand the Schrödinger equation $(H_0 + V)|E> = E|E>$. This procedure yields the coupled integral equations

$$(\varepsilon_1 + <\varepsilon_1|V|\varepsilon_1>)<\varepsilon_1|E> + \int_S d\varepsilon \; <\varepsilon_1|V|\varepsilon><\varepsilon|E> = E<\varepsilon_1|E> \tag{8}$$

and

$$<\varepsilon|V|\varepsilon_1><\varepsilon_1|E> + \varepsilon<\varepsilon|E> + \int_S d\varepsilon' \; <\varepsilon|V|\varepsilon'><\varepsilon'|E> = E<\varepsilon|E>. \tag{9}$$

Together with a normalization condition, these equations determine allowed eigenvalues and the elements $<\varepsilon_1|E>$ and $<\varepsilon|E>$ of the eigenvectors of H. We solve eqs. (8) and (9) for the functions $<\varepsilon_1|E>$ and $<\varepsilon|E>$ instead of using eq. (1) to calculate them. There is no need to introduce resonant wavefunctions $<r|E>$ which diverge for large r. For specific models in the energy representation of eq. (7), well-defined resonance eigenvectors $|z_i^R>$ may be determined by analytically continuing the solutions $<\varepsilon_1|E_n>$ and $<\varepsilon|E_n>$ which characterize a bound state of H.

We illustrate these ideas with the solution to a one-dimensional potential scattering problem. The Hamiltonian has the form $H = H_0 + V(x)$ where H_0 includes the free particle Hamiltonian and any attractive potential which yields one bound state $|\varepsilon_1>$ and a non-degenerate continuum of scattering states $|\varepsilon>$ of H_0. To obtain a solvable form for eqs. (8) and (9), we choose $V(x) = \lambda\delta(x - x_0)$ with λ positive real. This repulsive scattering center will decrease the binding energy of $|\varepsilon_1>$, and for sufficiently large λ, this state may become anti-bound or resonant.

Matrix elements of H are written

$$\begin{aligned}
<\varepsilon_1|H|\varepsilon_1> &= \varepsilon_1 + \lambda|<x_0|\varepsilon_1>|^2 \equiv h_0 \\
<\varepsilon|H|\varepsilon_1> &= \lambda<\varepsilon|x_0><x_0|\varepsilon_1> \equiv V(\varepsilon) \\
<\varepsilon|H|\varepsilon'> &= \varepsilon\,\delta(\varepsilon-\varepsilon') + kV(\varepsilon)\bar{V}(\varepsilon')
\end{aligned} \tag{10}$$

where we have introduced the H_0 wavefunctions $<x|\varepsilon_1>$ and $<x|\varepsilon>$, and the constant $k = (\lambda<x_0|\varepsilon_1><\varepsilon_1|x_0>)^{-1}$. Details of the solution to eqs. (8) and (9) for this special case may be found in reference [6]. Bound state eigenvectors may be written

$$|E_n> = \begin{pmatrix} <\varepsilon_1|E_n> \\ <\varepsilon|E_n> \end{pmatrix} = \begin{pmatrix} [\eta'(E_n)]^{-1/2} \\ V(\varepsilon)(E_n-\varepsilon)^{-1}\beta^{-1}(E_n)[\eta'(E_n)]^{-1/2} \end{pmatrix} \tag{11}$$

with E_n a solution to the eigenvalue equation

$$\eta(z) \equiv z - h_0 + \beta^{-1}(z)\int_S |V(\varepsilon)|^2(\varepsilon - z)^{-1}\,d\varepsilon = 0 \tag{12}$$

and

$$\beta(z) \equiv 1 + k\int_S |V(\varepsilon)|^2(\varepsilon - z)^{-1}\,d\varepsilon. \tag{13}$$

It is easy to show that $\eta^{-1}(z) = <\varepsilon_0|(z-H)^{-1}|\varepsilon_0>$ and that $\eta(z)$ can have a resonance zero z_i in the lower half plane when the Cauchy integrals in its definition are analytically continued through the cut to the region Im $z < 0$. If $V(\varepsilon)$ and $|V(\varepsilon)|^2$ can be analytically continued to z_i, then $V(z_i)$, $\beta(z_i)$ and $\eta'(z_i)$ are well defined in the unphysical sheet, and can be used in eq. (11) to define the right eigenvector $|z_i^R>$.

For E any element of the contour S in eqs. (9) and (10), the Hamiltonian has a continuum solution $|E>$ defined by

$$<\varepsilon_1|E> = \bar{V}(E)\beta^{-1}(E+i0)\eta^{-1}(E+i0) \tag{14}$$

and

$$<\varepsilon|E> = \delta(E-\varepsilon) + V(\varepsilon)[1+k(E-h_0)](E-\varepsilon+i0)^{-1}<\varepsilon_1|E>. \tag{15}$$

Our justification of eq. (4) required that we be able to analytically continue the integrand of eq. (2). For our non-degenerate model, this integrand is $<\psi|E><E|\omega>$. Using eq. (1) and the notation $<\psi|\varepsilon_1> = \psi_1$ and $<\psi|\varepsilon> = \psi(\varepsilon)$, we may write

$$<\psi|E> = \frac{\psi_1\bar{V}(E)}{\beta(E+i0)\,\eta(E+i0)} + \psi(E) + \frac{\bar{V}(E)[1+k(E-h_0)]}{\eta(E+i0)\,\beta(E+i0)}\int_S \frac{\psi(\varepsilon)V(\varepsilon)\,d\varepsilon}{E-\varepsilon+i0}\,. \tag{16}$$

The function $<E|\omega>$ may be obtained from eq. (16) by taking its complex conjugate and replacing ψ_1 by $\omega_1 = <\varepsilon_1|\omega>$ and $\psi(\varepsilon)$ by $\omega(\varepsilon) = <\varepsilon|\omega>$. The contour deformation procedure used in deriving eq. (4) is appropriate if the only singularities encountered in continuing $<\psi|E><E|\omega>$ are those due to the resonant zeroes in the continuation of $\eta(E+i0)$. This is the case when the functions $\psi(E)$, $\omega(E)$, $V(E)$ and $\overline{V}(E)$ are analytic in the region between S and the deformed contour Γ. When these conditions are satisfied, the functions defining $|E>$ in eqs. (14) and (15) can be continued to Γ to define $|z^R>$ in the unphysical energy region. We show in reference [6] that the complex energy eigenvectors described above form a complete, orthonormal set.

References:

1. M. L. Goldberger, K. M. Watson, Collision Theory, John Wiley & Sons, New York.
2. W. J. Romo, Nucl. Phys. A116, 618 (1968).
3. T. Berggren, Nucl. Phys. A109, 265 (1968).
4. R. C. Fuller, Phys. Rev. 188, 1649 (1969).
5. E. C. G. Sudarshan, C. B. Chiu, V. Gorini, Phys. Rev. D, to appear.
6. T. Bailey, W. C. Schieve, Nuovo Cimento, to appear.

P11.

On the Inverse Problem of the Abstract Relativistic Scattering Theory.

H. Baumgärtel

Let $U(x)$, $U_0(x)$ be unitary representations of R^n (translation group) in Hilbert spaces $\mathfrak{H}$, $\mathfrak{H}_0$. Let P^{ac}, P_0^{ac} be the orthoprojections onto the subspaces of absolute continuity (with regard to the Lebesgue measure of R^n). Let J be a bounded (identification) operator from $\mathfrak{H}_0$ into $\mathfrak{H}$. Let $a \in R^n$, $|a| = 1$. With regard to the direction a one can introduce wave operators $W_\pm(U,U_0,J,a) = \text{s-lim}_{t\to\pm\infty} U(ta)JU_0(-ta)P_0^{ac}$. Desirable properties of $W_\pm$ are (I) $W_\pm^* W_\pm = P_0^{ac}$ (partial isometry); (II) $W_+(a)$ (resp. $W_-(a)$) should be independent of a, if a is varying within a certain fixed cone K. Further, if U_0, K are given, one can introduce the "wave morphism" $\mu_\pm(A) = \text{s-lim}_{t\to\pm\infty} U_0(ta)AU_0(-ta)P_0^{ac}$ ($\mu_\pm(A)$ independent of a, if $a \in K$). dom $\mu_\pm$ is called the wave algebra or the algebra of asymptotic constants. If $W_\pm$ exists, then the scattering operator is defined by $S = W_+^* W_-$. It is unitary on $P_0^{ac}\mathfrak{H}_0$ and commutes with $U_0(x)$. The forward problem of the scattering theory consists in the investigation of $W_\pm$ and S, if $\{U,U_0,J,K\}$ is given. The inverse problem is the problem to find the class of all U, such that $W_\pm(U,U_0,J,K)$ exist and $W_+^* W_- = S$, if $\{U_0,J,K,S\}$ is given, where S is unitary in $P_0^{ac}\mathfrak{H}_0$ and commutes with U_0.

THEOREM (Wollenberg, Rehberg). *The inverse problem of scattering theory is soluble if and only if there is a partial isometry* V *such that* $J - V \in \ker \mu_\pm$. If $\mathfrak{H} = \mathfrak{H}_0$ and $J=1$, then as a corollary one obtains: The inverse problem is soluble for every S, which is unitary in $P_0^{ac}\mathfrak{H}$ and commuting with regard to U_0. The proof of the theorem is a constructive one.

There is a certain abstract characterization of the class of all solutions. There are some applications to multichannel scattering (the problem is to prove, that the identification operators, which occur in the multichannel case, satisfy the condition of the theorem).

References

[1] M. Wollenberg, On the Inverse Problem in the Abstract Theory of Scattering, preprint, Berlin 1977.

[2] J. Rehberg, Dissertation, Berlin 1978.

THE RIGGED HILBERT SPACE AND DECAYING STATES*

A. Bohm
Center for Particle Theory, University of Texas, Austin, Texas 78712

The rigged Hilbert space (RHS) $\Phi \subset H \subset \Phi^{\times}$, introduced into physics around 1965,[1] has already displayed so many features which make it ideally suited for the description of quantum mechanics[2] that one wonders why it has not yet beem more generally accepted by physicists. More recently, around 1975, a new attribute of the RHS was uncovered: It occurred that decay phenomena are most naturally described[3] using generalised eigenvectors of the self-adjoint energy operator with complex eigenvalue.[5] Such a description would establish the link between the S-matrix description of a resonance state as a pole and the usual description of states as vectors in a linear space; this is probably the reason why the complex energy eigenvectors have immediately caught the fancy of physicists working on decaying systems.[6,7] However, there are two points which were the actual motivation for the introduction of these generalised eigenvectors[4] and which have not been adequately mentioned in reference 6: 1) Their justification from the physical production process of decaying states, and 2) their application in the derivation of the decay rate formula.

1) Unstable systems are prepared by scattering experiments in which the delay time is large. The intermediate quasistationary system, when it is considered as an isolated decaying system ignoring the mode of formation, must therefore have the properties of a resonance which means that the state vector ϕ^R, which is to represent it, must have a Breit-Wigner energy distribution, i.e., if we write

$$<\phi^R| = \int dE <E| f^*(E-E_R) \tag{1}$$

where $|E>$ are the generalised eigenvectors of the essentially self-adjoint H with E belonging to the spectrum of $\bar{H}$, then

$$|f(E-E_R)|^2 = \frac{1}{\pi} \frac{\Gamma/2}{(E_R-E)^2 + (\frac{\Gamma}{2})^2} \tag{2}$$

with (E_R,Γ) the resonance parameters. Consequently

$$f(E-E_R) = \sqrt{\frac{\Gamma}{2\pi}} \frac{1}{z_R - E}; \qquad z_R = E_R - i\frac{\Gamma}{2} \tag{3}$$

We take the "scalar product" of (1) with $\varphi \in \Phi \subset H \subset \Phi^{\times}$, which has the property that $<E|\varphi>$ is the limit of a function $<z|\varphi>$ analytic in the upper half-plane (precisely, the value of the functional $<\phi^R|$ at $\varphi \in \Phi \cap H_+$ where H_+ is the space of Hardy class functions with respect to the upper plane). Then

*) This paper has been added to supplement and clarify the origin and development of the ideas discussed in the paper by Gorini and Parravicini.

$$<\phi^R|\varphi> = \sqrt{\frac{\Gamma}{2\pi}} \int dE <E|\varphi> \frac{1}{z_R^* - E} = \sqrt{\frac{\Gamma}{2\pi}} (-2\pi i) <z_R^*|\varphi> \tag{4}$$

In the last equality we have used a well known mathematical theorem[10] after extending the integration over $-\infty < E < +\infty$ whereby we have made a small error which goes to zero as $E_R/\Gamma \to \infty$.

In the same way one calculates

$$<\phi^R|H|\varphi> = z_R^* <z_R^*|\varphi> (-i\sqrt{2\pi\Gamma}) \tag{5}$$

and

$$<\phi^R|e^{iHt}|\varphi> = e^{iE_R t} e^{-\frac{\Gamma}{2}t} <\phi^R|\varphi> \tag{6}$$

Omitting again the arbitrary $\varphi \in \Phi \cap H_+$, we see that

$$<\phi^R| = -i\sqrt{2\pi\Gamma} <z_R^*| = -i\sqrt{2\pi\Gamma} \frac{1}{2\pi i} \int dE <E| \frac{1}{E - z_R^*} \tag{7}$$

is a vector which is a) normalized (element of H_+), b) not in the domain of $\bar{H}$, c) a generalised eigenvector of H with eigenvalues $z_R = E_R - i\frac{\Gamma}{2}$ and d) a decaying state vector.[8] Properties a) (which may be true only in the limit of the lower bound of the spectrum going to $-\infty$.) and b) are easily established and c) and d) are given by (5) and (6), respectively. A mathematical generalization of this physically motivated definition (7) is given by equations (6') of reference 6a.

2) The usefulness of describing decaying states by (7) is seen if one calculates the decay rate. The transition rate from a state ϕ into a subspace ΛH is given by (omitting all other quantum numbers but the energy):

$$\begin{aligned} \dot{P}(t) &= \frac{d}{dt} (\mathrm{Tr}\, \Lambda|\phi(t)><\phi(t)|) \\ &= -i \int dE_b \int dE dE' e^{-i(E-E')t} <b|V|E><E'|V|b><E|\phi><\phi|E'> \left(\frac{1}{E' - E_b - i\varepsilon} - \frac{1}{E - E_b + i\varepsilon} \right) \end{aligned} \tag{8}$$

where $\phi = \phi(t=0)$, $V = H - H_0$, $<b|V|E>$ is the T-matrix, $|b> = |E_b,0>$ is a basis of eigenvectors of H_0 for the subspace ΛH and the integration $\int dE_b$ runs over this subspace, and where generalised eigenvectors $|E>$ of H and $|E,0>$ of H_0 are connected by the Lippman-Schwinger equation[9]

$$|E> = |E,0> + \frac{1}{E - H_0 + i\varepsilon} V|E> \quad .$$

Though ϕ^R is not exactly a physically preparable state (described by elements of Φ)[2] because a resonance is always accompanied by background, the decay rate is the transition rate from the decaying state $\phi = \phi^R$. Inserting (7) into (8) one obtains for the decay rate

$$\begin{aligned} \dot{P}(t) = -i \int dE_b \int dE dE' <b|V|E><E'|V|b> \frac{\Gamma}{2\pi} \frac{e^{-iEt}}{E - (E_R + i\frac{\Gamma}{2})} \frac{e^{iE't}}{E' - (E_R - i\frac{\Gamma}{2})} \times \\ \times \left(\frac{1}{E' - (E_b + i\varepsilon)} - \frac{1}{E - (E_b - i\varepsilon)} \right) = \dot{P}_1 + \dot{P}_2 \end{aligned} \tag{9}$$

The integration can now be carried out for $\varepsilon > 0$ using again the well known mathematical theorem[10] and assuming that $\langle E|V|b\rangle$ is such that its conditions are fulfilled. For the first term $\dot{P}_1$ of (9) one integrates first over E' then takes the complex conjugate, integrates over E and takes again the complex conjugate. The result is:

$$\dot{P}_1(t) = 2\pi\Gamma i\int dE_b \, \langle b|V|E_R - i\tfrac{\Gamma}{2}\rangle\langle E_b + i\varepsilon|V|b\rangle \, \frac{e^{-iE_R t} \, e^{-\frac{\Gamma}{2}t} \, e^{iE_b t} \, e^{-\varepsilon t}}{E_b - E_R + i(\varepsilon + \frac{\Gamma}{2})} \tag{10}$$

where we have used $\langle b|V|E+i\frac{\Gamma}{2}\rangle = -\langle b|V|E-i\frac{\Gamma}{2}\rangle$ for the T-matrix, which is an immediate consequence of the well known symmetry relation for the S-matrix.[11] For the initial decay rate $\dot{P}(0)$ one obtains from this for $\varepsilon \to \frac{\Gamma}{2} \to 0$

$$\dot{P}_1(0) = 2\pi\Gamma i\int dE_b \, \langle b|V|E_R - i\tfrac{\Gamma}{2}\rangle\langle E_b + i\tfrac{\Gamma}{2}|V|b\rangle \, \frac{1}{E_b - E_R + i0} \tag{11}$$

A similar expression is obtained for $\dot{P}_2(0)$ with $+i0$ replaced by $-i0$. If one then uses the well known relation between distributions

$$\frac{1}{E_b - E_R + i0} - \frac{1}{E_b - E_R - i0} = -2\pi i\delta(E_b - E_R) \tag{12}$$

and re-inserts (7) one obtains the well known expression for the initial decay rate

$$\dot{\tau}(0) = 2\pi\int dE_b \, \langle b|V|\phi^R\rangle\langle\phi^R|V|b\rangle \, \delta(E_b - E_R) \tag{13}$$

If the T-matrix is a slowly varying function of the complex energy z, $\langle E_b + i\varepsilon|V|b\rangle = \langle E_R + i\frac{\Gamma}{2}|V|b\rangle$ for $E_b \approx E_R$, then one can use (10) also for Γ which are larger than the energy resolution of the detector, $\Delta E_b \ll \Gamma$, and obtains $(\varepsilon \to +0)$:

$$\begin{aligned}\dot{P}(0) &= 2\pi\Gamma i\int dE_b \left\{ \frac{\langle b|V|E_R - i\frac{\Gamma}{2}\rangle\langle E_b + i\varepsilon|V|b\rangle}{E_b - E_R + i(\varepsilon + \frac{\Gamma}{2})} - \frac{\langle b|V|E_b - i\varepsilon\rangle\langle E_R + i\frac{\Gamma}{2}|V|b\rangle}{E_b - E_R - i(\varepsilon + \frac{\Gamma}{2})} \right\} \\ &\approx 2\pi\Gamma i\int dE_b \, \langle b|V|E_R - i\tfrac{\Gamma}{2}\rangle\langle E_R + i\tfrac{\Gamma}{2}|V|b\rangle \left\{ \frac{-i\Gamma}{(E_b - E_R)^2 + (\frac{\Gamma}{2})^2} \right\} \\ &= 2\pi\int dE_b \, \langle b|V|\phi^R\rangle\langle\phi^R|V|b\rangle \left\{ \frac{\Gamma/2\pi}{(E_b - E_R)^2 + (\frac{\Gamma}{2})^2} \right\}\end{aligned} \tag{14}$$

where the last factor is the well known natural line width. If one compares these simple straightforward calculations with the conventional procedure[9] one will appreciate the usefulness of our new vectors (7).

The question we have addressed ourselves to is the description of a decaying state with the resonance parameter (E_R, Γ); our suggestion is (7). We have not discussed the question how one calculates the position of the resonance. That is a completely different problem whose answer depends upon the particular property of the energy operator H. If one has a particular energy operator with a particular set of resonance parameters $(E_R^{(n)},\Gamma^n)$, n = 1,2,..., then one would like to have only those generalised eigenvectors with complex eigenvalue contained in ϕ^x which have these resonance parameters. The work of Napiorkowski[12] would suggest to do this by constructing the topology in ϕ such that all unwanted generalised eigenvectors are excluded from ϕ^x; one readily sees that this method does not work for the generalised eigenvectors (7). Also, the question of the topology for ϕ has not yet been treated.

Acknowledgement: I am particularly grateful: to B. Nagel, who taught me the properties of the generalised eigenvectors with complex eigenvalue many years ago and with whom I discussed the vectors defined by (7); to K. Napiorkowski with whom I discussed (7) and who referred me to the work of Baumgärtel; to H. Baumgärtel for his extensive correspondence on the subject of this paper over the past two years. I also gratefully acknowledge discussions of this subject with L. Horwitz, E.C.G. Sudarshan, I. Prigogine and V. Gorini in Fall 1976 and with W.C. Schieve, A. Grecos, V. Gorini and G. Parravicini more recently.

Footnotes and References

1. J.E. Roberts, Journal Math. Phys. 7, 1097 (1966); A. Böhm, Boulder Lectures in Theoretical Physics, Vol. 9A, 255 (1966).
2. A. Böhm, The Rigged HilbertSpace and Quantum Mechanics, Springer Lecture Note 78 (1978); J.E. Roberts, Comm. Math. Phys. 3, 98 (1966);J.P. Antoine, Journal Math. Phys. 10, 53 (1969); O. Melsheimer, Journal Math. Phys. 15,902 (1974). G. Lassner, Wiss. Z. Karl-Marx-Univ. Leipzig Math.-Naturwiss. R. 22 (1973) H.Z.; A. Böhm, Lectures at the Istambul Summer Institute 1970 A.O. Barut, editor, Reidel Publishing Co, 1973
3. Two distinct suggestions have been made, the first (which constructs a RHS that contains exactly those complex energy eigenvectors which are related to the discrete eigenvalues of $H_0 = H - V$) by H. Baumgärtel in Math. Nachr. 75, 133 (1976) and the second (which defines complex energy eigenvectors that have a Breit-Wigner energy distribution) by A. Böhm in a manuscript distributed to colleagues in 1976 which will appear as Chapter XXI of reference 4.
4. A. Böhm, Quantum Mechanics, Springer, New York (1979)
5. That such eigenvectors exist was known to physicists for a long time from their careless manipulation with generators of non-compact subgroups in representations of Lie algebras. The first detaile rigorous discussion of such vectors of SU (1,1) was given by G. Lindblad, B. Nagel, Ann. Inst. Henri Poincaré XIII, 27, 1970. See also G.J. Iverson, Phys. Letters 26B, 229 (1968). Usually one tries to exclude from ϕ^x all generalised eigenvectors with eigenvalues that do not belong to the spectrum by an appropriate choice of the nuclear topology of ϕ; see reference 12.
6. a) V. Gorini, G. Parravicini, Proceedings of the VIIth International Group Theory Colloquium.
b) E.C.G. Sudarshan, C.B. Chiu, V. Gorini, Phys. Rev. D, to appear. How much the use of the rigged Hilbert space simplifies the description can be seen from a comparison of these two papers. G. Parravicini, V. Gorini, E.C.G. Sudarshan, C.B. Chiu, Journal Math. Phys., to be submitted.

7. The association of complex energies with resonances is of course almost as old as quantum mechanics, and other recent suggestions include the use of non-self-adjoint Hamiltonian operators. See e.g. J.S. Howland, J. Math. Anal.Appl. 50, 415 (1975); B. Simon, Annals of Math. 97, (1973), 247; J. Aguilar and J.M. Combes, Comm. Math. Phys. 22, 280 (1971); L.P. Horwitz and I. Sigal, Helv. Phys. Acta, to be published. T.K. Bailey, W.C. Schieve, Nuovo Cimento, Dec. 1978; C. George, F. Henin, F. Mayne, I. Prigogine, Hadron Journal, vol. 1 (1978); A.P. Grecos, I. Prigogine, Proc. Nat. Acad. Sci., USA, 69, 1629 (1972). A. Grossmann, J. Math. Phys. 5, 1025, (1964).
8. F. Fonda, G.C. Ghirardi, A. Rimini, Rep. Prog. Phys. 41, 587 (1978), describe the well known deviations from the exponential decay law which however do not apply for ϕ^R because ϕ^R is not in the domain of the Hilbert space operator $\bar{H}$ (small times) and ϕ^R was defined in the approximation $E_R/\Gamma \to \infty$ with the lower bound of the spectrum going to $-\infty$ (large times).
9. M.L. Goldgerger, K.M. Watson, "Collision Theory," Wiley, 1964.
10. See e.g. H.M. Nussenzveig, "Causality and Dispersion Relations," Academic Press, 1972, Equation (1.6.6.).
11. e.g. H.M. Nussenzveig, "Causality and Dispersion Relations," Academic Press, 1972, Equation (2.8.16).
12. K. Napiorkowski, Bulletin of the Polish Academy of Sciences 22, 1215 (1974); 23, 251 (1975).

Excitation and Decay of a Multilevel Atom*

L. Davidovich
Departamento de Física, Pontifícia Universidade Católica, Rio de Janeiro, Brazil
and
H.M. Nussenzveig
Instituto de Física, Universidade de São Paulo, Brazil

The time evolution of a model system consisting of a nonrelativistic hydrogen atom interacting with the quantized radiation field is analysed. The treatment employs Van Hove's resolvent operator method, but not the Van Hove limit. With a truncated Hamiltonian, which eliminates persistent perturbation effects, the model becomes exactly soluble and allows a complete discussion of the excitation process and the decay law, including to some extent the effects of transitions between excited states. In order to apply a similar procedure to the full Hamiltonian, the persistent effects are first removed by means of a dressing transformation, defined order-by-order in perturbation theory. Van Hove's formalism is then applied to the Hamiltonian obtained by carrying out the transformation up to second order, and the implications for the decay problem are discussed.

*Work partially supported by the National Research Council of Brazil.

Dynamical Spin Spreading of Unstable Particles in Four Models; G. N. Fleming, Davey Lab., Penna. State Univ., University Park, Pa. 16802

We show that in four models of unstable particles the matrix elements of the Pauli-Lubanski invariant are not compatible with a sharp spin spectrum (1). The models are:

(i) The interaction responsible for the instability is separated off from the Poincare generators and the unstable particle states lie in an irreducible representation space of the remaining generators at a definite time.

(ii) The same as (i) except that the generator separation occurs on a definite null plane rather than at a definite time (thus escaping Haag's theorem).

(iii) The unstable particle states are constructed by applying a nonconserved charge operator to the states of an irreducible representation space of the full Poincare generators. The charge operator is evaluated at a definite time.

(iv) The same as (iii) except that the charge operator is defined on a definite null plane (thus escaping Coleman's theorem).

The results are:

$$\text{(i)}\quad \langle\vec{p}',t|W^2|\vec{p},t\rangle = -2p_0\delta^3(\vec{p}'-\vec{p})\left\{\frac{(2\pi\hbar)^6}{2p_0}\oint dn\ \delta^3(\vec{p}-\vec{p}_n)|\langle n|\mathcal{L}_{int}(0)\vec{L}|\vec{p},0\rangle|^2\right\}$$

where $\vec{L}|\vec{p},0\rangle \equiv ih\left(\vec{p}\times\frac{\partial}{\partial\vec{p}}\right)|\vec{p},0\rangle$ and the $\langle n|$ form a complete sets of states.

$$\text{(ii)}\quad \langle\tilde{p}',\tau|W^2|\tilde{p},\tau\rangle = -2p_-\delta^2(\bar{p}'-p)\delta(p'_--p_-)$$

$$\left\{(2\pi h)^6\frac{\hbar^2}{2}\oint dn\ p_-\delta^2(\bar{p}-\bar{p}_n)\delta(p_--P_{n-})|\langle n|\mathcal{L}_{int}(0)\ \frac{\partial}{\partial\bar{p}}\ |\tilde{p},0\rangle|^2\right\}$$

$$\text{(iii)}\quad \langle\vec{p}',t|W^2|\vec{p},t\rangle = -2p_0\delta^3(\vec{p}'-\vec{p})\left\{(2\pi\hbar)^6\frac{\hbar^2}{2p_0}\oint dn\delta^3(\vec{p}-\vec{p}_n)|\langle n|\partial^\mu j_\mu(0)\vec{L}|p\rangle|^2\right\}$$

$$\text{(iv)}\quad \langle\tilde{p}',\tau|W^2|\tilde{p},\tau\rangle = -2p_-\delta^3(\tilde{p}'-\tilde{p})\left\{\frac{\hbar^4}{2}(2\pi\hbar)^6\oint dnp_-\delta^3(\tilde{p}-\tilde{p}_n)|\langle n|\partial^\mu j_\mu(0)\ \frac{\partial}{\partial\bar{p}}|p\rangle|^2\right\}$$

---

[1] G. N. Fleming, J. Math. Phys. 13, 626 (1972).

SEQUENTIAL DECAYS OF OPEN QUANTUM SYSTEMS

L. Fonda[+*], G.C. Ghirardi[+*], C. Omero[*], A. Rimini[*], T. Weber[*]
+ International Center for Theoretical Physics, Trieste, Italy
* Istituto di Fisica Teorica, Università di Trieste, Italy

The same treatment which takes into account the occurence of system-environment interactions as the one developed for ordinary decay[1] can be applied to the case of sequential decays, yielding the classical exponential-type laws for the occupation numbers of the various levels[2]. We give here a summary of the procedure.

We assume that the Hilbert space $\mathcal{H}$ of the system is the direct sum of sectors $\mathcal{H}^{(i)}$ identified by the states $|j,\alpha>_{(i)}$ obtained by Hamiltonian evolution from the state $|i>$ characterizing the i-th excited level of the system. j denotes that the system is in the j-th level, while α specifies the number, type, momenta, etc. of the decay products as obtained from the process $i \to j$ even through intermediate steps. We next choose our apparatuses so that, when the emitted decay fragments corresponding to the transition $i \longrightarrow j$ are found, they are absorbed. The global effect on the ϱ-operator of an interaction with the environment is then expressed as

$$\varrho(t) \longrightarrow \varrho^*(t) = \sum_i \sum_{j\alpha} |j>_{(i)} <j,\alpha|\, \varrho(t)\, |j,\alpha>_{(i)} <j| \qquad (1)$$

If λ is the mean frequency of the randomly distributed measurements, it is easy to obtain an evolution equation for the ϱ-operator[2]. If an additional measurement is performed at time t, it is even possible to get an equation for the occupation number $N_j^*(t)$ for the j-th level (one takes of course $\varrho(0) = |1><1|$):

$$N_j^*(t) = e^{-\lambda t} P_{1\to j}(t) + \lambda \int_0^t d\delta\, e^{-\lambda\delta} \sum_{i=1}^{j} P_{i\to j}(\delta) N_i^*(t-\delta) \qquad (2)$$

where $P_{i\ j}(t) = \sum_\alpha |_{(i)}<j,\alpha|\exp(-iHt)\,|i>|^2$ and H is the self-adjoint Hamiltonian. Eq. 2 has a simple diagrammatic representation:

1 ═══ j (0 to t) = 1 ─── j (0 to t) + 1 ═══ i ─── j (0, t−δ, t)

where single (double) line represents evolution without (with) measurements.

One readily gets for the Laplace transform of Eq. (2):

$$n_j^*(s) = p_{1\to j}(s+\lambda) + \lambda \sum_{i=1}^{j} p_{i\to j}(s+\lambda)\, n_i^*(s) \qquad (3)$$

which is solved by iteration:

$$n_j^*(s) = \left\{ 1-\lambda p_{j\to j}(s+\lambda) \right\}^{-1} \left\{ p_{1\to j}(s+\lambda) + \lambda \sum_{k=1}^{j-1} p_{k\to j}(s+\lambda)\, n_k^*(s) \right\} \qquad (4)$$

The inverse Laplace transforms of (4) are dominated by the poles occuring for s real $= -\gamma_j$ where the denominators $1-\lambda p_{j\to j}(s+\lambda)$ vanish. One has that $0 < \gamma_j \ll \lambda$. The contributions to $N_j^*(t)$ not associated with these poles are then negligeable after some reductions have taken place, so that we finally get

$$N_j^*(t) = \sum_{k=1}^{j} q_{kj}^* \exp(-\gamma_k t) \qquad (5)$$

which is just the classical formula one obtains for sequential decays. Therefore also in this case the measurements wipe out the deviations from the exponential-type law. The numbers N_j^* are simply related to the numbers N_j obtained when no extra measurement is performed at time t(see Ref. 2). One has only to substitute q_{kj}^* with $q_{kj}^*(1-\gamma_k/\lambda)^{-1}$. Since $\gamma_k/\lambda \ll 1$, the two numbers differ very little.

---

1. See L. Fonda, G.C. Ghirardi, A. Rimini and T. Weber, these Proceedings
2. L. Fonda, G.C. Ghirardi, C. Omero, A. Rimini and T. Weber, ICTP/78/35 (Trieste), in press on Phys. Rev. and invited paper at "Mathematical Problems in Irreversible Quantum Processes",Naples, 1978

QUANTIZATION VIA THE IMPRIMITIVITY SYSTEMS AND SUPERSELECTION RULES

N. Giovannini and C. Piron
University of Geneva (Switzerland)

In usual classical physics, states are represented by points in the phase space Ω or, better said, by point-valued functions on Ω. Observables are themselves functions on these states with values in subsets of Ω. Hence Ω can be identified with a set $\Gamma=\{\Gamma_A\}$ of values of a collection of observables $\{A\}$. In quantum physics, if states are now rays in Hilbert spaces, they also can be characterized by the same set of physical observables, that is by a collection of spectral families defined on the same Γ. Our point of view is thus to start from this real space Γ, to define on it a kinematical symmetry group by usual physical equivalence postulates and to search for the various possible realizations of the physical system in the following sense. An elementary system is assumed to be an irreducible representation in a state space K that admits covariant projections defined for each Borel subset E of each Γ_A corresponding to the subset of states having as property a value within E for the observable A.

As two examples we consider the non relativistic and relativistic particles, without specifying the interaction, and the observables $\{\vec{p},\vec{q},t\}$ and $\{p^{\mu},q^{\mu}\}$ respectively. The symmetry groups are the Newton group (translations in $\vec{p}$, in $\vec{q}$, in t and rotations) and the Einstein group (translations in p^{μ}, in q^{μ} and Lorentz rotations) respectively.

If now K is assumed to be a separable Hilbert space $\mathcal{H}$ the above projections turn out to be the usual covariant projection-valued measures on Γ, i.e. the problem can be solved in terms of Mackey's systems of imprimitivity (in particular in what concerns the admissible observables and the compatibilities of various observables). We show however that this is not enough to recover, in the Newton case, neither quantum mechanics (lack of time observable), neither of course classical mechanics, even though the above scheme was independent of any realization. The frame needs thus to be extended and we propose, on the basis of a property of systems of propositions, to consider families of Hilbert spaces parametrized by a set of, possibly continuous, superselection rules (s.s.r) that turns out to correspond to those observables that are compatible with all the other ones.

Generalizing thus the notions of imprimitivity systems and of induced representations to direct unions of Hilbert spaces, we find, for the above examples, two Newton elementary spinless particles : a quantal one that corresponds to the direct union, over the time as s.s.r, of Hilbert spaces all isomorphic to $\mathcal{L}^2(\mathbb{R}^3)$ and a classical one that corresponds to the direct union over phase space of 1-dim. Hilbert spaces. We also find two elementary spinless Einstein particles, a quantal one that does not contain any s.s.r and a classical one of 1-dim. Hilbert spaces over phase space.

A CLASS OF SOLUBLE ONE PARTICLE MODELS

A. Grossmann and R. Hoegh-Krohn

We study point interactions with several centers in two and three dimensions, and obtain explicit formulas for the resolvent. This provides useful models for various question in molecular and solid state physics.

RESONANCES IN QUANTUM MECHANICS OVER PHASE SPACE

A. Grossmann

Every reference frame in complexified phase space defines an irreducible representation space for canonical commutation relations. Quantum mechanical operators, defined by a quantification formula restrict to these representation spaces. The natural spectral theory in this context includes the concept of resonance.

COMPLEX MASS AND FIELD OPERATOR FOR UNSTABLE PARTICLE

Jerzy Lukierski

Institute of Theoretical Physics, University of Wroclaw

Wroclaw, Poland

Let us consider the model (see e.g. [1])

$$(\Box - m_0^2)B(x) = g_0 :A^2(x): \qquad (\Box - \mu_0^2)A(x) = g_0 :A(x)B(x): \tag{1}$$

The fields satisfy the following LSZ asymptotic condition

$$A(x) \underset{t\to\pm\infty}{\longrightarrow} A_{\substack{in\\out}}(x) \qquad B(x) \underset{t\to\pm\infty}{\longrightarrow} 0 \tag{2}$$

Using (2) one can write eq.(1) in Yang - Feldman form. After infinite number of iterations one obtains <u>the exact equation for the unstable particle field operator B</u>(x)

$$(\Box - \hat{M}_0^2)B(x) = g_0 :A_{in}^2(x): + g_0^3 F[x,g_0;A_{in},B] \tag{3}$$

where the nonlocal mass operator has a form

$$\hat{M}_0^2(x,x^1) = m_0^2\delta(x-x^1) + g_0^2\Delta_R(x-x^1;\mu_0^2)\{A_{in}(x),A_{in}(x^1)\} \tag{4}$$

and F is nonlocal and nonlinear functional of the field B. One can introduce the retarded <u>Green operator</u> satisfying the operator equation

$$(\Box - \hat{M}_0^2)\hat{\Delta}_R[g_0; A_{in}] = \delta \tag{5}$$

The unstable field operator can be written as follows

$$B(x) = B_{in}(x) + g_0^3 \int d^4x^1 \hat{\Delta}_R[x,x^1;g_0;A_{in}]F[x^1;g_0;A_{in};B] \tag{6}$$

where

$$B_{in}(x) = g_0^2 \int d^4x^1 \hat{\Delta}_R[x,x^1;g_0;A_{in}] :A_{in}^2(x^1): \tag{7}$$

In perturbation theory one can write

$$\hat{\Delta}_R = \Delta_R - \Delta_R(\hat{M}_0^2 - m_0^2)\Delta_R + \cdots \tag{8}$$

If we discuss unstable particles it is however important <u>not to expand $\hat{\Delta}$</u> in perturbative serie.

One can approximate the mass operator $\hat{M}_o$ by a function (distribution) if we assume in (7)

$$\hat{M}_o^2(x,x^1) \to \langle 0[\hat{M}_o^2(x,x^1)]0\rangle = M_o^2(x-x^1) \tag{9}$$

where the Fourier transform of M_o^2 can be written as follows

$$\tilde{M}_o^2(p) = m_o^2 + g_o^2 \int_{4\mu_o^2}^{\infty} ds \frac{\sigma_2(s)}{p^2-s+i0p_o} \qquad \sigma_2(s) = \frac{1}{16\pi^2}\sqrt{\frac{s-4\mu_o^2}{s}} \tag{10}$$

The introduction of complex mass parameter is based on the formula

$$M^2 - m_o^2 = g_o^2 \int_{4\mu_o^2}^{\infty} ds \frac{\sigma_2(s)}{m^2-s+i0} \qquad M^2 = m^2 - i\Gamma \tag{11}$$

The substitution (9) leads to the following equation (see also [2])

$$(\Box - M_{\pm}^2(\Box))B_{in}^{(\pm)}(x) = g_o : A_m^{(\pm)2}(x) \tag{12}$$

where $B_{in}^{(+)}$ ($B_{in}^{(-)}$) describes the positive (negative) frequency part of unstable field operator and one can prove that

$$M_{\pm}^2(p^2) = M_{\pm}^2 \pm \frac{1}{\pi i}(p^2-m^2)\int_{4\mu_o^2}^{\infty} ds \frac{\operatorname{Im} M^2\pm(s)}{(p^2-s\pm i0)(m^2-s+i0)} \tag{13}$$

where $M_{\pm}^2 = M^2 \mp i\Gamma$.

We see clearly that the mass operator for unstable particle <u>can never be represented by purely numerical complex mass parameter because in local QFT the real and imaginary parts of $M_{\pm}^2(p^2)$ are related by a Hilbert transform</u>. The formula (11) can be used for the discussion of deviations from pure "Breit - Wigner" formula which are due to local nature of decay interaction.

The full text of our lecture will appear under the same title as ICTP Trieste preprint.

References:

1. M.Veltman, Physica 29, 186 (1963).

2. A.Brzeski and J.Lukierski, Nuovo Cimento Lett.9, 205 (1974).

A canonical description of semigroup law for unstable systems

G. Vitiello[†]

Institute for Theoretical Physics, University of Alberta, Edmonton, Canada

Unstable systems are described by taking advantage of the existence of unitarily inequivalent representations of the canonical commutation relations (CCR) in Quantum Field Theory (QFT)[(1)]. Let $V_{\bar{k}}^{\dagger}(V_{\bar{k}})$ be the creation (annihilation) operator for the unstable $V_{\bar{k}}$-particle at time t = 0. Assume the volume V finite and suppose the $V_{\bar{k}}$-particle is a fermion (extension to boson is possible). The anticommutation relations for $V_{\bar{k}}^{\dagger}$, $V_{\bar{k}}$ are the usual ones. Let $|0\rangle_V$ denote the vacuum state at $t=0$: $V_{\bar{k}}|0\rangle_V = 0$, ${}_V\langle 0|0\rangle_V = 1$. Our task is the construction of a state $|0(\theta)\rangle_V$ such that

$$ {}_V\langle 0(\theta)|V_{\bar{k}}^{\dagger}V_{\bar{k}}|0(\theta)\rangle_V = \exp(-\Gamma_{\bar{k}}t) \qquad \underline{\text{at the time } t} \quad , \tag{1}$$

where $\Gamma_{\bar{k}}$ is the inverse of the life-time and $\theta \equiv \theta(t)$. By a construction which is the analog of the GNS construction, we introduce a fermionic operator $\tilde{V}_{\bar{k}}^{\dagger}(\tilde{V}_{\bar{k}})$ with usual anticommutation relations and anticommuting with $V_{\bar{k}}^{\dagger}$, $V_{\bar{k}}$. The "tilde" vacuum at t = 0 is $\tilde{V}_k|\tilde{0}\rangle_{\tilde{V}} = 0$, ${}_{\tilde{V}}\langle\tilde{0}|\tilde{0}\rangle_{\tilde{V}} = 1$. The notation $|0\rangle_V \equiv |0,\tilde{0}\rangle_V$ for the direct product of $|0\rangle_V$ and $|\tilde{0}\rangle_{\tilde{V}}$ is used. Next we consider the Bogoliubov transformation

$$V_{\bar{k}}(\theta) = V_{\bar{k}}\cos\theta_{\bar{k}} - \tilde{V}_{\bar{k}}^{\dagger}\sin\theta_{\bar{k}} \quad , \quad \tilde{V}_{\bar{k}}(\theta) = \tilde{V}_{\bar{k}}\cos\theta_{\bar{k}} + V_{\bar{k}}^{\dagger}\sin\theta_{\bar{k}} \tag{2}$$

and h.c. which preserves the canonical anticommutation relations. The generator of (2) and the "vacuum" $|0(\theta)\rangle_V$ are

$$G_V(\theta) = i\sum_{\bar{k}}\theta_{\bar{k}}(V_{\bar{k}}^{\dagger}\tilde{V}_{\bar{k}}^{\dagger} - \tilde{V}_{\bar{k}}V_{\bar{k}}) = G_V^{\dagger}(\theta) \tag{3}$$

$$|0(\theta)\rangle_V = e^{-iG}|0\rangle_V = \prod_{\bar{k}}(\cos\theta_{\bar{k}} + \sin\theta_{\bar{k}}V_{\bar{k}}^{\dagger}\tilde{V}_{\bar{k}}^{\dagger})|0\rangle_V \quad , \quad {}_V\langle 0(\theta)|0(\theta)\rangle_V = 1 \quad . \tag{4}$$

The fictitious tilde-system plays the role of a "heat-bath". Our requirement, eg. (1), is now ${}_V\langle 0(\theta)|V_{\bar{k}}^{\dagger}V_{\bar{k}}|0(\theta)\rangle_V = \sin^2\theta_{\bar{k}} \equiv \exp(-\Gamma_{\bar{k}}t)$, which determines $\theta_{\bar{k}}$ as a function of time. From eq. (4) we see that $|0(\theta_{\bar{k}}(0))\rangle_V = V_{\bar{k}}^{\dagger}\tilde{V}_{\bar{k}}^{\dagger}|0\rangle_V$ and $|0(\theta_{\bar{k}}(+\infty))\rangle_V = |0\rangle_V$, i.e. the one-$V_{\bar{k}}$-particle state at t = 0 and the no-particle state at t = +∞, respectively. The decay products $B_{\bar{k}}^{\dagger}(B_{\bar{k}})$ can be included in the description and eq. (4) is written as

$$|0(\theta)\rangle = \exp(-K)\exp(\sum_{\bar{p}} B_{\bar{p}}^{\dagger}\tilde{B}_{\bar{p}}^{\dagger})\exp(\sum_{\bar{p}} V_{\bar{k}}^{\dagger}\tilde{V}_{\bar{k}}^{\dagger})|0\rangle \quad , \tag{5}$$

$$K = \frac{1}{2}\sum_{\bar{k}}(V_{\bar{k}}^{\dagger}V_{\bar{k}}\log\sin^2\theta_{\bar{k}} + V_{\bar{k}}V_{\bar{k}}^{\dagger}\log\cos^2\theta_{\bar{k}}) - \frac{1}{2}\sum_{\bar{p}}(B_{\bar{p}}^{\dagger}B_{\bar{p}}\log\cos^2\theta_{\bar{p}} + B_{\bar{p}}B_{\bar{p}}^{\dagger}\log\sin^2\theta_{\bar{p}}) \; . \tag{6}$$

† Permanent address: Istituto di Fisica, Via Vernieri 42, 84100 Salerno, Italia.

Eq. (5) shows that the time evolution, namely transitions among the states $|0(\theta(t))>$, is controlled by a semigroup law. The probability of finding the state $|0(\theta_{\bar{k}}(0))>$, i.e. the $V_{\bar{k}}$-particle state at $t=0$, at the time t is $<0(\theta_{\bar{k}}(0))|\exp(-K)|0(\theta_{\bar{k}}(0))> = \exp(-\Gamma_{\bar{k}}t)$. Eq. (6) suggest that K can be regarded as the "entropy" for the unstable system. In fact, $<0(\theta_{\bar{k}}(0))|K|0(\theta_{\bar{k}}(0))> = -\log\exp(-\Gamma_{\bar{k}}t)$, which is zero at $t=0$ and $+\infty$ at $t=+\infty$: the "entropy" increases as the system evolves towards a stability condition at $t=+\infty$. It is interesting to note that the possibility of referring to K as to the "entropy" and, at the same time, as to the operator which controls the time evolution, reflects the irreversibility of the time evolution of an unstable system. The statistical nature of a decaying process thus naturally emerges. In the infinite-volume limit the states $|0(\theta(t))>$ with different values of the time t become orthogonal each other and equations as (3) and (4) become only formal: the above construction ends up in a parametrization by the time t of the unitarily inequivalent representations of the CCR: physically, this means that the expectation values of the number operator $V^{\dagger}V$ in the "vacua" $|0(\theta)>$ are different at different values of time t. The non-unitary character of the time evolution of an unstable system is thus recovered as the non-unitary equivalence among the representations of the CCR: an unstable system is a system whose time evolution is described by the running over the unitarily inequivalent representations of the CCR. Application to the time evolution of a N-soliton is given in ref. 2. The parametrization of the representations of the CCR by a proper-time parameter has been used to handle with the non-uniqueness of the vacuum state in the quantization of a free matter field in curved space-time. A particle creation process is described, which, in the case of black hole matric, has many properties of the Hawking process. Quantum fluctuations from inequivalent representations play a crucial role in the one-loop renormalizability of the theory.

1. S. De Filippo and G. Vitiello, Lettere al Nuovo Cimento 19, 92 (1977).
2. P. Garbaczewski and G. Vitiello, Nuovo Cimento 44A, 108 (1978).
3. M. Martellini, P. Sodano and G. Vitiello, Vacuum structure for a quantum field theory in curved space-time, preprint 1978.

DEFORMATION OF SYMPLECTIC STRUCTURE AND QUANTIZATION

François BAYEN
Département de Mathématiques
Université de Paris 6
4 Place Jussieu
75230 - PARIS Cedex 05

A/ MECHANICS WITH TWISTED PRODUCTS

1. Introduction

The basic mathematical structures of classical mechanics are the symplectic structures attached to phase space : the algebra N of C^∞ functions on phase space W (a symplectic, C^∞ , paracompact manifold) under ordinary multiplication of functions, and the Lie algebra structure induced on N by the Poisson bracket P of W. In ref. [1] the formal differentiable deformations of these two algebras were examined. In particular we have introduced an associative algebra ($*$-product algebra) that is a deformation of the algebra N of functions with ordinary multiplication. For $f,g \in N$ we write the new (deformed) product as $(f,g) \mapsto f*g$. The corresponding Lie algebra defined by $(f,g) \mapsto [f*g] \equiv (f*g - g*f)/i\hbar$ is a deformation of the Poisson Lie algebra. A particular case of this type of deformation of classical mechanics is known as the Moyal product and associated Moyal bracket.

It is our intention to show in the parts A and B of this review that quantum mechanics can be replaced by a deformation of classical mechanics. A description of quantum phenomena in terms of ordinary functions on phase space, including a complete and autonomous physical interpretation can in fact be performed. This alternative formulation of quantum theory will include some features that are not usually associated with phase space. In order to know what to expect it is worthwhile to recall the elements of the theory of Weyl [6], Wigner [7] and Moyal [8].

2. Weyl application [2]

A Weyl *application*, $\Omega : f \to F$ maps a family of functions on phase space W bijectively to a family of operators in the Hilbert space H of quantum mechanics. The inverse map $\Omega^{-1} : F \to f$ of a Weyl application is called a Wigner application. If $W = \mathbb{R}^{2\ell}$ one has :

$$f(x) = (2\pi\hbar)^{-\ell}\int_{\mathbb{R}^{2\ell}} \tilde{f}(y)\ \omega(y)\ \exp(i\,\frac{x.y}{\hbar})\ dy \qquad (1)$$

$$F = (2\pi\hbar)^{-\ell}\int_{\mathbb{R}^{2\ell}} f(y)\ \exp(i\,\hat{x}.y/\hbar)\ dy \qquad (2)$$

Here $x,y = (q,p)\in\mathbb{R}^{2\ell}$; $q,p\in\mathbb{R}^{\ell}$ are natural coordinates for W, $\hat{x} = (X^1,\dots,X^{\ell}, -i\hbar\partial/\partial X^1,\dots,-i\hbar\partial/\partial X^{\ell})$ are the corresponding operators in $H = L^2(\mathbb{R}^{\ell})$ and $x.y = \sum_{j=1}^{2\ell} x^j\, y^j$, $\hat{x}.y = \sum_{j=1}^{2\ell} \hat{x}^j\, y^j$. The function ω is an entire function with constant term equal to 1 and with no linear term. (Different choices of ω correspond to different ordering conventions). The measure dy is the Lebesgue measure on $\mathbb{R}^{2\ell}$. If $\omega \equiv 1$ one gets the original Weyl application [6].

The Weyl applications and their inverses have been given several alternative forms, and their domain of validity have been studied by various authors [9]-[11]. We mention a few facts in the case $\omega = 1$. On the domain of differentiable vectors for the representation of the $2\ell+1$ dimensional Heisenberg group, the operator defined by eq. (2) makes sense for functions $f \in \mathcal{O}_M$ (the space of slowly increasing functions). If $f\in L^2(\mathbb{R}^{2\ell})$, $F = \Omega f$ is a Hilbert-Schmidt operator the kernel $\mathcal{F}$ of which may be written :

$$\mathcal{F}(X,Y) = (2\pi\hbar)^{-\ell}\int_{\mathbb{R}^{\ell}} \exp(-i(X-Y).p/\hbar)\ f((X+Y)/2,p)dp \qquad (3)$$

Here $X,Y\ \mathbb{R}^{\ell}$, $X.p = \sum_{j=1}^{\ell} X^j\, p^j$ and if $\phi\in L^2(\mathbb{R}^{\ell})$, $F\,\phi(X) = \int_{\mathbb{R}^{\ell}} \mathcal{F}(Y,X)\ \phi(Y)dY$.

The inverse map in this case may be written :

$$f(q,p) = \hbar^{\ell}\int_{\mathbb{R}^{\ell}} e^{-iX.p}\ \mathcal{F}(q-\hbar/2\ X,q+\hbar/2\ X)dX \qquad (4)$$

(The previous equalities hold almost everywhere). Finally if f is a polynomial one has :

$$F = \exp(-i\ \hbar/2\ \mathcal{D})\ f|q = Q,\ p = P\ ,\ Q \text{ before } P.$$

Where $\mathcal{D} = \sum_{j=1}^{\ell} \partial^2/\partial q^j\, \partial p^j$ and $\hat{x} = (Q,P)$.

3. Twisted products of functions defined on a symplectic manifold

An associative $*$-product is induced in N in a natural way. Let $f,g\in N$ and $F,G = \Omega g$ be the corresponding operators, then

$$f * g = \Omega^{-1}(FG) = \Omega^{-1}((\Omega f)(\Omega g)) \qquad (5)$$

A formal direct calculation gives in the case where $W = \mathbb{R}^{2\ell}$ and $\omega = 1$:

$$f * g = \sum_{n=0}^{\infty} \frac{\lambda^n}{n!} P^n(f,g) \qquad (\lambda = i\hbar/2) \tag{6}$$

Here P^n is the nth power of the Poisson bracket P, $P^0(f,g) = fg$ and $P^1 = P$. P^n is a bidifferential operator acting on the pair (f,g) and may be computed as follows :

$$P^n(f,g) = \Lambda^{i_1 j_1} \dots \Lambda^{i_n j_n} \; D^n_{i_1 \dots i_n} f \; D^n_{j_1 \dots j_n} g \tag{7} \quad (n \geq 1)$$

$$P(f,g) = \partial f/\partial q^i \; \partial g/\partial p^i - \partial f/\partial p^i \; \partial g/\partial q^i$$

where $D^n_{i_1 \dots i_n} = \partial^n/\partial x^{i_1} \dots \partial x^{i_n}$. The $*$-product defined by eq.(6) is called the Moyal $*$-product. The Moyal bracket is defined by :

$$[f \;\; g]_M = (f * g - g * f)/i\hbar = (2/\hbar) \sin(\hbar/2P)(f,g) \tag{8}$$

We emphasize that it is this bracket and not the Poisson bracket that corresponds to the quantum commutator. The series which occurs in eq.(6) may diverge and this equality is interpreted formally i.e. in the space $E(N,\lambda)$ of formal series in $\lambda \in \mathbb{C}$ with coefficients in N [1]. The fact that one has to do with a formal deformation of classical mechanics, with $\hbar$ playing the role of deformation parameter, is brought out by eq.(6).

Other Weyl applications have been considered in the literature (in the case $W = \mathbb{R}^{2\ell}$). They are such that $\omega = 1 + \Sigma \hbar^r \omega_{2r}$ where ω_{2r} is a homogeneous polynomial. It can be shown [1] that the corresponding deformations of N are, in precise sense, cohomologically equivalent to the deformation defined by (6). For example assume $\ell = 1$, take $\omega(x) = \exp(i[\hbar^{-1}/2]qp)$ and denote by $*'$ the corresponding product. One has $(T f) * (T g) = T (f *' g)$ where $T = \exp(i\hbar/2\mathcal{D})$ and $*$ is Moyal twisted product. This corresponds to the so called mathematical quantization [11] and the corresponding multiplication $*'$ may be written :

$$f *' g = \sum_{n=0}^{\infty} \frac{(-i\hbar)^n}{n!} \frac{\partial^n f}{\partial p^n} \frac{\partial^n g}{\partial q^n} \tag{9}$$

For completeness let us mention that for a large class of functions Moyal $*$-product may be written in integral form :

$$(f * g)(x) = (\pi\hbar)^{-2\ell} \int_{\mathbb{R}^{4\ell}} f(y) \, g(z) \exp\left(\frac{2i}{\hbar} [x-y, \, x-Z]\right) dy \, dZ \tag{10}$$

where $x,y = \sum_{j=1}^{\ell} (x^j y^{j+\ell} - x^{j+\ell} y^j)$; $x, y, z \in \mathbb{R}^{2\ell}$.

4. Trajectories and motions

Suppose now that $h \in N$ is a classical hamiltonian without explicit time dependence. The equation of motion for a fonction $f \in N$ will be :

$$\begin{cases} \partial/\partial t\ f_t = [f_t * h] \equiv (f_t * h - h * f_t)/i\hbar \\ f_o = f \end{cases} \quad (11)$$

It may be called a twisted equation of motion and it is the "classical" analogous of the Schrödinger equation (apply Ω to eq.(11)). The previous bracket defines a derivation of the $*$-product algebra :

$$[(f * g) * h] = f * [g * h] + [f * h] * g \quad (12)$$

whence $\partial/\partial t(f_t * g_t) = (\partial/\partial t\ f_t) * g_t + f_t * (\partial/\partial t\ g_t)$ (13)

However, unless h is particular (a polynomial of degree ≤ 2 in case of the Moyal $*$-product), we have for almost all f,g :

$$[(fg) * h] \neq [f * h]\ g + f\ [g * h] \quad (14)$$

In other words, one solves the equations of motion (11) for f,g and fg and discovers that, in general $(fg)_t \neq f_t\ g_t$. The impression that this result is paradoxical must be dispelled by a proper interpretation.

To begin with we shall sketch a formulation of classical mechanics that makes a fundamental distinction between the p's and q's as observables (elements of N) and the coordinates of points on a trajectory in phase space. This formulation is made here for the sake of simplicity in the case where $W = \mathbb{R}^{2\ell}$. The idea is to isolate the two principal elements of the theory, equations of motion and initial conditions, from each other and to associate the former with observables and the latter with states.

The Hamiltonian equation of motion for an observable f :

$$\partial/\partial t\ f = - P(h,f) \quad (15)$$

will be interpreted as defining a derivation of the algebra N.

The solution $f_t = \exp(-t\ P(h,.)).f$ defines a map of $\mathbb{R} \times N$ into N, $(t,f) \mapsto f_t$. Thus it is true that

$$\begin{cases} \partial/\partial t\ f_t = P(f_t,h) \\ f_o = f \end{cases} \quad (16)$$

and the map $t \mapsto f_t$ is a <u>trajectory in N through f</u>. In particular $q^i(t)$ and $p^i(t)$ are functions on W and there is not yet any reference to trajectories in W.

To any set of initial conditions one can associate a real (pseudo probability) distribution ρ on phase space, normalized so that

$$\int_{\mathbb{R}^{2\ell}} \rho(y)\, dy = 1 \tag{17}$$

We refer to such a distrubution as a state. In most problems of classical mechanics it is enough to consider $\rho(y) = \delta^{2\ell}(x-y)$ with $x = (q,p)$. The result of a measurement of the observable f at the time t on the state ρ is

$$\langle f \rangle_t = \int_{\mathbb{R}^{2\ell}} f_t(y)\, \rho(y)\, dy \tag{18}$$

This reformulation of classical mechanics is valid for mechanics whose equations of motion are given by the (nonlocal) eq.(11). In this case eq.(13) shows that $(f * g)_t = f_t * g_t$. The paradox means that trajectories in W have no invariant meaning. Determinism in a narrow sense is lost and roughly speaking using different coordinate systems one gets a fuzzy trajectory. This is related to Heisenberg uncertainty principle.

The analogue of the formula (18) for the measured value $\langle f \rangle_t$ of the observable $f \in N$ at time t in the state ρ will be :

$$\langle f \rangle_t = \int_{\mathbb{R}^{2\ell}} f_t(y) * \rho(y)\, dy \tag{19}$$

In this case $\Omega\rho$ is a density matrix. The equality (19) appears in the literature in the equivalent form (18). This comes from the equality valid for Moyal $*$-product :

$$\int_{\mathbb{R}^{2\ell}} \exp(i\,[x,y])(f * g)(y)dy = \int_{\mathbb{R}^{2\ell}} \exp(i\,[x,y])\; f(y + \frac{\hbar}{2}\,x)\; g(y)\; dy \tag{20}$$

Fore more general $*$-products we shall adopt (19).

5. <u>Spectral theory in phase space</u>

This part will be only outlined. The examples given in part B show that these considerations apply to various situations. Let $h \in N$ and consider the formal series :

$$\mathrm{Exp}(ht) = \sum_{n=0}^{\infty} \frac{1}{n!}\,(t/i\hbar)^n\,(h*)^n \tag{21}$$

where $(h*)^n = h* \ldots * h$ n times. Assume that there exists $R > 0$ such that for $|t| < R$ the power series converges to a distribution on W. Suppose also that this quantity or an analytic continuation (with respect to the variable t) of it has a Fourier-Dirichlet expansion :

$$\mathrm{Exp}(ht) = \sum_{\lambda \in I} \pi_\lambda \exp(\lambda t/i\hbar) \qquad (22)$$

where $I \subset \mathbb{C}$ is a sequence and $\pi_\lambda \in N$. In the case of the Moyal ✱-product, if $\Omega h = H$ is a normal operator in $\mathcal{H}$, then I is the spectrum of H and the $\Omega\pi_\lambda$ are the projectors for the spectral decomposition. This motivates the :

<u>Def</u>. If h N satisfies (22) we call I the spectrum of h. From (21) one obtains

$$i\hbar(d/dt)\,\mathrm{Exp}(ht) = h \ast \mathrm{Exp}(ht) \qquad (23)$$

and from this it follows that $h \ast \pi_\lambda = \pi_\lambda \ast h = \lambda\, \pi_\lambda$, $h = \sum_{\lambda \in I} \lambda\, \pi_\lambda$, $\pi_\lambda \ast \pi_{\lambda'} = \delta\lambda\lambda' \pi_\lambda$

In case of Moyal ✱-product the multiplicity $n_\lambda \in \mathbb{N} \cup \{+\infty\}$ of an eigenvalue $\lambda \in I$ will be

$$n_\lambda = (2\pi\hbar)^{-\ell} \int_{\mathbb{R}^{2\ell}} \pi_\lambda(x)\, dx \qquad (24)$$

In the general case it was shown in ref. [1] that formula (24) is valid if dx is replaced by a suitably normalized Liouville measure on W.

Finally if the spectrum of the operator H has a continuous part, a similar treatment is possible (see part B).

B/ CALCULATION OF SPECTRA. EXAMPLES.

1 - <u>Harmonic oscillator</u>. In this case we utilize Moyal ✱-product. We denote the Hamiltonian by h :

$$h(q,p) = 1/2(p^2 + q^2) \qquad (25)$$

and we begin with the one-dimensional oscillator. Suppose $f \in C^\infty(\mathbb{R},\mathbb{C})$ then $f \circ h \in N$ and one gets easily :

$$h \ast f(h) = h\, f(h) - \frac{\hbar^2}{4} (f'(h) + h\, f''(h)) \qquad (26)$$

Thus $(h\ast)^n = K_n(h)$ is a polynomial of degree n. The series (21) has a radius of convergence equal to π and its sum is :

$$\mathrm{Exp}(ht) = (\cos t/2)^{-1} \exp\left(\frac{2h}{i\hbar} \tan \frac{t}{2}\right) \qquad (27)$$

Fourier analysis is easily carried out and leads to

Proposition : For fixed $t \in \mathbb{C}$ with $\mathrm{Im}\, t \leqslant 0$ and $t \neq (2K+1)\pi$

$$\mathrm{Exp}(ht) \equiv (\cos t/2)^{-1} \exp\left(\frac{2h}{i\hbar} \tan \frac{t}{2}\right) = \sum_{n=o}^{\infty} \pi_n\, e^{-i(n+1/2)t} \tag{28}$$

The series converging in $\mathcal{S}'(\mathbb{R}^2)$ for the weak topology. Moreover, if $t = \pm\pi$ the series converges (in $\mathcal{S}'(\mathbb{R}^2)$) to $\mp i\pi\hbar\delta$.

The projectors π_n satisfy $\int_{\mathbb{R}^2} \pi_n(q,p)\, dq\, dp = 2\pi\hbar$. They are given by

$$\pi_n = 2 \exp\left(-\frac{2}{\hbar} h\right) (-1)^n L_n\left(\frac{4}{\hbar} h\right) \tag{29}$$

where L_n denotes the Laguerre polynomial of degree n.

Remarks : a) Similar results may be obtained with other twisted products e.g. the product defined by eq. (9), in this case the computations are more complicated. In fact one finds

$$\mathrm{Exp}(ht) = (\cos t)^{-1} \exp\left\{\frac{h \tan t}{i\hbar} + \frac{2i \sin^2 t/2}{\hbar \cos t}\, qp\right\}$$

However there exists a type of quantization where it is easier to get the spectrum than with Moyal $*$-product. It is the so called holomorphic quantization [12], [13]. In this case the observables are functions of two complex variables Z, Z' and the twisted product is

$$(f * g)(Z,Z') = \sum_{n=o}^{\infty} \frac{\hbar^n}{n!} \frac{\partial^n}{\partial Z'^n} f(Z,Z') \frac{\partial^n}{\partial Z^n} g(Z,Z')$$

In this framework for the harmonic oscillator $h(Z,Z') = ZZ' + 1/2$ and one obtains :

$$\mathrm{Exp}(ht) = e^{-it/2} \exp\left(\frac{ZZ'}{\hbar}(e^{-it}-1)\right)$$

Fourier analysis is then trivially performed :

$$\mathrm{Exp}(ht) = \exp\left(-\frac{ZZ'}{\hbar}\right) \sum_{n=o}^{\infty} \frac{1}{n!} \left(\frac{ZZ'}{\hbar}\right)^n e^{-i(n+1/2)t}$$

b) The hamiltonian $h(q,p) = p^2/2 + q$ may be treated in the same manner. With Moyal star product one has

$$\mathrm{Exp}(ht) = \exp((t^3/6 + ht)/i\hbar)$$

and the spectrum of h is the whole real line.

The ℓ-dimensional case is analogous to the one-dimensional case. One finds if $t \in \mathbb{C}$, $\mathrm{Im}\, t \leqslant 0$ and $t \neq (2K+1)\pi$:

$$\mathrm{Exp}(ht) = (\cos t/2)^{-\ell} \exp\left(\frac{2h}{i\hbar} \tan \frac{t}{2}\right) = \sum_{n=o}^{\infty} \pi_n^{(\ell)}\, e^{-i(n+\ell/2)t} \tag{30}$$

The series converging in $\mathcal{S}'(\mathbb{R}^{2\ell})$. If $t = \pm\pi$ (30), converges (in $\mathcal{S}'(\mathbb{R}^{2\ell})$) to $(\mp i\pi h)^{\ell}\delta$. This occurence of δ is not surprising. In fact if P is the parity operator in $L^2(\mathbb{R}^{\ell})$ defined by $P\,\phi(X) = \phi(-X)$ one has $P = \Omega(\pi^{\ell}\delta)$ and $\exp(\frac{H\pi}{i\hbar}) = (-i)^{\ell}P$.

The study of the harmonic oscillator is instructive by the appearance of the sl(2,$\mathbb{R}$) Lie algebra. Indeed we have a sl(2,$\mathbb{R}$) Lie algebra (for both Poisson and Moyal brackets) generated by $A = \{X = \alpha\,p^2 + 2\beta\,p.q + \gamma\,q^2 \mid \alpha, \beta, \gamma \in \mathbb{R}\}$ where $p.q = \sum_{j=1}^{\ell} p^j.q^j$, $p^2 = p.p$. For such an X equation (26) is replaced by :

$$X\ f(X) = X\,f(X) - \ell\,d\,\hbar^2\,f'(X) - d\,\hbar^2 X f''(X)$$

where $d = \alpha\gamma - \beta^2$. For an hyperbolic element such that $X = p.q$ one finds an expression similar to (28) with trigonometric functions replaced by hyperbolic functions. The continuous spectrum is obtained with the help of some generalized projectors (cf. [1] for their expressions) $\pi(\lambda, X)$:

$$\mathrm{Exp}(Xt) = \int_{-\infty}^{+\infty} \exp\left(\frac{\lambda t}{i\hbar}\right)\,\pi(\lambda, X)\,d\lambda \qquad (31)$$

Finally the functions $\{\mathrm{Exp}(Xt) \mid X \in A\}$ generate by $*$-multiplications a $*$-representation of the metaplectic group (the twofold covering of SL(2,$\mathbb{R}$). For $\ell = 1$ one may show that one gets a sum of two irreducible $*$-representations [1].

We have seen here in a specific case, SL(2,$\mathbb{R}$), that $*$-products and representations of Lie groups are strongly related. In fact, a part of $*$-products theory can be viewed as a "representation theory without operators" [2] in the same way we perform an autonomous quantum mechanics without operators.

Remark. This is a technical one. Suppose $\ell = 1$, consider $X = qp$ and take $\hbar = 1$ for simplicity. Then $(X*)^n$ is a polynomial G_n in the variable X, of degree n. A natural question to ask is : are the polynomials G_n known in the literature ? The answer is the following. Consider the Hilbert space $\mathcal{E} = L^2(\mathbb{R}, (\mathrm{ch}\pi\lambda)^{-1}d\lambda)$. A sequence $(E_n)_{n\geq 0}$ of polynomials of degree n, orthonormal in $\mathcal{E}$, is uniquely specified by the requirements that the coefficient of the term of degree n be positive. These polynomials were studied by Hardy [14]. Then one may check that :

$$G_n(X) = \sum_{m=0}^{n} \frac{(2X)^m}{m!}\,(\lambda^n, E_m)$$

$$X^n = \frac{n!}{2^n}\,E_n(X*)$$

where (,) is the scalar product in $\mathcal{E}$. The last relation is called in special functions theory a symbolic relation [15] and may be written $X^n = n!\,2^{-n}\,E_n(G(X))$.

2. Angular momentum

Let $M_{jK} = q_j p_K - q_K p_j$ $(1 \leq j, K \leq \ell, \ell > 2)$ be generators of a so(ℓ) Lie algebra (for P and $[*]_M$). The first Casimir is here (we utilize Moyal $*$-product) :

$$C = \sum_{j<K} (M_{jK} *)^2 = g(p,q)^2 - \ell(\ell-1) \hbar^2/4 \tag{32}$$

where $g^2(p,q) = p^2 q^2 - (p.q)^2$. One then gets a formula of the type (26) for $(g^2) * f(g^2)$ but more involved (with fourth-order derivatives). Then a detailed analysis shows that $\gamma = g - (\ell-2)\hbar^2/4g$, defined where $g \neq 0$, satisfies $(\gamma\ )^2 = g^2 - \frac{(3\ell-4)\hbar^2}{4}$; and that more generally for $n \geq 1$:

$$(\gamma *)^n = g^{-1} J_{n+1}(g) \tag{33}$$

where the polynomials J_n of degree n are defined by :

$$(\cos \frac{t}{2})^{2-\ell} \exp(\frac{2h}{i\hbar} \tan \frac{t}{2}) = \sum_{n=0}^{\infty} n!^{-1} (t/i\hbar)^n J_n(h) \tag{34}$$

From this one deduces, using the preceding section, that for fixed $s \in \mathbb{C}$ with Ims < 0 one has (weakly in $\mathcal{S}'(\mathbb{R}^{2\ell})$) functions Π_n (of g) such that

$$\mathrm{Exp}(\gamma s) = \sum_{n=0}^{\infty} \Pi_n(g) \exp(-is(n + \frac{1}{2}(\ell - 2))) \tag{35}$$

The spectrum of γ is thus $\hbar(n + \frac{1}{2}(\ell - 2))$ and that of $C = (\gamma *)^2 - (\ell - 2)\hbar^2/4$ is $\{n(n + \ell - 2) \mid n \in \mathbb{N}\}$.

Remark : The spectrum of the generator M_{jK} may also be found [1].

3. Hydrogen atom

Here we take p,q $\mathbb{R}^\ell$ with $h = 1/2\, p^2 - r^{-1}$, $r = \sqrt{q^2}$, for $\ell = 3$ we have the physical case. A treatment with Moyal $*$-product seems hopeless. However the physical space is in fact $W = T^* S^\ell$ with momentum in S^ℓ and the SO(ℓ+1) symmetry discovered by Fock.

We shall therefore construct explicitely ([1] and [3]) a $*$-product on the cotangent bundle to the ℓ-dimensional sphere. This star product will be so(ℓ+1) invariant in the sense of **ref [1] , [2]**.

The manifold $W = T^* S^\ell$ may be embedded in $\mathbb{R}^{2\ell+2}$ by the two conjugate constraints $\frac{1}{2}\pi^2 \equiv \frac{1}{2}\sum_{\alpha=1}^{\ell+1} (\pi^\alpha)^2 = \frac{1}{2}$ and $\frac{\pi}{|\pi|^2}\xi = \sum_{\alpha=1}^{\ell+1} \frac{\pi^\alpha}{|\pi|^2}\xi_\alpha = 0$. $\mathbb{R}^{2\ell+2} = T^*\mathbb{R}^{\ell+1} = \hat{W}$ being endowed with the usual symplectic structure and Poisson bracket $\hat{P}$). It is known [16] that the Dirac bracket [17] corresponding to this situation is nothing but the Poisson bracket P on W. We define a thickening of W to an open subset $\hat{W}_o \subset \hat{W}$ by a group G of symplectic transformations, so that on G-invariant functions $\hat{P}$ and P coïncide. Furthermore $\hat{\mathbf{P}}^n$ restricted to G invariant functions will still be a G-invariant functions and will enable us to define a $*$-product on N = N(W). More precisely let G be the two-dimensional solvable group acting on $\mathbb{R}^{2\ell+2}$ according to $(\pi,\xi) \to (\rho\pi, \rho^{-1}\xi+\sigma\pi)$ $(\rho > 0, \sigma \in \mathbb{R})$.

The space of orbits of G in $W_o = (\mathbb{R}^{\ell+1}-\{0\}) \times \mathbb{R}^{\ell+1}$ is diffeomorphic to W the projection ϕ being defined by :

$$\phi(\pi,\xi) = (\pi/|\pi|,\ |\pi|\xi - (\pi.\xi)/|\pi|\pi) \tag{36}$$

ϕ defines an isomorphism between the space N_G of differentiable G-invariant functions on $\hat{W}_o$ and N by the formulae :

$$\tilde{\phi}\hat{u} = \hat{u}|W \qquad \hat{u} = \tilde{\phi}^{-1} u = u \circ \phi \qquad u \in N\ ,\ \hat{u} \in N_G \tag{37}$$

This enables one to define $P^n(u,v) = \hat{P}^n(\hat{u},\hat{v})|W$ and a global associative $*$-product on N by :

$$u *' v = \sum_{n=o}^{\infty} \frac{\lambda^n}{n!} P^n(u,v) \tag{38}$$

where $u, v \in N$. It is clear that $u *' v = (\hat{u} * \hat{v})|W$ where the $*$-product in the right hand side is Moyal $*$-product in W_o. To make computations effective one will utilize a symplectic chart guaranteed by Darboux's theorem and then determine with the help of ϕ the corresponding curvilinear coordinates on $\hat{W}_o$.

We consider the generators of the SO(ℓ+1) symmetry :

$M_{jK} = q_j\, p_K - q_K\, p_j$, $M_{j,\ell+1} = -\frac{1}{2}(p^2 - 1)q_j + (p.q)\, p_j$, $1 \leq j, K \leq \ell$

Their expressions on W and on $\hat{W}_o$ are $M_{\alpha\beta} = \xi_\alpha \pi^\beta - \xi_\beta \pi^\alpha$. These generators are G-invariant functions. Using the results of the preceding section, the spectrum of the Casimir (expressed with $*'$) $C = \Gamma^2 - \ell(\ell+1)\hbar^2/4$ where $\Gamma = 1/2\ r(p^2 + 1)$ is found to be $n(n+\ell-1)\hbar^2$, and the $*'$-square root of $C + (\ell-1)^2\hbar^2/4$ is $\gamma = \Gamma-(\ell-1)\hbar^2/(4\Gamma)$. The $*'$-spectrum of γ is given by the following expansion of the $*'$-exponential :

$$\operatorname{Exp}(\gamma s) = \sum_{n=o}^{\infty} \pi'_n(\Gamma) \exp(-is(n + \tfrac{1}{2}(\ell-1))) \tag{39}$$

with multiplicity $(2\pi\hbar)^{-\ell}\int_{\mathbb{R}^{2\ell}} \pi'_n\, dp\, dq = (2n+\ell-1)\,\frac{(n+\ell-2)!}{n!(\ell-1)!}$, the number of spherical harmonics of degree n on S^ℓ.

In order to find the eigenvalue of h one utilizes the procedure of energy twist. The spectral problem

$$(h - E) * \phi = 0 \qquad \text{with} \quad \phi = \bar{\phi} \qquad \text{and} \quad E < 0$$

is transformed by Exp(-Ts), where $e^s = (-2E)^{1/2}$, T = pq, and becomes

$$(1 - e^{-s}(\gamma_o *)^{-1}) * \tilde{\phi} = 0$$

where $\gamma_o = \frac{1}{2}(p^2 + 1)^{1/2} * r * (p^2 + 1)^{1/2}$. Explicitely one has :

$$\mathrm{Exp}(-Ts) * (h - E) * \mathrm{Exp}(Ts) = \frac{1}{2} e^{2s}(p^2+1)^{1/2} * (1 - e^{-s}(\gamma_o *)^{-1}) * (p^2+1)^{1/2} .$$

Now the map τ which transforms Moyal $*$-product algebra to the $*'$-product algebra transforms $(\gamma_o *)^2$ into $(\gamma *')^2$. These two functions have then the same spectrum, wherefrom one proves [1] that the $*$-spectrum of γ_o is identical with the $*'$-spectrum of γ. It follows that $(-2E)^{1/2} = e^{-s} = (n + \frac{\ell-1}{2})\hbar$ $(n \in \mathbb{N})$. In particular, for $\ell = 3$ one finds the usual levels of the hydrogen atom $E = -1/2\, n^2\hbar^2$ with multiplicity n^2.

A similar treatment with the same "twist" is possible for the continuous spectrum $(E > 0)$.

C/ GEOMETRIC THEORY OF CONTRACTIONS AND TWISTED PRODUCTS [4]

This part will be a very brief survey of the question. In part B, star representations of some Lie groups on symplectic manifolds were encountered. ($SL(2,\mathbb{R})$ and $SO(\ell)$ on $\mathbb{R}^{2\ell}$ and $SO(\ell+1)$ on T^*S^ℓ). It seems therefore interesting to study contractions of representations of Lie groups or Lie algebras in terms of twisted products. The contractions will thus have a natural geometric interpretation in phase space.

We consider an example : The contraction of the Lie algebra $so(\ell+1)$ to $\varepsilon(\ell)$ the Lie algebra of the inhomogeneous rotation group $E(\ell)$ in $\mathbb{R}^\ell$. Consider the following realization of these two Lie algebra in $N(\mathbb{R}^{2\ell+2})$ and $N(\mathbb{R}^{2\ell})$ respectively :

$$M'_{jK} = q_j\, p_K - q_K\, p_j \qquad 1 \leq j < K \leq \ell+1 \qquad (40)$$

$$\begin{cases} M_{jK} = q_j\, p_K - q_K\, p_j & 1 \leq j < K \leq \ell \\ K_j = c\, p_j & c \in \mathbb{R} \end{cases} \qquad (41)$$

The generators M_{jK} and K_j are formally obtained from the generators M'_{jK} by the conjugate constraints $k_1 = q_{\ell+1} + c$, $k_2 = p_{\ell+1} = 0$. In ref. [16] the Dirac bracket [17] is interpreted as an instant of a formal deformation of the Poisson bracket. Let us recall the expression of the Dirac bracket :

$$[u,v]_D = P(u,v) + [P(u,k_1)\, P(k_2,v) - P(u,k_2)\, P(k_1,v)]\, P(k_1,k_2)^{-1}$$

In our case it is trivial to prove the

Proposition. The contraction of the Lie algebra $so(\ell+1)$ to $\varepsilon(\ell)$ is given by a formal deformation of the Poisson bracket on $\mathbb{R}^{2\ell+2}$ to the Dirac bracket relative to the constraints $k_1 = q_{\ell+1} + C$, $k_2 = p_{\ell+1}$.

A deformation of the envelopping algebra $\mathcal{U}(so(\ell+1))$ to $\mathcal{U}(\varepsilon(\ell))$ may then be performed and it extends the contraction map defined by the proposition. (The star product is Moyal $*$-product).

Now we consider the problem of reducing the $*$-representation of $E(\ell)$ which is defined by formulas (41). The spectral analysis of the Casimir $\mathcal{C} = \sum_{j=1}^{\ell} (K_j *)^2$ is easily performed [4].

It is clear that on the cotangent bundle $W_R^{\ell-1}$ to the sphere, defined by the equations $\sum_{j=1}^{\ell} p_j^2 = R^2$, $\mathcal{C}$ is a scalar. On W_R^{ℓ} the functions M'_{jK} are generators with respect to the Poisson bracket or to the bracket $[*']$ defined by (38) of a representation of $so(\ell+1)$. With the notations of part B §3, $\widehat{M}'_{jK} = M'_{jK}$. On $W_R^{\ell-1}$ the generators M_{jK} and K_j are generators with respect to the Poisson bracket or to the bracket defined by (38) of a representation of $\varepsilon(\ell)$. Here $\widehat{M}_{jK} = M_{jK}$ and $\widehat{K}_j = c\, p_j/|p|$. We are now in position to formulate the following.

Proposition. The restriction of the Dirac deformation from T^*S^{ℓ} to the submanifold $T^*S^{\ell-1}$ defined by the constraints $k_1 = q_{\ell+1} + c$ and $k_2 = p_{\ell+1}$ realizes the contraction of the representation of $so(\ell+1)$ on T^*S^{ℓ} to a representation of $\varepsilon(\ell)$ on $T^*S^{\ell-1}$ with scalar Casimir element $\mathcal{C}$.

REFERENCES

1. F. BAYEN, M. FLATO, C. FRONSDAL, A. LICHNEROWICZ and D. STERNHEIMER : Deformation theory and quantization. Ann.Phys. (N.Y.) 111 (1978) 61-151. Cf. also Lett. Math. Phys. 1 (1977), 521.

2. C. FRONSDAL : Some ideas about quantization. UCLA/77/TEP/18. (Preprint October 1977) To be published in Reports of Math. Phys.

3. A. LICHNEROWICZ : Construction of twisted products for cotangent bundles of classical groups and Stiefel manifolds. Lett. Math. Phys. 2 (1977), 133.

4. D. ARNAL and J.C. CORTET : Geometrical theory of contractions of groups and representations. Dijon 805 (prepring May 1978). To be published in J.Math. Phys.

5. M. GERSTENHABER : Ann. Math. 79 (1964), 59.

6. H. WEYL : The theory of groups and quantum mechanics, Dover, New York, 1931.

7. E.P. WIGNER : Phys. Rev. 40 (1932), 749.

8. J.E. MOYAL : Proc. Cambridge Phil. Soc. 45 (1949), 99.

9. E.S. AGARWAL and E. WOLF : Phys. Rev. D2, 2161 (1970)

10. K.C. LIU : J. Math. Phys. 17, 859 (1976).

11. A. VOROS : Developpements semi-classiques (thèse Université de Paris-Sud,n°1843, may 1977).

12. D. BABBITT : "Hilbert spaces of analytic functions" in Studies in Mathematical Physics, Princeton University Press, Princeton 1976.

13. I. DAUBECHIES : An application of hyperdifferential operators to holomorphic quantization (1978) to be published in Lett. Math.Phys.

14. G.H. HARDY : Proceedings of the Cambridge Philosophical Society 36, (1940), 1.

15. E.D. RAINVILLE, Special functions, Chealsea publishing co. (1960).

16. M. FLATO, A. LICHNEROWICZ and D. STERNHEIMER : J. Math. Phys. 17 (1976) 1754.

17. P.A.M. DIRAC : Lectures on Quantum Mechanics, Belfer Graduate School.

PRESYMPLECTIC HAMILTON AND LAGRANGE SYSTEMS, GAUGE TRANSFORMATIONS AND THE DIRAC THEORY OF CONSTRAINTS

Mark J. Gotay

Center for Theoretical Physics

University of Maryland, College Park, Maryland 20742

James M. Nester

Department of Physics

University of Saskatchewan, Saskatoon, Canada S7N OWO

1. <u>INTRODUCTION</u> Differential equations of the form

$$i_X\omega = \alpha \tag{1}$$

where α is a closed 1 form (Hamiltonian) and ω is a closed 2 form (Lagrange bracket) frequently arise in physics. If ω is *symplectic* such equations have unique solutions.[1] However ω is often *pre*symplectic (i.e. $\beta : X \rightarrow i_X\omega$ is <u>not</u> an isomorphism). This occurs, e.g., (i) for many infinite dimensional systems, (ii) *a priori* - as in Künzle's[2] treatment of a spinning particle in curved space-time, (iii) for systems derived from degenerate Lagrangians (i.e., the fiber derivative[1] $FL : TQ \rightarrow T^*Q$ is not a diffeomorphism) as in gravity and electromagnetism. If ω is presymplectic neither existence nor uniqueness of solutions to the system (1) is guaranteed.

We present a geometric constraint algorithm which provides the necessary and sufficient conditions for the existence of solutions to the generalized Hamilton equations (1) on a (finite <u>or</u> infinite dimensional) *pre*symplectic manifold (M,ω). We discuss the application of our algorithm to Lagrangian and Hamiltonian systems and indicate its relationship to the Dirac theory of constraints. We also discuss the non-uniqueness of the solutions of (1) and present an algorithm for determining the associated "space of true degrees of freedom".

2. <u>THE CONSTRAINT ALGORITHM</u>[3] Our constraint algorithm finds the *unique maximal* submanifold $N \xrightarrow{i_N} M$ along which (1) possesses solutions *tangent* to N. This final

constraint submanifold is the limit $N \equiv \cap M_\ell$ of a string of sequentially constructed constraint submanifolds[4]

$$M_{\ell+1} \equiv \{m \in M_\ell \mid \alpha_m \in \beta(T_m M_\ell)\} \tag{2}$$

(with $M_1 \equiv M$) which follow from applying consistency conditions[5] to (1).

Theorem The presymplectic Hamiltonian system (1) has solutions if N is non-empty, in which case we have consistent equations of motion along the final constraint submanifold N

$$0 = (i_X\omega - \alpha)|_N \in T^*_N M \tag{3}$$

with solutions $X \in TN$.

If β is closed and TM is reflexive Banach (which we henceforth assume) an equivalent but much more convenient characterization of the constraint submanifolds is

$$M_{\ell+1} \equiv \{m \in M_\ell \mid \langle TM_\ell^\perp|\alpha\rangle_m = 0\} \tag{4}$$

where $TW^\perp \equiv \{Z \in TM \mid \omega(Y,Z) = 0 \;\forall\; Y \in TW\}$ is the presymplectic orthogonal complement.

3. DIRAC CONSTRAINT THEORY When applied to systems described by degenerate Lagrangians our algorithm significantly generalizes as well as globalizes the local Dirac-Bergmann (DB) theory of constraints.[6,7] In this case $M \equiv FL(TQ) \xrightarrow{j} T^*Q$ is Dirac's primary constraint submanifold, $\omega \equiv j^*\Omega$ with Ω being the canonical symplectic structure on T^*Q, and $\alpha = dH$ where the Hamiltonian H satisfies

$$H \circ FL = E_L \tag{5}$$

with $E_L(v) \equiv \langle v|FL(v)\rangle - L(v)$.

The DB constraint theory works on all of phasespace (T^*Q) and proceeds to find (analytically) an extension of H to T^*Q and secondary constraints which will guarantee preservation of all constraints. Our (geometric) algorithm works directly on the primary constraint submanifold M. The constraint functions $\langle TM_\ell^\perp|\alpha\rangle$ in (4) correspond to Dirac's secondary constraints.

In recent years there has been some nice work done in giving a symplectic geometric description of the results of the DB constraint theory. (See the works of Śniatycki,[8] Tulczjew,[9] Lichnerowicz,[10] Marsden and Weinstein[11] and ref. 3.) Our work (following the presymplectic approach of Hinds[12]) geometrizes the constraint algorithm itself and thereby complements and extends this description.

In constrast to the Dirac method, our algorithm does not *a priori* require M to be a submanifold of some given symplectic manifold (e.g. T*Q). This fact is responsible for the wide range of applicability of the algorithm.

4. LAGRANGIAN SYSTEMS Our constraint algorithm can be applied directly to the inherently presymplectic Lagrange equations

$$i_X \Omega_L = dE_L \tag{6}$$

on velocity phase space TQ. The presymplectic form Ω_L is $FL^*\omega$ or equivalently $- dJ^*dL$ where J^* is the adjoint of Klein's[13] *almost tangent structure*

$$J : TTQ \longrightarrow TTQ$$

which satisfies

$$\ker J = \operatorname{Im} J = V(TQ) \equiv \ker T\tau_Q .$$

The ability of our algorithm to deal directly with Lagrangian systems on TQ is of particular importance because the fiber derivative FL may be too pathological for a Hamiltonian formalism to exist in T*Q. (For example (5) may not define a single valued function H.) There are two special cases of interest.

Definition L is *almost regular*[14] if FL is a submersion onto its image and the fibers of FL are connected.

Definition L is *quasi-regular* if the leaf space $TQ/(\ker FL_*)$ of the foliation of TQ generated by the involutive distribution $\ker FL_*$ is a manifold such that the canonical projection is a submersion. (Note almost regular implies quasi-regular.)

Theorem[15] If L is almost regular there exists a corresponding Hamiltonian formulation of the dynamics on a submanifold of T*Q and the Lagrangian and Hamiltonian formalisms are *equivalent* (i.e. each constraint submanifold of the Lagrangian system maps onto a corresponding constraint submanifold for the Hamiltonian system).

Theorem If L is quasi-regular there is a corresponding equivalent generalized Hamiltonian system; however it is not given on a submanifold of T*Q but rather on the leaf space $TQ/(\ker FL_*)$.

4A. Second Order Equation Problem Lagrangian systems differ from Hamiltonian systems in that from both physical[1] and variational[16] viewpoints the Lagrange equations (6) must be supplemented[17] by the 2nd order equation condition

$$T\tau_Q X = \tau_{TQ} X \quad . \tag{7}$$

This condition was first investigated for homogeneous degenerate Lagrangians by Künzle[18] who considered only degeneracies which arise from the homogeneity. He developed an algorithm which in principle solves the problem, but he gave no specific results in the general case. We have taken a different approach and considered the general degenerate Lagrangian, finding that (7) is compatible with our constraint algorithm.

Theorem[19] If L is quasi-regular there exists a 1 to 1 correspondence between solutions to (6,7) and solutions to the corresponding generalized Hamiltonian system on $TQ/(\ker FL_*)$.

5. NON-UNIQUENESS OF SOLUTIONS Presymplectic Hamiltonian systems in general do not have unique solutions. To any vector field $X \in TN$ satisfying (3) along the final constraint submanifold $N \xrightarrow{N} M_1$ one can add any element Y in

$$K_1 \equiv TN \cap TM_1^{\perp} \quad . \tag{8}$$

(These vector fields are generated by Dirac's primary 1st class constraints.) The presence of these "gauge" vector fields indicates that N contains redundant information. For certain purposes (e.g. quantization) one would like to eliminate this redundant information by constructing a *reduced phase space* (RPS) i.e. a "space of true degrees of freedom". One would like the RPS to be a symplectic manifold with a unique evolution determined by a Hamiltonian system.

5A. Standard RPS A reduced phase space can be constructed as follows: pull back (3) to N obtaining

$$i_X\omega_N = \alpha_N \tag{9}$$

where $\alpha_N \equiv i_N^*\alpha$ and $\omega_N \equiv i_N^*\omega$. The solutions to (9) are unique up to elements in

$$K_N \equiv TN \cap TN^{\perp} , \tag{10}$$

the characteristic vector fields of ω_N. These vector fields form an involutive distribution which determines a foliation of N. On the leaf space of this foliation $\overline{N} \equiv N/(K_N)$ there is a symplectic (hence unique solution) Hamiltonian system

$$i_{\overline{X}}\omega_{\overline{N}} = \alpha_{\overline{N}} \tag{11}$$

induced by (9) - i.e. our constraint algorithm guarantees that α is *admissible* for N in the sense of Lichnerowicz.[10] We call $\overline{N}$ the *standard* RPS. Regarding $\overline{N}$ as the "space of true degrees of freedom" of the system is equivalent to regarding all of K_N as gauge vector fields which corresponds to Dirac's extended Hamiltonian using all 1st class constraints (primary and secondary) as gauge generators. Dirac himself was uncertain as to whether this was correct - "I think it may be that all the first-class secondary constraints should be included among the transformations which don't change the physical state, but I haven't been able to prove it."[20]

5B. Gauge Vector Field Algorithm (GVA) RPS[21] In order to see how the manifest K_1 gauge invariance of (3) could imply K_N gauge invariance we have developed a gauge vector field algorithm (GVA) for extracting the "hidden" gauge transformations in (3). The associated GVA RPS is obtained by "moding out" the "gauge directions" in N: i.e. one wants to identify points connected by "gauge vector fields" (GVF's) since such points can be reached from the same initial data. The set of GVF's can be considerably larger than K_1 for if Z is a GVF then so is [X,Z] where X is any solution to (3). This follows from the diagram

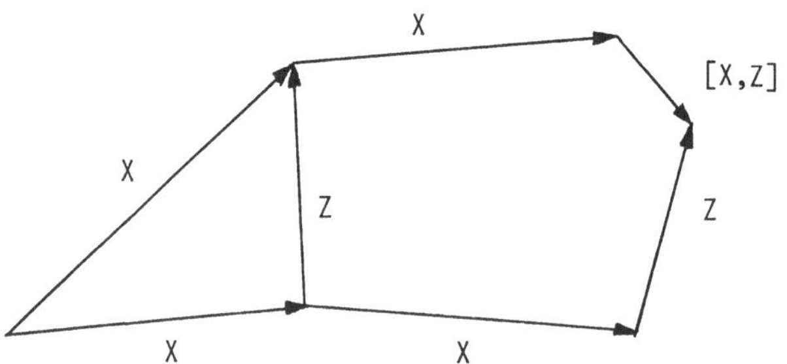

A similar construction shows that [Z,Y] is a GVF whenever Z and Y are GVF's. Starting from $G_1 \equiv K_1$ (the manifestly obvious GVF's) and iterating one obtains the *implicit* GVF's

$$G_{\ell+1} \equiv G_\ell + [G_\ell, G_\ell] + [X, G_\ell] \qquad (12)$$

whose limit is an involutive distribution G - the Lie module generated by $(\mathcal{L}_X)^\ell K_1$. The associated GVA RPS is the leaf space $\tilde{N} \equiv N/(G)$ of the foliation generated by G. Is the GVA RPS the same as the standard RPS?

Theorem $G = K_N$ iff the *first class constraints* $\langle K_\ell | \alpha \rangle$ are "good" constraints (i.e. their differentials do not vanish along N) where $K_\ell \equiv TN \cap TM_\ell^\perp$.

In general G is a proper Lie submodule of K_N (i.e. some 1st class secondary constraints do not generate gauge transformations of (3)) and consequently $\tilde{N}$ is larger than $\overline{N}$. In general $\tilde{N}$ is inherently presymplectic. (In fact $\tilde{N}$ can be odd dimensional!) Whenever $G \neq K_N$ there is no natural symplectic Hamiltonian system on $\tilde{N}$ for the *unique* evolution $\tilde{X} \varepsilon T\tilde{N}$. Consequently regarding $\tilde{N}$ as the space of true degrees of freedom has some undesirable features (e.g. How would one quantize such a system?). On the other hand $\tilde{N}$ has the virtue of literally respecting the equations of motion (4).

Which RPS is the "space of true degrees of freedom"?

5C. A Possible Resolution If the Hamiltonian contains a built in gauge condition then G will respect it while K_N will not. So it may be possible to interpret $G \neq K_N$ as a consequence of built in gauge conditions. Two examples may clarify this point. Consider the Lagrangian densities

$$\mathcal{L} = -\frac{1}{4} F^{\mu\nu}F_{\mu\nu} - \frac{1}{2\lambda}(\partial_\mu A^\mu)^2 \qquad (13)$$

$$\mathcal{L} = -\frac{1}{2}(\partial_\mu A_\nu)\partial^\mu A^\nu - A^\mu \partial_\mu \phi - \frac{\lambda}{2}\phi^2 \quad . \qquad (14)$$

where λ is a Lagrange multiplier and $F_{\mu\nu} \equiv \partial_\mu A_\nu - \partial_\nu A_\mu$. From (13) one obtains Maxwell's equations and the Lorentz condition $\partial_\mu A^\mu = 0$. If we interpret this system as electromagnetism with the Lorentz gauge condition imposed then it is easy to understand why $G \neq K_N$, for G respects the Lorentz condition while K_N does not. The Lagrangian density (14) in fact yields the same effective equations as (13), though now one may be tempted to interpret it as what it appears to be - not E&M but an entirely different (massless, spin 1, divergence free) field. This is the GVA interpretation of both (13) and (14). Thus whether or not Dirac's conjecture is correct (and which RPS is correct) depends on one's physical interpretation of the system. In general our algorithm permits a whole class of interpretations bounded by one (G) which extracts the gauge invariance of a given Lagrangian (and yields an inherently presymplectic RPS) and the standard one (K_N) which extracts the maximum possible invariance of the given system (and has a symplectic Hamiltonian system for the unique evolution on its RPS). The former violates Dirac's conjecture while the latter satisfies it by interpreting $G \neq K_N$ to be entirely a consequence of built-in gauge conditions.

1. R. Abraham and J. Marsden, Foundations of Mechanics (2nd Edition) Benjamin 1978.
   P. Chernoff and J. Marsden, Properties of Infinite Dimensional Hamiltonian Systems, Lecture notes in Math #425, Springer-Verlag 1974.
2. H.P. Künzle, J. Math. Phys. 13, 739 (1972).
3. M. Gotay, J. Nester, G. Hinds, "Presymplectic Manifolds and the Dirac-Bergmann Theory of Constraints", to appear J. Math. Phys. (1978).
4. We assume here and throughout that quotients and constraints yield sufficiently nice manifolds. For many interesting systems this is not true - one must then cut the system up into nice pieces. See refs. 8 and 12.
5. Actually this is a particular application of a general integrability algorithm that we have developed for systems of equations.
6. P.A.M. Dirac, Lectures on Quantum Mechanics, Academic Press 1965.
7. A. Hanson, T. Regge, C. Teitelboim, "Constrained Hamiltonian Systems", Accademia Nazionale dei Lincei #22, Rome (1976).
8. J. Śniatycki, Ann. Inst. Henri Poincaré A20, 365 (1974).
9. W. Tulczjew, Symposia Mathematica 14, 247 (1974).
10. A. Lichnerowicz, C.R. Acad. Sci. A280, 523 (1975).
11. J. Marsden, A. Weinstein, Rept. Math. Phys. 5, 121 (1974).
12. G. Hinds, "Foliations and the Dirac Theory of Constraints" thesis, Univ. of Maryland (1965).
13. J. Klein, Symposia Mathematica 14, 181 (1974).

14. This generalizes Śniatycki's definition.[8]

15. M. Gotay and J. Nester, "Presymplectic Lagrangian Systems I: The Constraint Algorithm and the Equivalence Theorem", submitted to Ann. Inst. Henri Poincaré.

16. J. Nester, "An Invariant Derivation of the Euler Lagrange Equations", in preparation.

17. If L is degenerate (7) is <u>not</u> a consequence of (6).

18. H.P. Künzle, Ann. Inst. Henri Poincaré <u>A11</u>, 393 (1969).

19. J. Nester and M. Gotay, "Presymplectic Lagrangian Systems II: The Second Order Equation Problem", in preparation for Ann. Inst. Henri Poincaré.

20. See ref. 5, pp. 23,24.

21. J. Nester and M. Gotay, "Presymplectic Hamilton Equations: Gauge Transformations, and the Reduced Phase Space", in preparation.

DEFORMATION THEORY AND QUANTIZATION

André LICHNEROWICZ (Collège de France)

It is possible to give a complete description of Classical Mechanics in terms of symplectic geometry and Poisson bracket. It is the essential of the hamiltonian formalism. In a common program with Bayen, Flato, Fronsdal, Sternheimer and J. Vey, we have study properties and applications of the <u>deformations</u> of the Poisson Lie algebra. Such deformations give a new approach for Quantum Mechanics. I consider here only dynamical systems with a finite number of degrees of freedom, but the approach and a significative part of our results can be extended to physical fields.

<u>1- Lie algebras associated to a symplectic manifold</u>.

a) Let (W,F) be a <u>smooth symplectic manifold</u> of dimension $2n$; F is a closed 2-form of rank $2n$. We denote by $\mu : TW \to T^{*}W$ the isomorphism of vector bundles defined by $\mu(X) = - i(X)F$ (where $i(.)$ is the interior product); this isomorphism can be extended to tensors in a natural way; we denote by Λ the skewsymmetrical contravariant 2-tensor $\mu^{-1}(F)$. We put $N = C^{\infty}(W;R)$.
A <u>symplectic infinitesimal transformation</u> (i.t) is defined by a vector field X such that $\mathcal{L}(X)F = 0$ (where $\mathcal{L}$ is the Lie derivative) it is an infinitesimal automorphism of the structure. We denote by L the (infinite dimensional) Lie algebra of the symplectic i.t.; $X \in L$ iff the 1-form $\mu(X)$ is closed. If $X,Y \in L$:

$$\mu([X,Y]) = d\, i(\Lambda)(\mu(X) \wedge \mu(Y)) \tag{1-1}$$

Let L^{*} be the subspace of L defined by the converse images of the exact 1-forms ($X_u = \mu^{-1}(du)$; $u \in N$). An element of L^{*} is a <u>hamiltonian vector field</u>. It is well-known (Arnold, myself) that L^{*} is exactly the commutator ideal $[L,L]$ of L (each element of $[L,L]$ is, by definition, a finite sum of brackets of elements of L); $\dim L/L^{*} = b_1(W)$, where $b_1(W)$ is the first Betti number for the homology of W with compact supports.

b) We denote by L^c (<u>Lie algebra of the conformal symplectic i.t.</u>) the Lie algebra of the vector fields X such that there exists a constant $k(X)$ for which :

$$\mathcal{L}(X)\, F + k\,(X)\, F = 0 \qquad \text{or} \qquad \mathcal{L}(X)\Lambda = k\,(X)\Lambda$$

L and L^{*} are ideals of L^c. If $X \in L^c$, we have $k(X)F = d\mu(X)$.
<u>If F is non exact</u> (in particular if W is compact), $k(X) = 0$ for every $X \in L^c$ and $L^c = L$.
<u>If F is exact</u>, $F = d\mu(Z)$ and $Z \in L^c$ with $k(Z) = 1$; we have $[L^c,L^c] = L$ and $\dim.\ L^c/L = 1$.

c) Let $\bar{N}$ be the space of the classes of elements of N modulo the additive constants; $\Pi : u \in N \to \bar{u} \in \bar{N}$ is the projection of N onto $\bar{N}$. The natural isomorphism between the spaces L^{*} and $\bar{N}$ induces on $\bar{N}$ a structure of Lie algebra defined in the following way : if $\bar{u}$, $\bar{v} \in \bar{N}$, it follows from (1-1) that the function

(1-2) $$w = i(\Lambda)\ (d\bar{u} \wedge d\bar{v})$$

defines a class $\bar{w}$ which is the bracket of $\bar{u}$, $\bar{v}$. The function (1-2) is the Poisson bracket of $\bar{u}$, $\bar{v}$, or of two representants u,v in N; we put $w = \{u,v\}$. The Poisson bracket defines on N a structure of Lie algebra; (N,{ , }) is the Poisson Lie algebra of the manifold and we have a homomorphism of the Poisson Lie algebra on the Lie algebra L^* of the hamiltonian vector fields.

2 - Classical Dynamics and Symplectic Manifolds.

a) Consider a dynamical system with time independent constraints and n degrees of freedom. The corresponding configuration space is an arbitrary differentiable manifold M of dimension n. It is well-known that the cotangent bundle T^*M admits a natural symplectic structure defined by the exact Liouville 2-form. For the hamiltonian formalism, a dynamical state of the system is nothing other as a point of $W = T^*M$, which corresponds to the usual phase space. The analysis of the equations of Mechanics have showed, from a long time, that it is essential to may introduce changes of the classical variables (q^α, p_α) which does not respect the cotangent structure of W. We are thus led to introduce as phase space a symplectic manifold (W,F) of dimension 2n. Dynamics is determined by a function $H \in N$, the hamiltonian of the system, which defines a hamiltonian vector field X_H. A motion of the dynamical system is given, by definition, by an integral curve c(t) of the hamiltonian vector field X_H, the parameter t being the time. Such is the geometrical meaning of the classical equations of Hamilton.

b) We can adopt another viewpoint. The space N admits the following two algebraic structures :

1) a structure of associative algebra defined by the usual product of functions (which is here commutative)

2) a structure of Lie algebra defined by the Poisson bracket.

The derivations of the product are given by the vector fields; in particular it follows from $\{u,v\} = \mathcal{L}(X_u)v$ that :

(2-1) $$\{w,uv\} = \{w,u\} \cdot v + u \cdot \{w,v\}$$

Consider a family u_t of elements of N satisfying the differential equation :

(2-2) $$du_t / dt = \{H,u_t\}$$

and taking the initial value u_o at t = 0. It follows from (2-1) that the evolution in the time of u_t process from the trajectories which appear in the first viewpoint;

(2-2) can be considered as the intrinsic equation of Classical Dynamics.

c) We have completely described Classical Mechanics in terms of the two laws of composition of N, connected by (2-1). It is natural to study if it is possible to deform in a suitable way these two algebraic laws so that we obtain a model isomorphic to the usual Quantum Mechanics. The first results are positive.

3- Chevalley cohomology and derivations.

a) The Chevalley cohomology of the Poisson Lie algebra (N,{ , }) is defined in the following way : a p-cochain C of N is an alternate p-linear map of N^p into N, the 0-cochains being identified with the elements of N. The coboundary of the p-cochain C is the (p+1)-cochain ∂C defined by :

$$(3\text{-}1)\quad \partial C(u_o,\ldots,u_p) = \varepsilon^{\lambda_o\ldots\lambda_p}_{o\ldots p}\left(\frac{1}{p!}\{u_{\lambda_o},C(u_{\lambda_1},\ldots,u_{\lambda_p})\} - \frac{1}{2(p-1)!}C(\{u_{\lambda_o},u_{\lambda_1}\},u_{\lambda_2},\ldots,u_{\lambda_p})\right)$$

where ε is the skewsymmetrical Kronecker indicator and where $u_\lambda \in N$. The space of the 1-cocycles of (N,{ , }) is the space of the <u>derivations</u> of the Lie algebra, the space of the exact 1-cocycles is the space of the <u>inner derivations</u>.

A p-cochain C is called <u>local</u> if, for each $u_1 \in N$ such that $u_1|_U = 0$ on a domain U, we have $C(u_1,\ldots, u_p)|_U = 0$. If C is local, ∂C is local.

A p-cochain C is called d-differential ($d \geqslant 1$) if it is defined by a multidifferential operator of maximum order d in each argument. If C is d-differential, ∂C is d-differential also.

b) I have proved the following non trivial theorems :

<u>Theorem 1 - If T is a local 1-cochain of N such that $C = \partial T$ is d-differential, T is d-differential itself.</u>

<u>If W is non compact, the assumption of locality for T can be suppressed.</u>

<u>If W is compact, each exact d-differential 2-cochain C is the coboundary of a d-differential 1-cochain T.</u>

<u>Theorem 2 - If W is non compact, each derivation of N is given by $\mathcal{D} = \mathcal{L}(X)+k(X)$, where $X \in L^c$. If W is compact, each derivation $\mathcal{D}$ of N is given by</u> :

$$\mathcal{D}u = \mathcal{L}(X)\, u + c\int_W u\,\eta$$

<u>where $X \in L$, $c \in \mathbb{R}$ and η is the symplectic volume element. There are non local derivations.</u>

4 - Formal deformations.

I will first recall and extend the main elements of the theory of Gerstenhaber concerning the deformations of the algebraic structures, in particular of the Lie algebras [2]

a) Let $E(N;\lambda)$ be the space of the formal functions of $\lambda \in \mathbb{C}$ with coefficients in N. Consider an alternate bilinear map $N \times N \to E(N;\lambda)$, which gives a formal series in

(4-1) $$[u,v]_\lambda = \sum_{r=0}^{\infty} \lambda^r C_r(u,v) = \{u,v\} + \sum_{r=1}^{\infty} \lambda^r C_r(u,v)$$

where the C_r's $(r \geqslant 1)$ are differential 2-cochains of $(N,\{\ ,\ \})$. These cochains can be extended to $E(N;\lambda)$ in a natural way. If $u,v,w \in N$, we have

$$S\left[[u,v]_\lambda,w\right]_\lambda = \sum_{t=1}^{\infty} \lambda^t D_t\,(u,v,w)$$

where S is the summation after circular permutation and where D_t is the 3-cochain :

$$D_t(u,v,w) = S \sum_{r+s=t} C_r(C_s(u,v),w) \qquad (r,s \geqslant 0)$$

We say that (4-1) defines <u>a formal deformation of the Poisson Lie algebra</u> if Jacobi's identity is satisfied formally, that is if $D_t = 0$ $(t = 1,2,\ldots)$. We put

$$E_t\,(u,v,w) = S \sum_{r+s=t} C_r\,(C_s(u,v),w) \qquad (r,s \geqslant 1)$$

and we have $D_t \equiv E_t - \partial C_t$. If (4-1) is limited to the order q, we have <u>a deformation of order q</u> if Jacobi's identity is satisfied up to the order (q+1). If such is the case, E_{q+1} is mechanically a 3-cocycle of N. We can find a 2-cochain C_{q+1} satisfying $D_{q+1} \equiv E_{q+1} - \partial C_{q+1} = 0$ iff E_{q+1} is exact; E_{q+1} defines a class which is <u>the obstruction at the order (q+1)</u> to the construction of a formal deformation.

A deformation of order 1 is called an <u>infinitesimal deformation</u>. We have $E_1 = 0$ and so only $\partial C_1 = 0$, that is a 2-cocycle of N.

b) Consider a formal series in λ :

(4-2) $$T_\lambda = \sum_{s=0}^{\infty} \lambda^s T_s = \mathrm{Id.} + \sum_{s=1}^{\infty} \lambda^s T_s$$

where the T_s's $(s \geqslant 1)$ are differential operators on N; T_λ acts naturally on $E(N;\lambda)$. Consider also another alternate bilinear map $N \times N \to E(N;\lambda)$ corresponding to the formal series

(4-3) $$[u,v]'_\lambda = \{u,v\} + \sum_{r=1}^{\infty} \lambda^r C'_r\,(u,v)$$

where the C'_r are differential 2-cochains again. Suppose (4-2), (4-3) such that we have formally the identity

$$T_\lambda \left[u,v\right]'_\lambda = \left[T_\lambda u, T_\lambda v\right]_\lambda \tag{4-4}$$

Using some universal formulas, we prove by recursion the following :

Proposition - The formal deformation (4-1) of the Poisson Lie algebra being given, each formal series (4-2) generates a unique bilinear map (4-3) satisfying (4-4). This map is a new formal deformation which is called equivalent to (4-1). In particular, a deformation is called trivial if it is equivalent to the identity deformation ($C_r = 0$ for every $r \geqslant 1$).

If two deformations are equivalent at the order q, there appears a 2-cocycle, element of $H^2(N,N)$, which is the obstruction to the equivalence for the order (q+1). In particular, two infinitesimal deformations defined by the 2-cocycles C_1 and C'_1 are equivalent if $(C'_1 - C_1)$ is exact.

c) We may give, for the associative deformations of the usual product on N, a definition similar to the definition of the deformations of the Poisson Lie algebra. Such a deformation is defined by a bilinear map $N \times N \to E(N;\nu)$ ($\nu \in \mathbb{C}$) given by :

$$u *_\nu v = u \cdot v + \sum_{r=1}^{\infty} \nu^r \Gamma_r (u,v) \tag{4-5}$$

where the Γ_r's are differentiable bilinear map $N \times N \to N$ such that we have the associtivity identity

$$(u *_\nu v) *_\nu w = u *_\nu (v *_\nu w) \tag{4-6}$$

There is a cohomology - the Hochschild cohomology - which is adapted to the problems of associative deformations. The Γ_r's are Hochschild 2-cochains. In the following part, we consider deformations (4-5) such that Γ_r are null on the constants. Moreover, we suppose that $\Gamma_r(u,v)$ is symmetric in u,v if r is even, skewsymmetric if r is odd. We obtain by skewsymmetrization, with $\lambda = \nu^2$

$$\left[u,v\right]_\lambda = (2\nu)^{-1} (u *_\nu v - v *_\nu u) = \{u,v\} + \sum_{r=1}^{\infty} \lambda^r \Gamma_{2r+1} (u,v) \tag{4-7}$$

which is a deformation of the Poisson Lie algebra, generated by (4-5).

Under these assumptions, I have proved that there exists at most one associative deformation which generates a given Lie deformation. Moreover we have

Theorem 3 - Consider the associative algebra defined on E(N;ν) by (4-5). If $b_1(W)=0$, all the derivations of this algebra are inner derivations; all the automorphisms $A_\nu = \sum_{r=0} \nu^2 A_r$, such that $A_o = Id.$, are inner automorphisms.

d) I consider here a very interesting example of deformation described recently by Jacques Vey. My viewpoint is different from the viewpoint of Vey. [5]

5 - The flat case.

a) Let (W,F) be a symplectic manifold. Such a manifold admits atlases of charts for which F(or Λ) have constant components (natural chart $\{x^i\}$ $(i,j,\ldots = 1,\ldots;2n)$).

A symplectic connection Γ is a linear connection without torsion such that $\nabla F = 0$, where ∇ is the operator of covariant differentiation defined by Γ. If $\{\Gamma^i_{jk}\}$ are the usual coefficients of a connection Γ in a natural chart $\{x^i\}$, we introduce the coefficients $\Gamma_{ijk} = F_{il}\,\Gamma^l_{jk}$. Such coefficients $\{\Gamma_{ijk}\}$ define a symplectic connection iff they are completely symmetrical for every natural chart. A symplectic manifold admits infinitely many symplectic connections; the difference between two symplectic connections is given by a symmetrical covariant 3-tensor.

b) Suppose that (W,Λ) admits a symplectic connection without curvature; if such is the case, the manifold (W,Λ,Γ) is called a flat symplectic manifold. The simplest example is the cotangent bundle of $\mathbb{R}^n$, that is $\mathbb{R}^n \times \mathbb{R}^n$. Introduce on a flat symplectic manifold the bidifferential operators P^r of maximum order r on each argument defined by the following expression on each domain U of an arbitrary chart $\{x^i\}$:

$$P^r(u,v)_{|U} = \Lambda^{i_1 j_1} \ldots \Lambda^{i_r j_r} \nabla_{i_1 \ldots i_r} u \, \nabla_{j_1 \ldots j_r} v \qquad (u,v \in N) \tag{5-1}$$

We put $P^o(u,v) = u$. For $r = 1$, we obtain the Poisson bracket operator P with $P(u,v) = \{u,v\}$.

Given a formal function f(z) with constant coefficients such that $f(o) = 1$, substitute P^r to z^r in the expansion of $f(\nu z)$; we obtain a bilinear map $(u,v) \in N \times N \to u *_\nu v = f(\nu P)(u,v) \in E(N;\nu)$. We will choose f so that we define thus a deformation of the associative algebra (N,.). The answer is given by the following :

Proposition - If (W,Λ,Γ) is a flat symplectic manifold, there is only one formal function of the Poisson bracket P (up to a constant factor and a linear change of the deformation parameter ν) that generates a formal deformation of the associative algebra (N,.) : it is the exponential function.

We have :

(5-2) $$u *_\nu v = \sum_{r=0}^{\infty} (\nu^r/r!)\, P^r(u,v) = \exp(\nu P)(u,v)$$

which generates the deformation of the Poisson Lie algebra ($\lambda = \nu^2$)

(5-3) $$[u,v]_\lambda = \sum_{r=0}^{\infty} (\lambda^r/(2r+1)!)P^{2r+1}(u,v) = \nu^{-1} \sin h(\nu P)(u,v)$$

It is remarkable that, for $\nu = i\hbar/2$, we deduce from (5-3) a bracket $\frac{2}{\hbar}\sin(\frac{\hbar}{2}P)$ given in 1949 by Moyal in the context of the Hermann Weyl-Wigner quantization. Consider the term P^3 of (5-3). If this cocycle were exact in the Chevalley cohomology, it would be the coboundary of a 1-cochain which can be assumed 3-differential, according to Theorem 1. But it is easy to see that such a coboundary has no term of bidifferential type (3,3). It is possible to prove that, for a flat symplectic manifold, the second space $H^2(N;N)$ of Chevalley cohomology has the dimension 1; P^3 defines a cohomology 2-class β which is a generator for this space. We see that the deformation (5-3) is non trivial even for the order 1. It is the same for the deformation (5-2).

6 - Generalizations.

It is natural to study if the deformations (5-2), (5-3) can be generalized to non flat symplectic manifolds. It is easy to see that we doe not obtain such generalizations if we extend the formula (5-1) to the case where ∇ corresponds to an arbitrary symplectic connection Γ .

a) if $u \in N$, denote by $\mathcal{L}(X_u)\Gamma$ the symmetric covariant 3-tensor defined by means of the Lie derivative of the symplectic connection Γ by the hamiltonian vector field X_u. The 2-cochain S^3_Γ defined by :

(6-1) $$S^3_\Gamma(u,v)_{|U} = \Lambda^{i_1 j_1}\Lambda^{i_2 j_2}\Lambda^{i_3 j_3}(\mathcal{L}(X_u)\Gamma)_{i_1 i_2 i_3}(\mathcal{L}(X_v)\Gamma)_{j_1 j_2 j_3}$$

admits the same principal symbol as P^3. We have $\partial S^3_\Gamma = 0$, according to the properties of the Lie derivatives. The same argument as for the flat case shows that the 2-cocycle S^3_Γ is non exact. If we change the symplectic connection, S^3_Γ is changed by addition of a coboundary. We see that the cohomology 2-class β of (N,P)defined by this 2-cocycle depends only upon the symplectic structure of the manifold.

b) Introduce now the following notations : we denote by Q^r a bidifferential operator of maximum order r on each argument, null on the constants and such that its principal symbol coincides with the principal symbol of P^r; Q^r is supposed symmetric in u,v if r is even, skewsymmetric if r is odd. We take in particular $Q^o(u,v) = u.v$, $Q^1 u,v) = P(u,v)$ and $Q^3 \in \beta$; J. Vey has recently proved by a long and fine cohomology study using Gelfand-Fuks results, the following

<u>Theorem 4 - (Vey). Let (W,F) be a symplectic manifold such that the third Betti number $b_3(W) = 0$. There exists formal deformations of the Poisson Lie algebra of the manifold such that</u>

$$[u,v]_\lambda = \sum_{r=0}^{\infty} (\lambda^r/(2r+1)!)\, Q^{2r+1}(u,v) \tag{6-2}$$

General explicit forms for Q^{2r+1} are not known. For the 2-cocycle Q^3, I have proved the following result : there is a unique connection Γ such that

$$Q^3 = S^3_\Gamma + \partial K \tag{6-3}$$

where K is a differential operator of order $\leqslant 2$, such that $K(1) = \text{const.}$

c) I shall say, that we have a $*_\nu$-product(or twisted product) on the symplectic manifold (W,F) if there are Q^r 's such that

$$u *_\nu v = \sum_{r=0}^{\infty} (\nu^r/r!)\, Q^r(u,v) \tag{6-4}$$

is <u>associative</u> . The general problem of the existence of a $*_\nu$-product on (W,F) is more difficult than the problem solved by Vey and the answer is unknown. I have obtained however construction processes of such $*_\nu$-products for large classes of cotangent of classical groups and homogeneous spaces.

I will limit myself to the simplest example. Consider the flat symplectic manifold defined by the cotangent bundle of the space $R^n-\{0\}$, that is the manifold $E = (\mathbb{R}^n - \{0\}) \times \mathbb{R}^n$.

The solvable group G_2 of dimension 2 acts on E in the following way

$$(x,y) \in E \leqq (\mathbb{R}^n - \{0\}) \times \mathbb{R}^n \to (x' = e^\rho x, y,\ y' = e^{-\rho}(y + \sigma x)) \qquad (\rho,\sigma \in \mathbb{R})$$

The group G^2 preserves the natural symplectic structure of E and the flat connection. It follows that it preserves the P^r's defined by (5-1). The space of the orbits of E by this group is isomorphic to T^*S^{n-1}, where $S^{n-1} = SO(n)/SO(n-1)$ is the sphere of dimension (n-1). We deduce from the $*_\nu$-product invariant under G_2, defined on E by the P^r's a natural $*_\nu$-product on T^*S^{n-1}; this product is invariant under SO(n). We may deduce from this method the existence of natural $*_\nu$-products for example for the cotangent bundles of the Stiefeld manifolds and of the Grassmann manifolds. Twisted products may be defined also on the symplectic manifolds determined by the orbits of a Lie group for the coadjoint representation, according to the classical theorem of Kirilov-Souriau-Kostant.

7 - Introduction to a spectral theory and quantization.

a) Come back to the flat symplectic manifold $\mathbb{R}^n \times \mathbb{R}^n$. Under suitable assumptions, Hermann Weyl has defined in this case, in terms of Fourier transform, a map Ω (the Weyl map) which associates to each element u of a large class of complex-valued functions or distributions a unitary operator $\hat{u}$ of a Hilbert space and conversely. The usual quantization processes in terms of these operators. But the $*_\nu$-product defined by (5-1) corresponds by Ω to the product of operators (for $\nu = i\hbar/2$). We note that if $u, v \in L^2$, we have

(7-1) $$\int_W (u * v)\, \eta = \int_W (u\, v)\eta$$

The change of ordering may be translated in terms of equivalent twisted products. It appears us possible to develop directly Quantum Mechanics in terms of ordinary functions or distributions and $*$-products, without reference to some Ω and to operators, in a complete and autonomous way.

b) Consider a symplectic manifold (W,F) admitting a $*_\nu$-product; we put $N^c = C^\infty(W;\mathbb{C})$. Let H be the classical hamiltonian of our problem. We are led to translate the dynamical Schrödinger equation by

(7-2) $$du_t/dt = [H, u_t]_{\nu^2}$$

Introduce the $*$-powers of $H(H^{(*)p} = H^{(*)p-1} * H)$ and the $*$-exponential of Ht. We put

(7-3) $$\mathrm{Exp}_*(Ht) = \sum_{p=0}^{\infty} (t/2\nu)^p (1/p!)\, H^{(*)p}$$

If $u_o \in N^c$, define u_t formally by :

(7-4 $$u_t = \mathrm{Exp}_*(Ht) *_\nu u_o *_\nu \mathrm{Exp}_*(-Ht)$$

(7-4) gives a formal solution of (7-2) taking the value u_o at $t = 0$.

c) Now consider the viewpoint of the mathematical analysis and give to ν the value $i\hbar/2$. Assume that H is such that, for t in a complex neighborhood of the origin, the right side of (7-3) converges to a distribution denotes by $Exp_*(Ht)$ again. Suppose for simplicity that $Exp_*(Ht)$ has a unique Fourier-Dirichlet expansion

$$(7\text{-}5) \qquad Exp_*(Ht) = \sum_{\lambda \in I} \Pi_\lambda \, e^{\lambda t/i\hbar}$$

where I is a subset of $\mathbb{C}$ and $\Pi_\lambda \in N^c$; I gives the spectrum of H and a trace formula gives the multiplicity. This expansion is similar to the spectral decomposition of an operator. It is easy to see that :

$$(7\text{-}6) \qquad \Pi_\lambda * \Pi_{\lambda'} = \delta_{\lambda\lambda'} \Pi_\lambda \ , \ \Sigma\Pi_\lambda = 1, \ \ H * \Pi_\lambda = \Pi_\lambda * H = \lambda\Pi_\lambda, \ H = \Sigma\lambda\Pi_\lambda$$

A $*$-product is sayed to be non degenerated if for any $u \in N^c \cap L^2$, $\bar{u} * u = 0$ implies $u = 0$. If such is the case, the spectrum of each real-valued function admitting a spectral expansion in the sense of (7-5) is real.

The Moyal product and the $*$-products deduced by quotient are non degenerated.

d) The previous algorithm, directly applied to the flat case, gives, for the harmonic oscillator of dimension n, energy levels and multiplicities. For the Hydrogen Atom, we consider T^*S^3 as the phase space and we introduce the corresponding $*$-product invariant under SO(4) (Fock). We obtain then the complete spectrum, that is the negative discrete spectrum and the positive continuous spectrum.

All the results of the usual Quantum Mechanics may be obtained by this way.

References.

[1] A. Avez and A. Lichnerowicz C.R. Acad. Sci. Paris t.275,A(1972), p.113.

[2] M. Gesrstenhaber Ann. of Math 79, (1964), 59-103.

[3] M. Flato, A. Lichnerowicz, D. Sternheimer Compos. Matem. 31 (1975), 47-82 ; C.R. Acad. Sci. Paris t.283, A (1976), 19-24.

[4] J.E. Moyal Proc. Cambridge Phil. Soc. 45, (1949), 99-124.

[5] J. Vey Comm. Math. Helv. 50, (1975), 421-454.

[6] A. Lichnerowicz Journ. Geom. Diff. Liège, déc 1976.

[7] F. Bayen, M. Flato, C. Fronsdal, A. Lichnerowicz, D. Sternheimer, Lett in Math. Phys.1(1977),521-530;Deformation Theory and Quantization Ann. of Phys.111(1978) 61-152.

[8] A. Lichnerowicz C.R. Acad.Sci. Paris t.286,A(1978),49-53; Sur les algèbres formelles associées par déformation à une variété symplectique (to appear)

ON CERTAIN EVENTS IN GEOMETRIC QUANTIZATION

J. Czyz
Institute of Mathematics, Technical University of Warsaw
Warsaw 00-661, Pl. Jednosci Robotniczej 1

Abstract: Certain cases which may appear in geometric quantization of a complex compact manifold are listed.

2. Geometric quantization of a complex manifold.

Throughout the paper cohomology classes in the sense of Čech and de Rham will not be distinguished. If we have a class $q \in H^2(M,\mathbb{Z})$ and $\varepsilon: H^2(M,\mathbb{Z}) \to H^2(M,\mathbb{R})$ is the canonical injection then we write also q instead of $\varepsilon(q)$.

Definition: Let us consider a triplet $\langle M^r,\omega,M\rangle$ where: - M is a complex manifold (i.e. complex and holomorphic), - M^r is a real smooth manifold having the same set of points as M and both structure M and M^r are equivalent in the sense of a real smooth diffeormorphism preserving orientation, - ω is a differential two-form which is complex with respect to M and real with respect to M^r.

Moreover, the following axioms hold

q1) The pair $\langle M,\omega\rangle$ is a Kähler manifold (hence $\langle M^r,\omega\rangle$ is a symplectic manifold).

q2) Let us denote $q:=[\omega]-\frac{1}{2}c_1(M)$, where $c_1(M)$ means the first Chern class of M. Then $q \in H^2(M,\mathbb{Z})$.

q3) $q \geq 0$, i.e. $q(a) \geq 0$ for any positive oriented two-cycle a of M.

Then a complex line bundle Q such that the class $c(Q)$ (i.e. the first Chern class) is equal to q will be called the quantum bundle. If all first order torsions of M vanish then the mapping $\varepsilon: H^2(M,\mathbb{Z}) \to H^2(M,\mathbb{R})$ is injective and the bundle Q is defined uniquely up to equivalence of bundles.

The above system of axioms may be simply enlarged onto a case almost complex symplectic manifold, see [3].

From now the manifold M will be compact.

By virtue of /q1/ and the Hodge-Kodaira theorem any cohomology class q contains the only harmonic form η $(\Delta\eta=0)$. Thus we may choose a connection ∇ such that curv $(\nabla,Q)=\eta$ and a ∇-invariant hermitian structure $(\cdot|\cdot)$ in the bundle Q. Then we may turn the set of holomorphic sections of Q into a Hilbert space $\mathcal{H}$ putting

$$(s_1,s_2): = \int_{M^r} (s_1|s_2)\omega^n \ , \ 2n = \dim M^r$$

If M is simply connected then the structure $(\cdot|\cdot)$ is unique up to a positive multiplicative constant c and the space $\mathcal{H}$ is determined uniquely up to an isomorphism given by multiplication by a factor $c\cdot\exp(if(\cdot))$ where the real function f depend only on a choice of the connection ∇ .

For an exhausting procedure of this quantization procedure see [3]. One can easily prove that for any complex compact manifold admitting a Hodge structure the above procedure may be carried out, see [1]. Let M be any complex compact manifold which admits a Hodge structure. Then the following four questions arise:

1. Does the geometric quantization procedure give the unique results / i.e. whether both class q and space $\mathcal{H}$ are determined uniquely or not / ?
2. Does exist quantum bundle Q such that $c_1(Q)=nc_1(M)$ for each natural n ?
3. Does exist quantum bundle Q such that $c_1(Q)=p$ for each non-negative cohomology class $p\in H^2(M,\mathbb{Z})$?
4. Does $\dim\mathcal{H}$ depend on the complex structure M only via $c_1(M)$ / i.e. on the same way as quantum bundle Q /?

The answers are as follows:

sort of manifold	question 1	2	3	4
M admits a Hodge structure	No	?[1]	?[1]	Yes, if $c(Q)-c_1(M)$ contains a Kähler form
$\dim H^2(M,\mathbb{R})=1$	No	Yes[2]	Yes[2]	"
M may be covered with an open bounded domain	?	Yes[2]	?	"
M is $\mathbb{C}$-homogeneous	Yes	Yes	Yes	Yes

1/ Specialists in complex analysis are convinced that the answer should be "yes" but I did not find any counterexample.

2/ This is true in a case of quantization procedure like in 2 or 3, in which ω or $-\omega$ is a Kähler form. For complete proofs see to [1].

Acknowledgement

I am deeply indebted to K. Maurin and N.M.J. Woodhouse for fruitful conversations.

References

[1] J. Czyz, "On geometric quantization of compact and complex manifolds", submitted to Comm. Math. Phys.

[2] J. Czyz, "On a modification of the Kostant-Souriau geometric quantization procedure", Bull. Polon. Acad. Sci., Ser. Sci. Math. Astronom. Phys. 26, 1978, 129-138

[3] J. Czyz, "On geometric quantization and its connections with the Maslow theory", to appear in Rept. Math. Phys.

KOSTANT-SOURIAU QUANTIZATION OF ROBERTSON-WALKER COSMOLOGIES WITH A SCALAR FIELD

MARK J. GOTAY *JAMES ISENBERG*

University of Maryland

INTRODUCTION

The application of the Kostant-Souriau method of quantization[1] to homogeneous cosmologies enables us to study two important problems of theoretical physics: (1) the compatibility of gravity and quantum theory, and (2) the usefulness of the K-S procedure for quantizing classical systems. The K-S technique is an important tool for studying quantum gravity since (as we show) it quantizes some gravitational systems which cannot be easily handled by traditional methods. The homogeneous cosmologies, on the other hand, are interesting systems for testing the K-S method as they are simple, yet nontrivial and in some sense physically realistic.

The homogeneous cosmologies[2] have long been used in studies of the quantization of gravity, since their phasespaces are finite-dimensional. By using them instead of general spacetime models, we can set aside the difficulties inherent to systems with an infinite number of degrees of freedom, and focus instead on problems associated with the choice of gauge, the constraints, and the nonlinearities of Einstein's theory. These problems are severe within the traditional quantization schemes: (a) If we make no choice of gauge, we have a vanishing Hamiltonian. It can only generate quantum evolution via some Klein-Gordon perspective with its attendant complications; (b) If we do choose a gauge then the Hamiltonian tends to be time-dependent, non-commuting for different times, and square root in form. The Schrodinger equation is then nearly impossible to solve; (c) Different choices of gauge give us various (usually inequivalent) quantizations; (d) Factor-ordering ambiguities also lead to different quantum systems.

While the K-S quantization procedure avoids some (and perhaps all) of these problems, it introduces a few new ones. In particular, for a given classical system there may exist more than one of each of the necessary prequantization line bundles, metaplectic frame bundles, and quantum Hilbert spaces. We thereby obtain inequivalent quantizations. Further ambiguities are introduced due to the freedom in the choice of polarization. However, unlike their counterparts in canonical quantization schemes, the inequivalent quantizations of the K-S method can be classified, at least partially, by specific Čech cohomology classes. This classification, however, does not apply to the alternate choices of polarization. Nor does it in any way indicate which one of the possible quantizations is "physically correct."

CLASSICAL RWϕ MODELS

The "RWϕ" model cosmologies -- Robertson-Walker spacetimes containing a Klein-Gordon scalar field coupled to gravity -- are of special interest as they are the simplest cosmologies which are dynamically nontrivial. The classical system corresponding to the RWϕ universe, described by the metric

$$ds^2 = -N^2(t)dt \otimes dt + R^2(t)g_{ij}\,\sigma^i \otimes \sigma^j,$$

consists of a 6-dimensional phasespace T^*R^3, a symplectic form

$$\omega = h^{-1}(d\pi_N \wedge dN + d\pi_R \wedge dR + d\pi_\phi \wedge d\phi), \tag{1}$$

a Hamiltonian

$$H = -N\kappa = -N\{\frac{1}{24R}\,\pi_R^2 - \frac{1}{2R^3}\,\pi_\phi^2 + 6KR - \frac{1}{2}m^2\phi^2R^3\}, \tag{2}$$

and the set of constraints

$$\pi_N = 0 \quad , \quad \kappa = 0. \tag{3a,b}$$

Here ϕ is the scalar field with mass m, and h is Planck's constant.

The vanishing Hamiltonian indicates that the system is in parameterized form, and therefore admits reduction via "choice of time."[2] This is effected by specifying t as a function of the canonical variables, solving the constraint (3b) in the form $\kappa = \pi_t + H = 0$, and then eliminating the redundant variables. One obtains a two-dimensional phasespace M with Hamiltonian $H = -\pi_t$. There are many possible choices of time, including $t = R$, $t = \pi_R$ and $t = \phi$. We are studying the geometric quantization of all of these, as well as that of the unreduced (parameterized) system itself. Here, we restrict our attention to one specific reduction.

GEOMETRIC QUANTIZATION OF INTRINSIC-TIME MODEL

If we choose an "intrinsic-time" gauge $t = R$, and carry through the standard reduction, we obtain a 2-dimensional phasespace M [coordinates (ϕ,π_π)]. The symplectic structure on M is $\omega = h^{-1}d\pi_\phi \wedge d\phi$, and the unconstrained Hamiltonian is

$$H = [24t(\frac{1}{2t^3}\,\pi_\phi^2 + \frac{1}{2}m^2t^3\phi^2 - 6Kt)]^{1/2}.$$

Choosing $K = -1$ or 0 to ensure that H is well-defined, and treating t simply as a parameter, we see that in the massless case this system resembles a free particle, while in the massive case it mimics a harmonic oscillator.

In the massless case, the phasespace can be taken to be R^2 and the simplest choice of the polarization P (which diagonalizes H) is the horizontal polarization. It follows that the prequantization line bundle L, the metaplectic frame bundle, and the bundle of 1/2-forms $N^{1/2}(P)$ must be trivial. The corresponding quantum pre-Hilbert space consists of compactly supported (modulo P) sections of $L \otimes N^{1/2}(P)$ of the form:

$$\psi = f(\pi_\phi)exp[(-i/\hbar)\pi_\phi\phi].\nu,$$

where f is arbitrary, and ν is the appropriate 1/2-form.

Since the polarization diagonalizes H, the Hamiltonian operator on our Hilbert space acts simply by multiplication $\check{\delta}_H\psi = H.\psi$. This operator commutes for different times, and we can solve the Schordinger equation by expanding in an evolving complete set of energy eigenfunctions $\{\psi_{E(t)}\}$. These eigenstates satisfy

$\psi_{E(t)} = exp[\frac{i}{\hbar}\int_{t_o}^{t} E(s)ds]\psi_{E(t_o)}$, where $\check{\delta}_H(t_o)\psi_{E(t_o)} = E(t_o)\psi_{E(t_o)}$. From (4), we find that

$$\psi_{E(t_0)} = f(\pi_\phi)exp[(-i/\hbar)\pi_\phi\phi]\delta(24t_0\ \{(\pi_\phi^{\ 2}/(2t_0^3) - 6Kt_0\} - E^2(t_0)).\nu$$

and that the spectrum of $H^2(t_0)$ is $(-144Kt_0^2\ ,\ \infty)$.

Blyth and Isham have quantized this system in the coordinate representation using canonical techniques.[3] We have demonstrated that the difference between their approach and ours lies in the choice of polarization only. Using the BKS transform we have shown that the K-S results are in exact agreement with theirs.

Blyth and Isham (or anybody else, to our knowledge) have not been able to quantize in the massive case because the canonically quantized Hamiltonian operator does not commute for different times. A major advantage of geometric quantization is that it allows one to quantize using a Hamiltonian polarization (that is, one which contains the Hamiltonian vectorfield of H), in which the K-S version of this operator *does* commute for all times.

Taking $M = R^2-\{0\}$ in the massive case (so that the Hamiltonian polarization is well-defined), the prequantization line bundle is still unique (and trivial), but now there are *two* metalinear frame bundles for every polarization. Following Simms,[1] we find that the quantum Hilbert space (in either case) must be identified with the cohomology group $H^1(M,S_P)$ [since $H^0(M,S_P) = 0$], where S_P is the sheaf of germs of covariantly constant sections of $L \otimes N_{1/2}(P)$. Explicitly, the state functions corresponding to the trivial metalinear frame bundle take the form $\psi = [\sum_{n=0}^{\infty} f(n)e^{in\theta}].\nu$, where n ranges through the positive integers, f is an arbitrary (ℓ^2-bounded) function of n, ν is the appropriate 1/2-form, and θ is related to the original coordinates on phasespace by

$$\theta = \frac{1}{2i}ln[(\pi_\phi-it^6m^2\phi)/(\pi_\phi+it^6m^2\phi)].$$

For the other metalinear structure, which is a bundle over $R^2-\{0\}$ with one "twist," the wave functions take a similar form; but now $n = 1/2\ ,\ 3/2, \cdots$. For both quantizations, the Hamiltonian is "diagonal". Thus the Hamiltonian operator commutes for different times. and the Schrodinger equation can be solved. Unlike the (evolving) energy eigenstates of the massless case, however, those in the massive case are energy quantized. For the trivial metalinear frame bundle

$$E_n(t) = \frac{2}{t}\sqrt{6}\ [\frac{nt^6m^2}{4\pi} - 6t^4K]^{1/2}.$$

The eigenvalues for the nontrivial structure are obtained by replacing n by $n + 1/2$.

The physical implications of these quantizations are presently under study. We are also comparing these quantizations with those obtained via different choices of polarization and time gauge.

REFERENCES

1. Simms, in Differential Geometric Methods in Mathematical Physics, Lecture Notes in Mathematics #570, (Springer, Berlin, 1977).
2. Ryan , Hamiltonian Cosmology, Lecture Notes in Physics #13, (Springer, Berlin, 1972).
3. Blyth and Isham, Phys. Rev. D, 11, 768 (1975).

INFINITE-DIMENSIONAL LIE GROUPS AND QUANTUM DYNAMICAL SYSTEMS

R.Vilela Mendes
CFMC, Inst. Nacional de Inv. Científica
Av. Gama Pinto 2, Lisboa 4, Portugal

A systematic study of infinite dimensional Lie groups as dynamical groups of quantum systems is proposed. Some results are presented for the case of the gauge group $\{\mathcal{P}, C^{\infty}(V_4, G)\}$

Infinite dimensional Lie groups (IDLG) emerge as the natural mathematical structures in many domains of physics (gauge theories, general relativity, hydrodynamics, etc.). One wonders why IDLG's do not provide a framework for dynamics in these theories as the Poincaré group does for relativistic quantum mechanics. The answer is to be found in the state of the mathematical theory itself,in the complexity of the functional analytic approachs wide enough to include Cartan's theory for pseudogroups of diffeomorphisms, but above all in the non--existence of a systematic representation theory

[1] However, progress in the last few years in the representation theory and the developments in the method of orbits and geometric quantization, lead us to suspect that the basic tools are now available for a full exploration of IDLG's as dynamical groups of physical systems [i.e. as groups defining(canonical)ω-preserving diffeomorphisms in the symplectic manifold (M,ω) of the states of motion]. The program has two steps. First one should classify all classical systems compatible with G, i.e.all manifolds with a G-invariant symplectic form. In practice this is achieved studying in the Lie algebra dual g^* the orbits generated by $\mu \to Ad^*h(\mu) + \theta(\mu)$ $\mu\epsilon g^*$ $h\epsilon g$ where $Ad^*h(\mu)$ is the coadjoint representation of G and $\theta(\mu)$ a symplectic cocycle. Finally, to characterize the quantum dynamical systems one quantizes the orbits (i.e. obtains the unitary representations).

Of particular importance for quantum physics are the symmetry groups of gauge theories $\{\mathcal{P}, C^{\infty}(V_4, G)\}$ and the general covariance groups for pure gravity $\mathcal{D}^{\infty}(M)$ and for gravity with matter $\{\mathcal{D}^{\infty}(M), C^{\infty}(M, SL(2,C))\}$. $\mathcal{P}$ is the Poincaré group, G a compact group and $\mathcal{D}^{\infty}(M)$ the group of local C^{∞}diffeomorphisms on a Riemannian manifold M.

Here, one summarizes some of the results of the application of our program to the gauge group $G = \{\mathcal{P}, C^{\infty}(V_4, G)\}$ of Lie algebra $g = \{P_\mu, M_{\mu\nu}, \phi^i(x)X_i\}$ and Lie algebra dual $g^* = \{P^\mu, M^{\mu\nu}, h_i(x)X^i\}$. $\phi^i(x)$ has compact support and $h_i(x)$ belongs to Schwartz's distribution space $\mathcal{D}'$. Full specification of G requires a definition of the operational relations between the Poincaré and the gauge parts. The usual assumption is commutativity of $P_\mu, M_{\mu\nu}$ with X_i and a representation

$$P_\mu \sim i\partial_\mu \quad ; \quad M_{\mu\nu} \sim i(x_\mu\partial_\nu - x_\nu\partial_\mu) + \Sigma_{\mu\nu} \tag{1}$$

for the action of the Poincaré group in the cross sections of a vector bundle over V_4.

The adjoint and coadjoint representations of G and g are obtained and one proves that the g^* 1-cohomology of G vanishes if the 1-cohomology of G vanishes. (If G is semisimple for example). In this case the classical dynamical manifolds are simply the orbits of the coadjoint representation. These are found not to be amenable to splitting into irreducible elementary Poincaré components, i.e. all masses and spins must exist in each classical dynamical system possessing G as a dynamical group. The unsplitable nature of the classical orbits is therefore identified as the true source of the classical difficulties previously found in the construction of unitary representations of the gauge group compatible with the Poincaré group.

To obtain orbit splitting the only possibility is to change the group structure namely the commutativity proprieties of the gauge and Poincaré parts, which amounts to replace the representation (1) by

$$\mathcal{P}_\mu \sim i\partial_\mu + A_\mu \qquad M_{\mu\nu} \sim i(x_\mu\partial_\nu - x_\nu\partial_\mu) + \Sigma_{\mu\nu} + R_{\mu\nu} \tag{2}$$

with properties

$$\begin{aligned} \mathrm{Ad}(\exp i\phi(x)X_i)A_\nu &= A_\nu - \partial_\nu\phi(x)X_i \\ \mathrm{Ad}(\exp i\phi(x)X_i)R_{\mu\nu} &= R_{\mu\nu} - (x_\mu\partial_\nu - x_\nu\partial_\mu)\phi(x)X_i \end{aligned} \tag{3}$$

A_μ and $R_{\mu\nu}$ are the abstract generalization of the well-known invariant connection and gauge field angular momentum in Lagrangian gauge theories. The conclusion is:

"Local gauge invariance is compatible with the splitting of the physical world into elementary systems if and only if the generators of the physical Poincaré group are the covariant momentum and angular momentum defined by Eqs. (2-3)".

The quantum systems (unitary representations) are obtained from direct product representations of the gauge and the "covariant" Poincaré group. Those that correspond to finite-dimensional orbits are found to correspond to functionals with support at a finite number of points and are classified through the construction of a family of finite dimensional Lie algebras. These unitary representations are exact solutions of theories having $\mathcal{G}$ as a dynamical group. However to extract from them results in the free field plus interaction form is probably as difficult as to obtain exact solutions from perturbation theory, because it amounts to split the unitary representations of $\mathcal{P}_\mu$ and $M_{\mu\nu}$ into the non unitary actions associated to $P_\mu, M_{\mu\nu}$ and $A_\mu, R_{\mu\nu}$.

In Lagrangian non-abelian gauge theory one checks explicitely the structure of Eqs. (2) and (3). However the Noether charge $\int d^3x\, J_0[\phi]$, that by exponentiation will correspond to an unitary operator in the representation of $C^\infty(V_4, G)$, vanishes identically, therefore the resulting unitary representation of $C^\infty(V_4, G)$ is the identity representation. The present results suggest therefore that there might be other (non-Lagrangian?) dynamical theories carrying non--trivial unitary representations of $C^\infty(V_4, G)$.

A detailed version of the results will be published elsewhere [2].

[1] R.S.Ismagilov; Funct.Anal.Appl.5,209(1971);9,154(1975);Math.URSS Izv.6,181(1972);Math.URSS Sborn.29,105(1976)
A.A.Kirillov; Sov.Math.Dokl.14,1355(1973);Moscow Univ.Math.Bull.29, 60(1974)
A.N.Rudakov, Math.USSR Izv.8,836(1974)
A.M.Vershik, I.M.Gelfand, M.I.Graev; Russian Math.Surv.30,1(1975)
A.B.Borisov; J.Phys.A11,1057(1978)

[2] Submitted to Journ.Math.Phys.

Geometric Quantization of Nuclear Collective Models

G. Rosensteel, Department of Physics, Tulane University,
New Orleans, Louisiana 70118

& E. Ihrig, Department of Applied Mathematics,
McMaster University, Hamilton, Ontario, Canada L8S4K1

The Kostant-Souriau theory of geometric quantization has made significant progress toward a clear understanding of the relationship between classical and quantum mechanics [1-3]. Perhaps the most impressive application of the theory has been to the relativistic free particle. One defines a classical model for a free particle as a phase space on which there is a transitive action of the Poincare group by canonical transformations. Since these phase spaces are naturally given as co-adjoint orbits of the Poincare group, all the possible classical models for a free relativistic particle are determined. It is found that there is a phase space for every possible mass and spin particle. This is in contrast to the conventional models, given by the cotangent bundles of orbits of the Lorentz group in Minkowski space, which describe only spinless particles.

The same physical principle which determines the classical models for a free particle was used by Wigner[4] in order to fix the quantum models. Thus, the irreducible unitary representations of the Poincare group yield all possible quantum models for a free relativistic particle. The relationship between the classical and quantum models is that the quantum irreps are given by quantization of the co-adjoint orbits. This quantization is the Kostant-Souriau construction suitable for arbitrary phase spaces meeting the Bohr-Sommerfeld quantization conditions.

In this note these same physical arguments are applied to the Bohr-Mottelson model of the spatial collective modes of a nucleus [5-7]. In the conventional quantum collective model, the wavefunctions are given by the complex-valued functions on the classical configuration space of the liquid drop. This configuration space may be taken to be an orbit of SL(3) in the space Q of three by three, real symmetric positive-definite matrices [8-10]. A point of Q is physically identified with the quadrupole moment of a classical fluid $q_{ij} = \int \rho(\underset{\sim}{x})\, \underset{\sim}{x}_i\, \underset{\sim}{x}_j\, d^3x$, where ρ is the density distribution of the fluid. Thus, the action of SL(3) on Q is given by $g \in SL(3)$: $Q \to Q$, $q \to g\, q\, {}^t g$. However, the quantum collective models given by quantization of these classical configuration spaces represent only a small fraction of the range of physical possibilities.

A more powerful algebraic characterization of the quantum collective model is given by observing that the Lie group CM(3) of collective observables may be used to define the collective model in the same fashion as the Poincare group determined the quantum models of a relativistic free particle. Thus, a quantum collective model is defined to be an irreducible unitary representation of CM(3).

The group CM(3) may be given as a subgroup of SP(3,R) according to

$$CM(3) = \{(\omega,g) \equiv \begin{pmatrix} g & 0 \\ \omega g & {}^{t}g^{-1} \end{pmatrix} \mid g \in SL(3),\ \omega \in R^6\}, \tag{1}$$

where R^6 denotes the abelian group (algebra) of 3x3 real symmetric matrices under addition (commutation). The Lie algebra is

$$cm(3) = \{(\Omega, X) \equiv \begin{pmatrix} X & 0 \\ \Omega & -{}^{t}X \end{pmatrix} \mid X \in sl(3),\ \Omega \in R^6\}. \tag{2}$$

The physics of this group and algebra is given by interpreting the subgroup SL(3) as a kinematical group on R^3, and interpreting the subalgebra R^6 as the algebra spanned by the quadrupole moment with respect to the Poisson bracket or commutator.

Since the group CM(3) is a semidirect product with an abelian normal subgroup R^6, CM(3) $\simeq [R^6]SL(3)$, its unitary irreducible representations are given immediately by the inducing construction [11]. The irreps are indexed by two parameters (λ, L) where $\lambda^3 = \det q$ measures the size of the nucleus (liquid drop) and $L(L+1)$ measures the total shear momentum. The $(\lambda, L = 0)$ irreps are identical to the quantum models constructed from the orbits of SL(3) in Q. In addition, we have obtained by this algebraic technique a large class of new models for non-zero, but positive integral L. This is to be compared with the Poincare group irreps which encompass the non-zero spin possibilities.

A classical collective model is now defined to be a symplectic manifold with a transitive, canonical CM(3) action. These phase spaces are exhausted by the co-adjoint orbits of CM(3). The orbits are indexed by two real parameters (λ, v), where again $\lambda^3 = \det q$, but v^2 is the total shear momentum. The singular orbit with $v = 0$ is 10 dimensional and is just the cotangent bundle of an orbit of SL(3) in Q. The generic orbits, $v > 0$, are 12 dimensional and may be identified with $[R^6]SL(3)/[R]\ SO(2)$. The collective models with nonzero v are new models for the description of classical fluids. While the $v = 0$ model is suitable for an irrotational water droplet, the models with nonzero v describe fluids with viscosity.

The Kostant-Souriau quantization construction may now be applied to the classical collective models. However, only the co-adjoint orbits with $2\pi v$ integral meet the Bohr-Sommerfeld quantization conditions. For these integral orbits one can always find a polarization. For the $v = 0$ phase space, a polarization is given by the subalgebra $[R^6]\ so(3)$ of cm(3). However, for nonzero v, a complex polarization is necessary, $[R^6]b$, where b is the complex subalgebra of $so(3)_C$ spanned by L_3 and $L_+ = (L_1 + iL_2)$. After applying the quantization construction, every unitary irrep of CM(3) is recovered. The co-adjoint orbit (λ, v) yields the irrep (λ, L), $L = 2\pi v$.

There is an additional feature to the CM(3) application, which does not appear in the Poincare group situation. For a non-relativistic A-body problem, the classical (R^{6A}) and quantum ($L^2(R^{3A})$) state spaces are known, to which we may now relate the CM(3) models. Thus, the quantum CM(3) model is realized by decomposing $L^2(R^{3A})$ into irreducible components of CM(3). It is found that every irrep of CM(3) occurs in this decomposition with countably infinite multiplicity [8,10]. In addition, the classical CM(3) model is realized by decomposing R^{6A} into orbits of CM(3) $\subset$ SP(3, R), which has a natural canonical action on R^6 and, hence, on R^{6A}. These orbits are not symplectic manifolds. Nevertheless, there is a natural map Φ from R^{6A} onto the dual of cm(3), which maps orbits of CM(3) in R^{6A} onto co-adjoint orbits. If the Hamiltonian is in the enveloping algebra of cm(3), then its integral curves factor through Φ. This effects an enormous simplification of the original Hamiltonian problem on R^{6A}, since the dimension of the co-adjoint orbits is much less than the dimension of R^{6A}, for A large.

Further details concerning the geometric quantization of CM(3) will be published elsewhere [12].

References.

1. B. Kostant, "Quantization and Unitary Representations," in Lecture Notes in Mathematics, Vol. 170, Springer, Berlin, 1970.
2. J-M. Souriau, "Structure des Systemes Dynamiques," Dunod, Paris, 1970.
3. D. J. Simms and N. M. J. Woodhouse, "Lectures on Geometric Quantization," Lecture Notes in Physics, Vol. 53, Springer, Berlin, 1976.
4. E. P. Wigner, Ann. Math. 40 (1939) 149-204.
5. A. Bohr, Dan Mat. Fys. Medd. 26, No. 14 (1952).
6. K. Alder, A. Bohr, T. Huus, B. Mottelson and A. Winther, Rev. Mod. Phys. 28 (1956), 432.
7. A. Bohr, B. Mottelson and J. Rainwater, Rev. Mod. Phys. 48 (1976), 365.
8. G. Rosensteel, "On the Algebraic Formulation of Collective Models," Ph.D. Thesis, Univ. of Toronto, 1975.
9. G. Rosensteel and E. Ihrig, Phys. Rev. C17 (1978), 1179.
10. G. Rosensteel and D. J. Rowe, "On the Algebraic Formulation of Collective Models, I: The Mass Quadrupole Collective Model," preprint.
11. G. Rosensteel and D. J. Rowe, Ann. Phys. 96 (1976), 1-42.
12. G. Rosensteel and E. Ihrig, "Geometric Quantization of the CM(3) Model," preprint.

IDEA AND APPLICATION OF SPECTRUM-GENERATING SU(3) AND SU(4)*

A. Bohm
Center for Particle Theory, University of Texas, Austin, Texas 78712

and

R. B. Teese
Max-Planck Institute für Physik, 8 München 40, Fed. Rep. of Germany

Abstract

We review the basic ideas of the spectrum-generating SU(n) approach in particle physics, and show the analogy between this and the spectrum-generating method in atomic and molecular physics. We outline the tests of this framework involving one-hadron processes and discuss two tests of a fundamental relation of this framework (the Werle relation).

I. Introduction

Historically there appear to be two distinct stages in the use of groups in quantum physics. The first and best known stage involves the use of groups to describe symmetry transformations. The first name which comes to mind is Wigner, but of course many other famous people, such as Weyl, van der Wearden, Hund and Bargmann, also contributed. In this use of groups, one begins with the fundamental assumptions of quantum mechanics and plausible properties of the symmetry transformations, and is led to unitary representations of the covering groups of the symmetry transformations, with the fundamental observables as the generators of the transformation group.[1] The space of physical states is then the unitary representation space of the symmetry group; consequently, if we know the mathematical properties of this representation space, then we know the physical properties of the physical system that it describes. We may distinguish two classes of properties of these structures that are used: First, there are those properties that are used for the classification of physical states--for example, the irreducible representations and their reduction with respect to subgroups, the spectra of the generators and of elements of the enveloping algebra. Second, there are those properties that are used for the calculation of transitions between subsystems--for example (with definite assumptions about the transformation properties of the observables) the Wigner-Eckart theorem and the Clebsch-Gordan coefficients.

The second stage in the application of group theory started around 1965. Its purpose was the same as that for symmetry groups; namely, the properties of the representation served to classify the states, and assumed transformation properties of the observables were used to calculate transitions. However, there was an

*Talk presented by R. B. Teese at the Integrative Conference on Group Theory and Mathematical Physics, September 11-16, 1978, Austin, Texas.

essential difference: The existence of these group representations could not be derived from a symmetry transformation of the physical system. The name which was first given to this concept was "dynamical group,"[2] although many other names have been used since then, such as "spectrum-generating group"[3] and "non-invariance group."[4]

The first application[2] of this use of groups was to the rotator. This gave a mathematical structure which can describe, for example, a diatomic molecule in a particular vibrational and electronic state. The rotator can have any integral value of angular momentum j, so its weight diagram is that of Fig. (1). The symmetry group of the rotator is $SO(3)_{Ji}$ (the subscript indicates that J_i are the generators). Each line in the weight diagram corresponds to an irreducible representation R^j of SO(3). However, if we add operators Q_i such that J_i, Q_i generate E(3), then the entire weight diagram of Fig. (1) corresponds to the irreducible representation space

$$R \xrightarrow[SO(3)]{} \sum_j \oplus R^j \tag{1}$$

of E(3). That is, the dynamical group E(3) contains operators Q_i which transform between different irreducible representation spaces R^j of the symmetry group $SO(3) \subset E(3)$.

A question which naturally arises is then, whether or not such considerations have any application to elementary particle physics. In fact, this question was already addressed in the original paper by generalizing the rotator to a relativistic rotator.[2] However, there is another possible application, which I want to discuss today, and that is the reinterpretation of the SU(n) of particle physics in terms of spectrum-generating groups. These groups, SU(2), SU(3),...,SU(n), where n is apparently limited only by the current experimental budgets, are well accepted as groups whose irreducible representations classify the observed particles and resonances. They have been customarily treated as symmetry groups, which, for SU(2)-isospin, was a very good approximation. As experimental budgets continue to rise, though, this approximation has been getting much worse. The increase of "symmetry breaking" with increasing n will, if it has not already done so, prevent us from being able to use the Wigner-Eckart theorem in the conventional way. It is in an effort to save this second aspect of the usefulness of group theory that we have investigated the reinterpretation of SU(n).

II. Spectrum-generating SU(n)

To demonstrate this approach, we shall use SU(3) as an example. For the representation space $H^{SU(3)}$ we choose a basis labelled by I, I_3, Y and any other quantum numbers which may be needed, with the notation

$$|\alpha\rangle = |I, I_3, Y, \ldots\rangle \tag{2}$$

In addition to these charges, the hadrons have properties coming from the spacetime symmetry group, the Poincaré group $\mathcal{P}P_\mu L_{\mu\nu}$. Consequently, each hadron is also described by an irreducible representation space $H(m,s)$ of $\mathcal{P}$. The basis vectors usually used for this space are the Wigner basis vectors $|pss_3\rangle$, which are generalized eigenvectors of the momentum operator P_μ. Letting H^α denote the space spanned by $|\alpha\rangle$, the combination of internal and spacetime properties (according to a fundamental assumption of quantum mechanics concerning the combination of physical systems[5]) has as its space of physical states for the hadron, the direct product space $H(m,s)\otimes H^\alpha$. Such a space is represented by a dot in the weight diagram. For example, the weight diagram for the pseudoscalar meson octet is shown in Fig. (2). The space of physical states for the whole octet of Fig. (2) is

$$H^{\{8\}} = \sum_\alpha \oplus H(m_\alpha,s)\otimes H^\alpha \ . \tag{3}$$

The basis system that is usually chosen for this direct product space is the direct product basis

$$|pss_3\alpha\rangle = |pss_3 m_\alpha\rangle \otimes |\alpha\rangle \ . \tag{4}$$

This basis may not exist, if for example the operators whose eigenvalues are α and the momentum operator do not commute. Nevertheless, even if the basis (4) does exist, it is not suitable if we take into account the fact that the SU(3) classification group is not a symmetry. To illustrate this, we shall compare this situation to that of the rotator in atomic physics.

For an atomic system we assume SO(3) rotational symmetry,

$$[H,J_i] = 0 \tag{5a}$$

where H is the Hamiltonian. The transitions between different angular momentum states take place through a triplet of operators Q_i having the property[1]

$$[Q_i,J_i] = i\epsilon_{ijk}Q_k \tag{5b}$$

The Wigner-Eckart theorem may then be applied to the transition matrix elements:

$$\langle E'j'j_3'|Q_{(K)}|j_3jE\rangle = C(j1j';j_3Kj_3')\langle E'j'\|Q\|jE\rangle \tag{5c}$$

where the reduced matrix elements $\langle E'j'\|Q\|jE\rangle$ do not depend upon j_3,K,j_3'.

The SU(3) which classifies the hadrons is, however, not a symmetry group. The mass operator, and therefore the 4-momentum operator cannot commute with all of the SU(3) generators E_α:

$$[P_\mu,E_\alpha] \neq 0 \ . \tag{6a}$$

To describe weak transitions from one hadron state to another, the algebra of observables must include the weak "current" operator. We assume that the transitions take place through a current-current term in the Hamiltonian

$$H_I \propto J_\mu J^\mu$$

where the current J^{μ} contains both leptonic and hadronic terms. For the hadronic term in $K_{\ell 3}$ and $\pi_{\ell 3}$ decays, one uses a Lorentz vector operator V^{μ}_{α} with the property

$$[E_{\alpha},V^{\mu}_{\beta}] = if_{\alpha\beta\gamma}V^{\mu}_{\gamma} \tag{6b}$$

and calculates transitions (decays) using the formula

$$<\alpha'p'|V^{\mu}_{\beta}|p\alpha> = \sum_{\gamma=F,D} C(\alpha\beta\alpha';\gamma)<p'\|V_{\mu}\|p>_{\gamma} \tag{6c}$$

Although this formula looks like the Wigner-Eckart theorem, it is not. Because of (6a), the quantities $<p'\|V_{\mu}\|p>_{\gamma}$ depend upon the particle masses through the momenta, so they are not independent of the SU(3) indices $\alpha\beta\alpha'$. One would expect that (6c) could be used as an approximation, to the extent that the mass differences in a multiplet may be negated. For SU(2), since $m_{\pi^+}/m_{\pi^0} \simeq 1$, the approximation is very good. For SU(3), since $m_K/m_{\pi} \simeq 4$, the approximation is highly questionable, and for SU(4), where $m_{\chi}/m_{\pi} \simeq 20$, the symmetry-breaking corrections to (6c) could be much larger than the effects of the Clebsch-Gordan coefficients.

It is clear from the above analogy that the problem with (6c) lies in (6a). In order to replace (6c) by an exact equation, one must assume that the SU(3) is a symmetry of something other than the momentum. A suggestion which was made many years ago by Werle[6] is that (6a) should be replaced by

$$[\hat{P}_{\mu},E_{\alpha}] = 0 \tag{7}$$

where $\hat{P}_{\mu} = P_{\mu}M^{-1}$ is the 4-velocity operator and M is the mass operator. Actually, (7) is more general than it at first appears. If we multiplied the momentum operator by a different function of M, it would lead to the unphysical relation $[M,E_{\alpha}] = 0$.

Under the assumption (7) that SU(3) is a symmetry of the velocity operator and the usual assumption

$$[L_{\mu\nu},E_{\alpha}] = 0 , \tag{8}$$

it is more convenient to use the velocity-Poincaré group $\mathcal{P}_{\hat{P}_{\mu}L_{\mu\nu}} = \hat{\mathcal{P}}$ rather than the physical spacetime symmetry $\mathcal{P}_{P_{\mu}L_{\mu\nu}}$. Neither this SU(3) nor $\hat{\mathcal{P}}$ are connected with physical symmetry transformations. Nevertheless, we may assume that $\hat{\mathcal{P}} \otimes SU(3)$ describes the spectrum of the physical system hadron octet, with each hadron of the octet being a different state of this physical system. The space of physical states is then

$$\mathcal{H}^{\hat{\mathcal{P}}}(1,s) \otimes \mathcal{H}^{\{8\}} \tag{9}$$

where $\mathcal{H}^{\{8\}}$ is the irreducible representation space of this SU(3). The physical Poincaré group $\mathcal{P}_{P_{\mu}L_{\mu\nu}}$ is still represented in this space, only P_{μ} cannot now be written in the direct product from $P_{\mu} \otimes 1$, due to (6a) and (7). As a basis for (9) one chooses

$$|\hat{p}ss_3,\alpha> = |\hat{p}ss_3> \otimes |\alpha> \tag{10}$$

where $|\hat{p}ss_3>$ are generalized eigenvectors of the 4-velocity operator $\hat{P}_{\mu}$ which span the space $\mathcal{H}^{\hat{\mathcal{P}}}(1,s)$. The mass operator acts only on $|\alpha>$.

Using (7) we can now write the Wigner-Eckart theorem for matrix elements of V_μ^α between the 4-velocity eigenvectors (10);

$$<\hat{p}'s's_3'\alpha'|V_\mu^\beta|\hat{p}ss_3\alpha> = \sum_{\gamma=F,D} C(\alpha\beta\alpha';\gamma)<\hat{p}'s's_3'\|V_\mu\|\hat{p}ss_3>_\gamma \tag{11}$$

The reduced matrix elements in (11), unlike in (6c), are SU(3)-invariant functions of the SU(3)-invariant 4-velocities. That is, (11) is not a symmetry-limit approximation, but rather an exact relationship. Continuing with the example of $K_{\ell 3}$ decays, (11) becomes[7]

$$<\pi\hat{p}_\pi|V_\mu^\beta|K\hat{p}_K> = C_F(\pi\beta K)[(\hat{p}_K+\hat{p}_\pi)_\mu F_+(\hat{q}^2) + (\hat{p}_K-\hat{p}_\pi)_\mu F_-(\hat{q}^2)] \tag{12}$$

The reduced matrix elements $F_\pm(\hat{q}^2)$ are functions of the SU(3)-invariant $\hat{q}^2 = (\hat{p}_K-\hat{p}_\pi)^2$. This is to be contrasted with the conventional expression which comes from (6c),

$$<\pi p_\pi|V_\mu^\beta|Kp_K> = C_F(\pi\beta K)[(p_K+p_\pi)_\mu f_+^{\pi K}(q^2) + (p_K-p_\pi)_\mu f_-^{\pi K}(q^2)] \tag{13}$$

in which the formfactors $f_\pm^{\pi K}(q^2)$ and the momenta p_K, p_π are not SU(3)-invariants.

We thus see that the basic idea behind spectrum-generating SU(n) is really very simple: Expression (6c) cannot be correct because of the mass differences, and it may not even be an acceptable approximation. By using the Werle relation, one obtains a formula which is in principle exact, and which can be tested. However, the task of testing this idea is not simple. Conventional formulas found in textbooks cannot be used, since they were derived using the momentum as the variable and not the 4-velocity. Quantities like the partial decay rates must be completely rederived, starting from the basic principles of quantum mechanics. The result of such rederivations is in general that the new formula differs from the conventional one by a factor (suppression factor), which is a well-defined function of the hadron mass ratios and differences.[8] The exact form of this function depends not only upon the Werle assumption (7) but also upon the assumptions made about the transformation property of the transition operators (weak and electromagnetic currents).

Although this approach has come to be known as "spectrum-generating SU(n)," we should point out that the name is not completely accurate. The transition operators V_μ^β (unlike the Q_i of E(3)) are not generators of the group, so SU(n) is not a spectrum-generating group in the same sense as E(3). What we are really using is a spectrum-generating algebra, which contains the algebra of an SU(n) that is not an invariance group of the momentum operator.

Before going on to discuss some applications of these ideas, we will quote van Dam and Biedenharn, who have independently studied the idea that some groups in particle physics should commute with the 4-velocity rather than 4-momentum operator. Referring to their "dynamical stability group of P_μ/M," before they knew of the spectrum-generating SU(3) results, they wrote,[9] "We suggest that the concept of a dynamical stability group is the proper concept to replace the unworkable concept of a global Lie group symmetry in relativistic quantum mechanics." We hope that the present results lend support to their suggestion.

III. Applications

Spectrum-generating SU(3) and SU(4) have been applied to five processes which involve no more than one hadron. Possible applications involving multi-hadron states will be discussed in the talk by Kielanowski.[10] Since this talk is a review, we will only briefly describe these five processes, and refer to the original literature for details.

1) $V \rightarrow e\bar{e}$. The leptonic decays of vector mesons (ρ,ω,ϕ,ψ) do not fulfill the ordinary quark model predictions which come from SU(4) symmetry (with mass differences taken into account in the phase space). Instead, they fulfill Yennie's empirical rule.[11] With a suitable assumption for the electromagnetic current operator within spectrum-generating SU(4), this rule can be derived.[7,12]

2) $V \rightarrow P\gamma$. The radiative decays of vector mesons $(\rho,\omega,\phi,\psi,K^{0*},D^*)$ into pseudoscalar mesons (π,η,K^0,χ,D) are difficult to explain with ordinary SU(3) and SU(4). In particular, the decays of the new particles $\psi \rightarrow \chi(2.8)\gamma$ and $D^{0*} \rightarrow D^0\gamma$ are strongly suppressed experimentally, although there is no principle such as the OZI rule to forbid them. Within spectrum-generating SU(4), this suppression is explained by the large hadron mass differences, since suppression factors arise naturally in this framework.[13]

3) Hyperon magnetic moments. Neither SU(3) symmetry nor quark model predictions[14] (including different "masses" of the quarks) can fit the present experimental magnetic moments. A suppression factor gives a slight improvement,[15] but the magnetic moments remain an unsolved problem.

4) Baryon semileptonic decays. This area will be discussed by Garcia in the next talk.[16]

5) Weak leptonic and semileptonic meson decays. We will discuss this in some detail below.

The original motivation for developing the spectrum-generating SU(3) approach was to explain the Cabibbo suppression (defined below) as a consequence of symmetry breaking.[17] That is, the Cabibbo angle was to be determined as a function of the hadron masses. Using the Werle assumption and a suitable form of the weak transition operator (hadronic current) it was in fact possible to obtain the Cabibbo suppression for some processes. However, it was not possible to find suitable assumptions which would give the correct suppression for all processes. Therefore, the Cabibbo angle had to be accepted as a phenomenological constant that cannot be expressed by a suppression factor. At present, we consider the spectrum-generating SU(n) approach to be rather a correction of the Cabibbo model, which takes the mass differences into account in a non-perturbative way.[18]

After much trial and error, we found that the following ansatz for the weak hadronic current gives the best results for all processes listed above:

$$J_\mu^{\text{Had.}} = \cos\theta(V_\mu^{-1} + A_\mu^{-1}) + \sin\theta(V_\mu^{-2} + A_\mu^{-2}) + \text{h.c.} \tag{14}$$

where

$$\begin{aligned} V_\mu^\beta &= \{M,\{M,\hat{V}_\mu^\beta\}\} \\ A_\mu^\beta &= \{M,\{M,\hat{A}_\mu^\beta\}\} \end{aligned} \tag{15}$$

and where $\hat{V}_\mu^\beta$ and $\hat{A}_\mu^\beta$ are SU(3)-octet operators which fulfill (6b). Cartan notation is used for the SU(3) indices,[7] and M is the mass operator. In the symmetry "limit," $[M,E_\alpha] = 0$, (14) is equivalent to the usual Cabibbo current, except for constant factors. One may consider (15) to be the phenomenologically-obtained formula which expresses the behavior of the currents away from the symmetry limit.

In an experimental test of the spectrum-generating SU(n) approach, it is important to separate effects which test the various assumptions individually. The ratios of decay rates (i.e., the suppression factors) depend upon both (15) and (7). However, there are other predictions which are independent of (15) and therefore provide a test of (7). These are:

a) Ratios of different formfactors for a single process. Using (6c) and (14), we find for the current in $\alpha \to \pi^0 \ell\nu$ decay ($\alpha = \pi^+$ or K^+)

$$\langle\hat{p}'\pi^0|J_\mu^{\text{Had.}}|\alpha\hat{p}\rangle = C_F(\pi^0,\{{}^{-1}_{-2}\},\alpha)\sin\theta(m_\alpha+m_\pi)^2F_+(\hat{t})(\hat{p}+\hat{p}')_\mu \tag{16}$$

where the velocity transfer $\hat{t}$ is related to the usual momentum transfer t^α by

$$\hat{t} \equiv (\hat{p}-\hat{p}')^2 = (m_\alpha m_\pi)^{-1}[t^\alpha - (m_\alpha-m_\pi)^2] \tag{17}$$

The factor $(m_\alpha+m_\pi)^2$ originates in the assumption (15). The formfactor F_- has been set equal to zero because it is "second class."[7] Expression (16) may be compared to the conventional form (13), using (17), to yield the formfactor ratio:

$$\xi(t) \equiv \frac{f_-^{\pi K}(t)}{f_+^{\pi K}(t)} = \frac{m_\pi-m_K}{m_\pi+m_K} = -0.57 \tag{18}$$

This prediction is independent of the factor in front of F_+ in (16), so it is a test of the Werle relation (7).

Prediction (18) is probably the single most important test of the spectrum-generating SU(n) approach, because it depends upon the fundamental assumption (7) rather than the detailed form of the current. If this prediction is not experimentally confirmed, then the whole approach would have to be rejected. For this reason, the experimental situation should be explained carefully. Since this has been done by Nieto,[19] however, we just give the results here: The data is statistically dominated by two experiments, one of which gives a value of $\xi(t)$ near zero, and the other of which gives a value compatible with (18). In short, the experimental situation is presently unclear, but (18) is certainly an experimentally favored value.[7]

b) Ratios of the Cabibbo suppression for leptonic and semileptonic processes. Phenomenologically, in the conventional Cabibbo model using (13), one needs two

Cabibbo angles:

$$J_\mu^{Cab.} = \cos\theta_V V_\mu^{-1} + \cos\theta_A A_\mu^{-1} + \sin\theta_V V_\mu^{-2} + \sin\theta_A A_\mu^{-2} + \cdots \tag{19}$$

Experimentally,[20]

$$\Gamma(K\to\mu\nu)/\Gamma(\pi\to\mu\nu) \text{ gives } \tan\theta_A^M = 0.276 \tag{20}$$

and

$$\Gamma(K\to\pi\ell\nu)/\Gamma(\pi\to\pi\ell\nu) \text{ gives } \tan\theta_V^M = 0.224 \tag{21}$$

with very small errors. The value of θ^B for baryon decays lies somewhere between θ_V^M and θ_A^M.

In the spectrum-generating SU(3) approach, the value determined for the $\tan\theta$ appearing in (14) is the same as $\tan\theta_A$ in (20), and is determined from the same data. However, the value determined from the semileptonic data is different from $\tan\theta_V$. The reason is the following[21]: The usual definition of the experimentally measured $\tan\theta_V$ is

$$\tan\theta_V = \frac{f_+^K(t^K=0)}{f_+^\pi(t^\pi=0)} = 0.224 \tag{22}$$

where both formfactors are evaluated at zero momentum transfer. In the spectrum-generating group approach, however, both formfactors should be evaluated at the same value of the SU(3)-invariant velocity transfer $\hat{t}$. Choosing the value $\hat{t} = 0$, which is always in the physical region, we have by (17) the following correspondence

$$\hat{t} = 0 \longleftrightarrow \begin{array}{l} t^\pi \simeq 0 \simeq t^\pi_{max} \\ t^K = (m_K-m_\pi)^2 = t^K_{max} \end{array} \tag{23}$$

so the suppression is

$$\tan\theta = \frac{f_+^K(t^K=t^K_{max})}{f_+^\pi(t^\pi=0)} = 0.276 \tag{24}$$

in agreement with the value determined from the leptonic decay. This result depends also on the formfactor parameter λ_+ in

$$f_+(t) = f_+(0)(1 + \lambda_+ \frac{t}{m_\pi^2}) \tag{25}$$

so this is not as clear a test of our approach as (18). Nevertheless, it is clear that this approach allows one to eliminate the use of different Cabibbo suppression factors for these different processes.

Although the equality of the vector and axialvector suppression factors is essentially a consequence of (7), the precise value of the Cabibbo angle depends upon the exact form of the operators appearing in (14) (e.g., on assumption (15)). That is, the Cabibbo suppression is explained not only by the presence of θ in (14) but also by the transformation property of the currents, as expressed for example by (15). The form (15) seems to be consistent with the data, but, as discussed by Garcia, the agreement for baryons is not impressive. Thus (15) may well have to be changed, but this will not change the value of tan very much because of the small baryon mass differences. One may therefore consider (24) to be a prediction of the model which is not very subject to change, and this prediction is very accurate.

Acknowledgements

We would like to thank our collaborators and colleagues for innumerable discussions about this subject.

This research was supported in part by NSF grant GF42060 and DOE grant E(40-1)3992.

References

1. E. P. Wigner, Nachr. Ges. Wiss. Gött., Math-Phys. Klasse (1932), p. 546. Reprinted in "Group Theory and Solid State Physics," Vol. I, P. H. Meijer, Editor (Gordon and Breach, New York, 1964), pp. 265-278. V. Bargmann, Annals of Math. 59, 1 (1954). G. Ludwig, Grundlagen der Quantum Mechanik (Springer, 1954), p. 101.
2. A. O. Barut and A. Bohm, Phys. Rev. 139B, 1107 (1965).
3. Y. Dothan, M. Gell-Mann and Y. Ne'eman, Phys. Rev. Lett. 17, 145 (1965).
4. N. Mukunda, L. O'Raifeartaigh, and E.C.G. Sudarshan, Phys. Rev. Lett. 15, 1041 (1965).
5. A. Bohm, "Quantum Mechanics," Chap. II (Springer, New York, 1979).
6. This relation has been suggested in 1965 by J. Werle, "On a symmetry scheme described by a non-Lie algebra," ICPT preprint, Trieste, 1965, unpublished. It has been incorporated in the spectrum-generating group approach for the mass and spin spectrum, A. Bohm, Phys. Rev. 175, 1767 (1968). Its application in connection with SU(3) has been developed through the following series of papers: A. Bohm, Phys. Rev. 158, 1408 (1967), Phys. Rev. D7, 2701 (1973); A. Bohm, E.G.C. Sudarshan, Phys. Rev. 178, 2264 (1969). The present scheme has been formulated in reference 7; A. Bohm, J. Werle, Nucl. Phys. B106, 165 (1976). As "dynamical stability group of P_μ/M" the same idea has also been arrived at by van Dam and Biedenharn (ref. 9).
7. A. Bohm, Phys. Rev. D13, 2110 (1976).
8. The existence of a theoretical permissibility of a correction factor depending upon the masses, at least for the electromagnetic interaction, has been noted several times before: M. Gourdin, in Symmetries and Quark Models, edited by R. Chand (Gordon and Breach, New York, 1970); R. P. Feynman, Photon Hadron Interaction (Benjamin, New York, 1972); the Duffin-Kemmer-Petian formalism: B. G. Kenney, D. C. Peaslee and M. M. Nieto, Phys. Rev. D13, 757 (1976) (see also the talk by Nieto in these Proceedings) for electromagnetic interaction in E. Fischbach, M. M. Nieto, M. Primakoff, C. K. Scott and Smith, Phys. Rev. Lett. 27, 1403 (1971) for $K_{\ell 3}$ decay; another way of describing symmetry-breaking effects by mass dependent correction factors in the effective coupling constant has been suggested by J. Schwinger in Proceedings of the 7th Hawaii Topical Conference on Particle Physics 1977, see also L. F. Urrutia, UCLA preprint (1977).
9. H. van Dam and L. C. Biedenharn, Phys. Rev. D14, 405 (1976).
10. P. Kielanowski (these Proceedings).
11. D. R. Yennie, Phys. Rev. Lett. 34, 239 (1975).
12. A. Bohm and R. B. Teese, Univ. of Texas preprint ORO 317 (to appear in Phys. Rev. D).
13. A. Bohm and R. B. Teese, Phys. Rev. Lett. 38, 629 (1977) and references therein. See also ref. (12).
14. R. Settles, private communication.
15. A. Bohm, Univ. of Texas preprint ORO 316 (to appear in Phys. Rev. D).
16. A. Garcia (these Proceedings).

17. The original suggestion comes, of course, from the nearness of m /m_K to the value or tan .
18. This does not mean that the Cabibbo angle cannot be derived from other hypotheses. For example, there have recently been suggestions that tan could be derived in terms of quark mass parameters. H. Fritzsch Phys. Lett. 70B, 436 (1977), A. Ebrahim Phys. Lett. 72B, 457, (1978), H. Terazawa Univ. of Tokyo preprint.
19. M. M. Nieto (these Proceedings).
20. M. Ross, Phys. Letters 36B, 130 (1971).
21. A. Bohm, M. Igarashi, and J. Werle, Phys. Rev. D15, 2461 (1977).

j = 3 • • • • • • •

j = 2 • • • • •

j = 1 • • •

j = 0 •

Fig. 1. The weight diagram of the rotator. The ordinate is j and the abscissa is j_3.

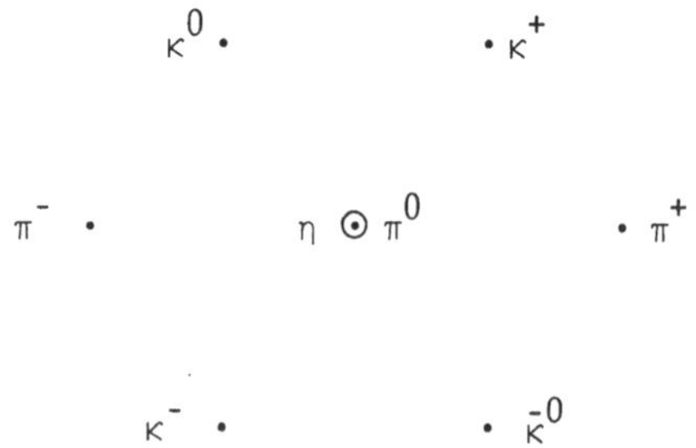

Fig. 2. The weight diagram of the pseudoscalar meson octet. The ordinate is Y and the abscissa is I_3.

Baryon Semileptonic Decays-Spectrum Generating SU(3) and the Cabibbo Model.

A. García*

Departamento de Física. Centro de Investigación y Estudios Avanzados, I.P.N. Apartado Postal 14-740, México 14, D.F., México.

Baryon semileptonic decays, in which a hyperon α decays into another one α' and a charged lepton-neutrino pair,

$$\alpha \rightarrow \alpha' + \ell + \nu$$

provide an excellent testing ground for strong interaction symmetries like SU(2) and SU(3). Our present approach to these decays consists of V-A theory [1], which provides a rather general framework, and of the requirement of basic symmetry properties of the hadronic part of the transition amplitude. We shall briefly review all of this, introducing the Cabibbo Model[1] and the Spectrum Generating SU(3) formulation[2]. We shall also discuss some important questions concerning the experimental information and point out what we think is the best way to use it.

V-A Theory.

The transition amplitude matrix element is given by

$$M = \frac{G}{\sqrt{2}} \langle p'\sigma'\alpha' | j^{\beta}_{\mu}(0) | p\sigma\alpha \rangle \, \bar{u}_{\ell}(p_{\ell}) \gamma^{\mu} (1+\gamma_5) v_{\nu}(p_{\nu})$$

where $j^{\beta}_{\mu} = V^{\beta}_{\mu} + A^{\beta}_{\mu}$; p , p' , ℓ , and p_{ν} are the four momenta of the particles involved, σ and σ' are spin labels of α and α', and β is an index. The hadronic part of M can be expressed, by use of Lorentz covariance, in terms of six form factors,

$$\langle p'\sigma'\alpha' | j^{\beta}_{\mu} | p\sigma\alpha \rangle = \bar{u}_{\alpha'}(p',\sigma') \Big\{ f_1^{\alpha'\beta\alpha}(q^2)\gamma_{\mu} + f_2^{\alpha'\beta\alpha}(q^2)\, i\sigma_{\mu\nu} q^{\nu} + f_3^{\alpha'\beta\alpha}(q^2)\, q_{\mu} + \big[g_1^{\alpha'\beta\alpha}(q^2)\gamma_{\mu} + g_2^{\alpha'\beta\alpha}(q^2)\, i\sigma_{\mu\nu} q^{\nu} + g_3^{\alpha'\beta\alpha}(q^2)\, q_{\mu} \big] \gamma_5 \Big\} u_{\alpha}(p,\sigma),$$

where $q_\mu = p'_\mu - p_\mu$ is the four-momentum transfer.

The above form factors can be classified under G-parity, the product of charge conjugation and a rotation of π degrees around the y-axis in isospace. A vector(axial -vector) current is said to be first class under G-parity if

$$G V_\mu G^{-1} = V_\mu \qquad (G A_\mu G^{-1} = - A_\mu) .$$

It turns out then that f_1, f_2, g_1, and g_3 are first class and f_3 and g_2 are second class.

The range of variation of q is small. It is essentially given by m_e and $\Delta m = m_\alpha - m_{\alpha'}$. This allows us to expand the form factors in a Taylors series and to keep the first two terms; for example,

$$f_1^{\alpha'\beta\alpha}(q^2) \simeq f_1^{\alpha'\beta\alpha}(0) \left(1 + \lambda_f^{\alpha'\beta\alpha} q^2 / m_\alpha \right)$$

The q^2-dependence is then characterized by some slope parameter λ_f. A similar expansion can be done for the other form factors. In practice it will turn out that only the q^2-dependence of f_1 and g_1 shall be noticeable. All other form factors can be taken to be constant.

In the case the emitted lepton is an electron then f_3 and g_3 will give very small contributions and can be safely ignored.

The Cabibbo Model.

This model consists of the following assumptions:(1) assume that currents are first class only, (2) assume that, except for mass differences, the SU(3) symmetry limit is good, (3) assume that V_μ^β and A_μ^β belong to SU(3) octets, (4) extend the conserved vector current hypothesis (CVC) to V_μ^β and (5) reformulate universality,

$$j_\mu(0) = \cos\theta_c \left(V_\mu + A_\mu\right)^{\Delta S=0} + \sin\theta_c \left(V_\mu + A_\mu\right)^{|\Delta S|=1}$$

The form factors are given, by use of the Wigner-Eckart Theorem, in terms of reduced form factors as

$$f_i^{\alpha'\beta\alpha} = \sum_{\gamma=1}^{2} C(\gamma; \alpha'\beta\alpha)\, f_i^\gamma(q^2) \quad ,$$

$$g_i^{\alpha'\beta\alpha} = \sum_{\gamma=1}^{2} C(\gamma; \alpha'\beta\alpha)\, g_i^\gamma(q^2) \quad , \quad i = 1, 2, 3 \quad ;$$

where $C(\gamma, \alpha'\beta\alpha)$ are Clebsch -Gordan coefficients ($\gamma = 1$ is antisymmetric and $\gamma = 2$ is symmetric). Under the above assumptions, we get $f_3(q^2) = 0$ and $g_2(q^2) = 0$ and

$$f_1^1 = \sqrt{6} \quad , \quad f_1^2(0) = 0 \quad , \quad f_2^1(0) = \sqrt{6}\left(\frac{\mu_p}{2} + \frac{\mu_n}{4}\right) \quad \text{and}$$

$f_2^2(0) = \mu_n \sqrt{5}\sqrt{6}/4$; where μ_p and μ_n are the anomalous magnetic moments of the proton and the neutron.

If the charged lepton is a muon, then $g_3(q^2)$ will contribute. it can be related to $g_1(q^2)$ by PCAC (partially conserved axial-vector current). We end up with three free parameters, θ_c, $g_1'(0)$ and $g_1^2(0)$. The q^2- dependence can be incorporated and fixed in terms of CVC and ν-scattering data.

Spectrum Generating SU(3)

Since SU(3) is not an exact symmetry because of the mass differences of the particles belonging to the same multiplet, we have that

$$[P_\mu , E_\alpha] \neq 0$$

where P_μ the four-momentum operator and E_α are the SU(3) generators. Instead, we can require that

$$[\hat{P}_\mu , E_\alpha] = 0$$

where $\hat{P}_\mu$ is the four-velocity operator. Since this operator does not depend on masses it can well commute with the SU(3) generators. This approach can be used to define new form factors that can be related to SU(3) with well defined transformation properties. The hadronic part of the transition amplitude can be expressed as

$$\hat{M} = \frac{g}{\sqrt{2}} \langle \hat{p}'\sigma'\alpha' | v_\mu^\beta + a_\mu^\beta | \hat{p}\sigma\alpha \rangle =$$

$$= \frac{g}{\sqrt{2}} \bar{u}_{\alpha'} \Big\{ \varphi_V^{\alpha'\alpha} \Big[F_1^{\alpha'\beta\alpha}(\hat{q}^2)\, \gamma_\mu + F_2^{\alpha'\beta\alpha}(\hat{q}^2)\, i\sigma_{\mu\nu}\hat{q}^\nu + F_3^{\alpha'\beta\alpha}\, \hat{q}_\mu \Big] +$$

$$+ \varphi_A^{\alpha'\alpha} \Big[G_1^{\alpha'\beta\alpha}(\hat{q}^2)\, \gamma_\mu + G_2^{\alpha'\beta\alpha}(\hat{q}^2)\, i\sigma_{\mu\nu}\hat{q}^\nu + G_3^{\alpha'\beta\alpha}(\hat{q}^2) \Big] \gamma_5 \Big\} u_\alpha$$

with $\hat{q}_\mu = \hat{p}'_\mu - \hat{p}_\mu$ being the four-velocity transfer. The factors $\varphi_V^{\alpha'\alpha}$ and $\varphi_A^{\alpha'\alpha}$ are overall constants to be determined by the transformation propertiesof v_μ and a_μ under SU(3) and by normalization and universality convention. For example, we can assume that v_μ and a_μ belong to octets. The connection between these form factors and the V-A ones is

$$f_1^{\alpha'\beta\alpha} = \frac{\varphi_V^{\alpha'\alpha}}{(m_\alpha m_{\alpha'})^{3/2}} \sum_{\gamma=1}^{2} C(\gamma,\alpha'\beta\alpha)\left[F_1^\gamma + \left(2 - \frac{(m_\alpha+m_{\alpha'})^2}{2m_\alpha m_{\alpha'}}\right)F_2^\gamma + \frac{m_\alpha^2 - m_{\alpha'}^2}{2m_\alpha m_{\alpha'}} F_3^\gamma\right],$$

$$f_2^{\alpha'\beta\alpha} = \frac{\varphi_V^{\alpha'\alpha}}{2(m_\alpha m_{\alpha'})^{5/2}} \sum_{\gamma=1}^{2} C(\gamma,\alpha'\beta\alpha)\left[(m_\alpha+m_{\alpha'})F_2^\gamma - (m_\alpha - m_{\alpha'})F_3^\gamma\right],$$

$$g_1^{\alpha'\beta\alpha} = \frac{\varphi_A^{\alpha'\alpha}}{(m_\alpha m_{\alpha'})^{3/2}} \sum_{\gamma=1}^{2} C(\gamma,\alpha'\beta\alpha)\left[G_1^\gamma + \frac{m_\alpha^2 - m_{\alpha'}^2}{2m_\alpha m_{\alpha'}} G_2^\gamma - \frac{(m_\alpha - m_{\alpha'})^2}{2m_\alpha m_{\alpha'}} G_3^\gamma\right],$$

$$g_2^{\alpha'\beta\alpha} = \frac{\varphi_A^{\alpha'\alpha}}{2(m_\alpha m_{\alpha'})^{5/2}} \sum_{\gamma=1}^{2} C(\gamma,\alpha'\beta\alpha)\left[(m_\alpha+m_{\alpha'})G_2^\gamma - (m_\alpha - m_{\alpha'})G_3^\gamma\right]$$

and similar expressions for f_3 and g_3 which we do not quote.

The present approach has some advantages. First, the result of the Cabibbo Model can be easily reproduced. Second, the symmetry breaking contributions due to mass differences are incorporated in a rigorous way. And, third, even in the absence of second class axial-vector currents, induced pseudo-tensor form factors g_2 are present. Let us notice that the corrections to f_1 are second order in symmetry breaking in accordance with the Ademollo-Gatto Theorem[1].

Another feature is that the q^2-dependence of f_1 and g_1 is more important now. For f_1 we get

$$f_1^{\alpha'\beta\alpha}(0) = F_1^{\alpha'\beta\alpha}(0)\left(1 - \frac{(m_\alpha - m_{\alpha'})^2}{m_\alpha m_{\alpha'}} b_f^{\alpha'\beta\alpha}\right) \qquad \text{and}$$

$$\lambda_f^{\alpha'\beta\alpha} = \frac{m_\alpha^2}{m_\alpha m_{\alpha'}} \frac{b_f}{1 - \frac{(m_\alpha - m_{\alpha'})^2}{m_\alpha m_{\alpha'}} b_f},$$

where $b_f^{\alpha'\beta\alpha}$ is the slope parameter of $F_1^{\alpha'\beta\alpha}(\hat{q}^2)$. An analogous expression is obtained for $g_1(q^2)$.

Experimental Information.

We want to make a few important remarks in order to be able to make a good comparison with experimental data. Very often experimental values for the ratios g_1/f_1 are quoted. This information is already biassed and it is dangerous to use it. Such values can only be obtained at present if theoretical assumptions are made and, thus, one runs into circular arguments.

A better way is to use quantities as directly related to experimental measurements as possible: rates and angular correlation coefficients. As an example, let us display the expression for the electron-neutrino angular correlation coefficient, assuming only V-A theory and recoil contributions up to second order,

$$a_{e\nu} = \frac{G^2 \Delta m^5}{60\pi^3} \Big\{ (1 - \tfrac{5}{2}\beta + \tfrac{11}{7}\beta^2)|f_1|^2 - \tfrac{2}{7}\beta^2|f_2|^2 + $$

$$+ (-1 - \tfrac{3}{2}\beta + \tfrac{25}{7}\beta^2)|g_1|^2 - 2\beta^2|g_2|^2 - \tfrac{3}{7}\beta^2 \, \mathrm{Re}\, f_1 f_2^* +$$

$$+ (4\beta - 2\beta^2)\, \mathrm{Re}\, g_1 g_2^* \Big\} / R$$

Here $\beta = (m_d - m_{d'})/m_d$ and R is the transition rate whose expression is also a quadratic function of the form factors. Now, in order to quote an experimental value for g_1/f_1, one must fix the values of f_2/f_1 and g_2/f_1. Usually, the assumptions of the Cabibbo Model are used to fix these last two ratios. And, therefore, the experimental value for g_1/f_1 thus obtained can not be used to test the Cabibbo Model. This would be a circular argument as mentioned before. Of course, a complete analysis should start by testing V-A theory to begin with but there are not enough data to do so and we must remain biassed in favor of V-A. Hopefully in a not too far future each of the hypothesis we would like to believe can be tested.

Another question is the radiative corrections. We shall only mention that if some approximations are accepted, then one can introduce effective form factors for f_1 and g_1.

$$f_1' = f_1 \left(1 + \frac{\alpha}{4\pi} c'\right) \quad \text{and} \quad g_1' = g_1 \left(1 + \frac{\alpha}{4\pi} d'\right)$$

in such a way that the V-A formulas for the rates and angular coefficients remain unchanged. Here α is the fine structure constant and c' and d' are two constants. These constants can be estimated using some model, the Weinberg-Salam model[3] for example. Roughly speaking, it turns out that the angular coefficients are not very much affected by radiative corrections and that the rates just get an overall correction constant, since $c' \approx d'$.

Conclusions.

Because of lack of space we can not give a detailed comparison with experimental data. It can be found in reference (4). We shall only sketch it. There are nineteen pieces of data available, nine rates and ten angular coefficients. We have used the latest data.

Assuming that the q^2- dependence of f_1 and g_1 is given by CVC and ν- scattering data and the absence of second class currents, we obtain for the Cabbibo Model a $\chi^2 = 39.7$, and for the Spectrum Generating SU(3) a $\chi^2 = 28.4$. If no q^2-dependence is included, then we get $\chi^2 = 36.6$ and $\chi^2 = 22.9$, respectively.

We shall now list our main conclusions.

1) The experimental evidence indicates that the Cabibbo Model should be improved.
2) The presence of the pseudo-tensor form factors g_2 is important and must be expected due to symmetry breaking contributions.
3) The Spectrum Generating SU(3) provides an interesting alternative to "perturbative" symmetry breaking corrections.
4) With present data, we obtain some improvement with Spectrum Generating SU(3) over the Cabibbo Model.

Let us finish with a word of caution. The improvement obtained is not impressive. If the experimental data were to be confirmed where they are now, it could well be that one ends with a major contradiction to the application of SU(3) to baryon leptonic decays. Both experimental information and the theoretical models should be further refined before a final conclusion can be reached.

Footnotes and References

*Supported in part by CONACyT (project No.936-A)

1) For a review see R.E. Marshak, Riazuddin and C.P. Ryan, The Theory of Weak Interactions in Particle Physics,(Wiley-Interscience, 1968).

2) A. Bohm and R.B. Teese, Jour.Math.Phys. 18,1434 (1977)

3) For a review see A. Sirlin, Rev. of Mod. Phys. 50, 573 (1978). A. García Phys. Lett. 73B, 299 (1978).

4) A. Bohm, R.B. Teese, A.García and J.S. Nilsson, Phys. Rev.D15, 689, (1977).

SPECTRUM-GENERATING SU(3) AND LOCAL CURRENT ALGEBRA*

Piotr Kielanowski, Institute of Theoretical Physics, Warsaw University, ul. Hoza 69, 00-681 Warszawa, Poland.

I. Introduction

The approach which considers the SU(3) group as the spectrum-generating group has been succesfully applied to the description of various phenomena for weak and electromagnetic interactions [1-8]. In contrast to the standard group theoretical methods this approach is free of many inconsistencies. On the other hand there exists a large number of the predictions of the conventional approach which have never been obtained or modified with the help of the SU(3) spectrum-generating group. These are the predictions of the algebra of currents [9]. The algebra of currents played a very important role in the field of group theoretical methods in particle physics. It first allowed one to give a precise meaning to the concept of the universality of weak interactions by fixing the scale of the currents due to non-linear structure of the commutators. Next Gell-Mann suggested that the SU(3) generators should be identified with the charges associated with the weak currents. He also assumed that the commutation relations of charges do not change even in the case of the symmetry breaking. The commutation relations can also be transformed into sum rules in analogy with atomic and nuclear physics. The current algebra formulated here shares some of these properties. In particular it fixes the scale of the currents and gives sum rules which are similar but not necessarily identical to the conventional results. In contrast to the current algebra of Gell-Mann we do not assume that the physical currents responsible for weak interactions have the same commutation relations as the SU(3) symmetric currents. We assume explicit symmetry breaking at the level of the currents. The treatment of the strangness-conserving and strangness-changing interactions is the same for both cases in our approach. We assume that the supression of the strangness-changing processes is the result of symmetry breaking and we do not have to introduce the Cabibbo angle.

*Supported in part by NSF grant # GF 42060.

II. Basic definitions and commutation relations

We shall consider here the SU(3) current algebra. We therefore introduce 8 vector and 8 axial vector currents $\hat{V}^a_\mu(x)$ and $\hat{A}^a_\mu(x)$. According to the standard formula they are:

$$\hat{V}^a_\mu(x) = e^{iPx}\,\hat{V}^a_\mu(0)e^{-iPx} , \tag{1}$$

$$\hat{A}^a_\mu(x) = e^{iPx}\,\hat{A}^a_\mu(0)e^{-iPx} , \tag{2}$$

where P is the four momentum operator in the hadron space. We assume that the operators $\hat{V}^a_\mu(0)$ and $\hat{A}^a_\mu(0)$ form octets with respect to the SU(3) spectrum generating group, i.e.,

$$\begin{aligned} [E^a,\hat{V}^b_\mu(0)] &= if^{ab}_c\hat{V}^c_\mu(0) , \\ [E^a,\hat{A}^b_\mu(0)] &= if^{ab}_c\hat{A}^c_\mu(0) . \end{aligned} \tag{3}$$

The operators E^a are the generators of the SU(3) spectrum generating group and f^{ab}_c are the structure constants of the SU(3) group. The operators E^a commute with the velocity operator $[E^a,\hat{P}^\mu] = 0$ [14]. We next assume that the time components of the currents $\hat{V}^a_o(x)$ and $\hat{A}^a_o(x)$ form the local Lie algebra with respect to equal time commutation:

$$\begin{aligned} [\hat{V}^a_o(x),\hat{V}^b_o(y)]_{x^o=y^o} &= if^{ab}_c\hat{V}^c_o(x)\delta_3(\vec{x}-\vec{y}) , \\ [\hat{V}^a_o(x),\hat{A}^b_o(y)]_{x^o=y^o} &= if^{ab}_c\hat{A}^c_o(x)\delta_3(\vec{x}-\vec{y}) , \\ [\hat{A}^a_o(x),\hat{A}^b_o(y)]_{x^o=y^o} &= if^{ab}_c\hat{V}^c_o(x)\delta_3(\vec{x}-\vec{y}) . \end{aligned} \tag{4}$$

One can derive the commutation relations (4) from the locality of the commutators of the currents and the commutation relations (3) after the application of the Jacobi identity.

One can also introduce the charges associated with the currents $\hat{V}^a_\mu(x)$:

$$\hat{E}^a(x^o) = \int d_3x\hat{V}^a_o(x) . \tag{5}$$

The charges $\hat{E}^a(x^o)$ have the following commutation relations with the currents

$$[\hat{E}^a(x^o),\hat{V}^b_o(y)]_{x^o=y^o} = if^{ab}_c\hat{V}^c_o(y) \quad ,$$
$$[\hat{E}^a(x^o),\hat{A}^b_o(y)]_{x^o=y^o} = if^{ab}_c\hat{A}^c_o(y) . \tag{6}$$

The commutation relations (6) are very similar to the previously introduced commutation relations (3), so one might be tempted to identify the charges $\hat{E}^a(x^o)$ with the generators of the spectrum-generating group E^a. This is, however, not true since, e.g., $\hat{E}^a(x^o)$ are time dependent and E^a are not. However, for further applications we assume that the diagonal, one-hadron matrix elements of the charges $\hat{E}^a(x^o)$ and E^a for a=3,8 are equal, so

$$\langle p|E^a|p'\rangle = \langle p|\hat{E}^a(x^o)|p'\rangle \quad . \tag{7}$$

We assume that the electromagnetic current is given by

$$J^{el}_\mu(x) = \hat{V}^3_\mu(x) + \frac{1}{\sqrt{3}}\hat{V}^8_\mu(x) \tag{8}$$

and the electromagnetic interactions are described in the conventional way with the help of the transition operator

$$H_{em} = e\int d_3x\; J^{el}_\mu(x)A^\mu(x) \quad . \tag{9}$$

The weak interactions cannot be described by the currents $\hat{V}^a_\mu(x)$ and $\hat{A}^a_\mu(x)$ due to the large symmetry breaking. We assume that the symmetry is broken by the mass operator only, so the physical currents responsible for interactions are functions of the SU(3) currents and the mass operator. We shall also assume that in the limit of exact symmetry the physical currents become the SU(3) currents. A very simple ansatz for physical currents satisfying these two requirements has the form ($V^a(x)$ and $A^a(x)$ are the physical currents):

$$V^a_\mu(x) = \hat{V}^a_\mu(x) + \alpha_1[M,\hat{V}^a_\mu(x)] + \alpha_2[M,[M,\hat{V}^a_\mu(x)]] + \text{higher commutators}$$
$$A^a_\mu(x) = \hat{A}^a_\mu(x) + \alpha_1[M,\hat{A}^a_\mu(x)] + \alpha_2[M,[M,\hat{A}^a_\mu(x)]] + \text{higher commutators} \tag{10}$$

The universal constants α_i should be determined from the weak interactions of hadrons. It can be shown that from time reversal invariance, all the coefficients α_i with i odd should vanish. The weak hadronic current is defined

$$J_\mu^{had} = V_\mu^{1+i2} + A_\mu^{k+i2} + V_\mu^{4+i5} + A_\mu^{4+i5} \tag{11}$$

and the transition operator is equal

$$H_{weak} = \frac{G}{\sqrt{2}} \int d_3x \, J_\mu(x) \mathcal{J}^{\mu+}(x) \quad , \tag{12}$$

where $J_\mu(x) = J_\mu^{lep}(x) + J_\mu^{had}(x)$. The leptonic part is constructed in the usual way.

In the rest of the paper we shall assume that the only non-vanishing coefficients α_i are α_2 and α_4. With that assumption one can satisfactorily describe the semileptonic decays of baryons and mesons.

III. Sum rules

We shall consider here two sum rules. Let us consider the commutator

$$[\hat{V}_o^{1+i2}(z/2), \hat{V}_o^{1-i2}(-z/2)]_{z^o=0} = 2\hat{V}_o^3(0)\delta_3(\vec{z}) . \tag{13}$$

If we sandwich this commutator between proton velocity states $\hat{p}$ then after standard manipulations one gets the sum rule

$$\frac{1}{\pi} \int_{-\infty}^{\infty} dq_o t^{oo} = -2\hat{p}^o m^3 \quad . \tag{14}$$

Here m is the mass of proton and $t^{\mu\nu}$ is the absorptive part of the amplitude for the scattering of "charged" photons with the momentum q. Now if we go to the limit of infinite proton velocity we get the Cabibbo-Radicati sum rule [10]

$$-F_1^V(o) + (F_2^V(0))^2 = \frac{1}{2\pi^2\alpha} \int \frac{d\hat{\nu}}{\hat{\nu}} (\sigma_{3/2}^V - \sigma_{1/2}^V) . \tag{15}$$

Here $\hat{\nu} = \hat{p}q$, F_1^V and F_2^V are conventional isotopic vector form factors and $\sigma_{3/2}$ and $\sigma_{1/2}$ are the cross sections for the scattering of the isotopic vector photons into the states with final isospin 3/2 and 1/2, respectively. This sum rule has been tested by Gilman and Schnitzer [11] and they found that it is fulfilled reasonably well. However, it should be stressed that the result is noteworthy since our currents are different from the conventional currents.

Let us consider the current

$$\hat{J}_\mu(x) = \hat{V}_\mu^{1+i2}(x) + \hat{A}_\mu^{1+i2}(x) + \hat{V}_\mu^{4+i5}(x) + \hat{A}_\mu^{4+i5}(x) \tag{16}$$

and the commutator

$$[\hat{J}_o(x), \hat{J}_o^+(y)]_{x^o=y^o} = \{4(\hat{V}_o^3(x) + \hat{A}_o^3(x)) + 2\sqrt{3}(\hat{V}_o^8(x) + \hat{A}_o^8(x)) - 4(\hat{V}_o^6(x) + \hat{A}_o^6(x))\}\times\delta_3(\vec{x}-\vec{y}) \tag{17}$$

Taking a Fourier transform of eq. (17) sandwiched between the proton or neutron states one gets:

$$(2\pi)^3\sum_N\{|\langle p|\hat{J}_o(0)|N\rangle|^2\delta_3(\vec{p}+\vec{q}-\vec{p}_N) - |\langle p|\hat{J}_o^+(0)|N\rangle|^2\delta_3(\vec{p}-\vec{q}-\vec{p}_N)\} = \left\{\begin{array}{ll} 5 & \text{for proton} \\ 1 & \text{for neutron} \end{array}\right. \tag{18}$$

After standard manipulations and taking into account the relation (10) between physical and SU(3) symmetric currents one gets:

$$\lim_{E_\nu\to\infty} \int_{thr}^{\infty} 1/\{1+\alpha_2(\sqrt{m^2+q^2+2m\xi}-m)^2+\alpha_4(\sqrt{m^2+q^2+2m\xi}-m)^4\}^2\times \left(\frac{d\sigma^\nu}{dq^2 d\xi} - \frac{d\sigma^{\bar{\nu}}}{dq^2 d\xi}\right)d\xi = \left\{\begin{array}{ll} \frac{5G^2}{2\pi} & \text{for } \nu p \text{ scatt.} \\ \frac{G^2}{2\pi} & \text{for } \nu n \text{ scatt.} \end{array}\right. \tag{19}$$

The notation is the following $q^2=(p_\nu-p_1)^2$, $\xi=(q\hat{p})$, p_ν is the initial neutrino momentum, p_1 is the final lepton momentum and $\hat{p}$ is the initial nucleon velocity. The sum rule (19) is different from Adler's original neutrino sum rule [12]. The difference is twofold. First, in the integration in the sum rule (19) there is an additional kinematical factor and second, we do not have any dependence on a Cabibbo angle. The sum rule (19) can be simplified if we introduce the variable w with the help of the equation

$$\frac{dw}{dq^2} = (1+\alpha_2(\sqrt{m^2+q^2+2m\xi}-m)^2+\alpha_4(\sqrt{m^2+q^2+2m\xi}-m)^4)^2 \tag{20}$$

After the introduction of the variable w the sum rule (19) reads

$$\lim_{E_\nu\to\infty}\left(\frac{d\sigma^\nu}{dw}-\frac{d\sigma^{\bar\nu}}{dw}\right) = \begin{cases} \frac{5G^2}{2\pi} & \text{for } \nu p \text{ scatt.} \\ \frac{G^2}{2\pi} & \text{for } \nu n \text{ scatt.} \end{cases} \tag{21}$$

The test of the sum rules (19) and (21) is at present impossible because of the large errors for neutrino-hadron cross sections [13].

IV. Conclusions

We have formulated the algebra of currents in the framework of the spectrum-generating group. Guided by the conventional current algebra methods we have obtained two sum rules for definite processes. Adler's neutrino sum rule has a different form from its original version. It can thus serve as a test of our approach. For further applications of our SU(3) current algebra it is necessary to formulate the analogue of PCAC, which is essential in the derivation of a large number of the current algebra results.

Acknowledgment: I want to thank Professor J. Werle for the suggestions, discussions and reading the manuscript. The hospitality of Professors A. Bohm and E.C.G. Sudarshan at the Center for Particle Theory, University of Texas, Austin is also acknowledged.

References:

[1] A. Bohm, Phys. Rev. D13, 2110(1976).
[2] A. Bohm and J. Werle, Nucl.Phys. B106,165(1976).
[3] A. Bohm et al., Phys. Rev. D15,348(1977).
[4] A. Bohm et al., Phys. Rev. D15,689(1977).
[5] A. Bohm, M. Igarashi and J. Werle, Phys. Rev. D15,2461(1977).
[6] A. Bohm and R.B. Teese, Phys. Rev. Lett. 38,629(1977).
[7] M. Igarashi, Institute of Theoretical Physics, Warsaw University Preprint IFT/77/9.
[8] A. Bohm and R.B. Teese, Phys. Rev. D18,331(1978).

[9] M. Gell-Mann, Physics 1,63(1964).

[10] N. Cabibbo and L. Radicati, Phys. Lett. 19,697(1966).

[11] F. Gilman and H. Schnitzer, Phys. Rev. 150,1562(1966).

[12] S.L. Adler, Phys. Rev. 143,1144(1966).

[13] H. Deden et al., Nucl. Phys. B85,269(1975).

[14] J. Werle, ICTP-Report, Trieste 1965.

THE EFFECT OF THE CHOICE OF WAVE FUNCTIONS ON THEORETICAL PREDICTIONS FOR SYMMETRY BREAKING PROCESSES: A VIEW FROM THE DKP FORMALISM*

Michael Martin Nieto
Theoretical Division, Los Alamos Scientific Laboratory
University of California, Los Alamos, New Mexico 87545/USA

Abstract

When considering an elementary particle matrix element, of necessity one must make an assumption, which often goes unnoticed, as to what formalism should be used for the wave functions. A current or interaction Lagrangian-density matrix-element is of the form $V = \overline{\psi}_{out}\Gamma\psi_{in}$, where ψ_{in} and $\overline{\psi}_{out}$ represent the physical ingoing and outgoing particles, and Γ represents the vertex function. A current must have the dimensions of $(\text{length})^{-3} = (\text{mass})^3$ in units of $\hbar = c = 1$. ψ_{in} and $\overline{\psi}_{out}$ must be described in terms of the physical on-shell masses or else one has no phase space. It is only the vertex function which can be symmetric in the internal symmetry under consideration.

The decision as to how much of the matrix element will be taken to be symmetric and how much of the matrix element will be taken to be associated with on-mass-shell wave functions is a fundamental assumption. Depending on how the assumption is made, different results will be predicted.

Normally first-order Dirac wave functions, with dimensions $(\text{length})^{-3/2}$ and second-order Klein-Gordon wave functions with dimensions $(\text{length})^{-1}$ are considered for spin-1/2 fermions and spin-0 bosons, respectively. We will discuss the types of new results which are obtained if, on the contrary, one chooses to consider bosons in the first-order Duffin-Kemmer-Petiau formalism. We will argue that the DKP formalism represents a complementary viewpoint to the spectrum generating approach. Both challenge the standard phenomenology: DKP by changing the wave function, spectrum generating by changing the vertex function.

*Work supported by the United States Department of Energy.

I. INTRODUCTION

The work I will discuss today results from a rather large collaboration[1-7] which proposed that one should consider using the Duffin-Kemmer-Petiau[8-10] first-order wave equation to describe mesons instead of the second-order Klein-Gordon equation. The main members of the collaboration were Ephraim Fischbach, C. Keith Scott, and myself, with a fourth major contributor being Henry Primakoff. There were also a number of others who joined us in this work, and they are credited in Refs. 1-7. These references and the works mentioned therein provide a guide to the literature.

This talk will differ from what would be my usual presentation since I have been asked to discuss our work with reference to the main topic of this session, "Spectrum Generating Algebras." Thus, the main thrust will be geared to answering the question, "What is the relation between the spectrum generating approach and the use of the DKP formalism?" I will argue that although there is not necessarily a one-to-one calculational equivalence, there is a complementary philosophical connection. The two ideas provide complementary alternatives to the standard ideas on symmetry breaking. The meaning of this assertion will become clear when I return to it in a few minutes. For now let me give a quick introduction into the DKP formalism.

II. DKP FORMALISM

Soon after the success of the original nonrelativistic Schrödinger equation, efforts began to generalize the Schrödinger theory to the relativistic domain. The first serious attempt was what we have come to call the "Klein-Gordon Equation," but which, however, was discovered by a number of people.[11]

The idea of the KG equation was a direct generalization of the Schrödinger idea that the classical nonrelativistic Hamiltonian equation be changed to an operator wave equation with the operator substitutions

$$E \to i\hbar \frac{d}{dt} \quad , \quad \vec{p} \to \frac{\hbar}{i} \frac{\vec{d}}{dx} \quad . \tag{2.1}$$

In particular, the KG equation takes the relativistic energy-momentum relation

$$E^2 = p^2c^2 + m^2c^4 \tag{2.2}$$

and changes it into a wave equation with the same substitutions, so one ends up with

$$\left[\hbar^2 \frac{d^2}{dt^2} - \hbar^2c^2 \left(\frac{\vec{d}}{dx}\right)^2 + m^2c^4\right]\phi = 0 \quad . \tag{2.3}$$

The Coulomb problem can be solved exactly for this system and it was discovered that the energy levels

$$\begin{aligned} E &= mc^2 \; [1 + \alpha^2/\nu^2]^{-1/2} \\ \nu &= n - (\ell + \tfrac{1}{2}) + [(\ell + \tfrac{1}{2})^2 - \alpha^2]^{1/2} \end{aligned} \tag{2.4}$$

do not correspond to the hydrogen atom. (Later it was realized that the KG equation could not work because of the spin of the electron.)

The great idea of Dirac was to extract the square-root of the energy-momentum relation (2.2). Dirac sought operators γ_μ which obey

$$(\vec{\gamma}\cdot\vec{p} - \gamma_0 p_0)^2 = (\gamma\cdot p)^2 = -m^2 \quad , \qquad m = \pm [-(\gamma\cdot p)^2]^{1/2} \quad . \tag{2.5}$$

(I will now usually take $\hbar = c = 1$.) The solutions are the now well-known Dirac matrices satisfying

$$\gamma_\mu\gamma_\nu + \gamma_\nu\gamma_\mu = 2\,\delta_{\mu\nu} \tag{2.6}$$

and which yield the first-order Dirac equation

$$[\partial\cdot\gamma + m]\psi = 0 \quad . \tag{2.7}$$

For the Coulomb problem, the solution to this equation yields the energy levels of the relativistic (but not second-quantized) hydrogen atom,

$$E = mc^2 \left\{ 1 + \frac{\alpha^2}{\{n - j - \frac{1}{2} - [(j+\frac{1}{2})^2 - \alpha^2]^{1/2}\}^2} \right\}^{-1/2} \quad . \tag{2.8}$$

Also due to its later successes, the use of this first-order formalism for electrons (and spin-1/2 particles in general) has gone on triumphantly for fifty years.

As to spin-0 particles, the story has been somewhat different. At the end of the 1930's, Duffin, Kemmer, and Petiau proposed what is a direct first-order generalization of the Dirac idea. (I refer the reader to Ref. 12 for a historical account of the work of these and other investigators.)

The DKP proposal can be put in terms of the following question: Are there operators β_μ which satisfy the energy-momentum relation

$$(p\cdot\beta)^3 = (p\cdot\beta)\,p^2 = -\,m^2(p\cdot\beta) \quad ? \tag{2.9}$$

The answer is, "Yes," and they have the algebra

$$\beta_\mu\beta_\lambda\beta_\nu + \beta_\nu\beta_\lambda\beta_\mu = \beta_\mu\,\delta_{\lambda\nu} + \beta_\nu\,\delta_{\mu\nu} \quad . \tag{2.10}$$

In particular there exists a (reducible) 16×16 representation of (2.10), which is given by the symmetric product of two Dirac spaces:

$$\beta_\lambda = \frac{1}{2}[I^{(1)}\,\gamma_\lambda^{(2)} + \gamma_\lambda^{(1)}\,I^{(2)}] \quad . \tag{2.11}$$

One can see, therefore, that in some sense one is combining internally two spin-1/2 (quark?) spaces. Spin-0 and spin-1 should come out, and indeed the 16×16 reducible representation can be decomposed into a trivial 1×1, a spin-0 5×5, and a spin-1 10×10 representation.

The spin-1 representation can be thought of as the equations for massive electrodynamics.[13] Six of the fields correspond to the $\vec{E}$ and $\vec{B}$ fields and four correspond to the physical A_μ fields (physical because the photon mass is not zero). For spin-1, this formalism is to be contrasted with the second-order formalism of Proca. For spin-0 a representation of the β's is[9]

$$\beta_1 = \begin{bmatrix} 0&0&0&0&0\\ 0&0&0&0&1\\ 0&0&0&0&0\\ 0&0&0&0&0\\ 0&1&0&0&0 \end{bmatrix} \qquad \beta_2 = \begin{bmatrix} 0&0&0&0&0\\ 0&0&0&0&0\\ 0&0&0&0&1\\ 0&0&0&0&0\\ 0&0&1&0&0 \end{bmatrix}$$

$$\beta_3 = \begin{bmatrix} 0&0&0&0&0\\ 0&0&0&0&0\\ 0&0&0&0&0\\ 0&0&0&0&1\\ 0&0&0&1&0 \end{bmatrix} \qquad \beta_4 = \begin{bmatrix} 0&0&0&0&-i\\ 0&0&0&0&0\\ 0&0&0&0&0\\ 0&0&0&0&0\\ i&0&0&0&0 \end{bmatrix} \quad . \tag{2.12}$$

The free particle solutions to the DKP spin-0 equation

$$(\partial\cdot\beta + m)\,\psi = 0 \tag{2.13}$$

are

$$\psi = \left[\frac{m}{p_0 V}\right]^{1/2} u(p)\, e^{ip\cdot x} \quad , \tag{2.14}$$

$$u = (2m^2)^{-1/2} \begin{pmatrix} i\,p_0 \\ i\,p_1 \\ i\,p_2 \\ i\,p_3 \\ -\,m \end{pmatrix} \quad . \tag{2.15}$$

The adjoint solutions are

$$\overline{\psi} = \psi^\dagger \eta_4 = \psi^\dagger (2\,\beta_4^2 - 1) \quad . \tag{2.16}$$

The dimensions of $(\text{mass})^{3/2} = (\text{length})^{-3/2}$ associated with (2.14) come from the Lagrangian derivable from (2.13). (Note that for the Dirac equation $\eta_4 = \gamma_4$, whereas here it is a polynomial of order two. This is a direct result of the fact that the algebra is three-fold vs. two-fold.)

Before proceeding to a comparison of the DKP and KG formalisms, let me mention that the procedure of DKP can be generalized, looking for operators satisfying a higher power energy-momentum relation. This was the idea of Bhabha, Lubański, and Mahadvarao, which resulted in what are known as the Bhabha first-order wave equations for arbitrary spin.[12,14] The matrices α_μ in the first-order wave equation

$$(\partial\cdot\alpha + \chi)\,\psi = 0 \tag{2.17}$$

are various representations (S,S) of the algebra so(5). A beauty is that the lowest representations, $(\frac{1}{2},\frac{1}{2})$ and (0,0), (1,0) and (1,1), are the Dirac and three DKP representation matrices. Thus, the Dirac and DKP equations are all special cases of the Bhabha set of equations,

These Bhabha equations are by no means the only first-order wave equations possible. However, since higher-spin equations would only give a detailed generalization of the physical points to be made today, this goes afield from the main topic under discussion. Therefore let me refer the interested members of the audience to Refs. 12, 14 and the review by Ginzburg and Man'ko[15] for references and an introduction to the subject.

III. DKP AND KG IN THE SYMMETRY LIMIT

Having obtained the formalism, one can ask if the two formalisms yield the same results. Since our answer is, "Not always," let's start by looking at where they do yield identical results.

a) Free particles. By deep theorems this must be true. Specifically, let

$$\psi_5 \equiv \phi \quad , \qquad \psi_{1,2,3,4} \equiv \phi_{4,1,2,3} \quad . \tag{3.1}$$

Putting this into Eq. (2.13) with the β-matrices of (2.12),

$$\begin{aligned} m\phi &= -\partial_\mu \phi_\mu \\ \partial_\lambda \phi &= -m\phi_\lambda \quad , \end{aligned} \tag{3.2}$$

which when combined yield

$$(-\Box + m^2)\phi = 0 \quad . \tag{3.3}$$

Thus, up to a (here) not important normalization, the two solutions are the same.

b) Pi-mesic atom. By the same procedure, one finds that the first-order DKP equation can be made into a second-order equation on ψ_5 which is identical to the KG equation for this problem. Thus, the same energy levels of Eq. (2.4) result.

c) QED. One can also show that 2nd-quantized QED processes involving mesons, such as $e^-e^+ \to \pi^-\pi^+$ and $e\pi$ scattering, yield identical results in the two formalisms. In fact, a lovely discussion of the QED of pi-mesons in the DKP formalism is contained in the book by Akhiezer and Berestetskii.[16] For people who like to play with the traces in Dirac QED, it can be a lot of fun to do these calculations. They are very similar to Dirac QED (except that there are modifications of the trace theorems). For example, the two diagrams in the DKP description of $\gamma\pi$ scattering yield the same results as the three diagrams in the KG description.

d) πNN coupling. Again the two formalisms agree, up to a normalization which can be absorbed in the coupling constant. This means that one will obtain the same predictions for πN scattering.

In the 1940's and 1950's a great deal of effort was expended proving the above results, and as a result of these efforts the KG formalism came to be preferred to the

DKP formalism, partially because of there not being a 5×5 matrix algebra associated with it.

However, our collaboration began with the realizations that the above discussions of the "equivalence" of the DKP and KG formulations were all done in the context of what we would now call the symmetry limit (before the discovery of the K meson), and that it was "folk lore" that this equivalence had been shown in the symmetry breaking case.

IV. DKP AND KG IN SYMMETRY BREAKING

Let us begin to make our point with a specific example: the case of non-conserved electrodynamics. With minimal electromagnetic substitution, the KG current is $(\partial_\lambda^\pm = \partial_\lambda \pm ieA_\lambda)$

$$j_\lambda^{KG} = -\, i[\phi_B^* \partial_\mu^- \phi_A - (\partial_\mu^+ \phi_B^*)\, \phi_A] \quad . \tag{4.1}$$

The DKP current is

$$j_\lambda^{DKP} = i\, \overline{\psi}_B\, \beta_\lambda \psi_A \quad . \tag{4.2}$$

By using the DKP wave equations one can rewrite (4.2) in terms of a properly normalized KG field ϕ. The result is

$$j_\mu^{DKP} = -\, i \left[(m_B^{1/2}\, \phi_B^*) \left\{ \frac{\partial_\mu^-}{m_A} (m_A^{1/2}\, \phi_A) \right\} - \left\{ \frac{\partial_\mu^+}{m_B} (m_B^{1/2}\, \phi_B^*) \right\} (m_A^{1/2}\, \phi_A) \right] \quad . \tag{4.3}$$

That is, the two currents are different when the current is not conserved, i.e., when there is symmetry breaking.

How can we understand this physically? On dimensional grounds, actually. Consider the $K_{\ell 3}$ decay $K \to \pi\ell\nu$, which usually is discussed in terms of a current-current interaction.

$$\langle \pi\ell\nu | H | K \rangle = \ell_\lambda \mathcal{H}_\lambda \quad , \tag{4.4}$$

$$\mathcal{H}_\lambda = g\overline{\psi}(p')\Gamma_\lambda \psi(p) \quad , \tag{4.5}$$

where ℓ_λ is the lepton current. The hadronic current $\mathcal{H}_\lambda$ has units of $(\text{length})^{-3} = (\text{mass})^3$, no matter what is the representation.

When one is doing a practical calculation, the wave functions must have physical (non-symmetric) masses or the phase space will be zero. Standardly, what is also done is to make the most important piece of vertex function Γ_λ be symmetric; in this case SU(3) symmetric. Now we come to the crux: IT IS AN ASSUMPTION to say what piece of the current is SU(3) symmetric, and which piece is physical. In this case, if one chooses the KG formulation, the wave functions $(1/2\ EV)^{1/2}$ $\exp(ip\cdot x)$ have dimensions of $(\text{length})^{-1}$ and the vertex function has units $(\text{length})^{-1}$. If one chooses the DKP wave functions of Eq. (2.14), then the vertex function will be dimensionless. In general one should expect different answers from a wave function of dimensions $(\mathrm{mass})^{3/2}$ which extrapolates to a new mass than from a wave function of dimensions (mass) which does a similar extrapolation.

Also observe that the same considerations apply when comparing Lagrangians of dimensions $(\mathrm{mass})^4$.

V. CONNECTION TO THE SPECTRUM GENERATING APPROACH

Members of the audience may already realize the point I am now trying to make. The DKP approach and the spectrum generating (SG) approach represent complementary viewpoints on the same problem.

Both approaches question the standard phenomenology which is used for mesons, that one should use wave functions of dimensions $(\mathrm{length})^{-1} = (\mathrm{mass})^1$ and standard vertex functions of dimensions $(\mathrm{length})^{-1}$. The way that DKP challenges the standard approach is to change the wave function. The physical wave functions have dimensions $(\mathrm{mass})^{3/2}$, leaving the vertex function dimensionless.

The spectrum generating approach, on the other hand, does not concentrate on the wave functions, but challenges the symmetry of the vertex function: like having the vertex function be changed from p_μ to p_μ/M where M is a mass operator.

An example which makes the similarity between DKP and SG more intuitive is to compare Eq. (4.1) for the KG current and Eq. (4.3) for the DKP current written in terms of KG wave functions. One sees that the current KG current operator $\overleftrightarrow{\partial}_\mu$ can be thought of as having been changed in DKP to a more complicated function with $\overleftrightarrow{\partial}_\mu$ and mass operators.

Whether a DKP approach will be calculationally the same as an SG approach will depend in detail on what DKP interaction and what SG interaction is used. But the common feature they have is that they provide alternatives to the standard KG approach with standard vertex functions.

At this point it is appropriate to observe that the questions about symmetry breaking which are raised by this discussion are not limited to mesons. The same type of questions can be raised about fermions. Indeed, if one looks back at the fundamental papers of Feynman on the path integral method[17] he preferred a second-order formalism for fermions. This point of view has been resurrected by Macrae[18] in his work on the unification of fermions and bosons in the fiber bundle picture. One could also change things by keeping a first-order formalism and putting in an SG interaction. In any case, the thing to be remembered is that different results can ensue. Which, if any, of these various possibilities is ultimately to be preferred depends, of course, on physical experiment.

VI. AN ILLUSTRATIVE EXAMPLE

In the references I have quoted,[1-7] applications have been made to the symmetry-breaking parameters of $K_{\ell 3}$ decay,[1] the Cabibbo angle,[2] pseudoscalar decays into two photons,[4] the PS meson-baryon-baryon coupling constants,[5] and vector-meson annihilation into e^+e^-.[7] Other processes also have been discussed in the literature.[19]

From these I wish to take a specific example and show explicitly how the DKP formalism approach yields deviations from the KG formalism, which deviations disappear in the symmetry limit. The example I will use was one of the results of our first paper. I choose it nonetheless first because it so nicely illustrates the DKP algebra and techniques, and second because the experimental situation is in such chaos.

In the standard discussion of $K_{\ell 3}$ decay, the KG vector current matrix element is

$$\langle \pi(p')|V_\lambda|K(p)\rangle = (4\, p_0 p_0'\, V^2)^{-1/2}[(p+p')_\lambda f_+(t) + q_\lambda f_-(t)] \quad . \tag{6.1}$$

The form factors $f_+(t)$ and $f_-(t)$ are expected to be slowly varying functions of t = momentum-transfer-squared, and if SU(3) is good, one expects the symmetry-breaking parameter

$$\xi \equiv f_-(0)/f_+(0) \tag{6.2}$$

to be small in magnitude compared to unity.

In fact, after first wandering all over the map, the Chounet-Gaillards review[20] summarized a growing consensus for a large negative value of order -1. Then two final grand experiments were done: a SLAC experiment with 1.6 million events which obtained[21] -0.11 ± 0.03 and a YALE[22] experiment with 2.2 million events which obtained -0.655 ± 0.127, and no one has ever understood the difference. So, taking in mind that a large negative value would be hard to understand, let's look at the DKP description.

The DKP current matrix element is ($m_K = m$, $m_\pi = \mu$)

$$\langle \pi(p')|V_\lambda(0)|K(p)\rangle = i\left(\frac{m\mu}{p_0 p_0'\, V^2}\right)^{1/2} \bar{u}_\pi(p')[\beta_\lambda g_V(t) + \frac{iq_\lambda}{m+\mu} g_S(t)]\, u_K(p) \tag{6.3}$$

$$\equiv [4 p_0 p_0'\, V^2]^{-1/2}[(p+p')_\lambda\, \hat{f}_+(t) + q_\lambda\, \hat{f}_-(t)] \quad , \tag{6.4}$$

$$\hat{f}_+(t) = \frac{m+\mu}{2(m\mu)^{1/2}}\, g_V(t) \quad , \tag{6.5}$$

$$\hat{f}_-(t) = \frac{m+\mu}{2(m\mu)^{1/2}}\, g_V(t) \left\{-\left(\frac{m-\mu}{m+\mu}\right) - \frac{g_S(t)}{g_V(t)}\left[1 - \frac{t}{(m+\mu)^2}\right]\right\} \quad . \tag{6.6}$$

This means the apparent "KG" symmetry breaking parameter is

$$\hat{\xi}(0) = \hat{f}_-(0)/\hat{f}_+(0) = -\left(\frac{m-\mu}{m+\mu}\right) - \frac{g_S(0)}{g_V(0)} = -0.57 - \frac{g_S(0)}{g_V(0)} \quad . \tag{6.7}$$

Thus, in the DKP formalism where one expects $g_S(0)/g_V(0)$ to be the quantity which is small in magnitude compared to unity, a small negative value would still give the apparent large symmetry breaking parameter ξ when written in the KG formalism. Thus, if the DKP formalism were correct, the apparent large symmetry breaking would only be due to having used the "incorrect" KG formalism.

A final comment on the SG analogy is that Böhm[23] recently reobtained the important factor -0.57 in Eq. (6.7) from the SG approach. It is instructive to follow this calculation exactly, and see the connection in this case. Let's start from the DKP approach, assuming that there is only the β_λ current, the "symmetry-breaking current" q_λ being zero. Then, as in a simpler version of (6.3), one finds that

$$i\bar{\psi}_\pi(p')\beta_\lambda\psi_K(p) = -i\left[(\mu^{1/2}\phi_\pi^*)\left\{\frac{\partial_\lambda}{m}(m^{1/2}\phi_K)\right\}-\left\{\frac{\partial_\lambda}{\mu}(\mu^{1/2}\phi_\pi^*)\right\}(m^{1/2}\phi_K)\right] \tag{6.8}$$

$$= (\mu^{1/2}\phi_\pi^*)\left[\frac{p'_\lambda}{\mu}+\frac{p_\lambda}{m}\right](m^{1/2}\phi_K) \tag{6.9}$$

$$= [\phi_\pi^*\phi_K]\frac{1}{2(m\mu)^{1/2}}[(m+\mu)(p+p')_\lambda + (-m+\mu)(p-p'_\lambda)] \, . \tag{6.10}$$

Equation (6.8) is a simpler version of the non-conserved current Eq. (4.3). In the intermediate equation (6.9) one has the form that SG would start with, the SG interaction p_λ/M combined with "velocity eigenstate" KG wave functions. Finally, simple algebra gives (6.10), which shows the relative normalizations of the $(p \pm p')_\lambda$ KG currents to be the factor $(-m+\mu)/(m+\mu)$.

ACKNOWLEDGEMENT

I would like to acknowledge the hospitality of the Aspen Center for Physics, where the first draft of this talk was written.

1. E. Fischbach, F. Iachello, A. Lande, M. M. Nieto, and C. K. Scott, Phys. Rev. Lett. 26, 1200 (1971).

2. E. Fischbach, M. M. Nieto, H. Primakoff, C. K. Scott, and J. Smith, Phys. Rev. Lett. 27, 1403 (1971).

3. E. Fischbach, M. M. Nieto, and C. K. Scott, Prog. Theor. Phys. (Kyoto) 48, 574 (1972).

4. E. Fischbach, M. M. Nieto, H. Primakoff, and C. K. Scott, Phys. Rev. Lett. 29, 1046 (1972).

5. F. T. Meiere, E. Fischbach, A. McDonald, M. M. Nieto, and C. K. Scott, Phys. Rev. D 8, 4209 (1973).

6. E. Fischbach, J. D. Louck, M. M. Nieto, and C. K. Scott, J. Math. Phys. 15, 60 (1974).

7. B. G. Kenny, D. C. Peaslee, and M. M. Nieto, Phys. Rev. D 13, 757 (1976).

8. R. J. Duffin, Phys. Rev. 54, 1114 (1938).

9. N. Kemmer, Proc. R. Soc. A 173, 91 (1939).

10. G. Petiau, Acad. R. Belg. C. Sci. Mem. Collect 8 16, No. 2 (1936).

11. O. Klein, Z. Phys. 37, 895 (1926);
V. Fock, Z. Phys. 38, 242 (1926); 39, 226 (1926);
E. Schrödinger, Ann. Phys. (Leipz.) 81, 109 (1926);
J. Kudar, Phys. Zeit. 27, 724 (1926);
W. Gordon, Z. Phys. 40, 117 (1926);
Th. de Donder, Compt. Rend. 182, 1380 (1926);
Th. de Donder and Fr. H. van den Dungen, ibid. 183, 22 (1926).

12. R. A. Krajcik and M. M. Nieto, Am. J. Phys. 45, 818 (1977).

13. A. S. Goldhaber and M. M. Nieto, Rev. Mod. Phys. 43, 277 (1971).

14. R. A. Krajcik and M. M. Nieto, Phys. Rev. D 15, 455 (1977). This is the seventh paper of a series on Bhabha first-order wave equations, giving a summary and conclusions.

15. V. L. Ginzburg and V. I. Man'ko, Fiz. Elem. Chastits At. Yadra 7, 3 (1976). [English trans. in Sov. J. Part. Nucl. 7, 1 (1976).]

16. A. I. Akhiezer and V. B. Berestetskii, Quantum Electrodynamics (Interscience, New York, 1965), Chap. IX.

17. See the discussion on p. 194 of R. P. Feynman and M. Gell-Mann, Phys. Rev. 109, 193 (1958).

18. K. Macrae, Phys. Rev. D (to be published).

19. E. Golowich and V. Kapila, Phys. Rev. D 8, 2180 (1973).

20. L. M. Chounet, J. M. Gaillard, and M. K. Gaillard, Phys. Reports 4C, 199 (1972).

21. G. Donaldson, et al., Phys. Rev. D 9, 2960 (1974).

22. J. Sandweiss, J. Sunderland, W. Turner, W. Willis, and L. Keller, Phys. Rev. Lett. 30, 1002 (1973).

23. A. Böhm, Phys. Rev. D 13, 2110 (1976).

ALGEBRAIC APPROACH TO HADRONS WITHOUT SEEING QUARKS

S. Oneda, Center for Theoretical Physics, Department of Physics and Astronomy University of Maryland, College Park, Maryland 20742. Dedicated to the memory of my mother, Kazu Oneda, January 21, 1896-July 22, 1978.

I. An approach to hadron physics through hidden quarks --

I wish to review the main idea of our algebraic approach and its recent development[1]. The approach deals solely with hadrons but it aims to synthesize the two successful notions of (perhaps confined) quarks. 1. The algebras of vector and axial-vector currents of quark fields are assumed to reflect the essential nature of quarks. No use of explicit properties of quarks is made. We assume that hadrons make up the entire Hilbert space but their masses, spectra and couplings are constrained by the presence of these algebras. 2. Hadrons are assumed to obey the level scheme of the symmetric $q\bar{q}$ and qqq quark model. In this model, multiplicity of hadrons associated with a level is automatically fixed, once the level is specified. Therefore, we may not need to introduce the notion of higher symmetry. The notion of levels may be more fundamental. One can, in fact, derive the SU(6) (SU(8),...) like constraints from a level realization requirement of the algebras. This is our prescription to try to synthesize the notions of current and constituent quarks. Although we are not after quark lines, the result usually permits an interpretation based on naive quark counting. Specifically the essential features of quark line rules[2] (QLR) are demonstrated within the theory. We do not need to impose QLR by hand.

II. Algebraic constraints (from quarks) in the world of hadrons

The starting algebras are Gell-Mann's chiral SU(3) $\otimes$ SU(3) (SU(4) $\otimes$ SU(4), ...) current algebras. In particular, the charge algebras are given by

$$[V_i,V_j] = if_{ijk}V_k,\ [V_i,A_j] = if_{ijk}A_k,\ [A_i,A_j] = if_{ijk}V_k\ . \tag{1}$$

(However, we do not use the concept of chiral SU(3) $\otimes$ SU(3) (SU(4) $\otimes$ SU(4),...) symmetry). SU(3) (SU(4),...) breaking will be characterized by the presence of exotic commutation relations involving $\dot{V}_\alpha = \frac{dV_\alpha}{dt}$,

$$[\dot{V}_\alpha,V_\beta] = 0 \text{ and } [\dot{V}_\alpha,A_\beta] = 0, \qquad (2) \text{ and } (3)$$

where (α,β) are the exotic combinations of physical SU(3)(SU(4),...) indices such as (K^o,K^o), (K^o,π^-), etc. Eq. (2) is an algebraic expression for the usual assumption of SU(3) (SU(4),...) breaking. Eq. (3) is a weaker assumption than the pure $(3,3^*) \oplus (3^*,3)((4,4^*) \oplus (4^*,4),...)$ chiral symmetry breaking. To study processes involving finite four momentum transfer or helicity changes, we also use the charge-charge density algebra such as

$$[[V_3^o(0),A_{\pi^+}],A_{\pi^-}]=2V_3^o(0),\ [[V_3^o(0),A_{\pi^+}],V_{\pi^-}] = 2A_3^o(0), \text{ etc.} \tag{4}$$

A distinct advantage of these algebras is that they could be valid in broken SU(3) (SU(4), ...) which enables us to make a non-perturbative approach[3] to broken SU(3)(SU(4),...).

III. Two Assumptions on quark dynamics-tools to extract information from algebras

1. Asymptotic SU(3)(SU(4), ...) symmetry - coping with broken SU(3)(SU(4), ...) symmetry.

We assume that quark dynamics permits the SU(3) (SU(4), ...) transformation to be still linear (including the possibility of SU(3)(SU(4), ...) particle mixing) in broken SU(3) (SU(4),...), but only in the infinite momentum limit, and that the non-linear terms vanish sufficiently fast as $\vec{p} \to \infty$. SU(3)(SU(4), ...) multiplet classification is thus carried out in this limit. SU(3)(SU(4),...) charges are assumed to annihilate the vacuum only in a segment of Hilbert space in which multiplets have infinite momenta. With asymptotic SU(3) (SU(4), ...) the implications of the algebras involving SU(3) (SU(4), ...) charges are:

(a) $[V,V] = V$, $[V,A] = A \to$ The asymptotic matrix elements of V_α and A_α can be parameterized, even in broken symmetry, by the usual prescription of exact SU(3) (SU(4), ...) plus mixing. Cabbibo's exact SU(3) analysis is then justified to the extent that we neglect SU(3) particle mixing. (IIIa)

(b) $[\dot{V}_\alpha,V_\beta] = 0 \to$ G-M-O mass formula with mixing. However, this formula is now derived as an exact constraint, not as a first order formula. (IIIb)

(c) $[\dot{V}_\alpha,A_\beta] = 0 \to$ Inter-multiplet mass relations ($K^2-\pi^2 = K^{*2} - \rho^2 = ...,$ $D^2 - \pi^2 = D^{*2} - \rho^2 = ...$, etc.) and inter-multiplet interplay among masses, mixing angles and asymptotic axial-vector matrix elements. (IIIc)

This interplay produces selection rules for the axial-vector matrix elements similar to the ones imposed by QLR.

2. Level realization of the algebras - Do we really need higher symmetry?

If the algebras involve axial charges and the intermediate states inserted among the factors of the commutators are not restricted by asymptotic SU(3) (SU(4), ...), the hypothesis of asymptotic level realization of these algebras is introduced. It produces SU(6) (SU(8), ...)-like constraints.

$[A,A] = V$, $[[V_3^0(0), A_{\pi^+}] = 2V_3^0(0)$ etc. } → hypothesis of asymptotic level realization of the algebra } → { Intermultiplet constraints among the couplings of groups of hadrons belonging to levels.

Insert these algebras between the states $\langle B_\ell(\vec{k},\lambda)|$ and $|B'_\ell(\vec{k}',\lambda')\rangle$ where B_ℓ and B'_ℓ are the hadrons belonging to the same level ℓ with helicity λ and λ' respectively ($\vec{k} \to \infty$ and $\vec{k}' \to \infty$). The r.h.s. is denoted as $g(B_\ell,\lambda;B'_\ell,\lambda')$. On the ℓ.h.s. we insert a complete set of single particle intermediate states between the factors of the commutators. We can then define a fractional contribution $f_i(B_\ell,\lambda;B'_\ell,\lambda')$ to the $g(B_\ell,\lambda;B'_\ell,\lambda')$ coming from all the intermediate states belonging to a level i. We now vary B_ℓ and B'_ℓ for given ℓ. The most stringent condition associated with the idea of level realization (which we call super realization) would be that $f_i(B_\ell,\lambda;B'_\ell,\lambda')$ depends only on λ and λ' but not on the choice of B_ℓ and B'_ℓ belonging to the same level ℓ. However, so far we have restricted to the weaker assumption, i.e., the external particles B_ℓ and B'_ℓ belong to the same SU(3) (SU(4), ...) multiplet.[1] We have called this realization asymptotic level realization of SU(3) (SU(4), ...) in the algebra under consideration. We have produced many good SU(6) (SU(8), ...) type constraints which are compatible with QLR. In V we discuss super realization.

IV. Constraints on Meson Multiplets -- Why are there two distinct nonet meson structures? Derivation of quark-line selection rules.

Sum rules described in III produce tight constraints among hadron multiplets. We have recently shown[4] that (i) They provide a resolution to the puzzling structure of SU(3) boson nonets. (ii) The qualitative feature of quark-line selection rules (QLR) involving charmed as well as strange quarks is demonstrated among these constraints.

Denote an SU(4) 16-plet as ($\pi_s, K_s, \eta_s, \eta_{cs}, D_s, F_s$ and η'_s), where s denotes J^{PC} and other quantum numbers. The η_s-η_{cs}-η'_s mixing angles defined in our asymptotic limit are θ^s, ϕ^s and ψ^s. The ideal configuration implies $\psi^s = 0$, $\phi^s = 30°$ and $\sin\theta^s = \sqrt{1/3}$. The procedure taken is as follows[4],

$[\dot{V},V] = [\dot{V},A] = 0$ plus asymptotic SU(4).
Asymptotic level realization of SU(4) in $[A,A]=V$ } → { Five mass-mixing angle sum rules for any 16-plet s.

elimination of mixing angles → { Two pure mass equations involving π_s, K_s, η_s, η'_s, η_{cs} and D_s.

In the limit of infinite charmed quark mass, i.e., $D_s^2 \to \infty$ and $\eta_{cs}^2 \to \infty$, one of the two pure mass equations yields $2D_2^2 = \eta_{cs}^2$, a statement of equal spacing. The other, however, produces, $N_S[N_S - 2(K_s^2-\pi_s^2)^2] = 0$. Therefore in our SU(3) limit, two (but <u>only</u> two) possible nonet mass splitting patterns are predicted. One of them is the well-known Schwinger's nonet mass relation $N_S = 0$, i.e.,

$$3N_S \equiv (3\eta_s^2+\pi_s^2-4K_s^2)\,(3\eta_s'^2+\pi_s^2-4K_s^2) + 8(K_s^2-\pi_s^2)^2 = 0\,, \tag{5}$$

well satisfied by the 1^{--} and 2^{++} nonets. It involves the ideal mass relation, $\pi_s^2 = \eta_s'^2$ and $\pi_s^2 + \eta_s^2 - 2K_s^2 = 0$, as its special case. The other, $N_S-2(K_s^2-\pi_s^2)^2 = 0$, is a new one[5]

$$(3\eta_s^2 + \pi_s^2 - 4K_s^2)(3\eta_s'^2 + \pi_s^2 - 4K_s^2) + 2(K_s^2 - \pi_s^2)^2 = 0\,, \tag{6}$$

well satisfied by the 0^{-+} nonet ($\pi, K, \eta, \eta' \equiv X(958)$). Eq. (6) predicts η' mass around 0.943 GeV. It is remarkable that the two theoretical possibilities are indeed realized in nature. In <u>pure</u> SU(3) Eq. (5) is the <u>only</u> possible nonet mass relation and cannot accommodate the 0^{-+} mesons. Apparently, three particle mixing in SU(4) provides more flexibility in choosing the $q\bar{q}$ structure of η'_s. The finiteness of the masses of η_c and D does not modify the above result significantly. The two pure mass equations permit us to compute D_s and η_{cs} (and also F_s) in terms of the nonet masses π_s, K_s, η_s and η'_s. For the 0^{-+} nonet, we obtain D = 1.65 and η_c = 2.39 GeV. The allowed domain of η' is <u>restricted</u> to $0.942 \leq \eta' \leq 0.980$ GeV, in which η_c becomes infinity at η' = 0.9425 GeV. Between η' = 0.960 and 0.951 GeV, η_c gradually increases from 2.26 to 3.12 GeV and at η' = 0.954, a shift of only 4 MeV from its measured value, we obtain D = 1.87, F = 1.94 and η_c = 2.69 GeV reasonably consistent with present experimental values. Let us now consider the asymptotic axial-vector matrix elements, $A^{st} \equiv \langle\eta_s|A_{\pi^-}|\pi_t^+(\vec{k})\rangle$, $B^{st} \equiv \langle\eta_s'|A_{\pi^-}|\pi_t^+(\vec{k})\rangle$ and $C^{st} \equiv \langle\eta_{cs}|A_{\pi^-}|\pi_t^+(\vec{k})\rangle$ with $C_sC_t = 1$ and $\vec{k}\to\infty$. In our SU(3) limit, two of the mixing angles become $\psi^s \to 0$ and $\phi^s \to 30°$ or $150°$. $\phi^s \to 30°$ is selected, since η_{cs} should have a structure close to a pure $c\bar{c}$ state. In this limit we obtain from the same set of sum rules which led to Eqs. (5) and (6),

$$\frac{A^{st}}{B^{st}} \equiv \frac{\langle\eta_s|A_{\pi^-}|\pi_t^+\rangle}{\langle\eta_s'|A_{\pi^-}|\pi_t^+\rangle} = -\tan(\theta^s-\theta_0), \quad \vec{k}\to\infty \tag{7}$$

where θ_0 is the magic angle. For given s, t is completely arbitrary provided $C_sC_t = 1$ and the r.h.s. of Eq. (7) depends only on θ^s, i.e. only on the nonet structure of s. When $\theta^s \to \theta_0$, η_s takes a pure $s\bar{s}$ structure. In this limit Eq. (7) exhibits the existence of a (hidden) dynamical selection rule $A^{st} = 0$ for any t. However, Eq. (7) equally prescribes the violation of the selection rule. θ^s can be computed from the G-M-O mass formula obtained from $[\dot{V},V] = 0$. Eq. (7) essentially realizes the QLR ansatz made by Okubo[3] and others. Our new perception of SU(3) also fixes the nonet structure. For the nonet satisfying Eq. (5), we obtain $\theta^s>0$ if $\eta_s'^2-\eta_s^2 < 0$. So we predict $\theta_{\omega\phi} \simeq +40°$ and $\theta_{ff'} \simeq +30°$ etc. For the nonet satisfying Eq. (6), Eq. (8) implies $\theta^s>0$ if $\eta_s'^2 > \eta_s^2$ and $\theta^s < 0$ if $\eta_s'^2<\eta_s^2$. For the 0^{-+} nonet it implies[6] $\theta_{\eta\eta'} \simeq +10°$. An example of our QLR: By using the hypothesis of level realization of SU(3) in the charge-charge density algebra $[[V_3^0(0),A_{\pi^+}],A_{\pi^-}] = 2V_3^0(0)$ and Eq. (7) we obtain

$$\frac{g_{\phi\pi\gamma}}{g_{\omega\pi\gamma}} = \frac{g_{\phi\rho\pi}}{g_{\omega\rho\pi}} = -\tan(\theta-\theta_0), \text{ where } \theta \text{ is the } \omega\text{-}\phi \text{ mixing angle.} \tag{8}$$

The sign as well as the magnitude of $g_{\phi\pi\gamma}/g_{\omega\pi\gamma}$ given by Eq. (8) is consistent with the recent experiment. (For details, see Ref. (4)).

In our theoretical framework η_{cs} approaches automatically to a pure $c\bar{c}$ state, if the mass of the charmed quark tends to infinity. In this limit, while the violation of QLR in SU(3) is described by Eq. (7), the quark-line rule selection rules involving charmed quarks become exact ($C^{st} = 0$ etc.). Experimentally this feature is preserved well for the (real) finite mass of η_{cs}.

V. Super realization -- Derivation of $g_A(0)$, nucleon anomalous magnetic moment relation, etc. from chiral SU(2) $\otimes$ SU(2) algebras

Oneda, Slaughter and Tanuma have recently applied[7] the idea of super realization to ground state baryons. Use of chiral SU(2) ⊗ SU(2) algebra is sufficient to derive tight intermultiplet constraints. Only the realization from the ground states (nucleons and Δ's) is studied. Consider, for example, the algebra $[A_{\pi^+},A_{\pi^-}] = 2V_{\pi^0}$. Insert this algebra between the ground state baryons with helicity $\frac{1}{2}$ and infinite momentum, i.e. between $\langle p,\frac{1}{2}|$ and $|p,\frac{1}{2}\rangle$, and $\langle \Delta^+,\frac{1}{2}|$ and $|\Delta^+,\frac{1}{2}\rangle$. Super realization in the ground state level then implies,

$$f^2 - 4h^2 = k \text{ and } g^2 + 4h^2 = 2k \quad (k > 0), \tag{9}$$

where k is the fractional contribution of the ground state and $\langle n,\frac{1}{2}|A_{\pi^-}|p,\frac{1}{2}\rangle \equiv f = g_A(0)$, $\langle \Delta^+,\frac{1}{2}|A_{\pi^-}|\Delta^{++},\frac{1}{2}\rangle \equiv -\sqrt{3/2}\, g$, $\langle p,\frac{1}{2}|A_{\pi^-}|\Delta^{++},\frac{1}{2}\rangle \equiv -\sqrt{6}\, h$. f and g are real. Super realization of the charge-charge density algebra $[[V_3^0(0),A_{\pi^+}],A_{\pi^-}] = 2A_3^0(0)$ inserted among all the possible ground state baryons $\langle B,-\frac{1}{2},\vec{s}|$ and $|B',\frac{1}{2},\vec{t}\rangle$ with $\vec{s} \to \infty$ and $\vec{t} \to \infty$, produces

$$5g + \sqrt{2}f = 0 \text{ and } a:b:c:d = 1:\tfrac{5}{2}:-\sqrt{2}:\sqrt{2}. \tag{10 and 11}$$

Here $\langle \Delta^+,-\frac{1}{2},\vec{s}|V_3^0(0)|\Delta^+,\frac{1}{2},\vec{t}\rangle \equiv a$, $\quad \langle p,-\frac{1}{2},\vec{s}|V_3^0(0)|p,\frac{1}{2},\vec{t}\rangle \equiv b$,
$\langle p,-\frac{1}{2},\vec{s}|V_3^0(0)|\Delta^+,\frac{1}{2},\vec{t}\rangle \equiv c$, $\quad \langle \Delta^+,-\frac{1}{2},\vec{s}|V_3^0(0)|p,\frac{1}{2},\vec{t}\rangle \equiv d$.

From Eqs. (9) and (10) we find a solution (choosing h to be real)

$$f = \frac{5}{3}\sqrt{k}, \quad g = -\frac{\sqrt{2}}{3}\sqrt{k}, \quad h = \pm\frac{2}{3}\sqrt{k}. \tag{12}$$

If k = 1 we recover the bad result of SU(6), $f = g_A(0) = 5/3$. However, by using PCAC, $|h|^2$ can be related to the width of $\Delta \to p\pi$ decay and k is found[8] to be around 0.6, producing $g_A(0) \simeq 1.2$. The same result on f and h, together with a prediction of SU(6), D/F = 3/2, was obtainted[8] previously without using super realization but assuming asymptotic level realization of SU(3) in the algebra [A,A] = V. We also find that the algebra $[[V_3^0(0), A_{\pi^+}],A_{\pi^-}] = 2V_3^0(0)$ is automatically super realized by the constraints obtained in Eqs. (9), (10), and (11). However, super realization of the algebra involving the electromagnetic current $J^\mu_{em} = V^\mu_3 + V^\mu_s$ (V^μ_s is SU(2) singlet), $[[J^0_{em}(0), A_{\pi^+}],A_{\pi^-}] = 2V_3^0(0)$, leads, among others, to $\langle p,-\frac{1}{2},\vec{s}|J^0_{em}(0)|p,\frac{1}{2},\vec{t}\rangle = -\langle n,-\frac{1}{2},\vec{s}|J^0_{em}(0)|p,\frac{1}{2},\vec{t}\rangle$ which yields a prediction $k_p = -k_n$ for the nuclcon anomalous moments. Experimentally it implies 1.79 = 1.91. Small discrepancy may be blamed for the mixing between the ground state baryons and their radially excited states. We have thus seen that the idea of super realization can explain (without introducing $SU(6)_W$ symmetry) the two important quantities associated with the nucleon, i.e., the value of $g_A(0)$ and the anomalous moment relation.

References

1. S. Oneda, in Proceedings of INS International Symposium on New Particles and the structure of Hadrons, Tokyo, 1977 edited by K. Fujikawa et. al. (Institute for Nuclear Study, University of Tokyo) p. 94. S. Oneda and Seisaku Matsuda, in Fundamental Interactions in physics, 1973, Coral Gables Conference, edited by B. Kursunoglu and A. Perlmutter (Plenum, New York, 1973, p. 175. Exhaustive references were given there.
2. S. Okubo, Phys. Rev. 16D, 2336 (1977) and the references cited there.
3. For somewhat similar non-perturbation theoretic approaches to broken SU(3) (SU(4),...), see for example, A. Bohm and J. Werle, Nucl. Phys. B106, 165 (1976) and A. Bohm and R. B. Teese, Phys. Rev. Lett. 38, 629 (1977).
4. H. L. Hallock and S. Oneda, Phys. Rev. D18, 841 (1978), and H. L. Hallock and S. Oneda, SU(3) in the world of more than three quarks — derivation of nonet structure and quark-line rule. Phys. Rev. in the press.
5. Actually Eq. (6) coincides with one of the three (two of them are useless) 15-plet mass formula derived in the early work on SU(4). However, the 15-plet assumption is certainly not realistic. B. J. Bjorken and S. L. Glashow Phys. Rev. Lett. 11, 255 (1964); D. Amati, H. Bacry, J. Nuyts and J. Prentki, Nuovo Cimento 34, 1732 (1964); R. E. Marshak, S. Okubo and J. H. Wojtaszek, Phys. Rev. Lett. 15, 464 (1965); G. Karl, Nuovo Cimento, 38A, 315 (1977).
6. This result is different from the usually favored choice $\theta \simeq 10°$.
7. S. Oneda, T. Tanuma and Milton D. Slaughter, to be published.
8. Seisaku Matsuda and S. Oneda, Phys. Rev. D5, 2287 (1972) and Milton D. Slaughter and S. Oneda, Phys. Rev. D14, 1314 (1976).

Gauge Invariance with Massive Gauge Bosons

R.L. Ingraham

Research Center, N.M. State University
Las Cruces, NM 88003, U.S.A.

We describe a theory in which the gauge bosons which mediate a local strong symmetry have mass ab initio, and yet the gauge-invariance is perfect. All this with a unique, invariant vacuum and without need of Higgs fields. The basic idea is therefore much at variance with the prevailing method of the last ten years or so to confer mass in some sense on initially massless gauge bosons. But as an introduction to this new idea, we must first describe a more basic theory, of which it is only one of many consequences. The detailed theory is appearing in Nuovo Cimento B; we restrict ourselves here to a qualitative sketch.

I. Conformal Relativity, a Theory of Mass

Building relativistic physics on the conformal group C as basic space-time symmetry automatically gives a theory of mass on an equal footing with four-momentum. But thereby one must use the correct base space, the 5-dimensional Riemannian one of space-time angle, not the "obvious" 4-dimensional one of space-time length.

This space V^5 has a Riemannian structure with the infinitesimal angle $d\theta$ between two nearby space-time spheres[1] of centers x^μ and $x^\mu+dx^\mu$ and radii $x^5 \equiv \lambda$ and $\lambda+d\lambda$ as the line element:

$$(1) \qquad d\theta^2 \equiv \gamma_{\alpha\beta}dx^\alpha dx^\beta = -\lambda^{-2}(dx^2+d\lambda^2) \quad , \quad dx^2 \equiv g_{\mu\nu}dx^\mu dx^\nu \equiv d\underset{\sim}{x}^2-dt^2 \quad .$$

(The proof, modulo signature questions, is a simple application of the Pythagorean Theorem.) The use of sphere space as the natural domain for the action of C follows from Liouville's Theorem on the conformal group. C of course also acts on points of space-time identified as null spheres (radii $\lambda=0$).

We are thus committed to finding a new degree of freedom, a new dimension, as a real feature of the physical world. The space-time "sphere", the locus[1]

$$(2) \qquad (\xi-x)^2 = -\lambda^2 \quad ,$$

of which λ is the radius, is physically a typical causal signal "wave front", which lies properly within the light cone. It follows

$$(3) \qquad t_{ret} = t - \sqrt{(\underset{\sim}{\xi}-\underset{\sim}{x})^2+\lambda^2} \quad ,$$

so that λ is a minimum causal signal time lag. This is its primary geometrical meaning, though the detailed theory reveals several other physical "meanings".

If we solve for the basis wave functions of the various C-IUR's (irreducible unitary representations of C) over this base space, we find massive as well as massless particle species. (On the other hand, space-time itself as base space cannot carry the massive C-IUR's. The non-realization of this important fact by most workers on the conformal group in physics has been the cause, in our opinion, of the fifty year hiatus in the development of relativity and a workable theory of particles, especially the hadrons.) In fact, λ is conjugate to mass just as x^μ is conjugate to four-momentum, so that mass is described by a differential operator in λ just as $p^\mu = -i\partial/\partial x^\mu$. This already makes it clear why one cannot expect a kinematical theory of mass from the study of C over space-time alone.

The wave functions f turn out to have the typical coordinate-dependence

(4a) $$f(x,\lambda) \propto \lambda^\alpha e^{ip\cdot x} \quad , \quad p^2 = 0 \qquad \text{(massless)},$$

(4b) $$f(x,\lambda) \propto \lambda^\beta J_\nu(m\lambda) e^{ip\cdot x} \quad , \quad p^2 + m^2 = 0 \qquad \text{(massive)},$$

(or a sum of such terms in (4b)), where the numbers α, β, ν depend on the invariants and other good quantum numbers. The wave functions (4b) describe massive resonance families with infinite discrete mass spectra. These are bare masses, of course.

One is forced by many physical and mathematical arguments to choose the domain of λ finite[2],

(5) $$0 < \lambda < \ell \quad .$$

Then the unitarity of the IUR's forces the discreteness of the bare mass spectra, which in fact must be the zeros of certain Bessel functions: $J_{\nu'}(m\ell) = 0$, $m \neq 0$.

II. Gauge Theory

The C-IUR's in general have quantum numbers beyond the Poincaré subgroup ones (particle-antiparticle and spin). In particular, they provide isospin labels, and the free Lagrangians, pure conformal group constructs, possess isospin SU(2) invariance as an accidental ("dynamical") symmetry. This interpretation survives many nontrivial tests, including the interaction theory, with complete success.

We can then turn this global SU(2) into a local symmetry by the introduction of gauge fields in the standard way[3] transposed to five dimensions. But great physical differences ensue because mass is treated so differently from nowadays.

For example (we treat a U(1) symmetry for simplicity), we introduce the covariant derivative

(6) $$\partial_\alpha \to \nabla_\alpha \equiv \partial_\alpha - ikB_\alpha(x,\lambda) \quad , \quad \partial_\alpha \equiv \partial/\partial x^\alpha \quad ,$$

in the free conformal spinor Lagrangian $\mathcal{L}_F$. Then under an infinitesimal local U(1) gauge transformation $\delta\chi(x,\lambda)$, spinor and gauge 5-vector transform

$$(7) \qquad \Psi \to (1-i\delta\chi)\Psi \quad , \quad B_\alpha \to B_\alpha - \frac{1}{k}\partial_\alpha\chi \quad ,$$

and $\mathcal{L}_F$ is gauge-invariant. The free boson Lagrangian

$$(8) \qquad \mathcal{L}_B \equiv -\tfrac{1}{4} B_{\alpha\beta}B^{\alpha\beta} \quad , \quad B_{\alpha\beta} \equiv \partial_\alpha B_\beta - \partial_\beta B_\alpha \quad , \quad B^{\alpha\beta} \equiv \gamma^{\alpha\gamma}\gamma^{\beta\delta}B_{\gamma\delta} \quad ,$$

is added on. $\mathcal{L}_B$ is obviously gauge invariant because the field strength $B_{\alpha\beta}$ is.

B_α may well be a massive solution, of type (4b), of the field equations derived from $\mathcal{L}_B$. A pure spin 1 solution has the simple form (as a second quantized free field)

$$(9) \qquad B_\mu(x,\lambda) = \sum_\kappa N(\kappa)\lambda J_1(\kappa\lambda)B_\mu(x;\kappa) \quad , \quad B_5 = 0 \quad ,$$

where $N(\kappa)$ is a certain normalizer and $B_\mu(x;\kappa)$ is a conventionally normalized, purely x-dependent, free spin 1 quantum field to mass κ. The sum goes over the mass spectrum, which may be $J_0(\kappa\ell) = 0$ or $J_1(\kappa\ell) = 0$. The point is, now, that even though the Lagrangian $\mathcal{L}_F + \mathcal{L}_B$ is manifestly gauge-invariant, the conventional mass terms are already contained in $\mathcal{L}_B$! We close by proving this.

The C-invariant action is defined

$$(10) \qquad I \equiv \int d\Omega\, \mathcal{L}(x,\lambda) \quad , \quad d\Omega \equiv \lambda^{-5}d^4x\, d\lambda \quad .$$

($d\Omega$ is the volume element of V^5.) We can express this in terms of an effective, four-dimensional, "reduced" Lagrangian density $\mathcal{L}(x)$ by

$$(11) \qquad I \equiv \int d^4x\, \mathcal{L}(x) \quad , \quad \mathcal{L}(x) \equiv \int d\lambda\, \lambda^{-5}\mathcal{L}(x,\lambda) \quad .$$

It remains now only to calculate $\mathcal{L}_B(x)$. From eq. (1) we see

$$B_{\alpha\beta}B^{\alpha\beta} = \lambda^4[B_{\mu\nu}\underline{B^{\mu\nu}} + 2\frac{\partial}{\partial\lambda}B_\mu \frac{\partial}{\partial\lambda}\underline{B^\mu}] \quad ,$$

where $\underline{B^\mu} \equiv g^{\mu\nu}B_\nu$ etc. i.e., index raised with the space-time metric. Now the λ-derivative brings out the mass according to

$$\frac{\partial}{\partial\lambda}[\lambda J_1(\kappa\lambda)] = \kappa\lambda J_0(\kappa\lambda) \quad .$$

Substitute expression (9) and do the λ-integral. But different mass modes are orthogonal, and for the direct terms, the λ-integrals are precisely cancelled by $N^2(\kappa)$:

$$(12) \qquad \int_0^\ell d\lambda\, \lambda\, J_1(\kappa'\lambda)J_1(\kappa\lambda) = \int_0^\ell d\lambda\, \lambda\, J_0(\kappa'\lambda)J_0(\kappa\lambda) = \delta_{\kappa'\kappa}N^{-2}(\kappa) \quad .$$

This follows from the very choice of the mass spectrum (eq. (12) is in fact equivalent to the unitarity of the C-IUR). Hence we get

$$\mathcal{L}_B(x) = \sum_{\kappa}[-\tfrac{1}{4} B_{\mu\nu}(x;\kappa)B^{\mu\nu}(x;\kappa) - \tfrac{1}{2}\kappa^2 B_\mu(x;\kappa)B^\mu(x;\kappa)] \quad , \tag{13}$$

which is just the usual massive boson free Lagrangian, Q.E.D.

A final remark: when the interaction Lagrangian

$$\mathcal{L}_{int}(x,\lambda) \equiv ik\overline{\Psi}\beta_{6'}\beta^{\alpha'}B_\alpha\Psi \quad ,$$

(where the β's are certain 8×8 "Dirac" matrices occurring in the spinor theory), is reduced, as in eq. (11), the standard SU(2)-invariant interaction of a "nucleon" family (J=T=½) and a T=0 spin 1 boson family is obtained. The same perfect isospin invariance results for all (strong) reduced interactions.

References

1. Indices μ, ν, ξ = 1 to 4 are space-time indices with $x^4 \equiv t \equiv$ time. Indices α, β, γ = 1 to 5 are V^5 indices with $x^\mu \equiv$ space-time and $x^5 \equiv \lambda$. Metric: $g_{\mu\nu} \equiv \text{diag}\ (+++-)1$. $(\xi-x)^2 \equiv g_{\mu\nu}(\xi^\mu-x^\mu)(\xi^\nu-x^\nu)$.
2. Then the "accelerations" must be dropped from the symmetry group.
3. R. Utiyama, Phys. Rev. **101**, 1597 (1956).

GROUP THEORETICAL INTERPRETATION OF RELATIVISTIC HADRONIC STRUCTURES

Y. S. Kim, University of Maryland, College Park, Maryland 20742, and
Marilyn E. Noz, New York University Medical Center, New York, New York 10016

Among various models of hadrons, the relativistic oscillator formalism is compatible with the existing rules of quantum mechanics and relativity, and provides a simple language for describing the basic hadronic phenomena including the mass spectra, form factor behaviors, and the parton phenomenon. The remaining question is whether this formalism corresponds to a representation of the Poincaré group. The purpose of this note is to point out that the answer to this question is "yes", and that the internal quark motion determines covariantly the mass and spin of the hadron.

We consider here a model hadron consisting of two quarks bound together by a harmonic oscillator potential of unit strength, and write down the wave function

$$\phi(X,x) = \psi(x,P)\ \exp(-iP\cdot X), \tag{1}$$

where X and P are the hadronic position and momentum variables respectively, and x measures the space-time separation between the quarks. The wave function $\psi(x,P)$ describes the internal quark motion, and satisfies the Lorentz-invariant differential equation

$$(\partial_\mu^2 - x_\mu^2)\psi(x,P) = 2\lambda\psi(x,P). \tag{2}$$

The P dependence of this wave function comes through the subsidiary condition

$$P^\mu a_\mu^\dagger\ \psi(x,P) = 0, \quad \text{where} \quad a_\mu^\dagger = x_\mu + \partial_\mu . \tag{3}$$

The point is that the operator $P^\mu a_\mu^\dagger$ commutes with the invariant Casimir operators of the Poincaré group for the total wave function. For this reason, the physical wave functions satisfying the subsidiary condition are diagonal in the Casimir operators of the Poincaré group.

The physical basis for the covariant oscillator formalism can be found in our review papers: Y. S. Kim and M. E. Noz, Am. J. Phys. 46, 480, 484 (1978). The detailed mathematical calculations are given in Y. S. Kim, M. E. Noz, and S. H. Oh, Univ. of Maryland CTP Tech. Rep. #78-097 (1978).

A SIMPLE BROKEN-SYMMETRY TREATMENT OF THE GROUND AND RADIALLY EXCITED STATE MESONS AND THE $\Delta^+ \to P\gamma$ DECAY

Milton Dean Slaughter

Theoretical Division, Los Alamos Scientific Laboratory, University of California, Los Alamos, New Mexico 87545/USA

I. INTRODUCTION

In this article, I wish to consider the ground state and radially excited state pseudoscalar and vector mesons and also the $\Delta^+ \to P\gamma$ decay in the context of a simple algebraic model which is of a non-perturbative form.[1] The model makes use of the well-known Gell-Mann SU(N) $\otimes$ SU(N) (N=2 or 4 in this article) algebras, assumed to be valid in broken symmetry. In addition, the dynamical concepts of algebraic (level) realization and asymptotic SU(N) are introduced and utilized in the infinite momentum frame limit.[1,2]

Special attention is given to deriving a number of mass-mixing angle constraints involving the 0^{-+} and the 1^{--} mesons in the ground state and radially excited states[3] and to relating the form factors G_M^* and G_E^* which describe the $\Delta^+ \to P\gamma$ decay[4] to $g_A(0)$ (the nucleon axial-vector coupling constant) and k_p and k_N (the proton and neutron anomalous magnetic moments).

II. A BRIEF REVIEW OF THE THEORETICAL FORMALISM USED

(A) Asymptotic SU(N)---In plain english, asymptotic SU(N) states that an SU(N) multiplet classification of physical particles can be accomplished in the infinite momentum frame with appropriate particle mixings.[1] Mathematically, if $a_j(\bar{k},\lambda)$ is a SU(N) representation annihilation operator, $a_\alpha(\bar{k},\lambda)$ is a physical particle annihilation operator and the Gell-Mann algebra holds, namely $[V_i, V_j] = i f_{ijk} V_k$, $[V_i, A_j] = i f_{ijk} A_k$, $[A_i, A_j] = i f_{ijk} V_k$ then

$$[V_i, a_j(\bar{k},\lambda)] = i f_{ijk}\, a_k(\bar{k},\lambda),$$

$$[V_i, a_\alpha(\bar{k},\lambda)] = i \sum_\beta u_{i\alpha\beta}(\bar{k},\lambda)\, a_\beta(\bar{k},\lambda) + \delta u_{i\alpha}(\bar{k}), \tag{1}$$

where $\bar{k} \equiv$ momentum, $|\bar{k}| \to \infty$, $\lambda \equiv$ helicity, and $\delta u_{i\alpha}(\bar{k}) \to |\bar{k}|^{-(1+\epsilon)}, \epsilon > 0$. Thus, the a_α are physical operators such that $a_\alpha^+|0\rangle = |\alpha\rangle$, where $\alpha = \pi, K, \Lambda, \Sigma$ etc. one finds that

$$a_\alpha(\bar{k},\lambda) = \sum_j C_{\alpha j}(\lambda)\, a_j(\bar{k},\lambda) \qquad |\bar{k}| \to \infty. \tag{2}$$

The $C_{\alpha j}(\lambda)$ will contain any necessary particle mixing parameters. The important thing to note about Eq.(2) is that the transformation between the "group" particle space and the "physical" particle space is linear in the infinite momentum frame limit. Note also that the consequences of Coleman's theorems[5] are avoided as $\langle B_1(\bar{k})|A_i|B_2(\bar{k})\rangle = \langle B_1(\bar{0})|O_{|\bar{k}|}^{-1} A_i O_{|\bar{k}|}|B_2(\bar{0})\rangle = \langle B_1(\bar{0})|\hat{A}_i|B_2(\bar{0})\rangle$, $|\bar{k}| \to \infty$ where, $O_{|\bar{k}|}$ is the boost operator in the $\bar{k}$ direction and $\hat{A}_i$ is now a null-plane charge. The linear relationship between the physical and group representation annihilation operators then implies that matrix elements of the vector and axial charges in the infinite momentum limit can be described (i.e. parametrized) as in exact SU(N) group symmetry. This in turn gives rise to the relations $[\dot{V}_\alpha, V_\beta] = [\dot{V}_\alpha, A_\beta] = 0$, the "exotic" commutation relations (C.R.'s) where $(\alpha,\beta) = \{(K^0,K^0), (D^0,\pi^-), (D^+,D^0), \text{etc.}\}$.[6]

(B) Algebraic (level) realization--- The concept of algebraic realization[1,2,3] utilizes the idea that hadrons occur in nature as members of multiplets, which are in turn grouped into levels. Thus, for instance, the algebras $[A_{\pi^+}, A_{\pi^-}] = 2V_{\pi^0}$ or $[[J^\mu(0), A_{\pi^+}], A_{\pi^-}] = 2J_3^\mu(0)$ ($J^\mu \equiv$ electromagnetic current) can be inserted between the hadronic states (at infinite momentum) $\langle \alpha_s(\bar{k},\lambda,\ell_e)|$ and $|\beta_t(\bar{k},\lambda',\ell_e)\rangle$, where s,t, = J^{PC} or J^P, λ, λ' = helicity, and ℓ_e specifies the level of the external states under consideration. The right hand side of the resulting equation is denoted by the quantity $g(\alpha_s,\lambda,\ell_t;\beta_t,\lambda',\ell_e)$. One can then sandwich a complete set of level states specified by ℓ_i (one-particle states) internal to the C.R.'s. The right hand side is now proportional to the quantity $f(\alpha_s,\lambda,\ell_e;\beta_t,\lambda',\ell_e;\ell_i,\lambda_i)$. One achieves "realization" for fixed s,t,ℓ_e,ℓ_i and λ_i by varying α and β but demanding that f be invariant. On the other hand, one obtains "super-realization" for fixed ℓ_e, ℓ_i, and λ_i by varying α, β, s, and t and demanding that f be invariant.

III. CONFIGURATION MIXING OF THE GROUND STATE AND RADIALLY EXCITED 0^{-+} AND 1^{--} MESONS IN SU(4)

I define $\langle\rho^0|A_{\pi^-}|\pi^+\rangle\equiv M_1$, $\langle\rho^0|A_{\pi^-}|\tilde{\pi}^+\rangle\equiv M_2$, $\langle\tilde{\rho}^0|A_{\pi^-}|\pi^+\rangle\equiv M_3$ and $\langle\tilde{\rho}^0|A_{\pi^-}|\tilde{\pi}^+\rangle\equiv M_4$ and similarly $\langle D^{*0}|A_{\pi^-}|D^+\rangle\equiv X_1$, $\langle D^{*0}|A_{\pi^-}|\tilde{D}^+\rangle\equiv X_2$ etc. where "$\sim$" denotes a radially excited state. In my model, I assume that: (1) SU(2) is a good symmetry, so that $\theta_{\rho^0\tilde{\rho}^0}=\theta_{\rho^-\tilde{\rho}^-}=\theta_{\rho^+\tilde{\rho}^+}\equiv\theta_{\rho\tilde{\rho}}$ and $\theta_{\pi^0\tilde{\pi}^0}=\theta_{\pi^-\tilde{\pi}^-}=\theta_{\pi^+\tilde{\pi}^+}\equiv\theta_{\pi\tilde{\pi}}$ etc., (2) $\theta_{D\tilde{D}}=\theta_{F\tilde{F}}\equiv\tilde{\theta}$; (3) $\theta_{D^*\tilde{D}^*}=\theta_{F^*\tilde{F}^*}\equiv\theta$; (4) $\theta_{\pi\tilde{\pi}}=\theta_{K\tilde{K}}\equiv\theta_1$, $\theta_{\rho\tilde{\rho}}=\theta_{K^*\tilde{K}^*}\equiv\theta_2$ and (5) the ground state and radially excited state mesons form two distinct levels. Using the C.R.'s $[V_{\pi^+},A_{\pi^-}]=2A_{\pi^0}$, $[V_{K^0},A_{\pi^-}]=0$ (and G-parity, i.e. SU(2) is good), it is easy to show that

$$
\begin{aligned}
&\langle K^{*+}|K^0\rangle=\langle K^{*0}|K^+\rangle=(\rho^+\rho^0)^{-1}\langle\rho^0|\pi^+\rangle,\\
&\langle K^{*+}|\tilde{K}^0\rangle=\langle K^{*0}|\tilde{K}^+\rangle=(\rho^+\rho^0)^{-1}\langle\rho^0|\tilde{\pi}^+\rangle,\\
&\langle D^{*0}|\tilde{D}^+\rangle=\langle D^{*+}|\tilde{D}^0\rangle\equiv X_1,\\
&\langle D^{*+}|\tilde{D}^0\rangle=\langle D^{*0}|\tilde{D}^+\rangle\equiv X_2,
\end{aligned}
\tag{3}
$$

where $\langle K^{*0}|\tilde{K}^+\rangle\equiv\langle K^{*0}|A_{\pi^-}|\tilde{K}^+\rangle$, etc. and $(\rho^+\rho^0)\equiv\langle\rho^+|V_{\pi^+}|\rho^0\rangle$ etc.. The C.R. $[V_{D^0},A_{\pi^-}]=0$ taken between the eight sets of states $\{\langle D^{*0}|,|\pi^+\rangle\}$, $\{\langle D^{*0}|,|\tilde{\pi}^+\rangle\}$, $\{\langle\tilde{D}^0|,|\rho^+\rangle\}$, $\{\langle\tilde{D}^0|,|\tilde{\rho}^+\rangle\}$, etc. yields the four equations

$$X_i=\sum_j A_{ij}M_j\ . \tag{4}$$

For instance,

$$X_1=(\tilde{D}^+\tilde{\pi}^+)(D^{*0}\rho^0)M_1-(\tilde{D}^+\tilde{\pi}^+)(D^{*0}\rho^0)M_2+(\tilde{D}^+\tilde{\pi}^+)(D^{*0}\tilde{\rho}^0)M_3-(\tilde{D}^+\pi^+)(D^{*0}\tilde{\rho}^0)M_4\ . \tag{5}$$

Realization of the algebra $[A_{\pi^+},A_{\pi^-}]=2V_{\pi^0}$ yields the four equations

$$M_i^2=2X_i^2\ ,\quad i=1,2,3,4\ , \tag{6}$$

Since in the limit of no configuration mixing we obtain $M_i=+\sqrt{2}\,X_i$, I choose $M_i=\sqrt{2}\,X_i$. Equations (4) and (6) then imply that $\sum B_{ij}M_j=0$. For a non-trivial solution, one has that $|B_{ij}|=0\Rightarrow|\tilde{\theta}'|\equiv|\tilde{\theta}-\theta_2|=|\theta-\theta_2|=|\theta'|$, $M_4=\varepsilon M_1$, $M_2=-\varepsilon M_3$, $\varepsilon=\pm1$. Defining $A\equiv(D^{*2}-D^2)+(\tilde{D}^{*2}-\tilde{D}^2)-(\tilde{\rho}^2-\tilde{\pi}^2)-(\rho^2-\pi^2)$, $R\equiv M_3/M_1$, $\varkappa\equiv\cos\theta'/\sin\theta'$ ($D^{*2}\equiv m^2_{D^*}$, etc.) and considering the exotic C.R. $[\dot{V}_{D^0},A_{\pi^-}]=0$ inserted between the states $\{\langle D^{*0}|,|\pi^+\rangle\}$, $\{\langle D^{*0}|,|\tilde{\pi}^+\rangle$, etc., I find that

$$A=0\ ,\quad R=\left[\{(D^{*2}-D^2)-(\rho^2-\pi^2)\}/\{(\tilde{D}^{*2}-D^2)-(\rho^2-\tilde{\pi}^2)\}\right]\varkappa\ . \tag{7}$$

Taking $(D^{*2}-D^2)-(\rho^2-\pi^2)\approx0$ from experiment and noting that $(\tilde{D}^{*2}-D^2)-(\rho^2-\tilde{\pi}^2)=0$ only if $\tilde{\pi}\lesssim0.19$ Gev (assuming that $\tilde{D}^*\gtrsim D^*\approx$ 2.01 Gev) we obtain $R\equiv M_3/M_1=0$ and $D^{*2}-D^2=\tilde{\rho}^2-\tilde{\pi}^2=\tilde{D}^{*2}-\tilde{D}^2$. Thus PCAC then implies that $\Gamma(\rho'\to\pi\pi)=0$, where $\rho'\equiv\tilde{\rho}$ and there is no restriction on the value of $\varkappa$ except that $\varkappa\neq\infty$ (i.e. $\theta'\neq0$).[7]

IV. THE $\Delta^+\to p\gamma$ DECAY

I consider realization of the C.R.'s $[[J^\mu(0),A_{\pi^+}],A_{\pi^-}]=2J_3^\mu(0)$ and $[J^\mu(0),A_{\pi^+}]=A^\mu_{\pi^+}$ in the hypercharge = 1 sector (SU(2) sector) of the ground state baryons.[1] So only the $\Delta^{++},\Delta^+,\Delta^0,\Delta^-,P$ and N are external and internal states. In this way, one mitigates the problem of mixing with higher lying states.[8] Defining $\langle P\uparrow|A_{\pi^+}|N\uparrow\rangle\equiv g_A(0)\equiv-\sqrt{1/2}\,(F-\sqrt{3}D)$, $\langle P\uparrow|A_{\pi^-}|\Delta^{++}\uparrow\rangle\equiv-\sqrt{6}\,H$, $\langle\Delta^+\uparrow|A_{\pi^-}|\Delta^{++}\uparrow\rangle\equiv-\sqrt{3/2}\,G$, $Z\equiv H/F$, $Y\equiv G/F$, $\langle P\rangle\equiv\langle P\uparrow|J^\mu(0)|P\downarrow\rangle$ $\langle N\rangle\equiv\langle N\uparrow|J^\mu(0)|N\downarrow\rangle$, $\chi_1\equiv Z\langle P\uparrow|J^\mu(0)|\Delta^+\downarrow\rangle$, $\chi_2\equiv Z\langle P\downarrow|J^\mu(0)|\Delta^+\uparrow\rangle$, $\langle\Delta^+\rangle\equiv\langle\Delta^+\uparrow|J^\mu(0)|\Delta^+\downarrow\rangle$, $\langle\Delta^0\rangle\equiv\langle\Delta^0\uparrow|J^\mu(0)|\Delta^0\downarrow\rangle$

where "$\uparrow$" denotes helicity $=+\frac{1}{2}$ and "$\downarrow$" denotes helicity $=-\frac{1}{2}$. Then eight equations are obtained. For example, $\langle p\uparrow|[[J^{M}(0),A_{\pi^+}],A_{\pi^-}]|p\downarrow\rangle = -\langle N\uparrow|[[J^{M}(0),A_{\pi^+}],A_{\pi^-}]|N\downarrow\rangle$ which reduces to $(\langle p\rangle+\langle N\rangle)(8z^2+\alpha^2)+(x_1+x_2)(5Y-\alpha)+(\langle\Delta^+\rangle+\langle\Delta^0\rangle)(-8z^2)=0$.
Solving these equations, I find that

$$\langle N\rangle = -\langle p\rangle \;,\; \langle\Delta^0\rangle = -\langle\Delta^+\rangle \;,\; x_2 \equiv -x_1 \;,$$

$$\langle\Delta^+\rangle = -(H/2G)\langle p|\Delta^+\rangle \;,\; \langle p|\Delta^+\rangle = [(-4HG)/(G^2+2H^2-\sqrt{2}\,g_A(0)\,G)]\langle p\rangle \equiv \delta\langle p\rangle \tag{8}$$

$$3G^3+(-4\sqrt{2}\,g_A(0))G^2+(2g_A^2(0)-4H^2)G+(4\sqrt{2}\,g_A(0)H^2)=0 .$$

In order to translate Eqs.(8) into statements about form factors, one writes
$T_\mu \equiv \langle p(\bar{p}),\lambda=+\frac{1}{2}|j_\mu(0)|\Delta^+(p^*),\lambda=-\frac{1}{2}\rangle = e\,\bar{u}(\bar{p},\tfrac{1}{2})\Gamma_{\beta\mu}\,u^\beta(\bar{p}^*,-\tfrac{1}{2})$ where

$$\Gamma_{\beta\mu} = (G_M^*-G_E^*)\,\varepsilon_{\beta\mu\sigma\tau}\,p^{*\sigma}p^\tau + i\,\frac{3(m^*+m)}{2mm^{*2}q_c^2}\,G_E^*\,\varepsilon_{\beta\sigma}(p^*p)\,\varepsilon_{\mu\sigma}(p^*p)\,\gamma_5$$

$m=m_p$, $m^*=m_\Delta$, q_c = center of mass momentum, and u^β is a Rarita-Schwinger spinor.
Write $\tilde{T}_\mu \equiv \langle p(\bar{p})\,\lambda=+\frac{1}{2}|j_\mu(0)|p(\tilde{p}^*)\,\lambda=-\frac{1}{2}\rangle = e\,\bar{u}(\bar{p},\tfrac{1}{2})[(F_1+k_pF_2)\gamma_\mu-(k_pF_2/2m)(p+\tilde{p}^*)_\mu]u(\tilde{p}^*,-\tfrac{1}{2})$
where $p^*=(p^{*0},0,0,|\bar{p}^*|)$, $p=(p^0,|\bar{p}|\sin\theta,0,|\bar{p}|\cos\theta)$, $\tilde{p}^*=(\tilde{p}^{*0},\bar{p}^*)$, $(\tilde{p}^{*0})^2=m^2+|\bar{p}^*|^2$.
Choose $\mu=0$ and take the non-collinear infinite momentum limit with $p^3=\lambda(p^*)^3$ such that $(p^*)^3\to\infty$, $p^1\to 0$, $\lambda\to 1$, $q^2=0$, $q\equiv p^*-p$.
Then
$$G_M^*-G_E^* = \delta\,[\sqrt{3}\,(\mu_p-1)/(\mu_p)]\,(2/3)\sqrt{2}\,\mu_p$$

To evaluate δ, I take $\Gamma(\Delta^{++}\to p\pi)=(6H^2q_c^3)/(4\pi f_\pi^2)=99$ Mev, $m_\Delta=1.211$ Gev, q_c = 246 Mev, $f_\pi=0.132$ Gev, and $g_A(0)$ = +1.25 .[9] Eqs.(8) then yield H^2 = 0.244, G = -0.347, and δ = 0.561. Thus,

$$G_M^*(1-G_E^*/G_M^*) \cong 0.624\,(2/3)\sqrt{2}\,\mu_p \tag{9}$$

Assuming that G_E^*/G_M^* = 0, one obtains[10]

$$G_M^* \cong 0.624\,\tfrac{2}{3}\sqrt{2}\,\mu_p . \tag{10}$$

Now $\mu^* \equiv \sqrt{3/2}\,\sqrt{m^*/m}\,G_M^*$ and thus $\mu^* = 0.868\,\frac{2}{3}\sqrt{2}\,\mu_p$

where μ^* is the conventional $\Delta^+ p$ transition magnetic moment. This should be compared to the experimental value of $\mu^*=(0.9\pm 0.1)(2/3)\sqrt{2}\,\mu_p$ obtained by considering the Δ contribution to the Cabibbo-Radicati sum rule.[11]

When one considers that mixing of the ground state baryons--$(56,0^+)$ in the SU(6) classification-- with their excited state counterpart has been neglected the results for the proton and neutron anomalous magnetic moments are good. The predictions for the Δ 's should be even better if the dominant mixing occurs with a $(70,0^+)$ supermultiplet since I treat only the SU(2) sector. Finally, one should recognize that the success of current algebra realization with the use of hadronic levels gives further impetus to the hope that the algebraic structure of the transformation between current and constituent quarks may one day be completely understood.

[1] S. Oneda, these proceedings and references therein.
[2] S. Oneda, Jung S. Rno, and Milton D. Slaughter, Phys. Rev. D17, 1389 (1978).
[3] Milton D. Slaughter and S. Oneda, Phys. Rev. Lett. 39, 309 (1977).
[4] R.C.E. Devenish, T.S. Eisenschitz, and J.G. Korner, DESY Report No. 75/48, (1975)
[5] S. Coleman, Phys. Lett. 19, 144 (1965);J. Math. Phys. 7, 787 (1966).
[6] H. Hallock, S. Oneda, and Milton D. Slaughter, Phys. Rev.D15, 884 (1977).
[7] Milton D. Slaughter,(to be published).
[8] S. Oneda, Milton D. Slaughter, T. Tanuma, (to be published).
[9] See S.R. Borenstein et al., Phys. Rev. D9 , 3006 (1974);S. Oneda and E. Takasugi, Phys. Rev. D10, 3113 (1974) and Milton D. Slaughter and S. Oneda, Phys. Rev. D14, 1314 (1976) for the broken symmetry decay reate formula values for and
[10] R.L. Walker, Phys. Rev. 182, 1729 (1969).
[11] Frederick J. Gilman and Inga Karliner, Phys. Rev. D10, 2194 (1974).

Self-Dual Gauge Fields and Space-Times

Joshua N. Goldberg

Syracuse University

Syracuse, New York 13210

1. Introduction

Self-dual solutions of gauge fields are of interest to field theorists mainly because of their possible application to the description of confined quark states. However, to workers in general relativity, these fields arose naturally from R. Penrose's twistor description of free fields.[1,2] In Minkowski space this led to Richard Ward's[3] encoding of the gauge fields into a line bundle structure over the projective twistor space PT $\cong$ CP^3. In general relativity this led to E. T. Newman's construction of H-space,[4,5] to J. Plebanski's construction of self-dual solutions of the complex Einstein equations,[6] and to Roger Penrose's derivation of self-dual solutions from the differentiable structure of a holomorphic deformation of twistor space.[7] Apart from the interesting mathematical properties being uncovered, the interest in these self-dual solutions is their possible use in an alternative approach to the quantization of general relativity.

This paper describes an approach to self-dual gauge fields which is different from what has been published. The goal is to obtain directly from the field equations the fiber bundle construction of gauge fields given by Richard Ward. As a by-product, one sees how to generalize E. T. Newman's generation of self-dual Maxwell fields from the electromagnetic news function on complex null infinity[8] $\mathcal{I}^+$ and a similar construction for gauge fields.[9] The application of these ideas to the Einstein equation gives the deformed twistor space ($\mathcal{T}$) constructed by Roger Penrose. However, the present approach allows one to understand the role of the differentiable structure of $\mathcal{J}$ and lets one define the holomorphic curves in $P\mathcal{J}$ corresponding to a point in the complex space-time manifold.

One finds that the gauge fields in Minkowski space may be obtained from the solution of a single differential equation on the complex sphere - what Newman[9] calls the "Sparling equation". The Sparling equation also arises in general relativity, but I have not shown that the equation is sufficient to yield a solution for the Einstein equations. However, an explicit coordinatization of the deformed twistor space $\mathcal{J}$ can be obtained. In the following we first outline the general construction for an arbitrary gauge field in Minkowski space. Details of this work will appear in a paper to be submitted for publication soon. The application of the argument to general relativity, as far as it has been carried out, is then presented.

2. Gauge-Fields in Complex Minkowski Space ($\mathcal{M}$)

The formalism works with a gauge field A_μ which arises from a local GL(n, $\mathbb{C}$) symmetry group $\mathcal{G}$. Thus, the A_μ are a set of n x n matrices which vary from point to point over the four dimensional complex manifold $\mathcal{M}$. They form the affine connection on the principal fiber bundle which locally is $\mathcal{M} \times \mathcal{G}$. To construct a self-dual field, up to a factor of i, one sets

$$F^{(-)\mu\nu} = F^{\mu\nu} + i\, F^{*\mu\nu} = 0$$

where $\quad F_{\mu\nu} = \partial_\nu A_\mu - \partial_\mu A_\nu + [A_\nu, A_\mu]$

and $F^{*\mu\nu}$ is the usual dual field.

Through each point of $\mathcal{M}$, there is a one parameter family of anti-self-dual planes defined by

$$V^{\mu\nu}(\tilde{\zeta}) = \ell^{\mu}(\zeta,\tilde{\zeta})\, m^{\nu}(\zeta,\tilde{\zeta}) - \ell^{\nu}(\zeta,\tilde{\zeta})\, m^{\mu}(\zeta,\tilde{\zeta})$$

where ζ and $\tilde{\zeta}$ are stereographic coordinates on the complex sphere and ℓ^{μ} and m^{ν} are null vectors as defined by Newman.[9,10] Then for fixed $\tilde{\zeta}$, the self duality condition implies that there exists $G(x,\zeta,\tilde{\zeta}) \in \mathcal{G}$ such that

$$A_{\mu}(x) = G^{-1}\,\partial_{\mu}G - B\ell_{\mu} - Cm_{\mu}$$

If $G(x,\zeta,\tilde{\zeta})$ is known as a function of ζ and $\tilde{\zeta}$ then B and C are determined by

$$C = \ell^{\mu}\,\tilde{\eth}(G^{-1}\,\partial_{\mu}\,G)$$

$$B = -m^{\mu}\,\tilde{\eth}(G^{-1}\,\partial_{\mu}\,G)$$

and we have the conditions

$$\ell^{\mu}\,\eth\,(G^{-1}\,\partial_{\mu}\,G) = m^{\mu}\,\eth\,(G^{-1}\,\partial_{\mu}\,G) = 0.$$

$\eth$ and $\tilde{\eth}$ are the usual spin-weighted differential operators on the sphere.[11] This can be shown to lead to the "Sparling equation"

$$G(x,\zeta\,\tilde{\zeta}) = V(u,v,\zeta,\zeta)\; G(x,\zeta,\tilde{\zeta})$$

with $u = x^{\mu}\ell_{\mu}$ and $v = x^{\mu}m_{\mu}$. Satisfaction of the Sparling equation is both necessary and sufficient for A_{μ} defined above to give a self-dual field. One can then apply these results to obtain the line bundle construction of R. Ward.[3]

3. General Relativity

Assume that through each point of space-time, $\mathcal{G}$, there is a one parameter family of anti-self-dual two surfaces as before. By requiring that the affine connection be integrable in all surface defined by $V^{\mu\nu}(\tilde{\zeta})$ one has that the Einstein equations must be satisfied and the conformal tensor must be self-dual. Furthermore, for the set of surfaces defined by a fixed value of $\tilde{\zeta}$, there exist coordinates $x^{\mu'}(x,\zeta,\tilde{\zeta})$ such that

$$\Gamma^{\mu}_{\rho\sigma} = \frac{\partial x^{\mu}}{\partial x^{\mu'}} \frac{\partial^2 x^{\mu'}}{\partial x^{\rho}\partial x^{\sigma}} + A^{\mu}\ell_{\rho}\ell_{\sigma} + B^{\mu}m_{\rho}m_{\sigma} + C^{\mu}(\ell_{\rho}m_{\sigma} + \ell_{\sigma}m_{\rho}).$$

Again if one assumes $x^{\mu'}$ is known as a function of ζ and $\tilde{\zeta}$, one can determine A^{μ}, B^{μ}, and C^{μ} in terms of $\tilde{\eth}$ acting on $\frac{\partial x^{\mu}}{\partial x^{\mu'}} \frac{\partial^2 x^{\mu'}}{\partial x^{\rho}\partial x^{\sigma}}$ in a manner similar to the gauge field example. Similarly one can show that the Sparling equation is satisfied by $\frac{\partial x^{\mu'}}{\partial x^{\rho}}$. One can then reconstruct the deformed projective twistor space with differentiable structure by which Penrose encodes the non-linear graviton.

Details of this work is being prepared in two papers which will be submitted for publication.

References

1. R. Penrose and M. A. H. MacCallum, Physics Reports 6C, 272 (1973).

2. R. Penrose, "Twistor Theory, Its Aims and Achievements", Quantum Gravity (eds. C. J. I. Shaw, R. Penrose, and D. W. Sciama, Oxford University Press, Oxford, 1975).

3. R. S. Ward, Phys. Lett. 61A, 81 (1977).

4. E. T. Newman, "The Bondi-Metzner-Sachs Group: Its Complexification and Some Related Curious Consequences", General Relativity and Gravitation, (eds. J. Rosen and G. Shaviv, John Wiley and Sons, New York, 1975).

5. M. Ko, E. T. Newman, and K. P. Tod, "H-Space - A New Approach", Asymptotic Structure of Space-Time (eds. F. P. Esposito and L. Witten, Plenum Press, New York, 1977).

6. J. F. Plebanski, J. Math. Phys. 16, 2395 (1975).

7. R. Penrose, GRG 7, 31 (1976).

8. M. Ko, M. Ludvigsen, E. T. Newman, and K. P. Tod, "The Theory of H-Space", preprint (1977).

9. E. T. Newman, "On Source-Free Yang-Mills Theories, preprint (1978).

10. E. T. Newman and R. Penrose, J. Math. Phys. 7, 863 (1966).

11. J. N. Goldberg, et. al., J. Math. Phys. 8, 2155 (1967).

TORSION AND QUANTUM GRAVITY

Andrew J. Hanson* and Tullio Regge**

Lawrence Berkeley Laboratory
University of California
Berkeley, CA 94720, U.S.A.

Istituto di Fisica
Universita di Torino
I-10125 Torino, ITALY
and
The Institute for Advanced Study
Princeton, New Jersey 08540, U.S.A.

Abstract: We suggest that the absence of torsion in conventional gravity could in fact be dynamical. A gravitational Meissner effect might produce instanton-like vortices of nonzero torsion concentrated at four-dimensional points; such torsion vortices would be the gravitational analogs of magnetic flux vortices in a type II superconductor. Ordinary torsion-free spacetime would correspond to the field-free superconducting region of a superconductor; a dense phase of "torsion foam" with vanishing metric but well-defined affine connection might be the analog of a normal conductor.

INTRODUCTION

The discovery of the instanton solution [1] to Euclidean SU(2) Yang-Mills theory has led to a new understanding of the Yang-Mills path integral quantization [2] and a new nonperturbative picture of the vacuum [3]. The instanton is characterized by finite action and by a self-dual field strength which is concentrated at a four-dimensional point and falls off rapidly in all directions.

The interesting properties of the Yang-Mills instanton naturally led to a search for analogs in Einstein's theory of gravitation. The most promising gravitational instanton seems to be the metric II of Eguchi and Hanson [4]. This Euclidean metric has self-dual Riemannian curvature concentrated at the origin of the manifold, is asymptotically flat, and has vanishing action. Furthermore, it can be shown to contribute to the spin 3/2 axial anomaly [5]. However, the manifold of this metric is not asymptotically Euclidean (S^3 at ∞), but is only asymptotically <u>locally</u> Euclidean (ALE); the 3-manifold at ∞ is S^3 with opposite points identified [6]. Gibbons and Hawking [7] have now found an entire series of "multiple gravitational instanton" solutions which asymptotically describe S^3 modulo the cyclic group of order k; the curvatures for these solutions are self-dual, have maxima in the interior of the manifold, and fall off rapidly at ∞. Hitchin [8] has in fact suggested the existence of an even larger set of manifolds admitting self-dual

* Research supported in part by the High Energy Physics Division of the United States Department of Energy.

** Research supported in part by the National Science Foundation Grant No. 40768X.

metrics which asymptotically describe all possible lens spaces of S^3: S^3 modulo the cyclic group of order k, the dihedral group of order k, the tetrahedral group, the cubic group and the icosahedral group. These manifolds may provide a complete classification of asymptotically locally Euclidean gravitational instantons with self-dual curvature.

In summary, we list the parallels between the Yang-Mills instanton solution of ref. [1] and the Einstein metric of ref. [4]:

Yang-Mills [1]	Einstein [4]
Self-dual field strength	Self-dual curvature
$A_\mu \rightarrow$ pure gauge at ∞	$g_{\mu\nu} \rightarrow$ ALE at ∞
No singularities	Geodesically complete
Finite action	Zero action
Gives spin 1/2 anomaly	Gives spin 3/2 anomaly.

GRAVITATIONAL MEISSNER EFFECT

The concept of gravitational instanton which we have just discussed involves classical Euclidean vacuum Einstein solutions with localized bumps in the curvature. We would now like to explore the idea that other interesting instanton-like objects might occur in a more general Euclidean Einstein-Cartan theory of gravity. The Cartan structure equations for a manifold with a metric are

$$ds^2 = \sum_{a=0}^{3} e^a e_a = g_{\mu\nu}(x)dx^\mu dx^\nu$$

$$\text{torsion} = T^a = de^a + \omega^a{}_b \wedge e^b$$

$$\text{curvature} = R^a{}_b = d\omega^a{}_b + \omega^a{}_c \wedge \omega^c{}_b , \tag{1}$$

where the e^a are vierbein one-forms and the $\omega^a{}_b$ are the connection one-forms. The standard Einstein theory is based upon the use of the Levi-Civita connection on a Riemannian manifold. The torsion-free condition and the metricity condition,

$$T^a = 0 , \qquad \omega^a{}_b = -\omega^b{}_a , \tag{2}$$

uniquely determine the Levi-Civita connection.

Now we would like to ask whether the condition that the torsion T^a vanish is necessarily fundamental. Is it possible that this could be a <u>dynamical effect</u> of a more general theory, much as the vanishing of the field in a superconductor is a dynamical effect of the Landau-Ginzberg theory?

In order to exhibit the possible parallels between gravity and superconductivity, we recall that the Landau-Ginzberg theory contains a Maxwell field coupled to a scalar field obeying the equation of motion

$$D_\mu D^\mu \varphi = g\,\varphi(|\varphi|^2 - \lambda^2). \tag{3}$$

Near the broken-symmetry vacuum, the magnitude of φ is constant,

$$|\varphi| = \lambda\,. \tag{4}$$

Far from walls and impurities, one finds in the static limit that φ is covariant constant,

$$D_\mu \varphi = (\partial_\mu + i\,e\,A_\mu)\varphi = 0. \tag{5}$$

Applying a second covariant differentiation to Eq. (5), we find

$$[D_\mu, D_\nu]\varphi = i\,e\,F_{\mu\nu}\,\varphi = 0\,, \tag{6}$$

so we conclude that either φ or $F_{\mu\nu}$ must vanish. In a type II superconductor, we obtain the Meissner effect: $F_{\mu\nu} = 0$ almost everywhere with the exception of Abrikosov vortices. Therefore we have

$$\begin{aligned} A_\mu &= -\frac{i}{e}\,\partial_\mu\,\theta \\ \varphi &= \lambda\, e^{i\theta}, \end{aligned} \tag{7}$$

where θ is a phase. Now consider a circular path parametrized by $0 \leqslant \theta \leqslant 2\pi$. Such a circle S^1 necessarily encloses a vortex line since as the circle shrinks to zero, the map $S^1 \to U(1)$ changes homotopy type; this is possibly only if $\varphi = 0$ somewhere inside the circle.

A gravitational analog of the Meissner effect might therefore arise from some object which is a <u>covariant constant</u> in a "superconducting region" where Einstein's theory of gravity is valid. We suggest that the appropriate object is the vierbein one-form itself, whose covariant constancy implies <u>vanishing torsion:</u>

$$T^a = D\,e^a = 0\,. \tag{8}$$

Requiring the exterior derivative of the torsion also to vanish results in the <u>cyclic identity</u>:

$$dT^a = 0 \rightarrow R^a{}_b \wedge e^b = 0 \rightarrow \epsilon_{ebcd} R^a{}_{bcd} = 0 . \qquad (9)$$

Hence ordinary spacetime, which is torsion-free and whose curvature satisfies the cyclic identity, would be analogous to the field-free region of a superconductor.

We are thus led to the following table showing the parallels between gravitation and superconductivity:

	Gravity	Superconductivity
Bundle Group	$SO(4) \subset GL(4, \mathbb{R})$	$U(1) \subset \mathbb{C}$
Connection 1-form	ω_{ab}	A
Field Strength	$R_{ab} = d\omega_{ab} + \omega_{ac} \wedge \omega_{cb}$	$F = dA$
Covariant Constant Object	$\left\{ \begin{array}{l} e^a{}_\mu \in GL(4, \mathbb{R}) \\ T^a = De^a = 0 \end{array} \right.$	$\left\{ \begin{array}{l} \varphi \in \mathbb{C} \\ D\varphi = 0 \end{array} \right.$
Meissner Effect	$R^a{}_b \wedge e^b = 0$	$F\varphi = 0$
Topological Object	Torsion Vortex (Pointlike)	Magnetic Vortex (Line)
Enclosing Manifold	S^3 = three-sphere	S^1 = circle
Inside Vortex	$T^a \neq 0, \quad R^a{}_b \wedge e^b \neq 0$	$F \neq 0$
Vortex Map	Iwasawa Decomposition of $e^a{}_\mu$	$U(1)$: $\varphi = \lvert\varphi\rvert e^{i\theta}$
Topological Charge	Euler and Pontrjagin numbers	Quantized Flux of F
Fundamental Equation	?	Landau-Ginzberg

It is very important to realize that $T^a = 0$ is the condition which locks together the $SO(4)$ gauge group and the group of coordinate transformations. $T^a = 0$ determines a rigid relation between the $SO(4)$ principal bundle and the tangent bundle of the underlying manifold. In this case, the topological invariants of the $SO(4)$ bundle are identifiable with the Euler characteristic and Hirzebruch signature of the base manifold. The latter correspondence may no longer be true if $T^a \neq 0$; as we shall now show, near a (point) torsion vortex, the $SO(4)$ principal bundle is unglued from the tangent bundle and the topological invariants of the two bundles no longer coincide. A torsion vortex in a localized region of Euclidean spacetime would thus combine essential features of both gravitational and Yang-Mills instantons.

CONNECTIONS WITH TORSION

We can make a simple example of a system with torsion by modifying the Levi-Civita connection of a flat Euclidean metric. Let $\{x_\mu\}$ be coordinates on $\mathbb{R}^4$. We let $\rho^2 = x_\mu x_\mu$ and introduce flat-space vierbeins in a nonstandard coordinate system,

$$
\begin{aligned}
e^0 &= 2\, x_\mu\, dx_\mu = d(\rho^2) \\
e^1 &= x_0\, dx_1 - x_1\, dx_0 + x_2\, dx_3 - x_3\, dx_2 \\
&= \rho^2\, \sigma_x\,, \qquad \text{cyclic.}
\end{aligned} \tag{10}
$$

The one-forms e^a obey the structure equations

$$
de^0 = 0, \qquad de^i = \rho^{-2}(e^0 \wedge e^i + \epsilon_{ijk}\, e^j \wedge e^k) \tag{11}
$$

and the σ_k can be expressed in polar coordinates on S^3 if desired. We remark that our coordinate system was chosen to give vierbeins e^a which are C^∞ (have infinitely differentiable coefficients). The Levi-Civita connection one-forms obtained by applying Eqs. (1) and (2) to Eq. (10) are self-dual and singular at the origin:

$$
\omega^i{}_0 = \omega^j{}_k = \rho^{-2}\, e^i\,, \qquad \text{cyclic.} \tag{12}
$$

The torsion vanishes by construction and the curvature vanishes because the metric was in fact flat.

Now we may introduce torsion by choosing a new regularized connection,

$$
\omega^i{}_0 = \omega^j{}_k = \varphi(\rho^2) e^i\,, \qquad \text{cyclic,} \tag{13}
$$

where $\varphi(\rho^2)$ is some C^∞ function regular at the origin and falling like $1/\rho^2$ at infinity:

$$
\varphi(\rho^2) \xrightarrow[\rho^2\to 0]{} \text{(regular)}, \qquad \varphi(\rho^2) \xrightarrow[\rho^2\to\infty]{} 1/\rho^2\,. \tag{14}
$$

The torsion two-forms become

$$T^0 = 0, \qquad T^i = (-\varphi + 1/\rho^2)[e^0 \wedge e^i + \epsilon_{ijk}\, e^j \wedge e^k] \tag{15}$$

and so fall off rapidly away from the origin. The curvatures are

$$R^1{}_0 = R^2{}_3 = (\varphi' + \varphi/\rho^2)e^0 \wedge e^1 + 2(\varphi/\rho^2 - \varphi^2)e^2 \wedge e^3 , \quad \text{cyclic.} \tag{16}$$

It can now be verified that both the torsion and the curvature are C^∞ forms everywhere, even at the origin. Clearly the connections with torsion do not provide a solution to the vacuum Einstein equations. The appearance of nonvanishing torsion is a necessary consequence of the presence of a <u>zero</u> in the vierbein and the regularization of the connection at the origin.

A typical choice for φ such as

$$\varphi = \rho^{-2}(1 - e^{-\rho^2/\lambda}) \tag{17}$$

in fact gives finite action,

$$\int R\, g^{\frac{1}{2}} d^4x = \frac{1}{4}\int_M R^{ab} \wedge e^c \wedge e^d\, \epsilon_{abcd} = \lambda^2/16 . \tag{18}$$

(If one chooses φ so that $\omega^a{}_b$ corresponds to the SU(2) Yang-Mills instanton connection, the action turns out to be infinite.) The topological invariants of our manifold and the bundle for which (13) is the connection are now drastically altered. So long as $\varphi \neq \rho^{-2}$, we find that the "Euler characteristic" volume term and surface corrections are [9]

$$\chi_V = -1 , \qquad \chi_S = 1 . \tag{19}$$

While our original flat $\mathbb{R}^4$ connection had $\chi = 1$, our new system has

$$\chi = \chi_V + \chi_S = 0 . \tag{20}$$

Similarly, we find for the new Pontrjagin number

$$P_1 = -\frac{1}{8\pi^2}\int_M \mathrm{Tr}\, R \wedge R = +2 . \tag{21}$$

The flat connection had Hirzebruch signature $\tau = P_1/3 = 0$. The altered topological invariants prove that our new SO(4) principal bundle is not isomorphic to the bundle of orthonormal tangent-space frames of Euclidean space.

CONCLUSIONS

We remark that there may be two phases of the system we have described. Since the presence of too many magnetic field vortices will cause a superconductor to undergo a phase transition to the normal state with $\langle \varphi \rangle = 0$, we can also conceive of a dense phase of torsion vortices with $\langle e^a{}_\mu \rangle = 0$ everywhere. In the presence of this "torsion foam," we may no longer have a sensible metric; only the SO(4) affine connection would remain well-defined. Physically, the "torsion foam" phase might dominate in the early universe or at very short distances [10].

The problem now is to develop a dynamical theory which has a stable torsion vortex as a solution. Such theories could of course include mechanisms (e.g. localized nonmetricity) more general than those discussed here. We would expect a theory with torsion vortices in the Euclidean regime to have a profound effect on our understanding of quantum gravity and its relation to elementary particle physics.

Acknowledgement: We are grateful to the Theory Group at CERN and to the members of the Center for Particle Theory of the University of Texas at Austin for their generous hospitality.

REFERENCES

1. A. A. Belavin, A. M. Polyakov, A. S. Schwarz and Yu. S. Tyupkin, Phys. Lett. 59B (1975) 85.
2. G. 't Hooft, Phys. Rev. Lett. 37 (1976) 8; Phys. Rev. D14 (1976) 3432.
3. R. Jackiw and C. Rebbi, Phys. Rev. Lett. 37 (1976) 172; C. Callan, R. Dashen and D. Gross, Phys. Lett. 63B (1976) 334.
4. T. Eguchi and A. J. Hanson, Phys. Lett. 74B (1978) 249; T. Eguchi and A. J. Hanson, submitted to Ann. Phys. (N. Y.).
5. S. W. Hawking and C. N. Pope, "Symmetry Breaking by Instantons in Supergravity" (DAMTP preprint); A. J. Hanson and H. Römer, "Gravitational Instanton Contribution to Spin 3/2 Axial Anomaly" (CERN, TH 2564).
6. V. A. Belinskii, G. W. Gibbons, D. N. Page and C. N. Pope, Phys. Lett. 76B (1978) 433.
7. G. W. Gibbons and S. W. Hawking, in preparation (private communication from S. W. Hawking).
8. N. Hitchin, private communication.
9. S. S. Chern, Ann. Math. 46 (1945) 674; see also T. Eguchi, P. B. Gilkey and A. J. Hanson, Phys. Rev. D17 (1978) 423.
10. See S. W. Hawking, "Spacetime Foam" (DAMTP preprint) for a similar proposal in the more restrictive context of Einstein's theory of gravity.

SURFACE DEFORMATIONS, THEIR SQUARE ROOT AND THE SIGNATURE OF SPACETIME

Claudio Teitelboim

The Institute for Advanced Study
Princeton, New Jersey 08540, U.S.A.

Abstract

The formulation of general relativity in terms of surface deformations and its extension to supergravity are briefly reviewed. The role of the spacetime signature ε is discussed. It is pointed out that ε may be used as a perturbation parameter. The "free" theory corresponds to $\varepsilon = 0$, which is halfway between hyperbolic ($\varepsilon = -1$) and Euclidean ($\varepsilon = +1$) spacetime.

1. Introduction

It is a great pleasure to report at this session dedicated to the memory of Alfred Schild, whom I admired not only for his ability as a physicist but also for his human qualities and especially for the encouragement he continuously gave to others.

I would like to review briefly(1) an approach to general relativity in which spacetime, in the sense of a four dimensional Riemannian manifold, is not regarded as a fundamental element of the theory, but it is rather considered as an object which could be modified, should the need arise.

In the scheme to be described attention is focused on three-dimensional space more than on spacetime, a point of view forcefully put forward by John Wheeler [2]. The main mathematical object in the formulation is not, however, three-space itself. Rather, it is the idea of a deformation of three-space which plays a central role. [3].

A key property of the deformations is that they obey a closed set of commutation relations. This property guarantees that the associated dynamical theory is "generally covariant" in the sense that,

(1) For a more extensive review including a more complete list of references than space allows here, see [1].

given the initial conditions, there appear four arbitrary functions of time for each space point in the corresponding solution of the equations of motion [4].

Only for one particular set of "structure constants" in the commutation relations of the deformations, can the different choices of arbitrary functions be interpreted as corresponding to different cuts through a single Riemannian manifold. In this sense the spacetime structure is contained in the algebra of the deformations. In other words, only for one choice of the algebra does spacetime exist.

However, that particular choice is by no means unique. Thus by shifting the emphasis from spacetime to the deformations of an embedded hypersurface, it becomes possible to modify the spacetime structure (by changing the algebra) without throwing overboard the whole mathematical structure of the theory.

There is in fact an interesting modification of the algebra which suggests itself and which will be discussed below under the at this point mysterious name, "the case of signature zero." In that case the generators of surface deformations are simplified by dropping a term which enters one of them as a "potential." That simplification changes the algebra of the generators and hence breaks the usual spacetime structure. Consequently the modification of general relativity thus obtained is vastly different from, and practically unthinkable in terms of, the usual changes in Einstein's theory which replace the Hilbert action by a different one without changing the underlying geometrical structure.

2. Surface Deformations and Many-time Field Theory

To start the analysis, one envisages a cloud of observers, one per space point, determining a three dimensional surface. The observers look at the evolution of the geometry of the surface (and also at the evolution of any fields defined on the surface) and make a record of it. Furthermore, the observers at each point may bifurcate and meet again in the future to compare their records. They may ask then whether their observations may be summarized as describing different slicings of a four dimensional Riemannian manifold. Not every law of evolution will permit such an interpretation and it is precisely the purpose of this section to write down the conditions which the Hamiltonian of a field theory must satisfy in order for the "embeddability interpretation" to be possible.

When looked upon from the standpoint of the enveloping spacetime, the requirement that the observations made by different families of observers should be amenable to an interpretation in terms of different cuts through that spacetime may be reformulated as follows: The idea of bifurcation and re-encounter of the observers is reexpressed by saying that the data on some given initial surface (where the bifurcation took place) are propagated to some given final surface (where the re-encounter arises) along different intermediate routes or "paths," each path being a one parameter family of surfaces that fills the "sandwich" in between the initial and the final surface. "Embeddability" becomes then synonymous with "path independence": the change in the field variables during the evolution from a given initial surface to a given final surface must be independent of the particular sequence of intermediate surfaces used in the actual evaluation of this change.

In order to formulate mathematically the consequences of path independence, we need at least a general form for the law of evolution from one surface to an infinitesimally altered one. We shall assume that such an evolution law may be given in Hamiltonian form. This is quite a strong assumption, but it looks like a reasonable one. Furthermore the concept of "rate of change as seen by the observer at the point x" translates, when looked upon from the enveloping spacetime, into the idea of "change induced by a deformation of the hypersurface localized at the point x."

Our starting point is then the dynamical law,

$$\delta F = \int dx\{[F,\mathcal{H}_{\perp}(x)]\delta\xi^{\perp}(x) + [F,\mathcal{H}_i(x)]\delta\xi^i(x)\} \equiv \int dx[F,\mathcal{H}_{\mu}]\delta\xi^{\mu} , \tag{2.1}$$

which gives the change in a generic functional F of the canonical variables of the theory under a deformation $\delta\xi$ of the hypersurface in terms of its Poisson bracket with the corresponding generators.

A generic deformation is characterized by giving its tangential components $\delta\xi^i$ (with respect to some coordinate system on the surface) and its normal component $\delta\xi^{\perp}$. Note that we leave open the signature of the time direction ($\varepsilon = n_{\alpha}n^{\alpha}$), for we want to see the role played by this quantity in the "embeddability conditions." The use of a coordinatized surface (with coordinates x) is not a point of principle, but it is a convenient mathematical technique; the conclusions will always be independent of the parametrization of the surface.

To obtain the conditions for path independence it is sufficient to consider the following chain of deformations: from the initial surface σ one goes to an infinitesimally altered surface σ_1 by a deformation $\delta\xi$. Next one goes from σ_1 to another surface σ' by a second deformation $\delta\eta$. If the two deformations are performed in reversed order, one arrives at a final hypersurface σ'' which is in general different from σ ; hence there is a non-zero deformation $\delta\zeta$ that directly connects σ' with σ'' .

Now, by a purely geometrical argument based on the assumption that all the hypersurfaces are embedded in a common four dimensional Reimannian manifold, the compensating deformation may be shown to be of the form:

$$\delta\zeta^{\gamma}(x'') = \int dx \int dx' \kappa_{\alpha\beta}{}^{\gamma}(x'';x,x')\,\delta\xi^{\alpha}(x)\,\delta\eta^{\beta}(x')$$

$$+ O[(\delta\xi)^2] + O[(\delta\eta)^2] \quad , \tag{2.2}$$

where the only nonvanishing κ's are

$$\kappa_{r\perp}{}^{\perp}(x'';x,x') = -\kappa_{\perp r}{}^{\perp}(x'';x',x) = \delta(x'',x)\,\delta_{,r}(x'',x') \quad , \tag{2.3a}$$

$$\kappa_{mn}{}^{r}(x'';x,x') = -\kappa_{nm}{}^{r}(x'';x',x) =$$

$$= \delta(x'',x)\,\delta_{,m}(x'',x')\,\delta^{r}_{n} - \delta(x'',x')\,\delta_{,n}(x'',x)\,\delta^{r}_{m} \quad , \tag{2.3b}$$

$$\kappa_{\perp\perp}{}^{r}(x'';x,x') = -\kappa_{\perp\perp}{}^{r}(x'';x',x) =$$

$$= \varepsilon g^{rs}(x'')\,[\delta(x'',x')\,\delta_{,s}(x'',x) - \delta(x'',x)\,\delta_{,s}(x'',x')]. \tag{2.3c}$$

These purely geometrical results become restrictions on the Hamiltonian generators of a field theory through the requirement of path independence of the dynamical evolution. In fact repeated application of (2.1) yields

$$\int dx \int dx' \left[F,[H_{\alpha}(x),H_{\beta}(x')]\right]\delta\xi^{\alpha}(x)\,\delta\eta^{\beta}(x')$$

$$= \int dx''\,[F,H_{\gamma}(x'')]\,\delta\zeta^{\gamma}(x'') \quad . \tag{2.4}$$

Now, if the "structure constants" $\kappa_{\alpha\beta}{}^{\gamma}$ had zero Poisson brackets with all the canonical variables, the condition

$$[H_{\alpha}(x),H_{\beta}(x')] = \int dx''\,\kappa_{\alpha\beta}{}^{\gamma}(x'';x,x')\,H_{\gamma}(x'') \quad . \tag{2.5}$$

would imply (2.4) and would thus guarantee "embeddability" by itself. However, $\kappa_{\perp\perp}{}^{r}$ given by (2.3c) depends on the metric g_{rs} of the surface, which is a dynamical variable in general relativity [5] and also in Dirac's parametrized field theories [4]. (Incidentally, for this same reason the $\kappa_{\alpha\beta}{}^{\gamma}$ cannot be regarded as being the structure constants of a group.)

Equations (2.5) have therefore to be supplemented by the constraints

$$H_r = 0 \quad . \tag{2.6a}$$

The constraints (2.6a) have in turn to be preserved under surface deformations. In particular, the change of H_r under a purely normal deformation has to vanish. We then obtain with the help of (2.1), (2.3) and (2.5),

$$H_{\perp} = 0 \quad . \tag{2.6b}$$

Thus, we arrive at the important conclusion that, once (2.5) is accepted, the path independence requirement implies that the Hamiltonian generators H_μ must be constrained to vanish whenever the metric of the surface is a dynamical variable.

3. Construction of the Hamiltonian

Equations (2.5) are the conditions for the dynamical evolution determined by the generators H_μ to be consistent with the geometrical structure associated with the algebra of deformations. If expressions (2.3) are used for the κ's, that structure corresponds to a Reimannian spacetime. In that case, when written in detail (2.5) read [3,4,6]:

$$[H_{\perp}(x), H_{\perp}(x')] = -\varepsilon(g^{rs}(x)H_s(x) + g^{rs}(x')H_s(x'))\delta,_r(x,x') \quad , \tag{3.1a}$$

$$[H_r(x), H_{\perp}(x')] = H_{\perp}(x)\delta,_r(x,x') \quad , \tag{3.1b}$$

$$[H_r(x), H_s(x')] = H_r(x')\delta,_s(x,x') + H_s(x)\delta,_r(x,x') \quad . \tag{3.1c}$$

Now, although our scheme is formally symmetric in all four generators, there is a great difference between $H_{\perp}$ and the H_r. In fact, the only generator of dynamical importance is $H_{\perp}$. The three H_r generate displacements which lie on the original surface and which amount merely to a change of spatial coordinates. Thus, once the behavior of the canonical variables under reparametrizations of three-space is specified, H_r can be written down. Hence, one may regard relations (3.1) as a set of equations for <u>the unknown</u> $H_{\perp}$. It turns out that (3.1b,c) are easy to satisfy. They simply say that $H_\perp$ and H_r must behave under changes of spatial coordinates as scalar and vector densities respectively. The really restrictive equation is (3.1a).

In order to determine $H_\perp$ from (3.1a) it is necessary to specify first the fields from which it will be constructed so that one can write down H_r. In any case, it is necessary that $H_\perp$ should depend on the metric g_{rs} of three-space. This is so because g^{rs} appears on the right hand side of (3.1a).

It is natural to look for an $H_\perp$ which is built solely from the g_{rs} regarded as canonical coordinates and of their canonically conjugate momenta π^{rs}. Such a solution indeed exists and it is unique under some reasonable assumptions. It is the Hamiltonian of Einstein's theory of gravitation [5] when no matter is present. The explicit expressions are:

$$H_i^{grav} = - 2 \pi_i{}^j{}_{;j} \tag{3.2}$$

and

$$H_\perp^{grav} = G_{ijk\ell}\pi^{ij}\pi^{k\ell} + \varepsilon\, g^{1/2}(R-2\lambda) \quad , \tag{3.3}$$

with

$$G_{ijk\ell} = \frac{1}{2}\, g^{-1/2}(g_{ik}g_{j\ell} + g_{i\ell}g_{jk} - g_{ij}g_{k\ell}) \quad . \tag{3.4}$$

Here ε is the signature, λ is the cosmological constant, R is the spatial curvature and the semicolon denotes covariant differentiation in g_{rs}.

The "embeddability conditions" (3.1) are thus enormously powerful. In fact, we have been able to deduce from them the Hamiltonian form of Einstein's theory directly, without recourse to a Lagrangian or any other four-dimensional quantity.

The approach also works well when matter fields are included. In the case of a vector field, which is among the most interesting examples, equations (3.1) lead naturally to gauge invariance and to the Yang-Mills Hamiltonian for any semi-simple Lie group:

$$H_r^{YM} = F_{rs}{}^a\, \pi_a{}^s \quad , \tag{3.5}$$

$$H_\perp^{YM} = \frac{1}{2}\, g^{-1/2}\, \pi_a{}^r \pi_{ar} - \frac{1}{4}\, \varepsilon\, g^{1/2}\, F_{rs}{}^a\, F^{rsa} \quad , \tag{3.6}$$

where a is a group index and

$$F_{rs}{}^{a} = A^{a}_{s,r} - A^{a}_{r,s} - C^{a}{}_{bc}\, A^{b}_{r}\, A^{c}_{s} \quad , \tag{3.7}$$

$C^{a}{}_{bc}$ being the structure constants of the gauge group.

The generators for the coupled theory are obtained by adding (3.5) to (3.2) and (3.6) to (3.3) respectively.

Lastly, the introduction of fermions is of special interest. The key point is the observation that the equations

$$\mathcal{H}_{\mu} = 0 \quad , \tag{3.8}$$

are the analog of

$$\mathcal{H} = p_{\mu}\, p^{\mu} + m^{2} = 0 \quad , \tag{3.9}$$

for a relativistic particle. The analogy is a precise one since the worldline of the particle may be regarded as the history of a space of dimension zero, whereas spacetime is the history of a space of dimension three. The function $\mathcal{H}$ defined by (3.9) generates deformations of the zero-dimensional space along the worldline, much in the same way as the $\mathcal{H}_{\mu}$ generate deformations of three-space along spacetime.

Now, spin degrees of freedom were introduced by Dirac in the particle case by taking the square root of (3.9), that is, by finding a function $\mathcal{S}$ which would yield $\mathcal{H}$ upon anticommutation with itself.

$$\{\mathcal{S},\mathcal{S}\} = \mathcal{H} \quad . \tag{3.10}$$

The form of $\mathcal{S}$ is well known; it is just the Dirac operator multiplied by γ_5 and divided by the square root of two:

$$\mathcal{S} = 2^{-1/2}\, \gamma_5 (i\gamma_{\mu} p^{\mu} + m) \quad . \tag{3.11}$$

The spin degrees of freedom show up through the new dynamical variables γ_{μ}, γ_5 which need to be introduced in order to satisfy (3.10) and which must obey anticommutation relations as a consequence.

A similar method may be followed here: one may extract the square root of the four generators $\mathcal{H}_{\mu}$ by introducing four functions $\mathcal{S}_{A}(x)$ ("supersymmetry generators") such that [7],

$$\{\mathcal{S}_{A}(x), \mathcal{S}_{B}(x')\} = \delta(x,x') \left(\delta_{AB}\, \mathcal{H}_{\perp} + \alpha^{i}\, \mathcal{H}_{i}\right) \quad , \tag{3.12}$$

where the α^i are the Dirac alpha matrices. The construction of the functions S_A requires the presence of new variables $\psi_i{}^A$ which obey anti-commutation rules and which may be thought of as the components of a massless spin three halves field. The S_A have the form [8]:

$$S = \alpha_i \psi_j \pi^{ij} + \text{"curl } \psi\text{"} \quad , \tag{3.13}$$

and the resulting theory is supergravity [9], obtained again directly in Hamiltonian form. Thus, in a literal sense, supergravity is the square root of general relativity.

After the square root is taken, the number of generators in the theory is enlarged. Besides the bosonic H_μ one now has the fermionic S_A . What has effectively been done is to enlarge the algebraic structure of the deformations by grading it.

4. The Role of the Signature

The signature ε appears explicitly in the closure relation (3.1a) and in the expressions (3.3,6) for $H_\perp$. This is an important property of general relativity. It implies, on account of the constraint $H_\perp = 0$, that the signature of the four dimensional manifold which emerges after solving the equations of motion is already determined by the initial value data. Therefore, initial data suitable for an Euclidean spacetime are not acceptable for a hyperbolic one and vice versa.

The parameter ε may take a continuous range of values. However, in what concerns the geometrical structure of the solutions of the equations of motion, there are only three essentially different cases: hyperbolic spacetime ($\varepsilon = -1$), Euclidean spacetime ($\varepsilon = +1$) and the case of signature zero ($\varepsilon = 0$). This last case is halfway between the previous two and is of a drastically different nature.

To see that there are just three basically different values of it is sufficient to observe that[(2)] if $\{g_{rs}(x,t),\ \pi^{rs}(x,t),\ N^\perp(x,t), N^s(x,t)\}$ is a solution of the equations of motion (2.1) and the constraints (3.8), then $\{g_{rs}(x,t), |\varepsilon|^{-1/2}\pi^{rs}(x,t)\ ,\ |\varepsilon|^{1/2}N^\perp(x,t), N^s(x,t)\}$, solves the equations of motion and the constraints with $\varepsilon = \pm 1$ depending on whether the

(2) We have employed here a more customary notation (Arnowitt, Deser and Misner, [5]) for the components of the deformation which connects two neighboring surfaces belonging to a one parameter family. If t is the parameter, then $\delta\xi^\mu = N^\mu \delta t$. It may be shown that $N^\perp = (\varepsilon g^{00})^{-1/2}$, $N^r = g^{rs} g_{s0}$.

the original ε was positive or negative. Thus the theories with $\varepsilon \neq 0$ can be brought to $\varepsilon = \pm 1$ by a (real) rescaling. The argument fails when $\varepsilon = 0$, a case which forms therefore a different class by itself.

Now, even though the actual physical world clearly does not have signature zero, the $\varepsilon = 0$ case might be a useful mathematical tool, for example as the starting point for a perturbation analysis of the full ($\varepsilon = -1$) theory. In fact, setting ε equal to zero in the gravitational generator (3.3) amounts to dropping the "potential term" $g^{1/2}R$. The Hamiltonian thus obtained appears to be considerably simpler than the original one, since it exhibits no coupling between different points of space (see in this context [10]). As a matter of fact, more than a field Hamiltonian, it resembles the one of a single particle moving in a manifold with curved metric $G_{ijk\ell}$ [11].

The simplifications inherent in the signature zero case are not restricted to the purely gravitational case. Putting ε equal to zero in the Yang-Mills generator (3.6) also has the effect of dropping the "potential" $\frac{1}{4}F_{rs}{}^{a}\,F^{rsa}$ and of reducing the problem to that of a particle moving on a curved manifold (the metric has this time extra diagonal entries $g^{-1/2}g_{rs}\delta_{ab}$ corresponding to the new "coordinates" A^{a}_{r}). Furthermore, the simplification still carries through in the supergravity case. Taking zero signature there amounts to eliminating the "curl ψ" term from the supersymmetry generator (3.12). The problem becomes then one of a spinning supersymmetric single particle moving on a curved geometry [12].

The unconventional feature of an approximation procedure in powers of the signature is that the starting point $\varepsilon = 0$ does not correspond to a Riemannian spacetime, but to a different sort of structure which lies halfway between hyperbolic and Euclidean spacetime.(3) Whether or not this possibility leads to something fruitful remains to be seen.

ACKNOWLEDGMENT

This work was supported in part by U.S. National Science Foundation Grant No. PHY77-20612 to the Institute for Advanced Study. The author holds an Alfred P. Sloan Research Fellowship.

(3) Note, incidentally, that when $\varepsilon = 0$, the metric g^{rs} drops out from the commutation relations (2.2) and, therefore in that case, the deformations form a group properly speaking. Furthermore, the "normal" deformations form then an Abelian subgroup.

Added Note: The idea of dropping the curvature term in $H_\perp$ and of using the resulting theory as the starting point for a perturbation expansion has been previously considered in an interesting paper by C.J. Isham [13]. No mention is made there, however, of the role of the spacetime signature in this context.

REFERENCES

[1] C. Teitelboim, in Einstein Centenary Volume (A. Held, ed., GRG Journal, in press).

[2] J.A. Wheeler, in Battelle Rencontres 1967 (C.M. DeWitt and J.A. Wheeler, eds., Benjamin, New York, 1968); also in Analytical Methods in Mathematical Physics (R.P. Gilbert and R. Newton, eds., Gordon and Breach, New York, 1970).

[3] C. Teitelboim, Ph.D. Thesis, Princeton University, 1973, unpublished. Ann. Phys. (N.Y.) 79, 542 (1973); S. Hojman, K. Kuchař and C. Teitelboim, Nature, Phys. Sci. 245, 97 (1973); Ann. Phys. (N.Y.) 96, 88 (1976).

[4] P.A.M. Dirac, Can. J. Math. 2, 129 (1950); 3, 1 (1951); Lectures on Quantum Mechanics (Academic, N.Y., 1964).

[5] P.A.M. Dirac, Proc. Roy. Soc. (London) A246, 326 (1958); R. Arnowitt, S. Deser and C.W. Misner, in Gravitation, an Introduction to Current Research (L. Witten, ed., Wiley, New York, 1962).

[6] P.A.M. Dirac, Phys. Rev. 73, 1092 (1948); J. Schwinger, Phys. Rev. 127, 324 (1962); J. Katz, C. R. Acad. Sci. (France) 254, 1386 (1962).

[7] C. Teitelboim, Phys. Rev. Lett. 38, 1106 (1977).

[8] R. Tabensky and C. Teitelboim, Phys. Lett. 69B , 453 (1977).

[9] S. Ferrara, D.Z. Freedman and P. van Nieuwenhuizen, Phys. Rev. D13, 3214 (1976); S. Deser and B. Zumino, Phys. Lett. 62B, 335 (1976).

[10] J.R. Klauder, Phys. Rev. D2, 272 (1970).

[11] B.S. DeWitt, Phys. Rev. 160, 1113 (1967).

[12] A. Barducci, R. Casalbuoni and L. Lusanna, Nuovo Cimento 35A, 377 (1976); F.A. Berezin and M.S. Marinov, Ann. Phys. (N.Y.) 104, 336 (1977). See also C. Teitelboim in Current Trends in the Theory of Fields (Proceedings of a conference held in Tallahassee, Florida in April 1977 on the 50th anniversary of the Dirac equation) and references therein.

[13] C.J. Isham, Proc. Roy. Soc. (London) A351, 209 (1976).

THE ROLE OF GROUP THEORY IN THE QUEST FOR EXACT SOLUTIONS OF EINSTEIN'S FIELD EQUATIONS: SOME RECENT DEVELOPMENTS

Frederick J. Ernst
Illinois Institute of Technology
Chicago, IL 60616

Abstract:

Recent discoveries suggest that now, sixty-two years after the birth of general relativity, the general solution to one physically significant problem, the spinning mass problem, may be close at hand.

INTRODUCTION

It should be a source of embarrassment to practitioners of exact solutions of Einstein's field equations of general relativity that even after the lapse of more than sixty years we still do not know the answers to some of the simplest physical problems which can be formulated. It is unnecessary to refer to the subtleties associated with the description of gravitational radiation; we can in fact underscore the paucity of our knowledge by considering the state of the spinning mass problem. Before 1963 one could have cited no exact solution of Einstein's vacuum field equations corresponding to the exterior gravitational field of a body spinning uniformly about its symmetry axis.[1] Even today essentially only one series of highly specialized spinning mass solutions is known.[2-4]

In spite of the frustration occasioned by the search for exceedingly elusive spinning mass solutions, a number of workers in the field have remained optimistic that one day the clouds of confusion will part and we shall in fact see, if not the general solution of this physical problem, at least the development of systematic procedures for solving particular cases of interest. The optimism of which I speak is to a large extent founded upon the realization that those solutions of the vacuum Einstein equations or the coupled Einstein-Maxwell equations which possess symmetries do not stand alone, but rather they are members of families of solutions related to one another by the action of a transformation group. It has even been conjectured that all stationary axially symmetric vacuum and electrovac solutions belong to one such family; i.e., that all stationary axially symmetric solutions, including the long sought spinning mass solutions, might be constructed from Minkowski space through the operation of a transformation group.[5]

No one has done more to enhance our understanding of the group structure of solution-generating transformations than has W. Kinnersley. In 1973 he showed the relevance of the group SU(2,1) to the problem of transforming one stationary electrovac spacetime into another.[6] In accomplishing this he did not work directly in terms of the metric tensor, but rather he took advantage of the complex potential formalism which I had introduced in 1968.[7] In this description of gravitational fields one employs a three dimensional metric γ_{ab}, a complex gravitational potential $\mathcal{E}$ and a complex electromagnetic potential Φ. The complex potentials are governed by the pair of differential equations

$$(\mathrm{Re}\,\mathcal{E} + \Phi^*\Phi)\, \gamma^{ab}\, \nabla_a \nabla_b \mathcal{E} = \gamma^{ab}\, (\nabla_a \mathcal{E} + 2\, \Phi^* \nabla_a \Phi)\, \nabla_b \mathcal{E} \ ,$$

$$(\mathrm{Re}\,\mathcal{E} + \Phi^*\Phi)\, \gamma^{ab}\, \nabla_a \nabla_b \Phi = \gamma^{ab}\, (\nabla_a \mathcal{E} + 2\, \Phi^* \nabla_a \Phi)\, \nabla_b \Phi \ .$$

Given any stationary electrovac spacetime, it is a straightforward matter to evaluate γ_{ab}, $\mathcal{E}$ and Φ. Conversely, if you have γ_{ab}, $\mathcal{E}$ and Φ, you can reconstruct the full four dimensional metric tensor of the spacetime. What Kinnersley accomplished was the identification of algebraic transformations

$$\mathcal{E}' = \mathcal{E}'(\mathcal{E},\Phi)\ , \quad \Phi' = \Phi'(\mathcal{E},\Phi)\ , \quad \gamma_{ab}' = \gamma_{ab}\ ,$$

under which the above equations for the complex potentials are left invariant. These transformations turned out to provide a nonlinear representation of the group SU(2,1). Key members of the group of algebraic transformations were the Ehlers transformation[8] and the Harrison transformation.[9] The transformation which I had employed in order to provide a logical derivation of the charged Kerr metric from the uncharged Kerr metric[10] was also included. Conspicuously absent was any transformation which would generate the Kerr metric from the Schwarzschild metric, or the Tomimatsu-Sato spinning mass solutions from their static counterparts, the Zipoy-Voorhees metrics (called Weyl metrics by Tomimatsu and Sato).

Because as long ago as 1917 H. Weyl provided the general solution of the static axially symmetric vacuum field equations[11], the discovery of a group of transformations which would map static asymptotically flat solutions into stationary asymptotically flat solutions would be a most significant step forward in the search for new spinning mass spacetimes. A recent series of papers by W. Kinnersley and D. M. Chitre suggests that the attainment of this objective may not be far off.[12] These authors have in fact succeeded in obtaining the Kerr metric from the Schwarzschild metric, and a generalization of the simplest Tomimatsu-Sato solution from the $\delta = 2$ Zipoy-Voorhees metric. Whether or not their method can be applied successfully to static metrics outside the Zipoy-Voorhees family is not yet clear. Nevertheless, I consider this to be one of the most exciting developments pertaining to exact solutions to have taken place during the last decade. Therefore, I should like to summarize for you the basic theory of complex potentials, and to introduce you to the concept of the Kinnersley-Chitre heirarchy of complex potentials, in terms of which the transformation group can be identified. The particular mode of description which I shall employ in this lecture is somewhat different from that employed by Kinnersley and Chitre, and is the result of a collaboration with my colleague, Isidore Hauser.[13]

COMPLEX POTENTIALS

As we explained in Ref. 14, complex potentials arise very naturally when one deals with spacetimes possessing isometries. One need only consider a complex 2-form W such that $dW = 0$ and $\mathcal{L}_{\underline{K}}(W) = 0$, where the latter is the Lie derivative of the 2-form W with respect to the Killing vector field $\underline{K}$. This can be expressed conveniently in terms of contractions, $K \lrcorner W$ and $K \lrcorner dW$, of the vector $\underline{K}$ with the 2-form W and the 3-form dW, viz.,

$$\mathcal{L}_{\underline{K}}(W) = K \lrcorner dW - d(K \lrcorner W) .$$

Under the stated conditions, it follows immediately that $d(K \lrcorner W) = 0$, so the complex 1-form $K \lrcorner W$ must be derivable from a scalar potential.

In the case of an electrovac field with a Killing vector field $\underline{K}$, an obvious candidate for the closed 2-form W is the self-dual part of the electromagnetic field 2-form $F = \frac{1}{2} f_{ab}\, dx^a\, dx^b$. Accordingly, we may identify W with $W^{(M)} := 2\,P\,F$, where P is a projection operator which yields the self-dual part. The complex electromagnetic potential Φ is defined so that $K \lrcorner W^{(M)} = d\Phi$.

The complex gravitational potential $\mathcal{E}$ arises from the consideration of another closed self-dual 2-form $W^{(E)}$ which can be constructed from $\omega = \frac{1}{2}\, dK$, where $K = K_a dx^a$ is the Killing 1-form. Using the field equations one can verify that $W^{(E)} := - 4\, P\, (\omega + \Phi^* F)$ is in fact closed, i.e., $dW^{(E)} = 0$. The potential $\mathcal{E}$ is defined so that $K \lrcorner W^{(E)} = d\mathcal{E}$.

These considerations do not, of course, fix the additive constants in the potentials. It has become customary to choose the additive constants in such a way that $\mathrm{Re}\,\mathcal{E} + \Phi^*\Phi = f := - \underline{K}\cdot\underline{K}$.

When one is concerned with stationary axially symmetric electrovac spacetimes, a matrix generalization of the complex potentials Φ and $\mathcal{E}$ may be introduced, for one then has a pair of commuting Killing vector fields $\underline{K}_A$ distinguished from one another by an index A. It is obvious that in this case there exist two associated complex electromagnetic potentials Φ_A, which may be regarded as components of a 2 x 1 matrix, and four complex gravitational potentials H_{AB}, which may be regarded as components of a 2 x 2 matrix. These are defined up to additive constants by $K_A \lrcorner W^{(M)} = d\Phi_A$ and $K_A \lrcorner W^{(E)}_B = dH_{AB}$. Kinnersley and Chitre restrict the additive constants so that the relation

$$\frac{1}{2}(H + H^\dagger) + \Phi\,\Phi^\dagger = f - i\,\varepsilon\,z \tag{1}$$

is satisfied. Here $\varepsilon = \begin{pmatrix} 0 & 1 \\ -1 & 0 \end{pmatrix}$ and f is a 2 x 2 matrix with components $f_{AB} := - \underline{K}_A\cdot\underline{K}_B$, while z is a real scalar field, which is defined as follows. Let $\rho^2 = - \det f$, and then solve the equation $dz = \circledast\, d\rho$, where $\circledast$ is the two dimensional duality operator, whose formal definition need not concern us in this brief introduction to the subject of complex potentials.

Finally, the equations which govern H and Φ arise directly from the self-duality condition which the W's satisfy. Therefore, we shall refer to these equations,

$$dH = - i \rho^{-1} f \varepsilon \circledast dH , \quad d\Phi = - i \rho^{-1} f \varepsilon \circledast d\Phi , \tag{2}$$

as the self-duality conditions. Kinnersley and Chitre refer to them as the field equations.

KINNERSLEY-CHITRE HEIRARCHY

I must admit that when Kinnersley first began talking about an infinite heirarchy of complex potentials, all of which satisfy the self-duality condition (2), I was very dubious that this was the proper road to new spinning mass solutions. However, the recent successes of Kinnersley and Chitre have forced me to revise my earlier assessment. After my colleague, I. Hauser, and I found that the whole formalism can be cast into a neat matrix language, I became distinctly enthusiastic about the infinite heirarchy idea.

The principal elements of our matrix formulation of the Kinnersley-Chitre theory are a 2 x ∞ matrix E consisting of 2 x 3 blocks $\overset{n}{E} = (\overset{n}{H}, \overset{n}{\Phi})$ and an ∞ x ∞ matrix S consisting of 3 x 3 blocks

$$\overset{mn}{S} = \begin{pmatrix} \overset{mn}{N} & \overset{mn}{M} \\ \overset{mn}{L} & \overset{mn}{K} \end{pmatrix} .$$

Here $\overset{mn}{N}$ is a 2 x 2 matrix, $\overset{mn}{M}$ is a 2 x 1 matrix, $\overset{mn}{L}$ is a 1 x 2 matrix, and $\overset{mn}{K}$ is a scalar field. The fields $\overset{1}{H}$ and $\overset{1}{\Phi}$ are identical with the potentials H and Φ introduced previously. Furthermore, $\overset{0}{H} = i\varepsilon$ and $\overset{0}{\Phi} = 0$, while $\overset{n}{E} = 0$ for all $n < 0$. On the other hand, $\overset{0n}{N} = - i \overset{n}{H}$, $\overset{0n}{M} = - i \overset{n}{\Phi}$, $\overset{0n}{L} = 0$ and $\overset{0n}{K} = \frac{1}{2} i \delta_{n,1}$, while all other elements with m or n less than or equal to 0 vanish, except for

$$\overset{1,-1}{N} = - \overset{-1,1}{N} = \overset{0,0}{N} = \varepsilon ,$$

$$\overset{1,0}{K} = - \overset{0,1}{K} = - \frac{1}{2} i .$$

It should be mentioned that this is one of the points at which we elected a different option from that elected by Kinnersley and Chitre, who defined

$$\overset{p,-p}{N} = - \overset{-p,p}{N} = \overset{0,0}{N} = \varepsilon ,$$

$$\overset{p,1-p}{K} = - \overset{1-p,p}{K} = - \frac{1}{2} i ,$$

for all $p \geqslant 1$, not just for $p = 1$.

The equations which govern the matrices E and S are the following:

$$dE = - i \rho^{-1} f \varepsilon \circledast dE , \tag{3}$$

$$\frac{1}{2i} E G E^\dagger = f - i \varepsilon z , \tag{4}$$

$$dS = E^\dagger \varepsilon dE , \tag{5}$$

$$E A = E G S . \tag{6}$$

The first of these is just the self-duality condition, which imposes no interrelationships among the various columns of the matrix E. The ∞ x ∞ constant matrix G in Eq. (4) consists of 3 x 3 blocks $\overset{mn}{G}$, all of which vanish except for

$$\overset{01}{G} = \overset{10}{G} = \begin{pmatrix} -\varepsilon & 0 \\ 0 & 0 \end{pmatrix}, \qquad \overset{11}{G} = \begin{pmatrix} 0 & 0 \\ 0 & 2i \end{pmatrix}.$$

Thus, Eq. (4) is nothing but Eq. (1) in disguise.

It is Eq. (5) which defines the matrix S except for additive constants. The integrability of this equation is guaranteed by the self-duality condition. Finally, using Eqs. (4) and (5), one can show that E G S satisfies the self-duality condition if E itself does. The matrix A is a unitary matrix which shifts the columns of E three columns to the left. Thus, Eq. (6) defines the potentials $\overset{n}{E}$ for $n > 1$.

Since $\overset{0n}{S}$ contains all the information contained in $\overset{n}{E}$, it is convenient to replace Eqs. (3) - (6) by a set of equations involving S alone; namely,

$$2\,(\rho - z \circledast)\, dS + A^{\dagger} \circledast dS = -\,(S - \mathfrak{E})\, G \circledast dS\,, \tag{7}$$

$$A^{\dagger} S - S A = -\,(S - \mathfrak{E})\, G S - \mathfrak{E} A + H\,, \tag{8}$$

$$dS = S^{\dagger}\, \mathfrak{E}\, dS\,, \tag{9}$$

$$S - S^{\dagger} - \mathfrak{E} = S^{\dagger}\, \mathfrak{E}\, S\,. \tag{10}$$

In these equations, $\mathfrak{E}$ and H are constant $\infty \times \infty$ matrices consisting of 3 x 3 blocks, all of which vanish except for the following:

$$\overset{0,0}{H} = -\overset{-1,1}{H} = -\overset{1,-1}{H} = \begin{pmatrix} 0 & 0 \\ 0 & \frac{1}{2}i \end{pmatrix},$$

$$\overset{-1,0}{H} = \overset{0,-1}{H} = -\overset{-2,1}{H} = -\overset{1,-2}{H} = \overset{0,0}{\mathfrak{E}} = \begin{pmatrix} \varepsilon & 0 \\ 0 & 0 \end{pmatrix}.$$

Certain constant terms in the Kinnersley-Chitre equations corresponding to our Eqs. (7) - (10) appear to have been left out inadvertently.

SOLUTION-GENERATING TRANSFORMATIONS

Tackling the solution of Eqs. (7) - (10) instead of Eqs. (1) and (2) would be absurd, if it were not for the fact that Eqs. (7) - (10) turn out to be invariant under certain algebraic transformations on the matrix S. At the present time only the infinitesimal transformations have been determined completely, although special examples of finite transformations are also known, e.g., the Ehlers and Harrison transformations.

Kinnersley and Chitre inferred from the infinitesimal versions of the Ehlers and Harrison transformations much more general infinitesimal transformations involving an infinite number of adjustable parameters. Under reasonably general assumptions Hauser and I have found that the Kinnersley-Chitre transformations are the only transformations of the form

$$\delta S = C^{(0)} + C^{(L)}\, S + S\, C^{(R)} + S\, C^{(2)}\, S\,, \tag{11}$$

with the C's constant matrices, which leave Eqs. (7) - (10) invariant.

The infinitesimal Kinnersley-Chitre transformations constitute a nonlinear representation of an infinite parameter Lie algebra, all the commutators for which were presented in their papers. In spite of the fact that the general exponentiation of the infinitesimal transformations (11) has not yet been accomplished, Kinnersley and Chitre observed that for certain specific transformations of type (11), using certain specific initial metrics, the effect of a finite transformation could actually be determined. Thus, for example, starting with a heirarchy appropriate to Minkowski space and centering attention upon certain specific transformations of type (11), they were able to show that all static axially symmetric vacuum spacetimes[11] can be obtained. As a result, Geroch's conjecture[5] appears very plausible indeed.

Kinnersley and Chitre also enjoyed success with the Zipoy-Voorhees metrics. In the case of these metrics all the higher potentials in the heirarchy can be expressed in terms of a small number of potentials by relations which are preserved by the transformation of type (11) which was considered. These constraints among the potentials were fed into Eq. (11), thereby reducing the set of equations to a small and manageable set. The final result was that from these static axially symmetric vacuum spacetimes one may obtain asymptotically flat spinning mass fields which are considerably more general than the Tomimatsu-Sato solutions, although of quite a similar character. Kinnersley and Chitre actually carried through the calculation for the $\delta = 2$ Zipoy-Voorhees metric, obtaining a five parameter generalization of the simplest Tomimatsu-Sato solution. The corresponding generalizations of the higher Tomimatsu-Sato solutions should involve an even greater number of parameters. Whether similar techniques will work in the case of metrics outside the Zipoy-Voorhees family is not yet known.

My colleague and I have spent a considerable amount of time reproducing the Kinnersley-Chitre theory in all its details. In the process we attempted to evaluate more carefully the role of the additive constants which arise when one calculates the matrix S for a given initial metric. We believe we have found inconsistencies in the treatment of these constants in the Kinnersley-Chitre papers. The existence of such anomalies in these otherwise excellent contributions probably can be attributed to the exhilaration which must have accompanied the discovery of a new approach to exact spinning mass solutions.

Of course, our primary motivation in developing a convenient matrix formulation of the Kinnersley-Chitre heirarchy has been our desire to initiate a systematic study of the exponentiation problem, for if this last step can be taken the determination of all spinning mass solutions will be within our grasp.

REFERENCES

1. R. P. Kerr, Phys. Rev. Lett. 11, 237 (1963).
2. A. Tomimatsu and H. Sato, Phys. Rev. Lett. 29, 1344 (1972); Progr. Theor. Phys. 50, 95 (1973).
3. M. Yamazaki and S. Hori, Progr. Theor. Phys. 57, 696 (1977).
4. C. M. Cosgrove, J. Phys. A10, 1481 (1977).
5. R. Geroch, J. Math. Phys. 12, 918 (1971); 13, 394 (1972).
6. W. Kinnersley, J. Math. Phys. 14, 651 (1973).
7. F. J. Ernst, Phys. Rev. 167, 1175 (1968); 168, 1415 (1968).
8. J. Ehlers, Les théories relativistes de la gravitation (CNRS, Paris, 1959).
9. B. K. Harrison, J. Math. Phys. 9, 1744 (1968).
10. F. J. Ernst, Phys. Rev. 168, 1415 (1968).
11. H. Weyl, Ann. Physik 54, 117 (1917).
12. W. Kinnersley, J. Math. Phys. 18, 1529 (1977); W. Kinnersley and D. M. Chitre, J. Math. Phys. 18, 1538 (1977); 19, 1926 (1978); 19, 2037 (1978); Phys. Rev. Lett. 40, 1608 (1978); D. M. Chitre, J. Math. Phys. 19, 1625 (1978).
13. We expect to submit a paper to the Journal of Mathematical Physics in the near future.
14. I. Hauser and F. J. Ernst, J. Math. Phys. 19, 1316 (1978).

On Group Covariant Physical Laws and Gravitation

Leopold Halpern
Dept. of Physics, Florida State University
Tallahassee, Florida 32306

Field equations covariant w.r.t. the De Sitter group and the conformal group were originally constructed by Dirac for the matter fields.(1) His method of imbedding of the space-time manifold was recently generalized to obtain a De Sitter covariant general relativistic theory of gravitation.(2) I present here first a general method to obtain the equations of matter fields covariant w.r.t. a semisimple Lie group of suitable signature of the Cartan-Killing metric and discuss conditions when such equations may be applicable to physics. I suggest then a generalization of the mathematical formalism which should lead to the inclusion of gravitational fields.(3)

We consider a semisimple group G_r of transformations of a V_n. The matrix of base vectors: (ξ^i_α) $(i=1\cdots n)$, $(\alpha=1\cdots r)$ is $4<n<r$ so that four-dimensional invariant varieties V_4 exist. The coördinates can be so chosen that $x^1\cdots x^4$ cover each of these V_4 and the $\xi^i_\alpha(x)\neq 0$ only for $i=1\cdots 4$. A metric on these V_4, which satisfies Killings equations is given by: $g^{ik}=\xi^i_\alpha\gamma^{\alpha\beta}\xi^k_\beta$ where $\gamma_{\alpha\beta}=C^\varepsilon_{\alpha\gamma}C^\gamma_{\beta\varepsilon}$ is the Cartan-Killing metric formed with the groups structure constants. Lagrangians of tensor fields can be formed which involve only Lie derivatives of the fields. E.g. for a scalar field ϕ:

$$\mathcal{L} = \sqrt{g}\ \gamma^{\alpha\beta}\xi^i_\alpha\left(\frac{\partial}{\partial x^i}\ \phi^*\right)\xi^k_\beta\left(\frac{\partial}{\partial x^k}\ \phi\right) \tag{1}$$

also g is formed only of ξ_α, $C^\gamma_{\alpha\beta}$.

The Lie derivative of Spinor fields ψ on the V_4 has been constructed in Ref. 3 with a Lie Spinor connection Γ^L_α defined by the equation:

$$\frac{\partial\gamma^i}{\partial x^\ell}\,\xi^\ell_\alpha - \frac{\partial\xi^i_\alpha}{\partial x^\ell}\,\gamma^\ell - [\gamma^i,\ \Gamma^L_\alpha] = 0 \tag{2}$$

where $\{\gamma^i,\gamma^k\}=2g^{ik}$. Γ^L_α is related to the metric Spinor connection Γ_k:

$$\Gamma^L_\alpha - \Gamma_k\ \xi^k_\alpha = \frac{1}{8}\ \sigma^{ik}\left(\frac{\partial}{\partial x^i}\ \xi_{\alpha k} - \frac{\partial}{\partial x^k}\ \xi_{\alpha i}\right) \tag{2a}$$

with $\sigma^{ik}=\frac{1}{2}[\gamma^i,\gamma^k]$. If the generators of the group Gr can be formed in spin space, one can also construct a spinning electron equation on V_4 with Lie derivative: $\psi_{|\alpha}=\xi^k_\alpha\frac{\partial}{\partial x^k}\psi+\Gamma^L_\alpha\psi$ out of group covariant expressions only, without explicit use of the metric.

An invariant variety V_4 of suitable extensions may be chosen as the space-time of a homogenous universe. To remain in agreement with experience, the law of motion of macroscopical bodies requires that every time like geodesics on V_4 should either coincide or be well approximated by a trajectory of the group. This is for example strictly fulfilled in case of the De Sitter group. We have the Theorem:

The trajectory of the symbol ξ_α of our group G_r coincides with a geodesic of V_4 for which G_r is a group of motion iff on every one of its points there exist four linear independent symbols ξ_δ (one of which may be ξ_α itself) such that

$$[\xi_{\alpha|}\xi_{\delta}]^{i}\, g_{ik}\, \xi^{k}_{\alpha} = C^{\gamma}_{\alpha\delta}\, \xi^{i}_{\gamma}\, g_{ik}\, \xi^{4}_{\alpha} = 0 \tag{3}$$

We shall consider here only groups which fulfill this condition.

There remains a family of other time-like trajectories of Gr which do not coincide with geodesics. Must we exclude such paths from the law of free motion? I don't think so, although we may never be able to observe them - just as we don't see a macroscopic system of decreasing entropy. One may explain it in a similar manner: the phase space available in the neighbourhood of a geodesic e.g. a maximal circle in the universe is so much larger than that of a circle of the size of our solar system that a limited number of samples will hardly be found in the latter state. In any case, all the generators are represented in the group covariant wave equations although some because of the large radius of the universe are locally less important. The best example for this speculation is again the situation in the De Sitter space.

The group covariant field equations on V_4 differ in general from the generally covariant equations on V_4, considered as a pseudo-Riemannian space of metric g_{ik}, only by a constant which because of the large extensions of the universe is very small.(1,2)

The formalism has to be generalized to the presence of localized mass inhomogenuities which we expect to perturb the metric. The example of the De Sitter group(2) shows that it is not always an impossible task to maintain the symmetry of the groups action in the local limit. The method considered here(3) introduces symmetry breaking and a general metric in a tempting way, but in its most natural form it results in torsion coupled to the metric, the applicability of which to physics has to be studied.

The generalization emerges from the covariance of the formalism of Lie groups of transformation w.r.t. constant linear transformations of the space of its base vectors.(4)

To the expressions $\xi^{i}_{\alpha}(x)$, $C^{\gamma}_{\alpha\beta}$ which enter in the physical equations one can apply a different linear transformation $C^{\alpha}_{\beta}(x)$ at every point x of V_4 without altering their meaning, if we only replace their derivatives by invariant derivatives:

$$\frac{\partial}{\partial x^{k}}\,\xi^{i}_{\alpha} \rightarrow \xi^{i}_{\alpha\cdot k} = \frac{\partial}{\partial x^{i}}\,\xi^{i}_{\alpha} + A^{\varepsilon}_{\alpha k}\xi^{i}_{\varepsilon} \tag{4}$$

so that

$$C\xi^{i}_{\alpha\cdot k} = (C\xi^{i}_{\alpha})\cdot k \qquad A'^{\beta}_{\alpha k} = C^{\xi}_{\alpha}(A^{\gamma}_{\varepsilon}C^{-1\,\beta}_{\gamma} + \frac{\partial}{\partial x^{k}}\, C^{-1\,\beta}_{\varepsilon}) \tag{4a}$$

This seems to do violence to the formalism in group space but we begin to view the ξ^{i}_{α}, A_k that occur in the physical equations, as physical quantities in a more general context - just as this was done with the metric. The generalized brackets

$$[\xi_{\alpha},\xi_{\beta}] = \xi^{i}_{\alpha\cdot\ell}\xi^{\ell}_{\beta} - \xi^{i}_{\beta\cdot\ell}\xi^{\ell}_{\alpha} = C^{\varepsilon}_{\alpha\beta}\xi^{i}_{\varepsilon}$$

and generalized Killing equation:

$$\frac{\partial g_{ik}}{\partial x^{\ell}} \xi^{\ell}_{\alpha} + g_{\ell k}\xi^{\ell}_{\alpha \cdot i} + g_{i\ell}\xi^{\ell}_{\alpha \cdot k} = 0 \tag{5}$$

remains valid if the modified expressions are inserted.

The new situation occurs if the A_k's are given such values that

$$F^{\beta}_{\alpha ik} = \frac{\partial}{\partial x^k} A^{\beta}_{\alpha i} - \frac{\partial}{\partial x^i} A^{\beta}_{\alpha k} + A^{\varepsilon}_{\alpha i}A^{\beta}_{\varepsilon k} - A^{\varepsilon}_{\alpha k}A^{\beta}_{\varepsilon i} \tag{6}$$

does not vanish. We obtain a generalized metric $g_{ik}(x)$ expressed by the A_k and ξ_{α}. We can form the curvature tensor and Field equations like in ref. 2 and obtain a theory which differs from general relativity w.r.t. the metric only by a cosmological constant.

We have, however, the other interesting possibility to consider the $\xi^{i}_{\alpha}(x)$, $A_k(x)$ together with the matter variables as fields and the matter Lagrangian plus the Lagrangian formed of the F_{ik}: $\sqrt{g}\, F_{ik}F^{ik}$, as our total Lagrangian, in which the matter fields are coupled to the $\xi^{'}_{\alpha}$, A_k fields which occur in the metric. One may request that the $C^{\gamma}_{\alpha\beta}$ are obtainable from those of the group G_r by a linear transformation with a matrix $C^{\beta}_{\alpha}(x)$ and introduce these as further variables. A Lagrangian multiplier can take account of their relation with the ξ_{α}, A_k. The resulting equations can then be written in a gauge where $\xi^{i}_{\alpha}=\delta^{i}_{\alpha}$ $\alpha = 1\cdots4$ and the remaining ξ_{β} vanish.

Restriction to a subgroup of the general linear group e.g. the adjoint group, can reduce redundant potentials. The theory establishes a relation between the broken group symmetry and the energy tensor of matter but it contains other variables besides the metris and only closer investigation will establish its physical content in the different cases.

References:

1. P.A.M. Dirac, Annals of Math. 30, p. 657 (1935) and (1936).
2. L. Halpern, Gen. Relat. & Gravit., V. 8, p. 623 (1977).
3. L. Halpern, "Gravitational Theories Generated by Groups of Transformation and Dirac's Large Number Hypothesis," FSU HEP 780713; SLAC-PUB-2166 July 1978. "Gravitation as the Theory of Broken Symmetry of Intransitive Groups of Transformation," submitted to Gen. Relat. 8 Gr.
4. L.P. Eisenhart, Continuous Groups of Transformation, Princeton (1933).

Acknowledgements

This research was supported in part by U.S. D.O.E. under grant number AT-(40-1)-3509. I thank Prof. P.A.M. Dirac and Prof. J. Lannutti for the possibility to work in their institute. Discussions with R. Parsons are acknowledged.

On the Exact Solutions of Tomimatsu-Sato Family for Stationary Axial-symmetric Gravitational Fields

Shoichi HORI

Department of Physics, Kanazawa University,
Kanazawa 920, Japan

Yamazaki and myself presumed the explicit form of the exact solution of Tomimatsu-Sato family generalized to the case of an arbitrary integral value of the deformation parameter δ. It is proved that the solution satisfies Ernst equation which is equivalent to Einstein equation in the case of stationary axial-symmetric gravitational fields. Actually it is easier to show that the solution satisfies intermediate integrals of Ernst equation. The intermediate integrals can most easily be derived with recourse to $SU(1,1) \cong SO(2,1)$ symmetry of Ernst equation. The explicit forms of metric functions are also obtained.

For details, see S. Hori, Prog. Theoret. Phys., 59 (1978), 1870.

Remarks to Homogeneous Solutions of Einstein's Field Equations*

by

István Ozsváth

The University of Texas at Dallas

I would like to report here about a method of obtaining homogeneous solutions and about some of the results, which justify this activity.

First of all the term "homogeneous" needs clarification. I should probably say "strictly homogeneous" instead. The four-dimensional space is strictly homogeneous, if it is a four-dimensional Lie group endowed with a left invariant metric. Or using the language of the transformation groups:

The four-dimensional space is strictly homogeneous if its maximal group of isometries has a four-dimensional simply transitive subgroup.

As a two-dimensional example:

The plane is strictly homogeneous but S_2 is not strictly homogeneous.

In the sequel I will say homogeneous in the sake of brevity but I always mean strictly homogeneous in the above sense.

A four-dimensional Lie group is a four-dimensional manifold carrying left invariant vector fields. One visualizes this as follows:

Take a vector in a point and carry it into any other point of the manifold by the translations of the group, generating a vector field, the invariant vector field. Since we can start with an arbitrary vector at our arbitrary origin, we see, that the left invariant vector fields form a vector space over the reals. More than that:

The theory says they form a Lie algebra:

Picking a base in the vector space of the invariant vector fields:

$$X_0, X_1, X_2 \quad X_3 \tag{1}$$

we can write

$$[X_a, X_b] = C^c{}_{ab} X_c \tag{2}$$

---

* In memory of Alfred Schild.

Where the structure constant tensor

$$C^c{}_{ab} = -C^c{}_{ba} \tag{3}$$

satisfies the Jacobi identities:

$$C^a{}_{fb}C^f{}_{cd} + C^a{}_{fc}C^f{}_{db} + C^a{}_{fd}C^f{}_{bc} = 0 \quad . \tag{4}$$

One uses base (1) combined with the dual base

$$\omega^0,\ \omega^1,\ \omega^2,\ \omega^3 \tag{5}$$

for the left invariant 1-forms defined by

$$\omega^a(X_b) = \delta^a{}_b \tag{6}$$

and satisfying

$$d\omega^a = -\tfrac{1}{2}C^a{}_{bc}\omega^b \wedge \omega^c \tag{7}$$

in order to span the tensor algebra of the tensor fields over our manifold.

We can impose a left invariant metric on this manifold by declaring that

$$g(X_a, X_b) = g_{ab} = \mathrm{diag}(-1, 1, 1, 1) \tag{8}$$

or alternatively

$$ds^2 = -(\omega^0)^2 + (\omega^1)^2 + (\omega^2)^2 + (\omega^3)^2 \tag{9}$$

this is our homogeneous space. Since the base is given up to Lorentz transformations, we have different homogeneous spaces for different structure constant tensors, tensors with respect to the Lorentz group.

The problem of the classification of the homogeneous spaces is the same as the problem of classifying Lie algebras with respect to the Lorentz group.

This is one part of the problem. The other is the field equations imposed on our metric.

We proceed as follows:

One introduces a connection by

$$\nabla_{X_a} X_b = \Gamma_{ab}{}^c X_c \tag{10}$$

and insists that it should be free of torsion and it should be metric: One obtains that

$$\Gamma_{abc} = \tfrac{1}{2}(C_{bca} + C_{cab} - C_{abc}) \quad , \tag{11}$$

where

$$C_{abc} = g_{af}C^f{}_{bc} \quad . \tag{12}$$

The components of the Riemann and Ricci tensor fields are given by

$$R^a{}_{bcd} = \Gamma_{cf}{}^a \Gamma_{db}{}^f - \Gamma_{df}{}^a \Gamma_{cb}{}^f - \Gamma_{fb}{}^a C^f{}_{cd} \tag{13}$$

and

$$R_{bc} = \Gamma_{fb}{}^g \Gamma_{gc}{}^f - \Gamma_{fg}{}^f \Gamma_{bc}{}^g \tag{14}$$

respectively. Therefore, if one imposes the field equations of Einstein

$$R_{bc} - \tfrac{1}{2} R g_{bc} - \Lambda\, g_{bc} = - \kappa T_{bc} \tag{15}$$

one has a system of quadratic equations, equations (4) and (15) for the 24 unknowns $C^a{}_{bc}$ to satisfy.

One proceeds by classifying the Lie algebras into broad classes by solving completely or partially the equations (4). As a second step, we impose the field equations in order to single out those algebras, which in turn generate the desired solution of Einstein's field equations. The basic ideas of this method originate with

E. L. Schücking, R. P. Kerr and I. Ozsváth.

(Conveniently forgetting about E. Cartan and others). The idea of looking for homogeneous solutions originates with Kurt Gödel.

There is a large body of results. I will mention a few less well-known ones, ignoring famous ones like the Einstein and the Gödel cosmos. The homogeneous solutions are simple; therefore, surveyable and excellent as examples or counter examples in settling basic questions of the theory.

For example Schücking and I wanted to construct a geodesically complete and singularity free vacuum solution in order to have a geniune gravitational field without sources. We called it the Anti Mach Metric (1962) [1].

This is the homogeneous space with the Lie algebra

$$\begin{gathered}[X_3, X_4] = 0,\ [X_4, X_2] = 0\ [X_2, X_3] = 0 \\ [X_2, X_1] = X_4,\ [X_3, X_1] = X_2,\ [X_4, X_1] = 0\end{gathered} \tag{14}$$

and the metric

$$g(X_a, X_b) = \begin{pmatrix} 1 & 0 & 0 & 0 \\ 0 & 1 & 0 & 0 \\ 0 & 0 & 0 & 1 \\ 0 & 0 & 1 & 0 \end{pmatrix} \tag{17}$$

the line element in a properly chosen coordinate system

$$ds^2 = dx^2 + (dy + xdz)^2 + 2dz(xdy + \frac{x^2}{2} dz + dt) \tag{18}$$

as far as I know this is the first example of a source free gravitational field.

This is a Robinson wave and the line element in the normal coordinates takes the form

$$ds^2 = dx^2 + dy^2 + 2du\,dv + [\tfrac{1}{2}(x^2 - y^2)\sin 2u + xy\cos 2u]du^2 \tag{19}$$

There are other vacuum solutions:

$$ds^2 = dx^2 + dy^2 + 2dudv + \{\tfrac{1}{2}(x^2 - y^2)\sin 2bu + xy\cos 2bu\}\,du^2 \tag{20}$$

1. Ozsváth and E. L. Schücking (1962) [1]

$$ds^2 = dx^2 + dy^2 + 2dudv + \frac{a}{(u)^2}\{\tfrac{1}{2}[x^2 - y^2]\sin 2b\ln u + xy\cos 2b\ln u\}du^2 \tag{21}$$

1. Ozsváth (1962) [2]

$$ds^2 = dx^2 + e^{2x}dy^2 + e^{2\omega x}dz^2 + e^{2\omega^2 x}d\bar{z}^2 \qquad \omega^2 + \omega + 1 = 0\,, \qquad z = u + i\,v \tag{22}$$

Petrov (1962) [3]

This list is complete (M. Cohen at all 1967 [4] see also R. E. Hiromoto and I. Ozsváth 1977 [5]) the first two are Robinson waves with a six parametric maximal group. The Petrow solution is of type I. There is an interesting dust solution constructed by Schücking and myself. The finite rotating universe [6].

The space sections are compact. The time like lines do not have the pathologies of the time like lines of the Gödel cosmos. These models show that there is a difference between a non rotating earth in a roatating universe and a rotating earth in a non rotating cosmos. We give these solutions in three different systems of coordinates exibiting the features mentioned above.

$$ds^2 = -dt^2 + R\sqrt{1-2k^2}\,\omega^3 dt + (\tfrac{R}{2})^2\{(1-k)(\omega^1)^2 + (1+k)(\omega^2)^2 + (1+2k^2)(\omega^3)^2\} \tag{23}$$

where $R>0$, $|k|<\frac{1}{2}$ are parameters and the 1-forms satisfy

$$d\omega^1 = \omega^2\wedge\omega^3, d\omega^2 = \omega^3\wedge\omega^1, d\omega^3 = \omega^1\wedge\omega^2$$

this form shows that our space time is homogeneous. The underlying Lie group is the direct product of a line with the three-dimensional sphere: $L \times S_3$

Using another coordinate system (we denote the coordinates with the same letters) we have

$$ds^2 = -dt^2 + \frac{4Rk^2}{\sqrt{2(1-4k^2)}}\,\omega^3 dt$$
$$+ (\tfrac{R}{2})^2\{(1-k)(\omega^1 \cos\tfrac{\alpha}{2}t + \omega^2 \sin\tfrac{\alpha}{2}t)^2 +$$
$$(1+k)(-\omega^1 \sin\tfrac{\alpha}{2}t + \omega^2\cos\tfrac{\alpha}{2}t)^2 + (1+2k^2)(\omega^3)^2\}$$

where

$$\frac{\alpha}{2} = \frac{1}{R}\sqrt{\frac{2}{1-4k^2}}\,,\ \frac{\kappa\rho}{2} = 1 - 4k^2,\ \Lambda = \frac{1}{R^2(1-k^2)}$$

in these coordinates the matter is at rest. The cross term indicate rotation. The motion of the matter is best studied in this system.

Still another coordinate system shows the global similarity to the Einstein cosmos

$$ds^2 = -dt^2+(\tfrac{R}{2})^2\{(1-k)(\omega^1 \cos\tfrac{\beta}{2}t + \omega^2 \sin\tfrac{\beta}{2}t)^2 + (1+k)(-\omega^1 \sin\tfrac{\beta}{2}t + \omega^2\cos\tfrac{\beta}{2}t)^2 + (1+2k^2)(\omega^3)^2\} \tag{25}$$

where

$$\tfrac{1}{2}\beta = \frac{1}{R}\sqrt{\frac{2(1-2k^2)}{1+2k^2}}$$

for further details consult [7].

There are two different families of dust solutions where the underlying group is the direct product of the real line with the three-dimensional hyperboloid: $L \times H_3$

$$ds^2 = dt^2 + R\sqrt{1-2k^2}\,\omega^3 dt + (\tfrac{R}{2})^2\{(1-k)(\omega^1)^2 + (1+k)(\omega^2)^2 - (1+2k^2)(\omega^3)^2\} \tag{26}$$
$$d\omega^1 = \omega^2\wedge\omega^3,\ d\omega^2 = \omega^3\wedge\omega^1,\ d\omega^3 = -\omega^1\wedge\omega^2$$

The formal similarity of (23) and (26) is striking but not surprising.

There exists another family of solutions for this group:

$$ds^2 = dt^2 + \frac{1}{\kappa\rho}\{(1+s)(\omega^1)^2 + (1-s)(\omega^2)^2 - (\omega^3)^2\} \tag{27}$$

where

$$\rho>0 \quad\text{and}\quad |s|<1$$

surprising is that there is no corresponding family in the case of $L \times S_3$.

In order to complete the list of dust solutions we mention that there is a tree parametric family of dust solutions given by

$$[X_1, X_2] = 0,\ [X_2, X_0] = 0,\ [X_0, X_1] = 0$$
$$[X_\alpha, X_3] = c^\beta{}_\alpha X_\beta \qquad \alpha,\beta = 0,1,2 \tag{28}$$
$$g(X_a, X_b) = \mathrm{diag}\,(-1,1,1,1)$$
$$U = -u\,X_0 + v\,X_1 + w\,X_2 \quad \text{(The velocity vector)}$$

Using the field equations one can write C^{β}_{α} and the components of U explicitely as functions of tree parameters. The cosmological constant is negative for this family. For the sake of simplicity I exhibit here a two parametric sub-family, which one obtains by a slight restriction on the motion of the matter:

The group is given by the Lie algebra

$$
\begin{aligned}
&[X_1, X_2] = 0, \ [X_2, X_0] = 0, \ [X_0, X_1] = 0 \\
&[X_0, X_3] = -a X_0 + (f-r)X_1 \\
&[X_1, X_3] = -(f+r) X_0 - a X_1 \\
&[X_2, X_3] = (2a+k)X_2
\end{aligned}
\tag{29}
$$

with the left invariant metric

$$g(X_a, X_b) = \mathrm{diag}\,(-1, 1, 1, 1)$$

and the time like unit vector

$$U = -aX_0 + v X_1$$

where

k, a, f, r, u, and v are given as

$$
\begin{aligned}
&k = \sqrt{3P-D}, \quad a = -\frac{P}{\sqrt{3P-D}}, \quad f = \sqrt{\frac{DP}{3P-D}}, \\
&r = \tfrac{1}{2}\sqrt{\frac{(P+D)(3P-D)}{P}} \\
&u = -\sqrt{\tfrac{1}{2}\left(\sqrt{\frac{P+D}{D}} + 1\right)}, \quad v = \sqrt{\tfrac{1}{2}\left(\sqrt{\frac{P+D}{D}} - 1\right)} .
\end{aligned}
\tag{30}
$$

P and D are arbitrary parameters subject to the only condition

$$P > \tfrac{1}{3} D > 0. \tag{31}$$

It is easy to see that the field equations

$$R_{ab} - \tfrac{1}{2} R\, g_{ab} = -2Du_a u_b - P\, g_{ab} \tag{32}$$

are satisfied.

If we interpret

$$P = -\Lambda \text{ and } D = \frac{\kappa\rho}{2} \tag{33}$$

then we have a dust solution with a negative Λ term and ρ is the density of the incoherent matter.

Since the Λ term is negative we could follow L. Shepley's suggestion and interpret in as a perfect fluid:

$$P = \kappa p \quad \text{and } D = \frac{\kappa(\epsilon + p)}{2} \tag{34}$$

where p is the pressure and ϵ is the proper energy density with the reality condition

$$p > \frac{1}{5}\epsilon > 0 \tag{35}$$

attempting to interpret the other models as fluid solutions one would have conditions which are even more unphysical.

This list is complete, contains all possible homogeneous dust solutions (see I. Ozsváth (1965) [8] and D. L. Farnsworth and R. P. Kerr (1966)) [9].

There is a family of homogeneous vacuum solutions with a non vanishing cosmological term:

$$ds^2 = -dt^2 + e^{2\sqrt{\frac{\Lambda}{3}}\,t}\{dx^2 + dy^2 + dz^2\}$$

de Sitter [1917] [10]

$$ds^2 = dx^2 + e^{2\sqrt{\frac{-\Lambda}{3}}\,x}\{-dt^2 + dy^2 + dz^2\}$$

anti de Sitter.

$$ds^2 = -e^{2\sqrt{\frac{-\Lambda}{3}}x}dt^2 + dx^2 + e^{2\sqrt{\frac{-\Lambda}{3}}z}dy^2 + dz^2$$

Bertotti (1959) [11]

$$ds^2 = dx^2 + e^{2\sqrt{\frac{-\Lambda}{3}}x}\{dy^2 + 2du\,dv\} + f\,e^{-\sqrt{\frac{-\Lambda}{3}}x}dv^2 \quad .$$

where f is an arbitrary parameter. The Weyl tensor of this space is of Typen.

I. Ozsvath (1962) [8] and M. Cahen (1964) [12]

$$ds^2 = dx^2 + e^{2\sqrt{\frac{-\Lambda}{3}}x}\{dy^2 + 2du\,dv\} +$$

$$f\,e^{-\sqrt{\frac{\Lambda}{-3}}x}\{-2\sqrt{2}\,dy + f\,e^{-3\sqrt{\frac{-\Lambda}{3}}x}dv\}\,dv,$$

where f is an arbitrary parameter. The Weyl tensor of this space is of type III.

I. Ozsváth (unpublished). [13]

This list is complete

I. Ozsváth to be published. See also M. Cahen (1964) [12]).

It is interesting that the maximal group of the type N solution is five parametric. Ivor Robinson suggests to interpret this solution as the plane fronted gravitational wave against the anti de Sitter background.

There are homobeneous solutions with electromagnetic field, with charged fluid, about which, however, I do not report here. An excellent bibliography is at the end of the book by M. P. Ryan and L. C. Shepley [14].

I would like to close with the following remark: I think one can find all homogeneous solutions for any given energy impulse tensor following the instructions explained in [5].

References

[1] I. Ozsváth and E. L. Schücking, An Anti Mach Metric, Recent Developments in Relativity, Pergamon Press (1962).

[2] I. Ozsváth, Lösungen der Einsteinschen Feldgleichungen mit Einfach Transitiver Bewegungsgruppe, Abhandlungen der Akademie in Mainz (1962).

[3] A. Z. Petrov, Recent Developments in General Relativity, Pergamon Press (1962), p. 379.

[4] M. Cohen, R. Debever and L. Defrise, A Complex Vectorial Formalism in General Relativity, J. of Math and Mech., 16, (1967), p. 761.

[5] R. E. Hiromoto and I. Ozsváth, General Relativity and Gravitation, 9, (1978), p. 299.

[6] I. Ozsváth and E. L. Schücking, Nature, 193, (1962), p. 1168.

[7] I. Ozsváth and E. L. Schücking, The Finite Rotating Universe, Annals of Physics, 55, (1969), p. 166.

[8] I. Ozsváth, New Homogeneous Solutions of Einstein's field Equations with Incoherent Matter, J. Math Phys., 6, (1965), p. 590.

[9] D. L. Farnsworth and R. P. Kerr, Homogeneous Dust Filled Cosmological Solutions, J. Math Phys. 7, (1966), p. 1625.

[10] W. de Sitter, On the Relativity of Inertia Proc. Kon. Ned. Akad. Wet. 19, (1917), p. 1217.

W. de Sitter, On the Curvature of Space Proc. Kon. Ned. Adad. Wet. 20, (1917), p. 229.

[11] B. Bertotti, Uniform Electromagnetic Field in the Theory of General Relativity Phys. Rev. 116, (1959), 1331.

[12] M. Cahen, On a Class of Homogeneous Spaces in General Relativity, Bull Acad. Roy. Belgique, 50, (1964), p. 972.

[13] I. Ozsváth, All Homogeneous Solution of Einstein's Vacuum Field Equations with a Non Vanising Cosmological Term (to be published).

[14] M. P. Ryan and L. C. Shepley, Homogeneous Relativistic Cosmologies, Princeton University Press, Princeton, N.J. (1975).

Description of particles with internal structure in General Relativity

G.E. Tauber and Y. Feldman

Dept. of Physics, Tel-Aviv University, Tel-Aviv, Israel

It can be shown that a (relativistic) particle is characterized by the space variables x^μ, momenta p_μ, and internal variables denoted generically by $Q^{\mu\cdot\cdot}$, which include at least a vector Q^μ and an antisymmetric tensor $S^{\mu\nu}$ generating infinitesimal transformations is given by

$$(p_\mu, x^\nu) = \delta^\nu_\mu \,, \; (x^\mu, Q^{\nu\cdot\cdot}) = 0, \quad (S^{\mu\nu}, Q^{\lambda\cdot\cdot}) = g^{\nu\lambda} Q^{\mu\cdot\cdot} - g^{\mu\lambda} Q^{\nu\cdot\cdot} + .. \qquad (1)$$

$$(p_\mu, Q^{\nu\cdot\cdot}) + \Gamma^\nu_{\mu\lambda} Q^{\lambda\cdot\cdot} + \ldots = 0, \quad (p_\mu, p_\nu) = \tfrac{1}{2} S^\lambda{}_\alpha R^\alpha{}_{\lambda\mu\nu}$$

However, internal and external variables are here intertwined, while to serve as coordinates of a phase-space they must be disentangled from one another, unless they are canonically conjugate. In our case the isomorphism between the internal gauge group and the geometrical symmetry group of the tangent space-time permits us to replace the ten gravitational potentials $g^{\mu\nu}$ by sixteen tetradic fields h^a_μ

$(a = 1,..4)$ satisfying completeness and orthogonality relations

$$h^\mu_a h^\nu_b \eta^{ab} = g^{\mu\nu} \quad \text{and} \quad h^\mu_a h^\nu_b g_{\mu\nu} = \eta_{ab} \text{ where } \eta_{ab} = \mathrm{diag}(-1,-1,-1,1) \qquad (2)$$

Upon defining gauge covariant momenta $\bar{p}_\mu$ by

$$\bar{p}_\mu = p_\mu + \tfrac{1}{2} \gamma_{\alpha\beta\mu} S^{\alpha\beta} \text{ where } \gamma_{\mu\nu\lambda} = h_{a\mu} h^a{}_{\nu;\lambda}$$

direct calculation yields

$$(\bar{p}_\mu, \bar{p}_\nu) = 0 \qquad (3)$$

Furthermore, upon projecting the internal vectors and tensors upon the tetrads h^a_μ

$$\Sigma^{ab} = S^{\mu\nu} h^a_\mu h^b_\nu \text{ and } q^{a\cdot\cdot} = Q^{\mu\cdot\cdot} h^{a\cdot\cdot}_\mu$$

the remaining brackets become (4)

$$(\Sigma^{ab}, q^{c\cdot\cdot}) = \Sigma^{bc} q^{a\cdot\cdot} - \Sigma^{ac} q^{b\cdot\cdot} + \ldots \text{ as well as } (\bar{p}_\mu, q^{c\cdot\cdot}) = 0$$

resulting in the desired separation of internal and external variables.

These commutation relations can be satisfied by the introduction of a spin-aligned orthonormal tetrad q^a_A (A = 1,..4) in a Minkowski frame

$$q^a_A\, q^b_B\, \eta_{ab} = \eta_{AB}\ ,\ q^a_A\, q^b_B\, \eta_{AB} = \eta_{ab} \tag{5}$$

in terms of which Σ^{ab} is expressed as

$$\Sigma^{ab} = \sigma(q^a_3\, q^b_4 - q^b_3\, q^a_4) - \sigma^*(q^a_1\, q^b_2 - q^b_1\, q^a_2) \tag{6}$$

Here, σ and σ^* are two scalars related to Σ^{ab} and its dual Σ^{ab*} through

$$\Sigma^2 - \Sigma^{*2} = \sigma^2 - \sigma^{*2},\ \Sigma^*\Sigma = \sigma\sigma^* \tag{7}$$

The introduction of two additional antisymmetric tensors Σ^{ab}_1 and Σ^{ab}_2 formed from (6) by cyclic permutation of the indices A = 1,2,3 then completes the Lie algebra which is seen to consist of 36 functions Σ^{ab}, Σ^{ab}_1, Σ^{ab}_2, q^a_A,σ and σ^*; and turns out to be Sp(8), while we are dealing with its real Sp(8,r) realization . However, only 8 of the internal variables are independent, viz. the two scalars and six components of the vectors q^a_A. These have to be adduced to the external x^μ and $\bar{p}_\mu$ to form a 16-dimensional phase space.

Two special cases are of interest. Suppressing the distinction of the vectors q^a_1 and q^a_2 and also the scalar σ^* an algebra spanned by the 15 generators σ, $\rho^a_B = \sigma q^a_B$ (B = 1,2) and Σ^{ab} (a,b = 1,..4) is obtained. This algebra is seen from its canonical form to be SO(6) and the real structure is again the least compact one, SO(3,3). In this case the internal bracket algebra is

$$\begin{aligned}
&(\sigma,\ \overset{*}{\rho}{}^a_3) = \rho^a_4,\quad (\sigma,\ \rho^a_4) = \rho^a_3,\quad (\sigma,\ \Sigma^{ab}) = 0\\
&(\rho^a_3,\ \rho^b_3) = \Sigma^{ab},\quad (\rho^a_4,\ \rho^b_4) = -\Sigma^{ab},\quad (\rho^a_3,\ \rho^b_4) = -\sigma\eta^{ab}\\
&(\Sigma^{ab},\ \rho^c_A) = \eta^{bc}\ \rho^a_A - \eta^{ac}\ \rho^b_A\\
&(\Sigma^{ab},\ \Sigma^{cd}) = \eta^{bc}\ \Sigma^{ad} - \eta^{ac}\ \Sigma^{bd} + \eta^{bd}\ \Sigma^{ca} - \eta^{ad}\ \Sigma^{cb}
\end{aligned} \tag{8}$$

The second even simpler case consists of suppressing σ as well and reducing the internal structure to the minimal SL(2,c) necessary to admit spin. The functionally independent generators are 6 for SO(6) and four SL(2,c).

The canonical variables are then found by taking linear (complex) combinations of the independent functions and satisfying

$$(\Theta_i, \Theta_j) = 0, \quad (\Psi_i, \Psi_j) = 0, \quad (\Theta_i, \Psi_j) = \delta_{ij} \quad (i,j = 1,..4) \tag{9}$$

Taking as the eight independent functions the following generators

$$H_1 = \tfrac{1}{2}(\sigma + i\sigma^* + i\Sigma^{12} + \Sigma^{34}) \qquad H_3 = \tfrac{1}{2}(\sigma - i\sigma^* + i\Sigma^{12} + \Sigma^{34})$$

$$H_2 = \tfrac{1}{2}(\sigma + i\sigma^* - i\Sigma^{12} - \Sigma^{34}) \qquad H_4 = \tfrac{1}{2}(\sigma - i\sigma^1 + i\Sigma^{12} - \Sigma^{34})$$

$$E_1 = \tfrac{1}{2}(\rho_3^3 + \rho_3^4 + \rho_4^3 + \rho_4^4) \qquad E_7 = \tfrac{1}{2}(i\rho_1^3 - \rho_2^3 + i\rho_1^4 - \rho_2^4)$$

$$E_5 = \tfrac{1}{2}(\Sigma^{13} + i\Sigma^{23} + \Sigma^{14} + i\Sigma^{24}) \qquad E_6 = \tfrac{1}{2}(-\Sigma^{13} + i\Sigma^{23} - \Sigma^{14} + i\Sigma^{24})$$

the canonical basis for the internal space is

$$\begin{aligned} &\Phi_1 = E_5, \quad \Phi_2 = E_6, \quad \Phi_3 = E_1/E_5, \quad \Phi_4 = E_7/E_6 \\ &\Psi_1 = H_1/E_5, \quad \Psi_2 = -H_4/E_6, \quad \Psi_3 = E_5/E_1, \quad \Psi_4 = i \ E_6/E_7 \end{aligned} \tag{10}$$

For SO(6) E_7 vanishes and therefore the basis reduces to Θ_j, Ψ_j (j = 1, 2, 3) with $\Theta_4 \Psi_4$ a constant of the motion. For SL(2,c) E_1 is not defined as well and the basis becomes Θ_k, Ψ_k (k = 1, 2) with $\Theta_3 \Psi_3$ an additional constant of the motion.

The equations of motions are generated from a Hamiltonian H according to

$$d\chi/ds = (H,\chi) \tag{11}$$

where χ is any function of the dynamical variables and s is a monotonic function of the time. For minimal coupling H can depend only on a linited number of scalars and in the general case is given by

$$H = H(N_A, \sigma, \sigma^*) \quad \text{where} \quad N_A = p_\mu p_A^\mu \quad (A = 1,..4) \tag{12}$$

For SO(6) N_1 and N_2 are not defined, but the mass $M^2 = g^{\mu\nu} p_\mu p_\nu$ may appear in the Hamiltonian instead. Finally, for SL(2,c) the Hamiltonian is given by

$$H = H(M,L,\sigma,\sigma^*) \quad \text{where} \quad L^2 = p_\mu \, p_\nu \, (S^{*2})^{\mu\nu} \tag{12a}$$

with both σ and σ^* being constants of the motion.

Upon defining an invariant distribution function, depending on x^μ, p^μ, as well as the internal variables (and scalars) and solving Liouville's equation one obtains the energy momentum tensor, entropy and number density by integrating over the momenta and internal variables. The imposition of thermal equilibrium as guaranteed by the H-theorem will limit the possible forms of the distribution function and make explicit calculations possible. The above results can be generalized to supergravity by adding anti-commuting Fermi coordinates θ^α as well as their conjugates.

SYMMETRIES AND STATISTICAL BEHAVIOR IN FERMION SYSTEMS+

J. B. French†

Department of Physics and Astronomy, University of Rochester
Rochester, New York 14627

and

J. P. Draayer

Department of Physics and Astronomy, Louisiana State University
Baton Rouge, Louisiana 70803

1. Introduction

The subject is the interplay between statistical behavior and symmetries in nuclei, as revealed for example by spectra and by distributions for various kinds of excitations. Much of the argument and general procedure may be applied also to other many-fermion spaces which have (as do nuclear shell-model spaces) a *direct-product* structure, being representable by distributing a certain number, m, of fermions over N single-particle states ($0 \leq m \leq N$). I shall talk about methods and general results because that seems appropriate here, rather than about specific applications, of which however there have been many.

Most people think of statistical behavior as being a relevant way to describe things only at high excitation, whereas the natural domain of symmetries, with the obvious exception of those which are exactly (or almost exactly) conserved*, is near the nuclear ground state. But even rather badly broken symmetries are useful in roughly partitioning an energy spectrum and play thereby an important role in determining locally averaged strengths, etc; thus symmetry effects extend *upward* in energy away from the ground-state region. On the other hand, if we introduce Hamiltonian ensembles to describe the statistical behavior (which is necessary for fluctuations, though not for *locally averaged* eigenvalue densities, etc.), it turns out that appropriate ensembles gives results which are *ergodic* and, when

---

+Supported in part by the U.S. Department of Energy and National Science Foundation.

†John Simon Guggenheim Memorial Fellow, 1977-78

*We speak of a symmetry as conserved if each wave function belong to a single irreducible representation

expressed in an energy scale supplied by the local eigenvalue density, are *stationary* in energy.[1] Thus statistical behavior extends *downward* into the ground-state domain so that the interplay which we discuss exists in all parts of the spectrum.

As I have indicated, statistical behavior can be discussed in terms of secular averages and of fluctuations about the average.* A finer decomposition is in terms of the sequence of ℓ-point correlation functions ($\ell = 1,2,\ldots$), of which $\ell = 1$ gives the density, while $\ell = 2$ dominates all of the measures which have been introduced (and perhaps all which can be experimentally measured). There are interesting connections between fluctuations and symmetries, for, while fluctuations are "small" and carry little information, the little which they do carry is closely connected with the existence or non-existence of a conserved or almost conserved symmetry; this is because the existence of such a symmetry moderates the "level repulsion" which dominates the fluctuation behavior. Methods for studying these connections have only recently been developed. I shall therefore say nothing more about fluctuations but focus instead on the (secular) behavior of locally averaged quantities.

2. Elementary Principles

The m-fermion wave functions for fixed m (=0,1,...,N) form an irreducible (anti-symmetric) tensor, of dimensionality $d(m) = \binom{N}{m}$, with respect to the group U(N) of unitary transformations in the single-fermion space. To each state $\Psi_\alpha(m)$ there exists a *complementary* one $\Psi_{\alpha_c}(N-m)$ which forms a contragradient tensor. The system has two vacuum states $|0\rangle$ and $|N\rangle$ and any state $\Psi_\alpha(m)$ can be described in terms of either:

$$\Psi_\alpha(m) = \psi_\alpha(m)|0\rangle = \tilde{\psi}_{\alpha_c}(N-m)|N\rangle \tag{1}$$

in which the operation ($\sim$) replaces every creation and destruction operator by its adjoint so that $\tilde{\psi}_\alpha(m) = \psi_\alpha(m{:}a_i^+ \rightleftarrows a_i)$. Since holes are also fermions (the fundamental commutation rules being invariant under ($\sim$)) we have that for any real

*In nuclear model systems, at least, there is a remarkable (well-understood) separation between the two phenomena which allows us to discuss them separately

operator, F

$$[\text{Trace }(F)\text{ in m-particle space}] \equiv \langle\langle F\rangle\rangle^m \equiv d(m)\langle F\rangle^m = \langle\langle \tilde{F}\rangle\rangle^{N-m} \tag{2}$$

where, as indicated, $\langle\langle F\rangle\rangle^m$ is the trace and $\langle F\rangle^m$ the average eigenvalue (or expectation value).

A k-body operator, $F(k) = \Sigma\langle k\beta'|F(k)|k\beta\rangle\psi_{\beta'}(k)\psi^{+}_{\beta}(k)$, transforms according to (k+1) irreps of U(N); for example, with n the number operator,

$$F(2) \sim \binom{n}{2}\left([1^{N}]\,\Big|\,N\right)+\binom{n-1}{1}\left([21^{N-2}]\,\Big|\,N-1\right)+\left([2^{2}1^{N-4}]\,\Big|\,N-2\right) = \binom{n}{2}F(0)+\binom{n-1}{1}F(1)+F(2) = \sum_{\nu=0}^{2} F^{(\nu)} \tag{3}$$

where $\binom{n}{2}F(0)$ measures the spectrally-averaged eigenvalue (which then "propagates" with particle number as $\binom{m}{2}$), $\binom{n-1}{1}F(1)$ (in which $F(1)$ is an SU(N) generator) gives the modification of the single-particle energies, and $F(2)$ may be regarded, in the context of the U(N) decomposition, as the "true" residual interaction.

There is, for each particle number m, a natural geometry for traceless operators in which

$$(\text{norm } F)^2 = d^{-1}(m)\langle\langle F^{+}F\rangle\rangle^m = \langle F^{+}F\rangle^m \tag{4}$$

The unitary decomposition of F is orthogonal in this geometry (true also for operators which do not conserve particle number). With this geometry we can consider such questions as "how much Q·Q is contained in a given H, acting in an m-particle space?" (answer: a large amount; the scalar product between the unit vectors for $H-\langle H\rangle^m$ and Q·Q (in statistical terms the "correlation coefficient" between H and Q·Q) is negative and, for light nuclei at least, ~ -0.5). Note that Q·Q is a sum of squares of traceless 1-body operators; it appears that realistic H's are rather well representable in this way.

There are central-limit theorems (CLT's) which ensure that, if N is large enough the smoothed spectrum converges, as we increase the particle number m, to a characteristic density describable by a few parameters. For example with non-interacting particles (H=H(k=1)) the single-particle spectrum convolutes and rapidly becomes Gaussian as we increase the particle number. The real function of the CLT is to make the unitary geometry an *effective* one. There are two aspects of this; the unitary norm for an operator might well be criticized on physical grounds because it treats all states on the same footing and therefore concerns itself mostly with states which lie high in energy (where the density is highest), even though most of these are not of direct physical significance and our interest might well be in the lowest ones. This difficulty disappears when the energy variation of the eigenvalue density is closely specified by a CLT, for what happens then at all energies is determined by the trace of H and the norms of low powers of H. At the same time we avoid the hopeless technical problem of evaluating high-order moments.

3. A Simple Example: Single-state Occupancies

The Strutinsky explanation of shell effects in fission may be taken as demonstrating the "survival" of single-particle information in the presence of strong interactions. As a simpler example of this let us ask whether the ground-state sum rule for the single-nucleon-transfer strength (which measures single-particle information) can tell us anything about the effective interaction.

The sum rule for transfer in an orbit j determines the ground-state occupancy $\langle W_g|n_j|W_g\rangle$. In general, for any H-eigenstate $|W\rangle$ of eigenenergy W, we have, with $\rho(W)$ the eigenvalue density function, $P_\nu(W)$ the orthonormal polynomials defined with $\rho(W)$ as weight function, and $\langle\rangle^m$ the m-particle (or model-space) average with $E \equiv E(m) = \langle H\rangle^m$ the eigenvalue centroid, that

$$\langle W|K|W\rangle = \rho^{-1}(W)\langle\delta(H-W)\rangle^m = \sum_\nu \langle K\,P_\nu(H)\rangle^m P_\nu(W) = \langle K\rangle^m + \frac{\langle K(H-E)\rangle^m}{\langle(H-E)^2\rangle^m}(W-E) + \ldots \tag{5}$$

It is easy to see that the ν-th order term would arise from the ν'th order deformations of $\rho(W)$ under the infinitesimal change of the Hamiltonian, $H \to H + \delta\alpha\times K$. But then, in the CLT limit, the terms of order larger than 1 must vanish (except for very high order, involving "wavelengths" comparable with the level spacing, for which the CLT smoothing is ineffective). Thus we have a linear form for the smoothed $\langle W|K|W\rangle$, the slope of which is proportional to the scalar product $\langle K(H-E)\rangle^m$. For any one-body operator, $K = n_j$ for example, only the $[2,1^{N-2}]$ part of H contributes (by orthogonality); the part of this deriving from the two-body H measures the Hartree-Fock-like contribution to the effective single-particle energy, proportional to $\{\sum_s W_{sjsj} - N^{-1}\sum_{sj} W_{sjsj}\}$. Observe how the CLT in (5) and the unitary geometry here have combined to modify a single-particle property of a complicated system. The modification has been measured in the ds shell, and used in determining which model interactions are acceptable.[2] Occupancies can of course be used in many other ways also.

4. Subgroup Structures, Spectra and Characteristic Forms

If, in the example above, we focus our attention on an N_1-dimensional subset of the single-particle states we are led naturally to think about the $U(N_1)$ subgroup of U(N), and more generally about "additive" decompositions of the U(N) algebra. If we consider neutron and proton transfer separately we are led to $U(N/2) \otimes U(2)$ and a direct-product decomposition. A great variety of other subgroups are encountered in other circumstances and so we must consider the subgroup problem more generally.

A given representation of a subgroup G, as with $SU_2(T)$ or $O_3(J)$, may occur very many times in a large space. We then naturally ask what representations of G are found and how often. This question is answered to some extent by considering the spectrum, in the m-particle space, of the bilinear Casimir operator C_2, which is, of course discrete with degeneracies equal to the number of times that the appropriate* representation occurs, multiplied by its dimensionality.

For any U(N) subgroup, C_2 may be written as a sum of squares of (non-commuting) traceless one-body operators (SU(N) generators) which have the same norm in all m-particle spaces, and which may be taken as orthogonal in the 1-particle space (and hence via (3) in m-particle spaces). Thus $C_2 = \sum_{\alpha=1}^{\ell} h_\alpha^2$ with $\langle h_\alpha \rangle^1 = 0$ and $\langle h_\alpha h_\beta \rangle^1 = \sigma^2 \delta_{\alpha\beta}$. Let us consider the moments of C_2 in the limit of a large dilute multi-particle space ($N\to\infty$, $m\to\infty$, $\frac{m}{N}\to 0$). For a traceless h we can decompose $\langle h^{2\nu} \rangle^m$ in terms of correlation patterns defined by the partition of 2ν into integers ≥ 2; then for example the partition $[2^\nu]$ gives a term proportional to $\{\langle h^2 \rangle^m\}^\nu = \{m(N-m)(N-1)^{-1}\sigma^2\}^\nu$. From the m-dependence it is easy to see that only binary correlations survive in the asymptotic limit.[3] Extending this to the sum of squares and observing that no cross-terms contribute we see that, in the present limit, C_2 is equivalent to a sum

*C_2 does not by itself distinguish all irreducible representations. We may say that it classes together all "C_2-equivalent" irreps.

of *commuting* squares so that its distribution is an ℓ-fold convolution; in other words only the dimensionality, ℓ, of the space of generators matters, nothing further about the nature of the group or about the way in which it is realized in the space. Thus the asymptotic spectrum of C_2 has a χ^2_ℓ form,

$$\rho_{C_2}(x) = \{2^{\ell/2}\sigma^{\ell-2}\Gamma(\ell/2)\}^{-1}\, x^{\ell/2-1}\, \exp(-x/2\sigma^2) \qquad (6)$$

in which, as above, σ^2 is the one-particle variance of a generator. An SU(3) example (ℓ=8) is shown in Fig. 1. For ℓ=3 (and thus for angular momentum) the distribution is Maxwellian, a result given in 1936 by Bethe[4] for non-interacting particles and used ever since, with minor corrections, for the J-dependence of the level density. For a quadrupole interaction (ℓ=5) Nomura[5] evaluated the moments to order four, and stressed that the asymptotic spectrum is non-Gaussian. From the argument above we see that the distribution is χ^2_5.

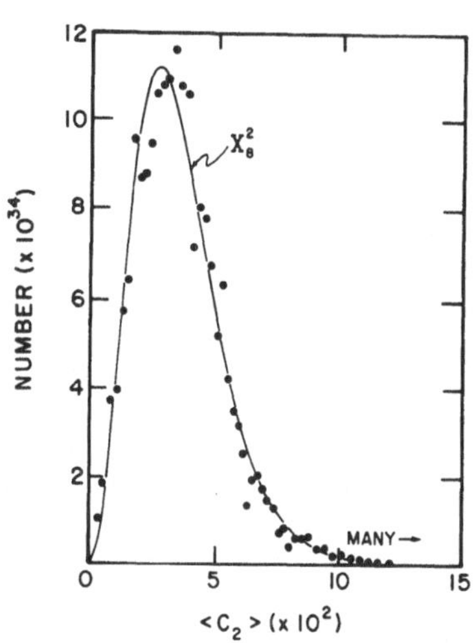

Fig. 1 The spectrum of C_2 for SU(3) realized as a direct-product factor of U(N=300). The particle number is 25. A running 5-point average has been used to reduce fluctuations. The smooth curve is χ^2_8.

This distribution (6) can be extended to finite particle number by explicit evaluation of the low-order moments (and, much less easily, to finite N as well). In general terms we can expect that an approximate χ^2_ℓ distribution will obtain as long as we have a half-dozen particles or more. For large ℓ, by the standard CLT, the χ^2_ℓ distribution is close to Gaussian. Since, as we have said, the Hamiltonian itself is often reasonably representable as a sum of squares we see that *its* asymptotic spectrum likewise should be approximately χ^2_ℓ (ℓ being now the number of squares of comparable magnitude) and in many cases then close to Gaussian, as is observed in detailed calculations.

Spectral distributions, as used in practice, involve the distribution in energy of states of a given symmetry, the distribution being a Hamiltonian-eigenvalue distribution if the symmetry is a good one, but otherwise an "intensity" distribution (giving for each Hamiltonian eigenvalue, the contribution to $<\Psi^+\Psi> = 1$ which arises from basis states of the specified symmetry). The moments of such a distribution follow from the simplest moments, $<H^p>^m$, by the model-space decomposition,* $\underset{\sim}{m} \to \sum_\Gamma (\underset{\sim}{m},\Gamma)$ where Γ defines the symmetry irrep; we come to that in Section 5. For now observe that if the C_2-H correlation coefficient, ζ, is small it follows from (5) that a given G irrep will distribute more or less uniformly over the whole energy spectrum, the decomposition (6) being then locally valid (this goes far towards justifying the use of Bethe's J-decomposition in level density studies). At the other extreme ($|\zeta|\sim 1$), the H spectrum will be close to that of C_2. In many cases the Hamiltonian centroids are determined by a single group invariant, C_2; in these cases, if we start with $H=C_2$ (to within a sign) and allow it to move away from C_2 ($|\zeta|$ going gradually to 0) the overall χ^2_ℓ distribution will be maintained for at least part of the range of ζ) even while the degenerate C_2 "spikes" are broadened by the movement of the individual levels and their admixing with levels of other symmetries. The line shape (which one would guess should be Gaussian, except perhaps near the extremes of the spectrum) and more generally the energy distribution of the fixed-symmetry states can be studied by the polynomial methods referred to in Section 3.

5. Propagation of Information

The asymptotic distribution (in energy) for a fixed symmetry is defined by three quantities only, the dimensionality, centroid and variance. For distributions which are not asymptotic the same characteristic form will obtain, but with m^{-1} and m/N corrections for which we need a few more pieces of information. Since the Hamiltonian and the other interesting operators of the system are *defined* in few-particle subspaces "embedded" in the m-particle space the general problem of understanding what goes on in a many-particle subspace may be thought of in terms of propagation of information from the simple subspaces to the more complicated ones. Alternatively if we measure things in a many-particle space we are "probing" the

---

*We use $\underset{\sim}{m}$ to denote the m-particle space.

structure of the physical few-particle spaces. Because of the filtering action of the CLT much of the information is lost* and conversely the many-particle probe is sensitive to only a few properties of the system.

Intuitively we might guess for a k-body operator, averaged over m, that

$$\langle F(k)\rangle^{m} = \langle F(k)\rangle^{k} \;\{\text{weight with which } \underset{\sim}{k} \text{ is found in } \underset{\sim}{m}\} \tag{7}$$

in which the natural definition of the weight is

$$\langle \sum_{\alpha} \psi_{\alpha}(k)\psi_{\alpha}^{+}(k)\rangle^{m} = \langle \rho(k)\rangle^{m} = \langle \binom{n}{k}\rangle^{m} = \binom{m}{k} \tag{8}$$

This result is easily seen to be correct; schematically (with particle number increasing upward so that $\psi_{\beta'}(k) \equiv \beta'\uparrow$ and $\psi_{\beta'}(k)\psi_{\beta}^{+}(k) \equiv \beta'\vee\beta)$ $F(k) = \sum F_{\beta'\beta}(\beta'\vee\beta)$ and then

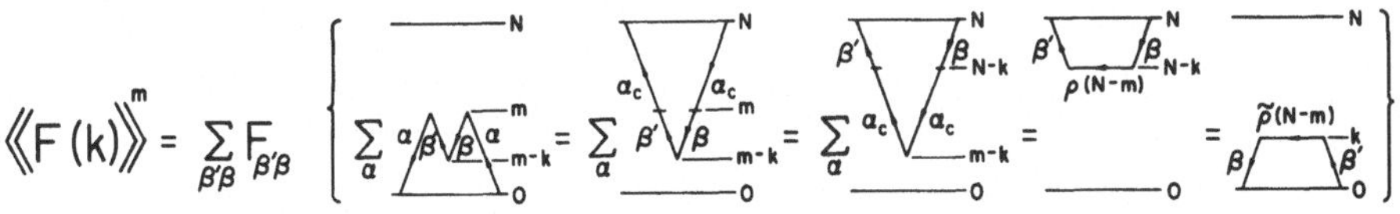

$$\langle\langle F(k)\rangle\rangle^{m} = \sum_{\beta'\beta} F_{\beta'\beta} \left\{ \sum_{\alpha} \alpha\,\beta\,\beta'\,\alpha \Big|_{0}^{m-k,\,m,\,N} = \sum_{\alpha} \beta'\,\alpha_c\,\alpha_c\,\beta \Big|_{0}^{m-k,\,m,\,N} = \sum_{\alpha} \beta'\,\alpha_c\,\alpha_c\,\beta \Big|_{0}^{m-k,\,N-k,\,N} = \beta'\,\rho(N-m)\,\beta \Big|_{0}^{N-k,\,N} = \beta\,\tilde{\rho}(N-m)\,\beta' \Big|_{0}^{k,\,N} \right\} \tag{9}$$

with ρ the eigenvalue density operator defined in (8). In going from the first to the second forms we have used the contragradient representation of the m-particle states, this followed by a commutation of state operators (which would not be valid for the first form) and a use of (2). Note that $\rho_{c}(N-m) = \rho(N-m)$; ρ is of course diagonal in β in the present case but not always so for its symmetry extensions.

Carrying out the sums we have

$$\langle\langle F(k)\rangle\rangle^{m} = \langle\langle \tilde{\rho}(N-m)F(k)\rangle\rangle^{k} = \langle\langle \rho(N-m)\tilde{F}(k)\rangle\rangle^{N-k} \tag{10}$$

which, with (8), reproduces the elementary result (7). The intuitive interpretation of (10) follows from the fact that $\tilde{\rho}(N-m)$ measures the number of holes in the complementary space $(\underset{\sim}{N}-\underset{\sim}{m})$. It is easy enough to see that (10) extends to any subspace Γ of $\underset{\sim}{m}$ (even to single states) so that, with $\langle\langle F\rangle\rangle^{m\Gamma}$ the trace over all m-

*The CLT does not smooth out the fluctuations but the "lost information is not to be found there. Fluctuations, as we have said, carry almost no information.

particle states belonging to all irreducible representations Γ of a group G, a subgroup of U(N), (involving therefore a sum over *all* irreps Γ)

$$<<F(k)>>^{m\Gamma} = <<\tilde{\rho}_{\Gamma_c}(N-m)F(k)>>^{k} = \sum_{\Gamma'} <<\tilde{\rho}_{\Gamma_c}(N-m)F(k)>>^{k\Gamma'}$$

$$\xrightarrow{(s)} \sum_{\Gamma'} <<\tilde{\rho}_{\Gamma_c}(N-m)>>^{k\Gamma'} <F(k)>^{k\Gamma'} = \sum_{\Gamma'} <<\rho_{\Gamma'}(k)>>^{m\Gamma} <F(k)>^{k\Gamma'} \tag{11}$$

the last two forms of which (the second of these following from the first via the same operations used in (9)) are valid when $\rho_{\Gamma_c}(N-m)$ behaves as a multiple of unity in each subspace, in which case we have simple (s) propagation, from the space in which F is defined, to the more complicated spaces of interest to us. This form displays beautifully how the trace information propagates throughout all the subspaces, the propagation coefficient being dependent only on the pair of representations involved, in fact on the weight with which one representation space is found in the other. The nature of simple propagation (in this case for "non-scalar" information) will become clearer on inspection of Fig. 2, for which the density operators are also available.[6]

By construction the operators $\rho_\Gamma(k)$ are k-body G scalars which satisfy $\rho_\Gamma(k)\psi_\alpha^{\Gamma'}(k) = \delta_{\Gamma\Gamma'}\psi_\alpha^{\Gamma'}(k)$. These properties completely define the operators and show that the set of them for fixed k forms a Green's Function, on the m=k subspace, by means of which we can calculate traces away from the defining "surface", i.e. for m>k. We can obviously construct one operator for each set of G irreps Γ. But there are also off-diagonal G scalars that can be constructed by coupling states from two equivalent irreps. Suppose that G is a subgroup of a factor U(k) where U(N) $\supset$ U(N/k) $\otimes$ U(k) and we specify also the

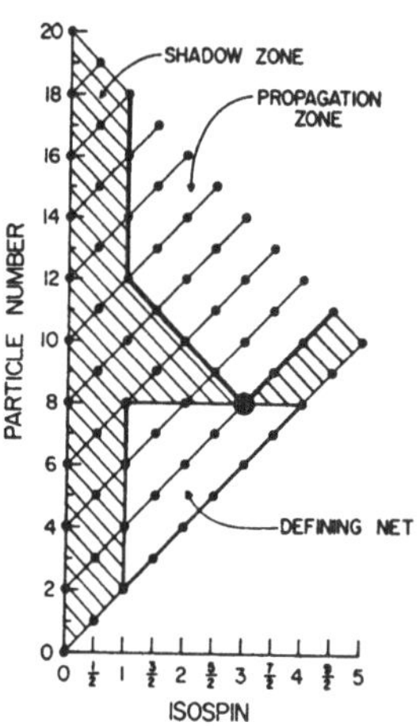

Fig. 2 Propagation on the isospin lattice for an operator with isospin-rank 2 and maximum particle rank 8. The triangle shows the effective defining lattice and the cross-hatching the "shadow zone" into which no propagation occurs for the part of the operator defined in the (m=8,T=3) subspace.

(single) irrep Λ of U(k); if now in the reduction $\Lambda \to \Gamma$ there is a multiplicity κ we have κ diagonal G scalars and $\frac{1}{2}\kappa(\kappa-1)$ off-diagonal ones, the total being $\frac{1}{2}\kappa(\kappa+1)$. Similarly if G is the last member of a chain, as in $U(N) \supset K \supset G$ and we also specify a single irrep Λ of κ. In these circumstances for each (Λ, Γ) pair there is a matrix of density operators which should enter into the propagation of information; we shall see that they do.

For U(N) itself (G=U(N)), Γ is of course unique and $\rho_\Gamma(k) = \binom{n}{k}$ as we have seen. For G=U(k) a factor of $U(N) \supset U(N/k) \otimes U(k)$ (and similarly for an additive decomposition) Γ is not unique but one sees easily that ρ_Γ is a function only of the generators of G and hence of its Casimir invariants. The total number of G irreps for $k \leq m$ particles then equals the number of terms in a polynomial, of maximum particle rank m, involving the U(N) invariant, n, and the G Casimir operators, this being true as long as N is large enough so that all the irreps can in fact be generated. In this case, which includes the nuclear physicist's isospin and spin-isospin SU(4), the propagation is simple in the sense of (11). We should mention here that the first systematic discussion in group theoretical terms of simple propagation and some of its extensions is given in an important paper by Quesne.[7]

Going beyond simple propagation, consider[8,9] $U(3) \supset O(3)$. The density operators are O(3) scalars constructed from the generators of U(3). By a simple counting of polynomial terms and representations one finds immediately that the polynomials are not constructible from the Casimir invariants of U(3) and O(3) alone; one new invariant is needed* for m=3 and one for m=4. We make contact here with the cataloging of invariants by means of the "integrity basis", that basis in the present case consisting of n, $C_2(O_3)$, $C_2(U(3))$, $C_3(U(3))$, X_3, X_4. As a demonstration that there are no higher ones we have for m=50 (and N large enough to generate the irreps) a polynomial of 35,993 terms and the same number of density operators (including the $(\kappa(\kappa-1)/2)$ off-diagonal ones which enter when there is a κ-fold U(3)-O(3) multiplicity). It should be noted that X_4 is redundant for the state-

*The first four operators give 5 polynomial terms for $m \leq 2$ and 9 for $m \leq 3$; there are 5 irreps $(\lambda\mu;L)$ for $m \leq 2$ and 10 for $m \leq 3$. In constructing the polynomials all of the scalars including the new ones can be taken as effectively commuting since commutation reduces the particle rank by unity giving scalars which have already been counted.

labeling problem, and thus the propagation of information, for which X_4 is necessary, makes more demands on the group structure than does the simpler problem.

For the decomposition[10] $U(N) \supset U(N/4) \otimes U(4)$ with $U(4) \supset SU(2) \otimes SU(2)$, which is of considerable interest to nuclear physicists, we find as the integrity basis (by counting the $U(4) \supset SU(2) \otimes SU(2)$ irreps) n, G_2, G_3, G_4, S^2, T^2 and non-Casimir invariants X_3, X_4, Y_4, Z_4, X_5, X_6, Y_6, explicit forms for which have in fact been given by Miller.[11] But here the integrity-basis counting gives numbers which grow slightly more quickly than the number of density operators, for m=7,9,25,50 the numbers are (73,201,57125,11766281) for the former, and (73,200,48653,6541506) for the latter. The source of the difference cannot be that we have too many basic invariants; it must therefore be that not all products and powers of the basic invariants are independent. The redundancies here could be resolved once again by counting irreps. Alternatively these redundancies can be anticipated from generating-function forms[9,12] for the number of scalars of given polynomial order. In this case we find that Z_4 ($=S \cdot E^2 \cdot T$ where E is the Gamow-Teller operator) and X_5 ($=S \cdot E^3 \cdot T$) can only occur linearly while no products of X_6 ($=(S \times S)^2 \cdot (E^4)^2$) and Y_6($=(E^4)^2 \cdot (T \times T)^2$) can occur at all. With these restrictions the polynomial counting precisely reproduces the irrep counting above.

As another example we mention Elliott's SU(3) which appears in the decomposition $U(N) \supset U(4) \otimes U(N/4)$; $U(N/4) \supset SU(3)$. Here one finds easily that when N=12,24 the invariants up to m=2 have a Casimir basis so that SU(3) centroids propagate simply in the p and ds-shell spaces; not so however for N=40 (fp shell), for which another invariant of particle rank 2 is needed.

We shall not discuss the explicit construction of the density operators for the non-Casimir case. We remark however that, in general the final forms of the propagation equation (11) will not apply, while, for the other forms we might have to resort to an explicit evaluation for each irrep (not just for the sum over all equivalent ones).

Finally we remark that we should be interested also in "non-scalar" information, as in Fig. 2. As a trivial example the quadrupole moment operator Q_0 has a zero average expectation value in the substates of a J-level but nonetheless carries information (about the intrinsic quadrupole moment) via the "double-barred" or tensor-coupled matrix element. Then there enter naturally the non-scalar density operators $\rho_\Gamma^\nu(m) \sim \sum_{\alpha\varepsilon\Gamma} (\psi_\alpha^\Gamma(m) \times \bar{\psi}_\alpha^\Gamma(m))^\nu$ (in which $(\times)$ denotes tensor coupling and $\bar{\psi}^\Gamma$ the tensor adjoint) and the corresponding extension of the integrity basis. This is particularly important for strength distributions.[12]

6. Information Propagation in Dilute Spaces

Looking at things from the standpoint of many-body physics we might well ask for the source of the great complications which show up in constructing the integrity basis. We recall that one type of unpleasantness arises in many-body-theory from the Pauli Principle which forbids two fermions to be in the same state. The (blocking) corrections involved disappear in dilute systems, those in which the single-state occupancies are much less than one. Thus $m \ll N$; the more detailed requirement arises from the need to avoid "condensations" in subsets of single-particle states whose occupancies then would be comparable with unity. There's no trouble calculating the occupancies (just go back to section 3) so that whether or not a system is dilute can in fact be determined.

If we first use completeness to write $\langle\langle \psi_\alpha(k)\psi_\beta^+(k)\rangle\rangle^m = \langle\langle \psi_\beta^+(k)\psi_\alpha(k)\rangle\rangle^{m-k}$, and then use (10) we find different propagation forms,

$$\begin{array}{l} \langle\langle F(k)\rangle\rangle^m = \langle\langle \tilde{F}(k)\rangle\rangle^{m-k} = \langle\langle F(k)\rangle\rangle^{N-m+k} = \langle\langle \tilde{\rho}(m-k)F(k)\rangle\rangle^k \\ \qquad\qquad\quad \downarrow \\ \qquad\qquad \displaystyle\sum_{p=0}^{k} \langle\langle (\tilde{F}(k))_p\rangle\rangle^{m-k} = \sum_{p=0}^{k} \langle\langle \tilde{\rho}(m-k-p)(\tilde{F}(k))_p\rangle\rangle^p \end{array} \qquad (12)$$

In the second branch of this equation $(\tilde{F}(k))_p$ is the p-body part of the k-hole operator $\tilde{F}(k)$ which of course has particle ranks $0,1,\ldots k$. The p-decomposition given here is important for dilute systems ($N\to\infty$, $m/N\to 0$) because in such systems $\langle\langle \tilde{F}(k)\rangle\rangle^{m-k}$ is dominated by its $p=0$ part; this comes about because the system has a great many holes, which are essentially counted by $\tilde{F}(k)$, but very few particles to inhibit the counting. The p'th term in fact is proportional to $(m/N)^p$, as one sees by noting that for the prototype operator $\binom{n}{k}$ we have* $\widetilde{\binom{n}{k}}_p = (-1)^p \binom{N-p}{k-p}\binom{n}{p}$.

Equation (12) is trivial as it stands; we need its decomposition according to a group G, producing thereby the equivalent of (11). But things are quite different now because any decomposition $\underset{\sim}{m} \to \sum_{\Gamma} (\underset{\sim}{m},\Gamma)$ will, for most operators, destroy the completeness used in the first step of (12); we get around this**, just as one does with multipole sum rules, by making use of the non-scalar density operators $\rho_{\Gamma}^{\nu}(m)$ introduced above and using recoupling algebra to isolate single intermediate Γ spaces. We end up then with a tensorial extension of (12) which we can use for propagating scalar or non-scalar information, and which is particularly appropriate for dilute systems.

In p=0 approximation we find for a k-body operator, $F^0(k:\lambda,\lambda)$, whose tensorial structure is $(\psi^{\lambda}(k)\times\bar{\psi}^{\lambda}(k))^0$, and which therefore is defined in the (k,λ) subspace, that

$$\langle\langle F^0 \rangle\rangle^{m\Gamma} = d(\Gamma) d^{-1}(\lambda) \langle\langle F^0 \rangle\rangle^{k} \sum_{\Gamma_1} d_{\ell}(m-k,\Gamma_1)\delta_{\Gamma\lambda\Gamma_1} \qquad (13)$$

where d_{ℓ} is a multiplicity, $d(\Gamma)$ a dimensionality, and $\delta_{\Gamma\lambda\Gamma_1}$ the triangle function. For the coupled trace of a non-scalar operator we would find not a sum over multiplicities but instead a Racah transform of the multiplicity function. The structure revealed here is as simple as could be; F^0 is measured in terms of its defining trace and the contribution to any $(\underset{\sim}{m},\Gamma)$ irrep is simply proportional to the number of parents which $(\underset{\sim}{m},\Gamma)$ has in the space $(\underset{\sim}{m}-\underset{\sim}{k})$. By taking F^0 as the (k,λ) density operator we evaluate the propagation coefficient precisely in terms of the weight of one space in the other. The complexities of propagation then disappear in the dilute-system limit. They gradually re-enter of course as we take into account terms of higher p.

---

*For scalar traces $F(k) \equiv \langle F(k) \rangle^k \binom{n}{k}$ so that the result is general. Note that the higher-p terms in (12) can also be expanded similarly, so that we have an expansion in (m/N). The error made in truncating at any stage is of course calculable in this simple case, but is significant because it can be applied to group decompositions as well.

**In defining ρ^{ν} above we have assumed that the tensor coupling is multiplicity free, but this restriction is in fact not necessary.

7. Conclusion

Our subject has been spectral distribution and group symmetries though we have been able to refer to only a few papers in each domain. We have stressed that group theory constructions enter in a particularly elegant way when many-body problems are treated via spectral distributions; and by the same token, many-body results shed some new light on problems in group theory.

References

1) A. Pandey, Ann. Phys. (N.Y.), in press.

2) V. Potbbare and S. P. Pandya, Nucl. Phys. A256 (1976) 253.

3) K. K. Mon and J. B. French, Ann. Phys. (N.Y.) 95 (1975) 90.

4) H. A. Bethe, Phys. Rev. 50 (1936) 332; J. G. Cleary and B. G. Wybourne, J. Math. Phys. 12 (1971) 45; T. R. Halemane, unpublished.

5) M. Nomura, Progr. Theoret. Phys. 48 (1972) 442.

6) J. B. French, in Isospin in Nuclear Physics, D. H. Wilkinson editor, (North-Holland, Amsterdam, 1969).

7) C. Quesne, J. Math. Phys. 16 (1975) 2427.

8) J. P. Elliott, Proc. Roy. Soc. A245 (1958) 128, 562; V. Bargmann and M. Moshinsky, Nucl. Phys. 18 (1960) 697, 23 (1961) 177; G. Racah, in Group Theoretical Concepts and Methods in Elementary Particle Physics, F. Gürsey editor (Gordon and Breach, New York, 1964).

9) B. R. Judd, W. Miller, Jr., J. Patera and P. Winternitz, J. Math. Phys. 15, (1974) 1787.

10) R. U. Haq and J. C. Parikh, Nucl. Phys. A220 (1974) 349; C. Quesne, J. Math. Phys. 17 (1976) 1452.

11) W. Miller, Jr., as quoted in Ref. 11.

12) R. Gaskell, A. Peccia and R. T. Sharp, J. Math. Phys. 19 (1978) 727.

13) J. P. Draayer, J. B. French and S. S. M. Wong, Ann. Phys. (N.Y.) 106 (1977) 472, 503.

SU_3 SYMMETRY AND INTEGRAL KERNELS FOR NUCLEAR CLUSTER PROBLEMS

K. T. HECHT and W. ZAHN
Department of Physics, University of Michigan, Ann Arbor, Mich., USA

1. Introduction

The theory of continuous groups has played an important role in our understanding of nuclei from the earliest studies of nuclear spectra. Wigner and Racah laid down their fundamental group theoretical program for spectroscopy with both nuclear and atomic spectra in mind. The full realization of this program has, however, been limited to very special cases. In practical nuclear physics applications the Wigner-Racah calculus is exploited to its fullest extent only in ordinary angular momentum theory. The reason for this is clear. The group chains of physical interest are hardly ever the "mathematically natural" group chains, the so-called canonical group chains for which the irreducible representations of the full subgroup chain give a complete solution to the labeling problem. Despite the fact that a fractional parentage coefficient, (to give one example), is a reduced Wigner coefficient, or (even better) can be factored into a product of several such reduced Wigner coefficients, cfp's are not actually calculated by the elegant methods which have been developed in recent years (by Biedenharn and Louck,[1)] e.g.) to calculate the Wigner coefficients for the canonical group chains such as $U(n)\supset U(n-1)\supset\ldots\supset U(1)$.

The simplest nontrivial continuous group, SU_3, plays a special role in nuclear shell theory: It is the symmetry group of the harmonic oscillator. Since the orbital angular momentum subgroup does not fit "naturally" into SU_3, group theoretical techniques have until recently not been used to their fullest even in this simple case. However, for SU_3 the technology needed for a detailed application of the Wigner-Racah calculus has now been developed to a stage that it can be exploited in challenging, practical problems in nuclear theory. SU_3 recoupling techniques, in particular, can be used to good advantage. Three points are important:

(1) The difficulties associated with the fact that the angular momentum quantum number does not fit "naturally" into the SU_3 scheme can be avoided until the very last step of a calculation by working in a basis in which the various components of a wave function are SU_3-coupled, successively, to resultants of good SU_3 symmetry. In such a so-called SU_3 strong-coupled basis results are given largely in terms of (subgroup label independent) Racah or recoupling-coefficients.

(2) By using a generalized Wigner-Eckart theorem to factor matrix elements into SU_3-reduced parts and SU_3 Wigner coefficients, any convenient basis, such as a Cartesian oscillator basis, can be used to calculate the SU_3-reduced matrix elements. Matrix elements in the physically meaningful basis of good angular momentum can then be constructed trivially by combining the SU_3-reduced matrix elements with

available $SU_3 \supset R_3$ coefficients.

(3) The needed Wigner and Racah coefficients are now readily available through efficient computer codes, the Wigner coefficient in both the $SU_3 \supset R_3$ and the $SU_3 \supset SU_2 \times U_1$ basis through the code of Akiyama and Draayer,[2)] the Racah coefficients and SU_3 analogs of 9-j recoupling coefficients also through this code and its generalization by Millener.[3)]

To illustrate the power of SU_3 recoupling techniques, we have chosen the special field of nuclear cluster-model physics to show how such techniques can facilitate calculations. The integral kernels which arise in a resonating group method used in detailed microscopic nuclear cluster model calculations lead to challenging technical problems, particularly in nuclear cluster systems made up of many fragments, or in systems with heavy fragments other than closed shell nuclei such as ^{16}O.

2. The Nuclear Cluster Model

The nuclear cluster model presupposes that a nucleus can be described by wave functions built from (n+1) cluster components or fragments, with no internal excitations. The cluster model basis is to be built from fully antisymmetrized wave functions of the type

$$\Psi_{\text{cluster}} = A \prod_{i=1}^{n+1} \phi_i(\xi_i)^{(00)} [[\chi(\vec{R}_1)^{(Q_1 0)} \times \chi(\vec{R}_2)^{(Q_2 0)}]^{(\lambda_{12}\mu_{12})} \ldots \times \chi(\vec{R}_n)^{(Q_n 0)}]^{(\lambda\mu)}_{\kappa LM} .$$

We will assume that the internal wave function of the ith fragment, ϕ_i, is built from 0s oscillator wave functions in the internal degrees of freedom (ξ_i) and hence must have mass number $A_i \leq 4$ and SU_3 symmetry $(\lambda\mu)=(00)$ in Elliott's notation. It is also assumed that the ϕ_i include the spin-isospin function for each fragment. The relative motion functions $\chi(\vec{R}_i)^{(Q_i 0)}$ are first assumed to be harmonic oscillator functions of equal oscillator frequency in the Jacobi vectors $\vec{R}_i$ which describe the relative positions of the various fragments. The assumption of equal size can be relaxed subsequently. For our purposes it will be extremely useful if these relative motion functions, each carrying Q_i oscillator quanta, are SU_3 coupled successively, to resultant SU_3 symmetry $(\lambda\mu)$. The square brackets denote SU_3 coupling. The above wave function can also be made to include cluster systems composed of heavier fragments. If some of the Q_i are limited to their lowest Pauli-allowed values, the antisymmetrizer, A, assures that the wave function of the heavy fragment, made up of a particular aggregate of this kind, is itself the internal wave function corresponding to a ground state shell model configuration of definite SU_3 symmetry of the corresponding nucleus. E.g., if $\vec{R}_1$ is the relative position vector for two α-particles and $\vec{R}_2$ is the position vector for a third α-particle relative to the center of mass of the first two, then with $Q_1=Q_2=4$, $(\lambda_{12}\mu_{12})=(04)$, the antisymmetrizer assures that this 12-particle fragment is in the lowest p-shell configuration of highest space symmetry, a good approximation for the ground state of ^{12}C. The

12-particle function with $(\lambda_{12}\mu_{12})=(04)$ also includes the 2^+ and 4^+ excitations which together with 0^+ make the ground state rotational band.

Two cluster systems will be used as illustrative examples: (1) the 3-cluster system $^{12}C=\alpha+\alpha+\alpha$ with arbitrary excitations Q_1, Q_2 for the two relative motion functions, for which we will discuss the full dynamics in terms of realistic interactions. There have been many cluster studies of this nucleus. Our purpose here is to illustrate a new method using SU_3 recoupling techniques which may have wider applicability in more challenging problems. (2) The 2-component cluster system made up of an α-cluster and a heavy fragment. The norm of the cluster function for such a system has useful applications in α-transfer spectroscopy or in the study of α-widths if the norms can be calculated for relative motion functions up to the very high oscillator excitations which may be needed to construct radial functions of realistic shape.

If the cluster wave function Ψ is abbreviated by

$$\Psi_{\text{cluster}} = A \prod_{i=1} \phi_i(\xi_i)\chi(\vec{R}),$$

where $\vec{R}$ stands collectively for $\vec{R}_1,\ldots,\vec{R}_n$, the kernel $K(\vec{\bar{R}},\vec{R})$ for an operator, H, acting on this Ψ is defined by

$$\int d\vec{\xi} \prod_{i=1} \phi_i^* \, H \, A \prod_{i=1} \phi_i \chi(\vec{\bar{R}}) = \int d\vec{R}\; K(\vec{\bar{R}},\vec{R})\chi(\vec{R}) \; .$$

The detailed solution of the nuclear cluster problem by a resonating group method is reduced to manageable proportions with the development of efficient techniques for the evaluation of integrals of the type $\int d\vec{\bar{R}}\, d\vec{R}\;\; \chi^*(\vec{\bar{R}})K(\vec{\bar{R}},\vec{R})\chi(\vec{R})$. If the kernel has complicated SU_3 or spherical tensor rank these equations are highly schematic since they imply considerable SU_3 or angular momentum coupling.

3. The Bargmann-Segal Transform

Integral transform techniques have proved to be very useful in the evaluation of such integrals. Of the many integral transforms used in microscopic cluster model calculations the Bargmann-Segal (B-S) transform[4)] is ideally suited to the exploitation of SU_3 coupling techniques since oscillator functions have very simple SU_3 properties in Bargmann space. It will therefore be advantageous to introduce the B-S transform of the kernel K:

$$H(\vec{\bar{K}},\vec{K}) = \int d\vec{\bar{R}} d\vec{R}\; A(\vec{\bar{K}},\vec{\bar{R}})K(\vec{\bar{R}},\vec{R})\; A^*(\vec{K},\vec{R}),$$

where $\vec{K}$ stands collectively for $\vec{K}_1\ldots\vec{K}_n$, the Bargmann space variables for the n relative motion vectors $\vec{R}_1\ldots\vec{R}_n$, and $A(\vec{K},\vec{R})$ generates the B-S transform

$$F(\vec{K}) = \int d\vec{R}\; A(\vec{K},\vec{R})\chi(\vec{R})$$

specifically $$A(\vec{K},\vec{R}) = \prod_{i=1}^{n} \prod_{\gamma=x,y,z} A(K_{i_\gamma},X_{i_\gamma})$$

where the 1-dimensional factor is defined by

$$A(K_x,X) = \sum_{n=0}^{\infty} \Psi_n^*(X)K_x{}^n/\sqrt{n!} \; .$$

The B-S transform of the normalized 1-dimensional harmonic oscillator function $\Psi_n(X)$, viz. $K_x{}^n/\sqrt{n!}$, is the normalized 1-dimensional oscillator function in Bargmann space. The fact that the oscillator ground state function is the simple number 1 leads to much of the simplicity of Bargmann space. The Bargmann space function

$$P(\vec{K})^{(Q,0)} = K_x^{n_x}/\sqrt{n_x!} \cdot K_y^{n_y}/\sqrt{n_y!} \cdot K_z^{n_z}/\sqrt{n_z!}$$

given here in a Cartesian oscillator basis, with $Q = n_x+n_y+n_z$, has SU_3 irreducible tensor character (Q0). The corresponding polynomial in $\vec{K}^*$ has SU_3 irreducible tensor character (0Q). An SU_3-coupled Bargmann space polynomial can be defined by

$$[P(\vec{K}_1)^{(Q_10)} \times P(\vec{K}_2)^{(Q_20)}]_\alpha^{(\lambda\mu)} = \sum_{\alpha_1\alpha_2} \langle (Q_10)\alpha_1 (Q_20)\alpha_2 | (\lambda\mu)\alpha \rangle P(\vec{K}_1)_{\alpha_1}^{(Q_10)} P(\vec{K}_2)_{\alpha_2}^{(Q_20)} ,$$

where the SU_3 subgroup labels, α, can be chosen in a convenient basis; including the angular momentum basis $\alpha = \kappa LM$. For many steps in the calculation a detailed knowledge of the SU_3 Wigner coefficients will not be necessary.

For a simple 3-cluster nucleus with two relative motion functions, the generating function can be expanded in terms of SU_3-coupled functions

$$A(\vec{K}_1,\vec{R}_1)\, A(\vec{K}_2,\vec{R}_2) = \sum_{Q_1Q_2} \sum_{(\lambda\mu)\alpha} [P(\vec{K}_1)^{(Q_10)} \times P(\vec{K}_2)^{(Q_20)}]_\alpha^{(\lambda\mu)} [\chi(\vec{R}_1)^{(Q_10)} \times \chi(\vec{R}_2)^{(Q_20)}]_\alpha^{(\lambda\mu)*} ,$$

with obvious generalizations to multi-cluster systems. The B-S transform of the kernel K can then be expanded in terms of SU_3-coupled $\vec{K}$-space polynomials

$$H(\vec{\bar{K}},\vec{K}) = \sum [P(\vec{\bar{K}}_1)^{(\bar{Q}_10)} \times P(\vec{\bar{K}}_2)^{(\bar{Q}_20)}]_{\bar{\alpha}}^{(\bar{\lambda}\,\bar{\mu})} [P(\vec{K}_1^*)^{(0Q_1)} \times P(\vec{K}_2^*)^{(0Q_2)}]_\alpha^{(\mu\lambda)}$$
$$\times \int d\vec{\bar{R}}\, d\vec{R}\, [\chi(\vec{\bar{R}}_1)^{(\bar{Q}_10)} \times \chi(\vec{\bar{R}}_2)^{(\bar{Q}_20)}]_{\bar{\alpha}}^{(\bar{\lambda}\,\bar{\mu})*} K [\chi(\vec{R}_1)^{(Q_10)} \times \chi(\vec{R}_2)^{(Q_20)}]_\alpha^{(\lambda\mu)} .$$

If we now imagine that the kernel (for a two-body interaction) is expanded in terms of SU_3 irreducible tensor components

$$K = \sum_{(\lambda_o\mu_o)} K_{L_o=0}^{(\lambda_o\mu_o)} ,$$

we can use the Wigner-Eckart theorem for the $(\lambda_o\mu_o)^{th}$ component of the $\vec{\bar{R}},\vec{R}$-space integrals to express these integrals in terms of SU_3-reduced (double-barred) matrix elements and SU_3 Wigner coefficients by

$$\sum_{\rho_o} \langle \bar{Q}_1\bar{Q}_2(\bar{\lambda}\,\bar{\mu}) || K^{(\lambda_o\mu_o)} || Q_1Q_2(\lambda\mu) \rangle_{\rho_o} \langle (\bar{\lambda}\,\bar{\mu})\bar{\alpha}(\mu\lambda)\alpha^* | (\lambda_o\mu_o)L_o=0 \rangle_{\rho_o} .$$

(The outer multiplicity label ρ_o is needed when the Kronecker product $(\bar{\lambda}\,\bar{\mu})\times(\mu\lambda)$ contains $(\lambda_o\mu_o)$ with a d-fold multiplicity, with $d>1$.) The SU_3 Wigner coefficients carry all dependence on SU_3 subgroup labels α. Using this form for the $\vec{\bar{R}},\vec{R}$-space integrals, the $\bar{\alpha},\alpha$ sums can be carried out; and the B-S transform of the kernel is given by the expansion

$$H(\vec{\bar{K}},\vec{K}) = \sum_{\bar{Q}_1\bar{Q}_2Q_1Q_2} \sum_{(\bar{\lambda}\,\bar{\mu})(\lambda\mu)} \sum_{(\lambda_o\mu_o)\rho_o} \langle \bar{Q}_1\bar{Q}_2(\bar{\lambda}\,\bar{\mu}) || K^{(\lambda_o\mu_o)} || Q_1Q_2(\lambda\mu) \rangle_{\rho_o}$$
$$[[P(\vec{\bar{K}}_1)^{(\bar{Q}_10)} \times P(\vec{\bar{K}}_2)^{(\bar{Q}_20)}]^{(\bar{\lambda}\,\bar{\mu})} \times [P(\vec{K}_1^*)^{(0Q_1)} \times P(\vec{K}_2^*)^{(0Q_2)}]^{(\mu\lambda)}]_{L=0}^{(\lambda_o\mu_o)\rho_o} .$$

By using the orthonormality of the $[P(\vec{K}_1)^{(Q_1 0)} \times P(\vec{K}_2)^{(Q_2 0)}]_\alpha^{(\lambda\mu)}$ in $\vec{K}$-space it is, in principle, possible to integrate over the complex variables $\vec{K}_1 \ldots, \vec{K}_2^*$ with the Bargmann $\vec{K}$-space measure[4)] to select a specific SU_3-reduced matrix element. In practice, a particular SU_3-reduced matrix element can be read directly from an expansion of $H(\vec{\bar{K}},\vec{K})$ in terms of the SU_3 irreducible tensor $\vec{K}$-space polynomials, SU_3-coupled successively to resultant $(\lambda_o\mu_o)$, in the form illustrated above. Hence no laborious irreducible tensor decomposition of the kernel $K(\vec{\bar{R}},\vec{R})$ needs to be carried out in $\vec{R}$-space. The SU_3-reduced matrix element will fall out of the B-S transform of this kernel quite naturally, if all factors in this transform are expressed in SU_3-coupled form.

The kernel of the unit operator is particularly simple since this is an SU_3-scalar, $(\lambda\mu)=(00)$-operator. The $\vec{R}$-space integrals give the norm, N, of the fully antisymmetrized cluster wave function. In this case

$$H(\vec{\bar{K}},\vec{K}) = \sum_{\bar{Q}_1\bar{Q}_2 Q_1 Q_2(\lambda\mu)} \frac{1}{N^2}[\dim(\lambda\mu)]^{\frac{1}{2}}[[P(\vec{\bar{K}}_1)^{(\bar{Q}_1 0)} \times P(\vec{\bar{K}}_2)^{(\bar{Q}_2 0)}]^{(\lambda\mu)} \times [P(\vec{K}_1^*)^{(0Q_1)} \times P(\vec{K}_2^*)^{(0Q_2)}]^{(\mu\lambda)}]^{(00)},$$

with straightforward generalizations to more complicated cluster systems: If the oscillator excitations, $Q_1 \ldots Q_{n-1}$ of the first (n-1) relative motion variables $\vec{K}_1, \ldots \vec{K}_{n-1}$ are restricted to their lowest Pauli-allowed values and SU_3-coupled to resultant $(\lambda_c\mu_c)$, the

$$[\ldots[P(\vec{K}_1)^{(Q_1 0)} \times P(\vec{K}_2)^{(Q_2 0)}]^{(\lambda_{12}\mu_{12})} \times \ldots \times P(\vec{K}_{n-1})]^{(\lambda_c\mu_c)} \equiv P(\vec{K}_1, \ldots, \vec{K}_{n-1})^{(\lambda_c\mu_c)}$$

become the B-S transforms of a heavy-fragment internal wave function, and, for fixed $\bar{Q}_1 = Q_1, \ldots, \bar{Q}_{n-1} = Q_{n-1}$, the generalization of the above equation gives the norm, N, of a [heavy-fragment + (n+1)[st] cluster component] - wave function with quantum numbers $(\lambda_c\mu_c)$, $(Q_n 0)$, $(\lambda\mu)$. For the 3-cluster system with arbitrary excitations $\bar{Q}_1, \bar{Q}_2$, Q_1, Q_2, the above $1/N^2$ is shorthand notation for a matrix which factors into submatrices for each $(\lambda\mu)$, with $\bar{Q}_1+\bar{Q}_2=Q_1+Q_2=\lambda+2\mu$. The eigenvectors for non-zero eigenvalues of $1/N^2$ give the Pauli-allowed states of this cluster system. Pauli-forbidden states have zero eigenvalues for $1/N^2$.

4. The $^{12}C = \alpha+\alpha+\alpha$ Cluster System

To show how the above program can be implemented in a practical cluster model calculation let us consider the simple 3-cluster system, $^{12}C = \alpha+\alpha+\alpha$. For such a cluster system the B-S transform of the kernel for the full interaction Hamiltonian is of simple Gaussian form if the two-body interaction can be expanded in terms of Gaussians. That is, in this case[4)]

$$H(\vec{\bar{K}},\vec{K}) = \sum_\beta a_\beta \exp\{\rho(\beta)\} \exp\{\sigma(\beta)\} \exp\{\tau(\beta)\} .$$

The β-sum arises from the antisymmetrizer, which can be expressed in terms of a double coset decomposition. The β-sum runs over the full set of double coset generators. The weighting factors a_β are calculated by the techniques of Kramer and Seligman.[5] The exponentials are abbreviations for

$$\exp\{\sigma(\beta)\} \equiv \exp\{\sum_{i,j=1}^{2} \sigma_{ij}(\beta)(\bar{\vec{K}}_i\cdot\vec{K}_j^*)\} ,$$

$$\exp\{\rho(\beta)\} \equiv \exp\{\sum_{i,j=1}^{2} \rho_{ij}(\beta)(\bar{\vec{K}}_i\cdot\bar{\vec{K}}_j)\} ,$$

$$\exp\{\tau(\beta)\} \equiv \exp\{\sum_{i,j=1}^{2} \tau_{ij}(\beta)(\vec{K}_i^*\cdot\vec{K}_j^*)\} .$$

An expansion of these exponentials followed by a succession of SU_3-recoupling transformations of the $\vec{K}$-space polynomials will lead to the needed final form in which $H(\bar{\vec{K}},\vec{K})$ is expanded in terms of the SU_3-coupled tensors

$$[[P(\bar{\vec{K}}_1)^{(\bar{Q}_1 0)}\times P(\bar{\vec{K}}_2)^{(\bar{Q}_2 0)}]^{(\bar{\lambda}\,\bar{\mu})}\times[P(\vec{K}_1^*)^{(0Q_1)}\times P(\vec{K}_2^*)^{(0Q_2)}]^{(\mu\lambda)}]_0^{(\lambda_o\mu_o)\rho_o}.$$

The basic building blocks for this expansion are

$$\frac{(\bar{\vec{K}}_i\cdot\vec{K}_j^*)^n}{n!} = [P(\bar{\vec{K}}_i)^{(n0)}\times P(\vec{K}_j^*)^{(0n)}]_{L=0}^{(00)}[\dim(n0)]^{\frac{1}{2}}$$

and, with i=j

$$\frac{(\bar{\vec{K}}_i\cdot\bar{\vec{K}}_i)^n}{n!} = \frac{[(2n+1)!]^{\frac{1}{2}}}{n!}\, P(\bar{\vec{K}}_i)_{L=0}^{(2n,0)}$$

while, with i≠j:

$$\frac{(\bar{\vec{K}}_i\cdot\bar{\vec{K}}_j)^n}{n!} = \sum_{(\lambda\mu)} [\lambda+1]^{\frac{1}{2}}(-1)^{\mu/2}[P(\bar{\vec{K}}_i)^{(n0)}\times P(\bar{\vec{K}}_j)^{(n0)}]_{L=0}^{(\lambda\mu)} ,$$

where $(\lambda\mu)$ runs over the values (2n,0), (2n-4,2), (2n-8,4),...(0n) for n=even,... (2,n-1) for n=odd.

With SU_3-recoupling transformations which take us from

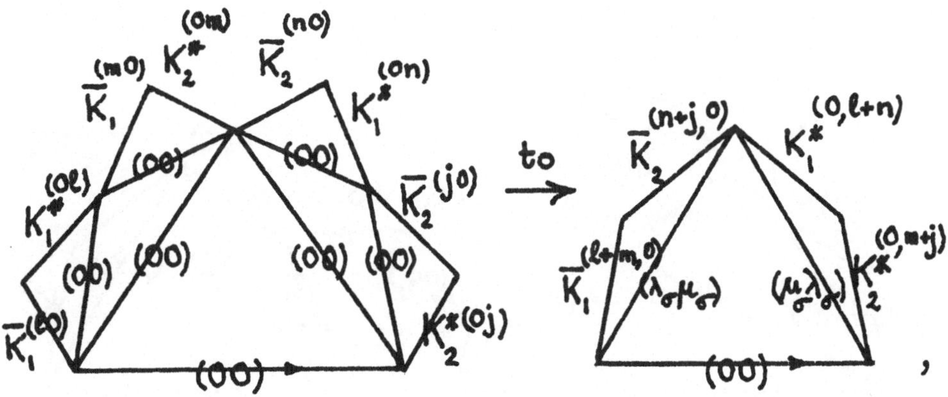

,

the exp{σ} factor can be expanded by

$$\exp\{\sigma\} = \sum_{(\lambda_\sigma\mu_\sigma)} \sum_{\substack{\ell+m,n+j\\ \ell+n,m+j}} C_\sigma\big(\ell+m,n+j,\ell+n,m+j;(\lambda_\sigma\mu_\sigma)\big)$$

$$[[P(\vec{\bar{K}}_1)^{(\ell+m,0)}\times P(\vec{\bar{K}}_2)^{(n+j,0)}]^{(\lambda_\sigma\mu_\sigma)}\times[P(\vec{K}_1^*)^{(0,\ell+n)}\ P(\vec{K}_2^*)^{(0,m+j)}]^{(\mu_\sigma\lambda_\sigma)}]_0^{(00)}$$

with

$$C_\sigma = \sum_{\substack{\ell\\(m,n,j)}} \sigma_{11}^{\ell}\sigma_{12}^{m}\sigma_{21}^{n}\sigma_{22}^{j}[\dim(\lambda_\sigma\mu_\sigma)]^{\frac{1}{2}}\left[\binom{\ell+m}{\ell}\binom{n+j}{j}\binom{\ell+n}{\ell}\binom{m+j}{j}\right]^{\frac{1}{2}}\begin{bmatrix}(\ell 0) & (m0) & (\ell+m,0)\\ (n0) & (j0) & (n+j,0)\\ (\ell+n,0) & (m+j,0) & (\lambda_\sigma\mu_\sigma)\end{bmatrix},$$

where the U_3 9-(λμ) coefficient contains at most 2-rowed representations so that it is equivalent to a very simple SU_2 9-j coefficient with 4 "stretched" couplings.

The expansion of the exp{ρ} factor in terms of SU_3-coupled $\vec{K}$-space polynomials is given by

$$\exp\{\rho\} = \sum_{\bar{q}_1\bar{q}_2}\sum_{(\lambda_\rho\mu_\rho)} B_\rho(\bar{q}_1\bar{q}_2;(\lambda_\rho\mu_\rho)) \times [P(\vec{\bar{K}}_1)^{(\bar{q}_1 0)}\times P(\vec{\bar{K}}_2)^{(\bar{q}_2 0)}]^{(\lambda_\rho\mu_\rho)}_{L_\rho=0},$$

where

$$B_\rho(\bar{q}_1\bar{q}_2;(\lambda_\rho\mu_\rho)) = \sum_{\substack{n,m\\ n+m=\frac{1}{2}(\bar{q}_1+\bar{q}_2)}} \rho_{aa}^{n}\rho_{bb}^{m}$$

$$\frac{[(2n+1)!(2m+1)!]^{\frac{1}{2}}}{n!m!}\Big\langle (2n,0)0;(2m,0)0\Big\|(\lambda_\rho\mu_\rho)0\Big\rangle\times d^{\frac{1}{2}\lambda\rho}_{\frac{1}{2}(\bar{q}_1-\bar{q}_2),(n-m)}(\beta).$$

The quadratic form $\{\Sigma\rho_{ij}\vec{\bar{K}}_i\cdot\vec{\bar{K}}_j\}$ has first been put into diagonal form: $\{\rho_{aa}(\vec{\bar{K}}_a\cdot\vec{\bar{K}}_a)+\rho_{bb}(\vec{\bar{K}}_b\cdot\vec{\bar{K}}_b)\}$. The SU_2 d-coefficient, with β given by $\tan\beta=2\rho_{12}/(\rho_{22}-\rho_{11})$, is the generalized Moshinsky bracket for SU_3-coupled oscillator functions which takes us from the $[P(\vec{\bar{K}}_a)\times P(\vec{\bar{K}}_b)]^{(\lambda\mu)}$ to the $[P(\vec{\bar{K}}_1)\times P(\vec{\bar{K}}_2)]^{(\lambda\mu)}$ basis. The double-barred coefficient is a very simple $SU_3\supset R_3$ Wigner coefficient with all L's equal to zero. A similar expansion holds for exp{τ}.

By coupling the expansions of the ρ, σ, and τ factors and combining these by SU_3 recoupling transformations which take us from

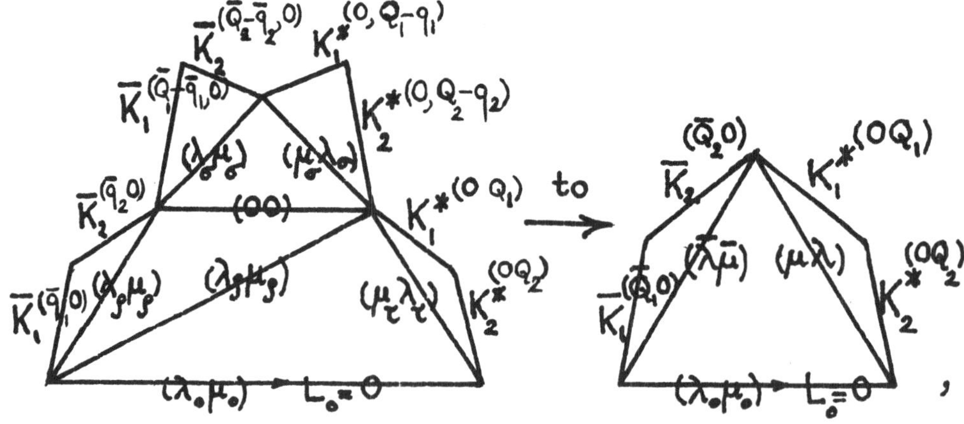

,

the resultant is expanded in a form needed to pick off the desired SU_3-reduced matrix elements

$$\exp\{\rho\}\ \exp\{\sigma\}\ \exp\{\tau\} = \Sigma\ C_\sigma(\bar{Q}_1-\bar{q}_1,\bar{Q}_2-\bar{q}_2,Q_1-q_1,Q_2-q_2;(\lambda_\sigma\mu_\sigma))\times B_\rho(\bar{q}_1,\bar{q}_2,(\lambda_\rho\mu_\rho))\times$$
$$B_\tau(q_1,q_2,(\lambda_\tau\mu_\tau))\times Z(\bar{Q}_1\bar{Q}_2Q_1Q_2\bar{q}_1\bar{q}_2q_1q_2(\lambda_\rho\mu_\rho)(\lambda_\sigma\mu_\sigma)(\lambda_\tau\mu_\tau)(\bar{\lambda}\ \bar{\mu})(\lambda\mu)(\lambda_o\mu_o)\rho_o)\times$$
$$[[P(\vec{\bar{K}}_1)^{(\bar{Q}_10)}\times P(\vec{\bar{K}}_2)^{(\bar{Q}_20)}]^{(\bar{\lambda}\bar{\mu})} \quad\times\quad [P(\vec{K}_1^*)^{(0Q_1)}\times P(\vec{K}_2^*)^{(0Q_2)}]^{(\mu\lambda)}]^{(\lambda_o\mu_o)\rho_o}_{L_o=0}\ ,$$

where

$$Z=U((\lambda_\rho\mu_\rho)(\lambda_\sigma\mu_\sigma)(\lambda_\rho\mu_\rho)(\mu_\sigma\lambda_\sigma);(\bar{\lambda}\ \bar{\mu});(00))\times\Big\{\sum_{\bar{\rho}_o}\Big\langle(\lambda_\rho\mu_\rho)L_\rho=0;(\mu_\tau\lambda_\tau)L_\tau=0\,||\,(\lambda_o\mu_o)L_o=0\Big\rangle_{\bar{\rho}_o}$$
$$\times U((\bar{\lambda}\ \bar{\mu})(\mu_\sigma\lambda_\sigma)(\lambda_o\mu_o)(\mu_\tau\lambda_\tau);(\lambda_\rho\mu_\rho)1\bar{\rho}_o;(\mu\lambda)1\rho_o)\Big\}$$

$$\left[\begin{pmatrix}\bar{Q}_1\\ \bar{q}_1\end{pmatrix}\begin{pmatrix}\bar{Q}_2\\ \bar{q}_2\end{pmatrix}\begin{pmatrix}Q_1\\ q_1\end{pmatrix}\begin{pmatrix}Q_2\\ q_2\end{pmatrix}\right]^{\frac{1}{2}}\left[\begin{matrix}(\bar{q}_10) & (\bar{q}_20) & (\lambda_\rho\mu_\rho)\\ (\bar{Q}_1-\bar{q}_1,0) & (\bar{Q}_2-\bar{q}_2,0) & (\lambda_\sigma\mu_\sigma)\\ (\bar{Q}_10) & (\bar{Q}_20) & (\bar{\lambda}\ \bar{\mu})\end{matrix}\right]\left[\begin{matrix}(Q_1-q_1,0) & (Q_2-q_2,0) & (\lambda_\sigma\mu_\sigma)\\ (q_10) & (q_20) & (\lambda_\tau\mu_\tau)\\ (Q_10) & (Q_20) & (\lambda\mu)\end{matrix}\right].$$

The U-coefficients are SU_3 Racah coefficients in unitary form.[2] The 9-($\lambda\mu$) coefficients contain at most 2-rowed U_3 representations so that they are again equivalent to simple SU_2 coefficients.

If the cluster wave function is expanded in a basis in which all internal and relative motion wave functions are harmonic oscillator wave functions of equal oscillator frequency, the lowest Pauli-allowed wave functions for ^{12}C must have $\bar{Q}_1=\bar{Q}_2=Q_1=Q_2=4$ and $(\bar{\lambda}\ \bar{\mu})=(\lambda\mu)=(04)$. In this restricted subspace of the cluster function basis the SU_3-irreducible tensor character of the interaction is restricted to components with $(\lambda_o\mu_o)=(00),(22),(44)$.

It is interesting to compare the relative magnitudes of these components. For this purpose the magnitudes of the relevant SU_3-reduced matrix elements for an interaction[6] of the type[7] used in earlier successful treatments of the $(\alpha+\alpha+\alpha)$-cluster system[8] are shown in the table below.

SU_3-Reduced Matrix Elements $\langle 44(04)\,||\,K^{(\lambda_o\mu_o)}\,||\,44(04)\rangle$

$(\lambda_o\mu_o)$	(00)	(22)	(44)	
Central Pot.	-799.9	51.9	-0.5	MeV
Kinetic E.	508.5	-29.5	0.07	
Coulomb Pot.	27.0	-1.5	-0.007	
Total	-264.4	20.9	-0.43	MeV

The SU_3-scalar component is by far the most important. The $(\lambda_o\mu_o)=(22)$-tensor is also significant. It gives rise to a nearly pure-rotational $0^+2^+4^+$ excitation spectrum. The $(\lambda_o\mu_o)=(44)$-tensor which, in principle, could have been responsible for large deviations from a pure L(L+1) excitation spectrum is seen to be almost completely negligible.

Clearly, a realistic calculation will require either a much larger pure oscillator basis, or, preferably, will stem from a variational treatment in which radial functions are expanded in terms of oscillator functions of several different frequencies.[8] In our treatment up to now it has seemingly been quietly assumed that all relative motion functions are oscillator functions with the same frequency, which may, however, differ from the frequencies of the internal wave functions of the cluster-components. Insofar as all of our techniques make use of oscillator functions $\chi(\vec{R})$ in dimensionless variables, $\vec{R}$, with unit weighting factors in the B-S transform $A(\vec{K},\vec{R})$, they apply with equal validity to the case where different relative motion functions have different frequencies. To apply SU_3 coupling techniques each $\vec{R}$ must be a dimensionless variable. A dimensionless $\vec{R}_i$ is equal to a physical $\vec{R}_i$, (measured in cm or fm), divided by the oscillator length parameter $[\hbar/M_i\omega_i]^{\frac{1}{2}}$ with appropriate reduced mass M_i and frequency ω_i of whatever size is convenient.

5. The Norm Problem for (α+Heavy-Fragment)-Cluster Systems

The technique outlined in section 4 handles antisymmetrization by a double coset expansion of the (n+1)-cluster problem. For heavy nuclei, with $A>16$, e.g., the number of terms in such an expansion becomes very large, and more indirect methods may be preferable. Such methods have been developed[9] for the norm problem, in particular. The 2-component cluster system made up of an α-particle and a heavy fragment, (e.g. $^{19}F=\alpha+^{15}N$), has been chosen as a specific example. Our aim here is to calculate the norm for arbitrary excitations of the α-heavy-fragment relative motion oscillator functions, since an expansion in an oscillator basis may have to include excitations up to high values to enable us to construct radial functions of realistic shape.

The B-S transform of the α+heavy-fragment cluster wave function has the form

$$[P(\vec{K}_1,\ldots\vec{K}_{n-1})^{(\lambda_c\mu_c)}\times P(\vec{K}_n)^{(Q_n0)}]_\alpha^{(\lambda\mu)} \quad .$$

The internal wave function of the heavy fragment is built from 0s internal wave functions of n cluster components and n-1 relative motion functions limited to their lowest Pauli-allowed excitations and coupled to resultant SU_3 symmetry $(\lambda_c\mu_c)$. In a basis of harmonic oscillator functions of equal frequency the norm is particularly simple since it is an SU_3-scalar. For an (n+1)-cluster system the B-S transform of the norm then has the general form

$$H(\vec{\bar{K}},\vec{K}) = \sum_\beta a_\beta \exp\{\sum_{i,j=1}^{n} \sigma_{ij}(\beta)(\vec{\bar{K}}_i\cdot\vec{K}_j^*)\} \quad ,$$

where the β sum runs over all double coset generators. For the restricted cluster system made up of a single α-particle and a heavy fragment frozen into its lowest Pauli-allowed state, the number of double coset generators is reduced to five; but the B-S transform of the norm is no longer of simple Gaussian form. The general form of the B-S transform for the unrestricted (n+1)-cluster function can be used to derive the form of the B-S transform of the restricted α+heavy-fragment cluster function by expanding the exponentials in the above and limiting the powers of $\vec{\bar{K}}_1,\ldots,\vec{\bar{K}}_{n-1};\vec{K}^*_1,\ldots,\vec{K}^*_{n-1}$ to their minimum Pauli-allowed values as dictated by the internal function of the heavy fragment. For this purpose it is convenient to split the exponentials into three factors:

(1) The first factor is to include only those terms in the sum with both $i,j \leq n-1$, that is, those terms which carry only oscillator excitations of internal degrees of freedom of the heavy fragment. In the restricted problem this factor leads to SU_3-coupled $\vec{K}$-space tensors of the form

$$[P(\vec{\bar{K}}_1,\ldots,\vec{\bar{K}}_{n-1})^{(\lambda'\mu')}\times P(\vec{K}^*_1,\ldots,\vec{K}^*_{n-1})^{(\mu'\lambda')}]^{(00)}_0$$

where the possible $(\lambda'\mu')$ are restricted to those Young tableaux which can be obtained by removing m squares from the Young tableau for $(\lambda_c\mu_c)$.

(2) The second factor is made up of the cross-terms of the form

$$\exp\left\{\sum_{i=1}^{n}\sigma_{in}(\vec{\bar{K}}_i\cdot\vec{K}^*_n)+\sum_{i=1}^{n}\sigma_{ni}(\vec{\bar{K}}_n\cdot\vec{K}^*_i)\right\}.$$

In the restricted problem it leads to contributions of the form

$$[P(\vec{\bar{K}}_1,\ldots\vec{\bar{K}}_{n-1})^{(m0)}\times P(\vec{K}^*_n)^{(0m)}]^{(00)}\times[P(\vec{\bar{K}}_n)^{(m0)}\times P(\vec{K}^*_1,\ldots\vec{K}^*_{n-1})^{(0m)}]^{(00)}.$$

These can be characterized by the single integer m, since each factor must be an SU_3 (00)-tensor, and both the $\vec{\bar{K}}_1,\vec{\bar{K}}_2,\ldots,\vec{\bar{K}}_{n-1}$-space and the $\vec{K}^*_1,\vec{K}^*_2,\ldots,\vec{K}^*_{n-1}$-space polynomials are restricted by the same coupling rule: $(\lambda'\mu')\times(m0)\rightarrow(\lambda_c\mu_c)$. If the heavy fragment is a p-shell nucleus, with $A\leq 16$, the Young tableaux for $(\lambda_c\mu_c)$ can have at most 4 columns so that m is restricted by $m\leq 4$ in this case.

(3) The third factor is left in its exponential form and leads to linear combinations of terms of the form

$$\exp\{\sigma_{nn}(p)(\vec{\bar{K}}_n\cdot\vec{K}^*_n)\} = \exp\left\{\frac{4(A-4)-pA}{4(A-4)}(\vec{\bar{K}}_n\cdot\vec{K}^*_n)\right\},$$

with $p=0,1,\ldots,4$ corresponding to the five DC generators of the α+(A-4)-particle fragment antisymmetrizer. The pth term arises from the exchange of p nucleons in the α-cluster with p nucleons in the heavy fragment.

For the case where the heavy fragment is a p-shell nucleus the B-S transform of the norm must thus have the general form

$$\begin{aligned}H(\vec{\bar{K}},\vec{K}) = \sum_{m=0}^{4}\sum_{(\lambda'\mu')} c_{m(\lambda'\mu')}&[P(\vec{\bar{K}}_1\ldots\vec{\bar{K}}_{n-1})^{(\lambda'\mu')}\times P(\vec{K}^*_1\ldots\vec{K}^*_{n-1})^{(\mu'\lambda')}]^{(00)}\\ \times&[P(\vec{\bar{K}}_1\ldots\vec{\bar{K}}_{n-1})^{(m0)}\times P(\vec{K}^*_n)^{(0m)}]^{(00)}\times[P(\vec{\bar{K}}_n)^{(m0)}\times P(\vec{K}^*_1\ldots\vec{K}^*_{n-1})^{(0m)}]^{(00)}\\ \times&\sum_{p=0}^{4}D_m(p)\exp\left\{\frac{4(A-4)-pA}{4(A-4)}(\vec{\bar{K}}_n\cdot\vec{K}^*_n)\right\},\end{aligned}$$

where we have assembled the three contributions with as yet undetermined coefficients, $c_{m(\lambda'\mu')}$. If the heavy fragment is a p-shell nucleus the coefficients

D(p) are functions of m only. This is related to the fact that the removal of m oscillator quanta from the heavy fragment function of symmetry $(\lambda_c\mu_c)$ to make a function of symmetry $(\lambda'\mu')$ requires the removal of m p-shell particles from the heavy fragment. Only exchange terms with $p\geq m$ can contribute to such terms in the sum, and $D_m(p)=0$ for $p<m$. These $D_m(p)$ have the simple form:

m \ p	0	1	2	3	4
0	1	-4	6	-4	1
1	0	-4	12	-12	4
2	0	0	6	-12	6
3	0	0	0	-4	4
4	0	0	0	0	1

The simplicity of these numbers is related to the fact that they are a group theoretical construct and can be calculated from properties of the permutation group. After expanding the exponentials in terms of $[P(\vec{K}_n)^{(Q_n-m,0)}\times P(\vec{K}_n^{*})^{(0,Q_n-m)}]^{(00)}$ the B-S transform of the α+(A-4)-particle-fragment norm can be put into the needed form

$$H(\vec{K}^{*},\vec{K}) = \sum_{(\lambda_c\mu_c)(Q_n0)(\bar{\lambda}\,\bar{\mu})} \frac{1}{[N(\lambda_c\mu_c)(Q_n0)(\bar{\lambda}\,\bar{\mu})]^2}\,[\dim(\bar{\lambda}\,\bar{\mu})]^{\frac{1}{2}}$$

$$\times\,[[P(\vec{K}_1,\ldots\vec{K}_{n-1})^{(\lambda_c\mu_c)}\times P(\vec{K}_n)^{(Q_n0)}]^{(\bar{\lambda}\,\bar{\mu})}\times[P(\vec{K}_1^{*},\ldots\vec{K}_{n-1}^{*})^{(\mu_c\lambda_c)}\times P(\vec{K}_n^{*})^{(0Q_n)}]^{(\bar{\mu}\,\bar{\lambda})}]^{(00)}$$

by straightforward SU_3-recoupling transformations which give the coefficient

$$\frac{1}{N^2} = \sum_{m=0}^{4}\sum_{(\lambda'\mu')} c_m(\lambda'\mu')\,\frac{(-1)^{\lambda_c+\mu_c+Q_n-\bar{\lambda}-\bar{\mu}}}{[\dim(m0)]^2}\left[\frac{\dim(\lambda_c\mu_c)\dim(Q_n0)}{\dim(\lambda'\mu')\dim(\bar{\lambda}\,\bar{\mu})}\right]^{\frac{1}{2}}$$

$$\times\left\{\sum_{p=0}^{4} D_m(p)\left[\frac{4(A-4)-pA}{4(A-4)}\right]^{Q_n-m}\right\}\begin{pmatrix}Q_n\\ m\end{pmatrix}$$

$$\times\left\{\sum_{(\lambda\lambda)=(00)}^{(mm)}\sum_{\rho}[\dim(\lambda\lambda)]^{\frac{1}{2}}U((\lambda_c\mu_c)(\mu'\lambda')(\lambda\lambda)(0m);(m0)_{--};(\mu_c\lambda_c)_{-}\rho)\right.$$

$$\left.\times U((Q_n0)(0,Q_n-m)(\lambda\lambda)(0m);(m0)_{--};(0Q_n)_{--})U((Q_n0)(\lambda_c\mu_c)(Q_n0)(\mu_c\lambda_c);(\bar{\lambda}\,\bar{\mu})_{--};(\lambda\lambda)\rho_{-})\right\}.$$

The expression in curly brackets, involving the sums over $(\lambda\lambda),\rho$, is akin to a 9-$(\lambda\mu)$ recoupling coefficient. The needed SU_3 Racah or U-coefficients are readily available.[2)] Where not needed, outer multiplicity labels, ρ, are replaced by a dash. For nuclei with $A\leq 20$, the number of m, $(\lambda'\mu')$ terms in the sum is ≤ 9; and the norm for arbitrary Q_n is determined with the evaluation of a relatively small number of coefficients $c_{m,(\lambda'\mu')}$. For nuclei with $A\leq 20$, these coefficients can be determined from the norms of the cluster functions with $Q_1,\ldots,Q_{n-1}$ and Q_n set equal to their minimum Pauli-allowed values. Such an antisymmetrized cluster function, however, is equivalent to a simple (valence) shell model wave function whose norm can be calculated by simple shell model techniques.[10),11)] For the $^{19}F=\alpha+^{15}N$ cluster system, e.g., the heavy fragment $(\lambda_c\mu_c)=(01)$ leads to eight possible m,$(\lambda'\mu')$ combinations: m=0,$(\lambda'\mu')$=(01); m=1,$(\lambda'\mu')$=(10),(02); m=2,$(\lambda'\mu')$=(11),(03); m=3,$(\lambda'\mu')$=(12),(04);

$m=4,(\lambda'\mu')=(13)$. The lowest Pauli-allowed Q-value is $Q_n=7$. The Pauli-forbidden terms with $Q_n\leq 3$ are automatically equal to zero via the structure of the $D_m(p)$. The eight $c_{m,(\lambda'\mu')}$ are evaluated from the Pauli-forbidden terms with $4\leq Q_n\leq 7$ and the single Pauli-allowed term with $Q_n=7$. As required, these are eight in number, corresponding to the $[(\lambda_c\mu_c)\times(Q_n0)](\bar{\lambda}\,\bar{\mu})=[(01)\times(Q_n0)](\bar{\lambda}\,\bar{\mu})$ possibilities with $Q_n=4$, $(\bar{\lambda}\,\bar{\mu})=(41),(30)$; $Q_n=5,(\bar{\lambda}\,\bar{\mu})=(51),(40)$; $Q_n=6,(\bar{\lambda}\,\bar{\mu})=(61),(50)$; $Q_n=7,(\bar{\lambda}\,\bar{\mu})=(71),(60)$. The coefficients, $1/N^2$, for the seven Pauli-forbidden terms must be equal to zero. The coefficient for the single Pauli-allowed term with $Q_n=7,(\bar{\lambda}\,\bar{\mu})=(60)$ has the value $1/N^2=[(3^4\times 5/2^{13})\times(19/15)^7]$ where N^2 for the shell model wave function $\Psi\,((0s)^4(0p)^{12}sd^3(60))$ has also been calculated by simple SU_3 recoupling techniques.[10)]

The $c_{m,(\lambda'\mu')}$ for a whole series of α+(A-4)-particle cluster systems with $12\leq A\leq 24$ are tabulated in ref.[9)] Once the $c_{m,(\lambda'\mu')}$ have been evaluated the norms for arbitrary Q_n are known. The technique is thus one which propagates information from the space of lowest Pauli-allowed excitations to aribtrarily high excitations of the α-heavy fragment relative motion functions of the cluster basis.

By merging the power of integral transform techniques with readily available SU_3-recoupling technology many of the technical problems in the detailed microscopic treatment of relatively complicated nuclear cluster structures may be greatly facilitated.

We gratefully acknowledge financial support by the National Science Foundation and the Deutsche Forschungsgemeinschaft, Bonn-Bad Godesberg, Germany.

References:

1. L. C. Biedenharn and J. D. Louck, Commun. Math. Phys. 8 (1968) 89.
2. Y. Akiyama and J. P. Draayer, J. Math. Phys. 14 (1973) 1904; Comp. Phys. Commun. 5 (1973) 405.
3. D. J. Millener, J. Math. Phys. 19 (1978) 1513.
4. T. H. Seligman and W. Zahn, J. Phys. G2 (1976) 79.
5. P. Kramer and T. H. Seligman, Nucl. Phys. A136 (1969) 545, A186 (1972) 49.
6. K. T. Hecht and W. Zahn, to be published.
7. H. Eikemeier and H. H. Hackenbroich, Z. Phys. 195 (1966) 412; Nucl. Phys. A169 (1971) 407.
8. H. Hutzelmeyer and H. H. Hackenbroich, Z. Phys. 232 (1970) 356;
W. Zahn, Burg Monographs in Science, Vol. 2, Burg Verlag, Basel, 1975;
H. H. Hackenbroich, T. H. Seligman and W. Zahn, Nucl. Phys. A259 (1976) 445.
9. K. T. Hecht and W. Zahn, to be published.
10. K. T. Hecht, Nucl. Phys. A283 (1977) 223.
11. M. Ichimura, A. Arima, E. C. Halbert and T. Terasawa, Nucl. Phys. A204 (1973) 225.

DYNAMICAL SYMMETRIES IN NUCLEI

F. Iachello
Kernfysisch Versneller Instituut, University of Groningen,
Groningen, The Netherlands

1. INTRODUCTION

Group theory has entered in nuclear physics in two ways. The first is a direct consequence of the shell model of the nucleus, where nucleons occupy a set of single particle levels characterized by quantum numbers, i_1, i_2, ... with degeneracies ν_1, ν_2, ... The states of t nucleons occupying the single particle level i_n with degeneracy ν_n can then be classified according to the irreducible representations of the group $U(\nu_n)$. Although this group allows to enumerate in a simple way all the states of the configuration i_n^t , it is not very useful in practice, especially in heavy nuclei, where both t and ν_n are large. The second, more interesting, way in which group theory has entered in nuclear physics is through the study of some particular property of the effective nucleon-nucleon interaction which gives rise to a <u>dynamical symmetry</u>. The existence of a dynamical symmetry, which is obviously only an approximate symmetry, implies that the states can be classified according to some group G and that the different representations of G are split but not admixed by the Hamiltonian H. Two examples of dynamical symmetries are known in nuclear physics, Wigner supermultiplet theory and Elliott model. Wigner theory[1] relies on the spin-isosopin structure of the nucleon-nucleon interaction and has found useful applications in light nuclei when both protons and neutrons occupy the same levels and the spin-orbit interaction does not play a dominant role. Elliott model[2] exploits the symmetries and degeneracies of the harmonic oscillator[3] and has found useful applications in nuclei at the beginning of the sd shell, $16 \leqslant A \leqslant 24$. Applications to heavier nuclei are spoiled by the presence of the spin-orbit potential which destroys the degeneracies of the harmonic oscillator, although several attempts have been made[4] to reintroduce Elliott model in heavy nuclei by studying the accidental degeneracies of simple particle levels in a Woods-Saxon well with a strong spin-orbit potential.

In this talk I will describe a new class of dynamical symmetries recently discovered in nuclei. These symmetries play an important role in the classification of medium mass and heavy even-even nuclei.

2. INTERACTING BOSONS

Heavy nuclei are characterized by the occurrence of low-lying collective states. However, the structure of these states changes from nucleus to nucleus. The changes are related to the presence of closed shells at nucleon number 50, 82, 126. The fact that the observed spectra change with mass number, A, makes the nuclear problem both difficult and interesting. We have developed a model which is able to account for the various situations encountered in nuclei, and the purpose of my talk is to discuss the group theoretical properties of this model.

We describe[5] an even-even nucleus as a system of N interacting boson. N is taken as the sum of the proton (π) and neutron (ν) pairs outside the closed shells, $N = N_\pi + N_\nu$. If more than half of the shell is full, we take N as the number of hole pairs. The bosons can have two values of the angular momentum, L=0, called s-boson, and L=2, called d-boson. The five components $\mu = 0, \pm 1, \pm 2$ of the d-boson and the single component of the s-boson span a six-dimensional space, yielding U(6) as the group structure of the problem. Introducing creation $(d^\dagger_\mu, s^\dagger)$ and annihilation (d_μ, s) operators, altogether denoted by $b^\dagger_{\ell\mu}$ (ℓ=0,2), which satisfy the commutation relations

$$[b_{\ell\mu}, b^\dagger_{\ell'\mu'}] = \delta_{\ell\ell'}\,\delta_{\mu\mu'} \quad , \tag{2.1}$$

one can write down the 36 generators of U(6) as

$$\begin{gathered}(d^\dagger \times \tilde{d})^{(0)}_0,\ (d^\dagger \times \tilde{d})^{(1)}_\kappa,\ (d^\dagger \times \tilde{d})^{(2)}_\kappa,\ (d^\dagger \times \tilde{d})^{(3)}_\kappa,\ (d^\dagger \times \tilde{d})^{(4)}_\kappa \\ (d^\dagger \times s)^{(2)}_\kappa,\ (s^\dagger \times \tilde{d})^{(2)}_\kappa,\ (s^\dagger \times s)^{(0)}_0 \quad ,\end{gathered} \tag{2.2}$$

where $\tilde{d}_\mu = (-)^\mu d_{-\mu}$ and the product sign denotes tensor products with respect to $0^+(3)$

$$(b^\dagger_\ell \times \tilde{b}_{\ell'})^{(k)}_\kappa = \sum_{\mu_1\mu_2} \langle \ell\,\mu_1\,\ell'\mu_2|k\,\kappa\rangle\, b^\dagger_{\ell\mu_1} (-)^{\mu_2}\, b_{\ell',-\mu_2} \quad . \tag{2.3}$$

The 36 generators can be denoted by $G^{(k)}_\kappa(\ell\ell')$, ($\ell,\ell'$=0,2). The most general Hamiltonian which includes one boson terms and boson-boson interaction can be written as

$$H = \sum_\ell \alpha_\ell(N)\, G^{(0)}_0(\ell,\ell) + \sum_{\substack{\ell\ell';\ell''\ell''' \\ k}} \beta^{(k)}_{\ell\ell';\ell''\ell'''}(N)\, G^{(k)}(\ell\ell')\cdot G^{(k)}(\ell''\ell''') . \tag{2.4}$$

Here the dot indicates scalar product with respect to $0^+(3)$ and the coefficients α, $\beta^{(k)}$ may only depend on N, which is obviously related to the linear Casimir operator of U(6). There are in all 2 coefficients of the type α and 7 coefficients of the type β.

3. DYNAMICAL SYMMETRIES

In general, given an even-even nucleus and thus N, in order to obtain eigenvalues and eigenstates, we should diagonalize the Hamiltonian (2.4) in an appropriate basis. Since we are dealing with a system of identical bosons, the appropriate basis is provided by the totally symmetric irreducible representations [N] of U(6) (for fixed boson number we can consider SU(6)). In the diagonalization, any chain of subgroups $SU(6) \supset G \supset G'$ ... may be used. In practical calculations we have preferred to use the chain (I) described below, $SU(6) \supset SU(5) \supset O^{+}(5) \supset O^{+}(3) \supset O^{+}(2)$. (Since the group $O^{+}(2)$ does not play any role at all, unless the nucleus is placed in an external magnetic field, it will not be written any longer). Transformation brackets from the other two chains II and III (also described below) to (I) have been recently constructed by Castaños, Chacon, Frank and Moshinski[6].

However, even more interesting is the possibility that dynamical symmetries will occur. As it is well known, this will happen whenever the Hamiltonian H can be written in terms only of invariants of a complete chain of subgroups of a given group $\tilde{G} \supset G \supset G' \supset \ldots$ Then the various representations are split but not admixed by H. In the present case, this will happen when the coefficients α, $\beta^{(k)}$, which depend on N, will assume some appropriate values. For example, if $\beta^{(0)}_{22,00}$ and $\beta^{(2)}_{22,20}$ vanish, the Hamiltonian H can be written in terms only of the Casimir operators of chain I, etc. Since the boson number N is related to the mass number A, we may encounter mass regions where approximate dynamical symmetries occur.

By inspecting (2.2) it is easy to see that there are three possible chains and thus three possible dynamical symmetries.

Chain I

$$SU(6) \supset SU(5) \supset O^{+}(5) \supset O^{+}(3) \quad , \tag{3.1}$$

Chain II

$$SU(6) \supset SU(3) \supset O^{+}(3) \quad , \tag{3.2}$$

Chain III

$$SU(6) \supset O^{+}(6) \supset O^{+}(5) \supset O^{+}(3) \quad . \tag{3.3}$$

The generators of the various subgroups of U(6) which appear in (3.1), (3.2) and (3.3) are

$$U(5):\ (d^{\dagger}\times\tilde{d})^{(0)}_{0},\ (d^{\dagger}\times\tilde{d})^{(1)}_{\kappa},\ (d^{\dagger}\times\tilde{d})^{(2)}_{\kappa},\ (d^{\dagger}\times\tilde{d})^{(3)}_{\kappa},\ (d^{\dagger}\times\tilde{d})^{(4)}_{\kappa}$$

$$O(5):\quad (d^{\dagger}\times\tilde{d})^{(1)}_{\kappa},\ (d^{\dagger}\times\tilde{d})^{(3)}_{\kappa}$$

$$O(3):\quad (d^{\dagger}\times\tilde{d})^{(1)}_{\kappa}$$

$$U(3):\ N=(s^{\dagger}\times s)^{(0)}_{0}+\sqrt{5}(d^{\dagger}\times\tilde{d})^{(0)}_{0}\ ;\ L^{(1)}_{\kappa} = (d^{\dagger}\times\tilde{d})^{(1)}_{\kappa};Q^{(2)}_{\kappa}=(d^{\dagger}\times s+s^{\dagger}\times\tilde{d})^{(2)}_{\kappa}-\frac{\sqrt{7}}{2}(d^{\dagger}\times\tilde{d})^{(2)}_{\kappa}$$

$O(6)$: $(d^\dagger \times \tilde{d})^{(1)}_\kappa$, $(d^\dagger \times \tilde{d})^{(3)}_\kappa$, $(d^\dagger \times s + s^\dagger \times \tilde{d})^{(2)}_\kappa$

Note that other chains are not possible, since we insist that G contains the physical angular momentum group $O^+(3)$. Also note that $O^+(6)$ is isomorphic to SU(4). The three possible dynamical symmetries can be labelled by the first subgroup G of SU(6) appearing in the chain, I) SU(5); II) SU(3); III) $O^+(6)$.

From the practical point of view, the importance of a dynamical symmetry is that one can provide, in closed form, a complete solution to the problem. For example, one can derive energy formulas. I quote, without derivation, the energy formulas corresponding to the three chains above:

I) $G \equiv SU(5)$[7]

$$E([N]\, n_d\ v\ n_\Delta\ L\ M) = \varepsilon\, n_d + \alpha\, \frac{1}{2}\, n_d(n_d-1) + \beta(n_d-v)(n_d+v+3) + \gamma[L(L+1)-6n_d] \qquad (3.4)$$

II) $G \equiv SU(3)$[8]

$$E([N]\,(\lambda,\mu)K\ L\ M) = (\tfrac{3}{4}\kappa+\kappa')L(L+1) - \kappa[\lambda^2+\mu^2+\lambda\mu+3(\lambda+\mu)] \qquad (3.5)$$

III) $G \equiv O^+(6)$[9]

$$E([N]\,\sigma\ \tau\ \nu_\Delta\ L\ M) = A\,\frac{1}{4}(N-\sigma)(N+\sigma+4) + B\ \tau(\tau+3) + C\ L(L+1) \qquad (3.6)$$

The symbols in parenthesis after E are the quantum numbers which are needed to specify uniquely the states and the constants ε, α, β, γ; κ, κ'; A, B, C are some appropriate linear combinations of the constants α_ℓ, $\beta^{(k)}_{\ell\ell'\ell''\ell'''}$ appearing in (2.4). Note also that each chain has a quantum number, n_Δ, K, ν_Δ respectively, not related to the eigenvalue of any of the Casimir operators of the groups appearing in (3.1), (3.2) and (3.3).

Typical spectra corresponding to (3.4), (3.5) and (3.6) are shown in Figs. 1, 2, 3. Selection rules for electromagnetic transition rates are also very different for the different symmetry types and are a characteristic feature of each coupling scheme.

In principle, there is no reason why actual nuclei should display any of these symmetry types. However, it turns out that <u>all three</u> are experimentally observed, although with some deviations from the corresponding energy formulas. Three examples are shown in Figs. 4, 5, 6. Others can be found in Refs. 7, 8 and 9.

In addition to suggesting the possible existence of dynamical symmetries in nuclei, the interacting boson model provides an interesting system in which broken symmetries can be studied in a straightforward way. There are three classes of broken symmetries: A) between I and II; B) between II and III; C) between III and I. The

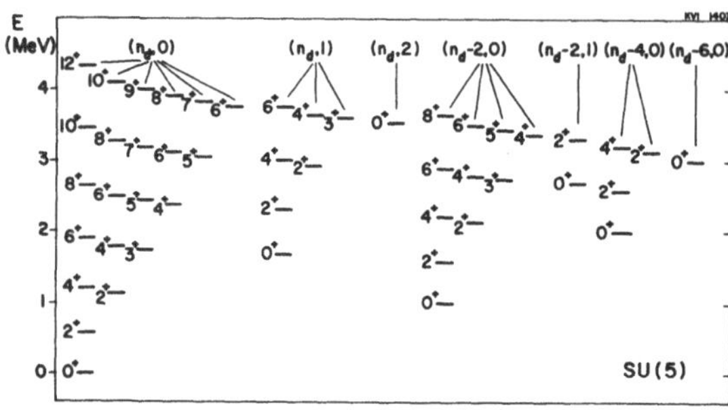

Fig. 1. A typical spectrum with SU(5) symmetry and N=6. In parenthesis are the values of v and n_Δ.

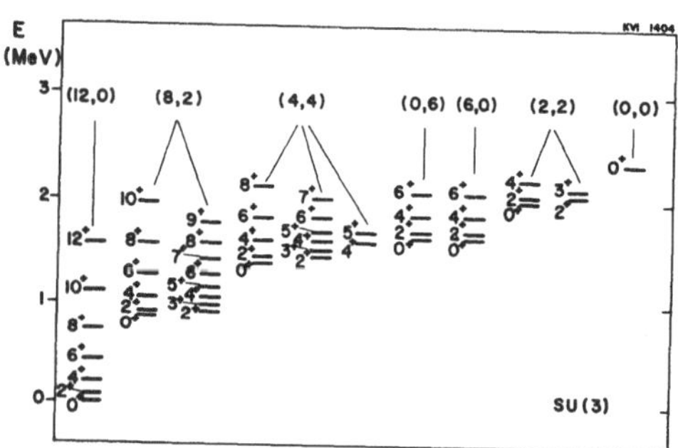

Fig. 2. A typical spectrum with SU(3) symmetry and N=6. In parenthesis are the values of λ and μ which label the SU(3) representations.

situation here is best shown by writing down the group lattice which corresponds to it, as suggested by Matsen and Plummer[11]. The two external sides of the lattice correspond to chains I and II respectively, while the internal line corresponds to chain III.

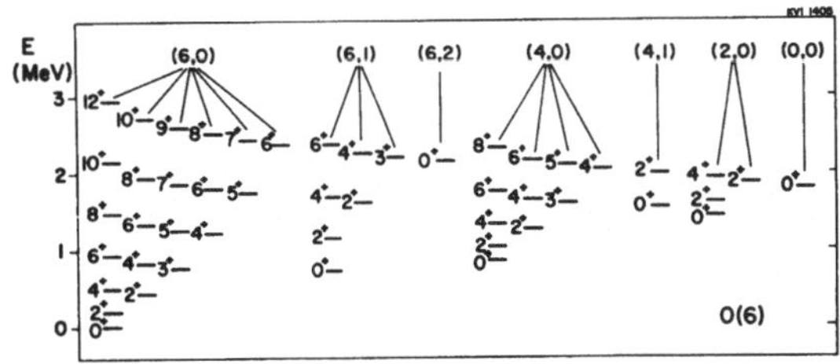

Fig. 3. A typical spectrum with O(6) symmetry and N=6. In parenthesis are the values of σ and ν_{Δ}.

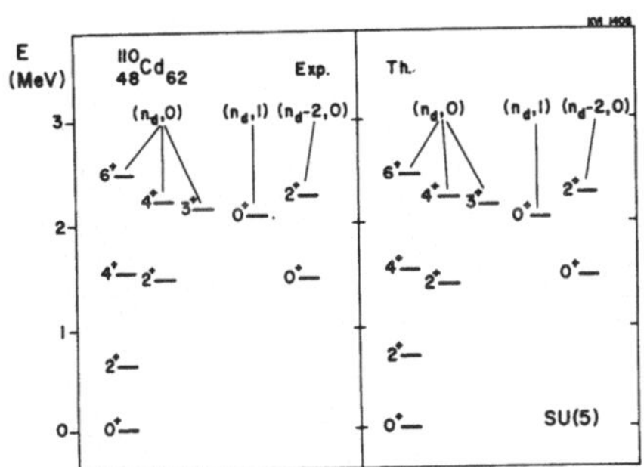

Fig. 4. An example[7] of a spectrum with SU(5) symmetry: $^{110}_{48}Cd_{62}$, (N=7).

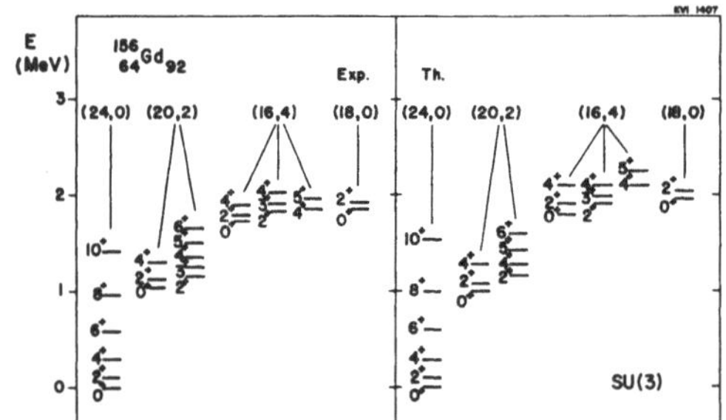

Fig. 5. An example[8] of a spectrum with SU(3) symmetry: $^{156}_{64}Gd_{92}$, (N=12).

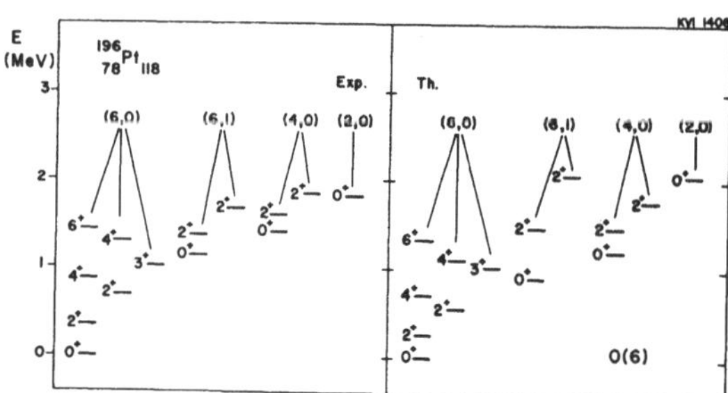

Fig. 6. An example[10] of a spectrum with O(6) symmetry: $^{196}_{78}Pt_{118}$, (N=6).

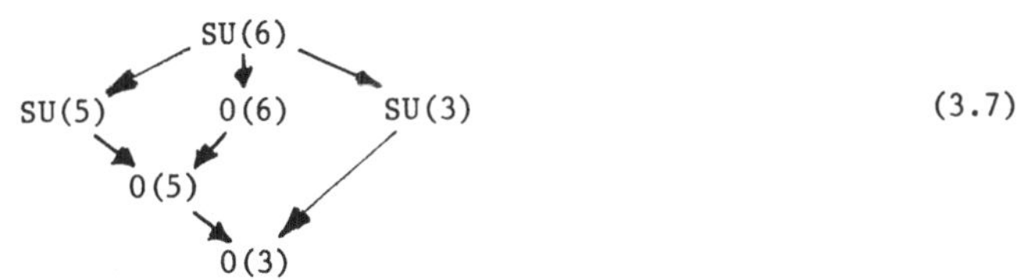

(3.7)

SU(6) is the head group and $O^{+}(3)$, the physical angular momentum group, is the tail group. As it has already been mentioned above, the three classes of broken symmetries A, B and C can be studied by diagonalizing H in the basis [N]. Although this is an interesting aspect of the model, I will not discuss it further.

4. PROTON AND NEUTRON BOSONS, F-SPIN

From the symmetry point of view, the dynamical origin of the bosons s and d which generate the collective spectrum in heavy even-even nuclei is not important. However, if one wants to understand why and where a particular symmetry occurs, one must in some way relate the bosons s and d to the original many-body system. Although this would require a long and detailed discussion of the effective nucleon-nucleon interaction, it is sufficient to say here that we have suggested[12] that the s and d bosons be identified with correlated nucleon pairs with L=0 and L=2. The L=0 pairs are similar to the Cooper pairs of the electron gas, the L=2 pairs are instead a new and important feature of the nuclear system.

The identification of the bosons with nucleon pairs opens the possibility of relating the model discussed in Sect. 3 to the original many-body problem. Although, again, this is a very important aspect of this model, I will not discuss it. Rather I will discuss another consequence of the identification of the bosons with nucleon pairs. The classification scheme discussed in Sect. 3 describes the properties of a system of N identical bosons. However, if we identify the bosons with nucleon pairs, we should actually introduce two kinds of bosons, corresponding to proton and neutron pairs respectively[12]. The simplest way to discuss the coupled system is then that of introducing a two-dimensional space. This space is analogous, but definitely not identical, to isospin, and it has been called F-spin. The group structure of the combined system is now SU(6) × SU(2). It is straightforward to see what is the effect of introducing this additional degree of freedom. The only major difference is that, while previously the only allowed representations were the totally symmetric [N], when introducing the proton-neutron degree of freedom, states with mixed symmetry [N-1,1], [N-2,2], etc. will appear. Little or no attention has been paid so far to these states. Their identification and the study of their effects on the totally symmetric states [N] is an important problem which has still to be analyzed in detail. This problem is related to the invariance of the proton (π) - neutron (ν) Hamiltonian $H = H_{\pi} + H_{\nu} + V_{\pi\nu}$ under proton-neutron interchange (F-spin invariance). Note that F-spin invariance is not directly related to isotopic spin invariance, since protons and neutrons occupy in heavy nuclei different major shells.

The three dynamical symmetries discussed in Sect. 3, and their generalization to the proton-neutron system, have been introduced within the framework of an algebraic description, in terms of some "elementary" boson variables s and d. Usually, the spectroscopy of collective states in nuclei is discussed in terms of the geometrical description of Bohr and Mottelson[13], in which the fundamental role is played by shape rather than algebraic variables. One may then inquire whether or not there is a connection between these two approaches. Although this connection is not straightforward, because the former describes a system with a finite number of degrees of freedom (the total number of active pairs) while the latter describes an infinite system (the liquid drop), I would like to mention that it is still possible to find some approximate correspondence. In this respect, an important development has been provided by Ref. 6, where the group structure of the model is analyzed in terms of the shape variables α_μ, $\bar{\alpha}$ and their conjugate momenta π_μ, $\bar{\pi}$. In the geometrical description, the three dynamical symmetries described above may then be viewed as:
I) anharmonic vibrations of the drop around a spherical shape; II) rigid rotations of the drop as a whole and III) vibrations of a γ-unstable drop.

The relationship between the dynamical symmetry II and the more general group structure of the rigid rotor, SL(3,R), discussed by Weaver and Biedenharn[14] has instead not yet been studied.

5. CONCLUSIONS

In conclusion, heavy even-even nuclei offer several examples of three dynamical symmetries. In addition they provide us with a system where broken symmetries can be studied in a straightforward way. The three new dynamical symmetries correspond to three different physical chains of subgroups of the group SU(6) (or its generalization SU(6) × SU(2)). This group appears to provide a complete and simple classification scheme for the low-lying collective states in medium and heavy mass nuclei, and it is generated by some "elementary bosons", which in turn represent correlated nucleon pairs with angular momentum L=0 and L=2.

REFERENCES

1. E. Wigner, Phys. Rev. 51, 106 (1937).
2. J.P. Elliott, Proc. Roy. Soc. (London) A245, 128 and 562 (1958).
3. For a review of the role of the harmonic oscillator in nuclear physics see, for example, P. Kramer and M. Moshinski, in "Group Theory and Its Applications", E.M. Loebl, ed., Academic Press, New York (1968), p. 339.
4. K.T. Hecht and A. Adler, Nucl. Phys. A137, 129 (1969).

5. A. Arima and F. Iachello, Phys. Rev. Lett. 35, 1069 (1975).
6. O. Castaños, E. Chacon, A. Frank and M. Moshinski, to be published in J. Math. Phys.
7. A. Arima and F. Iachello, Ann. Phys. (N.Y.) 99, 253 (1976).
8. A. Arima and F. Iachello, Ann. Phys. (N.Y.) 111, 201 (1978).
9. A. Arima and F. Iachello, Phys. Rev. Lett. 40, 385 (1978).
10. J.A. Cizewski, R.F. Casten, G.J. Smith, M.L. Stelts, W.R. Kane, H.G. Borner, and W.F. Davidson, Phys. Rev. Lett. 40, 167 (1978).
11. F.A. Matsen and O.R. Plummer in "Group Theory and Its Applications", E.M. Loebl, ed., Academic Press, New York (1968), p. 221.
12. A. Arima, T. Otsuka, F. Iachello and I. Talmi, Phys. Lett. 66B, 205 (1977).
T. Otsuka, A. Arima, F. Iachello and I. Talmi, Phys. Lett. 76B, 139 (1978).
13. A. Bohr and B.R. Mottelson, "Nuclear Structure", Vol. II, W.A. Benjamin, Reading, Mass. (1975).
14. L. Weaver and L.C. Biedenharn, Nucl. Phys. A185, 1 (1972).

Recent Work on Collective Motion

L. Weaver
Department of Physics, Kansas State University, Manhattan, KS 66506

I. Introduction

Representations of non-compact groups have been used for many years in attempts to describe the spectrum of the hadrons. In unpublished work Dothan, Gell-Mann, and Neeman suggested that the same techniques should work in nuclear physics. The specific suggestion they made was that for those nuclei in which quadrupole excitations are important the group generated by the time-derivatives of the mass quadrupole should be useful. "Useful" means several things: irreducible representations might span the spectrum of states; sum rules might be saturated by a few states; transition rates might be computable; the Hamiltonian might be simply expressed in terms of the group generators and invariants. So the time derivatives of the quadrupoles were computed using several model Hamiltonians. The groups thus generated were SL(3,R) and its relatives - we will discuss these below. This suggestion of Dothan, et al., prompted several investigations, some of which we will discuss this morning. These are all attempts to better understand some aspect of nuclear collective motion. This is a big problem and many other attempts have been made which we will not discuss. The most closely related work is that of Villars[1], Cusson[2], and Rowe[3] that uses canonical transformations to find collective degrees of freedom. Furthermore, although the suggestion of Dothan, et al., was the immediate cause of the investigations we discuss, a similar suggestion had been made much earlier by Lipkin and Goshen[4].

This report is organized as follows:

1) Physics - the algebra is roughly described, and sum rules, spectra, transition rates, and Hamiltonians are discussed.
2) Mathematics - the algebras are more carefully discussed.
3) ? - collective Hamiltonians suggested by the above work are discussed.

This is not a detective story, the conclusion may be revealed from the start: although the group theory makes several interesting suggestions along the way, its real payoff is in the final topic in the outline above - a collective Hamiltonian will be obtained using ideas inspired by group theory, but then this Hamiltonian will be properly derived from microscopic theory in an elegant coordinate-free way that makes no essential use of group theory.

II. Rough description of the algebraic framework; sum rules; transition rates.

A. Algebra. The rare earth nuclei 150<A<190 and the actinides 220<A exhibit electric quadrupole transition rates that are much larger than single particle values. The generally accepted view is that these are transitions mediated by the electric quadrupole moment of the entire nucleus. Our first assumption is that this electric quadrupole moment is proportional to the mass quadrupole

$$Q_{ij}^{el} \sim Q_{ij} = \sum_{n=1}^{A} m(x_{ni}x_{nj} - \frac{1}{3}\delta_{ij}\,\underline{x}_n\cdot\underline{x}_n)$$

The Q_{ij} all commute with one another. To find non-trivial algebraic relations we turn to their time derivatives:

$$\dot{Q}_{ij} = \frac{i}{\hbar}[H, Q_{ij}]$$

If the Hamiltonian H is the sum of the kinetic energy and a momentum-independent potential, then

$$\dot{Q}_{ij} = \sum_{n=1}^{A} (x_{ni}p_{nj} + p_{ni}x_{nj})$$

The $\dot{Q}_{ij}$ do not commute among themselves. For example

$$[\dot{Q}_{12}, \dot{Q}_{23}] = i\hbar \sum_{n=1}^{A} (x_{n1}p_{n3} - x_{n3}p_{n1}) = -i\hbar L_2$$

where L_2 is the orbital angular momentum of the nucleus about the 2-axis. The six $\dot{Q}_{ij}$ and the three L_i form a closed set on commutation. It is the algebra SL(3,R)[5]. Our second assumption is that the $\dot{Q}_{ij}$ do in fact generate this algebra. We will discuss it and its relatives after we see to what degree it is useful in physics.

B. Sum rules. The sum rules obtained from this algebra involve E2 transition rates weighted by the square of the transition energy. Explicitly

$$\sum_{\beta}(E_\beta-E_\alpha)^2\, B(E2{:}\alpha\to\beta)[J_\beta(J_\beta+1) - J_\alpha(J_\alpha+1) - 6] = \frac{150}{4\pi}\left(\frac{e\hbar^2}{m}\,\frac{Z}{A}\right)^2 J_\alpha(J_\alpha+1)$$

This cannot be directly evaluated at present because the absolute transition rates are not available. But we can estimate the left hand side from models. For ^{166}Er

Bohr and Mottelson[6] find that the low lying states exhaust about 1/3 of the sum rule. Thus, the sum rule suggests the existance of collective states at much higher energy - which need only couple weakly to the low-lying states because of the strong energy weighting. Furthermore, these states are constrained by another sum rule:

$$\sum_{\beta} (E_\beta - E_\alpha)^2 \, B(E2{:}\alpha\to\beta) \, (-)^{J_\beta - J_\alpha} W(J_\alpha 2 \, J_\alpha 2;\, J_\beta 3) = 0$$

This arises because the commutator of two L=2 operators couples to L=3, but the algebra says that this coupling must vanish.

The use of the algebra to find sum rules is thus not very encouraging. We turn to transition rates.

C. <u>Transition rates</u>. To compute the E2 transition rate from state α to state β we need the matrix element $\langle\beta|Q_{ij}|\alpha\rangle$. If we can find - from the algebra - the numerical values of $\langle\beta|\dot{Q}_{ij}|\alpha\rangle$ then we have

$$\langle\beta|Q_{ij}|\alpha\rangle = \frac{\hbar}{i} \, \frac{\langle\beta|\dot{Q}_{ij}|\alpha\rangle}{(E_\beta - E_\alpha)}$$

Now, finding the numerical values of $\langle\beta|\dot{Q}_{ij}|\alpha\rangle$ amounts to finding a representation of the algebra. To introduce the smallest number of parameters we make a <u>third assumption</u> that the actual nuclear states lie in an irreducible representation of the algebra.

Mathematicians have studied the representations of this algebra, notably Gelfand[7]. Irreducible representations in forms useful for physical applications have been found by many others, notably Gell-Mann[8] and D.W. Joseph[9]. Recently, Dj. Šijački[10] has given a particularly elegant presentation. Identifying physical states with the vectors on which the algebra acts, however, is quite difficult. We have here the same problem that SU(3) has in nuclear physics: The state labels in a representation are adapted to the subgroup chain U_1<SU(2)<SU(3) while nuclear physics deals with states labeled by O(2)<O(3)<SU(3). The solution we adopt is this: we represent the algebra on functions defined on the rotation group, expand these functions in terms of the D-matrices, $D^J_{KM}(r)$, and then use the quantum numbers JKM to characterize the states. The spectra that emerge are those of an <u>asymmetric rotator</u>. Transition rates can be computed and the results are generally similar to, though not identical with, those of the rotational model. There are two parameters in the algebraic approach, but for some cases these can be eliminated, specifically, the ratio of two $\Delta K=2$ transitions involves no free parameters:

$$\frac{B(E2:\ L2 \rightarrow L_1 0)}{B(E2:\ L2 \rightarrow L_2 0)} = \left(\frac{E_2}{E_1}\right)^2 \times \text{[leading order intensity in rotational model]}$$

Figure 1 shows a typical case. Notice that this approach is an improvement on leading order intensity rules in the rotational model, but is still usually outside the experimental errors. And while the rotational model suggests rather natural ways to improve its results - at the price of new parameters, of course - the algebraic approach is stuck. But the improvement over the rotational model is encouraging. Perhaps the discrepancies between theory and experiment could be overcome by a better solution to the state identification problem. The natural solution to this problem would be to find 'the' Hamiltonian and use its eigenstates. To do this successfully in this algebraic context the Hamiltonian must be a function of the operators in the algebra, $H = H(\dot{Q})$. But in fact the collective Hamiltonian is a function both of $\dot{Q}_{ij}$ and the Q_{ij} themselves. This is suggested by simple dimensional considerations, and also by the canonical transformations mentioned earlier[1-3]. So the question we should be asking is 'what is the algebra of the Q_{ij} and $\dot{Q}_{ij}$? and what is $H(Q,\dot{Q})$?'

III. <u>The algebra CM(3)</u>

The algebra obtained from the Q_{ij} and the $\dot{Q}_{ij}$ includes the operator $Q_o = \Sigma\, m\, \underset{\sim}{x}_n \cdot \underset{\sim}{x}_n$ and the angular momenta L_i. The Q_{ij} and Q_o from an abelian algebra T_6; the $\dot{Q}_{ij}$ and the L_i form SL(3,R). The combination is the semi-direct product $T_6 \circledS SL(3,R)$ which we call CM(3)[11,12,16]. Its structure is much like that of the Poincaré group, $P = T_4 \circledS SO(3,1)$. The representations of CM(3) contain a label $\Lambda = \det Q > 0$, characterizing the T_6 part, analogous to the label m^2 in P. Next, the little group of Q_o is found to be SO(3) just as the little group of (m,o,o,o) in P is SO(3). The CM(3) representation is thus also characterized by a spin we call v, for vorticity, analogous to the spin s in P. (There are also representations in CM(3) corresponding to $\Lambda = \det Q = 0$. This might describe a non-relativistic string.

The simplest such representations are labeled by a length and a minimum spin). The vorticity determines the response of a spherical quadrupole to rotation. Thus a non-zero value for v indicates the presence of some sort of rotational flow in the quadrupole. So we seek a Hamiltonian that contains such an internal flow.

IV. Collective Hamiltonians

A. Fastest route[13]. Assume the nuclear wave function depends explicitly only on the collective coordinates Q_{ij} which are themselves functions of the $\underset{\sim}{x}_n$. Now apply the kinetic energy T,

$$T = -\frac{h^2}{2m}\sum_{n=1}^{A}\nabla_n^2 ,$$

to ψ and use the chain rule. The result is that T is a function of Q_{ij} and $\partial/\partial Q_{ij}$. For small deformations Rosensteel and Ihrig[13] show that their function is exactly the Bohr Hamiltonian. The explicit form in two dimensions is

$$T = -\frac{h^2}{2m}\{4\partial_\sigma[A-1+\sigma\partial_\sigma+\delta\partial_\delta] + 4\sigma[\partial_\delta^2+\delta^{-1}\partial_\delta+(2\delta)^{-2}\partial_\theta^2]\},$$

where $\underset{\approx}{Q} = \begin{pmatrix}\cos\theta & +\sin\theta\\ -\sin\theta & \cos\theta\end{pmatrix}\begin{pmatrix}\frac{\sigma+\delta}{2} & 0\\ 0 & \frac{\sigma-\delta}{2}\end{pmatrix}\begin{pmatrix}\cos\theta & -\sin\theta\\ \sin\theta & \cos\theta\end{pmatrix}$.

B. A purely collective route[14]. This time let the positions of all particles be transformed by the same matrix: $\underset{\sim}{r}_n(t) = \underset{\approx}{M}(t)\cdot\underset{\sim}{r}_n(0)$. Then the velocity is $\dot{\underset{\sim}{r}}_n(t) = (\dot{\underset{\approx}{M}}M^{-1})\cdot\underset{\sim}{r}_n(t)$. The symmetric part of $\dot{\underset{\approx}{M}}M^{-1}$ provides irrotational flow, while the anti-symmetric part determines a rigid flow. Now express the kinetic energy $T = \frac{1}{2}m\sum\dot{\underset{\sim}{r}}_n\cdot\dot{\underset{\sim}{r}}_n$ in terms of the parameters of $\underset{\approx}{M}$, use many of the manipulations needed in CM(3), and find the result ($\partial_1 = \partial/\partial\mu_1$,etc.).

$$T = -\frac{h^2}{2m\Lambda}\{(\partial_1^2+\partial_2^2) + (\mu_1^2+\mu_2^2)^{-1}[2\mu_1\partial_1 - 2\mu_2\partial_2]$$

$$+\frac{1}{2}(\mu_1+\mu_2)^{-2}(\partial_\theta+\partial_\psi)^2 + \frac{1}{2}(\mu_1-\mu_2)^{-2}(\partial_\theta-\partial_\psi)^2\}$$

where $\underset{\approx}{M} = \underset{\approx}{R}(\theta)\begin{pmatrix}\mu_1 & 0\\ 0 & \mu_2\end{pmatrix}\underset{\approx}{R}(-\psi)$ and the quadrupole is $\underset{\approx}{Q} = \underset{\approx}{R}(\theta)\begin{pmatrix}\mu_1^2 & 0\\ 0 & \mu_1^2\end{pmatrix}\underset{\approx}{R}(-\theta)$. The parameter ψ seems ambiguous, for while θ determines the orientation of the quadrupole, ψ could be replaced by e.g. $\theta+\psi$ and leave $\underset{\approx}{Q}$ unaffected. Here group theory helps , for the vorticity invariant in the two dimensional case is proportional to ∂_ψ and not, e.g. $\partial_\psi + \partial_\theta$.

C. The BBC construction[15]. This is a simple, intrinsic, and exact separation of the kinetic energy into collective and internal motions. We first introduce three dyadics which simplify the subsequent discussion.

$$\underset{\approx}{D} = \sum_{n=1}^{A} \underset{\sim}{c}_n \underset{\sim}{r}_n, \quad \underset{\approx}{D}^T = \sum_{n=1}^{A} \underset{\sim}{r}_n \underset{\sim}{c}_n .$$

The c_n are orthonormal vectors in A-dimensional particle label space; the $\underset{\sim}{r}_n$ are ordinary vectors. Then the kinetic energy is

$$T = \tfrac{1}{2} m \Sigma \dot{\underset{\sim}{r}}_n \cdot \dot{\underset{\sim}{r}}_n = \tfrac{1}{2} m \mathrm{Tr} \dot{\underset{\approx}{D}}^T \cdot \dot{\underset{\approx}{D}} .$$

The first object is to find a simple expression for $\underset{\approx}{D}$. To this end BBC introduce the dyadics $\underset{\approx}{Q} = \underset{\approx}{D}^T \cdot \underset{\approx}{D}$ and $\underset{\approx}{\mathcal{Q}} = \underset{\approx}{D} \cdot \underset{\approx}{D}^T$. $\underset{\approx}{Q}$ is a 3x3 quadrupole and $\underset{\approx}{\mathcal{Q}}$ an AxA 'quadrupole'. They are both real, symmetric, positive semi-definite, and have the same trace. The eigenvalues and eigenvectors of $\underset{\approx}{Q}$ and $\underset{\approx}{\mathcal{Q}}$ are

$$\underset{\approx}{Q} \cdot \underset{\sim}{s}_\alpha = \lambda_\alpha \underset{\sim}{s}_\alpha , \quad \underset{\sim}{s}_\alpha \text{ a 3-vector}, \quad \lambda_\alpha \geq 0,$$

$$\underset{\approx}{\mathcal{Q}} \cdot \underset{\sim}{v}_q = \sigma_q \underset{\sim}{v}_q , \quad \underset{\sim}{v}_q \text{ an A-vector}, \quad \sigma_q \geq 0.$$

Then we have the remarkable theorem: Three of the σ_q are equal to the three λ_α; all others are zero. To prove it, apply $\underset{\approx}{\mathcal{Q}}$ to $\underset{\sim}{v}_\alpha = \lambda_\alpha^{-1/2} \sum_n \underset{\sim}{c}_n (\underset{\sim}{r}_n \cdot s_\alpha)$. Then use $\sigma_q \geq 0$ and Tr $\underset{\approx}{Q} = \mathrm{Tr} \underset{\approx}{\mathcal{Q}}$. Thus $\underset{\approx}{Q} = \sum_{\alpha=1}^{3} \underset{\sim}{s}_\alpha \lambda_\alpha \underset{\sim}{s}_\alpha$ and $\underset{\approx}{\mathcal{Q}} = \sum_{a=1}^{3} \underset{\sim}{v}_\alpha \lambda_\alpha \underset{\sim}{v}_\alpha$. Using this we find a very simple form for $\underset{\approx}{D}$:

$$D = \sum_{n=1}^{A} \underset{\sim}{c}_n \underset{\sim}{r}_n = \sum_{\alpha=1}^{3} \lambda_\alpha^{1/2} [\lambda_\alpha^{-1/2} \sum_{n=1}^{A} \underset{\sim}{c}_n (\underset{\sim}{r}_n \cdot \underset{\sim}{s}_\alpha)] \underset{\sim}{s}_\alpha = \sum_{\alpha=1}^{3} \underset{\sim}{v}_\alpha \mu_\alpha \underset{\sim}{s}_\alpha , \quad \mu_\alpha = \lambda_\alpha^{1/2}.$$

The motions allowed $\underset{\sim}{v}_\alpha$ and $\underset{\sim}{s}_\alpha$ are rotations in A-space and 3-space, respectively. This form for $\underset{\approx}{D}$ is the first main result.

BBC now introduce angular velocities through

$$\dot{\underset{\sim}{v}}_\alpha = \sum_{\beta=1}^{3} \dot{E}_{\alpha\beta} \underset{\sim}{v}_\beta + \sum_{k=4}^{3} \dot{E}_{\alpha k} \underset{\sim}{v}_k , \quad \dot{\underset{\sim}{s}}_\alpha = \sum_{\beta=1}^{3} \dot{\Omega}_{\alpha\beta} \underset{\sim}{s}_\beta$$

In terms of these they write the kinetic energy

$$T = T(\dot{\underset{\sim}{v}}, \dot{\mu}, \dot{\underset{\sim}{s}}, \mu) = \tfrac{1}{2} m \{ \sum_{\alpha=1}^{3} \dot{\mu}_\alpha^2 + \sum_{\alpha,k} \lambda_\alpha \dot{E}_{\alpha k}^2 + \sum_{\alpha<\beta} (\lambda_\alpha + \lambda_\beta)(\dot{E}_{\alpha\beta}^2 + \dot{\Omega}_{\alpha\beta}^2)$$

$$-4\sum_{\alpha<\beta} \mu_\alpha \mu_\beta \dot{E}_{\alpha\beta} \Omega_{\alpha\beta} \} .$$

This expression is exact, provides a clean separation of the rotational and vibrational terms from the internal motions (the only rotational-internal coupling is in the last term, and only the three internal angles $E_{\alpha\beta}$ are affected), and is 'intrinsic' in the sense that the coordinates used are all determined by the many-body system itself.

Quantization of T is carried out using the Pauli prescription and the Hamiltonian obtained[15]. Here we simply note that the momenta conjugate to $E_{\alpha\beta}$ and $\Omega_{\alpha\beta}$ are, in the two-dimensional case,

$$\begin{pmatrix} L = \partial T/\partial \dot{\Omega} \\ \mathcal{L} = \partial T/\partial \dot{E} \end{pmatrix} = \begin{pmatrix} \lambda_1 + \lambda_2 & -4\mu_1\mu_2 \\ -4\mu_1\mu_2 & \lambda_1 + \lambda_2 \end{pmatrix} \begin{pmatrix} \dot{\Omega} \\ \dot{E} \end{pmatrix} .$$

The eigenvalues of this moment of inertia matrix are just the rigid and liquid moments, and $\mathcal{L}$ is the CM(2) vorticity operator.

References

1. F.M. Villars, Nucl. Phys., 74 (1965).
2. R.Y. Cusson, Nucl. Phys., A114 (1968).
3. D.J. Rowe, Nucl. Phys., A152 (1970) and P. Gulshani and D.J. Rowe, Can.J.Phys. 54, (1976)
4. S. Goshen and H.J. Lipkin, Ann. Phys. (N.Y.), 6 (1959).
5. L. Weaver and L.C. Biedenharn, Nucl. Phys., A185, (1972).
6. A. Bohr and B.R. Mottelson, Nuclear Structure, Vol. II, W.A. Benjamin, Inc., Reading, Massachusetts (1975), p.411.
7. I.M. Gelfand and M.I. Graev, Am. Math. Soc. Trans. (2), 2 (1956).
8. M. Gell-Mann, unpublished (1965).
9. D.W. Joseph, unpublished (1970).
10. Dj. Sijacki, J. Math. Phys. 16 (1975).
11. L. Weaver, L.C. Biedenharn, and R.Y. Cusson, Ann. Phys. (N.Y.), 77 (1973).
12. G. Rosensteel and D.J. Rowe, Ann. Phys. (N.Y.), 96 (1976).
13. G. Rosensteel and E. Ihrig, Phys. Rev. C, 17 (1978).
14. Weaver et al. in Reference 16, Section III.
15. B. Buck, L.C. Biedenharn, R.Y. Cusson, to be published.
16. L. Weaver, R.Y. Cusson, and L.C. Biedenharn, Ann. Phys. (N.Y.), 102 (1976), especially Section IV.

Figure 1

Nucleus	$\frac{(L2 \to L'0)}{(L2 \to L''0)}$	B(E2) ratio Experiment	SL(3,R)	Rotator
Gd^{154}	$\frac{22 \to 00}{22 \to 20}$	0.43 ± 0.04	0.54	0.70
	$\frac{22 \to 20}{22 \to 40}$	6.9 ± 2	10.3	20
	$\frac{32 \to 20}{32 \to 40}$	1.03 ± 0.10	1.41	2.50
	$\frac{42 \to 20}{42 \to 40}$	0.15 ± 0.06	0.21	0.34
Er^{168}	$\frac{22 \to 00}{22 \to 20}$	0.56 ± 0.02	0.57	0.70
	$\frac{22 \to 20}{22 \to 40}$	12 ± 4	11.3	20
	$\frac{32 \to 20}{32 \to 40}$	1.56 ± 0.04	1.50	2.50
	$\frac{42 \to 20}{42 \to 40}$	0.15	0.22	0.34

The Calculation of 6j Symbols for Compact Groups

P.H. Butler
Physics Department, University of Canterbury, Christchurch, New Zealand.

For any finite or continuous compact group one can set up the Wigner-Racah algebra. We have previously reviewed the symmetry properties of generalized 3jm symbols, 6j symbols etc, the generalization of the Racah factorization lemma and the generalization of the Wigner-Eckart theorem. (P.H. Butler, Phil. Trans. Roy. Soc. (London) (1975) 277 545-585). The selection rules for a particular group, that is the Clebsch-Gordan series or Kronecker product decomposition, provides particular boundary values to the recursion relations within the general Wigner-Racah algebra. As a result the 6j and higher j symbols of the particular group may be calculated without recourse to ladder operators, projection operators or integration over group elements. Likewise the branching rules appropriate to restriction to a subgroup provide the additional boundary conditions to allow the calculation of 3jm symbols. (Sometimes 3jm symbols are called by the alternative names 3Γ, $3\lambda\mu$ symbols, isoscalar factors, V coefficients, fractional parentage coefficients, etc, depending upon the application the authors have in mind).

This method was first used to reproduce the standard formulas for $SO_3 \supset SO_2$ (P.H. Butler, Int. J. Quantum Chem (1976) 10 599-613) and to obtain tables for both true and spin irreps of the tetrahedral group, which is neither ambivalent nor multiplicity free (P.H. Butler, B.G. Wybourne, ibid, 581-598, 615-628). The method is being used to provide full (up to J = 8), computer generated and computer typeset tables of coefficients for all molecular point groups in all possible subgroup bases (P.H. Butler, Point Group Symmetry Applications, Methods and Tables, Plenum 1979, in press). Application of the method to the Lie group E_7 in the $SU_6 \times SU_3$ basis appropriate to a suggested six quark model provides the first results on the matrix elements of E_7 generators (P.H. Butler, R. Haase, B.G. Wybourne, Australian J. Phys. (1978) 31, 131-135; (1979) in press).

Whenever an irrep occurs more than once in a product or branching rule the free phase associated with the irrep (through Schur's lemmas) becomes a free unitary matrix of the appropriate size. Pseudoscalars of the group, if they exist, assist in separating this multiplicity (P.H. Butler, A.M. Ford, J. Phys. A, in press). M. Hamermesh (Group Theory and its Application to Physical Problems, Addison-Wesley, 1963) had shown that the alternating irrep $[1^n]$ of the symmetric group S_n leads to such symmetries but it was unclear how they were related to the reordering symmetries of the Wigner-Racah algebra. The octahedral point group (being the group S_4) provides, through its spin irreps, an interesting test case. The product $U' \times U'$ contains irreps T_1 and T_2 twice each, where T_2 is the pseudoscalar A_2 times T_1. The $U'U'T_2$ multiplicity is resolved by separation into symmetric and antisymmetric terms, but the $U'U'T_1$ multiplicity is not. However demanding a simple form the 6j symbol involving these two, the $T_1T_2A_2$ and the $U'U'A_2$ products leads to a separation of the $U'U'T_1$ multiplicity that has several nice properties. In particular $\sqrt{5}$ factors in previous tabulations are eliminated and many additional zeroes introduced.

Application of group theory to composite nuclear particle interactions.

G. John, P. Kramer, K.D. Hetzel
Institut für theoretische Physik der Universität Tübingen
German Federal Republic

Reactions between p-shell nuclei and ^{4}He can be used as models to investigate certain aspects of heavy ion interaction. The states of the ^{16}O and the ^{4}He nuclei are assumed to have s^4p^{12} and s^4 oscillator shell structure respectively. The system conveniently studied by solving an integral equation

$$\int [\mathcal{H}(t_1t_2,t_1't_2') - E\mathcal{N}(t_1t_2,t_1't_2')]\ u(t_1't_2')\ d\mu(t_1')d\mu(t_2') = 0$$

in Bargmann space, derived from the Schrödinger equation for the 20-particle system. The complex coordinates t_1t_2 refer to the positions and momenta of the two fragments. Let $|s^4p^{12}, s^4)$ be the orbital state of the two fragments and W_{τ_c} a Weyl operator which gives rise to a position and momentum translation t_1 and t_2 for each fragment. Further use the Young operator c(rfq) to establish the orbital partition f. Let H be the Hamiltonian of the 20-particle system and I the unit operator. It was shown in /1/ that

$$\{\mathcal{N},\mathcal{H}\}(t_1t_2,t_1't_2') = (s^4p^{12},s^4|W_{\tau_c}\circ c(\tilde{q}fr)\circ\{I,H\}\circ c(rfq)\circ W_{\tau_c'}|s^4p^{12},s^4)$$

Let us further introduce the matrix ε of overlap integrals between different single-particle states

$$\varepsilon_{i\ell} = (i|W_{\tau_c}\circ W_{\tau_c'}|\ell)$$

This matrix can be brought into the simple form

$$\varepsilon = \begin{bmatrix} \varepsilon_{11}\delta_{i\ell} & \varepsilon_{i5} \\ \varepsilon_{5\ell} & \varepsilon_{55} \end{bmatrix}$$

where for single-particle oscillator states up to the p-shell

i,ℓ: $/NLM) \rightarrow P^N_{LM}(x)$, NLM = 000,11-1, 110,111, we get

$$\varepsilon_{11} = \exp[t_1\cdot\bar{t}_1'] \quad \varepsilon_{i5} = P^N_{LM}(\bar{t}_2'-\bar{t}_1')\exp[t_2\cdot\bar{t}_1']$$

$$\varepsilon_{5\ell} = P^N_{LM}(t_2-t_1')\exp[t_2\cdot\bar{t}_1'] \quad \varepsilon_{55} = \exp[t_2\cdot\bar{t}_1']$$

The kernel $\mathcal{N}$ is given /2/ by a matrix element of the finite irreducible representation D^f of the group GL(j, $\mathbb{C}$) where j = 5 is the number of single-particle states,

$$\mathcal{N}(t_1t_2,t_1't_2') = D^f_{\tilde{q}q}(\varepsilon)$$

For the closed-shell system $s^4p^{12} + s^4$ and $f = [4^5]$ this expression becomes

$$\mathcal{N}(t_1t_2,t_1't_2') = (\det \varepsilon)^4$$

and after introduction of the relative vector $s_1 \sim t_1 - t_2$ and the c.m. vector s_2 one finds

$$\mathcal{N}(s_1s_2,s_1's_2') = \exp [s_2 \cdot \bar{s}_2'] \sum_{N=N_0}^{\infty} \eta_N (N!)^{-1} (s_1 \cdot \bar{s}_1')^N$$

The first part describes the unit operator with respect to the c.m. state, the second part yields as eigenfunctions of $\mathcal{N}$ the oscillator states of relative motion with eigenvalues η_N . The eigenvalues are zero for N below the first Pauli-allowed value N_0 and for large N tend towards the value 1, see the figure for $^{16}O + {}^4He$ and $^{40}Ca + {}^4He$.

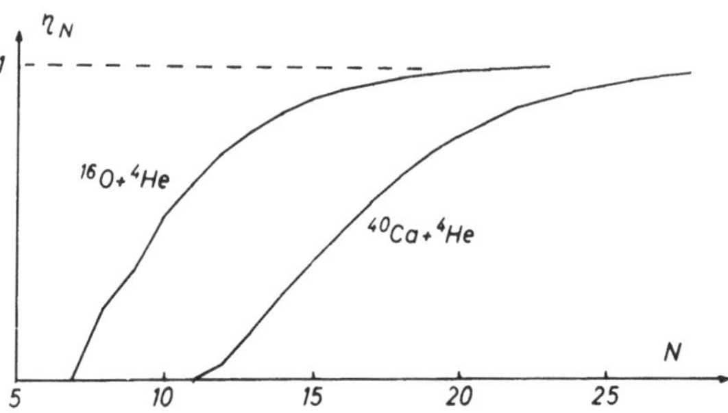

For the first fragment with open-shell configuration s^4p^m and orbital angular momentum LM, the second fragment s^4, and f = [44m] the computation /3/ yields

$$\mathcal{N}(t_1t_2,t_1't_2') = \sum_{askL'\tilde{L}''L''M'} \{(-1)^k (s!)^2 m! \, [(4-k-s)!(m-s)!k!]^{-1}$$

$$\langle [m]LM \,|\, [a]L'M'[s]\tilde{L}''M'' \rangle \langle [a]L'M'[s]L''M'' | [m]LM \rangle$$

$$(\varepsilon_{15}\varepsilon_{51})^{4-k-s} \, (\varepsilon_{55})^k \, P^S_{\tilde{L}''M''}(\varepsilon_{i5}) \, P^S_{L''-M''}(\varepsilon_{5\ell}) \, (-1)^{M''}\}$$

Here the brackets denote one-row Wigner coefficients of SU(3) in the chain SU(3) > SO(3,$\mathbb{R}$). A more elaborate analysis /3/ yields closed analytic expressions for the interaction kernels $\mathcal{H}$.

/1/ P. Kramer in: Proc. VI. Int. Coll. on Group Theoretical Methods in Physics, Springer Lecture Notes in Physics vol. 79, 1978

/2/ P. Kramer in: Group Theoretical Methods in Physics, Academic Press, New York 1977

/3/ P. Kramer, G. John and D. Schenzle, Group Theory and the Interaction of Composite Nucleon Systems, to be published.

Contravariant form for infinite-dimensional Lie algebras and superalgebras

V. G. Kac

Department of Mathematics, M.I.T., Cambridge, Mass. 02139

In this article a criterion of irreducibility of a representation with vacuum vector of the Lie superalgebras $\mathfrak{g}(A,\tau)$ (introduced in [1]) and of the string algebra and superalgebra (Virasoro and Neveu-Schwartz models [2]) is found. The method consists in the computation of the determinant of the so-called contravariant form. As a consequence we obtain a description of the composition series of the representation with vacuum vector (cf. [3] - [5]) and formulas for the characters of some irreducible representations with vacuum vector (cf. [4] - [6]). The well-known theorems of Šapovalov, Bernstein-Gelfand-Gelfand and Weyl are special cases of Theorems 1 and 2 when $\mathfrak{g}$ is a finite-dimensional Lie algebra.

1. The Setting. Let Γ be a free abelian group of rank n with a fixed linear ordering. Let $\mathfrak{g}$ be a complex (generally infinite-dimensional) Lie superalgebra with a consistent Γ-gradation: $\mathfrak{g} = \bigoplus_{\alpha} \mathfrak{g}_{\alpha}$ (i.e. $\mathfrak{g}_{\alpha}$ is finite-dimensional and entirely even or odd, and $[\mathfrak{g}_{\alpha}, \mathfrak{g}_{\beta}] \subset \mathfrak{g}_{\alpha+\beta}$). We suppose that $[h,x] = \hat{\alpha}(h)x$ for any $h \in \mathfrak{g}_0$, $x \in \mathfrak{g}_{\alpha}$, where $\hat{\alpha} \in \mathfrak{g}_0^*$. We set $\mathfrak{h} = \mathfrak{g}_0$, $\mathfrak{n}_+ = \bigoplus_{\alpha>0} \mathfrak{g}_{\alpha}$, $\mathfrak{n}_- = \bigoplus_{\alpha>0} \mathfrak{g}_{-\alpha}$; $\Delta^+ = \{\alpha \in \Gamma | \alpha > 0,\ \mathfrak{g}_{\alpha} \neq 0\}$, $\Delta^+_{\nu} = \{\alpha \in \Delta^+ | \deg \alpha = \nu\}$, $\nu = \bar{0}, \bar{1}$, $\tilde{\Delta}^+ = \{\alpha \in \Gamma | \alpha = n\beta,\ n \in \mathbb{N},\ \beta \in \Delta^+\}$. We set $\operatorname{sign} \alpha = (-1)^{\deg \alpha}$.

We call a representation of $\mathfrak{g}$ in the space V a representation with highest weight $\lambda \in \mathfrak{h}^*$ if: a) $V = \bigoplus_{\eta \geq 0} V_{-\eta}$ and $\mathfrak{g}_{\alpha}(V_{-\eta}) \subset V_{\alpha-\eta}$ and b) $V_0 = \mathbb{C}v_0$, $\deg v_0 = \bar{0}$, $h(v_0) = \lambda(h)v_0$ for $h \in \mathfrak{h}$ and $V = U(\mathfrak{n}_-)v_0$. (v_0 is the vacuum vector: $\mathfrak{n}_+(v_0) = 0$). We set $\operatorname{ch} V = \bigoplus_{\eta \geq 0} (\dim V_{-\eta}) e^{\eta}$ (e^{η} is a "formal" exponential). For each λ there exists a unique "maximal" representation $M(\lambda) = \oplus M_{-\eta}$ such that all the representations with highest weight λ are the quotients of $M(\lambda)$, and unique "minimal" representation $L(\lambda)$, which is irreducible: $L(\lambda) = M(\lambda)/I(\lambda)$. We set $P(\eta) = \dim M_{-\eta}$, $\eta \geq 0$. This function does not depend on $\lambda : \operatorname{ch} M(\lambda) = Q^{-1}$, where $Q = \prod_{\alpha \in \Delta^+} (1-(\operatorname{sign} \alpha)e^{\alpha})^{(\operatorname{sign} \alpha)(\dim \mathfrak{g}_{\alpha})}$.

Suppose that $\mathfrak{g}$ admits an involutive antiautomorphism ω such that $\omega|\mathfrak{h} = \mathrm{id}$, $\omega(\mathfrak{g}_{\alpha}) = \mathfrak{g}_{-\alpha}$. Then $M(\lambda)$ carries a bilinear symmetric form F, called a contravariant form which is uniquely defined by the properties: $F(v_0,v_0) = 1$ and $F(g(x),y) = F(x,\omega(g)y)$, for $x,y \in M(\lambda)$, $g \in \mathfrak{g}$. Clearly, $\operatorname{Ker} F = I(\lambda)$ and $F(M_{-\eta}, M_{-\eta'}) = 0$ if $\eta \neq \eta'$. We set $F_{\eta} = F|_{M_{-\eta}}$. The function $(\det F_{\eta})(\lambda)$ is a polynomial on $\mathfrak{h}^*$, which does not depend on the choice of the basis in $M_{-\eta}$ up to a non-zero constant factor. Clearly $M(\lambda)$ is irreducible iff $\det F_{\eta} \neq 0$ for all $\eta > 0$.

2. Lie Superalgebras $\mathfrak{g}(A,\tau)$. Let A be a complex $(r\times r)$-matrix and τ be a subset in $I = \{1,\dots,r\}$. We fix a system of free generators α_i, $i \in I$, of Γ, such that $\alpha_i > 0$, $i \in I$. We define a (unique) Lie superalgebra with Γ-gradation $\mathfrak{g}(A,\tau) = \oplus_\alpha \mathfrak{g}_\alpha$ by the properties: a) every Γ-graded ideal which intersects $\mathfrak{g}_0 = \mathfrak{h}$ trivially is zero, b) the $\mathbb{Z}_2$-gradation of $\mathfrak{g}(A,\tau)$ is given by: $\deg e_i = \deg f_i = \bar{1}$, $i \in \tau$, $\deg e_i = \deg f_i = \bar{0}$, $i \notin \tau$, $\deg h_i = \bar{0}$, c) $\mathfrak{g}(A,\tau)$ is generated by elements e_i, f_i, h_i, $i \in I$, such that $\mathfrak{g}_{\alpha_i} = \mathbb{C}e_i$, $\mathfrak{g}_{-\alpha_i} = \mathbb{C}f_i$, h_i's form a basis of $\mathfrak{h}$ and: $[h_i,h_j] = 0$, $[e_i,f_j] = \delta_{ij}h_i$, $[h_i,e_j] = a_{ij}e_j$, $[h_i,f_j] = -a_{ij}f_j$. Note that $\mathfrak{g}(A,\tau)$ admits an involutive anti-automorphism ω, defined by: $\omega(h_i) = h_i$, $\omega(e_i) = f_i$, $\omega(f_i) = e_i$.

Examples. Let $\mathfrak{p}$ be one of the simple finite-dimensional Lie algebras or basic classical Lie superalgebras $A(m,n)\, m \neq n$, $B(m,n)$, $C(n)$, $D(m,n)$, $D(2,1;\alpha)$, $F(4)$ or $G(3)$ [1]. Let E_i, F_i, H_i, $i = 1,\dots,r$, be the canonical generators of $\mathfrak{p}$.

a) If A is the Cartan matrix of $\mathfrak{p}$, $\tau = \{i \mid \deg E_i = \bar{1}\}$, then $\mathfrak{g}(A,\tau) \simeq \mathfrak{p}$ and the Γ-gradation is the root decomposition $\mathfrak{p} = \oplus \mathfrak{p}_\alpha$.

b) Let σ be an automorphism of order s of the Lie superalgebra $\mathfrak{p}$, and let $\mathfrak{p} = \oplus \mathfrak{p}_i$ be the corresponding $\mathbb{Z}_s$-gradation. We set $L(\mathfrak{p},\mathrm{id}) = \mathbb{C}[x,x^{-1}]\otimes_{\mathbb{C}} \mathfrak{p}$, and consider in $L(\mathfrak{p},\mathrm{id})$ the following subalgebra: $L(\mathfrak{p},\sigma) = \oplus_i (x^i \otimes \mathfrak{p}_{i \bmod s})$ (here $\mathbb{C}[x,x^{-1}]$ is the algebra of Laurent polynomials in x). $L(\mathfrak{p},\sigma)$ is an infinite-dimensional Lie superalgebra with a $\mathbb{Z}_2$-gradation defined by the $\mathbb{Z}_2$-gradation of $\mathfrak{p}$ and $\deg x = \bar{0}$. The universal central extension $\tilde{L}(\mathfrak{p},\sigma)$ of $L(\mathfrak{p},\sigma)$ is one of the Lie superalgebras $\mathfrak{g}(A,\tau)$. I shall explain it for $\sigma = \mathrm{id}$. Let θ be the highest root of $\mathfrak{p}$ and $\tilde{A}$ be the extended Cartan matrix of $\mathfrak{p}$. We set $\tilde{\tau} = \tau$ if $\deg \theta = \bar{0}$ and $\tilde{\tau} = \tau \cup \{0\}$ if $\deg \theta = \bar{1}$. Then the homomorphism $\mathfrak{g}(\tilde{A},\tilde{\tau}) \to L(\mathfrak{p},\mathrm{id})$, defined by $e_i \mapsto 1 \otimes E_i$, $f_i \mapsto 1 \otimes F_i$, $i = 1,\dots,n$, $e_0 \mapsto x \otimes E_{-\theta}$, $f_0 \mapsto x^{-1} E_\theta$, is the universal central extension, rank $\Gamma = r + 1$ and Γ-gradation of $\mathfrak{g}(\tilde{A},\tilde{\tau})$ is the pull-back of the decomposition $L(\mathfrak{p},\mathrm{id}) = \oplus_{\alpha,k} (x^k \otimes \mathfrak{p}_\alpha)$.

The finite-dimensional Lie superalgebras $\mathfrak{g}(A,\tau)$ are classified in [1]. One can show that provided that A is an indecomposable matrix and $a_{ij} = 0$ implies $a_{ji} = 0$ the Lie superalgebra $\mathfrak{g}(A,\tau)$ is infinite dimensional and has finite Gelfand-Kirillov dimension if and only if it is isomorphic to one of the Lie superalgebras $\tilde{L}(\mathfrak{p},\sigma)$ (cf. [7]). All the pairs (A,τ) corresponding to these Lie superalgebras are classified in [8].

From now on we assume that A is a symmetrisable matrix, i.e. $A = DB$, where $D = \mathrm{diag}(d_1,\dots,d_n)$, $\det D \neq 0$, and $B = (b_{ij})$ is a symmetric matrix. To $\eta = \Sigma k_i \alpha_i \in \Gamma$ we assign a linear function $\hat{\eta}$ on $\mathfrak{h}$ by $\hat{\eta}(h_j) = \Sigma_i k_i a_{ji}$; we set $h_\eta = \Sigma_i k_i d_i^{-1} h_i$. We introduce a bilinear form $(\ ,\ )$ on Γ by $(\alpha_i,\alpha_j) = b_{ij}$ and define $\rho \in \mathfrak{h}^*$ by $\rho(h_i) = \frac{1}{2} a_{ii}$.

Theorem 1 (cf. [3], [4]). a) For the representation $M(\lambda)$ of $\mathfrak{g}(A,\tau)$ one has: $\det F_\eta = \prod_{\alpha\in\Delta^+}\prod_{n\in\mathbb{N}}((\lambda+\rho)(h_\alpha) - \frac{1}{2}n(\alpha,\alpha))^{(\operatorname{sign}\alpha)P(\eta-n\alpha)\dim\mathfrak{g}_\alpha}$. In particular, $M(\lambda)$ is irreducible iff $2(\lambda+\rho)(h_\alpha) \neq (\alpha,\alpha)$ for any $\alpha\in\tilde{\Delta}^+$.

b) Any irreducible subquotient of $M(\lambda)$ is of the form $L(\lambda-\hat{\eta})$, where η is such that there exist $\beta_1,\dots,\beta_k\in\tilde{\Delta}^+$ for which $\eta = \Sigma_{i=1}^k\beta_i$ and $2(\lambda+\rho-\hat{\beta}_1-\dots-\hat{\beta}_{i-1})(h_{\beta_i}) = (\beta_i,\beta_i)$, $i = 1,\dots,k$.

Let $\gamma_1,\gamma_2,\dots$ be the minimal system of generators of the semigroup generated by Δ_0^+. Suppose that (*) $(\operatorname{ad}\mathfrak{g}_{\gamma_s})^m e_i = (\operatorname{ad}\mathfrak{g}_{\gamma_s})^m f_i = 0$ for some m and all $i\in I$, $s = 1,2,\dots$ (this property holds for any finite-dimensional superalgebra or one of $\tilde{L}(\mathfrak{g},\sigma)$). Denote by W the subgroup of $GL(\mathfrak{h}^*)$ generated by all the reflections in γ_i's.

Theorem 2 (cf. [4] - [6]). Suppose that $\mathfrak{g}(A,\tau)$ satisfies (*) and $L(\lambda)$ satisfies a) $\mathfrak{g}^m_{-\gamma_s}(L_0)$ for some m and all $s = 1,2,\dots,$ and b) $(\lambda+\rho)(h_\alpha)\neq 0$ for any $\alpha\in\Delta_1^+$ such that $(\alpha,\alpha) = 0$. Then $\operatorname{ch} L(\lambda) = Q^{-1}\Sigma_{w\in W}(\det w)e^{\lambda+\rho-w(\lambda+\rho)}$.

Proof (cf. [6]). It is easy to see that $e^{-\lambda-\rho}Q\operatorname{ch}L(\lambda) = \Sigma c_\mu e^{-\mu}$, where $c_{\lambda+\rho} = 1$ and $c_\mu\neq 0$ only for the μ's such that $L(\mu-\rho)$ is a subquotient of $M(\lambda)$. From a), b) and (*) once can deduce that these μ have form $w(\lambda+\rho)$, $w\in W$. But $e^{-\rho}Q$ is W-antiinvariant by (*) and $e^{-\lambda}\operatorname{ch}L(\lambda)$ is W-invariant by a). This completes this proof.

3. The Dual Models. The Virasoro algebra $\mathbf{V}$ is a complex Lie algebra with the basis e_0', e_i, $i\in\mathbb{Z} = \Gamma$, with the following commutation relations: $[e_i,e_j] = (i-j)e_{i+j} + \frac{1}{12}(i^3-i)\delta_{i,-j}e_0'$, $[e_i,e_0'] = 0$. The Γ-gradation is defined by $V_k = \mathbb{C}e_k$, $k\neq 0$, and $V_0 = \mathfrak{h} = \mathbb{C}e_0\oplus\mathbb{C}e_0'$. The involutive antiautomorphism ω is defined by $\omega(e_k) = e_{-k}$, $\omega(e_0') = e_0'$. For $\lambda\in\mathfrak{h}^*$ we set $\lambda(e_0) = h$, $\lambda(e_0') = c$. For $k,s\in\mathbb{N}$, $k\neq s$, let $\phi_{k,s}(h,c)$ denote the quadratic polynomial in h with the roots $\frac{1}{48}((13-c)(k^2+s^2)\pm(c^2-26c+25)^{\frac{1}{2}}(k^2-s^2) - 24ks - 2 + 2c)$ and let $\phi_{k,k}(h,c) = h + \frac{1}{24}(k^2-1)(c-1)$. We set $\psi_n(h,c) = \prod_{s|n}\phi_{s,n/s}$.

Theorem 3 [5]. a) $\det F_n(h,c) = \prod_{i=1}^n\psi_i^{p(n-i)}$, where $p(s)$ is the classical partition function. In particular the representation $M(h,c)$ is irreducible iff $\phi_{k,s}(h,c)\neq 0$ for any $k,s\in\mathbb{N}$.

b) Any irreducible subquotient of $M(h,c)$ is of the form $L(h+n,c)$ where n is such that there exist $n_1,\dots,n_k\in\mathbb{N}$ for which $n = \Sigma_{i=1}^k n_i$ and $\psi_{n_i}(h+n_1+\dots+n_{i-1},c) = 0$ for $i = 1,\dots,k$.

The Lie algebra $\mathbf{V}$ acts by derivations on the Lie algebra $\mathfrak{g} = \tilde{L}(sl_2, id)$, which contains a subalgebra $\mathfrak{s} \simeq sl_2$, such that $\mathfrak{s}$ commutes with $\mathbf{V}$. Let $\tilde{\mathfrak{g}}$ be the semi-direct sum of $\mathfrak{g}$ and V. Any representation $L(\lambda)$ of $\mathfrak{g}$ can be uniquely extended to $\tilde{\mathfrak{g}}$ [9]. $L(\lambda)$ is decomposed with respect to $\mathfrak{s}$:

$L(\lambda) = \oplus_{m \in \mathbb{N}} P_m$, where P_m is a sum of all irreducible subrepresentations of $\mathfrak{s}$ in $L(\lambda)$ equivalent to the $(m+1)$-dimensional representation T_m.

$Q_m = (\oplus_k L_{k(\alpha_0+\alpha_1)}) \cap P_m$ is invariant with respect to $\mathbf{V}$. We set $\mathcal{P}(t) = \prod_{k=1}^{\infty}(1-t^k)$. The crucial point in the proof of Theorem 3 is,

<u>Proposition 1.</u> <u>Let</u> λ_ε, $\varepsilon = 0$, <u>or</u> 1, <u>be the fundamental weights of</u> $\mathfrak{g}$ <u>(i.e.</u> $\lambda_i(h_j) = \delta_{ij}$, $i,j = 0,1$). <u>Then with respect to</u> $\mathfrak{s} \oplus \mathbf{V}$ <u>one has:</u> $L(\lambda_\varepsilon) = \oplus_{m \in \varepsilon + 2\mathbb{Z}_+} (T_m \otimes Q_m)$, <u>the representation</u> Q_m <u>of</u> $\mathbf{V}$ <u>is irreducible,</u> $Q_m \simeq L(m^2/4, 1)$ <u>and</u> $\Sigma_k \dim(L_{k(\alpha_0+\alpha_1)} \cap Q_m) t^k = t^{\frac{1}{2}(m-\varepsilon)^2}(1 - t^{m+1})\mathcal{P}(t)^{-1}$.

<u>Corollary.</u> a) <u>The representation</u> $M(h,0)$ <u>of</u> $\mathbf{V}$ <u>is irreducible iff</u> $h \neq \frac{1}{24}(m^2-1)$, $m \in \mathbb{Z}_+$. b) (Goldstone conjecture). <u>The representation</u> $M(h,1)$ <u>of</u> V <u>is irreducible iff</u> $h \neq \frac{1}{4}m^2$. <u>One has:</u> $\operatorname{ch} L(\frac{1}{4}m^2, 1) = \mathcal{P}(e)^{-1}(1-e^{m+1})$. <u>For</u> $h,c \in \mathbb{R}$, $h > 0$, $c > 1$ <u>the representation</u> $M(h,c)$ <u>is irreducible.</u>

<u>Conjecture.</u> $\operatorname{ch} L\left(\frac{1}{24}(m^2-1), 0\right) = \mathcal{P}^{-1}(e)\Sigma_{k \in \mathbb{Z}}(-1)^k e^{\frac{1}{2}(3k^2+mk)}$ for odd m and $= \mathcal{P}(e)^{-1}\Sigma_{k \in \mathbb{Z}}(-1)^k e^{6k^2+mk}$ for even m.

The <u>Neveu-Schwartz superalgebra</u> $\mathbf{S}$ is a complex Lie superalgebra with the basis e_0', e_i, $i \in \frac{1}{2}\mathbb{Z} = \Gamma$, such that e_i with $i \in \mathbb{Z}$ is even and e_i with $i \in \frac{1}{2} + \mathbb{Z}$ is odd, and the commutation relations are as follows: $[e_i, e_j] = (i-j)e_{i+j} + \frac{1}{8}(i^3-i)\delta_{i,-j}e_0'$ for $i,j \in \mathbb{Z}$, $[e_i, e_j] = (\frac{1}{2}i-j)e_{i+j}$ for $i \in \mathbb{Z}$, $j \in \frac{1}{2} + \mathbb{Z}$, and $[e_i, e_j] = 2e_{i+j} + \frac{1}{2}(i^2 - \frac{1}{4})\delta_{i,-j}e_0'$ for $i,j \in \frac{1}{2} + \mathbb{Z}$. The Γ-gradation and the involution ω are defined in the same way as for $\mathbf{V}$. For $k,s \in \frac{1}{2}\mathbb{N}$, such that $k - s \in \mathbb{Z} - \{0\}$, let $\phi_{k,s}(h,c)$ denote the quadratic polynomial in h with the roots $\frac{1}{8}((5-c)(k^2+s^2) \pm (c^2-10c+9)^{\frac{1}{2}}(k^2-s^2) - 8ks - \frac{1}{2} + \frac{1}{2}c)$ and let $\phi_{k,k}(h,c) = h + \frac{1}{16}(k^2-1)(c-1)$. For $k,n \in \frac{1}{2}\mathbb{Z}$ we write $k|n$ if there exists $s \in \frac{1}{2}\mathbb{Z}$ such that $k - s \in \mathbb{Z}$ and $2ks = n$. We set $\psi_n(h,c) = \prod_{s|n} \phi_{s,n/s}(h,c)$. In particular $\psi_n(h,0) = \prod_{s|n} (h - 1/16((4s - 2n/s)^2 - 1))$ and $\psi_n(h,1) = \prod_{s|n} (h - \frac{1}{2}(s - n/s)^2)$.

Theorem 4. a) *det* $F_n(h,c) = \prod_i \psi_i^{\tilde{p}(n-i)}$, *where* $\tilde{p}(s)$ *is equal to the number of partitions of* $s \in \frac{1}{2}\mathbb{N}$ *into the parts belonging to* $\frac{1}{2}\mathbb{N}$, *the parts from* $-\frac{1}{2} + \mathbb{N}$ *being different. In particular the representation* $M(h,c)$ *is irreducible iff* $\phi_{k,s}(h,c) \neq 0$ *for any* $k,s \in \frac{1}{2}\mathbb{N}$ *such that* $k - s \in \mathbb{Z}$.

b) *Any irreducible subquotient of* $M(h,c)$ *is of the form* $L(h+n,c)$, *where* n *is such that there exists* $n_1,\ldots,n_k \in \frac{1}{2}\mathbb{N}$ *for which* $n = \Sigma_i n_i$ *and* $\psi_{n_i}(h + n_1 + \ldots + n_{i-1}, c) = 0$ *for* $i = 1,\ldots,k$.

We set $\psi(t) = \mathcal{S}(t)^2/\mathcal{S}(t^{\frac{1}{2}})$.

Corollary. a) *The representation* $M(h,0)$ *of* S *is irreducible iff* $h \neq \frac{1}{16}(m^2 - 1)$, $m \in \frac{1}{2}\mathbb{Z}_+$. *The representation* $M(h,1)$ *is irreducible iff* $h \neq \frac{1}{2}m^2$, $m \in \frac{1}{2}\mathbb{Z}_+$. *One has:* $ch(\frac{1}{2}m^2, 1) = \psi(e)^{-1}(1 - e^{m+\frac{1}{2}})$.

Conjecture. $ch\, L\left(\frac{1}{16}(m^2-1), 0\right) = \psi(e)^{-1}\Sigma_{k \in \frac{1}{2}\mathbb{Z}} e^{k^2+\frac{1}{2}mk}$.

I am grateful to D. Kazhdan, J. Goldstone and G. Segal for interesting and stimulating discussions.

REFERENCES

1. V. G. Kac, Lie superalgebras, Advances in Math. 26 (1977), 8-96.
2. J. K. Schwartz, Dual resonance theory, Physics reports 8c. (1973), 269-335.
3. V. G. Kac, D. A. Kazhdan, Structure of representations with highest weight of infinite-dimensional Lie algebras, to appear.
4. V. G. Kac, Representations of classical Lie superalgebras, Lecture Notes in Math. 676 (1978), 597-626.
5. V. G. Kac, Highest weight representations of infinite-dimensional Lie algebras, Proceedings of ICM, Helsinki (1978).
6. V. G. Kac, Infinite-dimensional algebras, Dedekind's η-function, classical Möbius function and the very strange formula, Advances in Math. 30 (1978).
7. V. G. Kac, Simple irreducible graded Lie algebras of finite growth, Math. USSR Izvestija, 2 (1968), 1271-1311.
8. V. G. Kac, Automorphisms of finite order of simple Lie superalgebras, to appear.
9. G. Segal, preprint (1978).

DUAL SETS IN ASSOCIATIVE ALGEBRAS AND GENERALIZED LIE ALGEBRAS

Yehiel Ilamed
Soreq Nuclear Research Centre, Yavne, Israel

DEFINITIONS AND NOTATIONS. I. Let F be a field of characteristic zero and let P be the associative algebra over F generated by $x_1, x_2, \ldots$, noncommutative variables.

II. Let $f(p,q)$: $P \times P$ into P, be a bilinear map and let $u(i), v(i) \in P$, $i=1,2,\ldots$. We say that $\{u(1),\ldots,u(h)\}$ is dual to $\{v(1),\ldots,v(h)\}$ relative to $f(p,q)$, if for each pair $i \neq j$, $i,j=1,\ldots,h$, $f(u(i),v(j))$ can be expressed as a sum of commutators.

III. Let $s_h \equiv s_h(x)$ and $R_{h+1}(x;y) \equiv R_{h+1}(x_1,\ldots,x_h;y_1,\ldots,y_{h+1})$ be defined by

$$s_h(x) \equiv s_h(x_1,\ldots,x_h) = \Sigma_\sigma(sg\sigma)x_{\sigma 1}\ldots x_{\sigma h}$$

$$R_{h+1}(x;y) = \Sigma_\sigma\Sigma_\gamma(sg\sigma)(sg\gamma)y_{\gamma 1}x_{\sigma 1}y_{\gamma 2}x_{\sigma 2}\cdots y_{\gamma h}x_{\sigma h}y_{\gamma(h+1)}$$

where Σ_σ means summation over the h! permutations of $1,\ldots,h$, Σ_γ means summation over the h+1 cyclic permutations of $1,\ldots,h+1$ and $sg\sigma$ means the sign of σ.

IV. Let $s_h \overline{\wedge} s_k \equiv s_h \overline{\wedge} s_k(x_1,\ldots,x_{h+k-1})$ be the composition product defined by [1], $s_h \overline{\wedge} s_k = \Sigma_\tau(sg\tau)s_h(s_k(x_{\tau 1},\ldots,x_{\tau k}),x_{\tau(k+1)},\ldots,x_{\tau(k+h-1)})$, where Σ_τ means summation over all the permutations τ of $1,\ldots,h+k-1$ that satisfy $\tau 1 < \tau 2 <,\ldots,< \tau k$ and $\tau(k+1) < \tau(k+2) <,\ldots,< \tau(k+h-1)$.

V. Let J_{2h-1} be defined by $J_{2h-1}(x_1,\ldots,x_{2h-1}) = s_h \overline{\wedge} s_h(x_1,\ldots,x_{2h-1})$ and let V be a vector subspace of an associative algebra over F. We say that V is i) a weak h-Lie algebra or ii) an h-Lie algebra over F, if i) $s_h(v_1,\ldots,v_h) \in V$ or ii) $J_{2h-1}(v_1,\ldots,v_{2h-1})=0$ and i) are satisfied respectively for any v_j in V.

VI. Let H be the Heisenberg Lie algebra over F spanned by $c.1, q_1, p_1, q_2, p_2, \ldots$ i.e., $[p_i,q_j]=\delta_{ij}c1$, and all other commutators are zero; $c \in F$.

THEOREMS. 1. Let $y_1,\ldots,y_h$ be any h given elements of P and let $v_i = R_h(u_1,\ldots,u_{i-1},u_{i+1},\ldots,u_h;y_1,\ldots,y_h)$, $u_i \in P$, $i=1,\ldots,h$. The set $\{v_1,\ldots,v_h\}$ is dual to $\{u_1,\ldots,u_h\}$, relative to $f(p,q) = pq+qp$.

2. Let $w_1,\ldots w_{2h-1}$ and $y_1,\ldots,y_m$ be any 2h+m-1 given elements of P and let $u_i = s_{2h}(x_i,w_1,\ldots,w_{2h-1})$, $v_i = R_m(u_1,\ldots,u_{i-1},u_{i+1},\ldots,u_m;y_1,\ldots,y_m)$, $i=1,\ldots,m$. Then $\{x_1,\ldots,x_m\}$ is dual to $\{v_1,\ldots,v_m\}$ relative to the alternating bilinear map $f(p,q) = s_{2h+1}(p,q,w_1,\ldots,w_{2h-1})$.

3. H is a weak m-Lie algebra and a k-Lie algebra for $m \geq 2$ and even k respectively.

4. A weak m-Lie algebra is an m-Lie algebra if m is even.

5. If V is a weak m-Lie algebra for m=2n,2n+1,...,4n-1 or for m=2n+1,2n+2, ...,4n , then V is a weak m-Lie algebra for $m \geqslant 4n$ or for $m \geqslant 4n+1$ respectively.

6. Let M_n be the algebra of $n \times n$ matrices with entries in F . Then M_n is an m-Lie algbra for m = n+1,n+2,... .

7. Let L be the vector space over F spanned by $s_1, s_2, \ldots$. Then L with the multiplication law $s_h \overline{\wedge} s_k$ is a nonassociative algebra with the multiplication table $s_m \overline{\wedge} s_n = (m\delta_{n,odd} + \delta_{n,even} - \delta_{m,even}) s_{m+n-1}$.

8. Let L_1 and L_2 be vector spaces over F spanned respectively by $e_1, e_2, \ldots$ and $f_{1/2}, f_{3/2}, \ldots$, where $2e_n = s_{2n+1}$, n=0,1,2,... and $2f_j = s_{2j+1}$, j = 1/2 , 3/2 ,... . Then L_1 and $L_1 \oplus L_2$ are Lie algebras with the commutation relations $[e_m, e_n] = (m-n)e_{m+n}$, $[e_m, f_j] = -jf_{m+j}$, $[f_j, f_k] = 0$; where $[s_h, s_k]$ is defined by $[s_h, s_k] = s_h \overline{\wedge} s_k - s_k \overline{\wedge} s_h$.

NOTES. i) This research note is a continuation of the study started in Rfs. [2,3].

ii) The proofs of theorems 2 and 7 may be obtained using Rfs. [2] and [4] respectively, and using them the other theorems follows.

iii) Let us define $q_{-h} = p_h$ for $h > 0$ and $a_n = e_n + \alpha_n q_n$, $\alpha_n \in F$. Let us extend the commutation relations in theorem 8 to all the integers as subscripts. For appropriate values of α_n the algebra over F with basis $a_0, a_1, a_{-1}, a_2, a_{-2}, \ldots$ and $f_{1/2}, f_{-1/2}, f_{3/2}, f_{-3/2}, \ldots$ (an extended algebra of L_1, L_2 and H) is a realization of some string algebras [5].

iv) For h=2 , J_{2h-1} is the usual Jacobi identity. In general for even h, as for h=2 , J_{2h-1} is an identity due to the assumption A is associative.

v) The nonassociative algebra L is a Lie-admissible algebra.

vi) The problem of constructing a dual set to a set of h elements (vectors) of an associative algebra is connected to the problem of constructing a vector product [6] of h-1 vectors of the algebra, to what extent such a vector product exists.

References

[1] A. Nijenhuis, On a class of common properties of some different types of algebras, Nieuw Arch. Wisk., 17(1969) 17-46, 87-108 .

[2] Y. Ilamed, On identities for matrix rings and polynomials orthogonal to their arguments, Adv. Study Inst.,U.of Antwerpen,August 1978, Preprint.

[3] Heisenberg algebras, n-Lie algebras and alternating polynomials that are central, Notices Amer. Math. Soc. 25(1978) A-424 (78T-A111)

[4] A. Kovacs, The n-center of a ring, J. Algebra 40(1976)107-124 .

[5] J. Scherk, An introduction to the theory of dual models and strings, Rev. Modern Phys.,47(1975)123-164.

[6] B. Eckmann, Continuous solutions of linear equations-some exceptional dimensions in topology, Battelle Rencontres,1967 Lectures in Mathematics and Physics, Ed.C.M. DeWitt and J.A. Wheeler, p.516-526,Benjamin Inc. 1968.

MOLIEN FUNCTION FOR A SYMMETRIC GROUP

N.R.RANGANATHAN and J.S.PRKASH, MATSCIENCE, MADRAS-600 020, INDIA.

1. Introduction

In a perturbative expansion of Greens functions in many body systems we require information on the twin problems concerning the analysis of Feynman diagrams: (a) the number of topologically distinct diagrams; (b) their corresponding algebraic expressions. Akyopan [1] and more recently Trainor et al [2,3,4] have investigated the problem (a). The latter employed the properties of symmetric group to characterize the topologically distinct diagrams in terms of equivalence classes of permutations and have obtained a formula for the number in terms of the order of the centralizer of the permutation which labels the interaction lines in a natural order. One of us [5] has obtained this formula using the concept of subclass due to Wigner [6]. It is to be remarked that this approach does not yield us neat prescriptions for the problem (b).

Caianiello and his collaborators [7] have developed basic algorithm which expresses an arbitrary order in the perturbative expansion of Greens function for fermions in the form of a determinant each term of which givesus a Feynman diagram and its algebraic expression. Hence we feel that by suitably applying the approach of Trainor et al to Caianiello determinants, we will be able to provide answers to both problems (a) and (b).

2. Molien Function for a Symmetric Group.

Caianiello determinant provides us the algebraic expressions for all Feynman diagrams in an arbitrary order of perturbation. In order to locate the topologically distinct diagrams, we observe that these may be related to the invariants of symmetric group. A powerful tool for investigation of the invariants of a finite group is the Molien function which is a generating function from which we can obtain the various linearly independent, homogeneous invariants of different degrees [8]. Molien function has been studied recently for the space groups [9]. In this paper we present our calculation of Molien function for a symmetric group.

The Molien function $\Phi(\lambda)$ for a finite group G is given by

$$\Phi_G(\lambda) = \frac{1}{|G|} \sum_{A \in G} \frac{1}{\mathrm{Det}(I-\lambda A)}$$

where A is a (mxm) matrix, I is a (mxm) unit matrix and λ is a parameter. If G is the symmetric group of degree n, we consider the natural representation for its elements. We first observe that $\det(I-\lambda A)$ is a class function. The class of any permutation is determined by its cycle structure. We now choose a suitable permutation for each class so that its matrix representation has a block form in such a way that blocks correspond to matrices of dimensions equal tothe length of the cycles occurring in the permutation. Thus we can at once write that

$$\det(I-\lambda A) = \det(I_1-\lambda A_1) \times \ldots\ldots \det(I_k-\lambda A_k)$$

where A_K refers to a k x k matrix corresponding to a k-cycle and I_k

refers to a k x k unit matrix. Since there is one non-zero element in each row and column, we find that

$$\det(I_p-\lambda A_p) = \det I_p - \det \lambda A_p = (1-\lambda^p)$$

Hence

$$\det(I-\lambda A) = (1-\lambda)^{n_1} (1-\lambda^2)^{n_2} \times \ldots (1-\lambda^n)^{n_m}$$

where

$$n_1+2n_2+\ldots+m^{n_m} = n!$$

If we put $x_p = (1-\lambda^p)^{-1}$, we can write the Molien function $\Phi_{S_n}(\lambda)$ as

$$\Phi_{S_n}(\lambda) = \frac{1}{n!}\sum_{\text{all partitions of } n} \frac{n!}{n_1!\,2^{n_2}n_2!\cdots m^{n_m}n_m!}\, x_1^{n_1}\cdots x_m^{n_m}$$

We recognize that the R.H.S. of the above expression is the cycle index of S_n [10] . We can extend this result to any abstract group G by embedding it in a symmetric group whose degree is equal to the order of the group G and then considering the natural or regular representation of the corresponding subgroup of S_n. Because of this equality between Molien function and cycle index of a group, the Molien function of a group G = H ⊗ T is

$$\Phi_G(\lambda) = \Phi_H(\lambda)\,\Phi_T(\lambda)$$

We can express $\Phi_{S_n}(\lambda)$ using MacMahon's identity [11] , as

$$\Phi_{S_n}(\lambda) = \frac{1}{(1-\lambda)(1-\lambda^2)\cdots(1-\lambda^n)}$$

From this we see that the number of linearly independent homogeneous polynomials of degree n which are invariant under S_n (i.e. coefficient of λ^n in L.H.S.) is equal to the number of partitions of n with at most n parts (i.e. coefficient of λ^n in R.H.S.). This combinatorial interpretation of $\Phi_{S_n}(\lambda)$ is being used in our attempt to obtain the topologically distinct diagrams and their expressions from Caianiello determinant.

We wish to thank Professor Alladi Ramakrishnan, Director, Matscience for encouragement and Dr.V.V.Rama Rao for the benefit of helpful discussion on the problem.

References

1. A.A.AKYOPAN: JETP 20, 1172, (1965)
2. G.ROSENSTEEL, E.IHRIG and L.E.H.TRAINOR: Proc. Roy.Soc.Lond. A344 387(1975)
3. E.IHRIG, G.ROSENSTEEL, H.CHOW and L.E.H.TRAINOR: Proc.Roy.Soc. Lond. A 348, 339 (1976)
4. M.B.WISE and L.E.H.TRAINOR: Proc. Roy. Soc.Lond. A359, 111, (1978)
5. J.S.PRAKASH: (to be published)
6. E.P.WIGNER: Proc. Roy. Soc. Lond. A322, 181 (1971).
7. E.R.CAIANIELLO: Combinatorics and Renormalization in Quantum Field Theory, W.A.Benjamin Inc. USA (1973).
8. N.J.A.SLOANE: Amer. Math. Month 84, 82 (1977).
9. M.W.JARIC, J.L.BIRMAN, J. Math. Phys. 18, 1456, 1459, 2085 (1977)
10. J.RIORDAN: An Introduction to Combinatorial Analysis, John Wiley (1958)
11. P.A.MACMAHON: Combinatory Analysis, Cambridge University Press London (1916)

Generalized Lie Algebras

M. Scheunert, Dublin Institute for Advanced Studies

In this note we describe a close relationship between a class of generalized Lie algebras (called colour (super)algebras in reference [1]) and graded Lie (super)algebras. For a more detailed discussion see reference [2].

Let Γ be an abelian group and let $\mathbb{C}_*$ be the multiplicative group of non-zero complex numbers. A map $\varepsilon : \Gamma \times \Gamma \longrightarrow \mathbb{C}_*$ is called a *commutation factor* on Γ if

$$\varepsilon(\alpha,\beta)\,\varepsilon(\beta,\alpha) = 1 \quad , \quad \varepsilon(\alpha,\beta+\gamma) = \varepsilon(\alpha,\beta)\,\varepsilon(\alpha,\gamma)$$

for all $\alpha , \beta , \gamma \in \Gamma$. Note that $\alpha \longmapsto \varepsilon(\alpha,\alpha)$ is a homomorphism of Γ into $\mathbb{C}_*$ and that $\varepsilon(\alpha,\alpha) = \pm 1$ for all $\alpha \in \Gamma$. We set

$$\Gamma_0 = \{\alpha \in \Gamma \mid \varepsilon(\alpha,\alpha) = 1\} \quad , \quad \Gamma_1 = \{\alpha \in \Gamma \mid \varepsilon(\alpha,\alpha) = -1\}$$

and define a map $\varepsilon_0 : \Gamma \times \Gamma \longrightarrow \mathbb{C}_*$ by

$$\varepsilon_0(\alpha,\beta) = \begin{cases} -1 & \text{if } \alpha , \beta \in \Gamma_1 \\ 1 & \text{otherwise .} \end{cases}$$

Obviously, ε_0 is a commutation factor on Γ .

Let L be a complex Γ - graded algebra whose multiplication is denoted by a pointed bracket. Thus $L = \bigoplus_{\alpha\in\Gamma} L_\alpha$ (as a vector space) and $\langle L_\alpha , L_\beta\rangle \subset L_{\alpha+\beta}$ for all $\alpha , \beta \in \Gamma$. We call L an ε - *Lie - algebra* if $\langle\ ,\ \rangle$ satisfies the conditions

$$\langle A , B\rangle = -\varepsilon(\alpha,\beta)\langle B , A\rangle \quad , \quad \varepsilon(\gamma,\alpha)\langle A ,\langle B , C\rangle\rangle + \text{cyclic} = 0$$

for all $A \in L_\alpha$, $B \in L_\beta$, $C \in L_\gamma$; $\alpha , \beta , \gamma \in \Gamma$.

Let $\sigma : \Gamma \times \Gamma \longrightarrow \mathbb{C}_*$ be an arbitrary map. On the Γ - graded vector space L we introduce a new multiplication $\langle\ ,\ \rangle^\sigma$ by setting

$$\langle A , B\rangle^\sigma = \sigma(\alpha,\beta)\langle A , B\rangle$$

for all $A \in L_\alpha$, $B \in L_\beta$; $\alpha , \beta \in \Gamma$. The Γ - graded algebra which is obtained will be denoted by L^σ. Define a map $\delta : \Gamma \times \Gamma \longrightarrow \mathbb{C}_*$ by

$$\delta(\alpha,\beta) = \sigma(\alpha,\beta)\,\sigma(\beta,\alpha)^{-1} \quad \text{for all } \alpha , \beta \in \Gamma .$$

Then δ is a commutation factor on Γ and L^σ is an $\varepsilon\delta$ - Lie - algebra if σ satisfies

$$\sigma(\alpha,\beta+\gamma)\,\sigma(\beta,\gamma) = \sigma(\alpha,\beta)\,\sigma(\alpha+\beta,\gamma) \quad \text{for all } \alpha , \beta , \gamma \in \Gamma .$$

If Γ is finitely generated we can find a map σ such that this condition as well as the equation $\varepsilon\delta = \varepsilon_0$ are fulfilled. Then $L \longmapsto L^\sigma$ is a bijective transformation between ε - Lie - algebras and ε_0- Lie - algebras. However, the latter are just the Γ - graded Lie superalgebras (with Γ_0 being the subgroup of even degrees). The *graded* representations of L and L^σ are related by a similar construction.

[1] V. Rittenberg, D. Wyler, Generalized superalgebras, Rockefeller Univ., COO-2232B-150

[2] M. Scheunert, Generalized Lie algebras, Dublin Inst. Adv. Studies, DIAS-TP-78-36

SUPERCONFORMAL GROUP AND QUARK-LIKE FERMIONIC COORDINATES

by

Jerzy Lukierski

Institute of Theoretical Physics, University of Wroclaw, Wroclaw, Poland

Fundamental representation of the conformal group is known as twistor. Using twistors with commuting components as primary geometric objects one can introduce space-time coordinates as composite (see e.g.[1]). We develop parallel idea that quarks are primary fermionic geometric objects; one-quark system is described by infinite family of twistors with anticommuting coordinates. It appears that in fermionic twistor space one can introduce the metric invariant under superconformal transformations . This metric we use for the derivation of the action for quark-twistor string which appears to be fermionic nonlinear superconformal-invariant σ-model.

Such an approach allows to unify and geometrize all the symmetries occurring in hadron physics:

a) space-time, flavour and colour are unified by introduction of colour extension of superconformal transformations [2],

b) dual gauge transformations, with Virasoro algebra of generators, are related with different ways of counting of infinite family of fermionic twistors.

For details see [3,4] .

1.R.Penrose and M.A.H.Mac Callum, Phys.Rep.6C, 242 (1972).

2.J.Lukierski and V.Rittenberg, Phys.Rev.18, 385 (1978).

3.J.Lukierski, Lecture at Liblice (CSRR), June 1978;Wrocław Univ. preprint n.436.

4.J.Lukierski ICTP preprint IC/78/82, July 1978, to be published.

PARAFIELDS AND SUPERGROUP TRANSFORMATIONS

M. Omote and S. Kamefuchi
Institute of Physics, University of Tsukuba
Ibaraki 300-31, Japan

§1 Introduction

The problem of supersymmetry or fermi-bose symmetry is usually discussed within the framework of ordinary field theory.[1] In our opinion, however, a deeper insight into the problem may be acquired by studying it in a wider context, i.e., quantum parafield theory. The main purpose of the present paper is to show that parafields are related in some essential manner to what we may call supergroup transformations.

For convenience of later discussions let us first summarize the characteristic features of quantum parafield theory.[2] A parafermi (parabose) field to obey parafermi (parabose) statistics of order $p(=1,2,\dots)$ is described by a set of operators, a_k and $a_k^\dagger$, which annihilate and create, respectively, a particle of mode $k(=1,2,\dots,f\to\infty)$ and which satisfy the trilinear commutation relations:

$$[a_k,[a_\ell^\dagger,a_m]_\mp] = 2\delta_{k\ell}a_m,$$
$$[a_k,[a_\ell^\dagger,a_m^\dagger]_\mp] = 2\delta_{k\ell}a_m^\dagger \mp 2\delta_{km}a_\ell^\dagger, \quad [a_k,[a_\ell,a_m]_\mp] = 0. \tag{1.1}$$

Throughout the present paper upper and lower signs correspond to the parafermi and parabose cases, respectively. Further the vacuum state $|0\rangle$ is defined by

$$a_k|0\rangle = 0, \quad a_k a_\ell^\dagger|0\rangle = p\delta_{k\ell}|0\rangle. \tag{1.2}$$

The case p=1 provides an ordinary fermi or bose field where (1.1) is simplified to the bilinear commutation relations $[a_k,a_\ell^\dagger]_\pm=\delta_{k\ell}$, $[a_k,a_\ell]_\pm=0$. It is often convenient to work with Green-component operators $a_k^{(\kappa)}$, $a_k^{(\kappa)\dagger}(\kappa=1,2,\dots,p)$ such that

$$a_k = \sum_{\kappa=1}^p a_k^{(\kappa)}, \quad a_k^\dagger = \sum_{\kappa=1}^p a_k^{(\kappa)\dagger},$$
$$[a_k^{(\kappa)},a_\ell^{(\kappa)\dagger}]_\pm = \delta_{k\ell}, \quad [a_k^{(\kappa)}, a_\ell^{(\kappa)}]_\pm = 0 \qquad \text{(no summation over } \kappa\text{)}, \tag{1.3}$$
$$[\hat{a}_k^{(\kappa)},\hat{a}_\ell^{(\lambda)}]_\mp = 0 \quad (\kappa \neq \lambda),$$

where, for example, $\hat{a}_k^{(\kappa)}$ stands for $a_k^{(\kappa)}$ or $a_k^{(\kappa)\dagger}$. As can be easily checked, a_k and $a_k^\dagger$ expressed in this way satisfy (1.1) and (1.2).

It is also known[2] that a general system of coexisting parafields can be classified into families as follows. (i) All fields $\alpha,\beta,\gamma,\dots$ belonging to one and the same family must have the same order and obey the trilinear commutation relations:

$$[a_k^\alpha,[a_\ell^{\beta\dagger},a_m^\gamma]_{(\beta\gamma)}]_{-(\alpha\beta)(\alpha\gamma)} = 2\delta_{\alpha\beta}\delta_{k\ell}a_m^\gamma,$$
$$[a_k^\alpha,[a_\ell^{\beta\dagger},a_m^{\gamma\dagger}]_{(\beta\gamma)}]_{-(\alpha\beta)(\alpha\gamma)} = 2\delta_{\alpha\beta}\delta_{k\ell}a_m^{\gamma\dagger} + 2(\beta\gamma)\delta_{\alpha\gamma}\delta_{km}a_\ell^{\beta\dagger}, \tag{1.4}$$
$$[a_k^\alpha,[a_\ell^\beta,a_m^\gamma]_{(\beta\gamma)}]_{-(\alpha\beta)(\alpha\gamma)} = 0.$$

Here the Greek superscripts $\alpha,\beta,\gamma,\dots$ are to label the species of parafields; the signature symbols $(\alpha\beta)=(\beta\alpha)=+$ or $-$ are defined for each pair of parafields, and $(\alpha\alpha)$ in particular is chosen to be $-$ $(+)$ for α being a parafermi (parabose) field. (For the case p=1 the bilinear commutation relations $[a_k^\alpha,a_\ell^{\beta\dagger}]_{-(\alpha\beta)}=\delta_{\alpha\beta}\delta_{k\ell}$, etc. are also satisfied). (ii) Any two fields belonging to different families must obey the bilinear commutation relations:

$$[\hat{a}_k^{\alpha}, \hat{a}_{\ell}^{\beta}]_{(\alpha\beta)} = 0. \tag{1.5}$$

The above commutation relations, (1.4) and (1.5), together with the locality condition impose a strong restriction on the structure of physical observables.[2] As a consequence there follows the so-called gauge or colour symmetry O(p), SO(p), U(p) or SU(p), that leads to the specific selection rules. These problems, however, are beyond the scope of the present paper.

§2 Supergroups and superalgebras

We begin by noticing the fact that Lie's fundamental theorems for Lie groups can formally be generalized[3] to the case when the group parameters $x^i (i=1,2,\ldots,r)$ of a group G satisfy 'commutation relations' such as

$$[x^i, x^j]_{-(ij)} = 0 \tag{2.1}$$

with the signature symbols $(ij)=(ji)=+$ or $-$ being defined for each pair of parameters x^i and x^j. So-called Z_2-graded Lie groups often discussed in the literature[1] correspond to a special case of the above where $(ij)=(-1)^{g_i g_j}$ with g_i and g_j being integers. In what follows groups of such a general kind and associated Lie algebras will be referred to as supergroups and superalgebras, respectively.

As in the case of Grassmann algebras[4] it is difficult (or impossible?), in general, to define topology for a functional space of x^i's. In order to avoid this difficulty let us adopt here a standpoint such that all quantities concerned are regarded as formal (finite or infinite) power series in x^i's, thereby refraining completely from asking questions as to convergence of the series. Differentiation and integration can also be defined formally.

As usual the product of two elements $x=(x^1, x^2,\ldots x^r)$ and $y=(y^1, y^2,\ldots, y^r)$ can be written as $xy=f(x,y)$, and its coordinates $f^k(x,y)$ are expanded as

$$f^k(x,y)= \sum_{a,b=0}^{\infty} x^{i_1} x^{i_2} \ldots x^{i_a} y^{j_1} y^{j_2} \ldots y^{j_b} A^k_{i_1 i_2 \ldots i_a; j_1 j_2 \ldots j_b}. \tag{2.2}$$

Here the coefficients $A^k_{\ldots;\ldots}$ should have some obvious symmetry properties with respect to suffixes.

The structure of G is completely determined by fixing all $A^k_{\ldots;\ldots}$'s. In fact, by slightly generalizing the usual arguments[5] we can prove that i) for a given supergroup G the structure constants C^k_{ij} can be introduced which satisfy certain algebraic relations, and conversely, (ii) given a set of C^k_{ij}'s satisfying such relations a supergroup G can be uniquely constructed.

The structure constants C^k_{ij} are defined by

$$C^k_{ij} \equiv A^k_{i;j} - (ij)A^k_{j;i}, \tag{2.3}$$

and satisfy

$$C^k_{ij} = -(ij)C^k_{ji}, \tag{2.4}$$

$$(k\ell)C^m_{jk}C^i_{\ell m} + (\ell j)C^m_{k\ell}C^i_{jm} + (jk)C^m_{\ell j}C^i_{km} = 0. \tag{2.5}$$

where the summation convention is applied only for those indices which are repeated in factors other than signature symbols.

As in the case of usual Lie groups the function $f^k(x,y)$ satisfies a certain differential equation. Thus, when a set of structure constants C^k_{ij} are given, the corresponding $f^k(x,y)$ can be determined by solving this equation successively as a series expansion. Some of the lower order coefficients are as follows:

$$A^k_{;}=0,\ A^k_{i;}=A^k_{;i}=\delta^k_i,\ A^k_{i;j}=\frac{1}{2}C^k_{ij},\quad A^k_{ij;\ell}=\frac{1}{24}\{(ij)C^m_{j\ell}C^k_{im}+C^m_{i\ell}C^k_{jm}\},$$

$$A^k_{i;j\ell}=-\frac{1}{24}\{(i\ell)(j\ell)C^m_{ij}C^k_{\ell m}+(ij)C^m_{j\ell}C^k_{jm}\}. \tag{2.6}$$

The Lie generators $L_i (i=1,2,\ldots,r)$ defined in an appropriate manner satisfy the Lie commutation relations

$$[L_i,L_j]_{-(ij)}=(ij)C^k_{ij}L_k, \tag{2.7}$$

and further

$$[L_i,L_j]_{-(ij)}=-(ij)[L_j,L_i]_{-(ij)}, \tag{2.8}$$

$$(ki)[L_i,[L_j,L_k]_{-(jk)}]_{-(ij)(ki)}+(ij)[L_j,[L_k,L_i]_{-(ki)}]_{-(jk)(ij)}$$
$$+(jk)[L_k,[L_i,L_j]_{-(ij)}]_{-(ki)(jk)}=0, \tag{2.9}$$

The relations (2.7)~(2.9) characterize the superalgebra associated with G.

One of the important consequences of the above results is that the entire structure of G is determined completely by the 'local properties' that are specified solely by the structure constants C^k_{ij}. And as far as such properties are concerned, it suffices to consider only the first three terms in the expansion (2.2). This implies in effect that in such expressions we may treat the group parameters x^i and y^j as if they were infinitesimal.

§3 Commutation relations of parafields

On the basis of the results given in §2 we shall now show[3)] that the commutation relations (1.4) and (1.5) valid for a general system of parafields can be obtained as a whole as the Lie commutation relations of a certain supergroup.

Let us suppose that our system consists of several families and denote by $a^{\alpha_i}_k (a^{\alpha_i\dagger}_k)$ the annihilation (creation) operator for mode k of the α-th field belonging to the i-th family. We then consider a supergroup G whose infinitesimal transformations are given by

$$a^{\alpha_i}_k\to a^{\alpha_i\,\prime}_k=a^{\alpha_i}_k-i\sum_{\substack{m\\ \beta\in i}}[\xi^{\alpha_i\beta_i}_{km}a^{\beta_i}_m+\eta^{\alpha_i\beta_i}_{km}a^{\beta_i\dagger}_m$$
$$-(\beta_i\beta_i)\bar\theta^{\beta_i}_m M^{\alpha_i\beta_i}_{km}+(\beta_i\beta_i)N^{\beta_i\alpha_i}_{mk}\theta^{\beta_i}_m],$$

$$a^{\alpha_i\dagger}_k\to a^{\alpha_i\dagger\,\prime}_k=a^{\alpha_i\dagger}_k+i\sum_{\substack{m\\ \beta\in i}}[a^{\beta_i\dagger}_m\xi^{\beta_i\alpha_i}_{mk}+a^{\beta_i}_m\zeta^{\beta_i\alpha_i}_{mk}$$
$$-(\beta_i\beta_i)L^{\beta_i\alpha_i}_{mk}\theta^{\beta_i}_m+(\alpha_i\alpha_i)\bar\theta^{\beta_i}_m N^{\alpha_i\beta_i}_{km}],$$

$$N^{\alpha_i\beta_i}_{k\ell}\to N^{\alpha_i\beta_i\,\prime}_{k\ell}=N^{\alpha_i\beta_i}_{k\ell}+i\sum_{\substack{m\\ \gamma\in i}}[(\alpha_i\alpha_i)(\gamma_i\gamma_i)N^{\gamma_i\beta_i}_{m\ell}\xi^{\gamma_i\alpha_i}_{mk}-\xi^{\beta_i\gamma_i}_{\ell m}N^{\alpha_i\gamma_i}_{km}-\eta^{\beta_i\gamma_i}_{\ell m}L^{\alpha_i\beta_i}_{km}$$
$$+(\alpha_i\gamma_i)(\beta_i\gamma_i)M^{\beta_i\gamma_i}_{\ell m}\zeta^{\alpha_i\gamma_i}_{km}]+\frac{i}{2}(\alpha_i\alpha_i)(\alpha_i\beta_i)(\bar\theta^{\alpha_i}_k a^{\beta_i}_\ell-a^{\alpha_i\dagger}_k\theta^{\beta_i}_\ell), \tag{3.1}$$

$$L_{k\ell}^{\alpha_i\beta_i}\to L_{k\ell}^{\alpha_i\beta_i'} = L_{k\ell}^{\alpha_i\beta_i}+i\sum_{\substack{m\\ \gamma\in i}} [(\alpha_i\alpha_i)(\gamma_i\gamma_i)L_{m\ell}^{\gamma_i\beta_i}\xi_{mk}^{\gamma_i\alpha_i}$$

$$+(\beta_i\gamma_i)(\gamma_i\gamma_i)\xi_{m\ell}^{\gamma_i\beta_i}L_{km}^{\alpha_i\gamma_i}+(\alpha_i\beta_i)(\beta_i\beta_i)\zeta_{km}^{\alpha_i\gamma_i}N_{\ell m}^{\beta_i\gamma_i}$$

$$+(\alpha_i\beta_i)(\alpha_i\gamma_i)N_{km}^{\alpha_i\gamma_i}\zeta_{m\ell}^{\gamma_i\beta_i}]+\frac{i}{2}(\alpha_i\alpha_i)(\alpha_i\beta_i)(\bar{\theta}_k^{\alpha_i}a_\ell^{\beta_i\dagger}-a_k^{\alpha_i\dagger}\bar{\theta}_\ell^{\beta_i}),$$

$$M_{k\ell}^{\alpha_i\beta_i}\to M_{k\ell}^{\alpha_i\beta_i'}=M_{k\ell}^{\alpha_i\beta_i}-i\sum_{\substack{m\\ \gamma\in i}} [(\alpha_i\gamma_i)(\gamma_i\gamma_i)M_{m\ell}^{\alpha_i\beta_i}\xi_{km}^{\alpha_i\gamma_i}$$

$$+(\beta_i\beta_i)(\gamma_i\gamma_i)\xi_{\ell m}^{\beta_i\gamma_i}M_{km}^{\alpha_i\gamma_i}+(\alpha_i\gamma_i)(\beta_i\beta_i)N_{m\ell}^{\gamma_i\beta_i}\eta_{km}^{\alpha_i\gamma_i}$$

$$+(\alpha_i\gamma_i)(\beta_i\gamma_i)\eta_{m\ell}^{\alpha_i\beta_i}N_{mk}^{\gamma_i\alpha_i}]-\frac{i}{2}(\alpha_i\beta_i)(\beta_i\beta_i)(a_k^{\alpha_i}\theta_\ell^{\beta_i}-\theta_k^{\alpha_i}a_\ell^{\beta_i}),$$

for all i. Here the operators N, L and M are subject to

$$N_{k\ell}^{\alpha_i\beta_i\dagger}= (\alpha_i\alpha_i)(\beta_i\beta_i)N_{\ell k}^{\beta_i\alpha_i},$$

$$L_{k\ell}^{\alpha_i\beta_i\dagger}=M_{\ell k}^{\beta_i\alpha_i},\ L_{k\ell}^{\alpha_i\beta_i}=(\alpha_i\alpha_i)(\alpha_i\beta_i)(\beta_i\beta_i)L_{\ell k}^{\beta_i\alpha_i}. \tag{3.2}$$

On the other hand, the group parameters ξ,η,ζ,θ and $\bar{\theta}$ (any parameters x^i are transformed to $\bar{x}^i$ under hermitian conjugation) are subject to

$$\bar{\xi}_{k\ell}^{\alpha_i\beta_i}=\xi_{\ell k}^{\beta_i\alpha_i},\ \bar{\eta}_{k\ell}^{\alpha_i\beta_i}=\zeta_{\ell k}^{\beta_i\alpha_i},\ \eta_{k\ell}^{\alpha_i\beta_i}=(\alpha_i\alpha_i)(\alpha_i\beta_i)(\beta_i\beta_i)\eta_{\ell k}^{\beta_i\alpha_i},$$

$$[\hat{a}_k^{\alpha_i},A^{\beta_j\gamma_j}]_{-(\alpha_i\beta_j)(\alpha_i\gamma_j)}=0,\ [A^{\alpha_i\beta_i},B^{\gamma_j\lambda_j}]_{-(\alpha_i\gamma_j)(\alpha_i\lambda_j)(\beta_i\gamma_j)(\beta_i\lambda_j)}=0, \tag{3.3}$$

$$[\hat{\theta}_k^{\alpha_i},\hat{\theta}^{\beta_i}]_{(\alpha_i\beta_j)}=[\hat{\theta}_k^{\alpha_i},\hat{a}^{\beta_j}]_{(\alpha_i\beta_j)}=[\hat{\theta}_k^{\alpha_i},B^{\beta_j\gamma_j}]_{-(\alpha_i\beta_j)(\alpha_i\gamma_j)}=0,$$

where $A^{\alpha_i\beta_i}(B^{\alpha_i\beta_i})$ stands for any parameters (parameters or operators) having superscripts $\alpha_i\beta_i$, and $\hat{\theta}_k^{\alpha_i}$ stands for $\theta_k^{\alpha_i}$ or $\bar{\theta}_k^{\alpha_i}$.

Assuming that any transformation (3.1) can be generated by a unitary operator

$$U\equiv 1-i\sum_{\substack{\ell,m\\ \alpha_i,\beta_i,i}} (\xi_{\ell m}^{\alpha_i\beta_i}N_{\ell m}^{\alpha_i\beta_i}+\frac{1}{2}\eta_{\ell m}^{\alpha_i\beta_i}L_{\ell m}^{\alpha_i\beta_i}+\frac{1}{2}M_{\ell m}^{\alpha_i\beta_i}\zeta_{\ell m}^{\alpha_i\beta_i})$$
$$+\frac{i}{2}\sum_{\ell,\alpha_i,i}(\bar{\theta}_\ell^{\alpha_i}a_\ell^{\alpha_i}+a_\ell^{\alpha_i\dagger}\theta_\ell^{\alpha_i}), \tag{3.4}$$

and employing the standard procedure but with some necessary modifications, we obtain the Lie commutation relations. It is then an easy matter to see that these relations now contain (1.4) and (1.5).

The situation for the case when the system consists of a single parafield is naturally very simple. For the case of a parafermi (parabose) field G=SO(2f+1)[6] (Z_2-graded $S_p(2f,R)$[3]) where the generators are given by $a_k, a_k^\dagger, N_{k\ell}, L_{k\ell}$ and $M_{k\ell}$. Further, the operators $N_{k\ell}$, $L_{k\ell}$ and $M_{k\ell}$ alone form the Lie algebra of SO(2f)[7] ($S_p(2f,R)$[7]).

§4 Superfields and paragrassmann algebras

The transformations (3.1) to unify parafields are, in general, highly nonlocal when expressed in configuration space. A corresponding local unification of parafields is afforded, however, by generalizing the method of superfields.[8] In the present section we shall illustrate this by considering a simple model.

Let us begin by assuming that our group G has 8 parameters: the first 4, i.e., $x^\mu(\mu=0,1,2,3)$, are ordinary variables and the rest, i.e., θ_i and $\bar{\theta}^i(i=1,2)$, form a paragrassmann algebra of order $p(=1,2,\ldots)$[9] defined by

$$[\hat{\theta}_i,[\hat{\theta}_j,\hat{\theta}_k]] = 0,\ [\hat{\theta}_{i_1},\hat{\theta}_{i_2},\ldots,\hat{\theta}_{i_m}]_+ = 0 \quad \text{only for } m > p+1, \tag{4.1}$$

where $\hat{\theta}_i$ stands for θ_i or $\bar{\theta}^i$ (the same notation will be used hereafter for similar quantities) and $[\ ,\ ,\ldots]_+$ is the completely symmetrized product. As in (1.3) the Green components $\hat{\theta}_i^{(\kappa)}(\kappa=1,2,\ldots,p)$ may be introduced for $\hat{\theta}_i$'s as well. The group G, when reinterpreted as the one with group parameters x^μ, $\theta_i^{(\kappa)}$ and $\bar{\theta}^{i(\kappa)}$, falls precisely in the category of supergroups considered in §2.

Next we introduce the operators P^μ, $\bar{S}^{i(\kappa)}$ and $S_i^{(\kappa)}$ as the generators corresponding to the parameters $x_\mu, \theta_i^{(\kappa)}$ and $\bar{\theta}^{i(\kappa)}$, respectively, and assume that their commutation relations are given by

$$[S_i^{(\kappa)},S_j^{(\kappa)}]_+=[\bar{S}^{i(\kappa)},\bar{S}^{j(\kappa)}]_+=0,$$

$$[\hat{S}_i^{(\kappa)},\hat{S}_j^{(\lambda)}]=[\hat{S}_i^{(\kappa)},P^\mu]=[P^\mu,P^\nu]=0 \quad (\kappa\neq\lambda), \tag{4.2}$$

$$[S_i^{(\kappa)},\bar{S}^{j(\kappa)}]_+=2(\sigma_\mu)_i^j P^\mu \qquad (\text{no summation over } \kappa),$$

with $\sigma_\mu\equiv(1,\vec{\sigma})$. As for the commutation relations between $\hat{\theta}_i^{(\kappa)}$ and the operators we assume

$$[\hat{\theta}_i^{(\kappa)},\hat{S}_j^{(\kappa)}]_+=[\hat{\theta}_i^{(\kappa)},\hat{S}_j^{(\lambda)}]=[\hat{\theta}_i^{(\kappa)},P^\mu] = 0 \quad (\kappa\neq\lambda). \tag{4.3}$$

As can be easily checked, the structure constants and the signature symbols that are associated with (4.2) satisfy (2.4) and (2.5), so that the algebra of $P^\mu, \bar{S}_i^{(\kappa)}$ and $S_i^{(\kappa)}$ can be a superalgebra to define G.

Introducing the paragrassmann operators S_i and $\bar{S}^i$ defined by $S_i=\Sigma_{\kappa=1}^p S_i^{(\kappa)}$ and $\bar{S}^i=\Sigma_{\kappa=1}^p \bar{S}_i^{(\kappa)}$, we can eliminate the Green indices κ,λ from (4.2) and (4.3) to obtain trilinear relations.

We are now in a position to construct superfields. Note first that the unitary operator $L(x,\theta,\bar{\theta})\equiv\exp(i[\bar{\theta},S]/2)\exp(i[\bar{S},\theta]/2)\exp(-ixP)$, where $[\bar{\theta},S]=[\bar{\theta}^i,S_i]$, etc. and $xP=x_\mu P^\mu$, is changed under the transformation $U(y,\xi,\bar{\xi})\ \exp(i[\bar{\xi},S]/2+i[\bar{S},\xi]/2-iyP]$ with $\hat{\xi}$ being paragrassmann variables to $UL(x,\theta,\bar{\theta})=L(x^\mu+y^\mu+i[\bar{\theta},\sigma^\mu\xi]+i[\bar{\xi},\sigma^\mu\xi]/2,\ \theta+\xi,\ \bar{\theta}+\bar{\xi})$. We define a scalar superfield $\Phi(x,\theta,\bar{\theta})$ by

$$\Phi(x,\theta,\bar{\theta}) \equiv L(x,\theta,\bar{\theta})\Phi_0 L^{-1}(x,\theta,\bar{\theta}) \tag{4.4}$$

with Φ_0 being an operator. This superfield is transformed under U as $\Phi\to\Phi'=U\Phi U^{-1}$. For $U=U(0,\xi,\bar{\xi})$ its variation $\delta\Phi=\Phi'-\Phi$ is given, up to the first order in $\hat{\xi}$, by

$\delta\Phi=i[[\bar{\xi},S]+[\bar{S},\xi],\Phi(x,\theta,\bar{\theta})]/2.$

From the transformation property it is clear that Φ being independent of θ_i's is an invariant restriction, and $\Phi(x,\theta,\bar{\theta})=\Phi(x,\bar{\theta})$ will be assumed hereafter. Further let us restrict ourselves to the case p=2 for simplicity. When (4.1) for p=2 is taken into account, $\Phi(x,\bar{\theta})$ can be expanded in $\bar{\theta}^i$'s as

$$\Phi(x,\bar{\theta})=\phi(x)+[\bar{\theta}^i,\bar{\psi}^j]C_{ij}+\bar{\theta}^i\bar{\theta}^j C_{ij}A(x) +[\bar{\theta}^i\bar{\theta}^j\bar{\theta}^k C_{ik},\chi_j(x)]+\bar{\theta}^i\bar{\theta}^j\bar{\theta}^k\bar{\theta}^\ell C_{ik}C_{j\ell}B(x), \tag{4.5}$$

where C is the charge conjugation matrix. Evidently, ϕ, A and B are scalar fields, whereas ψ_i and χ_i are parafermi fields of order 2. The variation $\delta\Phi$ now takes the form

$$\delta\Phi=\delta\phi+[\bar{\theta}^i,\delta\bar{\psi}^i]C_{ij}+\bar{\theta}^i\bar{\theta}^j C_{ij}\delta A +[\bar{\theta}^i\bar{\theta}^j\bar{\theta}^k C_{ik},\delta\chi_j]+\bar{\theta}^i\bar{\theta}^j\bar{\theta}^k\bar{\theta}^\ell C_{ik}C_{j\ell}\delta B+\Delta\Phi, \tag{4.6}$$

where the variation of any component field consists only of other component fields and $\hat{\xi}_i$, and $\Delta\Phi$ is a sum of terms quadratic or quartic in $\bar{\theta}^i$'s such as $\bar{\theta}\xi\bar{\theta}\partial_\mu\bar{\psi}$, $\chi\bar{\theta}\bar{\xi}\bar{\theta}$, $\xi\bar{\theta}\partial_\mu\chi\bar{\theta}\bar{\theta}\bar{\theta},\ldots$, whose $\hat{\xi}$-dependent parts cannot be factorized as variations of any component fields without introducing explicit dependence on Green indices.

In constructing a Lagrangian invariant under the transformation $U(0,\xi,\bar{\xi})$ we note that

$$\int\Phi^2(x,\bar{\theta})d\Theta=8\{4\phi(x)B(x)-2[\bar{\psi}^i(x),\chi_i(x)]+A^2(x)\} \text{ with } d\Theta=d^2\bar{\theta}_1 d^2\bar{\theta}_2, \tag{4.7}$$

and that

$$\delta\int\Phi^2(x,\bar{\theta})d\Theta=8\partial_\mu\{2[(c^{-1}\sigma^\mu\xi)^i\phi(x),\chi_i(x)]+[\bar{\psi}^i(x),(\sigma^\mu\xi)_i A(x)]\}, \tag{4.8}$$

to which, fortunately, $\Delta\Phi$ does not contribute. By considering the dimensions of the component fields we may make an ansatz: $\chi_i=-i(\sigma^\mu\partial_\mu\psi_i)/2$, $A=A^*$, and $\Box\phi^*=4B$, which are invariant relations since $\delta\chi_i=-i(\sigma^\mu\partial_\mu\delta\psi_i)/2$, $\delta A=\delta A^*$ and $\Box\delta\phi^*=4\delta B$. Consequently, as an invariant (free) Lagrangian we may adopt

$$L_0=\frac{1}{16}\int\Phi^2(x,\bar{\theta})d^4x d\Theta = \frac{1}{2}\int d^4x\{-\partial_\mu\phi(x)\partial_\mu\phi^*(x)+i[\bar{\psi}(x),\sigma^\mu\partial_\mu\psi(x)]+A^2(x)\}. \tag{4.9}$$

Although the model considered here is too simple to be of physical interest, yet it serves to make clear that parafields can be unified locally through superfields with paragrassmann variables.

References

1) See, for example, P.Fayet and S.Ferrara, Physics Report 32C (1977) No.5.
2) Y.Ohnuki and S.Kamefuchi, Quantum Field Theory and Parastatistics, Soryushiron Kenkyu (mimeographed circular in Japanese) vol.55 (1977) special issue.
3) M.Omote, Y.Ohnuki and S.Kamefuchi, Prog. Theor. Phys. 56 (1976) 1948.
4) F.A.Berezin, The Method of Second Quantization, Academic Press Inc, N. Y. (1966); Y.Ohnuki and T.Kashiwa, Prog. Theor. Phys. in press.
5) See, for example, L.S.Pontrjagin, Continuous Groups (in Russian) 2nd ed., Moscow (1954).
6) C.Ryan and E.C.G.Sudarshan, Nucl. Phys. 47 (1963) 207.
7) S.Kamefuchi and Y.Takahashi, Nucl. Phys. 36 (1962) 177.
8) Abdus Salam and J.Strathdee, Nucl. Phys. B76 (1974) 477; J.Wess, Lecture Notes in Physics 37, Springer, Berlin·Heidelberg·New York (1975), p.352.
9) M.Omote and S.Kamefuchi, to be published.

SUPERMANIFOLDS*

Marjorie Batchelor
Mathematics Department, M.I.T., Cambridge, Massachusetts

The increasing recognition of the importance of supersymmetry transformations and Lie superalgebras, or graded Lie algebras in general has been complemented by the recognition of superspace, a generalization of ordinary space to include anticommuting coordinates, as a useful object. More generally, supermanifolds, or graded manifolds were defined to fill the need for an object resembling a manifold whose algebra of functions would include anticommuting elements. The purpose of this paper is to describe two different approaches to the definition of such objects, and to describe in what sense the two approaches define the same objects.

By way of motivation, consider a smooth real manifold X. By habit, the manifold X is usually considered the primary object, and one regards the algebra of smooth functions on X as a secondary object. But in fact all the geometry of X is contained in the algebra $C^\infty(X)$. The manifold X itself can be recovered from $C^\infty(X)$ by observing that the evaluation map

$$X \to \{p: \ C^\infty(X) \to R; \ \ p \text{ is an algebra homomorphism}\}$$

given by $x(f) = f(x)$ for every f in $C^\infty(X)$ for a point x in X, is a bijection of sets.

The first approach then is to mimick the usual definition of smooth manifold replacing the real numbers by a large exterior algebra. This approach, implicit in Berezin [3] and given in detail in DeWitt [6], is described in section 2. The second approach, described in section 3, is simply to extend the ring of functions to include anticommuting elements. This approach is due to Berezin and Leites [4] and Kostant [9].

I wish to thank Professor Shlomo Sternberg for bringing DeWitt's work to my attention and for several very helpful discussions on the subject.

1. Preliminaries: Notation, conventions, and basic definitions

All (ordinary) manifolds will be real, smooth, paracompact, Hausdorff manifolds. All algebras will be algebras over the real numbers unless otherwise indicated.

1.1 Definition. A Z_2-graded algebra (or superalgebra) is an algebra A together with a decomposition of A as a direct sum of linear subspaces

$$A = A_0 \oplus A_1$$

*Paper presented at the Integrative Conference on Group Theory and Mathematical Physics, September 11-16, 1978, at the University of Texas at Austin, Austin, Texas.

such that if a is in A_i and b is in A_j then

$$ab = (-1)^{ij}ba \qquad A_{i+j} \bmod 2$$

Examples. Consider the exterior algebra on N-dimensional Euclidean space, ΛR^N. If $\{x_1,\ldots,x_N\}$ is a basis for R^N, ΛR^N has a basis $\{x_1^{j_1}\ldots x_N^{j_N}\}$ where each j_k is either 0 or 1. An element u in ΛR^N is said to be homogeneous of degree i if u is a linear combination of basis elements $x_1^{j_1}\ldots x_N^{j_N}$ where $\sum_{k=1}^{N} j_k = i$. Denote by $\Lambda^i R^N$ the set of homogeneous elements of degree i.

The algebra ΛR^N has the structure of a Z_2-graded algebra, setting

$$\Lambda R^N = (\Lambda R^N)_0 \oplus (\Lambda R^N)_1$$

where

$$(\Lambda R^N)_0 = \bigoplus_{i \text{ even}} \Lambda^i R^N$$

$$(\Lambda R^N)_0 = \bigoplus_{i \text{ odd}} \Lambda^i R^N$$

Similarly, if A is any commutative algebra, $A \otimes \Lambda R^N$ is a graded algebra, setting $(A \otimes \Lambda R^N) = A \otimes (\Lambda R^N)_i$ for i = 0, 1. Also, any commutative algebra A is a Z_2-graded algebra, setting $A = A_0$ and $A_1 = 0$.

1.2 Definition. Let X be a manifold. A sheaf of algebras A( ) on X associates to each open set U in X an algebra A(U), with the following properties.

i) If U, V are open sets in X such that $V < U$, then there is an algebra homomorphism, called the restriction from U to V

$$\rho_{UV}\colon\ A(U) \to A(V)$$

ii) If U, V, and W are open sets in X such that $W < V < U$, then

$$\rho_{UW} = \rho_{VW}\rho_{UV}$$

Examples. If X is a manifold, the smooth functions, $C^\infty(\ )$ for a sheaf of algebras on X. Let $\pi\colon\ E \to X$ be a real finite dimensional vector bundle over X. Let ΛE be the associated exterior bundle over X, formed by replacing each fibre of E by the exterior algebra on that fibre. Define the sheaf of smooth sections $\Gamma(\ ,\Lambda E)$ by

$$\Gamma(U,\Lambda E) = \{s\colon\ U \to E;\ \ \pi s(u) = u \text{ for each u in U, and s is smooth}\}$$

$\Gamma(\ ,\Lambda E)$ is a sheaf of Z_2-graded algebras.

Definition. Let A( ), B( ) be two sheaves of algebras over X. A map of sheaves $p\colon\ A(\ ) \to B(\ )$ is a collection of algebra homomorphisms $\{p_U\}$ where

$$p_U\colon\ A(U) \to B(U)$$

such that $\rho_{UV}p_U = p_V\rho_{UV}$.

Example. Let E_x denote the fibre of E over a point x in X. The augmentation map sending an element u_x of ΛE_x to its component in $\Lambda^0 E_x$ induces a map of sheaves of algebras $\Gamma(,\Lambda E) \to C^\infty( )$.

2. Supermanifolds

In attempting to mimmick the ordinary definition of a manifold, substituting ΛR^N for R, it is necessary first to define analogues of Euclidean space and smooth functions on Euclidean space.

2.1 Definitions. Define (r,s)-dimensional super-Euclidean space, $E^{r,s}$ to be the set

$$E^{r,s} = (\Lambda R^N)_0^r \times (\Lambda R^N)_1^s$$

Define the augmentation map

$$\varepsilon: \ E^{r,s} \to R^r$$

$$(u_1,\dots,u_r,v_1,\dots,v_s) = (\varepsilon(u_1),\dots,\varepsilon(u_r))$$

where $\varepsilon(u_i)$ is the component of u_i in $\Lambda^0 R^N = R$.

The set $E^{r,s}$ can be given a topology by defining a set of U contained in $E^{r,s}$ to be open if and only if $U = \varepsilon^{-1}(V)$ where V is an open set on R^r. This topology on $E^{r,s}$ is coarser than the topology that $E^{r,s}$ would usually inherit as a real vector space; in particular it is not even Hausdorff. The coarseness of the topology compensates for the absence of inverses for nilpotent elements in ΛR^N in that, if a and b are elements in $\Lambda R^N = E^{1,1}$ then a and b have disjoint open neighborhoods if and only if a-b is invertible.

2.2 Having defined super-Euclidean space, it remains to decide which functions on super-Euclidean space should be considered smooth. The method of obtaining smooth functions is due to Kostant [9], section 2.18.

If A and B are sets, denote by F(A,B) the set of functions from A to B. In $F(E^{r,s},\Lambda R^N)$ there are projections p_i and π_j given by

$$p_i(u_1,\dots,u_r,v_1,\dots,v_s) = u_i \quad \text{for } i = 1,\dots,r$$

$$p_j(u_1,\dots,u_r,v_1,\dots,v_s) = v_j \quad \text{for } j = 1,\dots,s$$

Let SP denote the algebra over the real numbers generated by the projections. The algebra SP will be called the algebra of superpolynomials.

Remarks. i) Let $Sym(p_i)$ denote the symmetric algebra on $p_1,\dots,p_r$, and let $\Lambda(\pi_j)$ denote the exterior algebra generated by $\pi_1,\dots,\pi_s$. Then SP is isomorphic to the algebra $Sym(p_i) \otimes \Lambda(\pi_j)$, provided that $N \geq s$. For the rest of the paper, N is assumed to be an integer greater than s.

ii) Let $\Lambda R^N SP$ denote the algebra over ΛR^N generated by the projections. Then $\Lambda R^N SP \approx \Lambda R^N \otimes SP$. The algebra $\Lambda R^N SP$ can be regarded as a "better" generalization of ordinary polynomials than SP in that $\Lambda R^N SP$ contains the constant maps from $E^{r,s}$ to ΛR^N. However, $\Lambda R^N SP$ has other drawbacks which will be discussed later in 2.5.

2.3 Let D be the set of superderivations of SP. Let $\partial/\partial p_i$ denote the superderivation determined by $\partial/\partial p_i p_h = \delta_{ih}$, and $\partial/\partial p_i \pi_j = 0$ for all $j = 1,\ldots,s$. Similarly, let $\partial/\partial\pi_j$ denote the superderivation corresponding to π_j. Then any element in D can be expressed uniquely as a sum

$$\sum_{i=1}^{r} a_i \frac{\partial}{\partial p_i} + \sum_{j=1}^{s} b_j \frac{\partial}{\partial \pi_j}$$

where the coefficients a_i and b_j are elements of SP.

Let $\bar{D}$ be the sub-algebra of the algebra of linear maps from SP to itself generated by D. Elements of $\bar{D}$ can be expressed uniquely as a sum

$$\sum_{i_1,\ldots,i_r,j_1,\ldots,j_s} a(i_1,\ldots,i_r,j_1,\ldots,j_s)\frac{\partial i_1}{\partial p_1}\cdots\frac{\partial i_r}{\partial p_r}\frac{\partial j_1}{\partial \pi_1}\cdots\frac{\partial j_s}{\partial \pi_s}$$

where each i_h is a non-negative integer, each j_k is in $\{0,1\}$, and the coefficients $a(i_1,\ldots,i_r,j_1,\ldots,j_s)$ are elements of SP such that only finitely many of them are non-zero.

2.4 The algebra of superpolynomials can now be extended to a larger sub-algebra in $F(E^{r,s},\ {}^N)$ on which D acts. Let

$$SP^* = \{\alpha:\ SP \to R:\ \alpha \text{ is linear, } \ker\alpha > I \text{ such that } SP/I \text{ is a finite dimensional algebra}\}$$

Let $\langle\alpha,q\rangle$ denote evaluation of α at q for α in SP* and q in SP. Let

$$SP^{*\prime} = \{f:\ SP^* \to R:\ f \text{ is linear}\}$$

Similarly, let $\langle\alpha,f\rangle$ denote evaluation of f at α for α in SP* and f in SP*'. The properties of SP*' are summarized in the following proposition.

Proposition.

i) SP*' is an algebra.

ii) SP*' contains SP.

iii) $\bar{D}$ acts on SP*.

iv) There is a map ϕ: $SP^{*\prime} \to F(R^r,R)$

v) $SP^{*\prime} \to F(E^{r,s},\Lambda R^N)$ and the composite $SP \to SP^{*\prime} \to F(E^{r,s},\Lambda R^N)$ is given by the usual evaluation of a superpolynomial on points in $E^{r,s}$.

Indications of proofs.

i) The condition on SP* that ker α must contain an ideal I such that the quotient SP/I is a finite dimensional algebra guarantees a map

$$SP^* \to SP^* \otimes SP^*$$

induced by multiplication in SP. The algebra structure on SP*',

$$SP^{*\prime} \otimes SP^{*\prime} \to SP^{*\prime}$$

is the dual to this map.

ii) If q is in SP, q determines a linear map from SP* to R via

$$\alpha \to \langle\alpha,q\rangle \text{ for } \quad \text{in SP*}$$

iii) The action of $\bar{D}$ on SP gives rise to an action of $\bar{D}$ on SP* via

$$\langle\alpha\cdot d,q\rangle = \langle\alpha,dq\rangle \text{ for } d \text{ in } \bar{D},\ \alpha \text{ in SP*},\ q \text{ in SP}.$$

Similarly, $\bar{D}$ acts on SP*' via

$$\langle\alpha,df\rangle = \langle\alpha\cdot d,f\rangle \text{ for } d \text{ in } \bar{D},\ \alpha \text{ in SP* and } f \text{ in SP*'}.$$

iv) Notice that R^r is contained in SP* since we can define evaluation at a point $(x_1,\ldots,x_r)$ in R^r via

$$\langle(x_1,\ldots,x_r),q\rangle = q(x_1,\ldots,x_r,0,\ldots,0) \text{ for } q \text{ in SP}$$

regarding x_i as an element of $\Lambda^0 R^N$. Moreover the points of R^r are linearly independent in SP*. The inclusion R^r SP* then induces the desired map by restriction.

v) The map is given by the following formula: if f is in SP*'

$$f(u_1,\ldots,u_r,v_1,\ldots,v_s) = \sum_{\Sigma i_h+\Sigma j_k\leq N} \frac{1}{i_1!\ldots i_r!}\,\phi\Big(\frac{\partial i_1}{\partial p_1}\cdots\frac{\partial i_r}{\partial p_r}\frac{\partial j_1}{\partial \pi_1}\cdots\frac{\partial j_s}{\partial \pi_s} f\Big)$$
$$\cdot(\varepsilon(u_1),\ldots,\varepsilon(u_r))(u_1-\varepsilon(u_1))^{i_1}\ldots(u_r-\varepsilon(u_r))^{i_r} v_1^{j_1}\ldots v_s^{j_s}$$

where $(u_1,\ldots,u_r,v_1,\ldots,v_s)$ is in $E^{r,s}$, i_k is a non-negative integer, j_h is in $\{0,1\}$.

2.5 Using the algebra SP*' smooth maps can be defined.

Definition. Define the set of smooth maps in $F(E^{r,s},\Lambda R^N)$ by

$$\text{SMaps}(E^{r,s},\Lambda R^N) = \{f\varepsilon \text{SP*'}:\ \phi(df) \text{ is smooth on } R^r \text{ for every } d \text{ in } \bar{D}\}$$

The properties of the smooth maps are summarized in the following theorem.

Theorem.

i) SMaps $(E^{r,s},\Lambda R^N) = \{f\varepsilon \text{SP*'} \subset F(E^{r,s},\Lambda R^N)$: df is continuous on $E^{r,s}$ for all d in $\bar{D}$ }

ii) There is an isomorphism of algebras ψ: $\text{SMaps}(E^{r,s},\Lambda R^N) \approx C^\infty(R^r)\otimes\Lambda(\pi_j)$ given by the formula

$$\psi(f) = \sum_{(j_1,\ldots,j_s)} \phi\Big(\frac{\partial j_1}{\partial \pi_1}\cdots\frac{\partial j_s}{\partial \pi_s} f\Big)_1^{j_1}\cdots {}_s^{j_s} \text{ where } j_k \text{ is in } \{0,1\}$$

iii) $\text{SMaps}(E^{r,s},\Lambda R^N)$ can be given a topology so that ψ is a homeomorphism when $C^\infty(R^r) \times \Lambda(\pi_j)$ is given the topology of a product of 2^s copies of $C^\infty(R^r)$ and $C^\infty(R^r)$ is equipped with its usual topology. Relative to this topology, $\text{SMaps}(E^{r,s},\Lambda R^N)$ is the closure of SP in $\text{SMaps}(E^{r,s},\Lambda R^N)$. For a proof see [2].

Remarks. i) If U is an open set in $E^{r,s}$, the set of smooth maps from U to ΛR^N, denoted $\text{SMaps}(U,\Lambda R^N)$ can be defined by a similar technique.

ii) The procedure described here for defining SMaps from SP can be applied to ΛR^NSP to yield an algebra, ΛR^NSMaps, which turns out to be isomorphic to $\Lambda R^N \otimes$ SMaps. The principal objections to this algebra are that the constant maps have non-trivial superderivations, and the algebra is dependent on N. It may be however that physical applications indicate the use of this algebra over SMaps.

2.6 Equipped with a definition of super-Euclidean space and smooth functions, it is a simple task to define supermanifolds.

Definitions. An element T in $F(E^{r,s},E^{p,q})$ will be called smooth if the composition of T with each projection p_i and π_j from $E^{p,q}$ to ΛR^N is smooth. A bijection S in $F(E^{r,s},E^{r,s})$ will be called a superdiffeomorphism if S and S^{-1} are both smooth.

Definition. A supermanifold S is a topological space such that S has an open cover $\{U_i\}$ such that for each i there is a homeomorphism ϕ_i from U_i to an open set in $E^{r,s}$ and the composite

$$\phi_i\phi_j^{-1}: \ \phi_j(U_i \cap U_j) \to \phi_i(U_i \cap U_j)$$

is a superdiffeomorphism for each i, j.

3. Graded Manifolds

3.1 Definition. A graded manifold (X, A) consists of a manifold X and a sheaf of Z_2-graded algebras A( ) such that

i) There exists a surjective map of sheaves of Z_2-graded algebras

$$A(\ ) \to C^\infty(\ )$$

ii) There is an open cover $\{V_i\}$ of X and isomorphisms of sheaves of Z_2-graded algebras

$$T_i: \ A(\ )|_{V_i} \to C^\infty(\ ) \otimes \Lambda R^S|_{V_i}$$

where $A(\ )|_{V_i}$ and $C^\infty(\ ) \otimes \Lambda R^S|_{V_i}$ denote the sheaves A( ) and $C^\infty(\ ) \otimes \Lambda R^S$ restricted to V_i.

Examples. i) Let X be any manifold. Then $(X, C^\infty(\ ) \otimes \Lambda R^S)$ is a graded manifold.

ii) Let X be any manifold, and let $\Omega(\ )$ be the sheaf of differentiable forms on X. Then $(X,\Omega(\ ))$ is a graded manifold.

iii) More generally let π: $E \to X$ be a real vector bundle over X with finite dimensional fibre. If ΛE denotes the associated exterior bundle over X, and $\Gamma(\ ,\Lambda E)$ denotes the sheaf of sections of ΛE, then $(X, \Gamma(\ ,\Lambda E))$ is a graded manifold.

Remark. Given any graded manifold (X, A( )), there exists a bundle E over X such that A( ) is isomorphic to $\Gamma(\ ,\Lambda E)$ as sheaves of Z_2-graded algebras. There is however no canonical choice of vector bundle, nor is the isomorphism of sheaves unique. The ability to choose a vector bundle E and an isomorphism of sheaves is comparable to the ability to choose a set of coordinate charts for an ordinary manifold. For details see [1].

4. A comparison of graded manifolds and supermanifolds

The relationship between supermanifolds and graded manifolds is just the relationship of ordinary manifolds to their sheaves of functions. The key to relating the two definitions is the observation that to each supermanifold S there is a related ordinary manifold called the underlying manifold $\overline{S}$.

4.1 Let S be a supermanifold and let $\{U_i\}$ be an open cover for S such that there are homeomorphisms ψ_i of U_i with an open set in $E^{r,s}$ for all i, and

$$\phi_i\phi_j^{-1}: \phi_j(U_i \cap U_j) \to \phi_i(U_j \cap U_i)$$

is a superdiffeomorphism.

Then, by the definition of open sets in $E^{r,s}$, $\phi_i(U_i)$ projects via the augmentation ε onto an open set V_i in R^r. The superdiffeomorphisms $\phi_i\phi_j^{-1}$ give rise to diffeomorphisms

$$T_{ij}: \varepsilon\phi_j(U_i \cap U_j) \to \varepsilon\phi_i(U_i \cap U_j)$$

We can define a manifold $\overline{S}$, called the underlying manifold of S to be the disjoint union of the open sets V_i subject to an equivalence relation

$$\overline{S} = \cup V_i / \sim$$

where if x is in V_i and y is in V_j then $x \sim y$ if and only if $T_{ij}y = x$.

The augmentation ε: $E^{r,s} \to R^r$ then can be generalized to a projection

$$\varepsilon_S: S \to \overline{S}$$

where $\varepsilon_S(s) = \varepsilon\psi_i(s)$ if s in in U_i. The equivalence relation defining $\overline{S}$ guarantees that the map ε_S does not depend on the choice of i and hence the map ε_S is well defined.

4.2 The relationship between graded manifolds and supermanifolds is then summarized in the following theorem.

Theorem.

i) Let S be a supermanifold. Then $(\overline{S}, A(\ ))$ is a graded manifold where A is the sheaf given by

$$A(V) = \mathrm{SMaps}(\varepsilon_S^{-1}(V), \Lambda R^N) \text{ for an open set } V \text{ in } \overline{S}.$$

ii) Let (X, A( )) be a graded manifold. Then the set of homomorphisms of Z_2-graded algebras from A(X) to ΛR^N, $\mathrm{Hom}_{Z_2\text{-alg.}}(A(X), \Lambda R^N)$ is a supermanifold.

For a more precise statement and proof of this theorem, see [2].

References

1. M. Batchelor, "The Structure of Supermanifolds," Trans. Amer. Math. Soc., to appear.
2. M. Batchelor, "The Categories of Supermanifolds and Graded Manifolds," in preparation.

3. M. Berezin, The Method of Second Quantisation, Academic Press, New York, 1966.
4. M. Berezin and D. Leites, "Supervarities," Sov. Math. Dokl. 16, 1218-1222 (1975).
5. L. Corwin, Y. Ne'eman, and S. Sternberg, "Graded Lie Algebras in Mathematics and Physics (Bose Fermi Symmetry)," Rev. Mod. Phys. 47, 573-604 (1975).
6. J. Dell and L. Smolin, "Graded Manifold Theory as the Geometry of Supersymmetry," Harvard Preprint in preparation.
7. B. DeWitt, Differential Supergeometry, in preparation.
8. V. Kac, "Lie Superalgebras," Adv. in Math. 26, 8-96 (1977).
9. B. Kostant, "Graded Manifolds, Graded Lie Theory and Prequantisation," Differential Geometric Methods in Mathematical Physics. Lecture Notes in Mathematics, Springer-Verlag, vol. 570 (1977), pp. 177-306.

GRADED G-STRUCTURES

Charles P. Boyer , IIMAS , Univ. Nac. de México

In this communication I would like to give a progress report on a program to construct graded geometric structures on graded differential manifolds. Most of the results so far are on the formal algebraic level and many important problems of analysis remain.

Kostant[1] has given a definition of a graded differential manifold of dimension (n,m) and this is discussed in these proceedings by M. Batchelor. I only mention here that it consists of a pair (M,A) where M is an ordinary C^∞ manifold of dimension n and A is a sheaf of Z_2-graded commutative algebras whose underlying graded vector space $V = V_0 + V_1$ has $\dim V_0 = n$, $\dim V_1 = m$. A graded Lie group (G,A) then is a pair where G is a Lie group and the restricted dual $A°$ (vanishes on some ideal of finite codimension) has the structure of a graded Hopf algebra and can be represented as $A° = R(G) \wedge E(\mathcal{G})$ where R(G) is the group ring, $E(\mathcal{G})$ is the universal enveloping algebra of the graded Lie algebra $\mathcal{G} = \mathcal{G}_0 + \mathcal{G}_1$ with $\mathcal{G}_0$ the Lie algebra of G, and $\wedge$ denotes semidirect product.

In analogy with the usual situation, graded frames can be introduced and the graded frame bundle constructed. This consists of a principal fibre bundle L(M,G) with group $G = GL(n)\times GL(m)$ such that the graded tangent bundle $T(M,A)$ defined by Kostant is an associated vector bundle for L(M,G). The graded frame bundle (not really a bundle) is then given by the graded manifold $(L(M,G), A \otimes H)$ where $H° = R(G) \wedge E(\mathcal{G})$, $C - GL(n)\times GL(m)$, $\mathcal{G} = gl(n,m) = \mathrm{End}\, V$. Now if $G' \subset G$ is a closed Lie subgroup of G with a subHopf algebra $H' \subset H$, then $(L(M,G'), A\otimes H')$ is called a graded <u>G-structure</u> if L(M,G') is reduced subbundle of L(M,G). There is a natural right action R_a of H' defined on $(L(M,G'), A\otimes H')$. Associated with the graded tangent bundle of a G-structure, there is the tangent sheaf derA + derH (der A = derivations of A), and this is a free $A\otimes H$ module. A complement H to derH is called a <u>horizontal subspace</u>, and as usual H defines a <u>connection</u> if $R_a{*}H = H$. As underlying graded vector spaces H is isomorphic to V.

In analogy with the ordinary case, we study graded G-structures by studying graded Spencer cohomology. Let $\mathcal{G} \subset g\ell(n,m) = V\otimes V^*$ be a graded Lie algebra. We define the first <u>prolongation</u> $\mathcal{G}^{(1)}$ of $\mathcal{G}$ by all $T \in \mathrm{Hom}(V,\mathcal{G})$ which satisfy $T(u)v = (-1)^{|u||v|}T(v)u$ where u,v are homogeneous elements in V of degree $|u|$, $|v|$ respectively. The kth prolongation is defined inductively by $\mathcal{G}^{(k)} = (\mathcal{G}^{(k-1)})^{(1)}$. We construct Z_2 graded vector spaces $C^{k+1,\ell} = \mathcal{G}^{(k)} \otimes \Lambda^\ell(V^*)$ where $\Lambda^\ell(V^*) = \Lambda^\ell(V_0^*) \otimes S^\ell(V_1^*)$ is the graded exterior algebra over V*. The cochain map $\partial : C^{k+1,\ell} \longrightarrow C^{k,\ell+1}$ is defined by

$$\partial S(u_1,\ldots,u_{\ell+1}) = \sum_i (-1)^{i+|u_i|(|u_{i+1}|+\ldots+|u_{\ell+1}|)}[S(u_1,\ldots\hat{u}_i,\ldots u_{\ell+1}),u_i]$$

where $\wedge$ means remove that element. The graded Spencer cohomology $H^{k,\ell}(\mathcal{G})$ is the cohomology of the sequence

$$0 \longrightarrow C^{k+1,0} \xrightarrow{\partial} C^{k,1} \xrightarrow{\partial} C^{k-1,2} \xrightarrow{\partial} \cdots$$

We have proved the following results[2] on the formal algebraic level: 1) the formal Poincaré lemma — for $G = g\ell(n,m)$, $H^{k,\ell}(G) = 0$.

2) The Levi-Civita theorem for graded (pseudo)-Riemannian geometry- there is a unique torsion free connection $H^{0,2}(G) = 0$, $G = os(p,q;2m)$ (orthosymplectic algebra, signature (p,q)).

3) For $G = os(p,q;2m)$ (graded pseudo-Riemannian), $H^{1,2}(G) = 0$ if and only if the structure is flat.

4) For graded conformal structures $G = cos(p,q,2m)$ there is a torsion free connection, $H^{0,2}(G) = 0$.

5) For Wess-Zumino theory ($G = o(3,1)$), dim $H^{0,2}(G) = 208$.

References

1. B. Kostant in Differential Geometrical Methods in Mathematical Physics, K. Bleuder, A. Reetz (Eds) (Springer, N.Y., 1977) pg. 177-306.

2. C. P. Boyer, Formal Models of Graded Differential Geometry, UNAM preprint.

SPACE-TIME PROPERTIES OF SUPERGRAVITY

S. Deser*

Department of Physics
Brandeis University
Waltham, Massachusetts 02154

The structure, goals and current status of supergravity are reviewed from a space-time, rather than superspace, point of view. Its relations to general relativity are emphasized.

INTRODUCTION

The purpose of this brief survey is to discuss the physical ideas underlying supergravity and some of the results and problems of current interest. In view of the many contributions at this conference on the mathematical aspects of graded systems and on superspace methods, we shall emphasize here the ordinary space-time approach and the relation of these new models to Einstein's theory, particularly the senses in which they are a natural generalization of the latter. Two of the prime aims of the theory, and the degree of their success to date, will be mentioned. We emphasize at the outset that it is not so much the expectation that supergravity will be a realistic theory which makes it so exciting, but rather that the possibilities it raises in this direction, and the entirely new local gauge properties it embodies, may lead us to a deeper understanding of this basic goal of physics.

The physical content of supergravity may be understood in a number of instructive ways. We begin with some general remarks on these. At the most prosaic level, supergravity [1,2] is simply the consistently coupled system of Einstein gravity and a massless spin 3/2 field (as well as lower spins in the "extended" versions), which avoids the usual difficulties of higher spin coupling to gauge theories, gravity in particular. The restoration of consistency may in turn be traced to the fact that supergravity possesses a (non-manifest) new local fermionic gauge invariance. It is

---

**Research supported in part by NSF Grant PHY 78-09644.*

therefore, at a deeper level, a unified gauge theory in which gravitons and lower spin fields ("matter") form a gauge multiplet; in this respect, it fulfills one of Einstein's dreams, although probably not in the way unification used to be envisaged. One profound conceptual consequence of this unification is that space-time is no longer an invariant notion, but depends on the choice of the new gauge. A third aspect of the theory is embodied in the fact that it is, in a precise technical sense, the Dirac square root of general relativity [3]. In this incarnation, the spin 3/2 field is just the analog of the γ matrices, required in order to perform the square root. We shall return shortly to a fourth characterization of the theory as the necessary gauging or self-coupling of an initially linear system of uncoupled spin $2+3/2$ fields.

The prime goals of supergravity are twofold. First, it aims to provide a model which unifies gravitation with the other fundamental interactions in nature, and correspondingly to put into a single supermultiplet gravitons and the basic constituents of matter. Second, it hopes to avoid the divergence problems which plague quantum gravity, particularly in its normal interactions with matter. This hope is based on the strong constraints imposed by the new local invariance on the possible counterterms describing the divergences. We shall review the extent to which these aims have been fulfilled at present.

GRAVITY AND SUPERGRAVITY

It is enlightening to compare the basic properties of these two theories in a systematic way.

General relativity is the gauge field of the Poincaré algebra, its field variables ($e_{\mu a}$-vierbein, $\omega_{\mu ab}$-connection) appearing in the Poincaré algebra-valued vector $e_{\mu a}(x)P^a+\omega_{\mu ab}J^{ab}$ where (P^a,J^{ab}) are the usual generators. Supergravity adds to this vector a part, $\psi_\mu\bar{J}^\mu$, corresponding to the graded extension of the Poincaré algebra which includes (fermionic) spinor generators $\bar{J}_\mu$, and a vector-spinor field variable $\psi_\mu(x)$. Alternately, relativity is the consistent self-coupled generalization of linear $s=2$, $m=0$ in which local abelian (linearized coordinate) and global Lorentz invariances are merged into the local non-abelian coordinate group. To this is added, in supergravity, a new initial abelian local group and a global supersymmetry associated with the spin 3/2 field; these also merge into a non-abelian local (super) symmetry.

The underlying mechanism which implements this transition from free to coupled systems is a very simple one in both cases. In gravity, the linearized field equation (which satisfies a Bianchi identity because of the local abelian invariance) is no longer consistent in the presence of sources; to restore consistency, one must include the Noether current associated with the global symmetry, namely the stress tensor $T_{\mu\nu}$ of the gravitational field $h_{\alpha\beta}$ itself as a source [4]. The field equations become

$$G^{L}_{\mu\nu}(h) = T_{\mu\nu}(h) \equiv -G^{NL}_{\mu\nu}(h) \tag{1}$$

where $G^{NL}_{\mu\nu}$ is just the nonlinear part of the Einstein tensor. (We remark parenthetically that the basic principles of special relativistic quantum mechanics together with a few qualitative observed properties of gravitational interactions uniquely specify Einstein theory and its coupling to matter [5].) For supergravity, the above mechanism involves simultaneous coupling of the full $T_{\mu\nu}$ of spin 2 plus 3/2 in (1) and the Noether current $J_{\mu}(x)$ associated with the global supersymmetry as the source of the linear spin 3/2 field equation. We write this step schematically as

$$R^{L}_{\mu}(\psi) = J_{\mu}(\psi) \tag{2}$$

where R^{L}_{μ} is the free spin 3/2 field equation (see below); this basically turns the ordinary derivative in this Dirac-like equation into the required covariant one. At the same time these couplings fulfill the general requirements set by gravity for correct interaction with matter systems (general covariance). The full implementation of these self-couplings (which are necessary for consistency) as embodied in (1) and (2) is precisely equivalent to the supergravity equations given below.

The Free Spin 3/2 Field

Before coming to supergravity itself, it may be helpful, because of its exotic nature, to summarize the basic properties of its spin 3/2, m = 0 component. The spin 3/2 field, represented by the real (Majorana) vector-spinor $\psi_{\mu}(x)$, must contain only helicity ±3/2, but not ±1/2 components. This is achieved by requiring local gauge invariance under

$$\delta\psi_{\mu} = \partial_{\mu}\alpha(x) \tag{3}$$

where $\alpha(x)$ is an arbitrary spinorial function, just as the Maxwell invariance under

$\delta A_\mu = \partial_\mu \Lambda(x)$ excludes helicity zero there. In both cases, the field strength

$$f_{\mu\nu} \equiv \partial_\mu \psi_\nu - \partial_\nu \psi_\mu \tag{4}$$

is gauge invariant, as is its dual $*f^{\mu\nu} \equiv \frac{1}{2}\varepsilon^{\mu\nu\alpha\beta} f_{\alpha\beta}$, with $\partial_\mu *f^{\mu\nu} \equiv 0$. It is then clear from the properties of $*f$ that the action

$$I^L_{3/2} = -\frac{i}{2}\int \bar{\psi}_\mu \gamma_5 \gamma_\nu {}^* f^{\mu\nu} \tag{5}$$

is invariant under (3). [The γ_5 is needed both for parity and to keep the Lagrangian from being a total divergence.] The action (5) does not look like the usual spin 1/2 Dirac action, any more than does the Maxwell action resemble that of a scalar field. However, when the gauge and constraint components are removed, the (gauge invariant) pure helicity 3/2 part ψ_i^{TT} does have Dirac form:

$$I^L_{3/2} = -\frac{i}{2}\int \bar{\psi}_i^{TT} \not\partial \psi_i^{TT}. \tag{6}$$

Here ψ_i^{TT} is the doubly transverse spatial component of ψ_μ,

$$\vec{\gamma}\cdot\vec{\psi}^{TT} \equiv 0 \equiv \vec{\nabla}\cdot\vec{\psi}^{TT} \tag{7}$$

and therefore has only one (Majorana spinor) component, corresponding to the correct two helicity 3/2 degrees of freedom; all this is exactly as in free electrodynamics, $I_{MAX} = \frac{1}{2}\int A_i^T \Box A_i^T$ or linearized gravity where

$$I^L_2 = \frac{1}{2}\int h_{ij}^{TT} \Box h_{ij}^{TT},\ h_{ii}^{TT} \equiv 0 \equiv \partial_i h_{ij}. \tag{8}$$

A complete Hamiltonian analysis of supergravity [6] also shows that the constraint structure and the degree of freedom count remain the same after coupling.

Supergravity

We now proceed to full supergravity, whose action has the form

$$I_{SG} \equiv I_E(e,\omega) + I_{3/2}(\psi;\ e,\omega) \tag{9}$$

where

$$I_E \equiv \int e_{\mu a} e_{\nu b} {}^{**}R^{\mu\nu ab},\quad {}^{**}R \equiv \varepsilon\varepsilon R$$

is the usual first order Einstein action in terms of $(e_{\mu a},\ \omega_{\mu ab})$ as independent fields and $I_{3/2}$ is just the generally covariant form of the free action (5) with the usual covariant replacements $\eta_{\mu a} \to e_{\mu a}$, $\partial_\mu \to D_\mu(\omega)$. Although they are not manifest in (9), there are two closely related novel properties of this action. The first is consistency - the usual problems of higher spin coupling to gravity are overcome thanks to the second, a new local invariance. Consistency may be exemplified by computing the divergence of the spin 3/2 (Rarita-Schwinger) field equation, $D_\mu R^\mu$. It turns out to be proportional to the remaining two (Einstein and torsion) field equations, and therefore does vanish - iff the coupling is precisely the minimal one of (9). Also, the usual difficulties with characteristic surfaces of higher spin are removed - ψ_μ excitations propagate on the light cone (in any α-gauge!). The associated invariance (in addition to the manifest general coordinate and local Lorentz symmetries) is the non-abelian generalization of supersymmetry referred to earlier, namely that under the transformations

$$\delta_S \psi_\mu = 2D_\mu(\omega)\,\alpha(x), \quad \delta_S e_{\mu a} = i\bar{\alpha}\gamma_a \psi_\mu. \tag{10}$$

Furtherfore, the commutation of two such transformations on any variable is essentially a general coordinate transformation, which is another aspect of the Dirac square root idea. Note that $\delta_S e_{\mu a}$ is not a coordinate transformation; this is why space-time is no longer invariant.

Unification

The first major aim of supergravity was the unification of gravity with matter into a single multiplet. The original, "simple", supergravity accomplished this for spin 3/2 matter. However, "extended" models [7] can include all lower spins, even 1/2 and 0, which are themselves not gauge fields. Further, the graded algebra (grading is what avoids the usual no-go theorems of ordinary Lie algebra "unifications") relates the maximal spin to the various possible lower spins and their multiplicities. There is even an associated new (global) internal symmetry, O(N), where $N \leq 4\lambda$ is the degree of extension and λ is the maximal helicity. The maximal spin is a singlet, the one 1/2 a unit lower occurs N times, the next $\frac{N(N-1)}{2}$, etc. For supergravity ($\lambda = 2$) explicit models exist for $N \leq 4$ and no one doubts the possibility of $N > 4$ models, though only parts of the maximal O(8) have been found to date [8]. One way to understand the

O(N) buildup is to consider first the coupling of simple supergravity to a matter source of spins (3/2, 1). The second 3/2 field enters indistinguishably from the original one, hence the resulting O(2) invariance (spin 2 and 1 are singlets here). Higher O(N) models may be understood in a similar way in terms of further couplings to supermatter.

The physical question may now be asked - is O(8) symmetry sufficiently large to accommodate the known symmetries of strong (SU_3), weak and electromagnetic ($SU_2 \times U_1$) interactions? The answer is no - quite apart from more detailed problems of symmetry breaking and realistic particle assignments; all masses are of course initially zero before symmetry breaking. Thus, supergravity is probably not (at least in its present state) a realistic particle model. There do exist two ways of increasing the internal group and particle content within the present framework, neither satisfactory as it stands. The first is to include spins higher than 2, thereby increasing the maximal N beyond 8; but this is beset with both formal and physical difficulties, if it can be done at all (spin 5/2 cannot couple to gravity in the usual way, and even if it did, there would be more than 8 gravitons). The second possibility is to permit higher derivative actions, which always yield more particle states. While some of these (the supersymmetric versions of the Weyl action, quadratic in curvature) have interesting formal properties [9], the price paid is as usual, loss of unitarity, which seems too high.

Renormalizability

Quantum gravity has long been suspected of being nonrenormalizable, on simple power counting grounds, based on the dimensionality of the gravitational constant. The equivalence principle is the villain here, because it implies that all coupling to gravitons (including that of gravitons themselves) occurs through the stress tensor which rises with momentum (as p^2 for bosons and p for fermions). This means that each higher loop (involving an additional virtual graviton exchange) diverges more strongly, implying (in terms of dimensional regularization) that the required counter terms involve higher and higher powers of the curvature. The only possible escape from this would require miraculous cancellations which would remove enough of these dangerous counter terms. In the coupling of gravity to the usual spin 0, 1/2 or 1 fields, no such miracles occurred, and unacceptable counter terms were needed already at one loop.

Since supergravity is after all also a (particular example of) gravity-matter coupling, why should improvement be expected here? The answer is that the new invariance might exclude counter terms otherwise permitted by general covariance alone. Thus it is extremely encouraging that all source-free O(N) supergravity models are one-loop finite (S-matrix finiteness rather than renormalizability is relevant here), as has also been verified by explicit calculation. But O(N) models include coupling to lower spins, which means that diagrams with spin 3/2 loops compensate purely gravitational ones. This situation persists at two loops as well; as for one loop, there are no invariants which fail to vanish on shell and have correct dimensions. However, at 3 or more loops, such invariants can be formulated, at least for the O(1) and O(2) cases [10] (so far) and unless there is a (still) hidden symmetry forbidding them, it seems likely that the infinites will be back from this level on. Unfortunately, 2 (let alone 3) loop calculations are almost impossible to perform explicitly, so the renormalizability situation is still not settled. Of course, if further symmetries are found, it may be possible to set limits on the number and types of new counter terms without explicit calculation.

Positive Energy

Although a more complicated system than pure gravity, supergravity has a number of properties which follow basically from its group structure, notably the fact that its energy operator is nonnegative [11], a property hitherto not fully established even for classical Einstein theory. This result follows from the fact that bounded solutions (and energy is definable only for bounded systems, of course) of supergravity admit an asymptotic global graded Poincaré algebra. But this implies the relation $E = Q^2 \geq 0$, where Q is the (Majorana) spinor charge (the generator of time translations is the square of the supersymmetry generators). One can even show that no tachyonic four-momentum is possible, so all bounded solutions are future and probably only time-like. It is ironic that once established for (quantum) supergravity, these results follow *a fortiori* for ordinary classical gravity [12], without any explicit calculation. One simply considers the expectation value of the formally nonnegative energy operator in states with no on-shell spin 3/2 particles, and in the tree ($\hbar \to 0$) limit which removes all closed loops. Thus, even if nature does not choose supergravity, its very potentiality is sufficient to ensure those basic properties of classical Einstein theory.

SUMMARY; OPEN PROBLEMS

Supergravity is a new gauge theory unifying space-time and matter into a single supermultiplet. Whether it can provide a realistic model of the fundamental particles and their interactions is not at all clear, but the unification it provides does not even seem "kinematically" large enough to accommodate the invariances of the other interactions, quite apart from questions of symmetry breaking and mass assignments. On the more formal side, the renormalizability properties of the theory are also not very encouraging at present. However, the beautiful way in which spin and geometry are unified in a "miraculously" consistent fashion through the new gauge warrants the more modest hope that study of these models will teach us how to build more nearly physical unified models. Historically, the previous gauge theory of Yang-Mills had a gestation period of almost two decades before coming into its own!

We conclude with a few open problems in which both mathematical methods and experience from relativity should lead to important progress. (1) Almost no examples are known of exact solutions of the coupled equations. (2) There are surely other overall consequences of the asymptotic graded algebra besides positive energy which would be useful also for the classical limit. (3) Topological invariants constructed from the spin 3/2 fields; it is known formally that there exists a supersymmetric analog of the Gauss-Bonnet identity in four dimensions - does this imply the existence of some companion to the usual Euler number? (4) Are there generalizations of instanton solutions with either Minkowski or Euclidean signature? (5) To end on a slightly more concrete level, it would be interesting to see whether the spin 3/2 particles could play an appreciable role in the early stages of the universe. Direct observability is far more difficult because their coupling, while universal (with gravitational strength), does not lead to both macroscopic (only exchange of two fermions does this) and long-range forces and so they cannot be detected through their static effects.

REFERENCES

1. D. Z. Freedman, P. van Nieuwenhuizen and S. Ferrara, Phys. Rev. D13, 3214 (1976).
2. S. Deser and B. Zumino, Phys. Lett. 62B, 335 (1976).
3. C. Teitelboim, Phys. Rev. Lett. 38, 1106 (1477); R. Tabensky and C. Teitelboim, Phys. Lett. 69B, 453 (1977).
4. S. Deser, Gen. Rel. Grav. 1, 9 (1970).
5. D. Boulware and S. Deser, Ann. Phys. 89, 193 (1975).
6. S. Deser, J. H. Kay and K. S. Stelle, Phys. Rev. D16, 2448 (1977); M. Pilati, Nucl. Phys. B132, 138 (1978).
7. S. Ferrara and P. van Nieuwenhuizen, Phys. Rev. Lett. 37, 1669 (1976); D. Z. Freedman and A. Das, Nucl. Phys. B120, 221 (1977); S. Ferrara, J. Scherk and B. Zumino, Phys. Lett. 66B, 35 (1977); D. Z. Freedman, Phys. Rev. Lett. 38, 105 (1976); A. Das, Phys. Rev. D15, 2805 (1977); E. Cremmer and J. Scherk, Nucl. Phys. B127, 259 (1977); with S. Ferrara, Phys. Lett. 68B, 234 (1977).
8. B. DeWit and D. Z. Freedman, Nucl. Phys. B130, 105 (1977).
9. See for example, M. Kaku's report.
10. S. Deser, J. H. Kay and K. S. Stelle, Phys. Rev. Lett. 38, 527 (1977); S. Deser and J. H. Kay, Phys. Lett. 76B, 400 (1978).
11. S. Deser and C. Teitelboim, Phys. Rev. Lett. 39, 249 (1977).
12. M. T. Grisaru, Phys. Lett. 73B, 207 (1978).

THE GAUGE GROUP AND GEOMETRY OF SUPERGRAVITY

Tullio Regge
Istituto di Fisica del'Universita di Torino

In this brief talk I would like to discuss some of the recent approaches and results in the theory of supergravity. The language of forms is particularly suited to deal with gravity and Yang-Mills fields. We first review ordinary gravity as reformulated to Utiyama-Sciama-Kibble through the use of tetrads. Here two sets of 1-forms $\omega^{ab} = -\omega^{ba}$ (the connection) and ω^a (vierbein or tetrad) are given and can be considered as a single set ω^A where A = "ab" or "a" runs over the Lie algebra of the Poincaré group P. In other words, the ω^A are considered as Yang-Mills potentials for a Poincaré group gauge theory. The familiar Einstein action now takes the form

$$8A = \int_{M^4\subset P} R^{ab} \wedge \omega^c \wedge \omega^d \varepsilon_{abcd} \tag{1}$$

where

$$\begin{aligned} R^{ab} &= d\omega^{ab} + \omega^{ac} \wedge \omega^{cb} \\ R^a &= d\omega^a + \omega^{ab} \wedge \omega^b \end{aligned} \tag{2}$$

or simply

$$R^A = d\omega^A + \frac{1}{2} C^A{}_{BF}\omega^B \wedge \omega^F \tag{3}$$

where the $C^A{}_{BF}$ are the structure constants of P. The field equations turn out to be:

$$\begin{aligned} R^a &= 0 \\ R^{ab}\omega^c\varepsilon_{abcd} &= 0 \qquad d = 0\cdots 3 \,. \end{aligned} \tag{4}$$

Clearly, although the theory starts out as a Poincaré gauge theory, the action is not P gauge-invariant but only SO(3,1) (or SO(4) in the euclidean case) gauge-invariant. We can also see this fact as follows. It pays off to write the theory directly on the group manifold of P instead of using R^4 = P/SO(3,1) or P/SO(4). Let ε^A be an infinitesimal parameter set and let D denote the P-covariant derivative. In components, an infinitesimal gauge transformation is then written as:

$$\begin{aligned} \delta\omega^{ab} &= d\varepsilon^{ab} + \omega^{ac} \wedge \varepsilon^{cb} - \omega^{cb} \wedge \varepsilon^{ac} \\ \delta\omega^a &= d\varepsilon^a + \omega^{ab} \wedge \varepsilon^b - \omega^b \wedge \varepsilon^{ab} \end{aligned} \tag{5}$$

or

$$\delta\omega^A = D\varepsilon^A \tag{6}$$

We can easily check that if $\varepsilon^a = 0$ then (5) and (6) leave (4) and (1) invariant but that this is not true in general if $\varepsilon^a \neq 0$. In particular, $R^a = 0$ is not a translationally invariant equation, since under $\varepsilon^{ab} = 0$ R^a changes by $R^{ab}\varepsilon^b$.

To obviate for this lack of translational covariance, one writes an amended translation by changing (5) into (we set $\varepsilon^{ab} = 0$):

$$\delta\omega^{ab} = -2R_c{}^{ab}\varepsilon^c$$
$$\delta\omega^a = d\varepsilon^a + \omega^{ab} \wedge \varepsilon^b \tag{7}$$

Here the $R_c{}^{ab}$ are 1-forms obtained by contracting R^{ab} with the tangent vector T_c of the dual frame to the ρ^a. According to ref. (1) eq. (7) is nothing but a glorified infinitesimal coordinate transformation and the invariance of (1) and (4) under (7) is mathematically a trivial consequence of the general covariance of the language of forms.

Therefore, (7) has no further dynamical content beyond this covariance plus the Noether theorem and universal coupling of the translation current as generated by its flat limit (when matter is present). On the group manifold (7) can be generalized to

$$\delta\omega^A = D\varepsilon^A - 2R^A{}_B\varepsilon^B \tag{8}$$

where now the $R^A{}_B$ are usually severely restricted by the equations of motion. If in particular $\varepsilon^a = 0$, $\varepsilon^{ab} \neq 0$, certain coordinate transformations and gauge transformations coincide and we retrieve the usual SO(3,1) (resp. SO(4)) gauge covariance.

In supergravity we have similarly potentials ω^A where A = ("ab", or a, or α) where α is a Majorana spinor index.

The action is now

$$8A = \int_{M^4 \subset GP} (R^{ab} \wedge \omega^c \wedge \omega^d \varepsilon_{abcd} + 4i\bar{R}\gamma_5\omega^a\gamma^a) \tag{9}$$

where $R = d\omega + \frac{1}{2}(\sigma^{ab}\omega^{ab})\omega$. ω is the Majorana component of ω^A. GP is here the graded Poincaré group (see ref. 1) which has now 4 Fermi parameters generating supersymmetries, besides the usual 10 parameters of P. As in (1) we require that δA vanish for any $\delta\omega^A$ on any $M^4 \in$ GP. The resulting supergravity equations of motion are:

$$R^a = 0$$
$$R^{ab}\omega^f\varepsilon_{abfe} - 2i\bar{R}\gamma_5\gamma_e\omega = 0 \tag{10}$$
$$\gamma^a\omega^a R = 0$$

Eq. (10) restricts the curvature component $R^A{}_B$ appearing in (8) (which is of course valid for any Lie group provided one suitably interprets the range of A,B). If eqs. (10) are used, eqs. (8) appear in a simpler form, but they are then valid on shell only, be definition. Given the SO(3,1) or SO(4) gauge invariance of the theory, the dependence of the fields on these groups can be factored out, so that the theory appears effectively written on superspace $R^{4/4}$ = GP/SO(3,1), having four "Bose" coordinates x^μ and four "Fermi" coordinates θ^α forming a Majorana spinor. Physics however requires specifying data on an M^4 manifold in $R^{4/4}$ given by an embedding $\theta^\alpha = f^\alpha(x)$. Without loss of generality, we may take $\theta^\alpha = 0$. The structure of eqs. (10) is such that once the ω^A are specified on any M^4, they can be used to extend ω^A to the whole $R^{4/4}$ and hence to any other $M'^4 \subset R^{4/4}$, given for instance by $\theta^\alpha = \xi^\alpha(x)$ where the ξ^α are infinitesimal.

We can now perform a coordinate change such that M'^4 appears written as $\theta'^{\alpha} = 0$ in the new coordinates and the original restriction of ω^A on M'^4 now appears as a new field configuration ω'^A on M^4, which still satisfies eqs. (10). The relation between ω'^A and ω^A can be obtained by using (8) and is usually referred to as a "local supersymmetry" transformation. In ordinary gravity if we use a generic $M^4 \subset P$ and displace it in the SO(3,1) direction, the procedure still works; but one trivially obtains ordinary SO(3,1) gauge transformations. In $R^{4/4}$, displacement of M along the θ direction results in a nontrivial change in the fields ω^A. Clearly, the theory achieves maximum clarity if we view it first as a coordinate transformation (8), but in order to do so we need to consider the forms ω^A on $R^{4/4}$ and not only on M^4. This implies adding components in the θ directions, and more fields for the $R^A{}_B$ of (8). These fields appear under various labels as auxiliary fields in some recent versions of the theory and it is not an easy task to relate them to similar quantities appearing in different theories. Also, quite normally what appears in (10) as an equation of motion is taken as an a priori constraint in other formalisms, as exemplified in Zumino's talk at this conference (for some identifications of auxiliary superfields with the $R^A{}_B$ components, see 2). Anyway, eq. (8), contrary to some statements, is clearly an off-shell relation and does not depend on the particular choice of matter fields interacting with supergravity, including even torsion-generating fields. A definite advantage of (8) is that it is also automatically valid for the larger groups of extended supergravity.

References

1. Y. Ne'eman and T. Regge, Rivista del Nuovo Cimento, Ser. III, 1, #5, 1-42.
2. B. Zumino, Proc. XIX-th International Conference on High Energy Physics (Tokyo, 1978), to be published.

RECENT DEVELOPMENTS IN SUPERCONFORMAL GRAVITY

By Michio Kaku, Physics Dept.
City College of New York

(This work was done in collaboration with P.K. Townsend and P. van Nieuwenhuizen)

There are two types of supergravities. This is because there are only two classes[1] of graded Lie algebras which include both the compact internal symmetry groups and non-compact space-time groups: the graded Poincare group and the graded conformal group. Thus, there are also two types of supergravities, one based on gauging OSp(4/N) (which contains the de Sitter group Sp(4) and the O(N) group as subgroups) which is the usual supergravity, and one based on gauging SU(2,2/N) (which contains the conformal group SU(2,2) and the unitary group U(N) as subgroups), which is called superconformal gravity.

The two types of supergravities have different advantages and disadvantages. The usual supergravity[2]:

1) admits locally supersymmetric counter-terms in the lagrangian at all odd loop orders higher than one; this casts some doubt on the renormalization program for supergravity.

2) has a cosmological term which is many orders of magnitude too large to be compatible with astronomical data

3) is quantized in de Sitter space, on a five-dimensional closed hypersphere, which yields a non-causal theory

4) only admits O(8) as the highest internal symmetry group, which is too small to include the minimal SU(3) x SU(2) x U(1) of particle physics.

Superconformal gravity[3], which lacks a dimensional coupling constant because it is locally scale invariant, enjoys several advantages over the usual supergravity:

1) it is renormalizable (up to questions of anomalies)

2) it has no bothersome cosmological term

3) it is quantized in Minkowski space, and is therefore causal

4) it admits U(N) internal symmetry groups, which may be large enough to include the minimal SU(3) x SU(2) x U(1)

On the other hand, superconformal gravity lacks the Einstein term in the low energy limit[4] (instead, we have the Weyl lagrangian) and there also is the problem of ghosts arising from a higher derivative theory[5] (which may or may not cancel due to the large number of symmetries in the theory).

We begin by describing the gauge algebra of the superconformal group. We have the six Lorentz generators J_{ab} and the four translations P_a, four conformal K_a, one scale transformation D, one axial transformation A, and two sets of supersymmetric transformations Q and S. For the group SU(2,2/1), this gives a 24 parameter algebra:

$$[S,P_a] = \gamma_a Q; \quad [Q,K_a] = -\gamma_a S; \quad [S,A] = (3/4)i\gamma_5 S; \quad [Q,A] = -(3/4)i\gamma_5 Q$$

$$[S,D] = -(1/2)S; \quad [Q,D] = (1/2)Q; \quad [Q,J_{ab}] = \sigma_{ab}Q; \quad [S,J_{ab}] = \sigma_{ab}S \;;$$

$$\{Q_\alpha,Q_\beta\} = -(1/2)(\gamma^a C)_{\alpha\beta}P_a; \quad \{S_\alpha,S_\beta\} = (1/2)(\gamma^a C)_{\alpha\beta}K_a$$

$$\{Q_\alpha, S_\beta\} = -(1/2)C_{\alpha\beta}D + (\sigma^{ab}C)_{\alpha\beta}J_{ab} + (i\gamma_5 C)_{\alpha\beta}A \quad (a>b)$$

Yang-Mills is the local gauge theory of the unitary symmetry groups. Supergravity is the local gauge theory of the graded Poincare group. And superconformal gravity is the local gauge theory of the graded conformal group. We proceed to construct the theory in the same way as in Yang-Mills theory, by forming curvatures[6] of the theory:

$$R^A_{\mu\nu} = \partial_\nu h^A_\mu - \partial_\mu h^A_\nu + f_{BC}{}^A h^B_\mu h^C_\nu$$

where the f's represent the structure constants of the algebra: $[X_A, X_B\} = f_{BA}{}^C X_C$ There is only one lagragian quadratic in curvatures, which is scale invariant, with the correct parity, with curvatures sewn together by constant tensors rather than by metric tensors (except for the axial term):

$$L = \varepsilon^{\alpha\beta\rho\sigma}\{\varepsilon_{abcd}\alpha R^{ab}_{\mu\nu}(J)\, R^{cd}_{\rho\sigma}(J) + \beta R_{\mu\nu}(Q)\gamma_5 \bar{R}_{\rho\sigma}(S) + \gamma R_{\mu\nu}(A)\, R_{\rho\sigma}(D)$$
$$+ \delta e\, R_{\mu\nu}(A)\, R^{\mu\nu}(A)\ \} \qquad (\beta = \delta = 2i\gamma = -8\alpha)$$

Similarly, the gauge theory of the complete SU(2,2/N) group[7] can be shown to be S-supersymmetric (but Q-supersymmetry for these extended theories has yet to be demonstrated).

There have been several new developments in superconformal gravity. First, we have been able to couple superconformal gravity to matter multiplets.[4] In particular, we have combined superconformal gravity with a superconformal multiplet of scalars. By choosing a gauge to break conformal invariance, we obtain an action which reduces to ordinary supergravity in the low energy limit, and Weyl supergravity in the high energy limit. In this way, we can make superconformal gravity compatible with observational astronomical data.

The easiest way to construct such scalar interactions is simply to make a redefinition of fields according to:

$$e'_{a\mu} = k(A^2 + B^2)^{1/2}\, e_{a\mu}\ ; \quad A'_\mu = A_\mu - (A^2 + B^2)^{-1}\, [2(A\partial_\mu B - A\partial_\mu A - (1/4)i\bar{\chi}\gamma_\mu\gamma_5\chi$$
$$+ i\bar{\psi}_\mu\gamma_5(A - i\gamma_5 B)\chi]; \quad \psi'_\mu = (A - i\gamma_5 B)^{3/2}\, k^{-1/2}(A^2 + B^2)^{-1/2}\, \psi_\mu$$
$$+ \quad (A + i\gamma_5 B)^{1/2} k^{-1/4}(A^2 + B^2)^{-1/2}\chi$$

In addition to propagator terms, the first few terms of the extended lagrangian are:

$$L = \ldots + -(e/12)(A^2 + B^2)\, R - (1/12)(A^2 + B^2)\, \varepsilon^{\mu\nu\rho\sigma}\bar{\psi}_\mu\gamma_5\gamma_\mu D_\rho\psi_\sigma$$
$$+(2e/3)\ \bar{\chi}(A - i\gamma_5 B)\sigma^{\mu\nu}D_\mu\psi_\nu \quad + \ldots$$

Similar to the way in which a scale transformation on ordinary gravity will produce a scalar coupling to gravity (which can be eliminated by chosing the gauge properly), here we have extended ordinary supergravity to yield a superconformally invariant lagrangian, which will reproduce the low-energy behavior in a superconformal theory.

Recently, there has been much progress in reformulating superconformal gravity with tensor calculus methods and superspace methods. Eventually, these methods will reduce the tedious algebra needed to prove the invariance of superconformal theories and to generate new theories.

The tensor calculus starts with the observation[8] that, like in ordinary global supersymmetry, we can multiply multiplets (A,B,X,F,G) and (a,b,χ,f,g) to produce another supersymmetric multiplet (Aa - Bb, Ab + aB, $(A+i\gamma_5 B)\chi + (a+ib\gamma_5)X$, $Af + aF - Bg - bG - \bar{X}\chi$, $Ag + aG + Bf + bF + i\bar{X}\gamma_5\chi$)

Using these methods, one can generate the superconformal theory of SU(2,2/1) with relative ease.

The next method yields even greater simplifications, because the superfield[9,10,11] $E_A{}^B$ (A,B run over bosonic and fermionic indices) contains all the required fields of supergravity. The lagrangian of supergravity is just the determinant of this superfield[9] (with suitable constraints on the torsion), and the lagrangian of superconformal gravity turns out to be just the square of a fully symmetric tensor $W^{\alpha\beta\gamma}$ [9,11]

$$L \sim W_{\alpha\beta\gamma} W^{\alpha\beta\gamma}$$

With these powerful methods, we hope to solve unanswered questions: 1)what is the spectrum of the extended theory SU(2,2/N)? 2) can one write down spontaneously broken theories which break conformal invariance? 3) can we overcome the O(8) problem? In addition, we hope to establish the particle spectrum of the theory, to see if ghost killing mechanisms in this higher derivative theory exist.

REFERENCES

1. R. Haag, J.T. Lopuszanski and M. Sohnius, Nucl. Phys. B88 (1975) 257; V.G.Kac, Functional analysis and its applications, 9 (1975) 91; P.G.O.Freud and I. Kaplansky, J. Math. Phys. 17 (1976) 228; W. Nahm, V. Rittenberg and M. Scheunert. Phys. Lett. B61 (1976) 383.
2. D.Z.Freedman, P. van Nieuwenhuizen, and S. Ferrara, Phys. Rev. D13 (1976) 3214; S. Deser and B. Zumino, Phys. Lett. B62, 335 (1976).
3. M. Kaku, P.K.Townsend and P. van Nieuwenhuizen, Phys. Lett. 69B (1977) 304; M. Kaku, P.K.Townsend, and P. van Nieuwenhuizen, Phys. Rev. Lett. 39 (1977),1109; M. Kaku, P.K. Townsend, and P. van Nieuwenhuizen, Phys. Rev. D17 (1978) 3179.
4. M. Kaku and P.K. Townsend, Phys. Lett. 76B (1978),54 and Phys. Rev. Lett. 40 (1978) 1215.
5. S. Ferrara and B. Zumino, Nucl. Phys. B134 (1978) 301.
6. S.W.MacDowell and F. Mansouri, Phys. Rev. Lett. 38 (1977) 739;38,1376(E) (1977)
7. S. Ferrara, M. Kaku, P.K. Townsend, and P. van Nieuwenhuizen, Nucl. Phys. B129 (1977) 125
8. S. Ferrara and P. van Nieuwenhuizen, Phys. Lett. 76B (1978) 404.
9. J. Wess, and B. Zumino, Phys. Lett. 66B (1977) 361; R. Grimm, J. Wess, and B. Zumino, Phys. Lett. 73B (1978) 415; J. Wess and B. Zumino, Phys. Lett. 74B (1978) 51; L. Brink, M. Gell-mann, P. Ramond and J.H. Schwarz, Phys. Lett. 74B (1978) 336.
10. R. Arnowitt, P. Nath, Phys. Lett. 56B (1975) 117
11. W. Siegel, Harvard preprint 78/A014.

GEOMETRIZATION IN SUPERSPACE AND LOCAL SUPERSYMMETRY

Pran Nath and R. Arnowitt

Department of Physics, Northeastern University
Boston, Massachusetts 02115

Abstract

A brief review of the recent progress in gauge supersymmetry and the development of its connection with superspace formulations of supergravity is given. The equivalence of the metric approach and the supervierbein approach for supergravity mass-shell spaces is given. Recent progress in gauge supersymmetry includes the establishment of a rigorous theorem (including all quantum connections) on the existence of a residual preserved group, field content of gauge supersymmetry and finiteness of one and two loop connections to propagators for $N \geq 2$, where 4N is the dimensionality of the fermionic part of superspace.

Gauge supersymmetry[1,2] provides the first example of a completely unified gauge theory. The dynamics of gauge supersymmetry is free of arbitrary parameters and the twin principles of supersymmetry and local gauge invariance determine all the interactions in the theory including the sources and the Higgs field couplings. In conventional gauge theories, sources and the Higgs interactions are not fully determined by the gauge principles and one needs additional assumptions such as the number of quark flavors and the type of Higgs multiplet for a complete specification of the dynamics. Gauge supersymmetry which embodies local supersymmetry is formulated as a geometry of superspace because one believes that superspace[3] is linked on a fundamental level to supersymmetry. One is thus led in a natural fashion to develop unified gauge theories as geometries of higher dimensional spaces where the higher dimensions appear as anti-commuting fermionic coordinates of superspace. Actually the idea of trying to achieve unification of interactions by formulation of a gauge theory in higher dimensions can be traced back to Klein and Kaluza.[4] However, the problems attendant with higher Bose dimensions there are automatically absent here in the superspace formulations since the higher dimensions are fermionic and the physical space time is still only four dimensional.

First we discuss briefly the development of gauge supersymmetry as a geometry of superspace. Next we shall discuss its relationship to the superspace formulation of supergravity,[5,6] and the equivalence of the supervierbein and the supermetric tensor formulations of superspace supergravity.[7] Finally we shall report on some new developments in gauge supersymmetry with regard to its properties in the quantum domain.

Gauge supersymmetry is described by the invariant action[1]

$$I_0[g_{\Lambda\Pi}] = \int dz\, \sqrt{-g}\, \{(-1)^{\Lambda} g^{\Lambda\Pi} R_{\Pi\Lambda} + (4N - 2)\lambda\} \quad , \tag{1}$$

where $g_{\Lambda\Pi}(z)$ is the fundamental gauge multiplet of the theory with $g^{\Lambda\Pi}g_{\Pi\Sigma} = \delta^{\Lambda}{}_{\Sigma}$ and $z^{\Lambda} = (x^{\mu}, \theta^{\alpha a})$ $\alpha = 1\ldots4$ and $a = 1\ldots N$ are the superspace coordinates. The action of Eq. (1) is invariant under the general coordinate transformations in superspace $z^{\Lambda} = z^{\Lambda'} + \xi^{\Lambda}(z)$ which produces the gauge change

$$\delta g_{\Lambda\Pi}(z) = g_{\Lambda\Sigma}\xi^{\Sigma}{}_{,\Pi} + (-1)^{\Lambda+\Lambda\Sigma}\xi^{\Sigma}{}_{,\Lambda}g_{\Sigma\Pi} + g_{\Lambda\Pi,\Sigma}\xi^{\Sigma} \quad . \tag{2}$$

The gauge group of Eq. (2) unifies Yang-Mills, Einstein and supersymmetry groups. Equations of motion that result from Eq. (1) have the form $R_{\Lambda\Pi} = \lambda g_{\Lambda\Pi}$ where λ is a constant with dimensions of $(\text{mass})^2$.

In gauge supersymmetry spontaneous symmetry breaking plays a very crucial role in that the physical content of the theory can be extracted only after spontaneous breaking.(2) Spontaneous breaking implies a non-zero vacuum expectation value of $g_{\Lambda\Pi}$:

$$g_{\Lambda\Pi}(z) = \overset{o}{g}_{\Lambda\Pi} + h_{\Lambda\Pi} \quad , \tag{3}$$

where $\overset{o}{g}_{\Lambda\Pi} \equiv \langle 0|g_{\Lambda\Pi}|0\rangle$. In what follows we shall assume the existence of a globally supersymmetric vacuum state which requires that $\delta\overset{o}{g}_{\Lambda\Pi} = 0$ under superspace global supersymmetry transformations generated by $\xi^{\mu} = i\bar{\lambda}\Gamma^{\mu}\theta$, $\xi^{\alpha} = \lambda^{\alpha}$ where λ^{α} is a constant spinor. The most general form of a globally supersymmetric $\overset{o}{g}_{\Lambda\Pi}(z)$ then is

$$(\overset{o}{g}_{\Lambda\Pi}) = \begin{pmatrix} 0 & 0 \\ 0 & (\eta K)_{\alpha\beta} \end{pmatrix} + \begin{pmatrix} \eta_{\mu\nu} & -i(\bar{\theta}\Gamma_{\mu})_{\alpha} \\ i(\bar{\theta}\Gamma_{\mu})_{\alpha} & (\bar{\theta}\Gamma_{\mu})_{\alpha}(\bar{\theta}\Gamma^{\mu})_{\beta} \end{pmatrix} \quad . \tag{4}$$

$\eta = -C^{-1}$ where C is the charge conjugation matrix in Dirac space and K and Γ_{μ} are matrices in the Dirac and internal symmetry space so that $(\eta K)^{\sim} = -(\eta K)$ and $(\eta\Gamma_{\mu})^{\sim} = \eta\Gamma_{\mu}$. For the case when $K \neq 0$ one has a Riemannian theory (since the inverse of $g_{\Lambda\Pi}$ then exists). [Here a linear transformation on the θ coordinates allows one to set $K = 1$ so that the vacuum state does not really involve an arbitrary parameter K.] On the other hand the limit $K \to 0$ produces a non-Riemannian theory and in this limit gauge supersymmetry gives rise to supergravity. We discuss this limiting case of gauge supersymmetry next.

The first hint of a possible connection between gauge supersymmetry and supergravity arises due to a theorem on spontaneous breaking of gauge supersymmetry which shall be stated fully later on. Here the pertinent element of that theorem is that the supergravity gauge arises as one of the residual preserved gauges in the theory after spontaneous breakdown. Thus one may construct all the components of $g_{\Lambda\Pi}(z)$ in terms of only the supergravity multiplet which for the case N = 1 consists of the gravitational vierbein $e^{m}{}_{\mu}(x)$ and the spin 3/2 field $\psi_{\mu}{}^{\alpha}(x)$. $g_{\Lambda\Pi} = g_{\Lambda\Pi}(e^{m}{}_{\mu}, \psi^{\alpha}{}_{\mu})$ is then constructed so that the tensor transformations of $g_{\Lambda\Pi}$ of Eq. (2) and the supergravity transformation laws for $e^{m}{}_{\mu}$ and $\psi^{\alpha}{}_{\mu}$ are obeyed. The procedure consists in integrating Eqs. (2) order by order in θ with initial values given by

$$\xi^{\mu} = i\bar{\lambda}\gamma^{\mu}\theta\ , \qquad \xi^{\alpha} = \lambda^{\alpha}(x)\ ,$$

$$g^{(o)}_{\mu\nu}(z) = g_{\mu\nu}(x)\ , \qquad g^{(o)}_{\mu\alpha}(z) = \tfrac{1}{2}K\bar{\psi}_{\mu\alpha}(x)\ , \qquad g^{(o)}_{\alpha\beta}(z) = K\eta_{\alpha\beta}\ . \tag{5}$$

The process of integrating Eq. (2) order by order in θ <u>automatically introduces Bose space supergravity torsion</u> in the theory even though gauge supersymmetry possesses no superspace torsion: supergravity field equations then arise in the limit $K \to 0$ of gauge supersymmetry

$$[R_{\Lambda\Pi} - \lambda g_{\Lambda\Pi}]_{K\to 0} = 0\ . \tag{6}$$

Alternate superspace formulations of supergravity have been given by Wess and Zumino(5) and by Brink, Gell-Mann, Ramond and Schwarz(6) which use the supervierbein $V^{A}{}_{\Lambda}$ and do not require the limiting procedure. The connection between the metric and the vierbein approaches can be understood by noting that one can always construct a Riemannian metric given a supervierbein

$$g_{\Lambda\Pi} = V^{A}{}_{\Lambda}\eta_{AB}(-1)^{(1+B)\Pi}V^{B}{}_{\Pi}\ , \tag{7}$$

where $\eta_{AB} = (\eta_{mn}, k\eta_{ab})$. The analysis of Ref. (6) uses the formalism of auxiliary Breitenlohner fields. If one eliminates these auxiliary fields from the supervierbein and uses the supergravity field equations (i.e., the mass-shell conditions) one has then a supervierbein which through Eq. (7) produces the Riemannian metric of Ref. (7) also on shell.(7) The above equivalence holds for arbitrary values of K and further the transformation functions ξ^{Λ} of the two formalisms are also identical on-shell.

Gauge supersymmetry can be quantized in a manifestly globally supersymmetric fashion using the functional integral formulation. The statement of spontaneous symmetry breaking equations with <u>full</u> quantum corrections included then takes on a very simple form if one assumes globally supersymmetric breakdown of gauge supersymmetry. The equations have the form

$$\Gamma_{\mu}\Gamma^{\mu} = -\tfrac{1}{2}(\lambda - K')\ ; \qquad \mathrm{tr}\ \Gamma_{\mu}\Gamma^{\mu} = -4(\lambda - \lambda')\ . \tag{8}$$

λ' and K' include all the quantum corrections and can be expressed in terms of functional integrals.

The following set of theorems may then be deduced from the equations of spontaneous breaking:

<u>Theorem I</u>: Globally supersymmetric solutions of spontaneous breaking must rigorously preserve an internal symmetry subgroup.

<u>Theorem II</u>: The full set of residual preserved gauges of the theory after spontaneous breakdown consist of the Einstein gauge, the supergravity gauge and a set of Yang-Mills gauges associated with transformations $\xi^{\mu} = 0$, $\xi^{\alpha} = (M\theta)^{\alpha}$ where the matrices $M = M_0 + i\gamma^5 M_1$ obey the condition

$$[\Gamma_{\mu}, M] = 0\ , \qquad \tilde{M}_{0,1} = -M_{0,1}\ . \tag{9}$$

All other gauges than those stated in theorem II are spontaneously broken.

Finally we discuss briefly the quantum structure of gauge supersymmetry. The quantum loop calculations in gauge supersymmetry are generally difficult because of the large number of components that enter in the tensor multiplet. Progress can be made, however, by use of the superfield notation. A remarkable new result has emerged in that the one and two loop calculations in gauge supersymmetry are seen to be ultraviolet finite and work in progress indicates that this result may indeed hold in higher loops as well.[†] We exhibit the phenomenon of finiteness by examining the one loop contribution to the mass operator in superspace. One has

$$M_{\Pi\Lambda\Sigma\Delta}(q;\theta_1,\theta_2) = \int d^4k\ \Gamma(q,k;\theta_1)\Delta(q+k;\theta_1\theta_2)\Gamma(q,k;\theta_2)\Delta(k;\theta_2,\theta_1)$$

where Γ and Δ are the vertices and the propagators (to zeroth order). Global supersymmetry allows one to write

$$\Delta(k;\theta_1,\theta_2) = e^{\bar{\omega}\Gamma\cdot k\xi}\ \Sigma\ F_{(n)}(k;\omega)P_{(n)}(\xi)$$

where $\omega = \theta_1 - \theta_2$ and $\xi = \frac{1}{2}(\theta_1 + \theta_2)$. $P_n(\xi)$ denotes a polynomial in ξ. The structure of $F_{(n)}$ for large k can be derived from the equations obeyed by Δ and one has

$$F_{(n)}(k,\omega) \sim \frac{(\omega)^0}{k^{2+4N}} + \ldots + \frac{(\omega)^{4N}}{k^2} \quad .$$

Insertion of the form factors then produces convergence for $N \geq 2$.

[†]Note added in proof.

More recently we have established that the result of finiteness of quantum loopsdoes indeed hold to all orders. See Northeastern preprint NUB#2376.

Acknowledgment: This work was in part supported by the National Science Foundation.

References

1. P. Nath and R. Arnowitt, Phys. Lett. 56B, 171 (1975); R. Arnowitt and P. Nath, Gen. Rel. Grav. 7, 89 (1975). R. Arnowitt, P. Nath and B. Zumino, Phys. Lett. 56B, 81 (1975).
2. R. Arnowitt and P. Nath, Phys. Rev. Lett. 36, 1526 (1976). P. Nath and R. Arnowitt, Phys. Rev. D, October 15, 1978.
3. A. Salam and J. Strathdee, Nucl. Phys. B80, 499 (1974).
4. T. Kaluza, Sitzer. Preuss. Akad. Wiss, 966 (1921). O. Klein, Zs. f. phys. 37, 895 (1926). Y. M. Cho and P. Freund, Phys. Rev. D12, 1711 (1975).
5. J. Wess and B. Zumino, Phys. Lett. 66B, 361 (1977).
6. L. Brink, M. Gell-Mann, P. Ramond and J. Schwarz, Phys. Lett. 74B, 336 (1978); 76B, 417 (1978).
7. R. Arnowitt and P. Nath, Phys. Lett., to be published, Northeastern preprint NUB#2361 (1978). P. Nath and R. Arnowitt, Phys. Lett. 65B, 73 (1976).
8. Local indices are denoted by A,B,C,..=(m,n,s,..., a,b,c,...) and global indices by $\Pi,\Lambda,\Sigma,\ldots = (\mu,\nu,\lambda,\ldots,\alpha,\beta,\gamma,\ldots)$.

SUPERFIELD SUPERGRAVITY

Warren Siegel*

Lyman Laboratory of Physics

Harvard University
Cambridge, Massachusetts 02138

This talk is a summary of a paper of the same title written in collaboration with Jim Gates [1]. Our method of describing supergravity with superfields is to work in close analogy to globally-supersymmetric gauge theories, using the supertranslation group as the gauge group (i.e., the group generators are the supercoordinate partial derivatives $i\partial_M$.) We are then able to find locally supersymmetric actions, with minimal or nonminimal auxiliary fields, for supergravity (with or without a cosmological term), conformal supergravity, and the chiral and vector matter multiplets.

As for nonsupersymmetric gauge theories, the group structure of a supersymmetric gauge theory is found by generalizaing the global symmetry of the simplest matter multiplet to a local symmetry. For globally-supersymmetric gauge theories we therefore take a chiral superfield χ $(\bar{\partial}_{\dot{\alpha}}\chi = 0)$ which transforms under some representation of the global group, and make the group local by generalizing the constant gauge parameters to chiral functions of the supercoordinates (chiral in order to maintain the chirality of χ). Similarly, for supergravity we have the local transformation law $\chi' = \exp(i\Lambda)\chi$, where $\Lambda = \Lambda^M i\partial_M$ has chiral Λ^m and Λ^μ, but an arbitrary $\Lambda^{\dot{\mu}}$, since $\bar{\partial}_{\dot{\mu}}$ annihilates χ (but $\Lambda^{\dot{\mu}}$ will have an effect on the supergravity gauge fields). The group thus obtained is the largest local supercoordinate group, the local superconformal group; the local supersymmetry group is the subgroup given by the restriction $(n+1)(\partial_m\Lambda^m - \partial_\mu\Lambda^\mu) = (3n+1)\bar{\partial}^{\dot{\mu}}\Lambda_{\dot{\mu}}$ for some nonzero real number n.

We then look at the kinetic term of the chiral multiplet's action to see how the derivatives are covariantized by the gauge superfield. Before introducing gauge

*Research supported in part by the National Science Foundation under Grant Number PHY77-22864.

fields the action is $\int d^4x d^4\theta \bar{\chi} \exp(2\langle U\rangle)\chi$, where $\langle U\rangle = \theta^\alpha \bar{\theta}^{\dot{\beta}} \sigma^a_{\alpha\dot{\beta}} i\partial_a$ contains all derivatives. For globally-supersymmetric gauge theories we covariantize by replacing $\langle U\rangle$ with $\langle U\rangle + V$, where $V = V^i G_i$ is a real superfield contracted with the group generators G_i. Therefore, for supergravity we replace $\langle U\rangle$ with $\langle U\rangle + V^M i\partial_M \equiv U^M i\partial_M$, so the supergravity superfield is a real supervector U^M. The transformation law $\exp[2(\langle U\rangle + V')] = \exp(i\bar{\Lambda}) \exp[2(\langle U\rangle + V)] \exp(-i\Lambda)$ becomes $\exp(2U') = \exp(-\bar{\Lambda}^M \partial_M) \exp(2U) \exp(\Lambda^M \partial_M)$.

In globally-supersymmetric gauge theories we can construct covariant gauge-field quantities only from $\exp[2(\langle U\rangle + V)]$ and ∂_A. Therefore, due to the chirality of Λ, we can define $\nabla_\alpha = \exp[-2(\langle U\rangle + V)]\partial_\alpha \exp[2(\langle U\rangle + V)]$ as a covariant derivative, with $\nabla'_\alpha = \exp(i\Lambda)\nabla_\alpha \exp(-i\Lambda)$. $\nabla_{\dot{\alpha}} = \bar{\partial}_{\dot{\alpha}}$ is invariant under the same transformation, and allows us to define $\nabla_a = \frac{i}{4}\sigma_a^{\alpha\dot{\beta}}\{\nabla_\alpha, \nabla_{\dot{\beta}}\}$ to give a complete supervector ∇_A. Replacing $\langle U\rangle + V$ with U, we obtain the "semi-covariant" derivatives $\hat{\nabla}_A = \hat{E}_A{}^M \partial_M$ for supergravity, which are not fully covariant due to the nonchirality of Λ^μ, but are covariant enough to enable us to find an action for supergravity (just as one can use $e_a{}^m \partial_m$ in ordinary gravity in terms of the vierbein $e_a{}^m$ to make general-coordinate invariance manifest, but must treat local Lorentz invariance separately). Since dimensional analysis shows that L in the action $\kappa^{-2}\int d^4x d^4\theta L$ must be dimensionless, we can only use $\hat{E}_A{}^M$ as a power of its determinant. We therefore have $L = (\det \hat{E}_A{}^M)^n [1 \cdot \exp(-2\overleftarrow{U})]^{(n+1)/2}$ for nonzero real n, where the second factor must be included to make the action real (where $i\overleftarrow{\partial}_M$ in $U^M i\overleftarrow{\partial}_M$ acts to the left till it hits the 1). This action is invariant under the n-dependent restriction given above.

The actions given above for various n differ only in their auxiliary field structure. For $n = -1/3$, the action in terms of component fields is $\int d^4x (\det e_a{}^m)^{-1}(L_0 - \frac{4}{3}A^a A_a - \frac{4}{3}\bar{B}B)$ as found by Stelle and West [2] and Ferrara and van Nieuwenhuizen [3] (where L_0 is the usual Lagrangian for the physical supergravity fields). For $n \neq -1/3$, we have the action $\int d^4x (\det e_a{}^m)^{-1}[L_0 - \frac{4}{3}A^a A_a + \frac{4}{n}(3n+1)^2 \bar{B}B + \frac{4}{n}(3n+1)^2(\beta^\alpha \rho_\alpha + \bar{\beta}_{\dot{\alpha}}\bar{\rho}^{\dot{\alpha}}) + \frac{4}{n}(3n+1)v^a v_a + 12(3n+1)w^a w_a]$, which is Breitenlohner's action [4] for $n = -1$.

$\hat{\nabla}_A$ is sufficient to construct the supergravity action and its coupling to the chiral multiplet, but more complicated actions are difficult to construct unless one uses the true covariant derivatives ∇_A. These can be constructed by modifying $\hat{\nabla}_A$ to make them fully covariant, or by solving the following constraints:

$$T_{\alpha\beta}{}^C = T_{\alpha\dot\beta}{}^C + 2i\sigma^c{}_{\alpha\dot\beta} = T_{ab}{}^c (\text{or } R_{\alpha\dot\beta}{}^{cd}) = T_{\alpha b}{}^c - \tfrac{1}{4}\sigma_{b\alpha\dot\gamma}\sigma^{c\beta\dot\gamma}T_{\beta d}{}^d = 0 ,$$

$$\begin{cases} n = -\tfrac{1}{3} : T_{\alpha b}{}^b = 0 \\ n \neq -\tfrac{1}{3} : \tfrac{1}{12}\sigma_a{}^{\gamma\dot\alpha}\sigma_{b\gamma}{}^{\dot\beta}R_{\dot\alpha\dot\beta}{}^{ab} = \frac{n+1}{3n+1}\nabla_{\dot\alpha}T^{\dot\alpha}{}_b{}^b - \left(\frac{n+1}{3n+1}\right)^2 (T_{\dot\alpha b}{}^b)^2 , \end{cases}$$

where $\nabla_A = E_A{}^M\partial_M + \tfrac{1}{2}\phi_A{}^{bc}M_{bc}$ and $[\nabla_A,\nabla_B\} = T_{AB}{}^C\nabla_C + \tfrac{1}{2}R_{AB}{}^{cd}M_{cd}$ (M_{ab} are the Lorentz generators). The action $n^{-1}\kappa^{-2}\int d^4x d^4\theta (\det E_A{}^M)^{-1}$ with the above constraints is the same as the action given earlier. The action and constraints for $n = -\tfrac{1}{3}$ were found by Wess and Zumino [5] (with $T_{ab}{}^c = 0$); the case $n = -1$ (with $R_{\alpha\dot\beta}{}^{cd} = 0$) is our interpretation of the second-order formulation of Brink, Gell-Mann, Ramond, and Schwarz [6]. The solution to the constraints expresses ∇_A in terms of U^M, and also a superconformally-extending scalar superfield which can be gauged to 1 in the absence of a cosmological term. This scalar superfield is chiral for $n = -\tfrac{1}{3}$, and linear ($\bar\partial^2 T = 0$) for $n \neq -\tfrac{1}{3}$. (For construction of other actions, see ref. [1].)

References

1. W. Siegel and S. J. Gates,Jr., "Superfield Supergravity", Harvard preprint HUTP-78/A019 (June 1978).
2. K. S. Stelle and P. C. West, Phys. Letters 74B (1978) 330.
3. S. Ferrara and P.van Nieuwenhuizen, Phys. Letters 74B (1978) 333.
4. P. Breitenlohner, Phys. Letters 67B (1977) 49; Nucl. Phys. B124 (1977) 500.
5. J. Wess and B. Zumino, Phys. Letters 74B (1978) 51.
6. L. Brink, M. Gell-Mann, P. Ramond, and J. H. Schwarz, Phys. Letters 74B (1978) 336.

HOW TO BUILD HADRON MULTIPLETS FROM STABLE PARTICLES

A. O. Barut
Department of Physics, The University of Colorado, Boulder, Colorado 80309

"I want to understand light as well as
I can, without introducing things that
we can understand even less of."

Lord Kelvin

Abstract

New realizations of SU(3)-hadron multiplets are given in terms of stable particles, proton, electron, neutrino, and μ-meson, considered as a magnetically excited state of electron. Magnetic interactions between these particles can produce pairings of leptons and further coupling of these pairs with the proton. A shift of 2/3 of the leptonic charge to the electric charge relates leptons to quarks. Internal quantum numbers of hadrons and many rules of hadron interactions acquire simple physical interpretations.

1. Physical Principles

We propose here a theory of hadron multiplets and hadron interactions in terms of stable particles interacting through the electromagnetic field. All other "particles" are unstable; they come and go, and must be dynamically explained.

A. Particles

Absolutely Stable Particles: The only absolutely stable particles are proton, electron and neutrino and their antiparticles (as long as no annihilation takes place).

The Electromagnetic Field is a more general concept than photons. Only the free electromagnetic field in a radiation zone behaves like photons.

The Electron-Positron System is described by the Dirac equation with the usual superselection rule between e^- and e^+.

<u>The Neutrino</u>: The limit of the Dirac equation when $e \to 0$, $m \to 0$, such that $e/m \to \kappa$, is a four-component neutrino which can have anomalous magnetic moment interactions. Only the free massless Dirac equation splits into two 2-component equations, and a superselection rule occurs between ν and $\bar{\nu}$.

<u>The Proton</u>: We shall take here the proton as a given stable building block. It is indeed part of all baryons. About the internal structure of proton we shall comment later.

<u>The Muon</u>: The nature of the μ-meson is one of the most important (forgotten) problems of theoretical physics. We shall view μ as a magnetic-excited state of the electron. Under certain assumptions, one can derive the mass formula [1]

$$m_\mu = m_e\left[\frac{3}{2}\frac{1}{\alpha} + 1 + O(\alpha)\right] \tag{1}$$

which works very well and which indicates an electromagnetic origin to the mass of μ. Electric (Coulomb) excitations decay with the emission of a photon: $A \to B + \gamma$. Magnetic excitations seem to be connected with neutrinos, and decay via $A \to B + \nu + \bar{\nu}$. In building up the hadrons the electron can be excited to a muon, which is rather long-lived compared to hadronic processes. Hence temporarily we can consider μ as a building block whose eventual decay will account also for the decay of the hadrons. The μ-meson will be related to the strangeness quantum number of hadrons.

<u>Higher Excitations of Leptons</u>: Magnetic self-energy should give rise to higher excitations of electron after the muon, for example, τ-lepton. Similarly the neutrino will have excitations ν', ν'', ... These excited leptons will lead to new temporary hadronic states as we shall see.

<u>B. Forces</u>

The electromagnetic interactions manifest themselves as (i) electric (Coulomb) forces leading to Coulomb levels, hence to atomic and molecular matter; (ii) magnetic forces leading to high mass narrow resonances at very short distances. This phenomenon has been demonstrated in classical, semi-classical [2] and in relativistic Dirac equations [3], at the same level of understanding as the Coulomb levels, although a demonstration in quantum electrodynamics can so far be given only approximately, because it is a non-perturbative phenomenon [4, 5]. Such resonances between e^+e^-, as well as e^-e^-, have been numerically located in certain models [5].

The fundamental duality between electricity and magnetism can thus explain the duality between atomic and nuclear matter.

Magnetic Moment of the Neutrino: We can assume that the four-component neutrino has an anomalous magnetic moment form factor $F_2(q^2)$, even though $\kappa = F_2(0)$ might be vanishingly small. There are upper limits on the value of κ [6]. The phenomenon of magnetic resonances between leptons persists even for zero lepton masses and vanishing κ as long as $F_2(q^2) \neq 0$ [7].

Effective Magnetic Potential: The effective electromagnetic radial potential between leptons shows, at large r, first the Coulomb well with the centrifugal barrier, then at smaller distances new maxima and minima in which positive energy magnetic resonances occur [3, 5]. The effective potential depends on energy and angular momentum. For Coulomb levels which are localized at $r \sim 1/\alpha m$, the magnetic potential can be treated as a perturbation, but for states localized at $r \sim \alpha/m$ or α^2/m the potential of course must be treated non-perturbatively. Thus, there seems to be a fundamental magnetic pairing of leptons.

2. Construction of Hadron Multiplets

We put the leptons (ν, e^-, μ^-) into a triplet and denote the lepton number by $L \equiv B_\ell = \begin{pmatrix} 1 & & \\ & 1 & \\ & & 1 \end{pmatrix}$ and the electric charge by $Q_\ell = \begin{pmatrix} 0 & & \\ & -1 & \\ & & -1 \end{pmatrix}$. Consider also the three quarks (u, d, s), with $B_q = \begin{pmatrix} 1/3 & & \\ & 1/3 & \\ & & 1/3 \end{pmatrix}$ and $Q_q = \begin{pmatrix} 2/3 & & \\ & -1/3 & \\ & & -1/3 \end{pmatrix}$. We observe that [8] the sum of B_ℓ and charge Q_ℓ for leptons is the same as the sum of B_q and Q_q for quarks:

$$B_\ell + Q_\ell = B_q + Q_q. \tag{2}$$

This equation can be rewritten as

$$\left(B_\ell - \frac{2}{3} B_\ell\right) + \left(Q_\ell + \frac{2}{3} B_\ell\right) = B_q + Q_q \,,$$

and we find indeed that

$$B_\ell - \frac{2}{3} B_\ell = \frac{1}{3} B_\ell = B_q \quad \text{and} \quad Q_\ell + \frac{2}{3} B_\ell = Q_q. \tag{3}$$

Thus, quarks are obtained from leptons by shifting 2/3 of the fermionic charge to the electric charge.

The quarks obey the relation $Q_q = I_3 + \frac{1}{2}Y_q$ with $I_3 = \begin{pmatrix} 1/2 & & \\ & -1/2 & \\ & & 0 \end{pmatrix}$ and $Y_q = \begin{pmatrix} 1/3 & & \\ & 1/3 & \\ & & -2/3 \end{pmatrix}$. Hence keeping the same I_3 assignments for leptons, we obtain

$$Q_\ell = I_3 + \frac{1}{2}Y_q - \frac{2}{3}B_\ell \equiv I_3 + \frac{1}{2}Y_\ell$$

or,

$$Y_\ell = Y_q - \frac{4}{3}B_\ell \equiv B_\ell + S_\ell = \begin{pmatrix} -1 & & \\ & -1 & \\ & & -2 \end{pmatrix}. \tag{4}$$

Meson Multiplets

We construct the meson states as magnetically bound $\ell\bar{\ell}$-systems, hence $B=0$, $L=0$. In this case there is a precise correspondence with the usual $q\bar{q}$-system according to the above shifting procedure. The pseudo-scalar meson octet becomes as shown in Fig. 1.

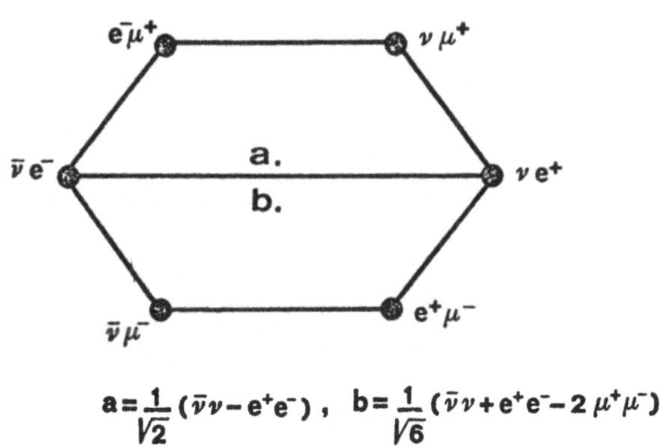

Fig. 1

Baryon Multiplets

The stable proton is surely the constituent of all baryons and gives them the baryonic (or better the protonic) number $B=1$. It further gives to the baryons their basic magnetic form factor. Therefore, the baryons with $B=1$, $L=0$, must be of the form $p\ell\bar{\ell}$ or $p\ell\bar{\ell}\ell\bar{\ell}$, ... [Note that the standard three-quark form of baryons, qqq, would go, under our shift-operation, into $\ell\ell\ell$ and would not give the correct total lepton number and charge to the baryons.] We have only to verify the additive quantum numbers Q, I_3 and Y for the baryons. These then determine uniquely the lowest $\ell\bar{\ell}$-content of the baryons shown in Fig. 2 for the spin $1/2^+$-octet, and in Fig. 3 for the spin $3/2^+$-decouplet. Any number of $\nu\bar{\nu}$, e^+e^- or $\mu^-\mu^+$-pairs can be added to each vertex without changing the additive quantum numbers.

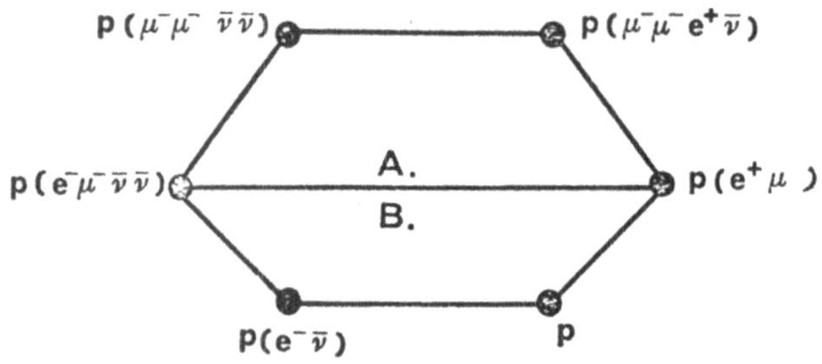

Fig. 2

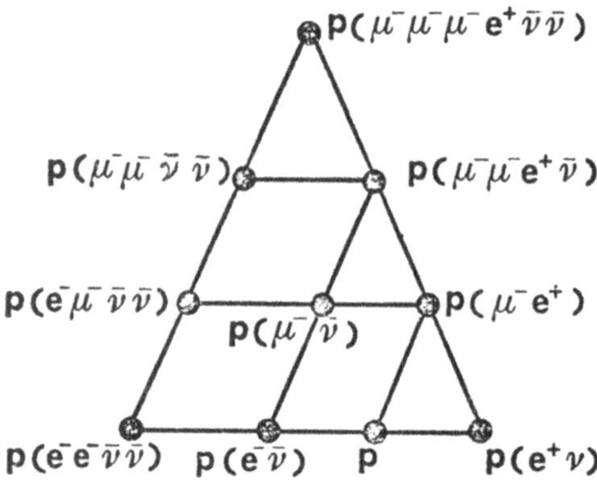

Fig. 3

Extensions to Other Leptonic Excitations

The shifting principle can be extended to other leptonic excitations. If a fourth lepton ν' is introduced as a magnetic excitation of ν, then the corresponding fourth quark would have charge +2/3, and corresponds to "charm." Similarly, higher excitations τ and ν_τ correspond to new quarks of charges -1/3 and +2/3, respectively.

A Prediction

Because magnetic forces also exist between two like-leptons, say e^-e^-, we predict sharp resonances of charge 2. The spectroscopy of charge 2 hadrons will be discussed elsewhere.

3. Physical Interpretation of Internal Quantum Numbers

The model of hadrons presented here provides a physical interpretation to the mysterious internal quantum numbers. This is a significant result of this work.

(i) "Strangeness" can be identified with the number of μ-constituents of hadrons, i.e., the number of "magnetically excited electrons":

$$S = N_{\mu^+} - N_{\mu^-}. \tag{5}$$

Strong interactions conserve the μ-number (strangeness), because these are of such short duration that μ-mesons do not have time to decay. A constituent μ can be exchanged between hadrons (associated production). Decays with $|\Delta S| = 1$ correspond to a de-excitation of μ, or an excitation of e into μ.

The number of higher excitations, ν', τ, ν_τ, . . . , similarly represent other new internal quantum numbers, like "charm."

(ii) The isotopic spin can be related to the number of absolutely stable particles (p, e, ν). The rising operator for isospin is related to the creation and annihilation of leptons by

$$I^+ = a^+_\nu a_{e^-} + a^+_{e^+} a_{\bar{\nu}} \quad , \tag{6}$$

while I_3 counts the absolutely stable particles as follows:

$$I_3 = \frac{1}{2}\left(N_p + N_{e^+} + N_\nu - N_{e^-} - N_{\bar{\nu}} \right). \tag{7}$$

Hence $[I^+, I^-] = 2I_3$. In this connection the physical proton might be thought to be associated with virtual lepton pairs in the form: $p \frac{1}{\sqrt{2}}(\nu\bar{\nu} - e^- e^+)$. Combining I_3 with the strangeness S, we derive the Gell-Mann Nishijima formula: $Q = I_3 + \frac{1}{2}(B + S)$.

The isospin conservation in hadron reaction also acquires an intuitive meaning. For example, the pure total isospin states $\frac{1}{2}(|pn\rangle \pm |np\rangle)$ correspond exactly to the states in which the lepton pair $(e^- \bar{\nu})$ is attached to one or the other proton, and one takes the symmetrized or antisymmetrized states, as in the H_2^+-ion of molecular physics.

Other SU(3) lowering and raising operators can be introduced similar to Eq. (6) from Figs. 1-3.

4. A Comparison

As compared to the quark model,

(1) there is no confinement problem; the constituent leptons do come out;

(2) there is no spin-statistics problem for baryons; the constituents have the correct Fermi-statistics;

(3) there is no need for a "colour" degree of freedom; [This follows from (2) and also from the following identity:

$$3(\text{colour}) \times \sum_q Q_q^2 = \sum_\ell Q_\ell^2 = 2 .$$

In π^o-decay, or in the ratio $R = \frac{\sigma(e^+e^- \to \text{hadrons})}{\sigma(e^+e^- \to \mu^+\mu^-)}$, the above factor enters, which would give a too-small value without the factor 3(colour).]

(4) there is no need for new "gluons"; the magnetic pairing provides the necessary force between the leptons.

5. Interactions of Hadrons

1. There are two fundamental strong processes:

(i) Rearrangement of constituents.

Examples:

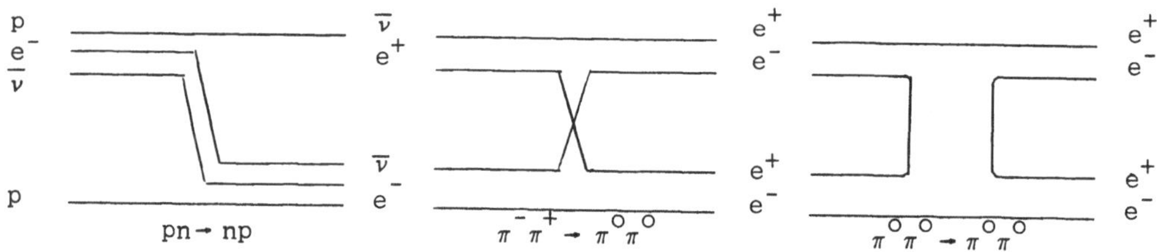

(ii) Pair production (or annihilation) of constituents.

Example:

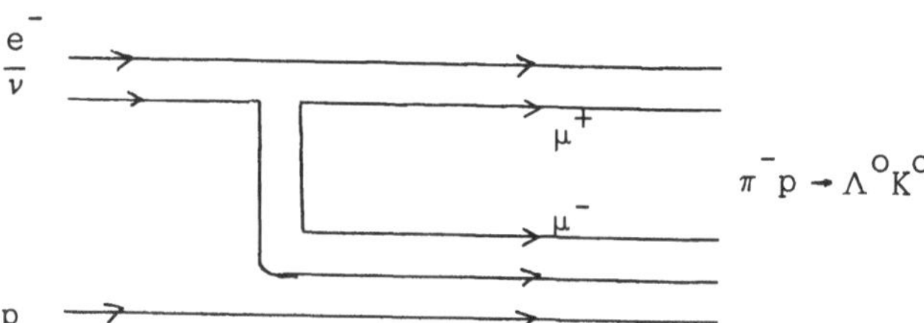

2. Electromagnetic decays correspond to annihilation of leptons.

Examples: π^o, η, Σ^o-decays.

3. There are two fundamental weak processes:

(i) μ-decay inside hadron.

(ii) Barrier penetration of constituents through the magnetic potential well.

In the first case, the μ-decay products may either combine with other leptons to form temporarily hadrons (nonleptonic decays), or the μ-decay products may penetrate the barrier (leptonic weak decays). Neutron decay can be interpreted as a pure barrier penetration as in α-decay. [Note that conversely, we could attribute the α-decay, if we wish, to a phenomenological interaction with some α-decay coupling constant.] If energetically possible, an electron can also be excited into a μ-meson during the barrier penetration. Example: $\pi \rightarrow \mu\nu$.

In this way, all decay schemes of particles can be intuitively pictured. Some mysterious concepts of particle physics are also amenable to an intuitive interpretation:

(i) In the neutral κ-mesons, the short and long κ-modes correspond to the superpositions $|e^-\mu^+\rangle \pm |e^+\mu^-\rangle$ and all the CP- and decay properties may be explicitly verified.

(ii) The Cabibbo angle measures the suppression factor of the magnetically bound μ-decay inside hadron relative to the free μ-decay, for example, in $\Lambda \rightarrow pe^-\bar{\nu}$.

(iii) The neutron-proton mass difference can now be understood as the sum of electron mass, and the (positive) magnetic energy of the $(e^-\bar{\nu})$-system around the proton.

(iv) The Zweig-rule can be stated as follows: A single lepton pair produced cannot come as a bound hadron, but the elements of the pair can form hadrons with other leptons present. The weak interaction $e\nu \rightarrow e\nu$ can be interpreted as due to the tail of the magnetic potential. There is also a difference, as we noted, between a four-component neutrino and a free two-component neutrino. In order to produce strong interactions in such reactions, the leptons must overcome the potential barrier, produce lepton pairs which then can combine into hadrons.

However, we leave as an open question, the problem of the μ-decay and weak interactions of the type $e\nu \rightarrow e\nu$, whether they might be interpreted as due to a new type of interaction (intermediate bosons), or due to magnetic interactions.

Finally, we comment about the structure of the proton and proton interactions. Part of the interactions between protons is due to the virtual lepton pairs (mesons) interacting magnetically. But the internal constitution and the mass of the proton is a problem on a different level than the other baryons of which it is a part. In the dyonium model of the proton [2] (extremely strongly bound state of two spin-zero monopoles, one charged, via the singular string giving the system a total spin 1/2 and Fermi statistics), we expect additional van der Waals forces between protons other than the meson exchange forces [9].

6. Conclusions

The theory presented here brings the high energy and particle physics back to the lines of traditional physics, and hopefully establishes continuity with nuclear and atomic physics. It is true that the evaluation of many numerical quantities is left to the dynamical theory of magnetic interactions of two or more particles. Of course, the same dynamical problem is present also in the usual quark model. However, the picture of fundamental particles is now considerably simplified. Not only no new particles are introduced, but also in principle no new interactions, no new coupling constants or parameters.

References

1. A. O. Barut, Physics Letters 73B, 310 (1978).

2. A. O. Barut, in The Structure of Matter, Proceedings Rutherford Centennial Conference, 1972, ed. B. G. Wybourne (Univ. of Canterbury Press, 1972), p. 22-80.

3. A. O. Barut and J. Kraus, Physics Letters 59B, 175 (1975) and J. Math. Phys. 17, 506 (1976).

4. A. O. Barut and J. Kraus, Phys. Rev. D16, 161 (1977).

5. A. O. Barut and R. Rączka, Nucl. Phys. (in publication).

6. J. Bernstein, G. Feinberg and M. Rudermann, Phys. Rev. 132, 1227 (1963).

7. A. O. Barut, to be published.

8. A. O. E. Animalu, Lett. N. C. 3, 729 (1972).

9. T. Sawada, Progr. Theor. Phys. 54, 149 (1978).

Structure of the Multiquark Meson States

S.K. Bose
Department of Physics
University of Notre Dame, Notre Dame, Indiana 46556

Before proceeding to discuss multiquark mesons, is is useful to give a brief survey of the nonet model.

The Nonet Scheme

The nonet scheme as developed by Okubo[1] is built upon two assumptions: 1) That the physical meson states are components of a (non-traceless) tensor G^{μ}_{α} and 2) That in no mathematical expression involving a nonet should the trace G^{λ}_{λ} appear _explicitly_. The resulting model makes two predictions. First, that the masses of physical mesons satisfy an equal spacing rule and, in addition, the mass of the state with I = 1, Y = 0 is degenerate with that of one of the two states with I = 0, Y = 0 (I = isospin, Y = hypercharge). Secondly, the model yields coupling schemes involving a nonet that are depictable graphically by means of the quark-line rule —— the Okubo-Zweig-Iizuka (OZI) rule.

The underlying group-structure of the nonet scheme was investigated by Iizuka[2] and by us[3,4]. Here one looks upon the nonet tensor G^{μ}_{α} to provide the representation $(\overline{3},3)$ of a non-chiral direct product group SW(3) $\equiv SU(3)_{\overline{q}} \times SU(3)_{q}$, where the factor $SU(3)_{\overline{q}}$ acts on the antiquark indices and $SU(3)_{\overline{q}}$ on the quark indices of the meson tensor. It was then found that the simplest extension of the GellMann-Okubo behavior for the SW(3) mass-splitting operator (see below for a more precise statement) leads to Okubo's results for the nonet. In regards to couplings involving a nonet it was found that most 3-particle couplings vanish in the limit of the non-chiral group. In fact, the only non-vanishing 3-particle coupling is the one involving 3-nonets with the final two mesons in a symmetric state, the corresponding coupling being[4]

$$g\, G_{\alpha}^{\lambda} \; \{G_{\mu}^{\beta}(1)\; G_{\nu}^{\gamma}(2)\; \varepsilon_{\alpha\beta\gamma}\; \varepsilon^{\lambda\mu\nu}\}^{*} + \text{H.C.} \tag{1}$$

The above coupling has the serious drawback that it cannot be extended to non-chiral groups with dimension greater than 3 (the number of quark flavors larger than 3), and for this reason will be discarded. The surviving symmetry is just SU(3) and consequently the OZI couplings must now be put in <u>by hand</u>.

<u>Multi-quark Meson States</u>

The multi-quark meson states that I shall now describe are built in close analogy with the $(\bar{q}q)$ nonet model, described above. The group SW(3) has a class of representations of the form $(\bar{d},d)$, where d is a symmetric tensor (triangular) representation of SU(3) and $\bar{d}$ the representation conjugate to d. I will call these the completely degenerate representations (CDR). Under the chain SW(3) $\supset$ SU(3), a CDR decomposes into a sequence of self-conjugate representations, each occurring once. For example, $(\bar{3},3) \to 1 + 8$, $(\bar{6},6) \to 1 + 8 + 27$ and so on. Thus, a CDR is the most natural and the most economic way of describing a meson multiplet. For this and for other (see below) reasons, I had advocated in 1966, the use of a CDR to describe multiquark meson multiplets.

Finally, let us note the form of CDR tensors. These are $G^{\mu\nu\lambda\ldots}_{\alpha\beta\gamma\ldots}$, with complete symmetry with respect to permutations of upper indices and of lower indices, separately.

Mass-splittings Within a CDR

The assumed transformation property

$$m_1 \sim (8,1) + (1,8) \tag{2}$$

of the mass-splitting operator m_1, leads[3,4,5] to the first order mass formula for physical masses (squared masses)

$$m = m_o + a(N_3 + N^3) \tag{3}$$

where a is a parameter and $N_3(N^3)$ the number operator for the lower (upper) index "3" (strange quark index) in a tensor component that describes a physical meson state. For instance, $N_3\, G^1_3 = G^1_3$, $N_3\, G^{12}_{12} = 0$. One can also write the mass formula on a basis provided by the (unphysical) SU(3) eigenstates. Then one obtains[3]

$$\langle n|m|n\rangle = m_o + \frac{2}{3}\, ar - \frac{2a\ (2r+3)}{(2n+1)\,(2n+3)} \left[I(I+1) - \frac{1}{4}\, Y^2 - \frac{1}{3}\, n(n+2) \right] \tag{4}$$

for the diagonal elements of the mass matrix. As for the off-diagonal components (transition masses), one finds that the only non-vanishing components are those connecting the adjacent SU(3) representations, such as $1 \leftrightarrow 8$, $8 \leftrightarrow 27$ (but not $1 \leftrightarrow 27$). The non-zero off-diagonal components of the mass-matrix are:

$$<n|m|n-1> = -\frac{a}{2(2n+1)}\left[(n-\frac{1}{2}Y)(n-\frac{1}{2}Y+1) - I(I+1)\right]^{1/2}$$

$$\times\left[(n+\frac{1}{2}Y)(n+\frac{1}{2}Y+1) - I(I+1)\right]^{1/2} \tag{4a}$$

$$\times\left[\frac{(2r+3)^2 - (2n+1)^2}{n(n+1)}\right]^{1/2}$$

Here, n is the label of SU(3) representation with dimensionality $(1+n)^3$ and r the label of SW(3) representation with dimensionality $\frac{1}{4}(r+1)^2 \times (r+3)^2$. If the matrix (4), (4a) is diagonalised, then eqn. (3) follows as a result[4]. Eqn. (3) can also be derived directly[5] from the stated transformation property eqn. (2). The following features of the mass-formula need to be noted: (1) From the two forms of the mass operator, given above, one can derive the connection between the physical states and the SU(3) eigenstates, that is, derive the representation mixing parameters. This was done explicitly for the representations $(\bar{3},3)$ and $(\bar{6},6)$ before[4]. Exactly same mixing parameters are derived from a study of the decomposition of the CDR tensor into SU(3) tensors[6] (symmetric and <u>traceless</u> tensors). Thus, the assumed behavior eqn. (2) of the mass operator is in complete accord with the requirement that the physical meson states be represented by components of CDR tensors. (2) Eqn. (3) leads to an equal spacing rule for the physical masses (squared masses). The level structure is thus the same as the harmonic oscillator energy levels. (3) All physical states with the same value of $(N_3 + N^3)$ are degenerate in mass. Included in this is the degeneracy of states with the same N_3 <u>and</u> N^3

i.e. states with the same hypercharge but differing in iso-spin. These iso-spin degeneracies have the same structure as the 'accidental degeneracies' of the hydrogen atom. Moreover, these iso-spin degeneracies remain true under arbitrary violations of the group SW(3), as long as the subgroup SW(2) is kept unbroken[4,7]. Thus, iso-spin degeneracy is a very general prediction of the present model. (4) The degeneracies and the equal spacing rule, detailed above, are also predicted by a naive quark model. But I must emphasize that eqn.(3) is a rigorous result.

The foregoing properties of the mass operator are rather attractive. This is another reason for selecting[3,4] the CDR to represent a meson multiplet. For the remainder of this talk, I shall concentrate on the representation $(\bar{6},6)$, specifically on its 3-particle couplings.

Couplings of the $(\bar{6},6)$ Multiplet

The $(\bar{6},6)$ is represented by the tensor $G^{\mu\nu}_{\alpha\beta}$, with $G^{\mu\nu}_{\alpha\beta} = G^{\nu\mu}_{\alpha\beta} = G^{\mu\nu}_{\beta\alpha} = G^{\nu\mu}_{\beta\alpha}$. This is thus a diquark-antidiquark state. Consider the coupling of a $(\bar{6},6)$ with a pair of nonets. The SW(3) invariant coupling exists only if the two nonets are in a symmetric state and is given by[4]:

$$g_1 \; G^{\mu\nu}_{\alpha\beta} \left[G^{\mu}_{\alpha}(1) \; G^{\nu}_{\beta}(2) + G^{\nu}_{\beta}(1) \; G^{\mu}_{\alpha}(2) \right]^{*} \tag{5}$$

In terms of quark-line diagrams the above represents the situation where the initial meson just falls apart into two ordinary ($\bar{q}q$) mesons . When the nonets are in an antisymmetric state, the SW(3) coupling is forbidden; but SU(3) together with OZI prescription for the nonets give the unique scheme:

$$g_2\, G^{\mu\lambda}_{\alpha\lambda} \left[G^{\mu}_{\sigma}(1)\, G^{\sigma}_{\alpha}(2) - G^{\sigma}_{\alpha}(1)\, G^{\mu}_{\sigma}(2) \right]^* \qquad (6)$$

The most striking feature of the above is that the entire 27-plet of the $(\bar{6},6)$ is now decoupled from the final states (does not decay into the final states). With $J^P = \frac{1}{2}^+$ baryons assigned to (1,8) (and antibaryons to (8,1)) of SW(3), the meson-baryon Yukawa coupling vanishes in the SW(3) limit. We now use SU(3) and, in addition, require (in direct analogy with the OZI rule for the nonet) that terms proportional to the contractions $G^{\mu\lambda}_{\nu\lambda}$, $G^{\nu\lambda}_{\nu\lambda}$ do not appear in the meson-baryon coupling. Then the following unique scheme is obtained[6].

$$g_3\, G^{\mu\nu}_{\alpha\beta} \left[\Psi^{\mu\nu\lambda}\, \Psi_{\alpha\beta\lambda} \right]^* \qquad (7)$$

where $\Psi_{\mu\nu\lambda}$ is the baryon tensor[6] (3 quark composite). The most interesting aspect of the above coupling is the following: of the two two states with I = 1 and Y = 0, one couples only to $\bar{\Sigma}\Sigma$ and $\bar{N}N$, while the other couples to $\bar{\Sigma}\Sigma$, $\bar{\Sigma}\Lambda$, $\bar{\Xi}\Xi$ but not to $\bar{N}N$ (N = nucleon). Of the three states with I = 0, Y = 0, one couples only to $\bar{\Xi}\Xi$, the other to $\bar{\Sigma}\Sigma$, $\bar{\Lambda}\Lambda$ and $\bar{\Xi}\Xi$ but not to $\bar{N}N$, while yet another is coupled to $\bar{N}N$, $\bar{\Sigma}\Sigma$ but not to $\bar{\Lambda}\Lambda$, $\bar{\Xi}\Xi$. Of the two $I = \frac{1}{2}$, $Y = \pm 1$ states only one can couple to a channel in which a nucleon is present. These results generalize the corresponding predictions of the scheme due to Sugawara and von Hippel[8] for the coupling of a meson nonet. Further discussions of the above couplings may be found in reference 6.

Models Based on SW(n)

When the number of quark flavors is greater than three, that is, when there are n quarks ($n > 3$) which provide a fundamental representation of SU(n), the meson group is SW(n). All results of SW(3), discussed thus far, can be generalised to SW(n), by letting the tensor indices run over the values 1,2,3,...,n. Note that eqn. (3) generalises to[5]

$$m = m_o + a_3 \ (N_3 + N^3) + a_4 \ (N_4 + N^4) + \ldots + a_n \ (N_n + N^n) \qquad (8)$$

Dynamical Models

I shall briefly review models of multiquark mesons based on specific, dynamical assumptions regarding the forces between quarks. First a general remark. A dynamical model must do two things. 1) It must reproduce the general features of the meson spectrum that I have detailed above, since these depend on, and only on, the underlying group structure. 2) It must, in addition, predict absolute values of masses and coupling constants, after the parameters of the model have been fixed in terms of appropriate inputs.

Probably the most interesting, certainly the most popular, of the dynamical models is the one based on the MIT bag model[9,10]. Here the quarks carrying indices 1 and 2 (up and down quarks) are treated as massless. The meson mass then depends on 4 parameters, which are fixed, once and for all, by supplying the low-lying hadron masses as input[11]. Then the model predicts multiquark meson masses. In addition to recovering our mass-formula eqn. (3), the model gives the value of parameters m_o and a, for each state of the meson's spin and parity. For instance, the $(\overline{6},6)$ multiplet with $J^P=2^+$ is found to appear in the range 1650-2250 MeV. Also the model says that the linear mass formula should be used. The bag model also predicts additional multiplets of the type $G^{[\mu\nu]}_{[\alpha\beta]}$, $G^{[\mu\nu]}_{\{\alpha\beta\}}$ and $G^{\{\mu\nu\}}_{[\alpha\beta]}$ ([] = anti-symmetric, {} = symmetric). These states, in general, do not belong to the class of CDR and hence, falls outside of the scope of the group-theoretic model developed in references 3 to 6. However, for for the special case of SW(3), $G^{[\mu\nu]}_{[\alpha\beta]}$ is a nonet because of the possibility $G^{[\mu\nu]}_{[\alpha\beta]} \equiv \varepsilon^{\mu\nu\lambda} \varepsilon_{\alpha\beta\gamma} G^{\gamma}_{\lambda}$ (note that we should now use coupling (1) if we wish to preserve simple quark-line graphs). Moreover, our mass formula eqn.(3) could be proved to be true for the multiplets $G^{[\mu\nu]}_{\{\alpha\beta\}}$, $G^{\{\mu\nu\}}_{[\alpha\beta]}$ as well . However, this procedure fails if the symmetry group is SW(n) with $n > 3$ and for that reason antisymmetric states have no natural place in the group-theoretic scheme based on the use of CDR[3-6].

A closely related model based on quark-gluon model combined with dual unitarization is due to Hong-Mo and Hogaasen[12]. Finally, multiquark mesons have been considered in the literature for various dynamical reasons. An incomplete list is cited in reference 13.

REFERENCES

1. S. Okubo, Phys. Letters 15, 165 (1963)
2. J. Iizuka, Prog. Theo. Phys. (suppl) 37, 21 (1966)
3. S.K. Bose, Phys. Rev. 150, 1231 (1966) and Nuove Cimento 46, 419 (1966)
4. S.K. Bose and E.C.G. Sudarshan, Phys. Rev. 162, 1398 (1967)
5. S.K. Bose, Prāmana 3, 491 (1977)
6. S.K. Bose and E.C.G. Sudarshan, CPT, Austin (Texas) preprint (1978)
7. See also, S.F. Tuan and T.T. Wu, Phys. Rev. Letters 18, 349 (1967)
8. H. Sugawara and F. von Hippel, Phys. Rev. 145, 1331 (1966)
9. R.L. Jaffe, Phys. Rev. D15, 267 (1977) and D15, 281 (1977) and D17, 1444 (1977)
10. K. Johnson and C. Thorn, Phys. Rev. D13, 1934 (1976). R.L. Jaffe and K. Johnson, Phys. Letters 60B, 201 (1976)
11. T. Degrand, R.L. Jaffe, K. Johnson and J. Kiskis, Phys. Rev. D12, 2060 (1975)
12. Chan Hong-Mo and H. Hogaasen, Phys. Letters 27B, 121 (1977)
13. J.L. Rosner, Phys. Rev. Letters 21, 950 (1968). D. Weingarten and S. Okubo, Phys. Rev. Letters 34, 1201 (1975). C. Kalman, Nuovo Cimento 21, 201, (1978). C. Rosenzweig, Phys. Rev. Letters 36, 695 (1976). M. Bander et al Phys. Rev. Letters 36, 697 (1976), R. Rossi and G. Veneziano, Nucl. Phys. B123, 507 (1977).

Octonionic Structures in Particle Physics*+

Feza Gürsey

Yale University, Physics Dept., New Haven, Conn. 06520

ABSTRACT

Octonionic extensions of Quantum Mechanics associated with finite exceptional Hilbert spaces are summarized. An interpretation of such spaces as the finite space of internal symmetries allows the introduction of color and flavor degrees of freedom for fundamental particles, color arising from the automorphism group of octonions. As an example the space admitting E_6 symmetry is discussed. The projection operators for various fermion states are constructed and shown to belong to a larger ternary algebra of hermitian operators. A possible application is to the grand unified gauge theory of leptons and quarks based on E_6. Another possibility involves the construction of an exceptional superspace by means of octonionic Grassman numbers.

* Research supported in part (Yale Report # COO-3075-218) by the U. S. Department of Energy under Contract No. EY-76-C-02-3075.

+ Invited talk presented at the 1978 Austin Conference on Group Theory and Mathematical Physics.

I. Introduction

Octonions were first introduced in Physics by Jordan, von Neumann and Wigner[1] during the turbulent years that followed the discovery of Quantum Mechanics, when daring physicists were seeking new mathematical structures for the description of baffling new phenomena connected with neutrons, nuclear forces, artificial radioactivity, β decay etc., that signalled the birth of nuclear physics. They discovered a new finite Hilbert space in which octonions replaced complex numbers. Octonions were already known in mathematics as the last entry of Hurwitz list of four composition algebras that were also division algebras consisting of R (real numbers), C (complex numbers), H (quaternions) and Ω (octonions or Cayley numbers). R and C are both associative and commutative, H associative but not commutative, Ω neither associative nor commutative. All four algebras are alternative, with totally antisymmetric associators. It later turned out that the symmetry group of the exceptional Hilbert space was the exceptional group F_4, as shown by Chevalley and Schafer[2] in 1950. The exciting discovery of Jordan et al soon dropped out of physics as Fermi and later Yukawa showed that ordinary Quantum Mechanics and Quantum Field Theory were, in principle, sufficient to explain β-decay (Fermi's theory of weak interactions) and nuclear forces (Yukawa's theory of meson exchange for strong interactions). On the other hand, Jordan's algebra of hermitian operators representing observables and its exceptional octonionic realization survived as a new branch of mathematics and led to profound developments in the 1950's linking algebra, Lie groups and geometry and leading to the discovery of new non Desarguesian geometries[3] that generalize the octonionic projective geometry of Moufang[4] which was born the same year as the exceptional Hilbert space. At the time there was no possible physical interpretation for octonionic algebras and geometries. However, with the advent of color and flavor degrees of freedom of quark and leptons regarded as fundamental constituents of matter, a new finite Hilbert space carrying the corresponding quantum numbers is required. It then becomes tempting to identify this generalized charge space of color and flavor with exceptional spaces. Since the symmetries of exceptional spaces are described by exceptional groups, and since all exceptional groups share SU(3) symmetry that comes from the automorphism group of 6 additional imaginary units that one has to add to the imaginary unit of the complex numbers in order to turn a complex number into an octonion, it follows that the SU(3) group of color is intrinsically built in all exceptional geometries and exceptional Hilbert spaces. The return of the prodigal octonionic space to its original physics home might well provide a fundamental explanation for the color degree of freedom and its modes of combination with flavor groups within the larger exceptional groups.

The context in which flavor and color groups are used in Particle Physics is that of a local gauge field theory. The fundamental fields are taken to be quarks and leptons interacting through the exchange of gauge bosons. The interaction of the fundamental fermions with the gauge bosons and that of the gauge bosons among themselves are completely determined by the gauge principle. The unbroken color group gives rise to the strong interactions of quarks through the exchange of color gluons, while the spontanteously broken flavor group generates the electromagnetic and weak interactions of quarks and leptons through the exchange of the gauge bosons of the flavor group (massless photon, massive weak bosons) and possibly other much weaker interactions through the exchange of even more massive flavor gauge bosons.

If color and flavor are unified within a simple gauge group G[5], then one has, in addition, ultra massive gauge bosons associated with transformations that mix leptons and quarks and the fundamental fermions with their antiparticles. In order to have a quasi stable proton in such grand unified gauge theories the mass scale M of the destabilizing gauge bosons must be of the order of the Planck mass in Quantum gravity, i.e. 10^{19} GeV. If the charge space is an exceptional Hilbert space invariant under F_4 or E_6, these simple groups would be good candidates for the group G of grand unified theories. In fact E_6 seems to be consistent with present phenomenology, with quarks and leptons fitting in two 27 dimensional representations of E_6. If this description is correct we have 3 fundamental problems:

1 - <u>Are there reasons that make the superheavy mass M acceptable</u>? Now the mass M gives rise to large renormalization effects[6] which shift the coupling constant parameter α to α_s for strong interactions. Using the logarithmic dependence of the

renormalization of α_s/α on the mass scale in asymptotically free theories, from the phenomenological value of $\alpha_s(0.2\text{-}0.3)$ we deduce for M a value comparable to the Planck mass. Another large renormalization shifts the unrenormalized Weinberg angle parameter $\sin^2\theta$ (typically 3/8 in such theories) to the observed value 0.2-0.25. Again the renormalization factor is consistent with the same M. Another reason can be sought in Cosmology. Assuming that the observed asymmetry in the baryon-antibaryon number of the universe (about 10^{-8} including the CP symmetric primodial cosmic black body radiation) originates in the CP violating decay of primordial heavy bosons(7,8), an estimate of their mass again leads to the same mass scale M. Hence, including the life-time of the proton we find the mass M from four different considerations which suggest that such a mass scale may exist after all.

2 - Unification with gravity within a supersymmetry group. This large mass M in grand unified theories makes it impossible to ignore gravitation. It also suggests a unification with gravitation since it coincides with the Planck mass which is the inverse square root of the gravitational constant. On the other hand the only known consistent, two loop renormalizable theory of the gravitation interacting with matter is provided by supergravity(9), and in the case of charged matter by SO(2) extended supergravity. The further extension of supergravity to a larger group including the color group SU(3) and the Weinberg-Salam flavor group SU(2)xU(1) would unify all interactions. Here there are two difficulties: the largest possible extension of the usual supergravity stretches the group SO(2) to SO(8) and no more(10). Clearly this is not large enough. Second, simple supersymmetry groups do not include F_4 or E_6 as their Lie subgroups. It seems therefore impossible to imbed the successful grand unified theory based on E_6 in an extended supergravity. A way out may be the reformulation of supergravity in superspace(11,12) consisting of space-time and a Grassmann manifold. Grassmann numbers are obtained by using direct product of quaternions. There are exceptional Grassmann numbers based on octonions(13). It is possible that the invariance properties of such exceptional superspaces will involve new supersymmetries that generalize the exceptional groups.

3 - The confinement problem of color states. Free quarks and free gluons, colored states in general are thought to be confined because of long range forces in chromodynamics (gauge theory of unbroken color) that tend to pull such states together to form colorless states. It follows that colored states behave as if they were unobservable and color degrees of freedom are like hidden variables. This means that we cannot make yes or no experiments involving colored states. In turn this means that Birkhoff and von Neumann's calculus of propositions is not representable by a projective geometry, or equivalently the usual quantum mechanics in Hilbert space may not hold for quarks and gluons. In an algebraic version of Quantum Mechanics this is equivalent to saying that colored "observables" may not be represented by a Jordan algebra. Since some properties of quarks are observable (their number, spin, charge, flavor quantum number, etc.) we would expect them to be associated with more general algebraic properties. Colored states may span a new space larger than the usual Hilbert space with geometrical properties more general that the projective space. As an example we can give the complex octonionic plane associated with E_6, which unlike the Moufang real octonionic plane is not a projective geometry. The Moufang plane itself, although it is a projective geometry has non Desarguesian properties. Thus it is entirely possible that the non observability of free localized quarks is not just due to a dynamical feature of non-abelian gauge theories but also stems from algebraic properties that transcend the calculus of propositions or the ordinary Jordan algebra of the usual observables. Hence new geometries associated with octonionic extensions of projective geometries may be good candidates for a generalized quantum mechanics involving colored states. Unusual properties of colored states may either be associated with non Desarguesian projective geometries (corresponding to a failure of the superposition principle) or with the still more unusual properties of octonionic geometries that are not even projective geometries. As a possible example we shall mention the algebraic properties of E_6 representations and the construction of projection operators within these larger algebras.

II. Quantum Mechanics in Jordan Form

The octonionic extension of standard Quantum Mechanics is made possible by its reformulation as the Jordan algebra of Hermitian matrices associated with observables. Observables are closed under the symmetric Jordan product

$$A\cdot B = 1/2\ (AB + BA) \tag{2.1}$$

This algebra is commutative but not associative. Its associator

$$[ABC] = (A\cdot B)\cdot C - (A\cdot B)\cdot C \tag{2.2}$$

has the properties

$$[ABA] = 0,\ [ABA^2] = 0 \quad . \tag{2.3}$$

The last identity (Jordan's identity) leads to power associativity. To complete the formulation of Q.M. states have to be introduced. States are represented by kets $|\alpha\rangle$ and bras $\langle\alpha|$ in Hilbert space. In the Jordan scheme they are associated with the corresponding Hermitian matrices

$$P = |\alpha\rangle\langle\alpha| \tag{2.4}$$

that represent projection operators for the states and obey

$$P_\alpha{}^2 = P_\alpha \quad , \qquad Tr\ P_\alpha = 1 \quad . \tag{2.5}$$

The connection with projective geometry is clear. If the n complex components of $|\alpha\rangle$ are regarded as the homogeneous coordinates of a point in (n-1) dimensional complex space, then $|\alpha\rangle$ and $\lambda|\alpha\rangle$ represent the same point for an arbitrary complex λ. Normalization of $|\alpha\rangle$ reduces the factor λ to a pure phase which in turn disappears in the expression of P_α. It follows that there is a one-to-one correspondence between a physical state in a n-dimensional Hilbert space and a point in the (n-1) dimensional projective space. The case n = 3 correspond to the complex projective plane. Superposition of states in 3-dim. Hilbert space corresponds to linear dependence of the corresponding points in the projective plane. Every theorem in the projective plane (like the Pappus Theorem or the Desargues theorem) involving intersection of lines and the joining of points by straight lines can be translated in quantum mechanical language involving superposition of states[13,14].

The physical interpretation of Quantum Mechanics involves transition probabilities between states $|\alpha\rangle$ and $|\beta\rangle$ that can be expressed either by the kets or their associated projectors

$$\Pi_{\alpha\beta} = |\langle\alpha|\beta\rangle|^2 = Tr(P_\alpha\cdot P_\beta) = (P_\alpha, P_\beta) \quad . \tag{2.6}$$

Note that the non observable amplitude $\langle\alpha|\beta\rangle$ cannot be expressed in terms of projection operators and their Jordan products. The invariance group of Q.M. is the group that leaves all transition probabilities invariant. This is the group SU(n) that is also the automorphism group of the Jordan algebra of n x n hermitian matrices of observables and states.

Transition probabilities also have a geometric interpretation. We can define a SU(n) invariant distance[15] $d_{\alpha\beta}$ in projective (n-1) space between points α and β by

$$\cos^2 d_{\alpha\beta} = (P_\alpha, P_\beta) = \Pi_{\alpha\beta} \quad . \tag{2.7}$$

This distance obeys all the required inequalities and allows statements about transition probabilities to be reexpressed in geometrical terms.

The extension of the notion of Hermitian operator requires the introduction of the Jacobson triple product[16]

$$\{ABC\} = (A\cdot B)\cdot C + A\cdot(B\cdot C) - (C\cdot A)\cdot B = \{CBA\} \quad . \tag{2.8}$$

We can define an operator U_H associated with the observable H and acting on states P_α by

$$U_H P_\alpha = \{H P_\alpha H\} \quad . \tag{2.9}$$

Then, if H is Hermitian we can write(14,17)

$$|\langle\alpha|H|\beta\rangle|^2 = Tr(P_\alpha \cdot U_H P_\alpha) = (P_\alpha, U_H P_\beta) = (U_H P_\alpha, P_\beta) \tag{2.10}$$

which can be compared with the hermiticity condition of ordinary quantum mechanics

$$\langle\alpha|H|\beta\rangle = (\alpha, H\beta) = (H\alpha, \beta) \quad . \tag{2.11}$$

The triple product also allows us to formulate the insertion of a complete set of states in the Jordan formalism. To the completeness relation

$$\langle\alpha|\beta\rangle = \sum_\gamma \langle\alpha|\gamma\rangle\langle\gamma|\beta\rangle \;, \qquad \sum_\gamma |\gamma\rangle\langle\gamma| = I \tag{2.12}$$

corresponds

$$Tr(P_\alpha \cdot P_\beta) = Tr \sum_\gamma \{P_\alpha P_\gamma P_\beta\} \;, \; \sum_\gamma P_\gamma = I \quad . \tag{2.13}$$

Another form is

$$Tr(P_\alpha \cdot P_\beta) = Tr\, P_\alpha \cdot \sum_\gamma \{P_\gamma P_\alpha P_\gamma\} \tag{2.14}$$

or, in terms of the Hermitian operator $U_\gamma = U_{P_\gamma}$,

$$(P_\alpha, P_\beta) = \sum_\gamma (P_\alpha, U_\gamma P_\beta) \quad . \tag{2.15}$$

Finally the infinitesimal SU(n) transformations of the Hermitian matrix C standing for either an observable or a state P_α is given by

$$\delta C = i[\Omega, C] \qquad , \; \Omega = \Omega^\dagger \quad . \tag{2.16}$$

To express this law in Jordan form we introduce two Hermitian matrices A and B such that

$$i\Omega = 1/2\, [A,B] \quad . \tag{2.17}$$

Then, the transformation law takes the form

$$\delta C = 1/2\, \big[[A,B], C\big] = \{ABC\} - \{BAC\} \quad . \tag{2.18}$$

In other words, defining the operator $T_{A,B}$ by

$$T_{A,B} C = \{ABC\} - \{BAC\} = \delta C \tag{2.19}$$

we can show directly that $T_{A,B}$ generate a Lie algebra that is the derivation algebra (associated with the infinitesimal automorphism transformations) of the Jordan algebra of observables.

To summarize, we have shown that Quantum Mechanics only requires Hermitian matrices (observables and states) and their Jordan algebra as long as we deal with observable transition probabilities. The associative but non commutative matrix algebra is only needed for the non-observable transition amplitudes and transformations of kets. It follows that we have a quantum mechanics (interpretable as a projective geometry) as soon as we have quantities that exhibit the Jordan properties (2.1) and (2.2).

III. Quaternionic and Octonionic Quantum Mechanics

The four Hurwitz algebras have an involution $x \to \bar{x}$ (conjugation) with which one defines a positive norm N(x) and an inverse of non zero elements by

$$(\bar{\bar{x}}) = x, \; N(x) = x\,\bar{x} = \bar{x}\,x, \; x^{-1} = \bar{x}/N(x) \quad . \tag{3.1}$$

The Hurwitz algebras C, H, have respectively one, three and seven imaginary units which change sign under conjugation. The imaginary units j_i ($i = 1,2,3$) and e_a ($a = 1,\ldots,7$) of quaternions and octonions obey respectively the relations

$$H: \quad j_m j_n = -\delta_{mn} + \varepsilon_{mnr} j_r \tag{3.2}$$

$$\Omega: \quad e_a e_b = -\delta_{ab} + \psi_{abc} e_c \tag{3.3}$$

where ψ_{abc} is antisymmetric and equal to one for the combinations (123), (246), (435), (367), (651), (572) and (714).

• The automorphism groups of H and Ω are respectively SO(3) and the exceptional group G_2. The subgroups that leave one imaginary unit invariant are respectively $SO(2) \sim U(1)$ and SU(3). It is remarkable that the only exact internal symmetry gauge groups in particle physics, namely the electromagnetic gauge group U(1) and the color gauge group $SU(3)$ coincide with the automorphism groups of the imaginary units additional to the one of ordinary complex Hilbert space.

We can now consider quaternionic Hermitian matrices H and quaternionic projection operators P_α associated with the quaternionic ket α. We have

$$H = \bar{H}^T \quad , \qquad P_\alpha = \alpha\,\bar{\alpha}^T \tag{3.4}$$

where T means transposed and the bar denotes quaternionic conjugation. Observables H and states P_α are elements of the quaternionic Jordan algebra. Hence we obtain a quaternionic Quantum Mechanics (Q.Q.M.). Transition probabilities, superposition of states, insertion of a complete set can now be defined as in the preceding section by means of the quaternionic Jordan algebra. There are only two differences with standard QM. First, the invariance group is no longer SU(n) but the symplectic group Sp(2n) (again generated by $T_{A,B}$ of Eq. (21.9)). The second difference is that the Pappus Theorem in projective geometry does not hold in Q.Q.M. Finally, we gain an intrinsic U(1) group compared with complex Q.M.

Turning now to real octonions, there exists a 3 x 3 octonionic Hermitian matrix J that obeys the Jordan laws, namely

$$J = \bar{J}^T = \begin{pmatrix} \alpha & c & \bar{b} \\ \bar{c} & \beta & a \\ b & \bar{a} & \gamma \end{pmatrix} \quad , \tag{3.5}$$

where α, β, γ are real numbers and a, b, c are real octonions. The element J of the exceptional Jordan algebra has 27 real components. The automorphism group of the algebra is F_4. $I_1 = \mathrm{Tr}J$ is invariant under F_4, so that the traceless part of J corresponds to the 26-dimensional representation of F_4. Its two invariants under F_4 are

$$I_2 = \mathrm{Tr}J^2 \quad \text{and } I_3 = \mathrm{Det}\,J \ , \tag{3.6}$$

$$\mathrm{Det}\,J = \alpha\beta\gamma - \alpha|a|^2 - \beta|b|^2 - \gamma|c|^2 + (ab)c + \bar{c}(\bar{b}\bar{a}) \ . \tag{3.7}$$

J can be diagonalized by an F_4 transformation, the diagonal elements being functions of I_1, I_2 and I_3.

The states are defined by Hermitian matrices of the form

$$P_\alpha = \alpha\,\bar{\alpha}^T \quad , \quad \mathrm{Tr}\, P_\alpha = 1,\ P_\alpha^2 = P_\alpha \tag{3.8}$$

having three octonionic components, one of them being a real number. If we define the Freudenthal product by

$$J \times J = J^{-1}\mathrm{Det}J = J^2 - J\,\mathrm{Tr}\,J - 1/2\ I(\mathrm{Tr}J^2 - (\mathrm{Tr}J)^2) \tag{3.9}$$

then, we have

$$P_\alpha \times P_\alpha = 0. \tag{3.10}$$

If P_i (i = 1,2,3) are the 3 diagonal matrices with the i'th element in the diagonal equal to one and the other elements equal to zero we have the completeness relation.

$$\sum_{i=1}^{3} P_i = I \tag{3.11}$$

and the orthogonality relations

$$P_i \cdot P_j = 0 \quad \text{for } i \neq j. \tag{3.12}$$

With a complete set of states thus defined and observables being represented by Hermitian matrices J that are elements of an exceptional algebra we can set up a 3-dimensional quantum mechanics as in the preceding section with each physical state being represented by a point in the octonionic Moufang plane. Günaydin, Piron and Ruegg(17) have recently shown that all postulates of axiomatic Q.M. are satisfied by this octonionic Q.M. The differences with complex Q. M. are the following: 1-The invariance group is F_4 instead of a unitary group. 2-The automorphism group SU(3) of the additional six imaginary units is a distinguished subgroup of F_4. 3-The Pappus Theorem does not hold in the Projective geometry of the Moufang plane. 4-The Desargues Theorem which holds in Q.Q.M. does not hold in the octonionic case. Hence we have a non Desarguesian Q.M.. If the SU(3) subgroup associated with octonions is identified with the color group, then color singlet projection operators (states) have both the Pappus and Desargues property as in ordinary Q. M. It follows that colored states have different superposition properties than color singlet states. This may be due to the presence of superselection rules associated with the U(1) group in Q.Q.M. and the SU(3) color group in octonionic Q.M. Examples of colored states with non Desarguesian properties can be found in the paper of Günaydin et al(17).

Two maximal subgroups of F_4 that contain the color group are of interest, namely SO(9) (or, more precisely its covering group spin 9) and SU(3) x $SU(3)^c$. The coset space F_4/SO(9) can be identified with the Moufang plane. Its dimension is 16, which corresponds to the spinor representation of SO(9). Under SO(8) the spinor splits into the two kinds of 8 dimensional spinor representations of SO(8), each one giving one octonionic component of a point in the Moufang plane. The remaining components of the 26-dimensional representation of F_4 transform like a singlet and a vector under SO(9). Under SU(3) x $SU(3)^c$ we have

$$26 = (8,1) + (3,3) + (\bar{3},\bar{3}). \tag{3.13}$$

By using the space of the Hermitian operators U_H defined by (2.9) it is possible to construct projection operators for all the 26 components of J. In this case (8,1) can be associated with color singlet lepton states, while (3,3) and $(\bar{3},\bar{3})$ represent respectively three color triplet quarks and their antiquarks.

Thus, the color SU(3) and the flavor SU(3) arise naturally in octonionic Q.M. Further, colored states are distinguished by their different geometric properties.

There are some problems that arise in connection with octonionic Q.M. If projection operators for states are assumed to belong to a Jordan algebra, then, according to Albert's theorem[18], the octonionic case only allows a three dimensional Hilbert space. The general projection operators are obtained from P_1, P_2, P_3 by applying to them the transformation (2.19) of the group F_4 that depends on two traceless Jordan matrices A and B. Since A and B have 26 parameters each, and since, because of non associativity, (2.19) cannot be rewritten as a double commutator, all components of A and B are independent and F_4 has 52 parameters. If P_α, P_β, P_γ are related to P_1, P_2, P_3 by the same F_4 transformation, then they remain idempotent and mutually orthogonal. Instead of the diagonal projection operators we can use these elements of the exceptional Jordan algebra to represent states. How do we construct projection operators for more states? Those must be associated with higher representations of F_4 that are elements of algebras larger than the Jordan algebra. In fact if we start from the triple product (2.8) instead of the binary Jordan product, then we can define elements like U_H that are elements of a more general algebra. In this way we can project out states that are in the direct product of two sets of P_i. An example of that will be given in the next section for the more realistic E_6, but the procedure is the same for F_4.

Another problem is the combination with the Poincaré group (or the conformal group) of space-time. The simplest way is to attach (as in the construction of a fibre bundle) an octonionic Hilbert space at each space-time point. Other possibilities will be discussed in the last section.

IV. Beyond Quantum Mechanics Associated with Projective Spaces.

Standard quantum mechanics, its quaternionic extension and its octonionic form corresponding to the Moufang plane are all based on the calculus of propositions which is represented by 4 Jordan algebras. However, colored states need not be observable in the usual sense and therefore need not be represented by Jordan matrices with entries from the four Hurwitz algebras. They may lie in a larger Hilbert space with new properties, the color singlet subspace being a Quantum Mechanical space consistent with the calculus of propositions as previously suggested[19]. It has also been argued[20] that colored "observables" need not even be power associative. In that case they would be represented by exotic algebras more general than Hurwitz algebras and Jordan algebras.

Projective spaces associated with Hilbert spaces are particular examples of Cartan's symmetric spaces or more generally homogeneous spaces labeled by representatives of the coset space K = G/H where G is a simple Lie group and H a maximal subgroup of G. The Moufang plane is the symmetric space $F_4/SO(9)$. The next octonionic symmetric space is $E_6/SO(10) \times SO(2)$ and is 78-45-1 = 32 dimensional. The coset representatives are points with two complex octonionic components. This space which is called the complex octonionic plane has been the object of numerous investigations[21]. It is not a projective geometry but includes the real Moufang plane that is a Non Desarguesian projective geometry as a subspace. It is the simplest generalization of octonionic Q. M.[13]. To construct states we start from the 27-dimensional representation of E_6. It is a complex 3 x 3 octonionic matrix of the form (3.5) with α, β, γ being complex numbers and a, b, c complex octonions. The bar denotes conjugation with respect to octonionic conjugation only. Such matrices are again closed under Jordan multiplication. The matrix J* where the star denotes complex conjugation is now distinct from J and corresponds to the 27* representation of E_6. Just like SU(3), the group E_6 has complex representations that are not equivalent to their complex conjugates. The Freudenthal product of two such matrices is E_6 covariant and is defined by

$$J_1 \times J_2 = J_2 \times J_1 = 1/2\,(J_1 + J_2)\times(J_1 + J_2) - 1/2\, J_1 \times J_1 - 1/2\, J_2 \times J_2 \tag{4.1}$$

corresponding to the composition 27 x 27 = 27*. One can form 4 invariants with J, namely

$$I_2 = 1/2\, \mathrm{Tr} J^*\cdot J, \quad I_3 + i I_3' = \mathrm{Det}\, J, \quad I_4 = 1/2\, \mathrm{Tr}(J\times J)\cdot(J^*\times J^*) \tag{4.2}$$

We now associate states with elements J that obey the E_6 invariant equation

$$S \times S = 0 \tag{4.3}$$

which is satisfied either by idempotent or nilpotent elements. Hence S, unlike the F_4 case is not a projection operator. Let us show that one can associate a projection operator with each S. As in the F_4 case we use the triple product (2.8). It can be shown that if A, B, C transform like (27), so does

$$F = \{A\ B^*\ C\} \tag{4.4}$$

We define operators U_{R^*} and V_{R,R^*} by

$$U_{R^*}J = \{R^*\ J\ R^*\} \quad , \qquad U_R J^* = \{R\ J^*\ R\} \quad , \tag{4.5}$$

$$V_{R,R^*}J = \{R\ R^*\ J\} \tag{4.6}$$

We have the identities

$$U_R J^* = R\ \mathrm{Tr}(J^*\cdot R) - 2(R\mathrm{x}R)\mathrm{x}J^* \ , \tag{4.7}$$

$$V^2_{R,R^*}J - 1/2\ V_{R,R^*}J = 1/2\ U_R U_{R^*}J \quad . \tag{4.8}$$

showing the covariance of these operators. Let R obey the conditions

$$R \times R = 0, \tag{4.9}$$

$$\mathrm{Tr}(R^*\cdot R) = 1 \quad . \tag{4.10}$$

Using (4.5) and (4.7) we obtain,

$$\Pi_R J = U_R U_{R^*} J = R\ \mathrm{Tr}(R^*\cdot J) \tag{4.11}$$

so that

$$\Pi^2_R\ J = R\ \mathrm{Tr}(R^*\cdot J)\mathrm{Tr}(R^*\cdot R) = \Pi_R J \quad , \tag{4.12}$$

and applied on an arbitrary J, we have

$$\Pi^2_R = \Pi_R \quad , \tag{4.13}$$

provided R satisfies (4.9) and (4.10). Note that we can also write

$$\Pi_R = 2V^2_{R,R^*} - V_{R,R^*} \quad . \tag{4.14}$$

We have succeeded in associating a projection operator Π_R with each element R of the complex Jordan algebra that satisfies (4.9) and is normalized according to (4.10). Such elements can be taken to represent states in the E_6 invariant Q.M., or points in the complex octonionic plane. They are covariant under the SO(10)xSO(2) subgroup of E_6 that acts as a stability group for states. Now, because of the identity

$$\mathrm{Det}\ J = 1/3\ \mathrm{Tr}[J\cdot(J\mathrm{x}J)], \tag{4.15}$$

we have Det R = 0. Another identity is

$$\mathrm{Det}\{K\ J^*\ K\} = (\mathrm{Det}\ K)^2\ \mathrm{Det}\ J^*. \tag{4.16}$$

It follows that

$$\mathrm{Det}(U_{R^*}J) = 0 \quad , \quad (U_{R^*}J) \times (U_{R^*}J) = 0 \quad , \tag{4.17}$$

and $U_{R^*}J$ is also a state.

It remains to define an invariant distance d(R,S) between two points R and S, or correspondingly, in analogy with (2.7), an E_6 invariant transition probability between states R and S. We write

$$(R,S) = \cos^2 d(R,S) = 1/2\ \mathrm{Tr}(R\cdot S^* + S\cdot R^*). \tag{4.18}$$

The states R, S will be orthogonal if (R,S) = 0. State S_3 is the superposition of S_1 and S_2 if

$$\mathrm{Tr}(S_3\cdot S_1 \times S_2) = 0 \tag{4.19}$$

Further, if (R,S) = 0, then the corresponding projection operators obey

$$\Pi_R \Pi_S J = 0 \tag{4.20}$$

for arbitrary J. H is an observable if the corresponding operator U_H is Hermitian so that

$$(R,\ U_H\ S) = (U_H\ R,\ S). \tag{4.21}$$

Now we have all the ingredients for setting up a quantum mechanical scheme based on the (non-projective) geometry of the octonionic planes. The invariance group of this exotic Q.M. space is E_6. The properties of this space are under investigation. Note that the projection operators Π_R are imbedded in an octonionic triple algebra defined by (4.4). The projection operators (e.g. those for SU(3)) in standard Q.M. are imbedded in a complex triple algebra. The triple algebra of Jacobson has been generalized by Kantor[22]. In the way that the Jordan algebra can be regarded as an algebra of observables, the Kantor algebra, which mathematically is an algebra of coset spaces, can be interpreted as the algebra of states. Thus, it is an ideal tool for generalized octonionic states associated with color. As shown recently by Bars and Günaydin[23], a Grassmann extension of the Kantor algebra can also be used to construct and classify superalgebras, the latter acting as the derivation algebra of the algebra of bosonic and fermionic states. This allows an algebraic approach to extended supergravity.

V. Is E_6 a Grand Unification Group or is the Standard Theory Embedded in E_6?

If the symmetries of the complex octonionic plane can be interpreted in terms of the color and flavor degrees of freedom of fundamental quarks and leptons, E_6 would be a good candidate for the unification of weak, electromagnetic and strong interactions through a simple gauge group. The advantage of this approach is to make the charge space essentially finite since E_6, unlike SU(n) cannot be extended to higher groups with a similar structure. In physical terms, this means that we cannot go on adding quarks and leptons at will. The only higher groups contained in the exceptional series are E_7 and E_8 which are groups with distinctly different properties from E_6. There are unique geometries associated with each of these groups. In the preceding section we sketched the properties of the complex octonionic plane. The associated symmetry is a good candidate for the classification of physical states. The physical labeling of quark and lepton states within E_6 which has been going on for a number of years[24,25] has now been pinned down to a unique assignment[13,26] that is compatible with present experiments. If this assignment is disproved by new data, charge space has to be described in terms of a different algebraic structure, possibly associated with supersymmetry. The disadvantage of the approach stems from the difficulty of embedding exceptional groups other than G_2 within a simple supergroup, as discussed in the introduction. The imbedding in E_6 within new octonionic triple algebras of the Kantor type might still be possible, since the automorphism superalgebras of simple Kantor algebras need not be simple.

To spell out the E_6 assignment of quark and lepton states R projected out by 27 mutually orthogonal projection operators Π_R introduced in the previous section, we identify the color group $SU(3)^c$ as the group leaving the octonionic unit invariant in the Jordan matrix J. Then, under the maximal subgroup $SU(3) \times SU(3) \times SU(3)^c$, J decomposes as

$$(27) = (\bar{3}, 3, 1) + (3, 1, 3) + (1, \bar{3}, \bar{3}). \tag{5.1}$$

If J, under the Lorentz group behaves like a left handed spinor, then this corresponds to 9 two-component leptons (color singlets), 3 left handed quarks (color triplets) and 3 antiquarks (color anti triplets). Using the notations

$$\psi_L = 1/2\ (1 + \gamma_5)\psi, \quad \hat{\psi}_R = 1/2\ (1 + \gamma_5)\gamma^c \tag{5.2}$$

we have the following lepton and quark assignments

$$L^{(e)} = \begin{pmatrix} \hat{N}^\tau_R & \hat{\tau}_R & \hat{e}_R \\ \tau^-_L & \nu^\tau_L & \hat{\alpha}^e_R \\ e^-_L & \nu^e_L & \beta^e_L \end{pmatrix}, \qquad q^i_L = \begin{pmatrix} u^i_L \\ d^i_L \\ b^i_L \end{pmatrix}, \quad \hat{q}^i_R = \begin{pmatrix} \hat{u}^i_R \\ \hat{d}^i_R \\ \hat{b}^i_R \end{pmatrix}, \tag{5.3}$$

Here i is the color index for quarks (i = 1, 2, 3), the Weinberg-Salam weak SU(2) group is the subgroup of one of the color singlet SU(3) groups of E_6, so that the 6 weak doublets are

$$\begin{pmatrix} \nu^e_L \\ e^-_L \end{pmatrix}, \quad \begin{pmatrix} \nu^\tau_L \\ \tau^-_L \end{pmatrix}, \quad \begin{pmatrix} \hat{\tau}_R \\ \hat{N}^\tau_R \end{pmatrix}, \quad \begin{pmatrix} u^i_L \\ d'^i_L \end{pmatrix} \tag{5.4}$$

the remaining states being weak singlets. Here the d'^i_L are Cabibbo mixtures of the down quark d with the strange quark s. We have omitted possible lepton mixtures. The τ is the heavy lepton of Perl et al[27] and ν^τ its neutrino. If the neutral lepton N^τ is heavy, then the above structure coincides with the universal V-A structure of the Weinberg-Salam Unified theory of weak and electromagnetic interactions[28]. b is the heavy bottom quark with charge -1/3 and is thought to be the constituent of the charmonium-like $\bar{b}b$ upsilon states discovered by Ledermann et al[29].

Another (27) representation of E_6 is needed to accomodate the muon family of leptons and the charm family of quarks. The corresponding new states are obtained from the old one by

$$e^- \to \mu^-, \ \tau^- \to M^-, \ N^e \to N^\mu, \ \nu^e \to \nu^\mu, \ \nu^\tau \to \nu^M, \ \alpha^e \to \alpha^\mu, \ \beta^e \to \beta^\mu, \tag{5.5}$$

$$u \to c, \ d \to s, \ b \to h \tag{5.6}$$

where M is a new charged heavy lepton, c is the charmed quark of charge 2/3, constituent of the charmonium and charmed mesons[30], s is the strange quark and h a new heavy quark, both of charge -1/3. Thus the scheme requires 6 quarks, the sixth (h) having charge -1/3 instead of the 6th top quark with charge 2/3 of the standard theory. Heavy neutral leptons N^e, N^μ as well as a new charged heavy lepton M^- are necessary to the scheme.

The unrenormalized Weinberg angle parameter is 3/8. As discussed in the introduction, the theory, like other grand unified theories, requires a superheavy mass scale M which causes a renormalization of the electromagnetic fine structure constant α to $\alpha_s \sim 0.3$ and leads to a renormalized Weinberg angle parameter [13,26] of $\sin^2\theta_W \sim 0.2$. Both these values and the above assignment of weak doublets are consistent with present weak interaction phenomenology including neutrino experiments, τ decays and parity violation in polarized electron scattering on nucleons.

The main theoretical problem, besides color confinement is the calculation of the lepton, quark and gauge boson mass spectrum from an effective Higgs mechanism (that may be dynamical in origin) in the E_6 gauge theory. Preliminary attempts [13,26] show that this may be achieved by using two Higgs fields belonging to the adjoint 78 dimensional representation and the 351 dimensional representation contained in the symmetrical part of the direct product $27 \otimes 27$. The latter contains another 27* that could also be used as an additional Higgs field. There are color singlet vacuum expectation values in (78) belonging to the (1,8) + (8,1) representations of SU(3) x SU(3) that give superheavy masses to the gauge bosons which destabilize the proton. Some color singlet gauge bosons also become superheavy so that E_6 is broken down to SU(2) x U(1) x SU(2) x U(1) x $SU(3)^c$. In this stage the fermions are still massless. The vacuum expectation values of 351 are color singlets that belong to the $(6,\bar{6})$ and $(\bar{3},3)$ representations of the color singlet subgroup SU(3) x SU(3) of E_6. A suitable subset of these splits fermion masses, making b, h, τ, M, N^e and N^μ heavy. They also give masses to weak gauge bosons $W^\pm$ and Z. Thus, a qualitatively acceptable mass spectrum is obtained in the zero approximation. It is not yet known whether these vacuum expectation values minimize the Higgs Lagrangian and whether the neutrinos remain naturally massless. The other interesting theoretical problems are: CP violation, mixing angles including the Cabibbo angle and mixing of d and s with b and h, and the effect of instanton solutions of E_6.

If E_6 turns out to be incompatible with future data, we may consider the alternative possibility of embedding the phenomenological SU(2) x U(1) x $SU(3)^c$ group in the exceptional (octonionic) supergroup structure F(4) instead of the exceptional Lie group E_6. It is also possible that, starting from a complex octonionic quantum mechanics based on E_6, and treating its stability group SO(10) x SO(2) as a local gauge group that generalizes the phase group, only the points in octonionic plane that have doublet structure (with two components that are complex octonions) would lead to acceptable states in Hilbert space. Then, those doublets would be identified with the combined lepton and quark doublets of the Weinberg-Salam theory. In this case we would recover the usual quark symmetry, leptons and quarks combining to form complex octonions. Three octonionic pairs could be associated with $(\nu^e, e; u,d)_L$, $(\nu^\mu, \mu; c,s)_L$, $(\nu^\tau, \tau; t,b)_L$ requiring the sixth quark to be the top quark t with charge 2/3.

At this stage it is difficult to foresee whether the octonionic structure in High Energy Physics will correspond to the 27 states of the small E_6 representation or to the coset E_6/SO(10) x SO(2) represented by octonionic doublets. The nature of the sixth quark will help us distinguish between an octonionic space based on projection operators for the sixteen states of the complex octonionic plane forming a spinor representation of the stability group SO(10) or the space spanned by the 27 projection operators constructed out of the triple algebra of complex Jordan matrices.

VI. Exceptional Superspaces and Annihilation Operators

In this final section we consider possible octonionic structures that may arise when we go beyond grand unification to superunification that brings together supergravity, weak, electromagnetic and strong interactions. We know that the algebraic structure needed for such a unification is a superalgebra (or, through exponentiation a supergroup). Supergroups act in superspaces which are vector spaces with some elements being complex numbers and some elements anticommuting Grassmann numbers. The supergroup leaves a quadratic form of these superspace vectors (supervectors) invariant. Now the usual Grassmann numbers are associative. As shown by Jordan and Wigner, they can be constructed out of the direct product of quaternions. To this end we use the split quaternion units

$$j = 1/2\ (j_1 + ij_2), \qquad j_0 = 1/2\ (1 + ij_3) \tag{6.1}$$

with the j_i defined by (3.2). Let us take commuting copies $j^{(r)}$, $j_3^{(r)}$ of the quaternionic units. Then the annihilation operators $a^{(r)}$ defined below form a Grassmann algebra

$$a^{(1)} = j^{(1)},\ a^{(2)} = j_3^{(1)} j^{(2)},\quad a^{(3)} = j_3^{(1)} j_3^{(2)} j^{(3)},\ \text{etc.} \tag{6.2}$$

$$[a^{(r)}\ a^{(s)}]_+ = 0. \tag{6.3}$$

Furthermore $a^{(r)}$, $a^{(s)}$, $a^{(t)}$ satisfy the Jacobi identity because of associativity.

Now we turn to the construction of exceptional annihilation operators (Grassmann) numbers by means of direct products of octonions[13]. Let

$$u_n = 1/2\ (e_n + ie_{n+3})\ ,\quad u_0 = 1/2\ (1 + ie_7)\ ,\quad (n = 1,\ 2,\ 3) \tag{6.4}$$

stand for octonionic split units with e_a defined by (3.3). Let $u_n^{(r)}$, $e_7^{(r)}$ denote copies of octonionic units. Then

$$\alpha_n^{(1)} = u_n^{(1)}\ ,\quad \alpha_n^{(2)} = e_7^{(1)} u_n^{(2)},\quad \alpha_n^{(3)} = e_7^{(1)} e_7^{(2)} u_n^{(3)},\ \text{etc.} \tag{6.5}$$

form a set of anticommuting annihilation operators (exceptional Grassmann algebra) that annihilate the vacuum

$$|0\rangle = u_0^{(1)} u_0^{(2)} u_0^{(3)} \ldots \tag{6.6}$$

but do not satisfy the Jacobi identity.

The Jordan Wigner annihilation operators have a built-in U(1) symmetry that changes all $j^{(r)}$ by the same phase. Similarly, the exceptional Grassmann algebra has an intrinsic $SU(3)^c$ symmetry that acts on $u_n^{(r)}$ and is the automorphism group of the 6 units $u_n^{(r)}$ and $u_n^{*(r)}$.

We can also construct hybrid Grassmann algebra by combining the quaternionic and the octonionic annihilation operators. The set consists of the $a^{(r)}$ and $\alpha_n^{(s)} j_3^{(1)} j_3^{(2)} \ldots$, with the new vacuum

$$|0\rangle = j_0^{(1)}\ j_0^{(2)} \ldots u_0^{(1)}\ u_0^{(2)}\ \ldots \tag{6.7}$$

The hybrid algebra admits an intrinsic U(1) x SU(3) group.

These Grassmann algebras can be used in two ways, either for building a local fermionic field by associating annihilation operators (and the complex conjugate creation operators for the opposite frequency) with each plane wave solution of a free relativistic wave equation, or for constructing vectors in superspace as kets of supergroups. The octonionic local fields will then have an intrinsic SU(3) symmetry that we can identify with the color group and the supergroups will admit $SU(3)^c$ as their subgroup. Note that the exceptional local fermionic fields will not be power associative except for the color singlet subspace of the associated Fock space. We have already seen that quark fields or the states they create by their actions on the vacuum need not be power associative. But the (qqq) and ($q\bar{q}$) subspaces of the Fock space would be power associative.

This approach seems to offer new possibilities for the construction of local color fields and for the introduction of new superspaces with color symmetry.

Acknowledgements I would like to express my thanks to many colleagues and friends who helped me formulate some of the ideas presented in this talk. Among them are M. Gell-Mann, M. Günaydin, I. Bars, H. Ruegg, C. Saçlioğlu, M. Serdaroğlu, G. Domokos,

S. Kövesi-Domokos, L. Michel, P. Ramond and P. Sikivie. The hospitality of the Aspen Institute where part of this work was done is also gratefully acknowledged.

References

1. P. Jordan, J. von Neumann and E. P. Wigner, Ann. Math. 35, 29 (1934).
2. C. Chevalley and R. D. Schafer, Proc. Natl. Acad. Sci. U. S. , 36, 137 (1950).
3. For a review see H. Freudenthal, Advances in Math. I., 145 (1965).
4. R. Moufang, Abh. Math. Sem. Univ. Hamburg 9, 207 (1933).
5. J. Pati and A. Salam, Phys. Rev. D8, 1240 (1973; H.Georgi and S. L. Glashow, Phys. Rev. Lett. 32, 433 (1974).
6. H. Georgi, H. Quinn and S. Weinberg, Phys. Rev. Lett. 33, 451 (1974); D. I. D'Jakonov, Leningrad Nucl. Phys. Institute preprint (1977).
7. M. Yoshimura, Phys. Rev. Lett. 41, 381 (1978); S. Dimopoulos, L. Susskind, SLAC-PUB-2126 (1978).
8. D. Toussaint, S. B. Treiman, F. Wilczek and A. Zee, "Matter-Antimatter Accounting, Thermodynamics and Black Hole Radiation," Princeton preprint (1978); D. Toussaint, F. Wilczek, "Elementary examples of baryon generation," Princeton preprint (1978).
9. D. Z. Freedman, P. van Nieuwenhuizen and S. Ferrara, Phys. Rev. D13, 3124 (1976); S. Deser and B. Zumino, Phys. Rev. Lett. 62B, 335 (1976).
10. M. Gell-Mann, P. Ramond and J. H. Schwarz, Phys. Lett. 76B, 417 (1978).
11. L. Brink, M. Gell-Mann, P. Ramond and J. Schwartz, Caltech preprint 68-644 (1978).
12. S. MacDowell, Yale preprint COO-3075-201 (1978).
13. F. Gürsey, in Second Workshop on Current Problems in High Energy Particle Theory, ed. G. Domokos and S. Kövesi-Domokos, p. 3 (Johns Hopkins U., Baltimore, Md. 1978).
14. F. Gürsey, "Non Associative Algebras in Quantum Mechanics and Particle Physics," in Proc. of the 1977 Charlottesville Conference on Non Associative Algebras.
15. For the hyperbolic case see G. D. Mostow, Strong Rigidity of Locally Symmetric Spaces, section 19, (Princeton, 1973). See also Ref. 14.
16. N. Jacobson, Structure and Representations of Jordan Algebras, p. 36 (Am. Math. Soc. 1968).
17. M. Günaydin, C. Piron and H. Ruegg, "Moufang Plane and Octonionic Quantum Mechanics," U. of Genève preprint (1977).
18. A. A. Albert, Ann. Math. 35, 65 (1934).
19. M. Günaydin and F. Gürsey, Phys. Rev. D9, 3387 (1974); F. Gürsey in "Johns Hopkins Workshop on Current Problems in High Energy Particle Theory" p. 15 (Johns Hopkins U., 1974).
20. M. Gell-Mann, private communication.
21. R. J. Faulkner, Memoirs of the Am. Math. Soc. No. 104 (Providence, R. I., 1970) nad references therein.
22. I. L. Kantor, Dokl. Akad. Nauk SSSR, 208, No.6 (1973), Translation Soviet Math. Dokl. 14, No.1 (1973).
23. I. Bars and M. Günaydin, Harvard preprint (1978).
24. F. Gürsey, in "Int. Symp. on Math. Problems in Th. Phys." ed. H. Araki, p. 189 (Springer, 1975).
25. F. Gürsey, P. Ramond and P. Sikivie, Phys. Lett. 60B, 177 (1976); F. Gürsey and M. Serdaroğlu, Lett. al Nuovo Cimento, 21, 28 (1978).
26. Y. Achiman and B. Stech, Physics Letters 77B, 389 (1978).
27. For a recent review, see M. L. Perl in Proc. Int. Symp. on Lepton and Photon Int. at High En., Hamburg 1977, ed. F. Gutbrod (Desy, 1977).
28. S. Weinberg, Phys. Rev. Lett. 19, 1264 (1967); A. Salam in Nobel Symp. No.8, ed. N. Svartholm (Wiley, N. Y. 1969).
29. S. W. Herb et al. Phys. Rev. Lett. 39, 252 (1977).
30. For a review see T. Appelquist, R. M. Barnett and K. Lane "Charm and Beyond" to be published in Ann. Rev. Nucl. and Particle Sci., vol. 28.

Multilocal Field Theory

Chinn-Chann Chiang
Institute of Physics
National Taiwan Normal University
Taipei, Taiwan, Republic of China

It had long been thought that the difficulty of divergences in quantum field theory was due to our over simplification of considering elementary particles as point particles. In an attempt to solve both the difficulty of divergences and the problem of intrinsic structure of elementary particles, Yukawa proposed a non-local theory in which an elementary particle was treated as an extended object described by multilocal field. This approach involves two main problems, namely to have a correct mass spectrum representing the observed particles and to construct a unitary S-matrix to deal with interaction problems. Furthermore, it has become clear that whether a theory with multi-local fields is finite or not largely depends on how the multi-local fields are quantized.

In recent years, it has been generally believed that a hadron is made up of quarks and has structure as an extended object. Due to the fact that free quarks have not been observed, many attempts have been made to confine quarks inside the hadron. However, if we disregard the dynamical origin of quark confinement, we may consider an elementary particle as an extended object and describe it by multi-local fields. This comes back to the original idea of Yukawa.

To incorporate with the idea of quark model, we use bilocal and trilocal field respectively to represent bosons and baryons. The field equations satisfied by these multi-local fields are chosen to be covariant and yield bound state solutions with a Chew-Frantschi plot. Furthermore, the "constituents" are supposed to be permanently confined. These requirements can be fulfilled if we have only the center of momentum of the constituents propagates and the relative coordinates bound via a simple harmonic-oscillator potential. In terms of the commutation relations of the fields we have also constructed various propagators.

To further develop the formalism we consider a "ϕ^3" vertex and construct a formally unitary S-matrix by making use of the method similar to that of Yang and Feldman, and Källen in local theory. The explicit form for the S-matrix to second order is obtained and the unitarity of the S-matrix to this order is also explicitly checked.

We further compute the two-body elastic scattering amplitude and the self-energy. The scattering amplitudes show Regge behavior and damp in a Gaussian manner in both the s and t channels. The second order self-energy is finite.

References:

(1) A. Z. Capri and C. C. Chiang, Prog. Theor. Phys. 59, 996 (1978).

(2) A. Z. Capri and C. C. Chiang, Prog. Theor. Phys. 59, 1376 (1978).

MULTIQUARK STATES

H. Høgäasen and P. Sorba

Recently, several narrow resonances above and even below threshold have been discovered in $p\bar{p}$ formation and production experiments; they have been called baryonium states. Moreover, encouraging results have been obtained in the search for dibaryonic resonances in Λp and pp systems, and also in $\Delta\Delta$ systems. The description of such resonances as $2q,2\bar{q}$ states has been proposed for baryonium states, while properties of 6q states (dibaryons) as well as $5q,1\bar{q}$ (mesobaryonium) have also been considered. Spectrum of such states as well as qualitative properties about their decays are studied in the quark gluon model, the mass splitting between states built with quarks of the same flavour being assumed to be due to colour gluon exchange between the quarks. In this brief note we will summarize some results by considering separately s-wave n-quark states and orbital momentum $L \neq 0$ states.

As far as an n-quark state is in an s-wave, calculations can be done in the framework of the MIT bag model where the quarks are assumed to be enclosed inside a spherical cavity. For $2q,2\bar{q}$ states, Jaffe and Johnson (Phys. Letters 60B (1976) 201) remarked that to the usual quark model 0^{+} nonet ($q\bar{q}$, L = 1) should be added a "crypto-exotic" nonet 0^{++} which lies lower in mass. In particular, possible identifications of these states with $\varepsilon(700)$, $S^{*}(993)$ and $\delta(976)$ can then be made. For s-wave $4q,1\bar{q}$, detailed calculations have also been done (H. Høgäasen and P. Sorba, Nuclear Physics, to be published) allowing, in particular, some predictions on $K^{-}n$ backward scattering from the study of Regge trajectory for exotic exchange. In the case of 6q-states, Jaffe (Phys. Rev. Letters 38 (1977) 195) predicted the existence of a stable dihyperon of the form (2u,2d,2s) at a mass around 2150 MeV. Generalizing the approach of Jaffe for a number k of flavours bigger than 3, it can be shown (H. Høgäasen and P. Sorba, CERN preprint TH 2531) that many more stable dibaryons can then be predicted, these states being classified in the representations  ,  and  of the SU(k) flavour group. In the case k = 4, 25 charmed states appear then to be stable. Now considering the relative mixtures of baryon-baryon contained in such dibaryonic states, it appears that a non-negligible part comes from two colour octet baryons. In the case of the deuteron, experimental results on the deuteron form factor at large q^2 suggested the possibility of tunnelling transition of a real deuteron into a 6q state with a probability of about 7%: an estimation of the net composition of a real deuteron would then be 93.8% of p-n component, 0.6% of $\Delta\Delta$ and 5.6% of "hidden colour" component (V.A. Matveev and P. Sorba, Lett. al Nuovo Cimento 20 (1977) 435; Nuovo Cimento 45A (1978) 257).

As for high spin multiquark states, it appears that colour singlet states of large stability may be found in configurations consisting of two subsystems with k and n-k quarks (antiquarks) respectively. Inside each cluster the quarks are in a relative s-wave, but an angular momentum between the two clusters creates a centrifugal barrier which prevents the quarks from recombining into unbound colour singlets. In the case of $2q,2\bar{q}$, two different configurations of states $2q\text{-}2\bar{q}$ corresponding to colour cluster $\bar{3}$-3 and $\bar{6}$-6 are particularly interesting.

Many resemblances with the experimental situation of baryonium states above threshold are found in the spectrum deriving from this model (H.M. Chan and H. Høgäasen, Phys. Letters 72B (1977) 121; Nuclear Phys. B136 (1978) 401; R.L. Jaffe, Phys. Rev. D7 (1978) 1444). When $4q,1\bar{q}$ states are considered, the configurations which are the most likely to be rather stable are $qqq\text{-}q\bar{q}$ where both clusters form a colour octet and $qq\text{-}qq\bar{q}$ in the colour 6-$\bar{6}$ representations. Some candidates for 5q states already exist. Very recently a narrow enhancement in the Σ^-K^+K mass spectrum at a mass of 2.58 GeV has been observed (M. Mazzucato et al., CERN/EP/PHYS 78-24): this resonance involving three strange quarks was predicted in our model at a mass around 2.61 GeV/c^2 (M. de Combrugghe, H. Høgäasen and P. Sorba, CERN preprint TH.2537). We emphasize that the existence of such states could be considered as a direct signature of the colour degree of freedom.

Let us conclude this note by mentioning that group theoretical techniques are essential tools in this multiquark states approach. A condition for progress is now the computation of more complete tables of SU(6) Clebsch-Gordan and fractional percentage coefficients.

ON THE ERIKSENLIKE FORM OF THE MELOSH TRANSFORMATION AND ITS GROUP THEORETICAL INTERPRETATION

M. JASPERS and J. BECKERS,
Physique Théorique et Mathématique, Institut de Physique,
Université de Liège, Liège, Belgique.

The Melosh transformation[1] has been interpreted in the group theoretical context by different authors[2]. Its similarity with the Foldy-Wouthuysen (F.W.) transformation[3] is very well known. At the level of ordinary relativistic quantum mechanics, the Melosh proposal reads :

$$U_M = \exp\,(iS_M), \quad S_M = \frac{1}{2}\,tg^{-1}\,\frac{\vec{\gamma}_\perp \cdot \vec{p}_\perp}{m}, \quad \vec{\gamma}_\perp \equiv (\gamma_1,\gamma_2). \tag{1}$$

Then, applied to the free Dirac Hamiltonian, it gives :

$$H_D' = U_M H_D U_M^{-1} = \alpha_3 p_3 + \kappa\beta, \quad \kappa = (m^2 + \vec{p}_\perp^{\,2})^{1/2}. \tag{2}$$

It eliminates the α_1 and α_2 terms as the F.W. transformation does for <u>all</u> the odd terms.

Taking into account of the Melosh and F.W. similarities <u>on the one hand</u> and of the Eriksen[4] - F.W. connection <u>on the other hand</u>, we propose here an <u>Eriksenlike form</u> of the Melosh transformation. Let us consider the Eriksen problem in connection with (1) and (2) in the <u>free case</u>. Simple considerations lead to point out the specific rôle of the matrix Σ_3 (in correspondence with the matrix β in the F.W. context) and to the determination of λ :

$$\lambda = \Sigma_3(m + i\vec{\gamma}_\perp \cdot \vec{p}_\perp)\kappa^{-1/2}. \tag{3}$$

The Eriksenlike form of the Melosh transformation writes :

$$U_M = \frac{1}{2}(1 + \Sigma_3\lambda)[1 + \frac{1}{4}(\Sigma_3\lambda + \lambda\Sigma_3 - 2)]^{-1/2}. \tag{4}$$

In the <u>interacting case</u>, if we introduce the usual correspondence:

$$\vec{p}_\perp \rightarrow \vec{\pi} \equiv \vec{p}_\perp - e\vec{A} \tag{5}$$

in Eqs.(3) and (4), we get the Eriksenlike form of the Melosh transformation associated with a spin $\frac{1}{2}$ -particle with anomalous magnetic moment in a homogeneous magnetic field. This new form can also be written in exponential form. So, we simply recover the Tsai transformation[5]. Finally, let us mention that the discussion[6] of such transformations in connection[7] with the SO(4,1)-de Sitter group is under study.

REFERENCES

(1) H.J. MELOSH, Quarks : Currents and Constituents (EFI 73/26, unpublished);Phys.Rev. D9,1095 (1974).

(2) F. BUCCELLA, C.A. SAVOY and P. SORBA, Lett.Nuovo Cim.25A,331(1974); N. MARINESCU and M. KUGLER, Phys.Rev. D12,3315(1975); V. ALDAYA and J.A. DE AZCARRAGA, Phys.Rev. D14,1049(1976).

(3) L.L. FOLDY and S.A. WOUTHUYSEN, Phys.Rev. 78,29(1950).

(4) E. ERIKSEN, Phys.Rev. 111;1011(1958).
(5) W.Y. TSAI, Phys.Rev. D7,1945(1973).
(6) J. BECKERS and M. JASPERS, to be published (1979).
(7) A.J. BRACKEN and H.A. COHEN, J.Math.Phys. 10,2024(1969); Progr. Theor.Phys. 41,816(1969).

WORKING WITH H.A. KRAMERS IN SU(4)

P. JASSELETTE,
Physique Théorique et Mathématique, Institut de Physique,
Université de Liège, Liège, Belgique

In the thirties H.A. Kramers[1] initiated a method of calculation that easily gives results of physical interest when the system is SU(2) or rotation invariant. This method is based on the notion of fundamental invariants and on expansions in powers of components of elementary spinors.

This method was shown[2] applicable in SU(3) provided the calculations be made in reducible representations and projected back on irreducible ones afterwards.

The purpose of the present communication is to show that the method is also applicable to SU(4), to display the adequate reducible representations and to define the required projectors.

As an example of application, the multiplicities of irreducible representations of the type (n,0,m) containing n quarks and m antiquarks appearing in direct products of similar representations are obtained in closed form.

Whole material including numerical values for the matrix elements of projectors will be published elsewhere.

REFERENCES :

(1) H.A. KRAMERS, Proc.Roy.Soc.Amsterdam 34,956 (1931); 33,953 (1930); M.C. BRINKMAN, " Application of Spinor Invariants in Atomic Physics ",North Holland (1956).

(2) P. JASSELETTE, Thesis Université de Liège (1967); Nucl.Phys. B1, 521 (1967); Nucl.Phys. B1,529 (1967); Bull.Soc.Roy.Sc.Liège 36, 654 (1967).

DYNAMICAL GROUPS AND THE QUARKONIUM PROBLEM*

C.S. Kalman
CONCORDIA UNIVERSITY-UNIVERSITE DU QUEBEC
ELEMENTARY PARTICLE PHYSICS GROUP
Montreal, P.Q. Canada H3G 1M8

Suppose that we observe a spectrum. Let us say that the particle states before and after emission of the field quanta correspond to the eigenfunctions of an operator H_0^{part}. Similarly the states of the quantized field after emission correspond to eigenfunctions of an operator H_0^{field}. We assume that it is meaningful to write $H = H_0^{part} + H_0^{field} + H'$ (1). We examine the observed spectrum for symmetries and interpret the symmetries as due to the conservation of some quantities in the physical model. For example, in the hydrogen atom the symmetries are due to the conservation of the angular momentum and Runge-Lenz vectors. To avoid contradiction we impose the condition that each operator corresponding to a conserved quantity commutes with H_0^{part}. We hypothesize that the closure of this set of operators under the addition and commutation is the Lie Algebra of some Lie group which we refer to as the Maximum Symmetry Group. Similarly the closure under addition and commutation of the generators of the maximum symmetry group with H' is the Lie algebra of the dynamical group. Such a dynamical group does not commute with H since the generators of the symmetry group do not cummute with H' and the additional generators of the dynamical group do not commute with H_0^{part}. The generators of the maximum symmetry group correspond to physical observables (the conserved quantities). The remaining generators of the dynamical group correspond to the interactions of the particles and can be used to calculate cross sections and masses of particles.

In the Kuriyan-Sudarshan strong coupling model[1]. $H' = \sum_{i<j} a_{ij}M_{ij}$ where the a_{ij} are creation operators for mesons and the M_{ij} are meson coupling metrices. If there are n possible types of quarks in the quarkonium problem under consideration, then SU(n) is the maximum symmetry group. H_0^{part}, H_0^{field}, and a_{ij} are SU(n) scalars and M_{ij} is an element of an n^2-1 operator representation of SU(n). M_{ij} can also be represented as a product of n^{plet} and anti n^{plet} SU(n) representations. Let q_k and $\bar{q}_k$, k=1, 2,...,n be elements of such representations. Then if A_{ij}; i,j= 1,2, ...,n are the generators of SU(n),

$$[A_{ij}, A_{km}] = \delta_{im} A_{kj} - \delta_{jk}A_{im} \quad \sum_{i=1}^{n} A_{ii}=0 \qquad i, j, k, m = 1,2,\ldots,n \qquad (2)$$

$$[A_{ij}, q_k] = \delta_{ik}q_j \quad [A_{ij}, \bar{q}_k] = \delta_{ij} \bar{q}_i \qquad i, j, k = 1,2,\ldots,n \qquad (3)$$

The algebra is closed if $[q_i, \bar{q}_j] = \theta(\delta_{ij}A_{n+1,\, n+1} - A_{ji})$ i, j= 1,2,....,n (4)

where θ= +1 corresponds to the Lie algebra of SU(n+1); θ= -1 corresponds to that of SU(1, n); θ= 0 corresponds to that of $T_K \otimes$ SU(n), k= $(n+1)^2 - n^2$ and $A_{n+1,\, n+1}$ is a "diagonal" generator in addition to A_{ii}; i- 1,2,...,n needed to complete the first two of the above Lie algebras. We do not consider the group $T_k \otimes$SU(4) because by its use one cannot predict transitions between elementary particles.

If $|m\rangle$ represents a hadron state, classified according to SU(1,n) or SU(n) then in first order perturbation theory, its mass M(m) is given by

$$M(m) = C_0 + \langle m| \sum_{i,j=1}^{n} a_{ij} M_{ij} \; |m\rangle = C_0 + \sum_{i=1}^{n} C_i \langle m| \; q_i \bar{q}_i \; |m\rangle \tag{5}$$

where C_i, i= 1,2...., n are all constants. The first two terms of the sum in eq. 5 produces isoplet splitting and a renormalization of the constant C_0. Succeeding terms reproduce effects due to strangeness, charm, etc. Gavrilik and Shirokov[2] have proposed a modification of eq. 5. Their suggestion amounts to setting $M_{ij} = q_i\bar{q}_j + q_j\bar{q}_i$ and thus $M'(m) = d_0 + \sum_{i=1}^{n} (d_i^{(1)} \langle m|q_i\bar{q}_i|m\rangle + d_i^{(2)} \langle m|\bar{q}_i q_i|m\rangle)$ (5')

Some Examples

1. SU(1,1); ψ and Υ. The maximum symmetry group is U(1), which labels all the states by the single internal quantum number k. Before the breaking of the symmetry, all states corresponds to the same mass. Suppose k is taken to correspond to the radial excitation quantum number n of a quarkonium system. The representations of SU(1,1) are characterized by an integer K>0. There are two discrete series of representations determined by either $\infty > k > K+1$ or $-1 > k > -\infty$. For our purposes it suffices to consider the lower-bound series. Eq. 5 then takes the form MASS= $C_0 - Ck\,(K-k+1)$. Kalman[3] used ρ meson data to argue that only 3 or 4 ψ particles would be found. Assuming, that C_0 takes on different values for different quarkonium families, but that the breaking of the U(1) quark symmetry by SU(1, 1) is independent of quark mass, the Υ spectrum was obtained[3].

2. SU(4):The maximum symmetry group is SU(3). Three towers of representations occur.[4] Representations in the first tower are suitable for a description of baryons. Each SU(4) representation in the other towers consist of n-2 products of $q\bar{q}$. Various mass formulae for 27-plet mesons and baryons have been given by Kalman and Hongoh and Kalman. Such formulae can be applied to the baryonium problem for baryoniums of the same J^P. Kalman and collaborators have discussed predictions of charmed particle masses, selection of representation,electromagnetic mass differences etc. For reference see "Selection of a Dynamical Group for the Charmed Baryons"[5], which also discusses the problem of choosing between SU(n+1) and SU(1,n) as a dynamical group.

*Research supported in part by the NSERC (Canada).

1. J. Kuriyan and E.C.G. Sudarshan, Phys. Rev. 162, 1650 (1966).
2. A.M. Gavrilik, V.A. Shirokov, Institute for Theoretical Physics, Academy of Sciences of the Ukranian S.S.R. publication ITP-77-99P, Kiev (1977).
3. C.S. Kalman, Lett. Nuovo Cimento 21, 145 (1978).
4. C.S. Kalman, Lett. Nuovo Cimento 21, 201 (1978).
5. C.S. Kalman, Lett. Nuovo Cimento 21, 291 (1978).

WEIGHT MULTIPLICITIES OF THE EXCEPTIONAL LIE GROUPS.

R.C. King and A.H.A. Qubanchi, Department of Mathematics, The University, Southampton, England.

Given a Lie group G of rank r and a subgroup H of the same rank the weight multiplicities of the irreducible representations $\underline{\lambda}_G$ of G may be determined from a knowledge of the branching rule: $G \rightarrow H \quad \underline{\lambda}_G \rightarrow \sum_{\mu_H} B_{\underline{\lambda}_G}^{\underline{\mu}_H} \underline{\mu}_H$, and the characters of the irreducible representations $\underline{\mu}_H$ of H: $\chi^{\underline{\mu}_H}(\underline{\phi}) = \sum_{\underline{m}} M_{\underline{m}}^{\underline{\mu}_H} \exp i\, \underline{m}.\underline{\phi}$. One simply uses the same parameters $\underline{\phi}$ to describe the classes of G and writes: $\chi^{\underline{\lambda}_G}(\underline{\phi}) = \sum_{\underline{\mu}_H, \underline{m}} B_{\underline{\lambda}_G}^{\underline{\mu}_H} M_{\underline{m}}^{\underline{\mu}_H} \exp i\, \underline{m}.\underline{\phi} = \sum_{\underline{m}} M_{\underline{m}}^{\underline{\lambda}_G} \exp i\, \underline{m}.\underline{\phi}$ where the multiplicity of the weight $\underline{m}$ in the representation $\underline{\lambda}_G$ of G is $M_{\underline{m}}^{\underline{\lambda}_G}$.

Of the set of vectors $\underline{m}$ which arise in this way it is only necessary to consider those weights $\underline{\sigma}_G$ which are dominant weights of G. These are a subset of the dominant weights, $\underline{\tau}_H$, of the subgroup H, several of which may be related to one particular dominant weight of G through the action of the Weyl group, W_G of G. The dominant weights $\underline{\sigma}_G$ and $\underline{\tau}_H$ of G and H are used to label the irreducible representations of G and H.

The weight multiplicities for the exceptional Lie groups may be found from the branching rules[1,2] appropriate to: $G_2 \supset SU(3)$, $F_4 \supset SO(9)$, $E_6 \supset SU(2) \otimes SU(6)$, $E_7 \supset SU(8)$ and $E_8 \supset SO(16)$ and the known multiplicities of the dominant weights of the classical groups[3]. Some results are given in the table in which the dimensions of the representations have been given to aid in their identification.

REFERENCES

1. Wybourne, B.G. and Bowick, M.J. Aust.J.Phys. 30 259 (1977).
2. King, R.C. and Qubanchi, A.H.A. J.Phys.A. Math.Gen. 11 1491 (1978).
3. King, R.C. and Plunkett, S.P.O. J.Phys.A. Math. Gen. 9 863 (1976).

Dim $\underline{\lambda}_G$	$\underline{\lambda}_G$	$\underline{\mu}_H$		$M^{\underline{\lambda}_G}_{\underline{\sigma}_G}$				
	$G_1 = G_2$	H = SU(3)	$\underline{\sigma}_G$	0	1	1^2	2	21
7	1	1^2+1+0		1	1			
14	1^2	$21+1^2+1$		2	1	1		
27	2	$2^2+21+2+1^2+1+0$		3	2	1	1	
64	21	$32+31+2^2+2(21)+2+1^2+1$		4	4	2	2	1
	$G = F_4$	H = SO(9)	$\underline{\sigma}_G$	0	1	1^2	$\Delta;1$	2
26	1	$1+\Delta+0$		2	1			
52	1^2	$1^2+\Delta$		4	1	1		
273	$\Delta;1$	$\Delta;1+1^3+1^2+1+\Delta$		9	5	2	1	
324	2	$2+\Delta;1+1^4+1+\Delta+0$		12	5	3	1	1
	$G = E_6$	$H = SU(2)\otimes SU(6)$	$\underline{\sigma}_G$	0	$1;1$	$1;1^5$	$2;0$	$2;1^2$
27	$1;1$	$1\otimes 1+0\otimes 1^4$		0	1			
27	$1;1^5$	$1\otimes 1^5+0\otimes 1^2$		0	0	1		
78	$2;0$	$2\otimes 0+1\otimes 1^3+0\otimes 21^4$		6	0	0	1	
351	$2;1^2$	$2\otimes 1^2+1\otimes 21^3+1\otimes 1^5+0\otimes 2^3 1^2+0\otimes 2$		0	0	5	0	1
	$G = E_7$	H = SU(8)	$\underline{\sigma}_G$	0	1^6	21^6	$2^5 1^2$	2^6
56	1^6	1^6+1^2		0	1			
133	21^6	21^6+1^4		7	0	1		
1539	$2^5 1^2$	$2^5 1^2+21^2+2^2 1^4+21^6$		27	0	6	1	
1463	2^6	$2^6+2^2+2^2 1^4+1^4+0$		21	0	5	1	1
	$G = E_8$	H = SO(16)	$\underline{\sigma}_G$	0	1^2	2	21^2	2^2
248	1^2	$1^2+\Delta_+$		8	1			
3875	2	$2+\Delta;1_-+1^4$		35	7	1		
30380	21^2	$21^2+\Delta;1^2_+ +\Delta;1_-+1^6+1^2$		140	35	7	1	
27000	2^2	$2^2+\Delta;1^2_+ +\Delta;0_+ +1^8_+ +1^4+0$		120	29	6	1	1

Conserved Currents and Symmetries of the S-Matrix

Helmut Reeh
Institut für Theoretische Physik der Universität Göttingen

Consider generators of symmetry transformations defined as integrals over the time component of a conserved local current density. In the general case, the current density need not be covariant under translations. We work within the Wightman framework of local quantum field theory and assume in addition: i) Existence of a mass gap above the vacuum. ii) There are only isolated one-particle hyperboloids of finite multiplicity. iii) The vacuum is invariant under the symmetry transformation. iv) Asymptotic completeness. v) There exists a local interpolating field for every particle.

The results are as follows. Denote the generator by Q. Then 1) Q is closable and uniquely extendable to a dense set of scattering states. 2) Q commutes with the S-matrix. 3) Let $\psi_\nu^{ex}(x)$ denote the asymptotic fields (ex = in or out) for the particle of type ν according to Haag-Ruelle scattering theory. Then

$$[Q, \psi_\nu^{ex}(x)] = \sum_\mu P_{\nu\mu}(x, \partial_x)\, \psi_\mu^{ex}(x)$$

where $P_{\nu\mu}$ are polynomials in x and derivatives with respect to x vanishing if ψ_ν^{ex} and ψ_μ^{ex} have unequal mass. 4) According to 3) Q acts additively on multiparticle states, operates only within mass multiplets and is on one particle states represented by polynomials $P_{\nu\mu}(x, \partial_x)$. Assume there is only a finite number of asymptotic particles and assume they are all scalar. Then, on one particle states, Q is a polynomial of the generators P_κ, $M_{\lambda\rho}$ of the Poincare group. 5) If the $P_{\nu\mu}$ are of degree zero in x, then Q is selfadjoint since on one particle states it is a polynomial in the P_κ. In case of higher degree Q may have no selfadjoint extension as can be shown by an explicit example in terms of free fields of an hermitean Q with deficiency indices $(0, \infty)$.

Remarks: i) The results presumably stay true without the additional assumption in 4). ii) The situation resembles that of classical mechanics: There one can show that every additive conservation law of multiparticle systems on one particle states is given by a function f of momentum and angular momentum only. In classical mechanics one can show more: In case there is an interaction between the particles allowing all scattering processes consistent with energy, momentum and angular momentum conservation, then f can only be a linear combination of energy, momentum and angular momentum. The same conclusion presumably holds also for the field theoretic framework.

Proofs and more details can be found in [1] where are also references to relevant papers by other authors.

[1] W. D. Garber, H. Reeh: J. Math. Phys. 19, 59 (1978), J. Math. Phys. 19, 985 (1978) and forthcoming preprint.

GROUP THEORETICAL ASPECTS OF MORE-THAN-FOUR QUARK MODELS

Zdravko Stipčević
Institut za fiziku, Univerzitet u Sarajevu
Jugoslavija

SUMMARY

We focus attention on the six-quark models, consider O(6) flavor symmetry and investigate possibilities of embedding: $SU(2)_W \times U(1) \subset SU(4)$. Charge quantization in the frame of O(6) allows two options for the quark charge assignments. In one case (charges of u,d,s,c,t,b are 2/3,-1/3,-1/3,2/3,2/3,-1/3, respectively) we are led to the six-quark, six-lepton sequential model [1-3] whose weak and electromagnetic currents however cannot be embedded into SU(4), except for physically unacceptable values of the Cabibbo angles θ_1, θ_2, θ_3 and the absence of CP violation. We conclude that natural symmetry here is O(6) which then implies, from the grand unification point of view, the requirement for an $SU(3)_C \times SU(6)$ subgroup of G and consequently gives support to the E_7 sheme[4]. In the other case (charges of u,d,s,c,t,b are 2/3, -1/3,-1/3,-1/3,-1/3,-4/3, respectively) the corresponding six-quark model [5,6] possesses weak and electromagnetic currents which can be embedded into SU(4). The embedding produces constraints which allow for only one Cabibbo angle. An arbitrary phase δ still remains unconstrained and consequently there is room for CP violation. The SU(4) constraints within this charge option have been utilised [7] for a crude computation of the mass of a new heavy quark-antiquark bound state (the $t\bar{t}$ as a companion to the known ϕ, ω, ρ_0, Ψ and Υ vector meson resonances) and a value in the vicinity of 13 GeV is indicated.

REFERENCES:

1. M.Kobayashi and K.Maskawa, Prog.Theor.Phys. 49 (1973) 652;
2. H.Harari, Phys.Lett. 57B (1975) 265; Ann. of Phys. 94 (1975) 391;
3. Z.Stipčević, seminar, International School of Elementary Particle Physics, Baško Polje, 1975;
4. F.Gürsey and P.Sikivie, Phys.Rev.Lett. 36 (1976) 775;
5. C.H.Albright and R.J.Oakes, FERMILAB-Pub-75/53-THY; R.J.Oakes, "New Directions in Hadron Spectroscopy", Berger and S.Kramer, eds. ANL-HEP-CP 75-58;
6. F.E.Close and E.W.Colglasier, CERN preprint TH-2065 (1975);
7. S.Fajfer, S.Blatnik and Z.Stipčević, Univ. of Tuzla preprint, TF-PPP-2/78.

INFINITE UNITARY GROUP, QUARKS, UNITARY SYMMETRY

Igor Szczyrba

Department of Mathematical Methods in Physics, University of Warsaw

We try to estabilish links among some basic postulates of Quantum Mechanics, the unitary flavour group (possibly infinite-dimensional), the postulate of indistinguishability and the colour group of quarks.

Let H be a (possibly infinite-dimensional) Hilbert space of one-quark flavour space. Let $\overline{H}$ be the space conjugate to H, and let $H^N \otimes \overline{H}^M := H\otimes \ldots \otimes H \otimes \overline{H} \otimes \ldots \otimes \overline{H}$ describe the states of the system composed of N quarks and M antiquarks. The natural unitary representation U of the unitary flavour group AutH acts in $H^N \otimes \overline{H}^M$. In this space acts also the natural representation S of Cartesian product of the permutation groups $P_N \times P_M$ which permute separately the positions of quarks and antiquarks. Let K be an U-irreducible subspace of $H^N \otimes \overline{H}^M$ (see [1]) and let B(H) (resp. B(K)) denote the set of bounded operators in H (resp. K). As the main tool we use the physically motivated notion of general number operators [3]. This notion encomposes the operators which are related to the number of particles of a system, the operator of isospin, hypercharge etc., and as we will show the mass operator. From [3] it follows that for a given K the set of all general number operators is in one-to-one correspondence with the vector space of all linear, ultra-weakly-weakly continuous mappings $\Lambda: B(H) \to B(K)$ satisfying the following condition:

$$\Lambda(vAv^{-1}) = U_v \Lambda(A) U_v^{-1} \quad \text{for any } A \in B(H) \text{ and } v \in \mathrm{Aut}H. \qquad (*)$$

This equation reflects the assumption that no one-particle state is distinguished among the others.

In the case considered here any general number operator can be obtained as the linear combinations of the mappings:

$$B(H) \ni A \to \Lambda^i(A) := K(I\otimes \ldots \otimes I \otimes A \otimes I \ldots \otimes I)K \in B(K) \qquad i=1,2,..N+M$$

where K denotes also the orthogonal projection onto the space K and I is the identity operator in H. For a given representation in K not all of the mappings Λ^i are linearly independent. The explicit solution for that problem was given in terms of Young diagrams in [4].

For an arbitrary orthonormal basis $\{x_j\}_{j=1}^{\dim H}$ in the space H and for an arbitrary vector $\Psi \in K$ the expresion:

$$(\Psi|\Lambda^i_j\Psi) := (\Psi|\Lambda^i(|x_j)(x_j|)\Psi)$$

is equal to the number of particles in the state x_j which are contained in the multi-particle state Ψ at the position labelled by the i.

We claim that the (generalized) Gell-Mann mass operator has the following form:

$$G = \sum_{j=1}^{\dim H} \sum_{i=1}^{N+M} m^i_j \Lambda^i_j$$

where the constants m^i_j have the interpretation of masses of quarks labelled by position. The infinite sum is convergent e.g. in the strong topology. By using the property (*) and the equation $\Lambda^i_j(I_H)=I_K$ one can see that if dimH=3 the operator G coincides with the Gell-Mann mass operator with electromagnetic corrections. So the mass of a hadron can be obtained as the sum of masses of labelled quarks.

The operator G does not commute with the representation S. Therefore one should treat quarks in the hadron as being distinguishable (see [2]). The agreement with experiment can be gotten only if the mass of the quark depends on its label. This label should be compared with the notion of colour for quarks.

The expresion given above for the operator G enables us to obtain mass formulae (analogous to the Gell-Mann formula) for any type of the multi-quark space with an arbitrary number of quarks and antiquarks. We can also prove that the spectrum of all operators Λ^i_j is rational and hence the coeficients in the mass formulae are always integers. The other proved properties of the spectrum of Λ^i_j let us get for an arbitrary multi-quark system mass formulae analogous to the Coleman-Glashow formula.

We can show also that in fact all formulae mentioned above depend only on the symmetry type of multi-quark space and are independent of the dimension of H. Therefore one can consider even an infinite-dimensional one-quark flavour space. The detailed results will be published elsewhere.

[1] A.A.Kirillov, Dokl. Akad. Nauk SSSR 212 (1973) 288.

[2] R.H.Stolt, J.R.Taylor, Nucl. Phys. B19 (1970) 1.

[3] I.Szczyrba, Rep. on Math. Phys. 7.2 (1975) 251.

[4] I.Szczyrba, Group Theor. Meth. in Phys., Proc. of the Six Int. Coll., Tubingen1977, Lecture Notes in Physics 79 (1978) 540.

LIST OF PARTICIPANTS

Aebersold, D.	Bennington College, U.S.A.
Agacy, R.L.	J. Cook University, North Queenland, Australia
Ali, S.T.	University of Toronto, Canada
Al-Qubanchi, A.H.A.	University of Southampton, England
Antillon, A.	University of Mexico, Mexico
Arms. J.	University of Utah, U.S.A.
Bacry, H.	C.N.R.S. Marseille, France
Balachandran, A.P.	Syracuse University, U.S.A.
Barut, A.O.	University of Colorado, U.S.A.
Bassein, K.	Mills College, U.S.A.
Batchelor, M.	Massachusetts Institute of Technology, U.S.A.
Baumgartel, H.	Akad. der Wissenshaften, Berlin, DDR
Bayen, F.	University of Paris VI, France
Beckers, J.	University of Liege, Belgium
Beiglboeck, W.	University of Heidelberg, Germany
Biedenharn, L.C.	Duke University, U.S.A.
Bincer, A.	University of Wisconsin-Madison, U.S.A.
Binz, E.	University of Mannheim, Germany
Birman, J.	City College of C.U. New York, U.S.A.
Bleuler, K.	University of Bonn, Germany
Bohm, A.	University of Texas, U.S.A.
Bose, A.K.	University of Montreal, Canada
Bose, S.K.	University of Notre Dam, U.S.A.
Boyer, C.P.	University of Mexico, Mexico
Brandt, H.E.	Harry Diamond Lab, U.S.A.
Burt, P.B.	Clemson University, U.S.A.
Butler, P.H.	University of Canterbury, New Zealand
Bystricky, J.	University of Montreal, Canada
Campbell, D.	Los Alamos Scientific Lab, U.S.A.
Capps, R.H.	Purdue University, U.S.A.
Cattanero, U.	University of Neuchatel, Switzerland
Chacon, E.	University of Mexico, Mexico
Chiang, C.-C.	National Taiwan Normal University, Republic of China
Combe, P.	C.N.R.S. Marseille, France
Czyz, J.	University of Warsaw, Poland
Davies, B.L.	University of North Wales, England
Deans, S.R.	University of South Florida, U.S.A.
De Muynck, W.M.	Eindhoven University of Technology, the Netherlands
Deser, S.	Brandeis University, U.S.A.
Demirlioglu, D.	Bogazici University, Turkey
Dirl, R.	University of Wien, Austria
Deobner, H.D.	University of Clausthal-Zellerfeld, Germany
Draayer, J.P.	Louisiana State University, U.S.A.
Dragt, A.J.	University of Maryland, U.S.A.
Draut, A., Jr.	General Dynamics, Fort Worth, U.S.A.
Drechsler, W.	Max-Planck-Institut, Munich, Germany
Dresden, M.	SUNY, Stony Brook, U.S.A.
Dudley, A.L.	University of New Mexico, U.S.A.
Ekstein, H.	C.N.R.S. Marseille, France
Emch, G.G.	University of Rochester, U.S.A.
Ernst, F.J.	Illinois Institute of Technology, U.S.A.
Feldman, F.A.	Suffolk University, U.S.A.
Findley, G.L.	Louisiana State University, U.S.A.
Fleming, G.	Pennsylvania State University, U.S.A.
Flodmark, S.	Stockholm, Sweden
Fonda, L.	ICTP, Trieste, Italy
French, J.B.	University of Rochester, U.S.A.
Fukutome, H.	University of Kyoto, Japan
Fulling, S.A.	Texas A & M University, U.S.A.
Garber, W.	Göttingen University, Germany

Garcia, A.	University of Mexico, Mexico
Gazeau, J.P.	University of Paris, France
Geheniau, J.	Universite de Libre de Brussels, Belgium
Ghirardi, G.	University of Trieste, Italy
Gilmore, R.	University of South Florida, U.S.A.
Ginocchio, J.	Los Alamos Scientific Lab, U.S.A.
Giovannini, N.	University of Geneva, Switzerland
Goldberg, J.N.	Syracuse University, U.S.A.
Goldharbor, G.S.	Brookhaven National Lab, U.S.A.
Gorini, V.	University of Milano, Italy
Gotay, M.J.	University of Maryland, U.S.A.
Grecos, A.	Universite de Libre de Brussels, Belgium
Grossman, A.	C.N.R.S. Marseille, France
Gruber, C.	E.P.F., Lausanne, Switzerland
Gunther, N.J.	University of Southampton, England
Gursey, F.	Yale University, U.S.A.
Gutkin, E.	University of Utah, U.S.A.
Halpern, L.	Florida State University, U.S.A.
Hanson, A.	Lawrence Berkeley Lab, U.S.A.
Harnad, J.P.	University of Montreal, Canada
Harter, W.G.	University of Colorado, U.S.A.
Hecht, K.T.	University of Michigan, U.S.A.
Henrich, C.J.	Interdata, Inc., U.S.A.
Herrick, C	University of Oregon, U.S.A.
Hori, S.	Kanazawa University, Japan
Horwitz, L.P.	University of Tel-Aviv, Israel
Iachello, F.	Rijksuniversiteit, Groningen, the Netherlands
Ilamed, Y.	University of Tel-Aviv, Israel
Ingraham, R.L.	New Mexico State University, Las Cruz, U.S.A.
Jackiw, R.	Massachusetts Institute of Technology, U.S.A.
Jacob, P.	Max-Planck-Institut, Starnberg, Germany
Janner, A.	University of Nijmegen, the Netherlands
Jansson, P.-O.	University of Stockholm, Sweden
Jaric, M.V.	City College of C.U. New York, U.S.A.
Jaspers, M.	University of Liege, Belgium
Jasselette, P.	University of Liege, Belgium
Joel, J.S.	University of Michigan, U.S.A.
John, G.	University of Tübingen, Germany
Joseph, D.	University of Nebraska, U.S.A.
Jung, C.	Technical University, Berlin, Germany
Kac, V.	Massachusetts Institute of Technology, U.S.A.
Kadyshevsky, V.	Dubna, USSR
Kaku, M.	City College of C.U. New York, U.S.A.
Kalman, C.S.	Concordia University, Canada
Kamefuchi, S.	University of Tsukuba, Japan
Kataznlson, E.	University of Tel-Aviv, Israel
Kerner, R.	University of Paris VI, France
Kielanowski, P.	University of Warsaw, Poland
Kim, Y.S.	University of Maryland, U.S.A.
Klauder, J.R.	Bell Laboratories, U.S.A.
Komy, S.R.	University of Riyadh, Saudi-Arabia
Kostant, B.	Massachusetts Institute of Technology, U.S.A.
Kramer, P.	University of Tübingen, Germany
Krausser, D.	University of Berlin, Germany
Kummerer, B.	University of Tübingen, Germany
Laskar, W.	University of Nantes, France
Laskar, E.	University of Nantes, France
Lemire, F.	University of Windsor, Canada
Lichnerowicz, A.	College de France, France
Lopez, G.	University of Mexico, Mexico
Louck, J.D.	Los Alamos Scientific Lab, U.S.A.
Lukierski, J.	University of Wroclaw, Poland

Lund, F.	University of Chile, Chile
Mackey, G.W.	Harvard University, U.S.A.
Mainland, B.	Ohio State University, Newark, U.S.A.
Malin, S.	Ben Gurion University, Israel
Mansouri, F.	Yale University, U.S.A.
Marchand, J.P.	University of Denver, U.S.A.
Marmo, G.	University of Naples, Italy
Matsen, F.A.	University of Texas, U.S.A.
Mendes, R.V.	University of Lisbon, Portugal
Michel, L.	Bures-sur-Yvette, France
Mondaini, R.P.	C.B.P.F., Rio de Janeiro, Brazil
Morgan, T.	University of Nebraska, U.S.A.
Moshinsky, M.	University of Mexico, Mexico
Moylan, P.	University of Texas, U.S.A.
Nakano, T.	Osaka City University, Japan
Narcowich, F.J.	Texas A & M University, U.S.A.
Nath, P.	Northeastern University, U.S.A.
Ne'eman, Y.	Tel-Aviv University, Israel
Nester, J.M.	University of Saskatchewan, Canada
Neumann, H.	University of Marburg, Germany
Nieto, M.M.	Los Alamos Scientific Lab. U.S.A.
Nilsson, J.S.	University of Chalmers, Sweden
Nottrot, R.	Technical University Twente, the Netherlands
Nussenzveig, H.M.	University of Sao Paulo, Brazil
Okubo, S.	University of Rochester, U.S.A.
Omote, M.	University of Tsukuba, Japan
Oneda, S.	University of Maryland, U.S.A.
Opechowdki, W.	University of British Columbia, Canada
O'Raifeartaigh, L.	Dublin Institute of Advanced Studies, Ireland
Ozsvath, I.	University of Texas, Dallas, U.S.A.
Parravicini, G.	University of Milano, Italy
Pascolini, A.	University of Padva, Italy
Pasupathy, J.	University of Bangalore, India
Patera, J.	University of Montreal, Canada
Patterson, C.W.	Los Alamos Scientific Lab., U.S.A.
Paldus, J.	University of Waterloo, Canada
Petry, H.R.	University of Bonn, Germany
Phippen, J.W.	Weber State College, U.S.A.
Piron, C.	University of Geneva, Switzerland
Pitre, R.	University of Texas, U.S.A.
Plebanski, J.	University of Mexico, Mexico
Pommaret, J.F.	College de France, France
Radicati, L.A.	University of Pisa, Italy
Ranganathan, N.R.	University of Madras, India
Reeh, H.	University of Gottingen, Germany
Regge, T.	Institute for Advanced Studies, U.S.A.
Rideau, G.	University of Paris VII, France
Rieckers, A.	University of Tubingen, Germany
Ringhofer, K.	University of Osnabruck, Germany
Rodgers, H.H.	Los Alamos Scientific Lab., U.S.A.
Rodriguez, F.P.	University Autonoma Metropolitan, Mexico
Romerio, M.V.	University of Neuchatel, Switzerland
Ronveaux, A.	University of Notre Dame de la Paix, Namur, Belgium
Rosensteel, G.	McMaster University, Canada
Ruck, H.M.	Duke University, U.S.A.
Ryman, A.	University of Toronto, Canada
Saller, H.	Max-Planck-Institut, Munich, Germany
Scheunert, M.	Dublin Institute of Advanced Studies, Ireland
Schieve, W.C.	University of Texas, U.S.A.
Schober, A.	Technischen University Berlin, Germany
Scott, D.M.	Ohio State University, Mansfield, U.S.A.
Seiler, E.	Princeton University, U.S.A.

Seligman, T.	University of Mexico, Mexico
Sen, R.N.	Ben Gurion University, Neger, Israel
Shaw, G.	University of Oregon, U.S.A.
Shnider, S.	McGill University, Canada
Siegel, W.	Harvard University, U.S.A.
Slaughter, M.D.	Los Alamos Scientific Lab. U.S.A.
Sniatycki, J.	University of Calgary, Canada
Sobczyk, A.	Clemson University, U.S.A.
Solomon, A.	Open University, England
Sorba, P.	CERN, TH1211, Switzerland
Staunton, L.P.	Drake University, U.S.A.
Steinberg, S.	University of New Mexico, U.S.A.
Sternberg, S.	Harvard University, U.S.A.
Stipcevic, Z.	University of Saraievo, Yugoslavia
Strasburger, A.	University of Warsaw, Poland
Strom, S.	Chalmers Institute of Technology, Sweden
Sudarshan, E.C.J.	University of Texas, U.S.A.
Sullivan, J.	University of New Orleans, U.S.A.
Szczyrba, I.	University of Warsaw, Poland
Takasugi, E.	University of Texas, U.S.A.
Tauber, G.E.	University of Tel-Aviv, Israel
Teese, R.	Max-Planck-Institut, Munich, Germany
Teitelboim, C.	Princeton University, U.S.A.
Trebin, H.R.	City College of C.U. New York
Urrutia, L.F.	University of California, Los Angeles, U.S.A.
Uzes, C.	University of Georgia, U.S.A.
Van Alstine, P.W.	University of Nebraska, U.S.A.
Van Dam, H.	University of North Carolina, U.S.A.
Villarroel, D.	University of Chile, Chile
Vitiello, G.	University of Salerno, Italy
Wald, R.	Enrico Fermi Institute, U.S.A.
Weaver, L.	Kansas State University, U.S.A.
Werle, J.	University of Warsaw, Poland
Wheeler, J.	University of Texas, U.S.A.
Wigner, E.P.	Princeton University, U.S.A.
Wolf, K.B.	University of Mexico, Mexico
Wulfman, C.E.	University of Pacific, Stockton, U.S.A.
Yasskin, P.B.	University of Maryland, U.S.A.
Zahn, W.	University of Michigan, U.S.A.
Zaidi, S.A.A.	University of Texas, U.S.A.
Zia, R.K.P.	Virginia Technical Institute, U.S.A.
Zumino, B.	CERN, Switzerland

Selected Issues from

Lecture Notes in Mathematics

Vol. 532: Théorie Ergodique. Proceedings 1973/1974. Edité par J.-P. Conze and M. S. Keane. VIII, 227 pages. 1976.

Vol. 538: G. Fischer, Complex Analytic Geometry. VII, 201 pages. 1976.

Vol. 543: Nonlinear Operators and the Calculus of Variations, Bruxelles 1975. Edited by J. P. Gossez, E. J. Lami Dozo, J. Mawhin, and L. Waelbroeck, VII, 237 pages. 1976.

Vol. 552: C. G. Gibson, K. Wirthmüller, A. A. du Plessis and E. J. N. Looijenga. Topological Stability of Smooth Mappings. V, 155 pages. 1976.

Vol. 556: Approximation Theory. Bonn 1976. Proceedings. Edited by R. Schaback and K. Scherer. VII, 466 pages. 1976.

Vol. 559: J.-P. Caubet, Le Mouvement Brownien Relativiste. IX, 212 pages. 1976.

Vol. 561: Function Theoretic Methods for Partial Differential Equations. Darmstadt 1976. Proceedings. Edited by V. E. Meister, N. Weck and W. L. Wendland. XVIII, 520 pages. 1976.

Vol. 564: Ordinary and Partial Differential Equations, Dundee 1976. Proceedings. Edited by W. N. Everitt and B. D. Sleeman. XVIII, 551 pages. 1976.

Vol. 565: Turbulence and Navier Stokes Equations. Proceedings 1975. Edited by R. Temam. IX, 194 pages. 1976.

Vol. 566: Empirical Distributions and Processes. Oberwolfach 1976. Proceedings. Edited by P. Gaenssler and P. Révész. VII, 146 pages. 1976.

Vol. 570: Differential Geometrical Methods in Mathematical Physics, Bonn 1975. Proceedings. Edited by K. Bleuler and A. Reetz. VIII, 576 pages. 1977.

Vol. 572: Sparse Matrix Techniques, Copenhagen 1976. Edited by V. A. Barker. V, 184 pages. 1977.

Vol. 579: Combinatoire et Représentation du Groupe Symétrique, Strasbourg 1976. Proceedings 1976. Edité par D. Foata. IV, 339 pages. 1977.

Vol. 587: Non-Commutative Harmonic Analysis. Proceedings 1976. Edited by J. Carmona and M. Vergne. IV, 240 pages. 1977.

Vol. 592: D. Voigt, Induzierte Darstellungen in der Theorie der endlichen, algebraischen Gruppen. V, 413 Seiten. 1977.

Vol. 594: Singular Perturbations and Boundary Layer Theory, Lyon 1976. Edited by C. M. Brauner, B. Gay, and J. Mathieu. VIII, 539 pages. 1977.

Vol. 596: K. Deimling, Ordinary Differential Equations in Banach Spaces. VI, 137 pages. 1977.

Vol. 605: Sario et al., Classification Theory of Riemannian Manifolds. XX, 498 pages. 1977.

Vol. 606: Mathematical Aspects of Finite Element Methods. Proceedings 1975. Edited by I. Galligani and E. Magenes. VI, 362 pages. 1977.

Vol. 607: M. Métivier, Reelle und Vektorwertige Quasimartingale und die Theorie der Stochastischen Integration. X, 310 Seiten. 1977.

Vol. 615: Turbulence Seminar, Proceedings 1976/77. Edited by P. Bernard and T. Ratiu. VI, 155 pages. 1977.

Vol. 618: I. I. Hirschman, Jr. and D. E. Hughes, Extreme Eigen Values of Toeplitz Operators. VI, 145 pages. 1977.

Vol. 623: I. Erdelyi and R. Lange, Spectral Decompositions on Banach Spaces. VIII, 122 pages. 1977.

Vol. 628: H. J. Baues, Obstruction Theory on the Homotopy Classification of Maps. XII, 387 pages. 1977.

Vol. 629: W. A. Coppel, Dichotomies in Stability Theory. VI, 98 pages. 1978.

Vol. 630: Numerical Analysis, Proceedings, Biennial Conference, Dundee 1977. Edited by G. A. Watson. XII, 199 pages. 1978.

Vol. 636: Journées de Statistique des Processus Stochastiques, Grenoble 1977, Proceedings. Edité par Didier Dacunha-Castelle et Bernard Van Cutsem. VII, 202 pages. 1978.

Vol. 638: P. Shanahan, The Atiyah-Singer Index Theorem, An Introduction. V, 224 pages. 1978.

Vol. 648: Nonlinear Partial Differential Equations and Applications, Proceedings, Indiana 1976–1977. Edited by J. M. Chadam. VI, 206 pages. 1978.

Vol. 650: C*-Algebras and Applications to Physics. Proceedings 1977. Edited by R. V. Kadison. V, 192 pages. 1978.

Vol. 656: Probability Theory on Vector Spaces. Proceedings, 1977. Edited by A. Weron. VIII, 274 pages. 1978.

Vol. 662: Akin, The Metric Theory of Banach Manifolds. XIX, 306 pages. 1978.

Vol. 665: Journées d'Analyse Non Linéaire. Proceedings, 1977. Edité par P. Bénilan et J. Robert. VIII, 256 pages. 1978.

Vol. 667: J. Gilewicz, Approximants de Padé. XIV, 511 pages. 1978.

Vol. 668: The Structure of Attractors in Dynamical Systems. Proceedings, 1977. Edited by J. C. Martin, N. G. Markley and W. Perrizo. VI, 264 pages. 1978.

Vol. 675: J. Galambos and S. Kotz, Characterizations of Probability Distributions. VIII, 169 pages. 1978.

Vol. 676: Differential Geometrical Methods in Mathematical Physics II, Proceedings, 1977. Edited by K. Bleuler, H. R. Petry and A. Reetz. VI, 626 pages. 1978.

Vol. 678: D. Dacunha-Castelle, H. Heyer et B. Roynette. Ecole d'Eté de Probabilités de Saint-Flour. VII-1977. Edité par P. L. Hennequin. IX, 379 pages. 1978.

Vol. 679: Numerical Treatment of Differential Equations in Applications, Proceedings, 1977. Edited by R. Ansorge and W. Törnig. IX, 163 pages. 1978.

Vol. 681: Séminaire de Théorie du Potentiel Paris, No. 3, Directeurs: M. Brelot, G. Choquet et J. Deny. Rédacteurs: F. Hirsch et G. Mokobodzki. VII, 294 pages. 1978.

Vol. 682: G. D. James, The Representation Theory of the Symmetric Groups. V, 156 pages. 1978.

Vol. 684: E. E. Rosinger, Distributions and Nonlinear Partial Differential Equations. XI, 146 pages. 1978.

Vol. 690: W. J. J. Rey, Robust Statistical Methods. VI, 128 pages. 1978.

Lecture Notes in Physics

Vol. 68: Y. V. Venkatesh, Energy Methods in Time-Varying System Stability and Instability Analyses. XII, 256 pages. 1977.

Vol. 69: K. Rohlfs, Lectures on Density Wave Theory. VI, 184 pages. 1977.

Vol. 70: Wave Propagation and Underwater Acoustics. Edited by J. Keller and J. Papadakis. VIII. 287 pages. 1977.

Vol. 71: Problems of Stellar Convection. Proceedings 1976. Edited by E. A. Spiegel and J. P. Zahn. VIII, 363 pages. 1977.

Vol. 72: Les instabilités hydrodynamiques en convection libre forcée et mixte. Edité par J. C. Legros et J. K. Platten. X, 202 pages. 1978.

Vol. 73: Invariant Wave Equations. Proceedings 1977. Edited by G. Velo and A. S. Wightman. VI, 416 pages. 1978.

Vol. 74: P. Collet and J.-P. Eckmann, A Renormalization Group Analysis of the Hierarchical Model in Statistical Mechanics. IV, 199 pages. 1978.

Vol. 75: Structure and Mechanisms of Turbulence I. Proceedinas 1977. Edited by H. Fiedler. XX, 295 pages. 1978.

Vol. 76: Structure and Mechanisms of Turbulence II. Proceedings 1977. Edited by H. Fiedler. XX, 406 pages. 1978.

Vol. 77: Topics in Quantum Field Theory and Gauge Theories. Proceedings, Salamanca 1977. Edited by J. A. de Azcárraga. X, 378 pages 1978.

Vol. 78: Böhm, The Rigged Hilbert Space and Quantum Mechanics. IX, 70 pages. 1978.

Vol. 79: Group Theoretical Methods in Physics. Proceedings, 1977. Edited by P. Kramer and A. Rieckers. XVIII, 546 pages. 1978.

Vol. 80: Mathematical Problems in Theoretical Physics. Proceedings, 1977. Edited by G. Dell'Antonio, S. Doplicher and G. Jona-Lasinio. VI, 438 pages. 1978.

Vol. 81: MacGregor, The Nature of the Elementary Particle. XXII, 482 pages. 1978.

Vol. 82: Few Body Systems and Nuclear Forces I. Proceedings, 1978. Edited by H. Zingl, M. Haftel and H. Zankel. XIX, 442 pages. 1978.

Vol. 83: Experimental Methods in Heavy Ion Physics. Edited by K. Bethge. V, 251 pages. 1978.

Vol. 84: Stochastic Processes in Nonequilibrium Systems, Proceedings, 1978. Edited by L. Garrido, P. Seglar and P. J. Shepherd. XI, 355 pages. 1978

Vol. 85: Applied Inverse Problems. Edited by P. C. Sabatier. V, 425 pages. 1978.

Vol. 86: Few Body Systems and Electromagnetic Interaction. Proceedings 1978. Edited by C. Ciofi degli Atti and E. De Sanctis. VI, 352 pages. 1978.

Vol. 87: Few Body Systems and Nuclear Forces II, Proceedings, 1978. Edited by H. Zingl, M. Haftel, and H. Zankel. X, 545 pages. 1978.

Vol. 88: K. Hutter and A. A. F. van de Ven, Field Matter Interactions in Thermoelastic Solids. VIII, 231 pages. 1978.

Vol. 89: Microscopic Optical Potentials, Proceedings, 1978. Edited by H. V. von Geramb. XI, 481 pages. 1979.

Vol. 90: Sixth International Conference on Numerical Methods in Fluid Dynamics. Proceedings, 1978. Edited by H. Cabannes, M. Holt and V. Rusanov. VIII, 620 pages. 1979.

Vol. 91: Computing Methods in Applied Sciences and Engineering, 1977, II. Proceedings, 1977. Edited by R. Glowinski and J. L. Lions. VI, 359 pages. 1979.

Vol. 92: Nuclear Interactions. Proceedings, 1978. Edited by B. A. Robson. XXIV, 507 pages. 1979.

Vol. 93: Stochastic Behavior in Classical and Quantum Hamiltonian Systems. Proceedings, 1977. Edited by G. Casati and J. Ford. VI, 375 pages. 1979.

Vol. 94: Group Theoretical Methods in Physics. Proceedings, 1978. Edited by W. Beiglböck, A. Böhm and E. Takasugi. XIII, 540 pages. 1979.